MW00592590

Building Heat Transfer

Building Heat Transfer

Morris Grenfell Davies
The University of Liverpool, UK

Property of Alumni Library
Wentworth Institute of Technology

John Wiley & Sons, Ltd

697
.D28
2004 774775

Copyright © 2004 John Wiley & Sons Ltd, The Atrium, Southern Gate, Chichester, West Sussex PO19 8SQ, England

Telephone (+44) 1243 779777

Email (for orders and customer service enquiries): cs-books@wiley.co.uk
Visit our Home Page on www.wileyeurope.com or www.wiley.com

All Rights Reserved. No part of this publication may be reproduced, stored in a retrieval system or transmitted in any form or by any means, electronic, mechanical, photocopying, recording, scanning or otherwise, except under the terms of the Copyright, Designs and Patents Act 1988 or under the terms of a licence issued by the Copyright Licensing Agency Ltd, 90 Tottenham Court Road, London W1T 4LP, UK, without the permission in writing of the Publisher. Requests to the Publisher should be addressed to the Permissions Department, John Wiley & Sons Ltd, The Atrium, Southern Gate, Chichester, West Sussex PO19 8SQ, England, or emailed to permreq@wiley.co.uk, or faxed to (+44) 1243 770620.

This publication is designed to provide accurate and authoritative information in regard to the subject matter covered. It is sold on the understanding that the Publisher is not engaged in rendering professional services. If professional advice or other expert assistance is required, the services of a competent professional should be sought.

Other Wiley Editorial Offices

John Wiley & Sons Inc., 111 River Street, Hoboken, NJ 07030, USA

Jossey-Bass, 989 Market Street, San Francisco, CA 94103-1741, USA

Wiley-VCH Verlag GmbH, Boschstr. 12, D-69469 Weinheim, Germany

John Wiley & Sons Australia Ltd, 33 Park Road, Milton, Queensland 4064, Australia

John Wiley & Sons (Asia) Pte Ltd, 2 Clementi Loop #02-01, Jin Xing Distripark, Singapore 129809

John Wiley & Sons Canada Ltd, 22 Worcester Road, Etobicoke, Ontario, Canada M9W 1L1

Wiley also publishes its books in a variety of electronic formats. Some content that appears in print may not be available in electronic books.

Library of Congress Cataloging-in-Publication Data

Davies, Morris G.
Building heat transfer / Morris G. Davies.
p. cm.
Includes bibliographical references and index.
ISBN 0-470-84731-X (alk. paper)
1. Heating. 2. Heat–Transmission. I. Title.

TH7463.D28 2004
697–dc22 2003066075

British Library Cataloguing in Publication Data

A catalogue record for this book is available from the British Library

ISBN 0-470-84731-X

Typeset in 10/12pt Times by Laserwords Private Limited, Chennai, India
Printed and bound in Great Britain by Antony Rowe Ltd, Chippenham, Wiltshire
This book is printed on acid-free paper responsibly manufactured from sustainable forestry in which at least two trees are planted for each one used for paper production.

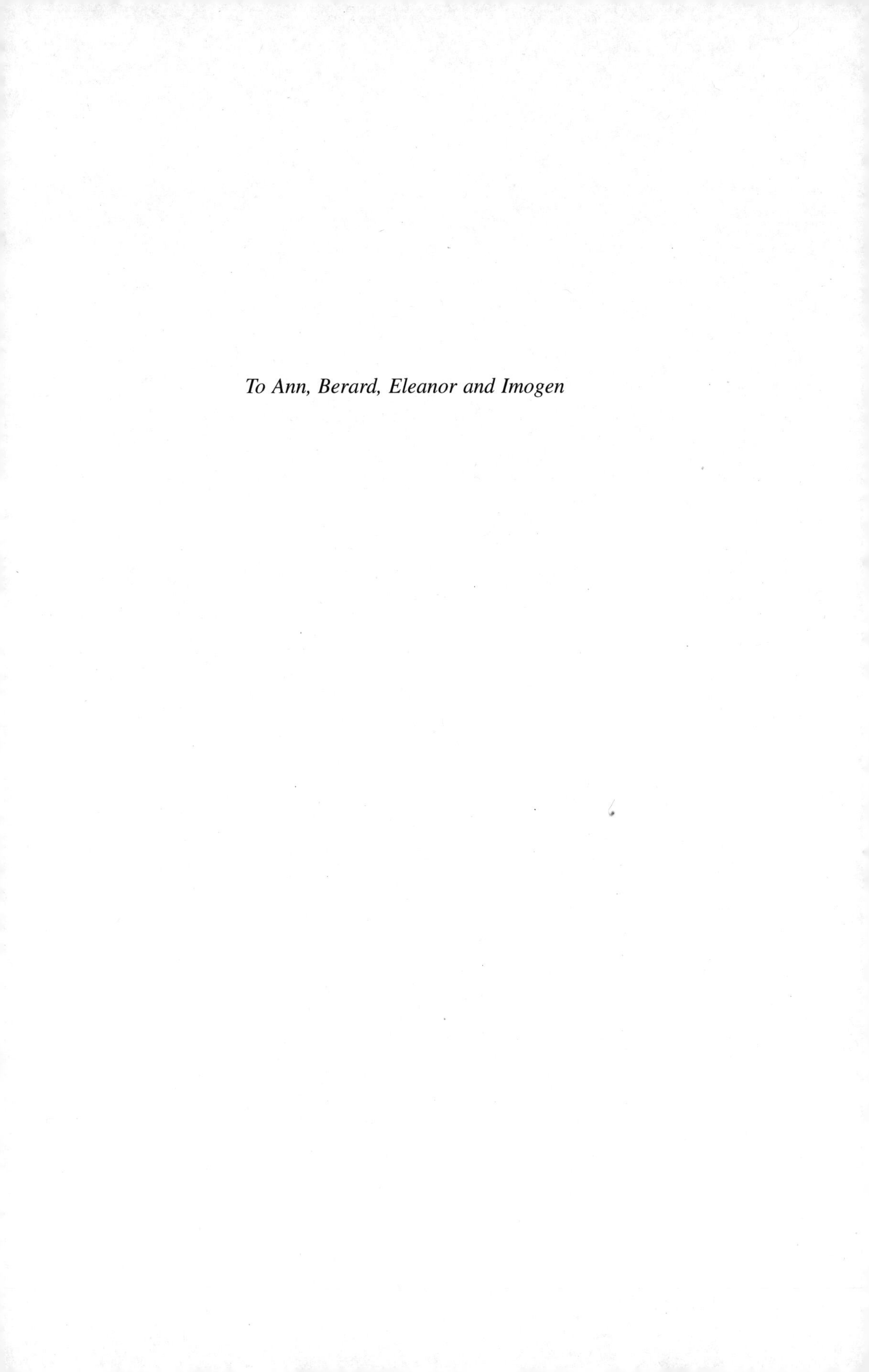

To Ann, Berard, Eleanor and Imogen

Contents

Preface

Building heat transfer is a gentle art. It does not involve low- or high-temperature technology, high-speed machinery, high pressures or a corrosive environment. Many of its elements are of long standing. Specific and latent heats, convection and radiation had been conceptionalised by the late eighteenth century. The Fourier continuity equation for wall conduction, the starting point for so much endeavour since the advent of electronic computation, first appeared in 1822 and the laws of thermodynamics evolved in the following decades. Box in his classic of 1868 provided values for conduction, convection and radiation exchange that do not differ seriously from those in present use.[1] Further, when estimates are to be made of indoor temperatures for human occupation, it is neither possible nor desirable to attempt high accuracy: uncertain assumptions have to be made for some of the physical parameters needed, the weather is variable, patterns of usage differ widely, individuals differ in their choice of conditions and some space and time variation in indoor temperature is generally welcomed.

However, one-third or more of the energy consumption of industrialised countries is expended on creating acceptable conditions of thermal comfort and lighting in their buildings (some 10% for air conditioning). Current estimates suggest that a little less than half the world's oil reserves have so far been used and that existing reserves can support current needs for more than 30 years.[2] There is no substitute for oil in sight however and the next generation will have to face the problem of supply, a problem compounded by the politics of some of the oil-producing countries. Mackay and Probert (1996) provide a comprehensive and readable account of energy and environmental policies in the UK. But consumption of fossil fuel contributes to global warming and ozone depletion, with problems which have to be faced now. Williams (1994: 97) remarks that 'to survive beyond the next 30 years, humanity's first priorities must be the conservation of resources and the preservation of the environment'. Following the Kyoto Climate Change Convention of 1997, worldwide endeavours have been initiated to reduce the emission of greenhouse gases to values of 10% or more below 1990 values. Much effort has been directed toward improving the efficiency of use of energy: environmentalists contend that this will lead

[1] *A Practical Treatise on Heat as Applied to the Useful Arts, for the Use of Engineers, Architects, Etc.*, by Thomas Box, London, E. and F. N. Spon, 1868. He used the still current symbol U to denote the time-averaged transmittance of a wall between the room index temperature and ambient temperature.

[2] Among the large literature on energy resources and the environment, I would mention the excellent coverage in the Journal *Applied Energy*.

to a reduction in energy consumption, although some economists have argued that the opposite may be the case (Herring 1999).

Expertise in the heat transfer processes that govern the distribution of energy flows in buildings, which has been accumulating for several decades through, among other means, development of programs to investigate the dynamic thermal behaviour of buildings, has thus become of special importance. Much of the corpus of classical heat transfer is relevant in a building context, and the number of well-established journals (more than 30 in English) devoted to publication of research results and experience in the energy field give clear and formidable indication of widespread concern for the environment in a broad sense.

Over a period of many years I gave a course of lectures on the fundamentals of building thermal response as an adjunct to the design of building thermal services, and this book brings together this material and more in the form of a monograph on heat transfer directed toward building thermal behaviour, providing in particular a logical basis for current design procedures. Chapter 1 treats elementary room response and Chapter 2 derives such basic thermal parameters whose physical explanation is simple. Chapter 3 discusses situations dominated by steady-state conduction. The thermal circuit (Chapter 4) provides a convenient means of visualising and analysing transfer by air movement (Chapter 5) and by long-wave radiant exchange (Chapter 6) and provides a means of merging these processes as is done in simple design procedures (Chapter 7). Basic aspects of moisture movement are discussed in Chapter 8 since, in their simplest form, the principles of moisture movement are similar to those for heat flow. In fact, moisture movement is much more complicated than heat flow. I have provided a sketch of some significant matters, with further topics at the end of Chapter 17, but moisture movement in buildings deserves a much more thorough presentation than can be given here. Chapter 9 deals with a few aspects of solar radiation, probably the most complex driver of heat flow in buildings (and one which has been well served by the literature since the 1970s). The later chapters are concerned with time-varying conduction and since the material relevant to building response is scarcely touched on in standard heat transfer texts, it is convenient to sketch developments in this area before indicating the sequence of chapters.

The design of heating systems remained for many years based on the time-averaged or steady state. Such mathematical interest as there was in Fourier's continuity equation for time-varying conditions was mainly concerned with error function solutions which have no bearing on unsteady flow in room walls, external or internal. More relevant solutions appeared in the 1910s and in 1925 Groeber published his much-quoted solution for the response of a slab, initially at uniform temperature and at time zero subjected to convective cooling to zero at both surfaces. But Groeber gave equal consideration to the response of a cylinder and a sphere and, sadly, presented somewhat meagre numerical results in a form that was difficult to interpret. The potential value of this analysis was not appreciated and in 1930 Krischer (with Esser) published an approach to wall dynamic response based on the fundamental idea that in a simple resistance/capacity circuit (thermal or electrical) the time constant or response time was the product resistance × capacity. The study was concerned with the response of a wall to a sudden (step) change in ambient temperature. Krischer was aware of the general form for the temperature profile in a slab but, unlike Groeber, proposed a flawed method for evaluating the constants. The theory was expounded at great length and was taken up by many workers over the next two decades. Out of it emerged the idea that a building response time might be evaluated as the 'storage/loss ratio', sometimes called the Q/U value. It could be regarded, however, as an unfortunate avenue of approach since the ratio is a steady-state construct and is inadequate to describe the more complicated time-varying behaviour.[3]

A more useful method of examining building dynamic response developed from changes in building design. Before the Second World War, heavyweight construction was the norm, with thick uninsulated walls and relatively small windows. Such constructions were slow to respond to internal sources of heat and to fluctuations in outside temperature. The post-war trend was towards lightweight framed buildings, sometimes with large areas of glazing. They were cheap and fast to erect and they looked good, but they proved disastrously uncomfortable. Areas near the windows tended to be so cold as to be unusable in winter and the solar gains, particularly large through south-facing facades on sunny days in spring and autumn, could lead to unacceptably high temperatures. Thus the need became apparent for measures of the thermal response of a (multilayer) wall to daily periodic excitation. Various parameters were proposed, including the heat flow into the wall due to sinusoidal variation in temperature on the same side (the admittance in UK nomenclature), numerically greater than the steady-state U value, together with the heat flow out of the wall from excitation on the opposite side (the dynamic or cyclic transmittance), which was less than U. The admittance was associated with a time lead and the transmittance with a time lag. Thus to express a wall response in terms of a steady-state component and a 24 h sinusoidal component, a set of five parameters was needed. This formed the basis for the UK 'admittance procedure', a very useful and simple if necessarily rather rough means of estimating manually the likely response of a building in its early stages of design. It first appeared in 1968. The procedure included the definition of an index, 'environmental temperature', which merged air and radiant measures of the room temperature. Similar measures were developed independently in the 1960s by other researchers. By including admittance and transmittance values for higher harmonics (12 h, 8 h, etc.), a more detailed response could be found but it was impractical to conduct the computations manually. The harmonic approach suffered the severe restriction that it could be used only when the building was in a steady-cyclic state. Its inputs consisted of predetermined values such as assumed values for the mean ambient temperature and its diurnal (sinusoidally varying) swing. It could not handle the response due to random ongoing excitation, expressed by hourly values of ambient temperature, for example.

About the time this simple steady-state/diurnal swings procedure was becoming accepted, Mitalas and Stephenson in Canada developed the means of estimating thermal response through the use of response factors – the flow of heat into or out of a wall following a triangular pulse of temperature excitation (1 K in height). In practice the base was of 2 h duration; the response factors then provided hourly heat flows and were intended for use with meteorological and user data specified at hourly intervals. A series of response factors is theoretically infinitely long, but the information they provide can be summarised in a remarkable and elegant manner through the use of transfer coefficients. The response of a typical wall due to data presented hourly could be provided by four series of 4–6 transfer coefficients each, some 20 values in total. (It may be mentioned however that while the values of admittance and transmittance for a multilayer wall could be evaluated and implemented using a hand-held calculator, evaluation of wall transfer coefficients was only

[3]The idea was current in the 1970s. It seemed to me to be ill-founded and I tested it using several simple models. It is not generally true.

possible using a sizeable computer program and again required computational facilities to implement it.)

Both the admittance and the transfer coefficient procedures were based on analytical solutions of the Fourier continuity equation in one dimension. In contrast, the finite difference procedure was based on a direct computational handling of the Fourier equation. The analytical methods refer to fluxes and temperature at the wall surfaces (or adjacent index temperature nodes), whereas the finite difference method supposes the wall to be composed of lumped resistances acting between nodes, together with lumped capacities at the nodes. At each hourly time step, the procedure recomputes the temperature distribution at nodes distributed through the thickness of the wall. It is conceptually very simple. Broadly speaking, European workers have used the finite difference approach whereas the Americans have favoured transfer coefficient methods.

The structure of the later chapters reflects this historical trend. Chapter 10 discusses the fundamentals of wall behaviour when the wall is represented as a chain of discrete resistance and capacity elements. It is comparatively easy to find the wall parameters and a time-domain method to find the transfer coefficients for such simple systems is given in Chapter 11. The Fourier continuity equation relates to a continuous distribution of resistance and capacity and Chapter 12 summarises the various possible forms of solution. Then come a few exact analytical models and some simplified models which have played a part in the thermal response field (Chapters 13 and 14.) The very important solution to the Fourier equation for sinusoidal excitation is developed in Chapter 15 to arrive at wall admittance and transmittance values. Chapter 16 discusses the means of implementing them in conjunction with the means of handling internal heat transfer (Chapter 7).

The thinking that led to transfer coefficients as a means of computing wall heat flows using time-domain driving data dates back at least to the early 1960s. The building needs at the time (problems associated with diurnal response) and the available mathematical tools, most notably advanced in Carslaw and Jaeger (1959), suggested the frequency-domain solution to the Fourier continuity equation as the appropriate starting point for evaluating these coefficients; it entailed Laplace and Z transforms and remains the favoured approach. They can, however, be found from time-domain solutions, using a simple development of standard Fourier analysis and without the use of transforms. This seems conceptually the simpler method and it is presented in Chapter 17. Chapter 18 examines the expected accuracy of heat flows estimated using transfer coefficients. Chapter 19, the final chapter, again bringing in internal transfer, shows how transfer coefficients can be used to find the response of an enclosure to ongoing excitation.

My own involvement in building thermal studies follows in part from a study of a building commonly called the Wallasey School, not far from Liverpool. It was designed in the late 1950s by Emslie Morgan, assistant borough architect to Wallasey Corporation, whose surviving notebooks show a strong, though not well-informed, interest in the possibilities of solar gains for heating purposes. He determined that suitable thermal conditions in the building should be attained in it by means of solar gains, the tungsten lighting system and the occupants' metabolic heat, without the use of a conventional hot-water heating system (although one was installed).[4] He patented the design (Morgan 1966). The school had a large area of nearly south-facing and completely unobstructed solar wall, mostly double

[4]He may very well have been encouraged by Billington's article on solar heat gain through windows (Billington 1947) to which he would have had ready access; his notebooks do not mention it however.

glazed, 125 mm of polystyrene insulation over all opaque areas (a remarkable thickness for its time), simple hand-operable ventilation control, time control of the lighting system and a large amount of internal thermal storage (concrete and brick) to absorb and counteract the effect of the large solar gains possible through the solar wall. More than a decade later, such a design came to be called 'passive'. It opened in 1961 and survived the bitterly cold and extended winter of 1963 in better shape than other schools in the area. It quickly achieved a certain fame, even notoriety, for its seeming ability to conjure heat from nowhere; thermal loss mechanisms were well understood by contemporary architects, but there was little understanding of the solar gain mechanism. It polarised opinion. Some said it was an excellent development because it 'worked'. Others condemned it: the school was a much-glazed building; much-glazed buildings were by then seen as environmentally unsuccessful, therefore the Wallasey School was unsuccessful. They complained of 'smell'.

Neither view was well argued. Adherents did not explain in what respect it worked; the opponents' doctrinaire view failed to take account of the many features the architect had incorporated to counter the problems of glazed areas. The headmaster flatly denied the existence of any smell.[5] In the early 1960s my former colleague, the late Dr C. B. Wilson, started an investigation of the building's thermal response and I took it over when he left in 1967.[6] Some of the problems the study threw up led me to developments reported in this book. (i) There was the basic problem: the solar wall readily admitted solar gains, even on dull days, but it also permitted large conduction losses; what was its net effect over a year? (ii) At that time very few admittance or transmittance values were available and they had been found using an electrical analogue computer for 24 h excitation; to estimate temperature swings in a building, a (digital) computer program was needed to find these parameters for a variety of constructions and for submultiples of 24 h (Chapter 15). (iii) Flaws in the argument leading to environmental temperature prompted an examination of how the complex radiant exchange in an enclosure could be optimally simplified so that it could be merged with convective exchange. This in turn led to the rad-air temperature as a room index (Chapters 6 and 7).[7]

Certain other related issues were addressed. The evaluation of diurnal temperature swings made clear that use of U, admittance and transmittance values (for periods of 24, 12, 8, etc., hours) did not afford a flexible method since the appropriate driving temperatures and fluxes had to be known in advance by preprocessing suitable hourly data: transfer coefficients would have allowed swings to be found for any hourly data. I later looked into alternative means of finding them (Chapter 17).[8]

[5]The smell certainly did exist. It was due to the serving of school dinners in the assembly room at the west end of the solar heated block and it drifted along the corridor to the classrooms. Visitors picked it up immediately since they normally entered the block through the assembly room, but it was not perceived as much of a problem by the staff and pupils.

[6]'Christopher Barrie Wilson: A man of many qualities', obituary by Edward Mathews in *Building and Environment* **30**(2), 155–156 (1995).

[7]It seemed to me from the time of a workshop on environmental temperature in 1969 that even though the formal argument could not be faulted as such, there was something amiss with the concept of environmental temperature to which it led. Environmental temperature purported to be an index to combine the radiant and convective components of room heat exchange so they could be handled as a single quantity. The argument was based on a cluster of misconceptions, but it took some time to articulate them.

[8]Wall 'response factors' are a series of numbers (units W/m^{2}K) which, when multiplied by say ambient temperature at a succession of hourly values, yield heat flows (W/m^2) into or out of the wall over the period

Other topics stem from courses given to students of building services engineering, whose needs have largely set the level of exposition. I have tried to ensure that the mathematical material remains simply a vehicle for the presentation of heat flow processes. Some knowledge of basic qualitative heat transfer is assumed. I have attempted a broadly uniform level of presentation, although important developments in the study of air movement both within a room and between adjoining rooms lie outside its scope. I have not ventured into aspects concerned with air conditioning and with control theory, or into practical aspects of construction, all of which were presented in other courses in the Liverpool department. These subjects are well covered by standard texts, handbooks and commercial literature. The topic of building heat transfer has not been so well served, (though Clarke (1985, 2001) puts material of this kind into a broader perspective.) Indeed, apart from studies of air movement and moisture transfer in walls, the main advances had been made by the 1970s; later work has consolidated and systematised its state. It has now reached a certain maturity. In 2000 the editor of the *Journal of Heat Transfer* (Howell 2000) reported the responses of some 36 heat transfer experts about where the field might be going in the twenty-first century. They threw up a wide variety of suggestions but no one mentioned a need for further experimental or numerical studies in building heat transfer as such: the fundamentals are in place, inviting application.[9] Although the journal literature is by now extensive, no book has appeared setting out its theory in a systematic way and I hope this monograph, primarily intended for building scientists and building services engineers, may provide a compact account, part survey (air and moisture

concerned. Their method of evaluation, set out in 1967 by Mitalas and Stephenson, starts with an expression for the response of a slab, initially at zero, and excited by a ramp in temperature applied to one surface. The expression, due to Churchill (1958), is quite simple in itself, but was an incidental product of a long and tortuous argument involving Laplace transforms, their inverse forms using contour integration, and complex quantities to evaluate residues at poles. Mitalas and Stephenson developed this approach to find the ramp response for a multilayer wall. Sometime in the 1970s I found that Churchill's expression for ramp response could be obtained in a straightforward way using elementary Fourier analysis, without the use of complex algebra. It entailed combining two solutions, one describing the response of the slab to a steadily increasing temperature and the other describing the form of transient or slump response when the slab surfaces were held at zero temperature. I set about using this approach to find the ramp response for a multilayer wall. I noted that a transmission matrix for a layer in a wall when the time variation had the form $\exp(-t/z)$ could be written as an expression closely similar to the well-known layer transmission matrix when excited by a signal varying in time as $\cos(\omega t)$. In both cases the overall properties of a multilayer wall could be found by multiplying the respective layer matrices. With suitable choice of isothermal and/or adiabatic boundary conditions and through a computer-based search, the transient response at some point within the wall could be found as sums of quantities whose variation with time was expressed as $\exp(-t/z_1)$, $\exp(-t/z_2)$, $\exp(-t/z_3)$, etc. ($z_1 > z_2 > z_3$), where z_1, etc., were a series of decay times or response times, similar in status to the time constant $z = RC$ of an elementary electrical RC circuit. The variation in time of the series of spacial sinusoidal temperature distributions through the wall associated with z_1, say, is independent of the variation associated with any other decay time (the Sturm–Liouville theorem, discovered in the 1830s but little applied in wall heat flows studies). A further form of transmission matrix was needed to find the response of a multilayer wall when excited by a steadily increasing temperature. With these means the ramp response of a wall could be found (hence the response factors) by elementary means and without the apparatus of the Laplace transform.

[9] Their comment simply accords with the fact that amid the huge mass of material appearing in heat transfer and similar journals in the 1990s, very little work was reported on building wall conduction, for example.

movement), part expository (radiative and conductive transfer), of the basis for current and future design methods.

Not long ago Doug Probert proposed the following word equation (Probert 1997):

$$\begin{pmatrix}\text{Overall benefit}\\ \text{achieved by}\\ \text{completing}\\ \text{the project}\end{pmatrix} = \underset{\text{(i)}}{\begin{pmatrix}\text{Worthwhileness of}\\ \text{its realistic and}\\ \text{achievable aims}\end{pmatrix}} \times \underset{\text{(ii)}}{\begin{pmatrix}\text{Effectiveness of planning}\\ \text{and of the means used to}\\ \text{achieve the aims}\end{pmatrix}}$$

$$\times \underset{\text{(iii)}}{\begin{pmatrix}\text{Effectiveness of}\\ \text{communicating the}\\ \text{worthwhile conclusions}\\ \text{of the project}\end{pmatrix}} \times \underset{\text{(iv)}}{\begin{pmatrix}\text{Effectiveness of}\\ \text{motivating those who}\\ \text{are capable of applying}\\ \text{the conclusions}\end{pmatrix}}$$

In this context, the overall benefit relates to provision of thermal comfort through efficient use of energy, which is just one consideration out of many in planning a building. The content of this book is part of item (ii) and so constitutes a part of a part of a part – a small element of the total enterprise. I hope that by sifting through a selection from a large literature, it will help to give that small element an ordered and coherent basis.

School of Architecture and Building Engineering
The University of Liverpool
February 2004

Acknowledgements

I have profited from discussions over the years with a large number of people, now beyond proper recollection. For more recent help of one kind or another I am indebted to Mr M.J. Barber, Dr G.C. Clifton, Professor J.A. Duffie, Dr I.W. Eames, Mrs J. Fazakerly, Professor B.M. Gibbs, Dr M. Gough, Mr I. Hunter, Mr J. Mitchell, Professor A.R. Millward, Dr J.F. Nicol, Professor J.K. Page, Dr F. Parand, Mr M. Reed, Dr K. Sedlbauer, Professor J.F. Spitler, Mr J.D. Tyson. Of the technical literature I have seen, I must single out my indebtedness to the book by Carslaw and Jaeger: my Chapter 12 is largely based on it. I am grateful to my former department for its support and provision of a range of facilities.

It may be that I have misunderstood or misleadingly or inaccurately reported work in this extensively studied field. I should be grateful if readers would pass on shortcomings of any kind or other comments to the publishers. I hope that it may prove possible to act upon them in some way.

The temperature in a building is the product of a number of factors: ambient air temperature and sunshine, the construction of the building, the casual heat inputs and ventilation rate due to occupation, and the heating or cooling from the plant. A 'passive' building largely or completely dispenses with the plant and the above school (at $53\frac{1}{2}^{\circ}$ N in Wallasey, near Liverpool, UK) is of this kind. It was designed in the late 1950s and was intended to function without a heating system; hot water radiators were installed but rarely used. The only mechanical control was a time-switch, set by the caretaker, which ensured that the lights (1100 W per classroom) could remain on when the school was unoccupied. The building was massively built to restrain daily variation in temperature due to the possibly large solar gains through the solar wall (mostly double glazed); the remainder of the envelope was covered with 125 mm of expanded polystyrene, a remarkable thickness for its time. High temperatures were avoided through use of simple ventilators and the cost of fuel (per child per year) proved to be less than for other schools in the area. I took part in a study of its thermal behaviour in the late 60s.

1

Elementary Steady-State Heat Transfer

From a thermal point of view a building can be looked upon as a modifier of the climate. For enjoyment and for many forms of manual and intellectual activity, a dry, draught-free and quiet space is needed with a much more restricted range of temperature than may be encountered out of doors. The fabric of the building serves as a filter or buffer generally and to this passive function is added that of its ability to impose a bias on ambient temperature through provision of heating or cooling. Much of this book will be taken up with internal temperature in relation to the fabric and supply of heat but it is convenient to present some brief and elementary account of the factors that lead to choice of room temperature, measures of severity of ambient temperature as it affects provision of comfort conditions, and the exchange of heat by ventilation and conduction between the internal and external environments: these factors determine the heating or cooling load.

1.1 HUMAN THERMAL COMFORT

Carefully devised heating appliances have been evident from early times: the Romans used warm-air heating in villas; one may note the flues built into the towers of the late thirteenth-century castles in Wales and the improvement to flue design urged by Count Rumford before 1800. (Rumford also installed a steam heating system in the Royal Institution in London.) The book by Roberts (1997) provides illustrations of heating, ventilating and refrigeration devices of earlier times. An article by Yunnie (1995) describes other early examples of climate control in the UK (including the system in St George's Hall, Liverpool, completed in 1845) and articles by Lewis (1995) and Greenberg (1995) give interesting accounts of the evolution of HVAC systems over 150 years in the United States.

Study of the relation between temperature, an objective measure, and perception of thermal comfort, a subjective measure, dates from the 1920s. There have been broadly two lines of enquiry: field observations and laboratory measurements. de Dear (1998) remarks: 'Both methods have their strengths and weaknesses. In the case of the climate chamber, precise measurements of, and control over, the main parameters in the comfort matrix are maximised. For example, the effects of inter-individual variations in clothing is typically eliminated by dressing subjects in a standard uniform, but the penalty

Building Heat Transfer Morris G. Davies
© 2004 John Wiley & Sons, Ltd ISBN: 0-470-84731-X

has been a reliance on small and possibly unrepresentative samples of human subjects, usually college students. Furthermore, the highly contrived setting of the climate chamber usually bears little resemblance to the complex environments within real buildings, raising doubts about the validity of generalizing from laboratory to "real" indoor environments. Field methods, on the other hand, involve real buildings, in use by large numbers of occupants going about their normal daily routine, and so retain the integrity of the person-environment relationship ... field study research ... seems most appropriate to the task of validating thermal comfort standards and models.' He goes on to describe a very large database of raw data on thermal comfort found from cross-sectional and longitudinal studies.

Here the response of the occupant is sought to temperature (or other physical variable); a good example is provided by the Bedford seven-point comfort scale (Bedford 1964: 94), where the subject in some location (e.g. a boot factory) is invited to say whether they are much too warm (coded as 1), too warm (2), comfortably warm (3), comfortable (4), comfortably cool (5), too cool (6) or much too cool (7). The wording was variable and a scoring scheme from +3 to −3 was also used.[1] A linear regression equation could then be evaluated, relating the ordinal measure with some measure of temperature; its success could be expressed through a correlation coefficient (values of around 0.5 were reported); additional variables (e.g. measures of humidity or air speed) could be included to improve the predictive value of the equation. This form of approach is still used; see for example Newsham and Tiller (1997).

In parallel with this empirical approach went enquiries into the physical and physiological mechanisms that might lead to perception of thermal comfort. This perception depends on control of deep body temperature, skin temperature and rate of loss of moisture by perspiration. The principal environmental factors affecting body heat loss are the room air temperature encountered, the radiant field the body intercepts, and the local humidity level and air speed. Heat loss is restrained by clothing, the amount and type of which may be determined by free choice or by custom. The body itself supplies two physiological mechanisms: vasoconstriction/vasodilation and perspiration or sweating. The involuntary aim of the mechanisms is to keep deep body temperature constant at around 36–37°C; at room temperatures lower than around 23°C most of the metabolically generated heat is lost by convection and radiation, but as the surrounding temperature increases, the loss by evaporation increases. In surroundings above 37°C the body *gains* heat by radiation or convection and sweating remains the only means of preventing a rise in deep body temperature. To counter conditions of extreme cold, the body has a further involuntary mechanism – shivering.

The interplay between the four physical heat flow mechanisms (convection, radiation, air speed and humidity), the voluntary choices of activity level and amount of clothing worn, together with the involuntary moisture loss mechanism, ensure that we can achieve thermal neutrality in a wide range of circumstances. Shivering is excluded since it is associated with discomfort. *Thermal comfort* itself has been defined as 'the condition of mind that expresses satisfaction with the thermal environment'.

These factors had been much studied since the 1920s (Bedford reviews them) but Fanger (1970) appears to have been the first worker to develop a comprehensive model to include them. Since then Fanger's model has been influential in the field of HVAC. It

[1]The ASHRAE wording is hot, warm, slightly warm, neutral, slightly cool, cool, cold.

is based on a steady-state continuity equation:

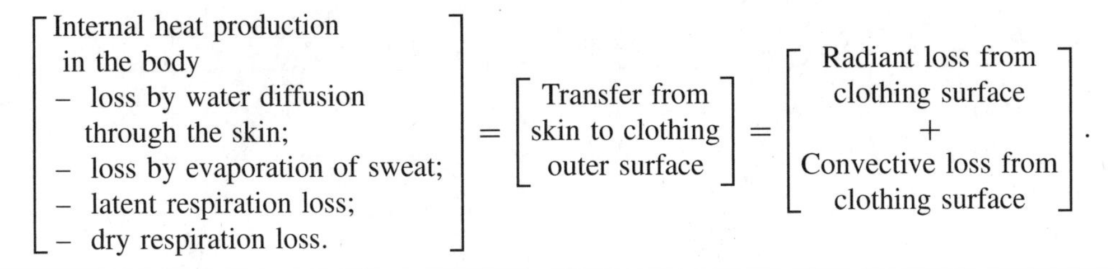
[Internal heat production in the body – loss by water diffusion through the skin; – loss by evaporation of sweat; – latent respiration loss; – dry respiration loss.] = [Transfer from skin to clothing outer surface] = [Radiant loss from clothing surface + Convective loss from clothing surface].

To this are added two empirical relations based on observations of subjects: mean skin temperature as a function of metabolic heat production per unit area (which decreases with increase in production) and evaporative heat loss from the body surface as a function of metabolic heat production; (the loss increases with metabolic rate). The pattern of exchanges can be represented approximately by a thermal circuit (Figure 1.1).

Fanger produced a series of charts indicating circumstances leading to thermal neutrality (Figure 1.2). The principal variables are air and radiant temperature. The parameters are the activity level and amount of clothing worn. A 'sedentary' activity describes for example a subject sitting quietly and corresponds to a metabolic heat output of about 60 W/m^2, 'medium activity' (120 W/m^2) corresponds to walking on the level at 3.2 kph and 'high activity' (175 W/m^2) corresponds to walking up a 5% slope at 3.2 kph. The thermal resistance of clothing is conventionally expressed in clo units and represents the resistance between the skin and the outer surface of the clothing. It is the resistance of the convective and radiative links between the skin and the clothing inner surface plus the resistance of the clothing itself. 1 clo $\equiv$ 0.155 m^2K/W. A light clothing ensemble of 0.5 clo might consist of long, lightweight trousers and an open-necked shirt with short sleeves. One clo unit is the resistance of a typical American business suit. The sloping solid lines in the figures relate to the surrounding air speed, however induced. The figures relate to a relative humidity (RH) of 50%.

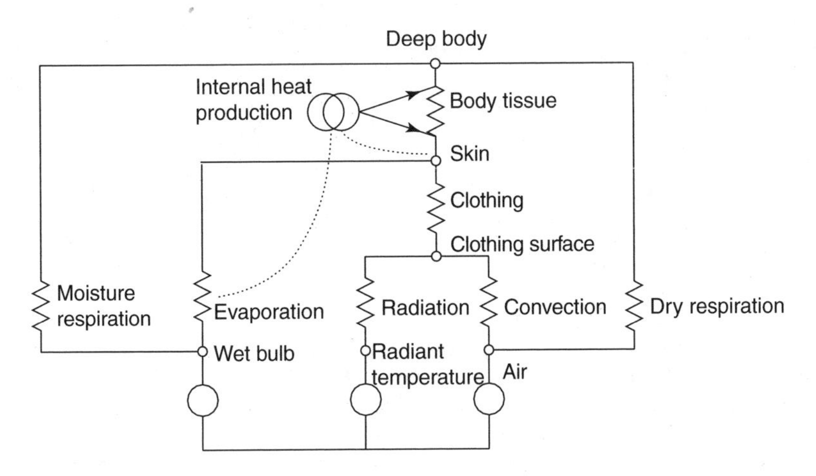

Figure 1.1 Simplified thermal circuit of Fanger's comfort model. The dotted lines indicate an imposed relation between the quantities linked

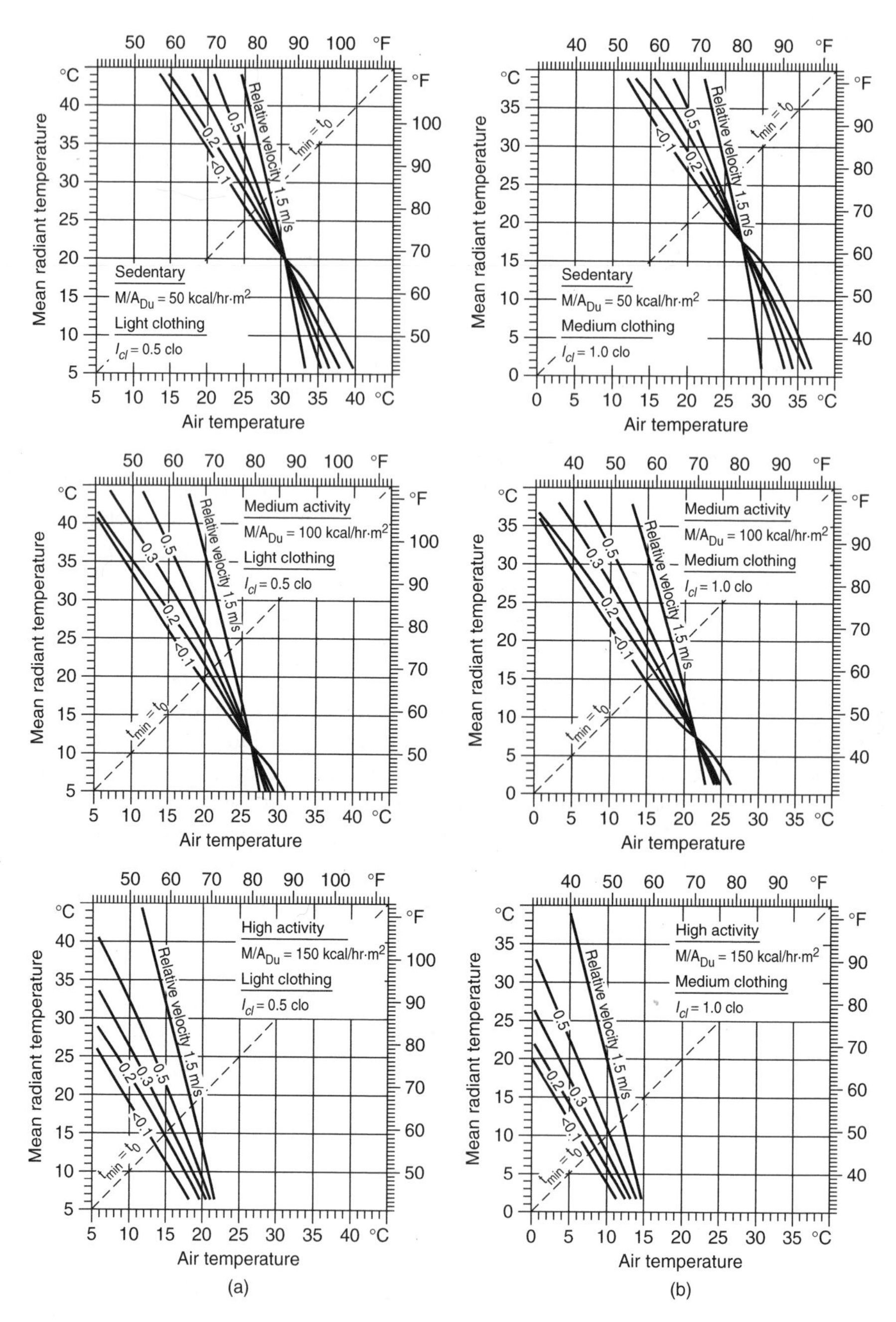

Figure 1.2 Comfort lines: air temperature versus mean radiant temperature with relative air velocity as parameter (a) for persons with light clothing ($I_{cl} = 0.5$ clo, $f_{cl} = 1.1$) and (b) for persons with medium clothing ($I_{cl} = 1.0$ clo, $f_{cl} = 1.15$). The relative humidity is 50% and plots are shown for three different activity levels (P.O. Fanger, *Thermal Comfort: Analysis and Applications in Environmental Engineering*, © 1972 McGraw-Hill, New York. Reproduced with permission of The McGraw-Hill Companies)

According to the Fanger model, if a lightly clothed subject engaged in medium activity is in a space at 50% RH and an airstream of 0.2 m/s, he/she will be in a thermally neutral state, neither wishing to be warmer or cooler anywhere between radiant temperature 35°C and air temperature 10°C and radiant temperature 11°C and air temperature 27°C. The dotted diagonal indicates the situation when radiant and air temperatures are equal. In this case a lightly clothed subject in still air conditions will be thermally neutral when *T* is about 26°C, 19°C and 13°C for sedentary, medium and high activity levels, respectively. With medium clothing the values are 23°C, 15°C and 7°C, respectively.

The results afford some justification for the preference of a temperature in the low 20s for office and domestic purposes where modest activity levels are the norm. When radiant and air temperatures are equal, RH = 50% and in still air conditions, the charts indicate the following values of temperature for thermal neutrality:

sedentary, light clothing	26°C	sedentary, heavy clothing	20°C	difference	6 K
high activity, light clothing	13°C	high activity, heavy clothing	2°C	difference	11 K

At a modest activity level in the 20s, one may be near thermal neutrality for a range of clothing levels; engaged in high activity, one's preferred temperature depends much more on the amount of clothing worn. The results also explain the common observation that while you may clothe yourself to withstand very low temperatures, you have to maintain an appropriate level of activity. If you pause, you will need extra clothing. A further factor favouring room temperatures in the 20s relates to manual dexterity; at low temperatures it becomes impaired and only coarser tasks may be undertaken. Gloves have limited utility.

Fanger's analysis is structured from the computational viewpoint. Skin temperature and moisture loss are not causally related to metabolic rate as Figure 1.1 suggests: all are controlled by the hypothalamus. The real independent variables are the activity level of the subject and the choice of clothing. Although the analysis has proved very valuable to building services engineers, it is a simplification of a very complicated process. See for example Griffiths and McIntyre (1974) and McIntyre (1980). We may sweat for reasons unrelated to thermal stress. The use of mean skin temperature obscures the fact that the face and hands are normally uncovered while other parts of the body may be under three layers of fabric. There is significant variation over the body surface in skin temperature and in the tendency to produce sweat. The ventilation action of clothing in motion makes it dubious to assess appropriate clo values.[2] The wide scatter of points relating body evaporative loss to metabolic rate is such that its representation as a straight line, while optimum, is not well founded. The same applies to the relation between skin temperature and metabolic rate, although there is less scatter. These empirical results were obtained from American college-age subjects. It is not clear whether they are also valid for an elderly population for whom matters of thermal comfort are more important.[3] Collins

[2]Ghali *et al.* (1995) have studied the modelling of heat and mass transfer in fabrics. There are four different forms of energy transport in the wicking of unevaporated sweat through a fabric that comes in contact with the skin. The main one is evaporation of the moisture to the atmosphere surrounding the fabric. Conduction, diffusion of moisture in the plane of the fabric, and convection of liquid in the plane of the fabric are less important.

[3]It is easy to find young subjects willing and able to participate in tests and surveys. It is much harder to gain access to elderly subjects and arrange that they undergo laboratory investigations.

and Hoinville (1980) describe a study to compare the thermal responses of elderly and young subjects with the following ages in years: 8 males 79 ± 7, 8 females 77 ± 6, 8 males 28 ± 5 and 8 females 27 ± 8. The elderly group wore clothing of about 1 clo and the young group wore clothing of around 0.87 clo. Subjects spent an initial period of 30 min at 22°C and then a further period of 2 h in a room controlled to 12, 15, 18, 21 and 24°C. Deep body temperature (urine temperature) decreased very little in the cool environment, but marked decreases occurred in mouth and ear temperatures, in the main with bigger falls for the elderly than for the young. Similar changes were noted for mean skin temperature, especially for the feet, but there was only a marginal difference for hand temperature. An ambient temperature of 21.1°C appeared to be satisfactory for old and young. The authors note how the vulnerability of elderly people in cold environments is essentially due to a lifestyle that involves a relatively low level of activity and an increased risk in the cold because of poor thermoregulatory responses and blunted perception of temperature changes.

Recent studies have taken account of much more physiological detail. The Fanger model (Figure 1.1) was based on single measures for skin and for clothing surfaces, but the model of Huizenga *et al.* (2001) includes a large range of detail. The body can be divided into an indefinitely large number of segments (typically 16). For each segment there are nodes corresponding to muscle, fat and skin temperature. Four types of link between the skin and the external environment are possible: bare skin in contact with solid surfaces (conduction loss), bare skin to the surrounding space (convective and radiant loss), clothed skin with conductive loss to solid surfaces, clothed skin with convective and radiant loss. Also included are mechanisms representing heat transfer by blood flow and the thermal and moisture capacity of clothing. Murakami *et al.* (2000) have used computational fluid dynamics and the k-ε turbulence model to examine sensible and latent heat transfer from the human body.

Analyses of thermal neutrality are based on a static model of thermal comfort: it views occupants as 'passive recipients of thermal stimuli driven by the physics of the body's thermal balance with its immediate environment and mediated by autonomic physiological responses' (de Dear and Brager 1998). It is taken to 'prescribe relatively constant indoor design temperatures... [which] have come to be regarded as universally applicable across all building types, climate zones and populations'. The model ignores the important cultural, climatic and social aspects of comfort and in treating the occupant as a passive element, it ignores his/her capacity to adapt to the environment. This includes behavioural adjustments – choice of clothing, local control of heating, cooling and ventilation, taking a siesta – psychological adaptation in that repeated exposure to a stimulus diminishes the response it evokes, and physiological adaptation whether intergenerational or by personal acclimatisation. Busch (1992) reported a study of comfort ratings of office workers in Bangkok, Thailand, in air-conditioned buildings and naturally ventilated buildings. According to widely adopted standards of thermal comfort, the upper limit for comfort is 26.1°C but Busch found values of 28°C for workers in cooled buildings and as much as 31°C for workers in naturally ventilated spaces. Imposition of American standards[4] results in a waste of energy.[5] See Santamouris and Wouters (1994). The adaptive approach to thermal comfort has been developed by Humphreys and Nicol (1998), also Nicol and

[4]See for example Chapter 8 of the 1993 *ASHRAE Handbook of Fundamentals*.

[5]Prins (1992) has written a provocative and highly critical discourse on the American rush to cooling. It is followed by a series of spirited rebuttals.

Humphreys (2002): *If a change occurs such as to produce discomfort, people react in ways which tend to restore their comfort.*[6] Humphreys and Nicol (2000: Figure 4) have summarised the results of empirical studies on how preferred comfort temperatures depend upon the mean outdoor temperature. For buildings that are heated or cooled, comfort temperatures lie in a band between about 18 and 24°C; for free-running buildings, numerous studies indicate preferred temperature of 26–29°C when ambient temperature is 25–29°C. Recognition of traditional preferences may lead to much reduced energy demand for comfort control.

The extensive literature on thermal adaptation in the built environment has been reviewed by Brager and de Dear (1998) and it is of interest to cite some of their conclusions:

> The adaptive approach to modeling thermal comfort acknowledges that thermal perception in 'real world' settings is influenced by the complexities of past thermal history, non-thermal factors and thermal expectations. Thermal adaptation in the built environment can be attributed to three different processes – behavioral adjustment, physiological acclimatization and psychological habituation or expectation. Evidence reviewed [in their article] indicates that the slower physiological process of acclimatization appears not to be so relevant to thermal adaptation in the relatively moderate conditions found in buildings, whereas behavioral adjustment and expectation have a much greater influence and should therefore be the focus of future research and development in this area.
>
> One of the most important findings from our review of field evidence was the distinction between thermal comfort responses in air-conditioned vs. naturally ventilated buildings. Analysis suggested that behavioral adaptation incorporated in conventional heat balance models could only partially explain these differences and that comfort was significantly influenced by people's expectations of the thermal environment. Occupants in naturally ventilated buildings had more relaxed expectations and were more tolerant of temperature swings, while also preferring temperatures that tracked the outdoor climatic trends. In contrast, occupants in closely controlled air-conditioned buildings had much more rigid expectations for a cool, uniform, thermal environment and were more sensitive to conditions that deviated from these constant setpoints.

Thus it appears that where cooling plant is used to achieve some value, it is desirable that it should indeed achieve it. Federspiel (1998) reports a survey of unsolicited complaints made by 23, 500 occupants in 690 commercial buildings in the US (millions of square metres area) and noted that the overwhelming majority of environmental complaints related to thermal sensation, mostly due to poor control performance and HVAC system faults. The neutral temperature was close to 23°C with minor variations for summer and

[6]Davies and Davies (1987VI) provide a compact illustration of adaptive response, found from the response of children in a passively solar-heated school to their thermal environment and expressed as correlation coefficients between various variables. There was a very high correlation between ambient and globe temperature, as indeed there must be in a passive building. There were moderately high correlations between globe temperature and the position of the windows, whether or not the lights were on (the lights were explicitly used for their heating potential), and the amount of clothing worn: the children had control over these factors. But the correlations between these control variables and the reported state of thermal comfort were low, showing that the controls had been used so as to reduce discomfort. Windows were more likely to be opened in warm weather, as is very clearly shown by Table 1 of Davies and Davies (1987VII).

winter, and for men and women. Countless studies have of course been conducted in which the respondent has been asked whether they are too warm, too cool, etc., in circumstances over which they had no control by way of personal action or by requesting some action. In this study, complaints were lodged and each complainant was assured that their complaint would be investigated. This might lead to a sharper sense of perception of the environment than passive respondents might have had.

1.2 AMBIENT TEMPERATURE

The heat loss from a building depends on ambient temperature, T_e. T_e can be recorded continuously but it is often reported as an average value over a period of an hour. T_e values over a period of months are needed to provide accurate (computer-based) estimates of the energy need for heating over the cold season or heating and cooling during warm or hot periods (Chapter 19). From the hourly values, other mean values may be derived such as daily mean and monthly mean values of T_e, which can be used to evaluate further measures that are suitable for manual use. Two will be presented briefly here.

1.2.1 Design Temperature

Consider the daily mean value $\overline{T}_e$ of T_e. Its frequency distribution over a year (365 values) is roughly binomial. A cumulative frequency distribution $\sum n(\overline{T}_e)$ starting from a value of T_e below the lowest value of $\overline{T}_e$ varies from 0 to 365 at the highest $\overline{T}_e$ value. In this way a value $\overline{T}'_e$ can be found below which $\overline{T}_e$ falls on some small number of occasions, say two per year. This is called a design temperature. The heating system is sized to cope with an ambient temperature of this value; the risk that it will not cope with the coldest day actually encountered is taken to be acceptable. Thus at some UK location, $\overline{T}_e$ may fall below -5.0°C just once per year on average and this serves as the design temperature for a lightweight building, one where the thermal capacity of the fabric is insufficient to prevent a rapid response to changing conditions. When the building is more massive, a less severe measure may be adopted based on the value of T_e averaged over a 48 h period, -3.5°C for the location concerned. Thanks to the Gulf Stream or, as has recently been suggested, the Rocky Mountains, the UK is much milder than most countries north of latitude 50°N. Much lower values than this are common.

1.2.2 Degree-Day Value

The heat requirement of a building may be taken as proportional to the difference between the comfort temperature T_c required and the external temperature T_e when lower than T_c. Assuming that a constant value of T_c is maintained, the total heat requirement over some period (a day, a month, the winter season) is proportional to $\int (T_c - T_e)\,\mathrm{d}t$ or $\sum (T_c - T_e)\,\delta t$, where δt is one hour or one day depending on the interval over which T_e is averaged. Some of the heat is supplied though casual gains – the sun, occupants, lighting and other equipment – and these gains are sufficient to maintain comfort conditions when T_e is above some value, say T_{base}. T_{base} rather than T_c serves as the temperature with which to estimate the plant output.

Neglecting the thermal capacity of the building and assuming steady conditions,

$$Q_{plant} + Q_{casual} = (T_c - T_e)K, \tag{1.1}$$

where K is the sum of the ventilation and conduction conductances (see below). The casual gains alone lead to an increment ΔT, equal to Q_{casual}/K. So

$$Q_{plant} = ((T_c - \Delta T) - T_e)K = (T_{base} - T_e)K, \tag{1.2a}$$

where

$$T_{base} = T_c - Q_{casual}/K. \tag{1.2b}$$

Karlsson *et al.* (2003) refer to T_{base} as the 'balance temperature' of a building. They are concerned with the solar contribution to Q_{casual}.

The heat load – the heat to be supplied by the heating system – is therefore proportional to $\sum(T_{base} - T_e)\,\delta t$, summed when positive. This is called the degree-hour or degree-day value for the location. It provides a compact means to summarise ambient temperature over a period of time as it relates to the need for heating in a building. Since Q_{casual} varies considerably from building to building, as does K, the value of T_c is a matter of choice; the value of T_{base} is arbitrary. A value of 15.5°C is taken in the UK, 18.3°C in the US and 18.0°C in parts of Europe. The degree-day value is accordingly

$$\mathrm{DD} = \sum(T_{base} - T_e)_+ \times (1\ \text{day}),\ \ \text{units K day}. \tag{1.3}$$

The subscript + denotes that only positive values are summed. Values for the heating season lie between about 1900 and 2900 K day in the UK and between 1000 and 5000 K day in the US. See for example Hitchin (1981). Thom (1954), Erbs *et al.* (1983), Hitchin (1983) and Schoenau and Kehrig (1990) provide means of converting values from to one base to another.

The quantity

$$\mathrm{DH} = \sum(T_{base} - T_e)_+ \times 1\ \text{hour}, \tag{1.4}$$

where T_e is the mean value of ambient temperature over a period of an hour, provides the most rigorous measure of severity, since little is gained through a finer time division. Waide and Norton (1995) discuss the degree-hour value as an index. DD is then simply DH/24. Degree-day values have been used since the 1930s. In the early days, data were most conveniently collected using a maximum-minimum thermometer to record T_e, reset daily, and DD values were evaluated from daily extremes of T_e rather than its continuous variation. For details of UK and US practice, see Day and Karayiannis (1998).

Degree-day values provide a satisfactory means of comparing temperature aspects of the severity of the weather on different sites; see Eto (1988). Hitchin (1990) has noted some possible improvement to their formulation but as noted below, they cannot give close estimates of the heat need for a particular building.

Attention should be drawn to the phenomenon of the urban heat island which is formed as urban areas expand and create their own climates. Air temperatures are higher than in the surrounding rural areas and this leads to increased cooling energy needs and accelerated smog formation in summer. See for example Meier (1997) and subsequent symposium papers.

See also the set of articles edited by Levermore (2002) which discuss the consequences of global warming for energy use in buildings as well as heat islands.

1.3 THE TRADITIONAL BUILDING HEATING MODEL

The total heat need in a room according to the traditional model is

$$Q = (T_i - T_e)\left(V + \sum AU\right). \tag{1.5}$$

T_i is the room index temperature, serving as the measure that drives the steady state heat loss to ambient T_e by the mechanisms of ventilation and conduction, and the temperature at which heat from the heating appliance and other sources is delivered. It also served as the measure of thermal comfort.[7] Since heat is input to a room and then distributed around the room by convection and radiation, two unlike mechanisms, this model provides a much simplified description of room internal heat transfer and the issue will be examined in more detail later on. It was, however, the main means of sizing heating plant up to about 1970 and may be expected to provide adequate estimates in simple situations. T_i and T_e values were discussed in Sections 1.1 and 1.2, respectively; we have to examine the ventilation and conduction loss terms, V and $\sum AU$.

The quantity $V + \sum AU$ is known as the heat loss factor or loss coefficient and is sometimes denoted by the single term *UA*. It is simple to measure: electric heaters are placed inside the building and room temperature is kept almost constant. Observations are made in stable conditions of ambient temperature and by night to avoid solar gains. *UA* is the ratio of heat input to temperature difference. Simmonds (1992) compares the details of its implementation in four European codes of practice.

1.3.1 Ventilation Loss

It is normally assumed that air at ambient temperature T_e enters a room, immediately becomes fully mixed with the room air and is lost again at room temperature T_i. The term 'natural ventilation' is often used to denote the exchange of air between the room and spaces external to it through architecturally designed openings such as open windows, vents and doorways. Infiltration is the uncontrolled movement of air through cracks of various kinds. Each is driven by a combination: wind forces bring about cross-ventilation due mainly to horizontal differences in pressure, and further flow may be generated by the thermal stack effect, which causes vertical differences in pressure. Forced ventilation implies an airflow driven by a fan, either simply installed in a wall or supplied through ductwork.

> Liddament (1998) has summarised the status of ventilation in buildings: Ventilation and air infiltration into buildings represents a substantial energy demand that can account for between 25–50% of a building's total space heating (or cooling) demand. As buildings become more thermally efficient, airborne energy loss

[7] T_i was an ad hoc index, not formally related to measurable temperatures although it was taken as corresponding roughly to the value found from a centrally placed thermometer. Figure 7.3 shows the relation between a formally defined index temperature and observed values.

is expected to become the dominant thermal transport mechanism. Unnecessary or excessive air change, therefore, can have an important impact on global energy loss. On the other hand, insufficient ventilation may result in poor indoor air quality and consequential health problems. Designing for optimum ventilation is therefore a vital part of building design to ensure energy efficiency and a healthy indoor environment. This task is made especially difficult, however, by the complexities of air flow behaviour, climatic influences, occupant characteristics and pollutant emission characteristics.

If v is the volume flow into the room (m^3/s) and s the volumetric specific heat of air (about 1200 J/m^3K), the difference in internal energy, $vs(T_i - T_e)$ must equal the heat gain Q_c to the air. Thus, ignoring the small decrease in density, the ventilation loss conductance V (W/K) is given as

$$V = Q_c/(T_i - T_e) = vs. \tag{1.6a}$$

It is common to express a required ventilation rate in terms of the supposed number of complete air changes per hour, so

$$\begin{aligned} V &= ((\text{number } N \text{ of air changes per hour}/3600)[\text{s}^{-1}]) \times (\text{room volume } V_r[\text{m}^3])(1200[\text{J/m}^3\text{K}]) \\ &= \tfrac{1}{3} N V_r [\text{W/K}]. \end{aligned} \tag{1.6b}$$

Recommended values of N for many classes of room lie between 1/2 and 1 volume air change per hour (CIBSE 1999: Table A4.10), although this can lead to excessive values for large spaces.

T_i is an ill-defined quantity and will be replaced later by T_a or T_{av} (6.55), the air temperature averaged over three-dimensional space and so a 3D construct. Strictly speaking, it should be written $T_{a,exit}$, which is the mean temperature of the air over the cross-sectional area of the duct or other opening through which the air leaves, and is therefore a 2D construct. The two are the same if the air is fully mixed but they may differ if there is significant short-circuiting between the points of entry and exit of the airflow.

1.3.2 Conduction Loss

The convective exchange between air and a solid surface is described by its convective heat transfer coefficient h_c, which the traditional model takes to be about 3 W/m^2K. The radiative exchange per unit area of a room surface, such as the floor, emissivity ε and the enveloping surfaces, supposedly black body, is εh_r, equal to about $0.9 \times 5.7 \approx 5$ W/m^2K.[8] The model merges these values to give an internal film coefficient of

[8]Most building materials have an emissivity ε of around 0.9. Dust collection, moisture condensation and corrosion lead to this value even though a clean new surface may have a lower value. But as noted by Goss and Miller (1989), it has been known since the 1930s that aluminium retains a high reflectivity (low emissivity) for radiant heat transfer due to a protective layer of transparent oxide. For the radiant exchange between surfaces, see equation (6.53).

$h_i = h_c + \varepsilon h_r \approx 8\,\mathrm{W/m^2K}$. Thus the heat flow from T_i to a surface bounding the room of area A and at T_n is

$$Q = (T_i - T_n)Ah_i, \tag{1.7}$$

where n denotes the nth layer in the wall, counting from outside. Similarly, the loss of heat by convection and radiation to ambient is

$$Q = (T_0 - T_e)Ah_e. \tag{1.8}$$

Subscript 0 denotes the interface between layers 0 and 1, where layer 0 here is the outer film and layer 1 is the outermost layer of the wall. Like h_i, the outer film coefficient has radiative and convective components but h_e is largely determined by the forced convection due to wind speed and is very variable; a value of $h_e = 18\,\mathrm{W/m^2K}$ is often assumed.

The one-dimensional heat flow by conduction through a slab of thickness X_1 and conductivity λ_1 and face temperatures T_0 and T_1 is

$$Q = (T_1 - T_0)A\lambda_1/X_1. \tag{1.9}$$

In steady-state conditions, the flow from inside at T_i through two such layers to ambient at T_e is

$$Q/A = \underset{\text{outer film}}{(T_0 - T_e)h_e} = \underset{\text{layer 1}}{(T_1 - T_0)\lambda_1/X_1} = \underset{\text{layer 2}}{(T_2 - T_1)\lambda_2/X_2} = \underset{\text{inner film}}{(T_i - T_2)h_i} \tag{1.10a}$$

$$= \frac{T_0 - T_e}{1/h_e} = \frac{T_1 - T_0}{X_1/\lambda_1} = \frac{T_2 - T_1}{X_2/\lambda_2} = \frac{T_i - T_2}{1/h_i} \tag{1.10b}$$

$$= \frac{T_i - T_e}{1/h_e + X_1/\lambda_1 + X_2/\lambda_2 + 1/h_i} \tag{1.10c}$$

$$= \frac{\text{temperature difference}}{\text{sum of the thermal resistances}}. \tag{1.10d}$$

The U value[9] or thermal transmittance of the wall is defined as

$$U = \frac{\text{heat flow per unit area in steady conditions}}{\text{temperature difference}} = \frac{Q/A}{T_i - T_e}, \tag{1.11}$$

so

$$1/U = 1/h_e + X_1/\lambda_1 + X_2/\lambda_2 + 1/h_i = \sum(\text{thermal resistances}). \tag{1.12}$$

If the wall includes a cavity, its resistance (around $0.18\,\mathrm{m^2K/W}$) must be included.

The overall behaviour of the wall can also be found by multiplication of the separate layer transmission matrices. Consider the flow through layer 1, the wall outer layer. Taking

[9] U is also called the U factor. The performance of the wall is also described by its resistance $R = 1/U$, with units $\mathrm{m^2K/W}$ or $\mathrm{h\,ft^2\,{}^\circ F/Btu}$. This has the merit that a high value of R denotes a well-insulated wall. The designation R-3, for example, denotes a resistance of $3\,\mathrm{h\,ft^2\,{}^\circ F/Btu}$ or $0.53\,\mathrm{m^2K/W}$.

T_e to be situated to the left of T_i and T_i to be higher than T_e, the heat flow q_0 at the left surface is $(T_1 - T_0)\lambda_1/X_1$ and in the negative x direction, so

$$-q_0 = (T_1 - T_0)\lambda_1/X_1 \quad \text{and} \quad q_1 = q_0. \tag{1.13a}$$

In matrix form this is

$$\begin{bmatrix} T_0 \\ q_0 \end{bmatrix} = \begin{bmatrix} 1 & X_1/\lambda_1 \\ 0 & 1 \end{bmatrix} \begin{bmatrix} T_i \\ q_i \end{bmatrix}. \tag{1.13b}$$

Since temperature and heat flux are continuous across the interface between two layers, $[\,T_1 \quad q_1\,]^T$ is given by a similar matrix involving the resistance X_2/λ_2, and similarly for the outside and inside films. Thus we can write

$$\begin{bmatrix} T_e \\ q_e \end{bmatrix} = \underbrace{\begin{bmatrix} 1 & 1/h_e \\ 0 & 1 \end{bmatrix}}_{\text{outer film}} \underbrace{\begin{bmatrix} 1 & X_1/\lambda_1 \\ 0 & 1 \end{bmatrix}}_{\text{layer 1}} \underbrace{\begin{bmatrix} 1 & X_2/\lambda_2 \\ 0 & 1 \end{bmatrix}}_{\text{layer 2}} \underbrace{\begin{bmatrix} 1 & 1/h_i \\ 0 & 1 \end{bmatrix}}_{\text{inner film}} \begin{bmatrix} T_i \\ q_i \end{bmatrix} \tag{1.14a}$$

$$= \begin{bmatrix} 1 & 1/h_e + X_1/\lambda_1 + X_2/\lambda_2 + 1/h_i \\ 0 & 1 \end{bmatrix} \begin{bmatrix} T_i \\ q_i \end{bmatrix}. \tag{1.14b}$$

The only significant term in the product matrix is the sum of the resistances, hence $U = q/(T_i - T_e)$ as before. Both these methods of arriving at U are trivial, but the matrix approach becomes essential in time-varying conditions when we must also take account of the thermal capacity of solid layers and all the elements become significant.

In the calculation it is assumed that the conductivity in some layer is constant and the temperature gradient is then uniform. However, λ values of some building materials increase with moisture content and in masonry materials λ may increase toward the exterior surface, either because moisture diffuses to cooler places or through wetting by rain. In this case the gradient decreases toward the outer surface.

A sheet of glass is so thin that its thermal resistance is negligible[10] and the U value for a window depends on the films alone. For single glazing (and any very thin wall), $U \approx (1/8 + 1/18)^{-1} = 5.5\,\mathrm{W/m^2K}$; for double glazing, $U \approx (1/8 + 0.18 + 1/18)^{-1} = 2.8\,\mathrm{W/m^2K}$. A value of $0.18\,\mathrm{m^2K/W}$ is usually taken for the resistance of an air cavity. Argon, krypton and xenon can replace air, and by using multiple glazing and low-emissivity coatings, transmission coefficients down to $0.5\,\mathrm{W/m^2K}$ can be achieved; see Muneer and Han (1996).

Bricks and blocks are sometimes provided with slots arranged in various ways which increase their face-to-face resistance and improve the thermal insulation they provide. Anderson (1981) shows how the resulting two-dimensional flow pattern can be analysed.

The ordering of the layers, in particular the position at which insulation is placed in the wall, does not affect the steady-state transmittance.[11] However, it becomes relevant for the dynamic behaviour of the wall: the combination of insulation inside/mass outside

[10] The resistance of 6 mm glass is $X/\lambda = 0.006/1.05 = 0.006\,\mathrm{m^2K/W}$, negligible compared with the inner film resistance of $0.12\,\mathrm{m^2K/W}$.

[11] At corners, a given thickness of insulation is most effective to reduce heat loss when placed inside, but then the structure is colder than a plane wall, with the risk of freezing.

results in a rapid response to heat input. This may be the desired outcome, but if the input is due to solar gain, it may lead to high room temperatures or a large cooling plant to restrain them. With mass inside/insulation outside the room is thermally more stable and solar gain may contribute usefully to the heat need, but the space may then require a large heat input to reach a comfortable temperature in reasonable time if the space has previously been unheated in cool conditions. Furthermore, the arrangement with insulation inside/mass outside may lead to interstitial condensation in the predominantly cold external structure. Sonderegger (1977) reaches these conclusions using the method of harmonic analysis presented in Chapter 15. Boji'c and Loveday (1997) describe a study comparing two wall constructions with the same U value, one of form masonry, insulation, masonry (MIM) and the other of form insulation, masonry, insulation (IMI). They confirm that if the building is to be intermittently heated, the IMI form is better but for intermittent cooling, the MIM form is better. For continuous cooling, the structure does not matter. The differences in energy needed are of order 30%.

Although much of this book is devoted to a study of wall behaviour in non-steady conditions, the simple U value or U factor of a wall remains its most important thermal descriptor. Methods to find the transmittance of building elements composed of bridged layers are given in Section 3.3.11 in Book A of the 1999 *CIBSE Guide*. Maximum permitted values are specified by the building regulations in many countries. Following the increased awareness in the 1970s of the amount of energy needed to heat and cool buildings, maximum permitted values have been progressively reduced, especially in Scandinavia.[12] By incorporating 300 mm of rock wool insulation, U values of around 0.1 W/m^{2}K are reached; a value of 0.09 W/m^{2}K has been reported in Finland.

A simple expression allows us to estimate the thickness X of insulation, conductivity λ, that might on economic grounds be added to a wall of basic U value U_0. X will increase with

- F, the cost of fuel, \$/J say;
- N degree-days per year, a measure of the severity of the climate;
- f, the proportion of the 24 h period during which comfort conditions are to be maintained.

X will decrease with

- P, the interest rate on the capital borrowed to purchase the insulating material;
- z, the cost of the insulating material, \$/m^3.

The optimal value of X is $(FNf\lambda/(Pz))^{1/2} - \lambda/U_0$. In effect, the optimal wall resistance is $(FNf/(Pz'))^{1/2}$, where z' is the cost of insulating material expressed as \$/m^2 per unit of added resistance. A closely similar expression is given in equation 13 of Hasan (1999). One would suppose that to conserve energy, a 'hot' surface requires thicker insulation than a 'cool' surface. Bejan's initial analysis (Bejan 1993) does not support this view. He considers insulating a surface whose temperature varies linearly from ambient to some high value using a certain fixed volume of insulation. The loss turns out to be the same when the insulation is applied uniformly and when its thickness is proportional to the

[12] Values for 2001 in the UK are walls 0.35, roofs 0.20, floors 0.30, glazing 2.20 W/m^{2}K.

temperature difference but Bejan's argument supposes that the insulation provides the only resistance to heat loss; it does not consider the outside film resistance.

Wall insulation can be viewed from a strictly economic standpoint: the saving in running costs. It can also be seen in relation to environmental pollution: the saving of running costs is concomitant with reduced pollution but the manufacture of insulating material together with its transport and installation entails increased pollution. Erlandsson *et al.* (1997) have made a life-cycle assessment for additional external wall insulation for Scandinavia; for economy, insulation thicknesses between 100 and 170 mm are appropriate but environmental considerations favour the greater thickness.

In steady-state conditions, the temperature gradient dT/dx through any one layer is constant but differs from layer to layer. If however temperature is plotted as a function of progressive resistance, the gradient $dT/d(x/\lambda)$ is uniform through the wall and the construction can be extended to include the surface films. (Strictly speaking, a profile cannot be traced through a film; the part associated with convection is unchanged in the bulk air and only changes within the boundary layer. The radiant component however cannot be displayed in this way.) In unsteady conditions, the temperature profile in any layer is curved, but when plotted against resistance, the gradient at an interface remains continuous.

Much work over a long period has been devoted to find experimental U values for a large range of wall types. This lies outside the scope of the book but observational values of wall and roof U are usually higher (i.e. worse) than the values computed from assumed h, λ and X values would suggest. Siviour (1982) reports that measurement of heat flow through a wall insulated with urea-formaldehyde corresponded to a U value of 0.65 W/m^{2}K while the calculated value was about 0.5 W/m^{2}K. Reasons for this include higher values of λ and h in practice than tabulated ones (since λ depends strongly on moisture content), evaporation of rain, thermal bridging due to wall ties or debris lodged in the cavity, ventilation of the cavity, thermal bridging at window frames and additional losses at corners. Errors may be made in the measurement of temperature itself; Bénard *et al.* (1990) report a detailed study of possible errors in measurement of surface temperature by a thermocouple.

Because of these factors and possible omission of insulation, one might suppose that the observed heat loss coefficient $UA = V + \sum AU$ should be larger than its calculated value, but this is not necessarily the case. Liu and Claridge (1995) summarise studies from the 1980s onwards which showed that the calculated value of UA could be double its observed value. They attributed this to neglect of air infiltration heat recovery and neglect of the heat discharge from thermal storage during the night.

Heat loss from a solid floor cast on earth is a three-dimensional flow problem which is much more complicated than that for simple wall losses. An estimate has been provided in the past by Macey's formula (Macey 1949). Consider a solid floor of length L (the major dimension) and breadth B, surrounded by a solid wall of thickness W (so that the external breadth is $B + 2W$). If the floor surface and surrounding land surfaces have temperatures T_{in} and T_{out}, the steady state flow outward is $(T_{in} - T_{out})UBL$, where

$$U = \frac{4\lambda}{\pi B} \operatorname{arctanh}\left(\frac{B}{B + W}\right) \exp\left(\frac{B}{2L}\right) \tag{1.15}$$

and λ is the soil conductivity. U is a *surface-to-surface* conductance, not an *index-to-index* conductance as usually defined. Although not misleading, Macey's expression has some logical flaws (see later).

Transparent insulating materials act as thermal insulation but simultaneously permit the transmission of solar energy. See Braun *et al.* (1992) and other articles in this issue. Wood and Jesch (1993) present a detailed account of transparent insulating materials. Affixed to the exterior of a massive wall with a dark exterior surface, a transparent insulating material acts as an insulator in the usual way but allows incident radiation to be absorbed by the wall. Most of the energy is transmitted with some phase lag to the space behind the wall. An analysis of the mechanism is somewhat like that indicated in Figure 9.14 (although these materials have optical properties which depend on the angle of incidence and on temperature.) Gorgolewski (1996) reports that in the Scottish climate (latitudes above about 55°N) a south-facing wall of this kind can reduce the annual heating load by 200 kWh/m^2. The material requires external protection and a movable blind must be supplied to prevent excessive gain in summer.

Hens (Hens and Fatin 1995) has listed a number of checks that relate to the performance of a cavity wall. Of these, the U value and thermal bridging have already been mentioned. Steady-state aspects of hygric stress and moisture balance are discussed in Chapter 8 with some mention of dynamic effects in Chapter 17 and Chapters 15 to 19 deal with unsteady heat conduction. Hens also discusses the permeance of a wall to airflow due to a pressure difference Δp_a (Pa) across it. He cites values such as $2.5 \times 10^{-3} \Delta p_a{}^{-0.5}$ kg/(m^{2}s Pa) and $10^{-6} \Delta p_a{}^{-0.28}$ kg/(m^{2}s Pa) but this consideration lies outside the scope of the book.

1.3.3 Loss from a Cylinder

If inside and outside temperatures remain constant, the effect of adding a layer of material to a plane wall is to reduce the heat loss. This is not necessarily the case if a layer of material is added to a cylinder at a fixed temperature. Consider a cylinder of radius R_0 and length L at temperature T_0. It loses heat Q (W) to the surroundings at T_2. The loss is $2\pi R_0 Lh\ (T_0 - T_2)$ where h (W/m^{2}K) is the combined convective/radiative film coefficient. Suppose now that a layer of material of conductivity λ is added to form a cylinder of radius R_1 with temperature T_1. From (3.17) and continuity of heat flow,

$$\frac{Q}{L} = \frac{2\pi\lambda(T_0 - T_1)}{\ln(R_1/R_0)} = 2\pi R_1 h(T_1 - T_2). \qquad (1.16)$$

It follows that

$$\frac{R_1 h/\lambda}{2\pi(T_0 - T_2)L}\left(\ln\frac{R_1}{R_0} + \frac{\lambda}{R_1 h}\right)^2 \frac{dQ}{dR_1} = h - \frac{\lambda}{R_1}. \qquad (1.17)$$

The expression is valid if $R_1 = R_0$. It shows that if material is added to the cylinder, it will lead to an *increase* in heat loss if $h > \lambda/R_0$. $R_0 h/\lambda$ has the form of a Biot number B (Chapter 13).[13] The heat loss is maximised when the perimeter of the insulation is $2\pi\lambda/h$ and Hsieh and Yang (1984) have shown that this is true too for a square section.

[13]Note that both the Biot number B and the Nusselt number Nu have the form, Xh/λ, where X is some characteristic length, h is a film coefficient (W/m^{2}K) and λ is conductivity (W/m K). The Biot number features in conduction-dominated problems; X is a layer thickness aligned in the direction of heat flow and λ is the conductivity of the layer material. The Nusselt number is used in convection-dominated problems; X can be chosen to be parallel or perpendicular to the flow direction and λ is the conductivity of the fluid.

1.4 SEASONAL HEAT NEED

By assuming some value for the hourly air change rate and summing over the various wall, window and roof elements of a building, the term $V + \sum AU$ can be found. According to the simple model, the energy to be supplied by the heating plant is

$$Q_{plant} = (T_{base} - T_e)\left(V + \sum AU\right), \tag{1.18}$$

so the energy need (J) is over some fixed period, a month say:

$$E = \mathrm{DD} \times 86\,400\left(V + \sum AU\right), \tag{1.19}$$

where DD is the degree-day value (equation 1.3) for the site for the relevant period and 86 400 is the number of seconds in a day; to convert to kilowatt-hours divide by 3.6×10^6.

This quantity is easy to evaluate but it represents a simplified approach to the problem of energy supply. The incremental temperature ΔT at some time t in fact depends on time-varying heat inputs. Further, the ventilation rate may well be higher by day than by night and when the internal temperature varies, the conduction loss conductance L (having the value $\sum AU$ in steady conditions but now including conduction into all bounding surfaces) in effect becomes a varying quantity. We have to write

$$\Delta T(t) = \frac{Q_{solar}(t) + Q_{occupants}(t) + Q_{lighting}(t) + Q_{equipment}(t)}{V(t) + L(t)}. \tag{1.20}$$

These elements have in varying degrees steady, cyclic and transient components and the sequence of values of ΔT at hourly intervals may be expected to show large variation and to differ from the sequence in a nearby room or building. Clearly, a degree-hour value

$$\mathrm{DH} = \sum (T_c - \Delta T(t) - T_e(t)) \times 1\ \text{hour}, \tag{1.21}$$

in which $T_e(t)$ too takes on hourly values, provides a coarse measure to estimate seasonal energy needs. Some indication of reliability is provided in a study by Day and Karayiannis (1999). They considered a particular model building with specified thermal capacity, fabric conductance, glazed area, infiltration rate, occupancy, casual gains and ten year weather data and they used advanced means to find its energy needs. With this as the 'truth' value, they found values based on various simplifications. They took a fixed value for the inside temperature T_i (the setpoint or T_c value), and also hourly values of T_i and its daily and monthly mean values. Similarly, hourly plus daily and monthly averaged values were taken for the casual gains. These assumptions lead to a series of base temperatures of form

$$T_{base} = T_i - Q_{casual}/\left(V + \sum AU\right). \tag{1.22}$$

See equation 1.2b. According to the definition of T_{base}, DD values ranged from 1117 to 2090 K days. The worst energy estimate was found with a combination of a fixed setpoint value and hourly gains, when the energy need was some 90% larger than its true value.

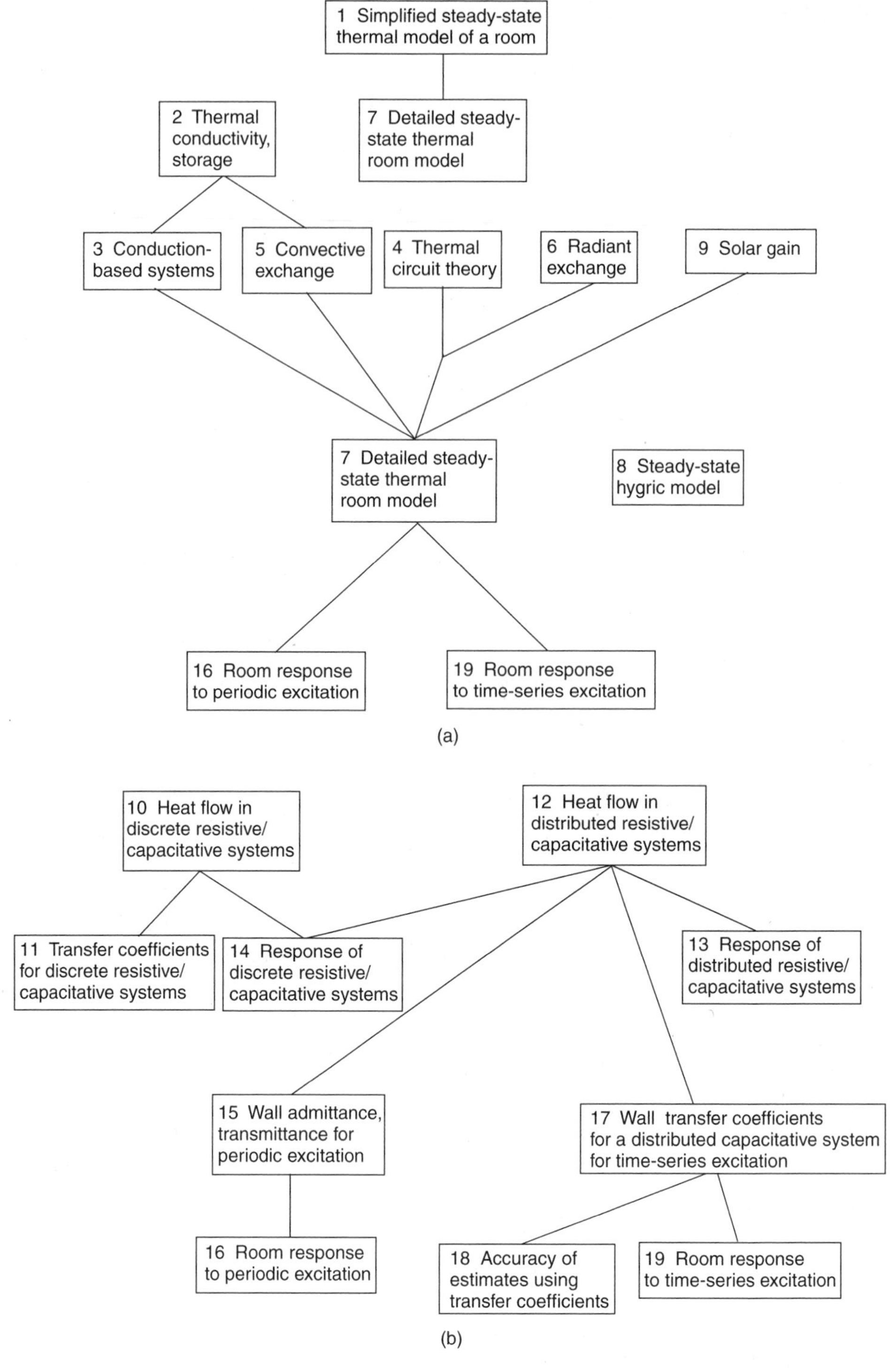

Figure 1.3 Plan of the book: (a) basic mechanisms and (b) wall conduction/storage and room thermal models

With actual values for T_i (rather than a supposed value T_c) and averaged over a day or a month, with similarly averaged values for the gains, the overestimate was reduced to less than 4%. The authors' worst case is based on hourly gains, and not even this information may be available to the building designer at an early stage of design. Their results therefore suggest that a seasonal energy estimate for a building based simply on its loss coefficient and the local degree-day value – the only information the designer may have early on – may be seriously in error. Better estimates involve more effort. The approach using transfer coefficients is given in Chapter 19.

1.5 PLAN OF THE BOOK

The principal question for building heat transfer studies is to find the amount of heating or cooling that the plant must deliver to maintain some specified level of temperature (and humidity), or the daily profile of temperature in a space if it responds to heat inputs in an uncontrolled manner, or perhaps some combination of fixed and floating conditions. Other texts, handbooks and technical literature cover details of plant design, operation and control, so the means of heating or cooling will be assumed without further discussion. The exception is the heating effect of the sun, discussed in Chapter 9. The theory of convection and conduction uses the conductivity λ, the density ρ and the mass specific heat at constant pressure c_p of the materials concerned. For gases these quantities can be found semiquantitatively by elementary kinetic theory, and this is outlined in Chapter 2. Chapter 3 considers the conduction-dominated three-dimensional heat flow from a floor slab. Convective and radiative studies can be treated separately (Chapters 5 and 6), and in combination they lead to a more detailed model for steady-state room heat transfer (Chapter 7). They are combined using some thermal circuit theory given in Chapter 4. The room elementary humidity model (Chapter 8) is formally similar to the thermal model.

In unsteady heat flow we must take account of the storage potential of wall materials in addition to their conductivities and this considerably complicates the calculation of wall response. A useful simplification is to suppose that the continuous distribution of storage and resistance can be represented as localised or discretised elements and Chapters 10 and 11 show how the conventional dynamic wall parameters, developments of the simple U value, can be found. Chapter 12 presents various forms of solution to the Fourier continuity equation then uses them to find the dynamic parameters when storage and resistance are considered as distributed properties, eventually to be used in Chapters 15 and 17. Before that, some classical solutions are presented for cases where a wall or a room is subjected to a step temperature excitation (Chapter 13) and for some room models so simplified that their thermal capacity is represented as one, two or possibly three lumped capacities (Chapter 14). Room models using period-based parameters are discussed in Chapter 16 and models using time-series parameters are covered in Chapter 19. The plan is illustrated in Figure 1.3.

Chapter 1 of Clarke (2001) gives a detailed justification for a study of these processes in the context of energy flow simulation in a building.

2

Physical Constants of Materials

A flow of heat is associated with a temperature difference (sensible heat flow) or a change of phase (latent heat flow). In steady conduction in a homogeneous medium, the flow from an isothermal area 1 to an isothermal area 2 depends on the intervening geometry, a single physical constant called the conductivity λ, and the difference in temperature $T_1 - T_2$. In time-varying conditions, the flow depends on two further constants: the density ρ and specific heat at constant pressure c_p, but only as their product ρc_p. For forced convection from area 1 to a fluid at temperature T_2, we need to consider one more property, the viscosity μ; the heat flow is proportional to $T_1 - T_2$. When the fluid movement is driven by a temperature difference, that is by buoyancy, the driving effect is represented by βg, where β is the coefficient of expansion of the fluid and g is the acceleration due to gravity. Thus the heat flow depends on the fluid properties, λ, ρ, c_p, μ and β and is now proportional to $(T_1 - T_2)^\alpha$ where α is a little larger than unity.

Room radiant exchange is more complicated, principally because it involves exchange between several surfaces forming a cavity. The heat flows are proportional to differences in the fourth power of the absolute temperatures of the surfaces and the relevant material property is the emissivity ε of each surface. Emissivity is a function of the atomic structure of the surface material and appropriate values will simply be assumed.

For practical calculations, appropriate observed values for thermal parameters would normally be used but the basic kinetic theory of gases provides a statistical model from which simple expressions for λ, ρ, c_p, μ and β can be derived for air. The values deviate somewhat from observational values, but the theory shows how they are interrelated and it will be sketched here. Parameter values will be assumed for liquids, as will the values of λ and ρc_p for solids.

2.1 THERMAL PARAMETERS FOR GASES: KINETIC THEORY

The principal laws governing the bulk behaviour of gases at low pressures were already determined by the beginning the nineteenth century:

- *Boyle's law*: the product of pressure p and volume V is constant when the temperature T is constant.

Building Heat Transfer Morris G. Davies
© 2004 John Wiley & Sons, Ltd ISBN: 0-470-84731-X

- *Charles' Law*: the product pV is proportional to temperature.
- *Dalton's law*: the total pressure of a mixture of gases is the sum of their individual pressures.
- *Gay-Lussac's law*: when two gases react chemically to produce a third gas (e.g. hydrogen and oxygen combining to form water), the volumes of the reacting gases and their (gaseous) product bear simple relations to each other when reduced to the same temperature and pressure. (Two volumes of hydrogen combine with one volume of oxygen to form two volumes of water vapour.)

The fact that these simple laws represent the behaviour of all gases at low pressures made it clear that gases have a common and simple structure. The kinetic theory gradually developed later in the century to explain these findings and to provide an explanation for parameters such as the conductivity and viscosity of a gas. A full and interesting account of its evolution is given by Brush (1976).[1]

The elementary theory assumes that a gas consists of spherical molecules that occupy a very small volume in comparison with the bulk volume and that are in random motion with a distribution of velocities. They make perfectly elastic collisions with each other and with surfaces they encounter. Except at impact, they exert no forces on each other.

To derive an expression for the pressure of the gas contained in a vessel of volume V, suppose there are n_1 particles in unit volume each of mass m and with velocity components u_1, v_1, w_1 parallel to the x, y, z axes. Then $n_1 u_1 \, \mathrm{d}S \, \mathrm{d}t$ particles pass through some area $\mathrm{d}S$ normal to the x axis during the interval $\mathrm{d}t$, each having a momentum mu_1, so the packet's momentum is $mn_1 u_1^2 \, \mathrm{d}S \, \mathrm{d}t$. Summing over all positively directed velocities, the total packet of momentum is $\frac{1}{2}m(n/V)\overline{u^2} \, \mathrm{d}S \, \mathrm{d}t$, where

$$\overline{u^2} = (n_1 u_1^2 + n_2 {u_2}^2 + \cdots)/(n_1 + n_2 + \cdots) \tag{2.1a}$$

and

$$n = (n_1 + n_2 + \cdots)V. \tag{2.1b}$$

The mean square velocity $c_{ms}^2 = \overline{u^2} + \overline{v^2} + \overline{w^2}$ and equals $3\overline{u^2}$ if there is no drift velocity, so $\overline{u^2} = \frac{1}{3}c_{ms}^2$.

Now if $\mathrm{d}S$ forms part of the surface of the containing vessel, the particles rebound from it in the negative direction and the change in momentum has twice this value. It corresponds to an impulse of $p \, \mathrm{d}S \, \mathrm{d}t$. Thus

$$pV = \tfrac{1}{3}mnc_{ms}^2 \quad \text{or} \quad \tfrac{2}{3} \times \tfrac{1}{2}mnc_{ms}^2 \tag{2.2}$$

The expression was first derived by Daniel Bernoulli in 1738. Now particle velocity, a microstate concept, corresponds to the bulk property of temperature. If c_{ms}^2 is constant, that is, if temperature is constant, then pV remains constant. Thus the assembly of particles exhibits Boyle's law. The pressure proves to be two-thirds of the total translational energy

[1]His book *The Kind of Motion We Call Heat* takes its title from an 1857 article by Clausius, an early contributor to the kinetic theory.

of the molecules in unit volume. In a mixture of gases the kinetic energies are additive and so too must be the pressures – Dalton's law.

Consider two gases A and B at the same temperature and pressure. Equality of temperature requires that the mean translational energies of their molecules are equal, (a result from the dynamical theory of gases), so $\frac{1}{2}m_A c_{ms,A}^2 = \frac{1}{2}m_B c_{ms,B}^2$. Equality of pressure requires that $\frac{1}{3}m_A(n_A/V)c_{ms,A}^2 = \frac{1}{3}m_B(n_B/V)c_{ms,B}^2$. Thus $n_A = n_B$, that is, equal volumes of all gases measured at the same temperature and pressure contain the same number of molecules. This is Avogadro's law and affords an explanation of Gay-Lussac's law. Thus the volume occupied by a kilogram-molecule (the mass in kilograms equal to the molecular weight of the gas) at the same temperature and pressure is the same for all gases. Avogadro's number N_0 is the number of molecules in 1 kilo-mole and has the value 6.0228×10^{26}. As applied to 1 kilomole, equation (2.2) becomes

$$pV = \tfrac{1}{3}mN_0c_{ms}^2. \tag{2.3}$$

Qualitatively speaking, temperature is a measure of the hotness or coldness of a body. To establish a quantitative scale, the values x of some physical quantity which varies with temperature are to be found at two fixed (reproducible) points: x_0 denotes the value at the melting point of ice, 0°C, and x_{100} at the boiling point of water, 100°C at 1 atmosphere pressure; the interval corresponds to 100 K. The temperature value of a body at some observed value of x, x' say, is then $t' = [(x' - x_0)/(x_{100} - x_0)] \times 100$.

The obvious temperature-varying quantity in principle is the value of pV for a gas, since the value of pV varies in the same way for all gases at low pressures: it is a general property, not specific to any particular gas.[2] With this definition, the temperature t(°C) varies incrementally with pV and the absolute temperature T (K) is proportional to pV. The observed constant of proportionality is $1/(273 + t)$. As applied to 1 kmol, pV can be written

$$pV = R(273 + t) = RT, \tag{2.4}$$

where $R = 8314$ J/(K kmol). This relation ceases to hold when the temperature falls to values near where the gas may liquefy. Comparing equations (2.3) and (2.4),

$$RT \equiv \tfrac{1}{3}mN_0c_{ms}^2. \tag{2.5}$$

Here mN_0 is the molecular weight M of the gas in kilograms; for example, oxygen has a molecular weight of 32 kg. Of the gas properties we wish to determine, two follow readily:

the coefficient of expansion $\beta = (\partial V/\partial T)_p/V = 1/T$, (2.6)

the density ρ is related to the pressure as $\rho = M/V = pM/RT$ (2.7)

and is around $10^5 \times 29.0/(8314 \times 293)$ or 1.19 kg/m³ for air at 20°C and 1 bar pressure.[3]

[2] Suppose instead that the saturated vapour pressure of water were chosen as the physical quantity varying with temperature; $p_0 = 0.66$ kPa and $p_{100} = 101.32$ kPa. This would be an unsuitable choice however since it is linked to a specific material, water, it is complicated by a phase change; compared with scales based on expansion or electrical resistance, a temperature scale defined this way is very non-linear and it would not serve at subzero temperatures.

[3] Dry air consists of 20.95% oxygen, 78.04% nitrogen, 0.93% argon, some carbon dioxide and traces of other gases (1993 ASHRAE Handbook of Fundamentals, p. 6.1). At the beginning of the industrial era, the CO_2 content was 0.028%; it had risen to 0.0365% by the late 1990s and is increasing at a rate of 0.00015% per

The kinetic theory provides a derivation of gas transport properties and a simplified account for conductivity and viscosity follows. It arrives at expressions which are qualitatively correct but which may differ by a factor of up to 2 from values found by observation or reached by more detailed reasoning. The classical account is given by Jeans (1954, 1962); see also Chapman and Cowling (1970). It is necessary to consider the distribution of velocities in a volume of gas, the mean distance between collisions and any directional aspects.

A simple argument gives qualitative justification for a distribution, that is, some molecules may have higher velocities than others. Consider two identical perfectly elastic spheres, A and B. B is at rest and A impacts upon it centrally with velocity v. After impact, A becomes stationary and B moves with the velocity v in magnitude and direction. If however A's velocity vector is not directed to the centre of B and impact is made non-centrally, after impact A and B each have forward components of velocity (the direction of v) and equal and opposite lateral components. Each has a speed less than v. If now the process is reversed, so that A and B approach each other, after impact B say is brought to rest while A assumes a velocity greater than either had before impact. An assembly of particles must have a range of velocities.

An expression for the distribution of velocity of the molecules was evaluated by Clerk Maxwell in the 1860s (see Section 2.4). The number of molecules in an assembly of N molecules having the velocity c in the range dc is

$$\mathrm{d}N_c = \frac{4N}{c\sqrt{\pi}}\left(\frac{Mc^2}{2RT}\right)^{3/2}\exp\left(-\frac{Mc^2}{2RT}\right)\mathrm{d}c; \tag{2.8}$$

see Roberts and Miller (1960: 81).

The most probable velocity (the value of c where $\mathrm{d}N_c$ has its maximum value) follows by differentiation. The mean velocity $\bar{c}$ by definition is $\int_0^\infty c\,\mathrm{d}N_c/N$ and the root mean square velocity has already been found (equation 2.5). To these measures may be added the velocity of sound. The quantity γ is defined below.

	sound velocity	most probable velocity	mean velocity	root mean square velocity
	$\sqrt{\gamma}\left(\frac{RT}{M}\right)^{1/2}$	$\sqrt{2}\left(\frac{RT}{M}\right)^{1/2}$	$\frac{2\sqrt{2}}{\sqrt{\pi}}\left(\frac{RT}{M}\right)^{1/2}$	$\sqrt{3}\left(\frac{RT}{M}\right)^{1/2}$
Coefficient of $\sqrt{RT/M}$ for air	1.18	1.41	1.60	1.73
Value for air at 0°C	331 m/s	396 m/s	446 m/s	485 m/s

See footnote 4.

year. It is the principal greenhouse gas. Weight for weight, methane, nitrous oxide and the chlorofluorocarbons are many times more powerful than carbon dioxide but they are only present in very small concentrations. The molecular weight of dry air is taken here to be $0.2095 \times 32 + 0.7804 \times 28.0 + 0.0093 \times 39.9 = 28.93$ or 29 kg. Oxygen has an atomic weight of 16 by definition.

[4]The probability that a molecule has a velocity between c and $c + \delta c$ decreases very rapidly with c. Now, if a projectile is to escape from the earth's gravitational field, it must have a velocity of some 11 000 m/s, which is many times greater than these representative values for air. The probability of an air molecule escaping from the earth's atmosphere is therefore exceedingly small.

It is a matter of simple observation that some foreign odour diffuses slowly through still air; its rate is orders of magnitude less than these measures since a molecule travels only a very short distance before colliding with another molecule. The average distance between encounters is the mean free path, Λ. Its value can be estimated by noting that if there are n molecules in unit volume, a molecule of diameter σ travelling a distance $\bar{c}t$ on average in time t (and tracing a volume $\frac{1}{4}\pi\sigma^2\bar{c}t$) will make encounters with the $n\pi\sigma^2\bar{c}t$ molecules (assumed for the moment to be static) within the volume $\pi\sigma^2\bar{c}t$. Thus the average distance between encounters is $\Lambda = 1/n\pi\sigma^2$. Taking the motion of the other molecules into account, the value becomes $\Lambda = 1/\sqrt{2}n\pi\sigma^2$.

Within the volume of the gas it is assumed that equal numbers of molecules are moving in all directions. It can be shown that an element of area $\mathrm{d}S$ receives a total of $(\mathrm{d}S/4)n\bar{c}$ collisions per second and that the number leaving $\mathrm{d}S$, making an angle θ with the normal to $\mathrm{d}S$ into the incremental angle $\mathrm{d}\theta$, is $(\mathrm{d}S/2)n\bar{c}\cos\theta\sin\theta\,\mathrm{d}\theta$. See for example Roberts and Miller (1960: 72, 222).

To derive an expression for the conductivity λ, consider the gas contained between two horizontal plates (in xy planes), the upper being at the higher temperature so there is a downward flow of heat by conduction and no transfer by convection. In the plane at a height z_0 (containing the horizontal surface $\mathrm{d}S$) the mean translational energy will be taken to be E joules per molecule. Molecules arriving at $\mathrm{d}S$ from above come on average from a plane $z = z_0 + \Lambda\cos\theta$ and their mean energy has the higher value $E + \Lambda\cos\theta\,\mathrm{d}E/\mathrm{d}z$. Thus the amount of energy transferred across $\mathrm{d}S$ in the upward direction is

$$-(E + \Lambda\cos\theta\,\mathrm{d}E/\mathrm{d}z)(\mathrm{d}S/2)n\,\bar{c}\cos\theta\sin\theta\;\mathrm{d}\theta,$$

and on integrating θ from 0 to π, the flux is $-\frac{1}{3}\,\mathrm{d}S\Lambda n\bar{c}\,\mathrm{d}E/\mathrm{d}z$ or $-\frac{1}{3}\,\mathrm{d}S\Lambda n\bar{c}(\mathrm{d}E/\mathrm{d}T)(\mathrm{d}T/\mathrm{d}z)$. By definition this is the heat flux $-\mathrm{d}S\lambda\,\mathrm{d}T/\mathrm{d}z$, (the Fourier law of conduction),[5] so the conductivity λ is $\frac{1}{3}\Lambda n\bar{c}(\mathrm{d}E/\mathrm{d}T)$.

Since E is the energy per molecule (J) and m is the mass of a molecule (kg), E/m is the energy of the gas in J/kg and $(1/m)(\mathrm{d}E/\mathrm{d}T)$ is the specific heat at constant volume, c_v (J/kg K). So the conductivity is

$$\lambda = \tfrac{1}{3}m\Lambda n\bar{c}c_v. \tag{2.9}$$

A similar argument leads to an expression for the viscosity μ. Suppose that the lower plane is stationary and the upper plane moves in the x direction such that at $z = z_0$, the horizontal velocity in the x direction is u_0, which is small compared with molecular velocities. In moving downward, each molecule from above z_0 carries with it on average a momentum $m(u_0 + \Lambda\cos\theta\,\mathrm{d}u/\mathrm{d}z)$. The total amount of momentum carried across $\mathrm{d}S$ in unit time is found by integrating the quantity

$$m(u_0 + \Lambda\cos\theta\,\mathrm{d}u/\mathrm{d}z)(\mathrm{d}S/2)n\bar{c}\cos\theta\sin\theta\,\mathrm{d}\theta$$

[5]The Fourier conduction law states that the heat flux density q(W/m^2) is proportional to the negative temperature gradient (K/m), so $q = -\lambda\nabla T$. Barletta and Zanchini (1997) state that this is associated with an increase of entropy. The Fourier continuity law relates a space change in gradient to a rate of change of temperature: $\lambda\Delta T = \rho c_p\partial T/\partial t$.

between $\theta = 0$ and π to give $\frac{1}{3}\,\mathrm{d}Sm\Lambda n\bar{c}\,\mathrm{d}u/\mathrm{d}z$. Now the shearing force on $\mathrm{d}S$ due to the fluid motion is $\mathrm{d}S\mu\,\mathrm{d}u/\mathrm{d}z$. So the viscosity is

$$\mu = \tfrac{1}{3}m\Lambda n\bar{c}. \tag{2.10}$$

(In 1866 Maxwell also derived expressions for the diffusion of one gas into another, as apparently Stefan did independently in 1871. They are cited by Mitrovic (1997). The diffusion of water vapour through air is central to a study of moisture in building walls. Fick's study of diffusion was published in 1855.)

Since Λ is inversely proportional to n, both λ and μ appear to be independent of pressure. This is counter-intuitive, but it proves to be true. It will be seen that according to this simplified approach,

$$\mu c_v/\lambda = 1. \tag{2.11}$$

The specific heat in this relation is at constant *volume*. Convective heat transfer, however, generally takes place under conditions of constant pressure and the Prandtl number Pr, a non-dimensional group fundamental to a study of convection, is defined in terms of the specific heat at constant *pressure*, c_p:

$$\mathrm{Pr} \equiv \mu c_p/\lambda. \tag{2.12}$$

There are two relations between c_v and c_p, from which we can evaluate Pr for gases. First, if a kilogram of a gas is heated at constant volume through 1 K, the amount of heat needed is by definition c_v. If it is then allowed to expand to its former pressure p, further heat is needed to maintain the 1 K increment as it does work against the surrounding pressure:

$$c_p = c_v + p(V_2 - V_1) = c_v + (R/M)(T_2 - T_1), \tag{2.13}$$

noting that the equation of state for a perfect gas is

$$pV = (R/M)T, \tag{2.14}$$

so

$$c_p - c_v = R/M, \tag{2.15}$$

since the change in temperature $T_2 - T_1$ is unity by definition.

The second relation derives from the atomicity of the gas – the number of atoms per molecule, two for each of the principal constituents of air. It depends on the number of degrees of freedom of a molecule, which is the number of independent squared terms in the energy expression of a system. For a monatomic gas this number is 3 since the energy of a molecule comprises the independent values of $\overline{u^2}$, $\overline{v^2}$ and $\overline{w^2}$. The molecule might have energy associated with spin but quantum theory rules this out. According to the principle of energy equipartition, the total energy of a system is equally divided between the degrees of freedom. Thus from equation (2.5) the mean energy per molecule is

$$\tfrac{1}{2}m\overline{u^2} = \tfrac{1}{2}m\overline{v^2} = \tfrac{1}{2}m\overline{w^2} = \tfrac{1}{2}RT/N_0, \tag{2.16}$$

so the energy per degree of freedom per kilomole is $\frac{1}{2}RT$ and per kilogram is $\frac{1}{2}RT/M$. For a monatomic gas with three degrees of freedom, the internal energy $U = \frac{3}{2}RT/M$ (J/kg) and so

$$c_v = (\partial U/\partial T)_v = \tfrac{3}{2}R/M \tag{2.17}$$

with units J/kg K

A molecule of a diatomic gas has three translational degrees of freedom. Additionally it has two degrees associated with the two orthogonal axes perpendicular to the axis through the two atoms. There are two further degrees associated with the kinetic and potential energy of vibration along this axis but they too are largely ruled out at moderate temperatures by quantum considerations. Thus $U = \frac{5}{2}RT/M$ and for air (assumed to be a diatomic gas) it has values

$$c_v = (\partial U/\partial T)_v = \tfrac{5}{2}R/M = \tfrac{5}{2}8314/29.0 = 717 \text{ J/kg K}, \tag{2.18a}$$

$$c_p = c_v + R/M = \tfrac{7}{2}R/M = 1003 \text{ J/kg K}. \tag{2.18b}$$

The value for c_p is in satisfactory agreement with the observed value of 1006 J/kg K at 0°C and the theoretical value for the ratio $\gamma = c_p/c_v$ of 1.4 agrees with the observed value of 1.403. However, according to the simple kinetic theory sketched above, $\mu c_v/\lambda = 1$ and with the Prandtl number defined as $\mathrm{Pr} = \mu c_p/\lambda$, Pr should be equal to γ or 1.4. In fact, using the observed values of λ and μ for air, Pr is about 0.71.

The value of c_p is not in doubt, but several assumptions have been made which can only be regarded as approximations to the real situation. The theory neglects intermolecular forces, yet these dominate the liquid and solid states. The dumb-bell form of a diatomic molecule can only crudely be represented as spherical, having some definite diameter σ. Further, molecules crossing the area dS do not arrive there from some precise locality typified by the mean free path Λ but from a spread of locations. The sketch, however, reflects the position in the mid nineteenth century and the subject was further advanced over a long period; see Chapman and Cowling (1970). The account here is intended to give a simple description, no more than semiquantitative, of the physical model for the transport parameters of gases. Its application lies in the theory of convection.

Parameters for dry air at 20°C and atmospheric pressure (Ede 1967b: 256): density $\rho = 1.20\,\mathrm{kg/m^3}$, specific heat $c_p = 1007$ J/kg K, viscosity $\mu = 1.81 \times 10^{-5}$ kg/m s, thermal conductivity $\lambda = 2.57 \times 10^{-2}$ W/m K, Prandtl number $\mathrm{Pr} = 0.71$, $\mathrm{Gr}/(L^3\Delta T) = g\beta\rho^2/\mu^2 = 1.47 \times 10^8 \mathrm{m^{-3}K^{-1}}$.

2.2 REPRESENTATIVE VALUES FOR SOLIDS

No corresponding simple theory can be advanced to explain the corresponding parameters for liquids and solids. Table 2.1 gives a selection of representative values for λ, ρ and c_p for solids (taken from the 1999 *CIBSE Guide A*, p. 3-44), together with derived values.

Table 2.1 A selection of thermal parameters for solids

	Thermal conductivity λ (W/m K)	Density ρ(kg/m^3)	Specific heat c_p (J/kg K)	Diffusivity $\kappa = \lambda/\rho c_p$ $\times 10^8$(m^2/s)	Effusivity $\sqrt{\lambda \rho c_p}$ (W s$^{1/2}$/m^{2}K)
Impermeable materials					
Asphalt	0.50	1 700	1 000	29	922
Glazed ceramic	1.40	2 500	840	67	1 715
Solid glass	1.05	2 500	840	50	1 485
Steel	45	7 800	480	1 202	12 980
Roofing felt	0.19	960	840	24	391
Non-hygroscopic materials					
Polyurethane foam	0.028	30	1 470	63	35
Glass fibre slab[a]	0.035	25	1 000	140	30
Loose-fill floor/roof screed	0.41	1 200	840	41	643
Preformed mineral fibreboard[b]	0.042	240	760	23	88
Acoustic tile	0.057	290	1 340	15	149
Expanded polystyrene	0.035	23	1 470	104	34
Rock wool	0.034	23–200	710		
Inorganic porous materials					
Aerated brick[c]	0.30	1 000	840	36	502
Brickwork, inner leaf	0.62	1 700	800	46	918
Brickwork outer leaf	0.84	1 700	800	62	1 069
Cement fibreboard	0.082	350	1 300	18	193
Cement mortar, dry	0.93	1 900	840	58	1 218
Cement mortar, moist	1.5	1 900	840	94	1 547
Dry ceramic tiles	1.20	2 000	850	71	1 428
Aerated concrete block	0.24	750	1 000	32	424
Lightweight concrete block	0.73	1 800	840	48	1 051
Heavyweight concrete block	1.31	2 240	840	70	1 570
Aerated cast concrete	0.29	850	840	41	455
Lightweight cast concrete, dry	0.23	770	840	36	386
Lightweight cast concrete, moist	0.38	770	840	59	496
Dense cast concrete	1.70	2 200	840	92	1 772
Lightweight masonry	0.22	570	840	46	324
Heavyweight masonry	0.90	1 850	840	58	1 183
Common earth	1.28	1 460	880	100	1 282
Granite	3.49	2 880	840	144	2 906
Sandstone[d]	1.83	2 200	710	117	1 691

Table 2.1 *(continued)*

	Thermal conductivity λ (W/m K)	Density ρ(kg/m^3)	Specific heat c_p (J/kg K)	Diffusivity $\kappa = \lambda/\rho c_p$ $\times 10^8$(m^2/s)	Effusivity $\sqrt{\lambda \rho c_p}$ (W s$^{1/2}$/m^{2}K)
Organic, hygroscopic materials					
Laminated paper	0.072	480	1 380	11	218
Wilton carpet	0.06	190	1 360	23	125
Wool felt underlay	0.04	160	1 360	18	93
Cork board	0.04	160	1 890	13	110
Pitchpine, dry	0.17	650	2 120	12	484
Pitchpine, moist	0.23	650	3 050	12	675
Hardboard	0.08	600	2 000	7	310
Hardboard	0.29	1 000	1 680	17	698
Woodwool roofing slabs	0.10	500	1 000	20	224

[a]For variation of the conductivity of fibreglass with moisture content, see Section 8.6.1.
[b]The thermal conductivity of fibrous composites is discussed by Pitchumani and Yao (1991).
[c]The effective conductivity of building bricks is discussed by Söylemez (1999); they consider its dependence on convection and radiation in the pores.
[d]Quoted values of λ vary between 1.3 and 5 W/m K.

The equilibrium moisture content (kg of moisture/kg of dry material) varies with relative humidity. The following values for wood are given by Luikov (1966: 498) for wood at 20°C for values of RH = 10, 20, ..., 90%: 0.03, 0.0482, 0.063, 0.077, 0.0917, 0.11, 0.13, 0.156, 0.205. It decreases with increasing temperature. The values at 50% relative humidity and −20, 0, 20, ..., 100°C are 0.1117, 0.102, 0.0917, 0.081, 0.071, 0.0634, 0.057.

2.3 DISCUSSION

According to the Fourier conduction equation, the steady heat flow Q_x(W) and q_x(W/m^2) in the x direction through a surface of area $y \times z$ into a rectangular block $x \times y \times z$ is

$$q_x \equiv \frac{Q_x}{yz} = -\lambda \frac{\partial T}{\partial x} = \lambda \frac{T_1 - T_2}{x}, \tag{2.19}$$

where T_1 and T_2 are the uniform temperatures of the front and back surfaces. A positive heat flow in the x direction is associated with a negative temperature gradient, or vice versa; physical constants such as λ are necessarily positive and the negative sign ensures that both sides of the equation have the same sign. The units of λ are $(\mathrm{W/m_y m_z})/(\mathrm{K/m_x})$ or simply W/m K when directions are ignored. Conductivity is, strictly speaking, a tensor quantity and some crystalline materials have distinct values λ_x, λ_y and λ_z but with the exception of wood, where λ(along the grain) > λ(across the grain), building materials do not have directional properties.[6]

[6]Equation (2.19) asserts that the *effect* (a conducted heat flow) is proportional to its *cause* (a temperature gradient); these may reverse their identities. A relation of this kind is to be expected and is similar to a number of simple relations: the rate of mass flow by diffusion (e.g. water vapour through air) is proportional

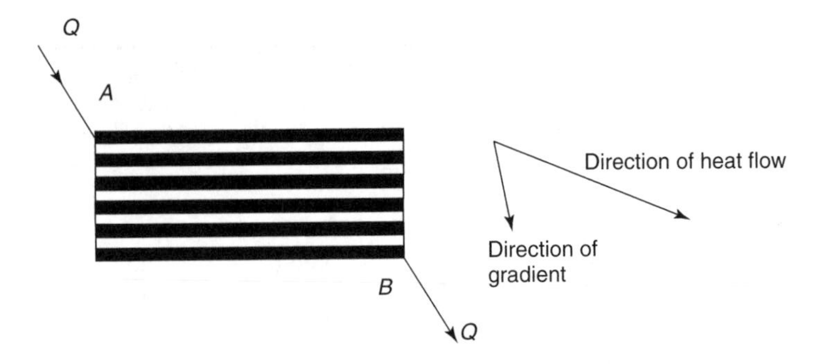

Figure 2.1 Conduction in a material having non-uniform conductivity

Although in principle it is very simple to measure the conductivity of building materials using a guarded hot plate rig, independent workers using this method on a single sample (with due precautions regarding conditioning, etc.) in an ostensibly identical manner do not obtain by any means the same values (Spooner 1980).

(In a homogeneous material, the directions of heat flow and temperature gradient are the same and the isotherms – lines or surfaces of equal temperature – are perpendicular to the heat flow vector. This can be formally demonstrated but the effect of non-homogeneity can be illustrated in a simple physical manner.

Figure 2.1 shows a block of material consisting of alternating layers of good (black) and bad (white) conducting material (uniform in depth). Heat is input at the top left edge at A and is withdrawn from the bottom right at B. On a macroscopic scale the flow is from A to B. However, each layer of good conductor must be at a near-uniform temperature: the components of the temperature difference between A and B are largely made up of the differences through the layers of poor conductor and the gradient is almost vertical. Thus in a non-homogeneous material, the directions of heat flow and temperature gradient may not be the same.[7])

to the concentration gradient (Fick's law); the flow of electricity in a conductor is proportional to the driving potential (Ohm's law); deformation of a solid is proportional to the stress applied (Hooke's law); the shear stress on a surface moving in a fluid is proportional to the velocity gradient at the surface; the flow of fluid is initially proportional to a pressure gradient (when viscosity provides the principal constraint, this is Darcy's law for flow in a porous material and it leads to Poiseuille's formula for flow through a cylindrical tube); the change in pressure of a gas at constant volume is proportional to a change in temperature (Charles' law); the transfers of heat by convection and radiation are approximately proportional to the temperature difference driving them. How far strict proportionality pertains is a matter for further consideration. In the conduction of heat and electricity and fluid stress at a surface, for most purposes the effect is strictly proportional to its cause. In other cases the proportionality is limited. Charles' law fails at low temperatures before the gas liquifies. As to Hooke's law, at sufficiently high stress there may be irreversible strain and possible rupture. Pipe flow is proportional to pressure difference Δp only when it is dominated by viscous restraint; when dominated by momentum change, it varies as $\sqrt{\Delta p}$ (Bernoulli's theorem). Convective transfer is proportional to around $\Delta T^{1.3}$, and radiant exchange is proportional to a difference of fourth powers of temperature.

[7]It is of no significance in a building context but it is interesting to note that a magnetic field modifies the relation between heat flow and temperature gradient in metals. Onsager (1931a: 425) remarks that 'if a transverse magnetic field is applied the temperature gradient [in an isotropic body] will have a component in the third direction perpendicular to flow and field.... If a circular metal plate is placed perpendicular to the magnetic field, heated in the middle and cooled at the edge, the heat will flow outward in spirals.' Onsager's reciprocal relations will be used later in connection with the simultaneous transport of heat, vapour and liquid water in a porous material.

The foregoing argument applies to certain crystalline materials where the transfer mechanism involves the oscillation of atoms about their lattice points. In electrically conducting materials, metals, the free electrons provide a further thermal transport mechanism and metals have higher thermal conductivities. Except for solid metals, most building materials, are porous, without directional properties, but having conductivities which depend on their porous nature. Convection and radiation contribute to the net transfer process; see Simpson and Stuckes (1990). In bone dry condition, such materials contain pockets of air which has a low conductivity, around 0.025 W/m K. Aerated materials clearly have low densities and low conductivities compared with the dense material (see Table 2.1). In insulating materials the conduction mechanism is largely air; there are minor contributions due to radiant transfer and fibre conduction of around 0.01 and 0.002 W/m K, respectively. Conductivity increases with density. Since radiant transfer increases rapidly with temperature, the radiant component in insulating materials increases and at higher temperatures the thermal conductivity rises again with decreasing bulk density.

In moist conditions the pores become partly filled with water which has a higher conductivity, around 0.6 W/m K, consequently the conductivity of porous materials varies strongly with moisture content (Table 2.1). A water content of 10% by volume can more than double the conductivity of porous building materials (Jakob 1949: 94).[8] See also Jesperson (1953). The uptake depends on the ambient vapour pressure and temperature. The mechanism is complicated by possible evaporation and condensation toward the cold surface and movement due to surface tension within the material. Equation 6 of Eckert and Faghri (1980) is an expression for the measured conductivity λ_s and its value in the bone dry state λ:

$$\lambda_s = \lambda + \rho_a D_s L \partial w / \partial T, \tag{2.20}$$

where ρ_a is the density of air, $D_a(\mathrm{m}^2/\mathrm{s})$ is the vapour diffusion coefficient in the solid, L (J/kg) is the latent heat of vaporisation and w is the vapour mass fraction. The apparent thermal conductivity of moist materials is discussed by Azizi *et al.* (1988). The thermal conductivity of ice, 2.2 W/m K, is much larger than that of water; the performance of an insulant deteriorates in subzero conditions.

To provide a qualitative description of the process, we note that the intermolecular attraction between molecules tends to restrain the escape of a water molecule from the liquid phase to the vapour phase. A molecule leaving a very small droplet is acted upon by a smaller volume of liquid than one leaving a plane surface and so has a relatively higher probability of escaping, thus the saturated vapour pressure outside such a convex surface is greater than the saturation pressure above a plane surface. Similarly, the saturation pressure just above a concave surface, such as a meniscus, is less than the plane surface value and the difference is $2\sigma/r$, where σ is surface tension and r the radius of the surface (Section 8.8). The difference only becomes significant when r is of molecular dimensions.

Figure 2.2 illustrates the situation.[9] It shows the relation between the saturated vapour pressure (SVP) of water and temperature above a plane surface, together with the relations for pressure above surfaces of small and large concave curvature. Condensation

[8]The water content by volume is the number of kilograms of water per cubic metre of the porous material (w). The content can also be described as the number of kilograms of water per kilogram of the dry material, w/ρ_{dry}.

[9]The build-up of liquid in the voids of porous materials is illustrated in Figure 1 of Ozaki *et al.* (2001).

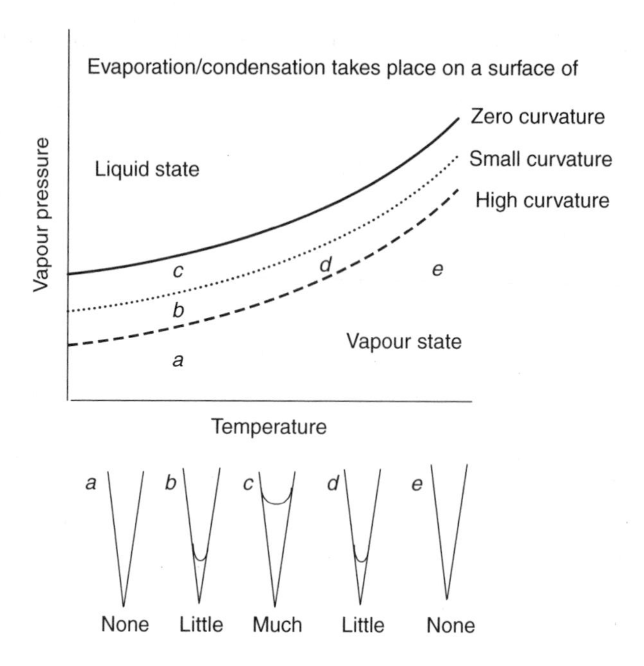

Figure 2.2 Condensation in a micro-wedge-shaped cavity

takes place on a surface if the ambient vapour pressure is greater than the SVP for that particular temperature and curvature. Consider the possibility of condensation taking place within a wedge- or conical-shaped cavity of very small dimensions. Condition (a) denotes low values of ambient temperature and vapour pressure and no condensation takes place. Condition (b) indicates a rather higher vapour pressure, above the SVP for a surface of high curvature, so a little condensation takes place to form a meniscus. With further increase in pressure, further condensation follows, condition (c). Suppose now that ambient temperature increases, with no further change in pressure, as shown by condition (d). Evaporation takes place until the curvature of the meniscus reaches its appropriate value and further temperature increase may drive out all moisture, condition (e).

Thus the uptake of water increases with vapour pressure at constant temperature and decreases with temperature at constant vapour pressure. In particular, the changes from (a) to (e) and from (b) to (d) lead to little change in moisture; they also describe a change of very roughly constant *relative* humidity. However, the change from the (a, e) conditions to the (b, d) conditions leads to an increase in moisture content and relative humidity.

The voids between the particles of masonry materials provide spaces for the progressive development of meniscuses, similar to those in the idealised wedge model. The relation between moisture content of some material and relative humidity at constant temperature is described as its sorption isotherm, illustrated in Figure 8.3. It is commonly remarked that moisture content increases with relative humidity but it is a concomitant relationship rather than a causal relationship.

The intermolecular forces which play only a minor role in determining the transport properties of gases dominate those of solids and liquids and there is no simple theory for values of ρ, λ and (for liquids) β and μ. However, some comments may be made about their specific heats. The value of c_p is some 1–5% larger than c_v The value of c_v at ordinary temperatures is given as

$$c_v = 3R/M; \tag{2.21}$$

this is Dulong and Petit's law of 1819. It comes about because an atom in a three-dimensional lattice has six degrees of freedom (the product of three directions with potential and kinetic energy in each); there is $\frac{1}{2}R/M$ associated with each degree of freedom. Materials with low molecular weights have high specific heats; the values of M and c_p for hydrogen and lead, two extreme examples, are $M = 2.0$, $c_p = 14\,200$ J/kg K and $M = 207$, $c_p = 129$ J/kg K, respectively As Table 2.1 shows, c_p for wood with a high hydrogen and carbon content is higher than c_p for masonry materials which consist of heavier elements.

In steady heat conduction only the conductivity is significant, but in unsteady conduction the density and specific heat are significant, as their product ρc_p. The product ρc_p combines with λ in two ways:

- *As the quotient* $\kappa = \lambda/\rho c_p$: this combination appears in the Fourier continuity equation. The diffusivity κ, sometimes called the thermometric conductivity, describes the rate of propagation of a thermal signal through a solid following a disturbance. Since λ is very roughly proportional to ρ, the value of diffusivity (metals apart) does not vary from substance to substance by large factors (Table 2.1).
- *As the product* $\lambda\rho c_p$: the heat flow per unit area into a large thickness of material is proportional to $\sqrt{\lambda\rho c_p/t}$, where t is some measure of time; $t = 24$ h if the material is excited sinusoidally with a period of 24 h. Effusivity $\sqrt{\lambda\rho c_p}$ is much larger for heavy materials than for insulating materials (Table 2.1).

2.4 APPENDIX: THE MAXWELLIAN DISTRIBUTION

After impact, two gas molecules may have different velocities, so there must be a distribution of velocities in a population. Its form was first derived by Maxwell in 1859, arguing along the following lines.[10]

Consider a population of N molecules. Their velocities can be represented as an assembly of vectors springing from some point O. The number having a velocity component u in the range $\mathrm{d}u$ in the x direction can be expressed as $\mathrm{d}N = Nf(u)\,\mathrm{d}u$, where the function $f(u)$ has to be determined. There are similar relations for the v and w components of velocity and since the x, y and z components of velocity of any molecule at some instant can be taken to be independent, the number of vectors terminating in the elementary volume $\mathrm{d}u\,\mathrm{d}v\,\mathrm{d}w$ is

$$\mathrm{d}^3N_c = Nf(u)f(v)f(w)\,\mathrm{d}u\,\mathrm{d}v\,\mathrm{d}w. \tag{2.22}$$

[10]Maxwell's ingenious and non-mechanistic argument gives no hint of the mechanistic quality behind his thinking about the distribution of velocities. In fact, he speculated on the possibility of a 'very observant and neat-fingered being', later dubbed Maxwell's demon, who could follow and manipulate individual molecules, admitting fast-moving molecules in one space through a kind of trapdoor into a second space, thereby causing air in the second space to increase in temperature without any work being done, so that heat travelled 'uphill'. The further fortunes of this gremlin have been described by von Baeyer (1998).

In spherical coordinates the elementary volume contained within the solid angle $d\omega$ is $c^2\,d\omega\,dc$, where $c^2 = u^2 + v^2 + w^2$, so the number of vectors terminating within it can be written in terms of some function $\phi(c)$:

$$d^3N_c = N\phi(c)c^2\,d\omega\,dc. \tag{2.23}$$

The number is the same in each case, so we can equate the two expressions for the number of terminations per unit volume:

$$f(u)f(v)f(w) = \phi(c) = \phi(\sqrt{u^2 + v^2 + w^2}). \tag{2.24}$$

The equation has the solution

$$f(u) = C\exp(-u^2/\alpha^2), \text{ etc.,} \tag{2.25a}$$

and

$$\phi(c) = C^3\exp(-c^2/\alpha^2), \tag{2.25b}$$

where C and α are constants to be determined.

Firstly, integrating u between $-\infty$ and $+\infty$, we find the total number of particles, so

$$N = \int_{-\infty}^{+\infty} NC\exp(-u^2/\alpha^2)\,du, \tag{2.26a}$$

which equals $NC\alpha\sqrt{\pi}$,[11] so

$$C\alpha\sqrt{\pi} = 1. \tag{2.26b}$$

The integral of $d\omega$ over all directions is simply 4π, so we can express the number of molecules with velocities between c and $c + dc$ as

$$dN_c = 4\pi NC^3\exp(-c^2/\alpha^2)c^2\,dc. \tag{2.27}$$

Now the quantity $3RT/M$ is the mean square velocity c^2_{ms} of the particles (equation 2.5) while the expression for the distribution gives c^2_{ms} as $\int_0^\infty [4\pi NC^3\exp(-c^2/\alpha^2)c^2]c^2\,dc/N$ or $4\pi C^3\left(\frac{3}{8}\alpha^5\sqrt{\pi}\right)$. Equating these two values for c^2_{ms} and doing some manipulation, we find equation (2.8) for the distribution:

$$dN_c = \frac{4N}{c\sqrt{\pi}}\left(\frac{Mc^2}{2RT}\right)^{3/2}\exp\left(-\frac{Mc^2}{2RT}\right)dc. \tag{2.28}$$

It can be shown that this distribution is stable: any other distribution goes over to this form. (On mechanistic grounds, some molecule might momentarily be in a state of rest ($c = 0$). According to the expression, the probability is zero of a molecule being momentarily at rest, but we must allow for short-term deviations from the most probable distribution.)

[11] This and the following integral are standard forms: $\int_0^\infty \exp(-x^2/\alpha^2)\,dx = \frac{1}{2}\sqrt{\pi}\alpha$ and $\int_0^\infty x^4\exp(-x^2/\alpha^2)\,dx = \frac{3}{8}\sqrt{\pi}\alpha^5$.

Although a relation for the frequency of occurrence of molecules of some velocity was derived in the middle of the nineteenth century, it was only observed experimentally in 1927 when Eldridge demonstrated a distribution in cadmium vapour. Discs with radial slots were mounted on a rotating axle. Molecules passing as a batch through the first slot were observed after they had passed through the second slot, now spread out in accordance with their velocity distribution.

3

Conduction-Dominated Systems

In this chapter we examine some systems relevant to building heat transfer where steady state conduction is the principal mechanism.

3.1 HEAT FLOW ALONG A FIN

In steady-state flow through a wall, the heat flow $q = -\lambda\, \mathrm{d}T/\mathrm{d}x$[1] and does not vary with x, so $\mathrm{d}^2T/\mathrm{d}x^2 = 0$. If however heat is lost laterally, as is the case for a fin forming part of a heat exchanger, more heat is conducted into an element δx than is conducted out, $\mathrm{d}^2T/\mathrm{d}x^2 \neq 0$ and the temperature profile is no longer linear. The solution $T(x)$ is useful directly in connection with heat exchanger design, and also serves as a simpler form for the solution when a wall undergoes a sinusoidal variation in time.

Consider a fin or bar of cross-sectional area A and periphery p; if the fin is circular with diameter d, $A = \frac{1}{4}\pi d^2$ and $p = \pi d$. The heat flowing into a section at $x = a$ is $-\lambda A(\mathrm{d}T_x/\mathrm{d}x)_a$. According to Taylor's theorem, the heat flowing out at $a + \delta x$ is

$$-\lambda A\left(\left(\frac{\mathrm{d}T_x}{\mathrm{d}x}\right)_a + \frac{\mathrm{d}}{\mathrm{d}x}\left(\frac{\mathrm{d}T_x}{\mathrm{d}x}\right)\delta x\right), \tag{3.1}$$

so (heat conducted in at a) – (heat conducted out at $a + \delta x$) is $(\lambda A\mathrm{d}^2T_x/\mathrm{d}x^2)\delta x$. This is balanced by the convective/radiative loss of heat $hp\,\delta x(T_x - T_e)$ to the surroundings at T_e. Heat balance requires that

$$\frac{\mathrm{d}^2(T_x - T_e)}{\mathrm{d}x^2} = \frac{hp}{\lambda A}(T_x - T_e) = \alpha^2(T_x - T_e) \quad \text{say}, \tag{3.2}$$

[1]The Fourier *conduction* law, noted in 1807, states that the component of heat flow q_x (W/m^2) in the x direction is proportional to the temperature gradient, $\partial T/\partial x$ (K/m), in the x direction. The constant of proportionality is the conductivity λ or k (W/mK), so $q_x = -\lambda\partial T/\partial x$; the negative sign is required by the sign convection that a positively directed flow results from a numerically negative temperature gradient. The Fourier *continuity* equation in one dimension (1822) expresses the requirement that if more heat flows into an elementary element at some instant with no lateral loss, it produces an increase in the stored energy and so an increasing temperature. $\lambda\partial^2T/\partial x^2 = \rho c_p\, \partial T/\partial t$, where ρc_p (J/m^{3}K) is the volumetric thermal capacity of the material.

Building Heat Transfer Morris G. Davies
© 2004 John Wiley & Sons, Ltd ISBN: 0-470-84731-X

where

$$\alpha = \sqrt{hp/\lambda A} \tag{3.3}$$

Taking $T_e = 0$, this has solutions

$$T_x = B_1 \exp(\alpha x) + B_2 \exp(-\alpha x) \tag{3.4a}$$

or

$$T_x = C_1 \cosh(\alpha x) + C_2 \sinh(\alpha x). \tag{3.4b}$$

The choice of form depends on the boundary conditions. If the length of the bar is effectively infinite, B_1 must be zero; for a bar of finite length the second form is suitable. Suppose the bar is of length X and suppose no heat flows at $x = X$, then $\mathrm{d}T_x/\mathrm{d}x = 0$, though T_x itself is not zero. The other boundary condition is that the temperature at $x = 0$ is T_0. Then the solution is

$$T_x = T_0 \cosh(\alpha(X - x))/\cosh(\alpha X). \tag{3.5}$$

The total heat loss Q from the bar can be found either as $-\lambda A(\mathrm{d}T_x/\mathrm{d}x)_0$ or as $\int_0^X T_x hp\ \mathrm{d}x$:

$$Q = T_0 \underbrace{\sqrt{hp \times \lambda A}}_{\text{product}} \tanh(\underbrace{\sqrt{hp/\lambda A}}_{\text{quotient}} X). \tag{3.6}$$

Notice that hp and λA enter in product and quotient form. There are three other points:

- The product $\sqrt{hp \times \lambda A}$ describes the ability of the bar to conduct and lose heat; an effective fin will have a high value.
- The quotient $\sqrt{hp/\lambda A}$ (defined as α) denotes (ability to lose heat)/(ability to conduct heat); a large value implies an initial steep gradient in the bar, making it ineffective as a heat exchanger element.
- The fin efficiency is defined as

$$\frac{\text{heat transfer by the fin to the surrounding fluid}}{\text{heat transfer if all the fin were at its base temperature}} = \frac{\tanh(\sqrt{hp/\lambda A}X)}{(\sqrt{hp/\lambda A}X)}, \tag{3.7}$$

  which is unity for small values of X and decreases to zero as X becomes large.

If the temperature and heat flow are known at two points 1 and 2, distance X apart along the bar, it can readily be shown that they are related as

$$\begin{bmatrix} T_1 \\ q_1 \end{bmatrix} = \begin{bmatrix} \cosh u & -(\sinh u)/v \\ -(\sinh u) \times v & \cosh u \end{bmatrix} \begin{bmatrix} T_2 \\ q_2 \end{bmatrix}, \tag{3.8}$$

where q is the heat flow per cross-sectional area, $u = X\sqrt{(hp/A)/\lambda}$ and $v = \sqrt{(hp/A) \times \lambda}$. The determinant of the matrix is unity. Forms not unlike these occur in a similar connection for transient flow through a wall (12.12), (12.13) and for sinusoidal flow (12.33), (12.34). The term hp/A denoting the lateral loss of heat per cross-sectional area (J/s m^{3}K) is replaced there by the term ρc_p/time (J/s m^{3}K), which denotes the effect of thermal capacity per cross-sectional area. Lombard (1995) has shown how matrices of this form can be used to examine combinations of fins.

We note too that

$$\frac{Q}{A} = -v\left(\frac{\mathrm{d}T}{\mathrm{d}u}\right)_{u=0} = v\int_{u=0}^{u} T\,\mathrm{d}u = T_0 v \tanh u. \tag{3.9}$$

Thus the fin property $v \tanh u$ can be found either by differentiation at the base of the fin or by integration along it. Now the heat flow from the inside surface of an exterior wall toward the building interior results from the ambient (or sol-air) temperature T_0 'now' and previous hourly values T_{-1}, T_{-2}, etc., and can be expressed as

$$(Q/A)_{now} = \phi_0 T_0 + \phi_1 T_{-1} + \phi_2 T_{-2} + \ldots, \tag{3.10}$$

where the ϕ values (W/m^{2}K) are known as response factors. The series is infinite but eventually ϕ_{n+1}/ϕ_n tends to a constant value less than 1, so the contribution of remote T values is negligible; this is shown in equation (17.63). Analysis of heat flow through a multilayer wall in time-varying conditions is much more complicated than this simple problem of steady-state heat loss from a fin but the values of ϕ, a kind of generalisation of v tanh u, can be found in two parallel ways: the frequency-domain approach rests on *differentiation* of a function of the thermal properties of a multilayer wall; the time-domain method follows from *integration* through the wall (Section 17.9).

Some forms of solar collector consist of a blackened absorbing plate thermally attached to a series of parallel tubes through which cooling water flows and which is provided with a cover to reduce losses to ambient. Solar gain q (W/m^2) absorbed at the collector is in part conducted toward the tubes. Neglecting two-dimensional heat conduction in the plate, the above theory may be used to examine the action of the collector: T_0 denotes the water temperature, supposed uniform; w denotes the width of the collector plate in the direction of flow of water and d is the plate thickness, so $A = wd$; X is the semi-spacing of the tubes, so $p = 2w$. To equation (3.5) we must add the incremental temperature q/h due to solar gain, where h is the overall loss to ambient (reduced by the presence of the cover). Collectors without covers may be used to heat swimming pools, a particularly suitable application of solar energy. Brinkworth (1997) cites the Foreword to BS 6785: 'The relatively low temperature requirements allow collectors (often simple unglazed types) to operate at high collection efficiencies. In addition the swimming pool itself provides a heat store reservoir while the pool filtration system offers a means of circulating the pool water through the collector circuit for a minimum extra cost.'

A further example of a continuity equation similar to equation (3.2) occurs in connection with wall 'dynamic insulation'; ventilation air enters the building by moving slowly through some porous insulating material and may reduce the conductive heat loss (Etheridge 1998). We define u (m/s) as the air velocity through the insulation, ρ and c_p as the density and specific heat of air, and λ as the conductivity of the wall material. Then

$$\mathrm{d}^2T/\mathrm{d}x^2 = (u\rho c_p/\lambda)\,\mathrm{d}T/\mathrm{d}x. \tag{3.11}$$

Etheridge comments on the potential of this form of heat recovery. Baker (2003) describes a study to examine its effectiveness.

3.2 HEAT LOSS FROM A SOLID FLOOR

While it is easy to find the steady state heat loss through a plane wall, calculation of the steady-state loss from a solid floor to the ground and ultimately to the exterior is a complicated undertaking involving three-dimensional heat flow and perhaps other factors such as possible placing of insulation in various ways, film coefficients, and imposition of a fixed bounding temperature through presence of a water table. Time variation in the external climate–sinusoidal for annual variation or a step change to idealise short-term changes–makes the problem even more complicated. Articles of interest have appeared since the 1940s and are summarised by Claesson and Hagentoft (1991). Some of the simpler issues are presented here.

We consider the heat loss from a solid rectangular floor, most probably of concrete laid on hardcore on earth and uninsulated, of length L, breadth B, surrounded by a solid wall of thickness W, so that the external length dimension is $L + 2W$. The inside surface is at T_i and the surface of surrounding land will be taken to be T_o. The steady-state heat loss can be written as

$$Q = \lambda(T_i - T_o)G = \lambda \Delta T G, \tag{3.12}$$

where G is some function of L, B and W and has the units of length. The floor U value is defined by

$$U = Q/(LB\Delta T) = \lambda G/(LB). \tag{3.13}$$

Forms for G can be found by taking the heat flow pattern in the earth variously to be determined by one-, two- and three-dimensional considerations.

3.2.1 One-Dimensional Heat Loss

Macey's expression (1.15) was derived in essence by a one-dimensional argument. Noting that

$$\text{arctanh}\, x \equiv \frac{1}{2} \ln\left(\frac{1+x}{1-x}\right), \tag{3.14}$$

the heat loss can be written as

$$Q = \lambda \Delta T L (2/\pi) \ln(2B/W + 1) \exp(B/2L). \tag{3.15}$$

Macey's argument applies to a floor of infinite length and leads to the $\ln(2B/W + 1)$ term; an argument leading to the $\ln(2B/W + 1)$ term is presented below, based on elementary properties of circles. The exponential term constitutes an ad hoc correction for length L. Although useful in its time, it has a flaw: for a square floor ($L = W$), $\partial Q/\partial L$ should equal $\partial Q/\partial B$ but in fact $\partial Q/\partial L$ may be less than $\frac{1}{2}\partial Q/\partial B$.

Consider the loss of heat from an indefinitely long, straight, thin cylinder (effectively a line source) radius R_0 at T_0 embedded in an infinite medium of conductivity λ. The heat flow q per unit length through some concentric cylindrical surface radius r is

$$q = -\lambda 2\pi r \; \mathrm{d}T/\mathrm{d}r. \tag{3.16}$$

After integration the temperature T_1 on a concentric cylindrical surface radius r_1 is

$$T_1 = T_0 - q \ln(r_1/R_0)/2\pi\lambda. \tag{3.17}$$

Suppose that a second identical cylinder is laid parallel to the first a horizontal distance $2d$ away and that its temperature is $-T_0$ so that it gains the entire output from cylinder 1. The temperature T_2 at distance r_2 due to the second cylinder alone is

$$T_2 = -T_0 + q \ln(r_2/R_0)/2\pi\lambda. \tag{3.18}$$

Thus the net temperature T_P at the field point P distances r_1 and r_2 from the two cylinder axes, respectively, is

$$T_P = T_1 + T_2 = q \ln(r_1/r_2)/2\pi\lambda. \tag{3.19}$$

If P moves subject to the constraint that r_1/r_2 has a constant value $\sqrt{\alpha}$, it traces out an isothermal loop (or surface in three dimensions.) It is readily shown (Section 3.4) that the loop is a circle of radius $c = 2d\sqrt{\alpha}/(\alpha - 1)$ and centre $x = \pm d(\alpha + 1)/(\alpha - 1)$, $y = 0$. The cylinder axes themselves are at $x = \pm d$, $y = 0$, so the values of α describe a family of circles (set 1) consisting of two subsets with centres lying outside $x = \pm d$. The vertical axis, $x = 0$, represents the case $r_1 = r_2$, $c = \infty$ and $T_P = 0$.

A second series of circles (set 2) can be constructed with centres on the vertical axis and passing through the cylinder centres, $x = \pm d$. It can be shown that any circle of set 1 intersects any circle of set 2 perpendicularly. Set 1 circles represent sections through isothermal surfaces, set 2 represent lines of flow from the cylinder at $+T_0$ to the cylinder at $-T_0$. (Figure 3.1a).

To adapt the model to find the heat loss from a floor, two qualitative transformations are needed: bounding surfaces and reversal of identities. To understand bounding surfaces, consider a limited region of the flow field, the region below the horizontal. It is a principle that temperature and flow in this region of the medium will remain unaltered

(i) if the region is terminated by a perfectly conducting surface, held at some fixed temperature, which coincides with an isothermal surface of that temperature in the parent system, or

(ii) if the region is terminated by a perfectly insulating surface which coincides with a heat flow surface in the parent system.

These ideas are self-evident. Suppose now that the two thin cylinders are replaced by cylinders of finite diameter W (emitting $\pm q$ W/m as before) so as to coincide with isothermal surfaces; the heat flow in the medium is unaltered and the volume within the cylinders does not form part of the flow field. Furthermore, the flow field below the horizontal remains unaltered if the horizontal surface between the cylinders plus the two infinitely extended horizontal surfaces outside the cylinders are replaced by perfectly insulating surfaces and the upper volume of the conducting medium is removed (Figure 3.1b).

Since the flow lines and isothermal surfaces form an orthogonal system, they can reverse their identities. An isotherm can be interpreted as a flow line, and a flow line can be interpreted as an isotherm.

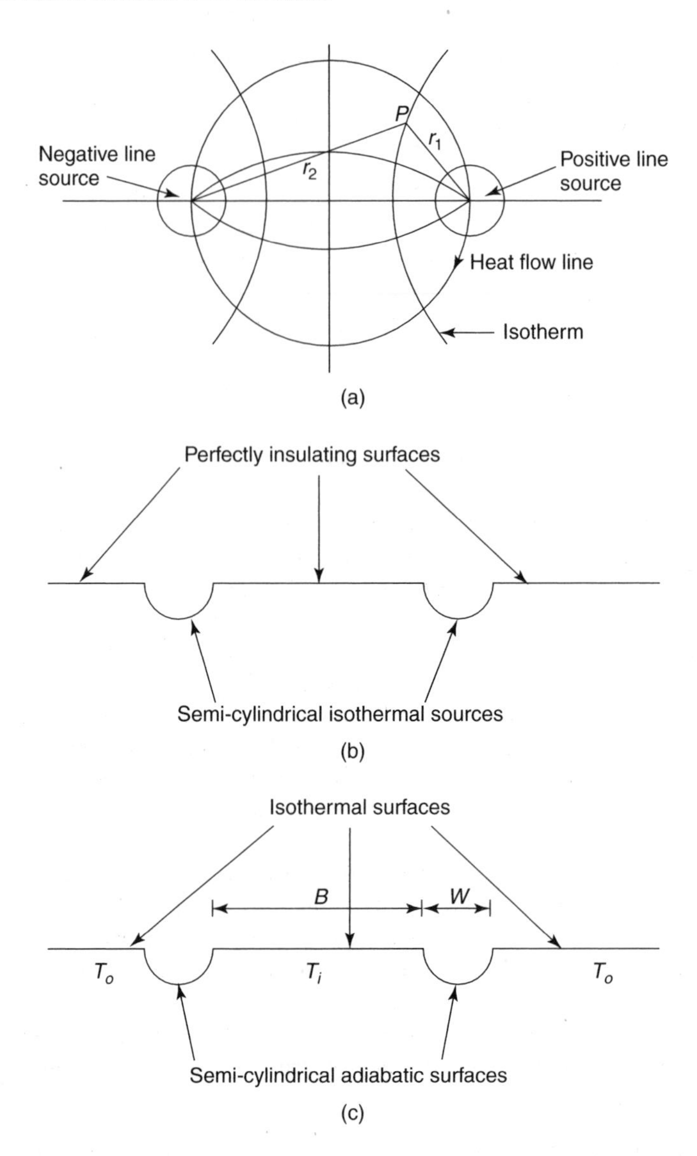

Figure 3.1 Macey's model for heat loss through a solid floor between isothermal surfaces separated by a gap: (a) isotherms and heat flow lines due to a line dipole source; (b) the model when the flow field is limited by isothermal and adiabatic surfaces; (c) the building model which results when surfaces of (b) reverse their identities. (Reprinted from *Building and Environment*, vol. 28, M.G. Davies, Heat loss from a solid ground floor, 347–359, © 1993, with permission from Elsevier Science)

Applying the transformation to Figure 3.1b, we arrive at Figure 3.1c, where the former horizontal insulating surfaces between and outside the semicylinders become isothermal surfaces at T_i and T_o. The former isothermal semicylindrical surfaces become adiabatic and must be composed of perfectly insulating material if we suppose that they are solid. Heat then flows from the floor of breadth B and at T_i to the infinitely extended land area

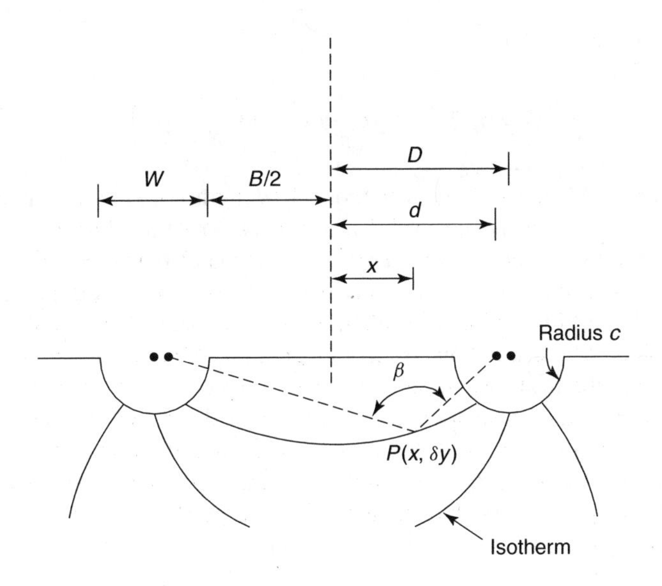

Figure 3.2 Diagram to help derive Macey's formula (Reprinted from *Building and Environment*, vol. 28, M.G. Davies, Heat loss from a solid ground floor, 347–359, © 1993, with permission from Elsevier Science)

at T_o along semicircular flow lines. The heat flow through the elementary area L dx is

$$dQ = -\lambda L\ dx(dT/dy) \text{ vertically downward} \tag{3.20a}$$

$$= -\lambda L\ dx(dT/d\beta)(d\beta/dy) \text{ say.} \tag{3.20b}$$

Consider the point $P(x, \delta y)$ on the circular isotherm just below the surface in Figure 3.2. It is specified by a constant value of the angle β. Suppose the field point P is moved from a point on the inner surface where $\beta = \pi$ to a point on the land surface T_o where $\beta = 0$. A change in β of π radians corresponds to the difference $T_i - T_o$. Also, the isotherms are spaced uniformly around the former thin-cylinder sources. So

$$dT/d\beta = (T_i - T_o)/\pi. \tag{3.21}$$

Further, β is given as

$$\beta = \pi - (\delta y/(d - x) + \delta y/(d + x)) \quad \text{so} \quad d\beta/dy = -2d/(d^2 - x^2). \tag{3.22}$$

On integrating equation (3.20) between $x = \pm\frac{1}{2}B$ and using equations (3.21) and (3.22), we have

$$Q = \lambda(T_i - T_o)\frac{2}{\pi}L\ln\left(\frac{\sqrt{1 + 2W/B} + 1}{\sqrt{1 + 2W/B} - 1}\right). \tag{3.23}$$

If the room is broad compared to wall thickness, as is usually the case, $W/B \ll 1$ and this approximates to

$$Q = \lambda(T_i - T_o)L\frac{2}{\pi}\ln\left(\frac{2B}{W} + 1\right) = \lambda(T_i - T_o)L(2/\pi)\ln(2x + 1), \tag{3.24a}$$

where $x = B/W$, or

$$Q = \lambda(T_i - T_o)L\frac{4}{\pi}\text{arctanh}\left(\frac{B}{B+W}\right). \tag{3.24b}$$

Equation (3.24b) is Macey's formula without the length correction. It is an approximation to an exact result that is only valid if any heat flow line starting from the floor at the wall itself follows a *semicircular path* to the ground outside, rather than the more direct path it must follow in reality. Thus the analysis tacitly assumes the presence of semicircular volumes of perfectly insulating material beneath the walls, blocking the flow at the points where the flow is in fact greatest. Macey's expression must therefore underestimate the heat loss, though Thomas and Rees (1999) report good agreement between the so-calculated heat loss from a floor and their own detailed measurements. Although the expression is flawed, the derivation is important since the simple geometrical construction provides physical insight into the pattern of heat flow from a slab; it is not made apparent by analytical methods. The associated error is less significant than the uncertainty in the value of soil conductivity which the designer may have to assume.

To see the implications of the expression, consider a room where the width $B = 5$ m, the wall thickness $W = 0.3$ m and the length L is indefinite. Suppose the soil conductivity is 1 W/m K. The expression gives a heat loss of 2.25 W/K per unit length in the long direction. This is the heat loss through a mat of insulating material, $\lambda = 0.04$ W/m K, width 5 m and thickness about 90 mm. Hagentoft (2002) points out that the resistance provided by the soil is equivalent to some 50–100 mm of mineral wool insulation.[2]

Although much research on heat loss through solid floors has since been published (see below), it has served in the main to extend the simple conclusions which follow from Macey's expression. It may be noted, however, that although the steady-state or time-averaged locations of the ground isotherms may be roughly semicircular, the annual variation in ground surface temperature[3] leads to ground isotherms at a particular time of year which are far from semicircular.

3.2.2 Two-Dimensional Heat Loss

Delsante *et al.* (1983: Section 3.3) and Anderson (1991a: equation 11) have independently supplied expressions to find the heat loss from a length L in a room of infinite length, breadth B and flanked by walls of thickness W:

$$Q = \lambda(T_i - T_o)L(2/\pi)[(x+1)\ln(x+1) - x\ln x] \quad \text{where} \quad x = B/W. \tag{3.25}$$

See also Delsante (1988) and Anderson (1991b).

[2]Thus, by traditional building standards, a solid floor would be regarded as inherently well insulated. However, the soil beneath a solid floor provides a large thermal capacity and it may take months or years before an almost steady-state condition is reached and during this period the heat losses will be larger than the value to which they tend. Insulation markedly reduces these losses.

[3]Thomas and Rees (1999) observed a variation of 10.3 K in the temperature of the ground at a depth of 0.25 m and 1 m outside the perimeter of their building.

The extent to which Macey's formula underestimates the loss can be seen from the difference between the non-dimensional parts of equations (3.24a) and (3.25).

$$\Delta = \underbrace{(2/\pi)[(x+1)\ln(x+1) - x\ln x]}_{\text{exact}} - \underbrace{(2/\pi)\ln(2x+1)}_{\text{Macey}}. \tag{3.26}$$

The value of Δ is around 0.19 and varies little with x.

An expression for the loss, together with details of the field in the earth below and outside the room can be found from an adaptation of a frequently presented solution to the Fourier continuity equation for steady state in two dimensions. Earlier heat was assumed to be lost by *convection* or *radiation* in a direction perpendicular to x. If instead heat is lost by *conduction* in the y direction, the field is determined by the continuity equation

$$\frac{\partial^2 T(x,y)}{\partial x^2} + \frac{\partial^2 T(x,y)}{\partial y^2} = 0. \tag{3.27}$$

Consider a lamina of length a in the x direction and c in the downward y direction. It is taken to be adiabatic on the left, right and lower edges and to have a temperature distribution

$$T(1,x,0) = T_1 \cos(\pi x/a) \tag{3.28}$$

imposed on the top edge (Figure 3.3a).[4] It can be readily checked that the $n = 1$ solution is

$$T(1,x,y) = \frac{T_1}{\cosh(\pi c/a)} \cos\frac{\pi x}{a} \cosh\frac{\pi(c-y)}{a}. \tag{3.29}$$

Further solutions are given by replacing a by a/n. This model supplies a solution to the problem where there is a uniform temperature T_i for the semi-breadth $0 < x < B/2$, a uniform fall in temperature beneath a wall of width W, and a uniform temperature $T_o = 0$ beyond (where $x > B/2 + W$). Using standard Fourier analysis, the temperature along the surface $y = 0$ is

$$T(x,0) = \sum T(n,x,0) = T_i \left[\frac{1}{2} c_0 + \sum_{n=1}^{\infty} c_n \cos(n\pi x/a) \right], \tag{3.30}$$

where

$$c_0 = (1/2a)\left(\tfrac{1}{2}B + \tfrac{1}{2}W\right)$$

and

$$c_n = \frac{4}{(n\pi)^2}\frac{a}{W}\left(\sin\frac{n\pi\frac{1}{2}W}{a} \sin\frac{n\pi\left(\frac{1}{2}B + \frac{1}{2}W\right)}{a} \right). \tag{3.31}$$

In theory a is infinite and a large value compared with B and W must be chosen for computation. As a result, a large value of n, 10^4 or 10^5, is needed to represent the $y = 0$ profile.

[4] In the more usual formulation, the lamina is isothermal on the three edges and has a temperature of $T(1,x,0) = T_1 \sin(\pi x/a)$ imposed on the top edge.

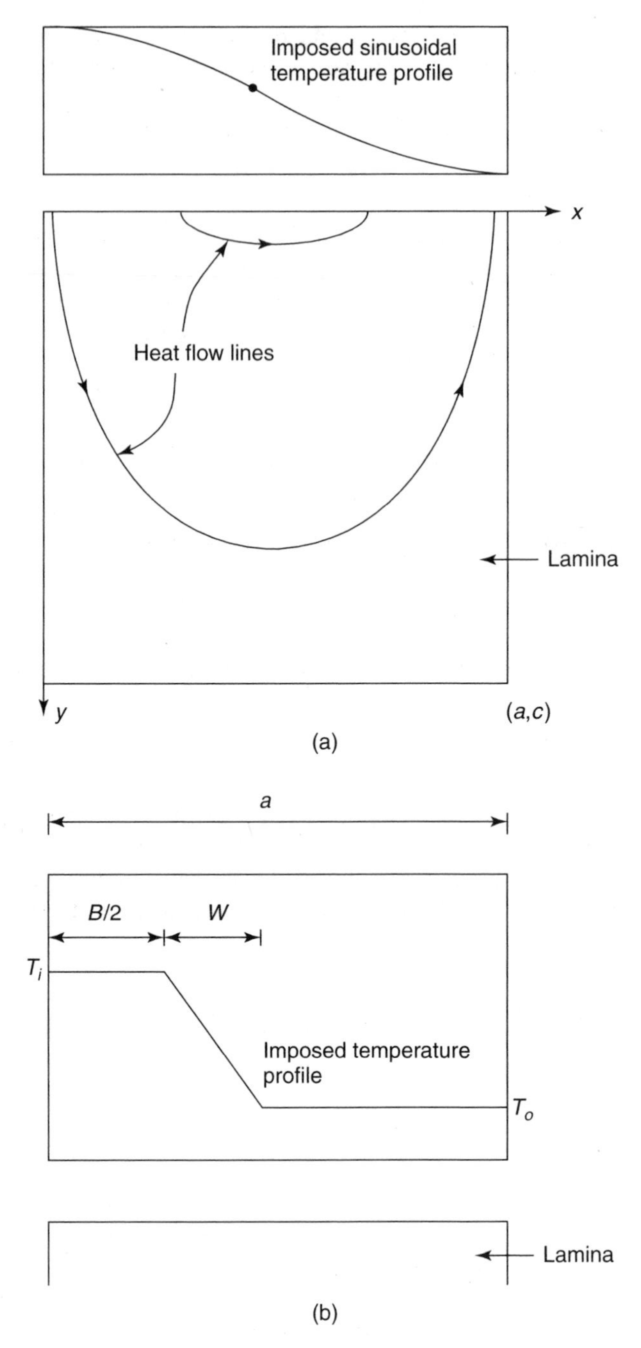

Figure 3.3 (a) A rectangular lamina with three adiabatic sides and a sinusoidally varying temperature distribution imposed on the fourth. The arrows show the direction of heat flow. (b) The temperature imposed on the lamina to approximate the temperature in the building model (Reprinted from *Building and Environment*, vol. 28, M.G. Davies, Heat loss from a solid ground floor, 347–359, © 1993, with permission from Elsevier Science)

The local vertical heat flow $q(1, x, 0)$ at the surface $y = 0$ is

$$q(1, x, 0) = -\lambda(\partial T(1, x, 0)/\partial y) = \lambda T_i \tanh(\pi c/a)\cos(\pi x/a). \tag{3.32}$$

The semi-total local heat input over the floor is found by integrating this expression over the floor semi-breadth $\frac{1}{2}B$:

$$Q = \lambda(T_i - T_o)L \sum c_n \tanh(n\pi c/a)\sin\left(n\pi \tfrac{1}{2}B/a\right). \tag{3.33}$$

Values found by this expression agree to 4 significant figures with those from equation (3.25). The temperature and flow field in the earth can thus be examined by summing terms similar to equation (3.29) with a/n replacing a.

If there is a water table at a depth below the building comparable with the building width and at the same temperature T_0, the table provides a heat sink in parallel with the ground surface and the heat loss is greater. Delsante (1993) has examined its effect and concludes that if the table is at a depth equal to one-fifth the width of the building, for example, the heat loss roughly doubles: the increase depends somewhat on the thickness of the wall, through which a linear fall in temperature is assumed.

3.2.3 Three-Dimensional Heat Loss

Delsante *et al.* (1983: Section 5) have also given an exact equation for the heat loss from a floor of breadth B and wall thickness W as above, but of finite length L:

$$Q = \lambda(T_i - T_o)(1/\pi)J(L, B, W), \tag{3.34}$$

where J is an extensive function of L, B and W with units of length. Defining $G = L - B$ and $D = L + B$, they obtain

$$\begin{aligned}
J(L, B, W) = {} & (2 + D/W)\sqrt{(L + W)^2 + (B + W)^2} - (D/W)\sqrt{L^2 + B^2} \\
& + ((L^2 + B^2)/W)(1 + \sqrt{2}\ln(\sqrt{2} - 1)) + 2W(\sqrt{2} + \ln(\sqrt{2} - 1)) \\
& - (\sqrt{2}G^2/W)\ln\{[\sqrt{G^2 + (D + 2W)^2} + D + 2W]/[\sqrt{G^2 + D^2} + D]\} \\
& - \sqrt{2}(1 + L/(2W))\sqrt{L^2 + (L + 2W)^2} \\
& - \sqrt{2}(1 + B/(2W))\sqrt{B^2 + (B + 2W)^2} \\
& + [(G^2 - (B + W)^2)/W]\ln\{[\sqrt{(L + W)^2 + (B + W)^2} + B + W]/[L + W]\} \\
& + [(G^2 - (L + W)^2)/W]\ln\{[\sqrt{(L + W)^2 + (B + W)^2} + L + W]/[B + W]\} \\
& + [L(2B - L)/W]\ln\{[\sqrt{L^2 + B^2} + B]/L\} \\
& + [B(2L - B)/W]\ln\{[\sqrt{L^2 + B^2} + L]/B\} \\
& - [(L^2 - W^2)/W]\ln\{[\sqrt{W^2 + (L + W)^2} + W]/[L + W]\} \\
& - [(B^2 - W^2)/W]\ln\{[\sqrt{W^2 + (B + W)^2} + W]/[B + W]\}
\end{aligned}$$

$$+ (2L + W)\ln\{[\sqrt{W^2 + (L + W)^2} + L + W]/W\}$$
$$+ (2B + W)\ln\{[\sqrt{W^2 + (B + W)^2} + B + W]/W\}$$
$$+ (\sqrt{2}L^2/W)\ln\{[\sqrt{L^2 + (L + 2W)^2} + L + 2W]/L\}$$
$$+ (\sqrt{2}B^2/W)\ln\{[\sqrt{B^2 + (B + 2W)^2} + B + 2W]/B\}. \quad (3.35)$$

The first five terms are symmetrical in L and B. The remaining terms are in pairs, with L and B interchanging positions.

L and B are normally large compared with W. Delsante *et al.* (1983) found an approximation to this expression which includes W terms as W^{-1} and W^0. Davies (1993) found a further approximation which includes terms in W^{+1} as well (equations 3 and 53); this provide acceptable approximations to the exact equation, but they are still cumbersome and a simple generalisation of equation (3.25) proves to be more useful.

In equation (3.25) $x = B/W$ served to non-dimensionalise the floor dimension. When L is finite, either L/B or B/L describes the *shape* of the floor and $\sqrt{L/B} + \sqrt{B/L}$ provides a symmetrical *shape* grouping. The relative *size* of the floor is $\sqrt{LB}/\text{W}$. Two parameters can be formed from the shape and size groups:

$$\underbrace{[\sqrt{LB}/W]}_{\text{size}} \times \underbrace{[\sqrt{L/B} + \sqrt{B/L}]}_{\text{shape}} = (L/W + B/W) \quad (3.36a)$$

and

$$\underbrace{[\sqrt{LB}/W]}_{\text{size}} \div \underbrace{[\sqrt{L/B} + \sqrt{B/L}]}_{\text{shape}} = (W/L + W/B)^{-1} = x, \text{ say,} \quad (3.36b)$$

since as L tends to infinity, x tends to B/W as formerly defined. A generalisation of equation (3.25) is then

$$Q = \lambda(T_i - T_o)(L + B)(2/\pi)[(x + 1)\ln(x + 1) - x\ln x] \quad \text{where} \quad x = LB/W(L + B). \quad (3.37)$$

We rewrite equation (3.34) as

$$Q = \lambda(T_i - T_o)(L + B)\{(2/\pi)J(L, B, W)/2(L + B)\}. \quad (3.38)$$

A comparison of the functions

$$\Phi_e = (2/\pi)J(L, B, W)/2(L + B) \quad (3.39a)$$

and

$$\Phi_a = (2/\pi)[(x + 1)\ln(x + 1) - x\ln x] \quad \text{where} \quad x = (W/L + W/B)^{-1} \quad (3.39b)$$

shows that the simple expression Φ_a is a remarkably good approximation to the complicated exact form Φ_e. Table 4 in Davies (1993) shows that Φ_a differs from Φ_e by little more than 1%. The heat loss from a solid floor can be reliably found from equation (3.37), assuming the model can be accepted.

3.2.4 Discussion of Floor Losses

Apart from exposed corridors linking buildings, and buildings having a small ground area, heat losses from solid floors are small. For a square floor with $L = B = 20$ m and $W = 0.3$ m, for example, the U value is $(\lambda/W) \times 0.086$ whereas the U value of the wall itself (neglecting the films) is λ/W. Although these values are not directly comparable, they demonstrate the relative unimportance of the ground loss.

It is intuitively obvious and easily demonstrated from the one-dimensional construction or the heat flow into the lamina that there is a concentration of heat loss through the floor near the walls, so it is best to place the insulation there.

Wall U values include the indoor and outdoor surface film resistances but these are ignored for a floor construction. However, the resistance of a flow tube linking an elementary area in the floor to the corresponding external area is large compared with the convective or radiative resistances, so they may normally be safely neglected.

Floor U values are proportional to λ, which is reported to vary between 0.7 and 2.8 W/m K (see page 22.13 of the 1993 *ASHRAE Handbook of Fundamentals*), and the designer may not have a reliable value. Furthermore, λ may vary with the wetness of the soil, variable near the surface because of evaporation, more so for grassed areas than paved areas.

The basis for long-term heat loss is more complicated than we have assumed since some heat is lost not to a time-varying ambient temperature, but to a constant value some metres below the surface. This may be further complicated by the presence of a water table of varying height (see below).

It would appear neither possible nor necessary to compute long-term floor losses to the same fractional accuracy as those associated with walls (particularly windows) and by ventilation. The short-term time-varying exchange of the floor with the room may be more significant and can be found using methods developed later. A review of the extensive work on earth-contact heat flows is given by Adjali *et al.* (1998).

3.2.5 Placement of Insulation

Any insulation may reduce floor losses, but the resistance to heat flow normally afforded by the soil itself is equivalent to 50 mm or more of insulating material and the thickness of added insulation should be of this order. Hagentoft and Claesson (1991) present graphs to help determine the steady-state loss from a rectangular slab on the ground with some thickness of insulating material beneath it. However, since the flux of heat (W/m^2) from an isothermal floor slab on earth is not uniform over the floor but is greatest near and in the foundation area of the outer walls, this is the best location for insulation if applied non-uniformly; it is immaterial, thermally speaking, whether the slabs are placed vertically (to a depth of 600 mm say, probably against the inner face of the foundation, backfilled) or horizontally inside or outside the wall. There are evident building problems: accommodation of building services, protection against damage and moisture. There is now an extensive literature on theoretical and observational studies of floor losses.

Landman and Delsante (1986, 1987a, 1987b) have examined such placement using an extension to the Fourier method sketched above. They considered a room of internal breadth B with external walls, each of width W. Vertical insulation was assumed, coplanar with the room inner wall and extending downward. They divided the ground area into several regions: region 1 was below the insulation and extending downward to infinity, region 2 was between the centreline and the insulation, region 3 was beyond the insulation and region 4 was the insulation itself. The continuity equation was taken to hold in each region and equations similar to equation (3.29) were set up with appropriate coefficients. The coefficients were determined by assuming continuity of temperature and flux at each boundary and the consequent solution of sets of simultaneous equations. The room was taken to be one of an infinite array of such rooms, well separated from each other. In this way, the number n of eigenfunction temperature distributions needed could be kept below 100. In their 1986 article they reported the reduction of heat loss which resulted from insertion of perfectly insulating material to depth d. When $d/B = 0.05$ the loss was 64% of the uninsulated value. In their first 1987 article they considered the insertion of a slab of finite thickness and conductivity. For a room of width 10 m with walls of thickness 0.2 m and vertical floor insulation of thickness 30 mm down to a depth of 0.5 m, losses of 70–80% of the uninsulated loss were found (i.e. reductions of 20–30%) depending somewhat on the ratio, (conductivity of insulation/soil conductivity). In their second 1987 article they discuss horizontally placed insulation. They express their results as a proportion of the loss of heat when some thickness of insulation is placed beneath the entire floor slab; most of the saving is achieved by widths of insulation of some 0.1 or 0.2 of the room width. Anderson's 1993 study of the effect of edge insulation concluded that if a slab of insulating material ($\lambda = 0.04$ W/m K) 1 m wide and 40 mm thick were placed at each edge and below a 10 m wide floor slab on soil of $\lambda = 2$ W/m K, it would lead to a reduction of heat loss from 2.45 to 2.05 W/K per metre run of floor.

Similar studies were conducted independently and about the same time by Hagentoft (1988) and Hagentoft and Claesson (1991), and by Krarti and his colleagues (Krarti 1989, 1993a, 1993b, Krarti *et al.* 1988a; Krarti and Choi 1995). In the above solutions, all the heat from the floor was taken to flow to the external surface at T_o which extended to infinity. The loss increases however if there is a water table beneath the floor. Krarti and his colleagues, placed a limit to the space beneath the floor slab by assuming a water table at some depth, (e.g. 5 m) which fixes the temperature. They did not make explicit the effect of wall thickness so that a step in temperature is implied between the temperatures of the slab and soil surfaces.[5] Placement of vertical insulation and horizontal insulation (specified simply as a ratio, λ/d) within and outside the wall was discussed and temperature profiles through the soil were plotted for assumed values of temperatures of the building air, the external soil surface and the water table. It was suggested that insulation may lead to reductions of heat loss of around 15%. Hagentoft (1996) noted that the presence of insulation reduced the relative increase in heat loss when a water table limited the flow field beneath a building.

Insulation may prove useful in warm climates as well as cold. Cleaveland and Akridge (1990) illustrate floor loads for Albany GA: the heating loads for December to March are comparable with the cooling loads for June to September. They examine the effect of various placements of layers having insulation of resistance 0.44, 0.88 and 1.76 m^2K/W, (thickness about 18, 36 and 72 mm).

[5] It is strictly invalid to juxtapose two surfaces with different temperatures. This assumption is routinely made in simple room models where it can be justified because the difference between the mean temperatures of a ceiling and some wall, say, may be expected to be fairly small. But the difference between slab and soil surface temperatures might be 15 K, significantly larger than internal temperature differences.

3.2.6 Heat Flow through Corners

Suppose that an outside wall of thickness a forms a corner with another outside wall of thickness b. At points in each wall remote from the corner, the isotherms run parallel to the surfaces and the heat flow lines are directed outward. In and near the cross-sectional area $a \times b$ the isotherms are curved and are determined by the two-dimensional continuity equation (3.27). The total flow pattern is further affected by the values of the inside and outside film coefficients (around 8 and 18 W/m^2K respectively). The temperature of each wall surface falls somewhat toward the corner, leading to higher relative humidities and higher probabilities of mould growth. This problem too has been studied by workers, sometimes ignoring the films and so assuming surfaces which are isothermal up to the corner. See for example Oreszczyn (1988). Tang (1997) reports a study which includes the films. The wall of thickness b affects flow in the other wall up to a distance of $3b$ on the other wall, measured from the inside edge of the corner. Mingfang and Qigao (1997) develop an approximate solution to the problem of corner flow by using the solution for heat flow through a cylinder equation (1.16); see also Sarkis and Letherman (1987).

3.3 SOLUTION USING THE SCHWARZ-CHRISTOFFEL TRANSFORMATION

The Schwarz–Christoffel transformation, found independently by Schwarz in 1869 and Christoffel in 1867, provides a method of finding the temperature field in some two-dimensional space from a knowledge of the field in some other space.[6] It has been used extensively in other disciplines, but not much in the thermal field. Thermal applications are described in Carslaw and Jaeger (1959: 441) and Kumar (1972: 45) and used by Letherman and Sarkis (1984), Krarti *et al.* (1994) and Hagentoft (2002).

Consider the polygon in Figure 3.4a. It is supposed that the temperatures of its sides are known, so the temperature field within it can be found. Since the point A, for example, cannot be at two different temperatures, we suppose that the two sides are not quite in contact. This implies a large local heat flow between the two sides concerned, but it does not affect the temperature field within the polygon. We now suppose that some side, AD for example, is broken at an arbitrary point E and the space within the polygon is stretched so that periphery $EABCDE$ is laid out on a ξ axis, as shown in Figure 3.4b. The four corners have corresponding locations as shown. An element $\mathrm{d}z$ of the field at P, an element of an isothermal for example, is transformed to an element $\mathrm{d}t$ at P' in the object plane, the plane of engineering interest.

It is convenient to refer the original polygon, the image plane, to axes x, y and to handle them in the complex plane $z = x + \mathrm{i}y$. The object plane is correspondingly referred to as $t = \xi + \mathrm{i}\eta$. The relation between an element $\mathrm{d}z$ in the image plane and an element $\mathrm{d}t$ in the object plane is

$$\frac{\mathrm{d}z}{\mathrm{d}t} = C(t-a)^{\alpha/\pi-1}(t-b)^{\beta/\pi-1}(t-c)^{\gamma/\pi-1}(t-d)^{\delta/\pi-1}, \qquad (3.40)$$

[6] A familiar example of a transformation is the image of, say, a square tile placed in front of a vertical convex cylindrical reflecting surface. The image appears laterally foreshortened with curved edges but with right-angled corners.

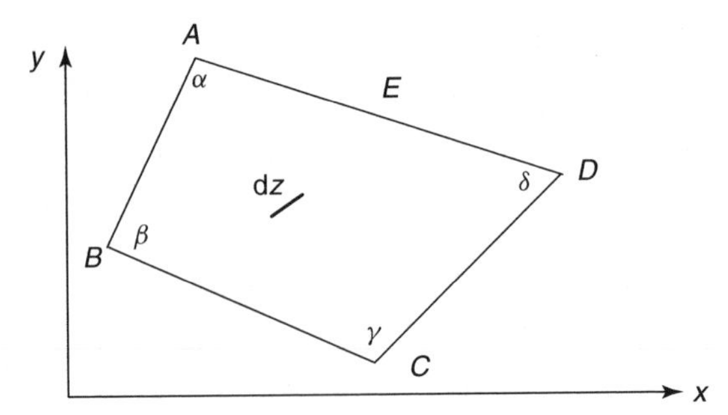

(a) Polygon within which the temperature field is known from the imposed temperatures at its sides.

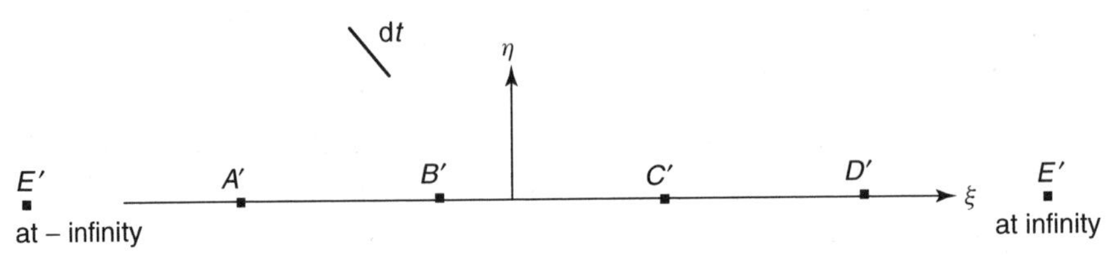

(b) Transform of the periphery of the polygon onto the ξ axis.

Figure 3.4 The Schwarz-Christoffel transformation (Reprinted from *Building and Environment*, vol. 28, M.G. Davies, Heat loss from a solid ground floor, 347–359, © 1993, with permission from Elsevier Science)

where C is a real constant, a is the location of A' on 0ξ, etc., three of which can be chosen arbitrarily, and α is the angle between AB and AD, etc.

As an elementary example, suppose we wish to determine the temperature field due to heat transmission between a semi-infinite plane at some temperature (the floor surface) to a coplanar plane at some other temperature (the soil surface) through the medium beneath them. This can be found from Figure 3.5a, representing flow between two parallel plates at floor and soil surface temperatures T_f and T_s extending between $x' = \pm\infty$, and a distance X apart. We write $x = x'/X$ and $y = y'/X$. Here y denotes the non-dimensional temperature, $(T(x, y) - T_s)/(T_f - T_s)$, and lies between 0 and 1. The surfaces form a degenerate two-sided polygon, the limit of Figure 3.5b as AD tends to zero, so that $\beta = 0$ and $\delta = \pi$.

$$\text{Thus} \qquad \frac{\mathrm{d}z}{\mathrm{d}t} = C(t-a)^{\beta/\pi-1}(t-b)^{\delta/\pi-1} = C\frac{1}{t}, \tag{3.41}$$

since a can be taken to be zero. So

$$z = C\ln t + k \tag{3.42a}$$

or

$$x + \mathrm{i}y = C\ln(\xi + \mathrm{i}\eta) + k = C((\ln\sqrt{\xi^2+\eta^2} + k/C) + \mathrm{i}C\arctan(\eta/\xi)). \tag{3.42b}$$

As ξ increases from $-\infty$ to $+\infty$, for any value of η, y increases from 0 to 1 and $\arctan(\eta/\xi)$ increases from $-\pi/2$ to $+\pi/2$, a total change of π, so $C = 1/\pi$. For $\xi = 0$, η may vary from 0 to ∞, so $\ln\eta$ varies from $-\infty$ to $+\infty$, as does x, hence the constant

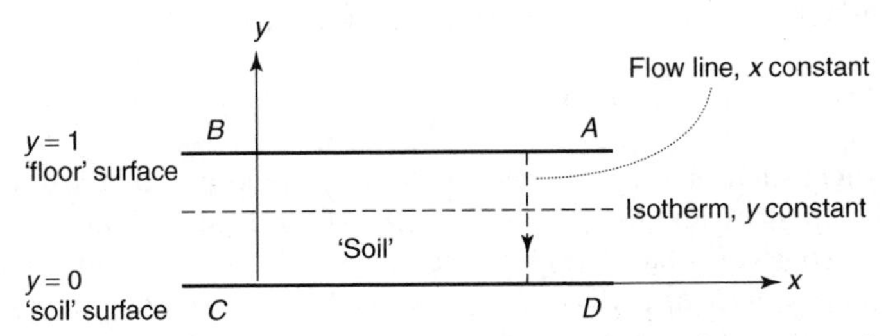

(a) The image plane: heat flow between infinite parallel plates ($-\infty < x < +\infty$, $0 < y < 1$).

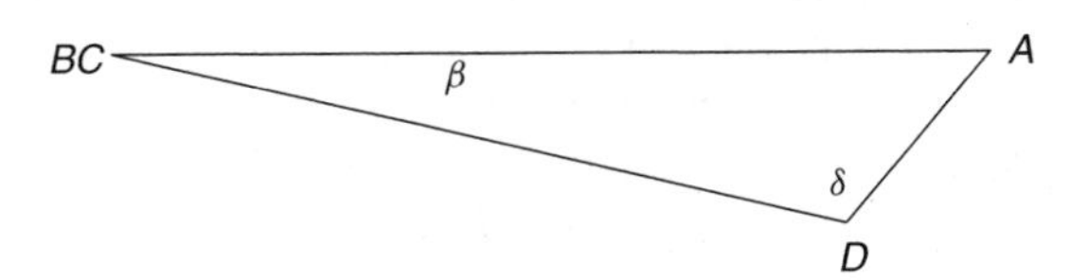

(b) A polygon which has (a) as its limit.

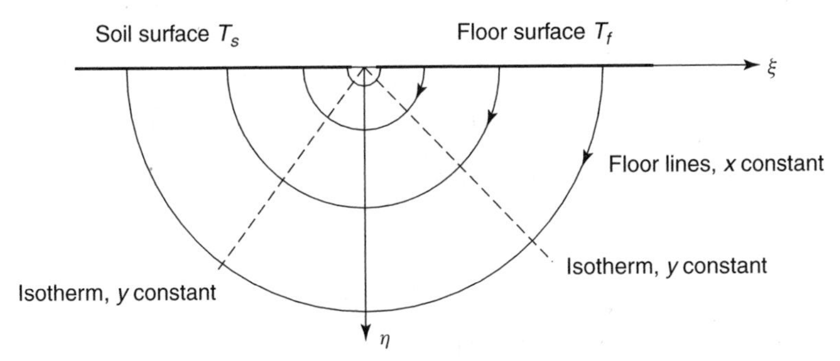

(c) The object plane: flow between two coplanar surfaces ($-\infty < \xi < +\infty$, $0 < \eta < \infty$).

Figure 3.5 Example of the Schwarz-Christoffel transformation

k is zero. Inverting and equating real and imaginary parts, we obtain

$$\mathrm{e}^{\pi x}\cos(\pi y) = \xi \quad \text{and} \quad \mathrm{e}^{\pi x}\sin(\pi y) = \eta, \tag{3.43}$$

thus

$$\mathrm{e}^{\pi x} = \sqrt{\xi^2 + \eta^2}. \tag{3.44}$$

The flow line x = constant in Figure 3.5a maps as the semicircle of radius $\mathrm{e}^{\pi x}$ in Figure 3.5c. The lines are uniformly distributed in Figure 3.5a between $x = \pm\infty$ but are increasingly clustered near $\xi = 0$ in Figure 3.5c. Further, we note that η is positive and that as ξ changes from large positive to large negative values, the temperature changes from near T_f to near T_s, so

$$y = 1 - (1/\pi)\arctan(\eta/\xi). \tag{3.45}$$

The isotherm y = constant in Figure 3.5a maps as a straight line in Figure 3.5c:

$$T(\xi, \eta) = T_s + (T_f - T_s)\pi^{-1}(\pi/2 + \arctan(\xi/\eta)). \tag{3.46}$$

The local heat flow from a strip of width $\mathrm{d}\xi$ from the floor into the soil is

$$\mathrm{d}q = -\lambda \mathrm{d}\xi (\mathrm{d}T/\mathrm{d}\eta)_{\eta=0} = -\lambda \mathrm{d}\xi (\mathrm{d}T/(\xi \mathrm{d}\theta)) = -\lambda(\mathrm{d}\xi/\xi)(T_{floor} - T_{soil})/\pi, \qquad (3.47)$$

where θ is the angle subtended by the element $\mathrm{d}\xi\mathrm{d}\eta$ at the origin; $\mathrm{d}T/\mathrm{d}\theta$ has a uniform value over a semicircular path of $(T_{floor} - T_{soil})/\pi$. Also q tends to infinity as ξ tends to zero, as must be. Thus equation (3.47) must be integrated between positive values of ξ. This implies a semicylinder of perfect insulator placed where the flow must in fact have its maximum value so that the flow is underestimated.

The field in the annular region between radii ξ_1 and ξ_2 is similar to the flow field from a cylindrical source in a medium of conductivity λ, described by equation (3.17), interchanging the roles of isotherms and flow lines.

This procedure can be used to examine the flow field of a floor slab of finite width. Figure 3.6a shows a semi-infinite soil surface at $T_s = -\frac{1}{2}$ and a semi-infinite slab surface at $T_f = +\frac{1}{2}$, together with a similar system at temperatures of $+\frac{1}{2}$ and $-\frac{1}{2}$, displaced by a distance $2a$. Superposed, they form a soil/floor/soil surface at temperatures $0 - 1 - 0$, (Figure 3.6b).

The temperature at the field point P due to the floor at unit temperature is

$$\begin{aligned} T(\xi, \eta) &= +\tfrac{1}{2} + (-1)\left(\tfrac{1}{2} + \pi^{-1}\arctan((\xi - a)/\eta)\right) \\ &\quad - \tfrac{1}{2} + (+1)\left(\tfrac{1}{2} + \pi^{-1}\arctan((\xi + a)/\eta)\right) \\ &= A/\pi. \end{aligned}$$

If P moves so that the angle A remains constant, then (i) $T(P)$ is constant, and (ii) P describes a circular arc. Thus the isotherms are circular. This is an alternative means of arriving at the result noted in Section 3.2.1. As Krarti *et al.* have shown, it is easy to use this construction to include several areas of different widths and temperatures.

These authors have extended the transform method to examine the temperature field around a basement with fixed surface temperatures, with soil of indefinite depth and of

$+^1/_2$ · $2a$ · $-^1/_2$

$-^1/_2$ $+^1/_2$

Figure 3.6(a) Two semi-infinite slab/soil surfaces superposed to form a floor of finite width

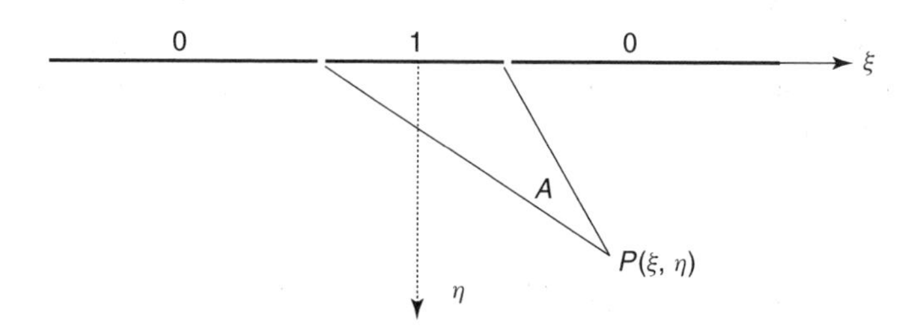

Figure 3.6(b) The field point P beneath a floor slab of width $2a$

finite depth with temperature being defined by a water table. The solution now involves Legendre elliptic functions and iteration is needed.

These studies indicate that although the Schwarz–Christoffel transform itself is elegant, it does not provide a ready means of examining heat losses. The simplest cases, the soil/slab and the soil/floor/soil situations, are more simply examined using the elementary methods of Section 3.2.1. More detailed cases involve two transformations and perhaps difficult solutions, although numerical packages are now being offered on the internet. The assumption that building surfaces are isothermal implies that heat flow lines meet the surfaces normally and this is physically unreasonable when it is no longer valid to assume one-dimensional heat flow. The method strictly implies high heat flows at some corners, or the implied location of perfectly insulating material. Internal convective and radiative exchange together with wall construction are ignored. Finally, representation of the disposition of isotherms well outside the building envelope, though visually attractive, is not of prime interest to designers; their concern is with the places of greatest heat loss.

3.4 APPENDIX: SYSTEMS OF ORTHOGONAL CIRCLES

Macey's approach to estimating heat loss from a solid floor is based on the simple geometrical properties of orthogonal circles. In Section 3.2.1 positive and negative sources were supposed placed at $(+d, 0)$ and $(-d, 0)$, respectively; the temperature T_P at the field point P is $q \ln(r_2/r_1)/2\pi\lambda$. If P moves subject to the condition that r_2/r_1 is constant, it traces an isothermal line in two dimensions. Now

$$\frac{r_1{}^2}{r_2{}^2} = \alpha = \frac{(x+d)^2 + y^2}{(x-d)^2 + y^2} \tag{3.48}$$

and this can be written as

$$\left(x - d\frac{\alpha+1}{\alpha-1}\right)^2 + y^2 = \left(\frac{2d\sqrt{\alpha}}{\alpha-1}\right)^2. \tag{3.49}$$

Now the equation of a circle radius c with centre on $x = D$ is

$$(x - D)^2 + y^2 = c^2, \tag{3.50}$$

so by identification, the locus of the field point P (describing an isothermal) is a circle with a centre at

$$x = D = d(\alpha+1)/(\alpha-1) \text{ and radius } c = 2d\sqrt{\alpha}/(\alpha-1).$$

The heat flow lines for the line source model are orthogonal to the isotherms, and one might guess that they too would be circular. To check this is so, consider the circle K (Figure 3.7) which passes through the points $\pm d$ on the axis, labelled M_1 and M_2; the point M_i at $x = D$ labels the centre of an isotherm. Now

$$\begin{aligned} M_iM_1 \times M_iM_2 &= (D-d)(D+d) = D^2 - d^2 \\ &= d^2\left(\frac{\alpha+1}{\alpha-1}\right)^2 - d^2 = \frac{4\alpha d^2}{(\alpha-1)^2} = c^2 = M_iP^2, \end{aligned} \tag{3.51}$$

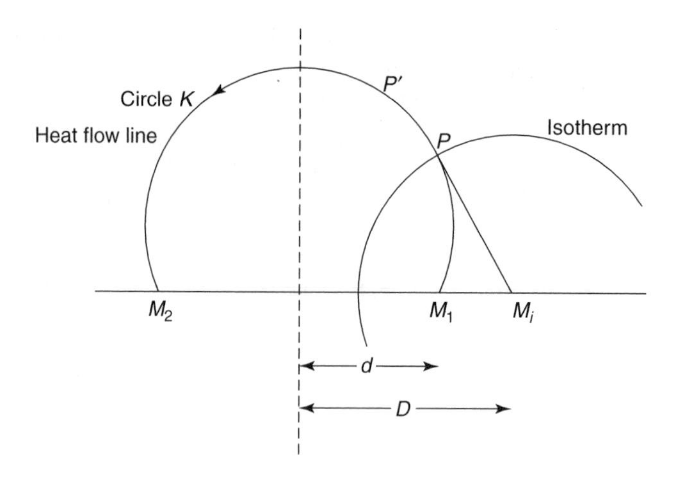

Figure 3.7 Orthogonal systems of circles are formed by the heat flow lines and thermal surfaces associated with conduction between a hot/cold pair of line sources (Reprinted from *Building and Environment*, vol. 28, M.G. Davies, Heat loss from a solid ground floor, 347–359, © 1993, with permission from Elsevier Science)

where P is some point on the isotherm centred at M_i. Thus, by a well-known property of circles, M_iP is a tangent to circle K and has the direction of the periphery of K at P; further, M_iP is perpendicular to the isotherm. Since the direction of heat flow is perpendicular to the isotherm, its direction must be that of circle K. The same is true of any point P' on K. Thus the heat flow lines too have circular paths.

Macey did not use the lines source argument to establish families of orthogonal circles. He started from an application of conjugate functions, based on Carslaw (1906). The approach is sketched in Section 2.1 of Davies (1993).

4

Thermal Circuit Theory

Circuit notation is a standard means of expressing the disposition of resistors and other elements in an electric circuit and the notation has come to be generally though not fully accepted as a means of describing building and other heat flow problems. Just as electric current flows due to a difference in potential between two points in a conductor, so heat flows through a conducting medium due to a temperature difference. Before the mid 1960s, thermal phenomena were often investigated by setting up an electrical analogue model so that a display in circuit form was routine; see for example Nottage and Parmlee (1954, 1955). The correspondence between thermal and electrical elements is presented in Section 10.5. A thermal circuit is sometimes known as an electrical analogue circuit but this is incorrect. Circuit notation exists in its own right in the thermal field.

4.1 BASIC THERMAL ELEMENTS

The basic elements of a thermal circuit are shown in Figure 4.1. Thermal capacity is introduced in Chapter 10.

4.1.1 Reference Temperature

All temperatures must be referred to a base or reference temperature, analogous to the earth potential of electrical theory. Its value is arbitrary but is most conveniently chosen as 0°C.

4.1.2 Temperature Node

The temperature of the air or of some surface will be represented here as a small circle. A black dot is often used. It normally denotes an average value, found from or implying uniformly distributed observations over two dimensions for a surface, or over three dimensions for the air or radiant temperature within an enclosure. Exceptionally it might denote the local value returned by a thermometer or other sensor.

Building Heat Transfer Morris G. Davies
© 2004 John Wiley & Sons, Ltd ISBN: 0-470-84731-X

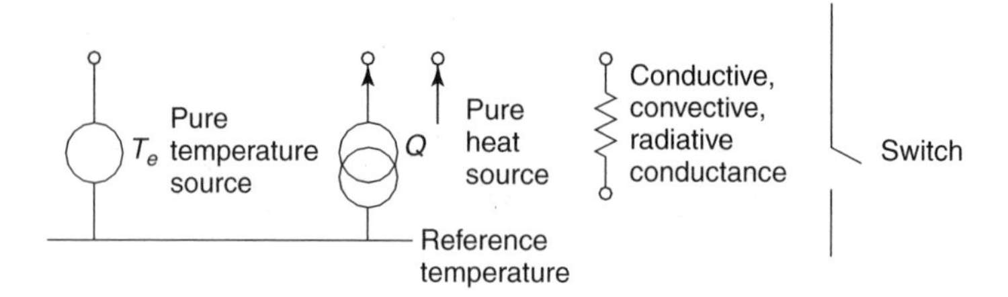

Figure 4.1 Thermal circuit elements

4.1.3 Pure Temperature Source

A pure temperature source represents some mechanism that maintains a certain prescribed temperature at some node, regardless of the associated heat flow. The source accordingly has zero impedance or infinite conductance and must be shown linked to the reference temperature. It is sometimes shown as a single circle, as in Figure 4.1. The main pure temperature source in a building context is ambient temperature T_e, which is taken to be a meteorological variable, independent of the heat flow from a building to it. Strictly speaking, ambient is not exactly pure since it is a little higher over large urban areas than in the surrounding country. For any individual building, however, it may be sufficient to model the local ambient temperature as a pure source.

Sky temperature constitutes a further temperature source.[1] In the absence of an atmosphere, surfaces would radiate to interplanetary space at a temperature of 2.7 K. The cloudless atmosphere of the earth emits long-wave radiation downward because of the various gaseous layers above the ground, which are heated to a range of temperatures. Thus a significant part of the short-wave radiation absorbed by the atmosphere is re-emitted radiatively, some downward and some outward to space. The wavelengths emitted depend on the emission characteristics of the various constituents of the atmosphere at various levels, and their temperatures. The downward flux onto a horizontal surface is the incoming long-wave radiation. It lies mainly in the wavelength band 3-μ m to 30-μ m. This is not black-body radiation and there are windows in the atmospheric spectrum between 8-μ m and 15-μ m (Scharmer and Greif 2000). The sky is effectively at a temperature near but lower than ambient by an amount which depends on the amount of moisture present. This accounts for the appearance of condensation on lightweight elements (e.g. car roofs) and ground frost on cold clear nights, when the ground and other exposed surfaces tend to a temperature which is intermediate between sky and ambient temperatures and which is also below dew point temperature.

Bliss (1961: Figure 19) presents the net long-wave radiant ground–sky exchange (flux from the ground to flux from the sky) in cloudless conditions as a function of air temperature near the ground and dew point (or relative humidity). It amounts to 45–100 W/m^2. Duffie and Beckman (1980: 123) have expressed the Bliss results in terms of sky, air and

[1]The exchange of radiation between the sky and the ground is often expressed as the difference between L_{sky} and L_{ground} (W/m^2), which are the downward and upward fluxes, respectively. However, the exchange takes place because the sky and the ground are in effect at different temperatures: the difference in temperature is not consequent on a radiant exchange. Thus it seems conceptually better to represent the effect of the atmosphere as a temperature rather than a flux source.

dew point temperatures as

$$T_{sky} = T_{air}(0.8 + (T_{dp} - 273)/250)^{1/4}. \quad (4.1)$$

McClellan and Pedersen (1997) also report an expression for sky temperature in clear sky conditions:

$$T_{sky} = (\varepsilon_s)^{1/4} T_{air}, \quad (4.2)$$

where the sky emissivity is

$$\varepsilon_s = 0.787 + 0.764 \ln(T_d/273) \quad (4.3)$$

and T_d (K) is dew point temperature. They give some further expressions, both simpler and more detailed.

Page 165 of the *European Solar Radiation Atlas*, together with results of Cole (1979) as adapted for use in the 1999 *CIBSE Guide Book A* (Section 2.A1.2.2), suggests that, taking the quantity $\sigma T_{sky}^4 (\mathrm{W/m^2})$ to be the long-wave irradiance on a horizontal surface, the effective sky temperature for radiation on a surface at an angle β from the horizontal is

$$T_{sky}(\beta) = T_{air}\{[0.904 - (0.304 - 0.0061 p_w{}^{1/2}) S_h - 0.0005 p_w{}^{1/2}] \cos^2(\beta/2) + 0.09 k_3[1 - (N/8)(0.00822 T_{air} - 1.5386)]\}^{1/4}. \quad (4.4)$$

S_h is the hourly sunshine fraction by day; by night it is $1 - N/8$ where N is the hourly cloud cover (oktas); p_w is the water vapour pressure (Pa). Also k_3 is a smoothly varying function of β with value zero at $\beta = 0°$ and value 0.3457 at $\beta = 90°$; it can be estimated as

$$k_3 = \frac{0.1826 \times 10^{-2} + 0.1795 \times 10^{-3}\beta^2 + 0.4751 \times 10^{-8}\beta^4}{1 + 0.4303 \times 10^{-3}\beta^2 - 0.9966 \times 10^{-8}\beta^4 + 0.2413 \times 10^{-11}\beta^6}. \quad (4.5a)$$

An alternative form for k_3 is given in the 1999 *CIBSE Guide*, page 2–71:

$$k_3 = 0.7629(0.01\beta)^4 - 2.2215(0.01\beta)^3 + 1.7483(0.01\beta)^2 + 0.054(0.01\beta). \quad (4.5b)$$

If moisture movement is included in a model, dew point temperature too constitutes a pure temperature source.

4.1.4 Pure Heat Source

A pure heat source similarly describes some mechanism that delivers a prescribed heat flow to some node, regardless of the conductance linked to the node. It has an infinite impedance or zero conductance. It can be shown (as overlapping circles or otherwise) linked to the reference temperature but that is unnecessary and it may lead to clearer diagrams to omit the link. There are several building examples: solar incidence absorbed at some surface, the lighting system and other electrical gains and heat from a burning gas jet. Metabolic heat production can be modelled more approximately as a pure source. The outputs from these sources split into a radiant fraction p which is absorbed at bounding

Table 4.1 Radiant fraction p (%) delivered by heating appliances

Forced warm air heaters	0
Natural convectors and convector radiators	10
Multicolumn radiators	20
Double-and treble-panel radiators and double-column radiators	30
Single-column radiators, floor warming systems, block storage heaters	50
Vertical and ceiling panel heaters	67
High-temperature radiant systems	90

and internal surfaces and a convective fraction $1 - p$ acting at the air temperature T_a. Values for heating appliances are suggested in Table 4.1, taken from Table A9 of the 1986 *CIBSE Guide*. A hot water radiator is not strictly a pure source since its output depends on room temperature; more detail is required to describe a radiator.

In a survey of heat output from a range of appliances and its split into the radiant/convective fraction, Hosni *et al.* (1999b) noted the following ranges of radiant output:

- office equipment, 15 items ranging from 9% to 41%, average 25%;
- laboratory equipment, 7 items from 5% to 57%, average 29%.

The convective component $100 - p$ of the heat transferred from heating appliances to the surroundings is an instantaneous load, added without time delay to air by convection. The radiant component is first absorbed at solid surfaces, mainly those of the room, and is re-emitted over time. This affects the sizing of air-conditioning equipment.

4.1.5 Conductance

The exchange of heat by conduction, convection and radiation between specified nodes can be expressed as the zigzag element in Figure 4.1; a rectangular box is often used instead. The value of the wall conduction conductance for a single layer in the wall of Chapter 1 is $A\lambda/X$ and Ah_i for the internal convective/radiative link. It has units W/K. Alternatively, a thermal circuit may apply to a wall only, in which case the area is unimportant: the elements, transmittances, then have units $\mathrm{W/m^2\,K}$.

The resistance is the inverse of the conductance, with units K/W, or the inverse of the transmittance, with units $\mathrm{m^2\,K/W}$. The context must show which is intended.

4.1.6 Switch

A switch can be incorporated with a heat source. It might be for infiltration of ambient air into a room; if at some time a ventilation or air-conditioning system is switched on, pressurising the space sufficiently to eliminate the infiltration effect, a switch might be appropriate. It is not appropriate for other elements except that in moisture movement, when evaporation from a surface ceases, its termination can be modelled by opening a switch in the link between the surface and room dew point.

4.1.7 Quasi Heat Source

Two further sources are needed when fluid motion is involved. Consider an elementary enclosure consisting of a six-sided space, five sides of which are adiabatic and perfectly reflecting and so have no thermal role. The sixth has an air-to-air conductance of L (equal to its U value times area). Air at ambient temperature T_e enters the enclosure, is fully mixed, is heated by a convective source Q and leaves at T_a. The thermal flow rate V (W/K) is the product of the mass flow rate (kg/s) and the specific heat (J/kg-K). The internal energies of the air on entry and exit are VT_e and VT_a and heat balance requires that

$$Q + VT_e = VT_a + L(T_a - T_e). \tag{4.6}$$

This can be represented as in Figure 4.2a, where VT_e and VT_a are shown as heat sources. They do not have the same status as the pure convective source Q but they have a similar effect and will be called quasi heat sources.[2]

The airflow leaving the room carrying VT_a with it may continue to another room, in which case the thermal flow is directed to the T_a node of the second room. An application is shown in Section 19.5. When the flow is directed back to ambient however, as in the present example, the two flows can be combined, representing V as a conductance which has the same status as the real conductance L, as shown in Figure 4.2b and equation (4.6) can be written as

$$(V + L)(T_a - T_e) = Q. \tag{4.7}$$

This is the way in which ventilation is usually modelled.

4.1.8 Quasi Temperature Source

Just as a difference in temperature and a fluid flow led to a form of *heat* source, so a real heat source together with fluid flow leads to a form of *temperature* source. If a fluid, flow

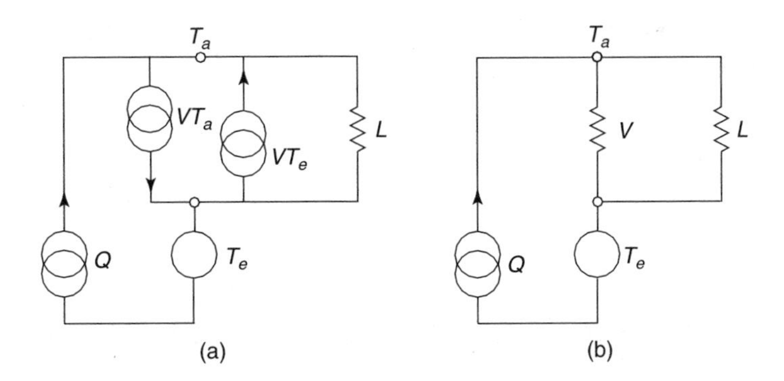

Figure 4.2 Ventilation modelled as (a) a pair of quasi heat elements and (b) a conductance

[2]Author's term. The 1986 edition of *Collins English Dictionary* defines *quasi* as 'resembling but not actually being'.

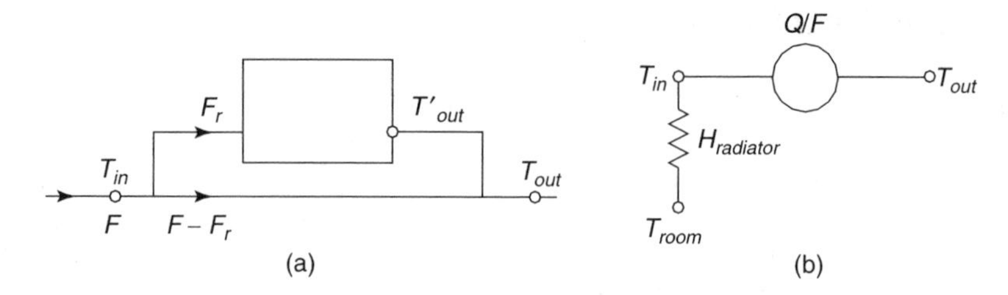

Figure 4.3 (a) Radiator in a one-pipe system and (b) its circuit representation

rate F (W/K), is heated in a pipe or duct, a temperature difference of $T_{out} - T_{in} = Q/F$ is maintained regardless of T_{in}.

A radiator can be modelled in this way. Consider the simple panel radiator in a one-pipe system (Figure 4.3a). The full thermal flow is F, of which F_r passes through the radiator. The heat output of the radiator is Q and as far as the hydronic system is concerned, it behaves as a temperature source Q/F, preceded and followed by similar sources.

To model the link with the room, suppose that the radiator has an area of p per unit length. To avoid the problem of heat emission to a wall, the radiator will be taken to be free-standing so that all its loss may be taken to act at the room temperature T_{room}; p therefore includes both sides. The heat loss δQ from the incremental length δx is

$$\delta Q = hp\ \delta x(T(x) - T_{room}) = F_r(T(x + \tfrac{1}{2}\delta x) - T(x - \tfrac{1}{2}\delta x)) = -F_r\,\delta T. \qquad (4.8)$$

Integrating between $x = 0$ and $x = x$,

$$T(x) - T_{room} = (T_{in} - T_{room})\exp(-hpx/F_r). \qquad (4.9)$$

The total loss from the radiator is

$$Q = \int_{x=0}^{x=X} (T(x) - T_{room})hp\,\mathrm{d}x = (T_{in} - T_{room})F_r(1 - \exp(-P/F_r)), \qquad (4.10)$$

where $P = hpX$. Thus the conductance H between the radiator inlet temperature and room temperature is

$$H = Q/(T_{in} - T_{room}) = F_r(1 - \exp(-P/F_r)). \qquad (4.11)$$

When F_r is small, so that there is a big temperature drop in the radiator, H tends to F_r. When F_r is large, so that the panel is at a near-uniform temperature, H tends to P.

4.2 THE HEAT CONTINUITY EQUATION IN AN ENCLOSURE

The temperature within a room is usually attributable to a single real temperature source, ambient temperature, and a number of heat sources acting at different nodes. There are two approaches to analysing the problem, one based on a closed configuration of resistances and associated with the name of G.R. Kirchhoff (1842–1887) and the other based on the temperature nodes. They are exactly equivalent and lead to a set of simultaneous equations. The mesh method leads to fewer equations than the nodal method and might be preferred for hand calculations on very simple circuits. The nodal method is the more convenient for computer formulation. Both approaches will be illustrated here.

We return to the simple enclosure of Section 4.1.7, an outer wall with insulating, reflecting inner surfaces, but we will include more detail in the outer wall. It is now decomposed into its inner film, the solid construction of thickness X and conductivity λ, and an outer film. A pure convective source Q_a continues to act at the air temperature T_a, but additionally a long-wave radiant source Q_r acts at the wall inner surface node, T_{wi}, and short-wave radiation (i.e. solar gain) Q_s acts at the outer surface node, T_{wo}.

4.2.1 The Mesh Approach

Formulated in thermal terms, Kirchhoff's laws are

(i) the algebraic sum of all heat flows arriving at a node is zero;

(ii) the sum of the products of heat flows multiplied by resistance taken round a closed loop equals the sum of the temperature sources acting in the loop (zero if the loop does not include a source).

In Figure 4.4a the inner film, wall, outer film and ventilation are represented in resistance form (K/W) as R_i, R_w, R_o and R_v respectively. The heat gains are shown. Q_a acts at T_a and splits into Q_1 and $Q_a - Q_1$ as shown, in accordance with the first law, and similarly at nodes T_{wi} and T_{wo}. Using the second law, starting from T_a and moving anticlockwise, we obtain

$$(Q_a - Q_1)R_i + (Q_a + Q_r - Q_1)R_w + (Q_a + Q_r + Q_s - Q_1)R_o - Q_1 R_v = 0, \quad (4.12)$$

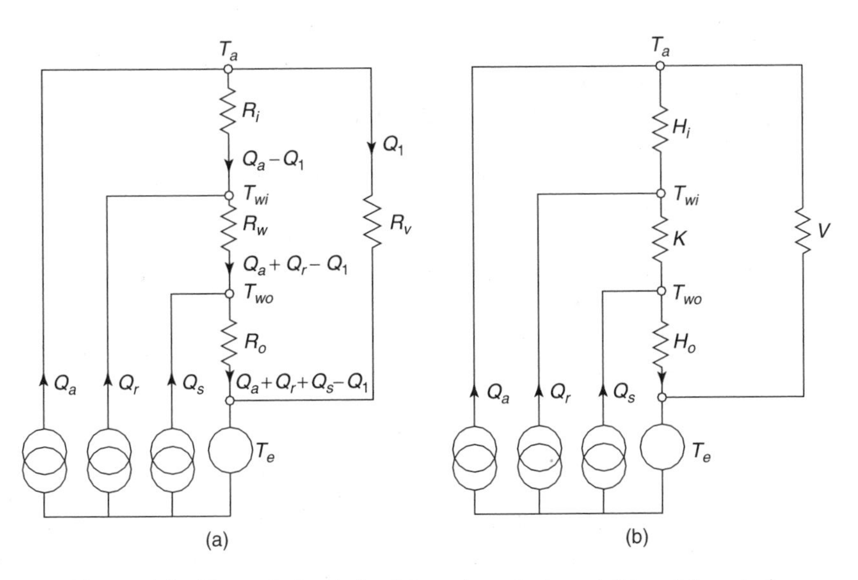

Figure 4.4 Thermal circuit for (a) mesh analysis and (b) nodal analysis

since the loop does not include the temperature source T_e. The last term is negative since we are progressing *against* the flow. So

$$Q_1 = \frac{Q_a R_i + (Q_a + Q_r)R_w + (Q_a + Q_r + Q_s)R_o}{R_i + R_w + R_o + R_v}. \tag{4.13}$$

The unknown temperatures follow:

$$T_a = T_e + Q_1 R_v \tag{4.14}$$

and T_{wo} and T_{wi} can be found similarly.

4.2.2 The Nodal Approach

Figure 4.4b shows the same circuit with the links in conductance form: $H_i (= Ah_i) = 1/R_i$, etc. A continuity equation must be written down at each node whose value of temperature is initially unknown:

$$\begin{aligned} &\text{at } T_a, \; Q_a = (T_a - T_{wi})H_i + (T_a - T_e)V; \\ &\text{at } T_{wi}, \; Q_r = (T_{wi} - T_a)H_i + (T_{wi} - T_{wo})K; \\ &\text{at } T_{wo}, \; Q_s = (T_{wo} - T_{wi})K + (T_{wo} - T_e)H_o. \end{aligned} \tag{4.15}$$

Any input of heat at T_e has no effect since T_e, as a temperature source, has zero impedance. So

$$\begin{bmatrix} H_i + V & -H_i & \\ -H_i & H_i + K & -K \\ & -K & H_o + K \end{bmatrix} \begin{bmatrix} T_a \\ T_{wi} \\ T_{wo} \end{bmatrix} = \begin{bmatrix} Q_a + T_e V \\ Q_r \\ Q_s + T_e H_o \end{bmatrix}. \tag{4.16}$$

This is of form $\mathbf{AT} = \mathbf{b}$. Thus we have three equations now while in the mesh method there was only one but this is unimportant for computer computation. The matrix $\mathbf{A}$ can be written down by inspection of the thermal circuit since (i) the conductances on the principal diagonal are the conductances linked to the node concerned (e.g. H_i and V for T_a) and (ii) the off-diagonal terms are symmetrical and are the negative of the conductance between the nodes concerned. If the nodes are not directly linked (as for T_a and T_{wo}), the entry is zero. (When the ventilation action is modelled as a source, the matrix is no longer symmetrical in V (Section 19.5).)

The vector $\mathbf{b}$ represents known, independent variables. The group $Q_s + T_e H_o$ leads to the so-called sol-air temperature T_{sa} (Mackey and Wright 1943, 1946a):

$$T_{sa} = q_s/h_o + T_e = (Q_s + T_e H_o)/Ah_o. \tag{4.17}$$

T_{sa} provides a useful combination of the independent time-varying values of solar incidence and ambient temperature, which is valid as a driver provided that h_o does not vary significantly. Variation of h_o is likely to be unimportant for a wall of high resistance but may be important for a thin wall or a window. Surfaces of different orientations or different short-wave absorptivities have different T_{sa} values. Other examples having this form are given later.[3]

4.3 EXAMPLES

4.3.1 The Ventilated Cavity

If a wall includes an unventilated cavity, the cavity resistance R_{cav} (often taken as $0.18\,\text{m}^2\,\text{K/W}$) forms a significant element in its transmission matrix and the overall wall transmittance is found by including the matrix in the sequence illustrated by equation (1.14a). If the cavity is ventilated to ambient, intentionally or otherwise, there is an additional heat loss mechanism. It can be examined using the ventilation conductance V of Section 4.1.7.

Consider a wall consisting of an outer and inner solid structure with a ventilated cavity. Each layer has a conduction resistance and two film resistances which are not of interest individually and whose sum will be denoted simply as R_o and R_i (K/W) as shown in Figure 4.5. (To avoid some complication, the radiant exchange between the surfaces will be taken as zero.[4]) The cavity air is represented as a two-terminal element with temperatures and heat flows as shown. From continuity,

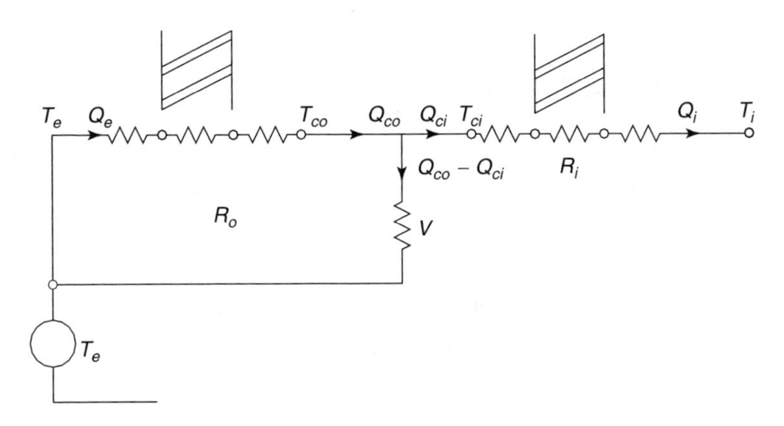

Figure 4.5 The ventilated cavity

[3]Mackey and Wright's definition of sol-air temperature only included a solar component. It was later broadened to include the long-wave exchange between a building surface and its surroundings. The derivation of T_{sa} in the 1986 *CIBSE Guide* (equation A5.102) however places the short-wave and long-wave exchanges on the same footing. This is not formally incorrect, but it is misleading since the short-wave solar flux is driven by a pure heat flow source, whereas the long-wave exchange is driven by a temperature difference, as noted earlier.

[4]If the radiation exchange is to be included, three matrices and the delta-star transform are needed. Take the temperature of the inner surface of the outer leaf as T_{oi} and the temperature of the surface it faces, as T_{io}. Take the radiation resistance between the surfaces as R_r and the convection resistances to the cavity air temperature T_{ca} as R_{oa} and R_{ia}. These three resistances form a delta pattern which can be transformed exactly to star form as shown below, to give resistances between the cavity index temperature T_{ci} and the real surface temperatures of R_{of}, R_{if}; these provide the resistive elements in matrices 1 and 3. The resistance between T_{ca} and T_{ci} is R_{af} and the e_{21} element of the centre matrix is $1/R_{af} + V$.

$$Q_{co} - Q_{ci} = VT_{co} \quad \text{and} \quad T_{co} = T_{ci},$$

so

$$\begin{bmatrix} T_{co} \\ Q_{co} \end{bmatrix} = \begin{bmatrix} 1 & 0 \\ V & 1 \end{bmatrix} \begin{bmatrix} T_{ci} \\ Q_{ci} \end{bmatrix}. \tag{4.18}$$

The outside and inside temperatures and flows are then related as

$$\begin{bmatrix} T_e \\ Q_e \end{bmatrix} = \begin{bmatrix} 1 & R_o \\ 0 & 1 \end{bmatrix} \begin{bmatrix} 1 & 0 \\ V & 1 \end{bmatrix} \begin{bmatrix} 1 & R_i \\ 0 & 1 \end{bmatrix} \begin{bmatrix} T_i \\ Q_i \end{bmatrix}. \tag{4.19}$$

On multiplying out and setting $T_e = 0$, the U value is found to be

$$U = \left(\frac{Q_i}{T_i}\right)_{T_e=0} = \frac{-1}{R_i + R_o/(1 + R_oV)}. \tag{4.20}$$

The negative sign denotes that the heat flow is directed outwards, to the left. If V is zero, U has its usual value. As V increases, $R_o/(1 + R_oV)$ decreases and U increases; in effect, the outer resistance is being bypassed. The U value can be readily found by elementary means; the matrix formulation only becomes advantageous when there are several elements. The elementary form for a capacity is similar to that for V; see equation (10.1). Cavity ventilation is said to be ineffective in drying out moisture from driving rain or condensation (Hens and Fatin 1995: 618).

4.3.2 A Basic Circuit for Thermal Response

Figure 4.6 illustrates an elementary circuit for heat transfer. A specific application is to identify T_1 with the rad-air temperature in a room (Chapter 7), and T_2 with the mean air temperature. L, X and V then denote respectively the conduction loss, internal exchange and ventilation loss conductances. L represents the sum of the conduction losses, $\sum L_j$ where $j = 3, 4, 5, \ldots$, (leaving subscripts 1 and 2 for rad-air and air respectively). If a heat input Q_1 acts at T_1, the temperatures (relative to the reference temperature) are readily found as

$$T_{11} = \frac{Q_1(V + X)}{VX + XL + LV}, \quad T_{21} = \frac{Q_1X}{VX + XL + LV}. \tag{4.21}$$

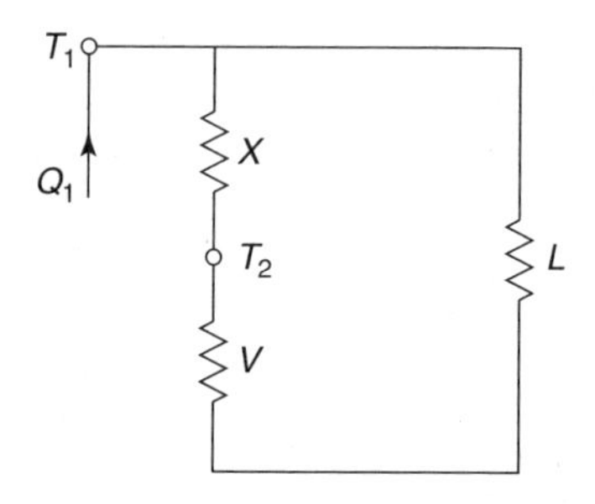

Figure 4.6 A basic circuit for thermal response

The first subscript denotes the temperature node, the second the node at which the heat flow acts.

Consider the application of a heat flow Q_2 at T_2. The temperature it generates at T_1 can readily be shown to be the same as application of the reduced flow $Q_2X/(X+V)$ at T_1, independent of L. Furthermore,

$$T_{12} = \frac{Q_2 X}{VX + XL + LV}, \tag{4.22}$$

so

$$T_{12}/Q_2 = T_{21}/Q_1. \tag{4.23}$$

Such a relation is known as a reciprocity relation.

Similarly, if a node T_3 were located on one of the components of L, say L_3, dividing it into conductances H_3 (the combined internal convective and radiant exchanges) and F_3 (the conductive and external film links) and a heat flow Q_3 (perhaps due to solar radiation) acted at T_3, its effect at T_1 is the same as the reduced flow $Q_3H_3/(H_3+F_3)$ acting at T_1, independently of X, V and the other components of L. This is the basis for the so-called 'surface factor' of the 1986 *CIBSE Guide*.[5]

Although the reduced flow $Q_2X/(X+V)$ applied at T_1 generates the same temperatures and heat flows in all parts of the network other than the link through T_2, the value it generates at T_2 itself is lower than the correct value by an amount $Q_2/(X+V)$; this deviation too is independent of L.

4.4 CIRCUIT TRANSFORMS

It is convenient to summarise a number of transformations for electrical and thermal circuits.

4.4.1 Thévenin's and Norton's Theorems

Thévenin's theorem states that, in thermal terms, as far as its effect at two nodes A and B is concerned, a circuit of complex form can be resolved into a temperature source together with a resistance in series with it (Figure 4.7a). Norton's theorem similarly resolves the complex circuit into a heat flow source and a resistance in parallel. It follows from Thévenin's theorem or Norton's theorem that the effects of a step in air temperature adjacent to a solid surface are the same as though the air temperature remained constant but an appropriate pure heat flow acted at the surface itself; this situation is discussed in Sections 12.3.1, 12.7 and 13.2. Norton's theorem is not of interest in a building context, but Thévenin's theorem formalises a result that has already appeared as equation (4.17) and can be generalised.

[5] Equation (A3.54) gives the expression for the surface factor $SF = 1 - R_{si}Y_c$ or $1 - (R_{si}/A)(Y_cA)$, where A is the area of the surface whose surface film coefficient is R_{si} (m^2 K/W) and whose admittance is Y_c(W/m^2 K), the relevant measure when the wall is excited sinusoidally with period 24 hs. In steady-state conditions, Y_cA is replaced by the wall UA value. In the present notation, $R_{si}/A \equiv 1/H_3$ and $UA \equiv 1/(1/H_3 + 1/F_3)$. Then SF is found to be $H_3/(H_3+F_3)$.

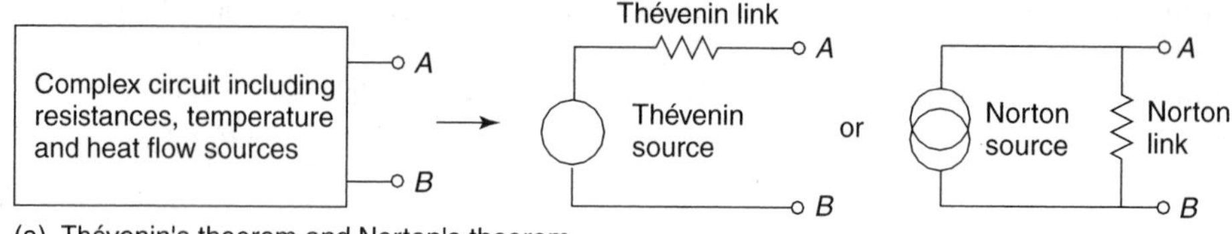

(a) Thévenin's theorem and Norton's theorem.

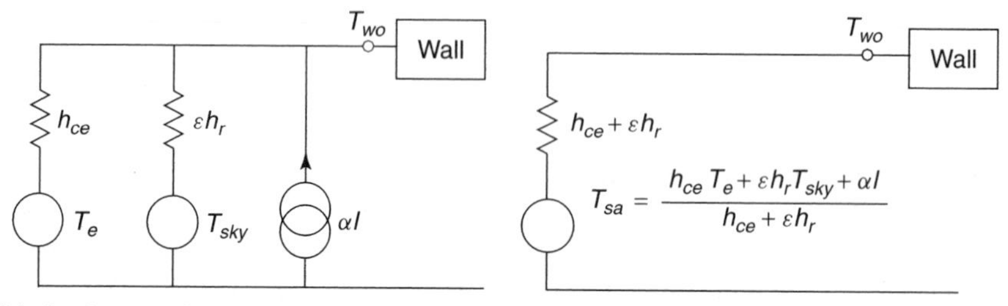

(b) Application of Thévenin's theorem to form sol-air temperature T_{sa}.

Figure 4.7 Equivalence theorems

The thermal forces acting on a building exterior at T_{wo} are ambient temperature T_e, which in extreme climates may lie outside ±40°C, a long-wave exchange, mainly a loss, of some tens of W/m^2 to the sky and surroundings and a gain of αI by solar radiation of order of hundreds of W/m^2. The heat flow q_{wo} into the outer surface of a flat roof is

$$q_{wo} = (T_e - T_{wo})h_{ce} + (T_{sky} - T_{wo})\varepsilon h_r + \alpha I, \tag{4.24}$$

where h_{ce} and εh_r are respectively the heat transfer coefficients by convection and long-wave radiation (see later). The flux can be written as

$$q_{wo} = (T_{sa} - T_{wo})(h_{ce} + \varepsilon h_r), \tag{4.25}$$

where

$$T_{sa} = \frac{T_e h_{ce}}{h_{ce} + \varepsilon h_r} + \frac{T_{sky} \varepsilon h_r}{h_{ce} + \varepsilon h_r} + \frac{\alpha I}{h_{ce} + \varepsilon h_r}. \tag{4.26}$$

A sloping or vertical surface may exchange long-wave radiation with surrounding buildings or other surfaces and the ground; long-wave radiation reflected from the ground may also be considered. This leads to complications in the details of the long-wave component of T_{sa} but not its general form. Values for T_{sa} for the United Kingdom are tabled in the 1999 edition of the *CIBSE Guide*. The following values (page 2-55) relate to a horizontal surface, London area, in July.

Time	T_e	T_{sa} (dark surface, $\alpha = 0.9$)	T_{sa} (light surface, $\alpha = 0.5$)
03 (before sunrise)	13.6	10.1	10.1
06 (after sunrise)	14.3	18.1	15.0

Before sunrise, T_{sa} is less than T_e due to the long-wave loss and is independent of the value of absorptivity to solar radiation. After sunrise, T_{sa} is greater than T_e and depends on absorptivity.

The sol-air form appears in connection with solar heating of internal spaces; see equations (9.26) and (9.28). Sol-air temperature, the external index temperature, and rad-air temperature, the internal index temperature, have the same structure; see equations (7.6) and (7.7). Each index is independent of the thermal details (conductivity, etc.) of the wall or roof.

4.4.2 Delta-Star Transformation

The delta-star transformation is illustrated in Figure 4.8. Consider the three conductances C_1, C_2 and C_3 directly linking the three temperatures T_1, T_2 and T_3. The configuration is supposed to form part of some larger circuit. For the larger circuit in which T_1, T_2, T_3 are enmeshed, the delta form can be replaced exactly by the star form, provided that

$$C_1K_1 = C_2K_2 = C_3K_3 = C_1C_2 + C_2C_3 + C_3C_1 = \frac{K_1K_2K_3}{K_1 + K_2 + K_3}. \tag{4.27}$$

The temperature at the centre of the star is

$$T_{star} = \frac{K_1T_1 + K_2T_2 + K_3T_3}{K_1 + K_2 + K_3}. \tag{4.28}$$

In applying this transform, the C conductances represent physical constructs and the K conductances their non-physical equivalents. T_{star} is clearly a weighted mean of T_1, T_2 and T_3 but has no immediate physical significance. In an electrical circuit, any potential is simply the potential at a point. In thermal circuits, temperature may denote the temperature at a point, the value returned by a thermometer or other sensor. But it may refer to the mean temperature of a surface, a *two-dimensional* construct, or to the mean temperature within a space, a *three-dimensional* construct. If T_1, T_2 and T_3 result from different forms of averaging, it is not valid to form the star temperature T_{star} (Section 7.3.3).

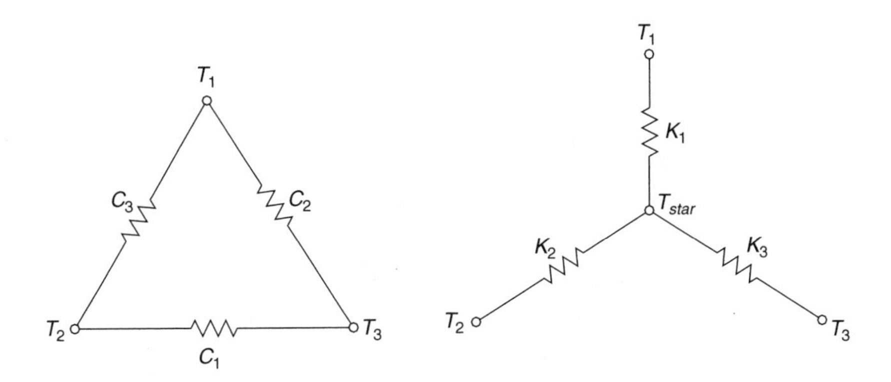

Figure 4.8 The delta-star transformation

4.4.3 Series-Parallel Transformation

A further transformation, the series-parallel transformation, will be used later to merge the convective and radiative exchange mechanisms in a room (see Figure 4.9). In the upper circuit, T_1, a surface temperature, is linked to T_a by the conductance C_1 and further to T_{rs} through S_1.[6] A heat flow Q_r acts at T_{rs}; the component Q_{f1} is lost by conduction from T_1 through the fabric and the remainder, Q_v, by ventilation from T_a. Then

$$T_1 = T_a + (Q_r - Q_{f1})/C_1 \tag{4.29a}$$

and

$$T_{rs} = T_a + (Q_r - Q_{f1})/C_1 + Q_r/S_1. \tag{4.29b}$$

T_a and T_1 have the same significance in Figure 4.9b, but a new node, T_{ra}, is located on C_1 defining segment conductances X and Y such that $X/Y = C_1/S_1 = \alpha_1$, say. Thus, although S_1 no longer appears explicitly, its value is not lost. Since $1/X + 1/Y = 1/C_1$,

$$X = (S_1 + C_1)C_1/S_1 = (1 + \alpha_1)C_1 \tag{4.30a}$$

and

$$Y = S_1 + C_1 = (1 + \alpha_1)S_1. \tag{4.30b}$$

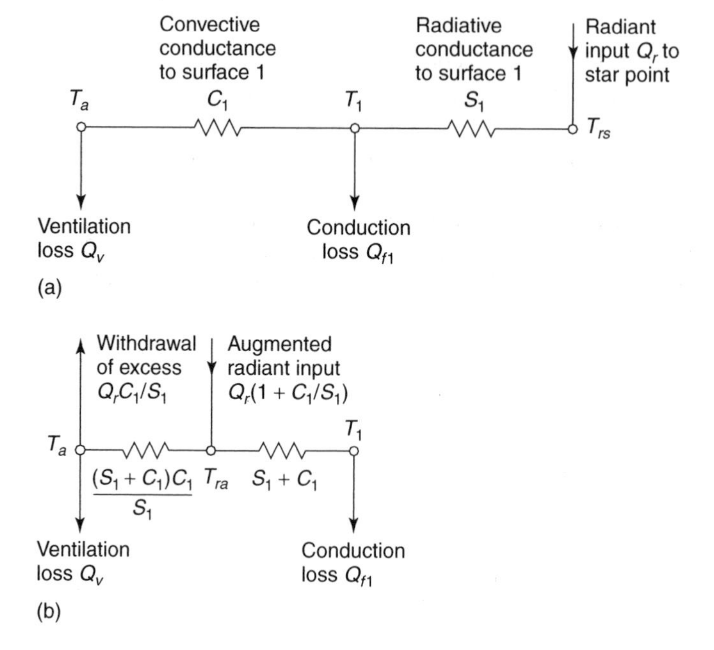

Figure 4.9 The series-parallel transformation

[6]The conductance S_1 and the radiant star node T_{rs} are equivalent constructs formed from the physical pattern of radiant transfer between the surfaces of a room. They are developed in Chapter 6.

The heat input is handled as an augmented flow $Q_r(1+\alpha_1)$ at T_{ra} with a withdrawal of the excess, $Q_r\alpha_1$, from T_a, so that the total input remains the same. Then

$$T_{ra} = T_a + [Q_r(1+\alpha_1) - Q_{f1}]/X. \tag{4.31a}$$

The value of temperature at the T_1 node, T_1' say, and possibly differing from the above value, is

$$T_1' = T_a + [Q_r(1+\alpha_1) - Q_{f1}]/X - Q_{f1}/Y. \tag{4.31b}$$

This in fact is found to equal T_1. Further, it will be found that T_{rs}, not included in the second circuit, can be found from it as

$$T_{rs} = T_{ra}(1+\alpha_1) - T_a\alpha_1 = T_{ra}(S_1+C_1)/S_1 - T_aC_1/S_1. \tag{4.32}$$

Thus the second circuit provides exactly the same information as the first: one is a transform of the other. In Figure 4.9a heat flows through S_1 and C_1 *in series*; in (Figure 4.9b) heat flows from T_{ra} to T_1 with S_1 and C_1 effectively *in parallel*, so that we have the series-parallel transformation.

When this transformation is used later on, the conductances have various physical interpretations.

- C_1 directly represents the convective exchange between surface 1 and the air.
- S_1 models the radiation falling on surface 1 from the remainder of the enclosure but since radiation is physically a surface-surface exchange, its representation as a surface-star point exchange involves some simplification of the exchange pattern.
- No physical significance can be attached to the conductance $X = (S_1+C_1)C_1/S_1$.
- The conductance $Y = S_1 + C_1$ has no strict physical interpretation but can be loosely regarded as representing the transfer of heat by both mechanisms from the enclosure as a whole, driven by an index temperature T_{ra} to one of its surfaces (T_1).

The transform can be extended to two surfaces at T_1 and T_2. Figure 4.10a is similar to Figure 4.9a and it can be shown that

$$T_{rs} = T_a + \frac{Q_r(S_1+C_1)(S_2+C_2) - Q_{f1}S_1(S_2+C_2) - Q_{f2}S_2(S_1+C_1)}{S_1C_1S_2 + S_1C_1C_2 + S_1S_2C_2 + C_1S_2C_2}. \tag{4.33}$$

If each surface is to be linked to a central node T_{ra}, as shown in Figure 4.10b, the links must be $S_1 + C_1$ and $S_2 + C_2$. The link X between T_{ra} and T_a, previously defined by equation (4.30a), must now take account of both surfaces and will be written as

$$X = (S_1+S_2+C_1+C_2)(C_1+C_2)/(S_1+S_2) \quad \text{or} \quad (S+C)C/S, \tag{4.34}$$

where

$$S = S_1 + S_2 \quad \text{and} \quad C = C_1 + C_2. \tag{4.35}$$

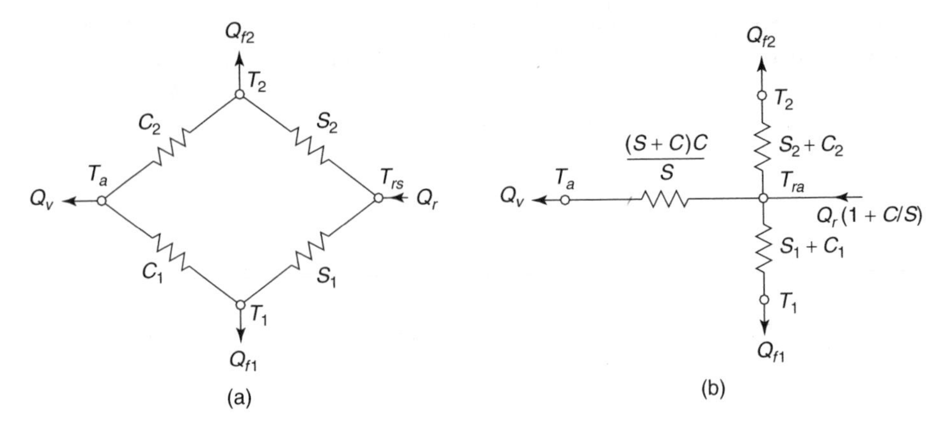

Figure 4.10 The series-parallel transformation for two surfaces

With an input at T_{ra} of $Q_r(1 + C/S)$,

$$T_{ra} = T_a + [Q_r(1 + C/S) - (Q_{f1} + Q_{f2})]/X. \tag{4.36}$$

The corresponding value of T_{rs} is given by the generalised form of equation (4.32), but written as T'_{rs} since it might not equal T_{rs} of equation (4.33):

$$T'_{rs} = T_{ra}(S + C)/S - T_a C/S, \tag{4.37a}$$

and on evaluation we obtain

$$T'_{rs} = T_a + \frac{Q_r(S_1 + S_2 + C_1 + C_2) - (Q_{f1} + Q_{f2})(S_1 + S_2)}{(S_1 + S_2)(C_1 + C_2)}. \tag{4.37b}$$

T'_{rs} clearly differs from T_{rs} in equation (4.33). However, if the conductances are such that

$$C_1/S_1 = C_2/S_2 (= \alpha \text{ say}), \tag{4.38}$$

then they assume the same value:

$$T'_{rs} = T_{rs} = T_a + \frac{Q_r(1 + \alpha) - (Q_{f1} + Q_{f2})}{C_1 + C_2}. \tag{4.39}$$

T_1 and T_2 also have the same values in the two formulations if $C_1/S_1 = C_2/S_2$. Thus, with this condition, heat flows expressed in terms of the two nodes T_a and T_{rs} can be computed exactly using the circuit centred on the node T_{ra}. The second formulation is less physical than the first but is simpler to handle.

The transformation remains true for an enclosure of three or more surfaces provided that

$$C_1/S_1 = C_2/S_2 = C_3/S_3 = \cdots = C_n/S_n. \tag{4.40}$$

This may be expected to be roughly true since convective and radiative conductances increase with surface area, though not in the same way. A mean value of α is given as

$$\alpha = \frac{C_1 + C_2 + C_3 + \cdots}{S_1 + S_2 + S_3 + \cdots} = \frac{\sum C_j}{\sum S_j} = \frac{C}{S}. \tag{4.41}$$

T_{ra} is a linear combination of T_{rs} and T_a:

$$T_{ra} = \frac{ST_{rs}}{S + C} + \frac{CT_a}{S + C}. \tag{4.42}$$

5

Heat Transfer by Air Movement

Air movement in a room is essential for the well-being of its occupants and it plays a major role in determining the thermal environment. The movement is associated with bulk transport due to ventilation or infiltration (Section 1.3.1). Section 19.5 notes the extensive work on air movement between rooms or between air-duct inlets and outlets. This chapter considers heat transfer between air and the solid surfaces bounding the room or the furnishings.

At a fundamental level, room heat transfer by air movement, convection, is much more complicated than transfer by conduction and radiation. Writing x_1, x_2 and x_3 for the Cartesian coordinates, x, y and z and u_1, u_2 and u_3 for the corresponding velocities, with p, ρ and μ for pressure, density and viscosity, respectively, the basic relationships, the Navier–Stokes equations for a fluid moving under the action of a body force F_i(m/s^2) and heat input Q (W/m^3) can be written as follows.

The conservation of mass (rate of change of mass per unit volume) is

$$\frac{\partial \rho}{\partial t} + \frac{\partial}{\partial x_j}(\rho u_j) = 0. \tag{5.1}$$

The conservation of momentum (rate of change of momentum per unit volume) is

$$\rho\left(\frac{\partial u_i}{\partial t} + u_j\frac{\partial u_i}{\partial x_j}\right) = \rho F_i + \frac{\partial}{\partial x_j}\left[\mu\left(\frac{\partial u_i}{\partial x_j} + \frac{\partial u_j}{\partial x_i}\right)\right] - \frac{2}{3}\frac{\partial}{\partial x_i}\left(\mu\frac{\partial u_j}{\partial x_j}\right) - \frac{\partial p}{\partial x_i}. \tag{5.2}$$

The conservation of energy (flow of energy per unit volume) is

$$\rho c_p\left(\frac{\partial \theta}{\partial t} + u_j\frac{\partial \theta}{\partial x_j}\right) = Q + \frac{\partial p}{\partial t} + u_i\frac{\partial p}{\partial x_j} + \frac{\partial}{\partial x_j}\left(\lambda\frac{\partial \theta}{\partial x_j}\right)$$
$$+ \mu\left[\frac{\partial u_i}{\partial x_j}\left(\frac{\partial u_i}{\partial x_j} + \frac{\partial u_j}{\partial x_i}\right) - \frac{2}{3}\left(\frac{\partial u_j}{\partial x_j}\right)^2\right]. \tag{5.3}$$

See for example Ede (1967a). To these must be added some model of turbulence. The k-ε model (Launder and Spalding 1974) has been used in the context of the built environment. It defines the turbulent exchange coefficient for momentum from two additional

Building Heat Transfer Morris G. Davies
© 2004 John Wiley & Sons, Ltd ISBN: 0-470-84731-X

variables, the turbulent kinetic energy k and its rate of dissipation ε, that characterise the local state of turbulence. Nielsen (1998) has discussed the use of this and other models to predict airflow patterns in rooms. For further details see Clarke (2001: Ch. 5). A representative example of their use – airflow driven by a radiator – is given by Lu *et al.* (1997). Solution by computer is slow, especially when the airflow is driven by weak buoyancy forces.

Such solutions may be needed to determine flow patterns in many circumstances and in special enclosures such as atria. The equations form the basis for several of the studies reported below.[1] Fortunately, the convective solutions required for determining the thermal behaviour of the room, in conjunction with radiation as the complementary mechanism for internal transport and conduction/storage for the wall, are relatively simple, consisting of expressions for heat flow through boundary layers. The topic can be addressed by adapting experimental findings presented through non-dimensional groupings or by a detailed examination of the action of buoyancy in a boundary layer or an enclosure as a whole.

5.1 LAMINAR AND TURBULENT FLOW

When a fluid moves relative to a surface, with or without associated heat transfer, the flow may be either laminar or turbulent. Provided the mean free path of the molecules is very small compared with the surface dimensions, invariably the case in building applications, the air in contact with the surface is stationary and the velocity vector of air particles close to the surface follows the broad contour of the surface in the direction of the bulk flow, the x direction say. These fluid layers move smoothly upon one another and the regime is described as laminar. Further away from the surface, instability may set in, the vector may have a steady x component $\overline{u}$ and randomly fluctuating x, y and z components u', v' and z'; then the flow is described as turbulent. The degree of turbulence can be described by the ratio $\sqrt{\frac{1}{3}(\overline{u'^2} + \overline{v'^2} + \overline{w'^2})}/\overline{u}$, and may be typically up to 10%. A classical illustration of the transition is the demonstration due to Reynolds in 1883. Water flows through a glass tube and an injection of dye shows the form of the flow. At low flow rates, the dye steam has the form of a thin thread moving downstream and only slowly diffusing sideways into the surrounding fluid. At a higher rate, the dye becomes rapidly dispersed to occupy the full tube section. Reynolds carried out experiments with different liquids and with different tube diameters. He found that the criterion governing the regime depends on the dimensionless grouping $\rho VD/\mu$ where the symbols denote fluid density, mean fluid velocity, pipe diameter and fluid viscosity, respectively. This is an early recognition of the value of similarity relations leading to dimensionless groups. $\rho VD/\mu$ is called the Reynolds number Re.[2] If $\mathrm{Re} < 2300$ the flow is laminar. Up to $\mathrm{Re} = 50\,000$ the flow may be laminar if sufficient care is taken, but the regime is unstable and a disturbance will bring on turbulence. In laminar conditions the velocity profile over a cross-section is

[1] There are many examples in the literature of their use to determine airflow patterns in rooms. See for example the special issue Numerical Solutions of Fluid Problems related to Buildings, Structures and the Environment in *Building and Environment* **24**(1) 1989 and further articles in *Building and Environment* **29**(3), presented at the Sixth International Conference on Indoor Air Quality, Helsinki, 1993.

[2] D here is transverse to the direction of the flow. A Reynolds number can be formed by taking a characteristic distance in the direction of the flow, perhaps the distance of some pipe location downstream of the entry point of the fluid into the pipe.

parabolic; in turbulent conditions it is much flatter, so transport takes place more in the form of a plug.

A study of fluid motion near a solid surface is conducted using the boundary layer concept introduced by Prandtl in about 1904.[3] The complete flow is divided into two regions: 'the thin boundary layer which develops very close to the solid wall, in which the frictional forces are as important as the inertia forces, and the external region in which the flow is practically frictionless' (Schlichting 1975). If a fluid is drawn into a pipe or across the surface of a plate, the flow adjacent to the pipe wall or plate is laminar in a boundary layer which increases in thickness with distance from the point of entry. The layer may eventually become turbulent, although the flow remains laminar in a thin layer in immediate contact with the solid surface. If the surface and bulk fluid temperatures are different, heat transfer will occur through molecular motion. In the region of laminar flow, heat is transferred between layers by molecular motion. Once the flow becomes turbulent, the eddying motion carries bulk particles of fluid across the main direction of flow, a more effective mixing mechanism, and heat transfer coefficients in turbulent flow are greater than those for laminar flow.

The fluid movement may be due to an external driver such as the wind, a fan or a pump; in this case the convective effect is described as 'forced'. At a vertical surface, however, the convective effect due to a difference in temperature and the consequent buoyancy is said to be 'natural'. Flow in the layer is initially laminar and may become turbulent. For a fuller discussion on boundary layers and turbulence, see Eckert and Drake (1972).

5.2 NATURAL CONVECTION: DIMENSIONAL APPROACH

5.2.1 Vertical Surface

The fluid and heat flow at the surface of a vertical plate immersed in a fluid at a lower bulk temperature is a fairly simple phenomenon. Fluid in immediate contact with the plate is heated, expands, becomes less dense than the surrounding fluid and thus experiences an upward thrust; the consequent upward motion is restrained by viscous drag at the plate. Two kinds of variable are involved:

- *Fluid properties*: conductivity λ[W/m K], mass specific heat at constant pressure c_p [J/kg K], density ρ[kg/m^3], viscosity μ[kg/m s], and coefficient of expansion β[1/K] of the fluid.
- *Situation variables*: the temperature difference ΔT [K] between the plate and bulk air, the distance x from the leading edge and the total height H of the plate, the acceleration g[m/s^2] due to gravity. Alternatively, the flow may be generated by a heat flow q[W/m^2] imposed at the vertical wall.

The outcome variables are the local and overall heat transfer coefficient h[W/m^{2}K], the boundary layer thickness δ, the velocity u (a function of x and y) and the total upward volume flow, a function of x. Interest usually centres on the overall heat transfer coefficient

[3] A tribute to Prandtl is given by Schlichting (1975).

h, which can be written as a function Φ:

$$h = \Phi[H^a, (\beta g)^b, (\Delta T)^c, \rho^d, c_p^e, \mu^f, \lambda^g]. \tag{5.4}$$

Only the product of β and g is significant.

Some progress can be made through examining the dimensions of the various terms: quantities on the two sides of an equation must have the same units; quantities with different units can be multiplied and divided, but not added or subtracted. These considerations are said to have been current at the time of Fourier. McAdams (1954) devotes a chapter to dimensional analysis.[4] The indices a to g form a series of seven unknowns. There are, however, five fundamental units: in addition to the usual three mechanical units of m, s and kg, the thermal units of J and K can be added. (This is allowable since although mechanical work done (J) has the mechanical units of kg m^2/s^2, in this particular problem the mechanical work is not converted into thermal energy, so thermal energy constitutes a further fundamental unit. The distinction cannot be made in equation (5.3).) According to the Buckingham Π theorem (1915), h, suitably non-dimensionalised, can be expressed as a function of $7 - 5 = 2$ further dimensionless groups on the right-hand side of the equation.[5] If the flow is generated by a temperature difference ΔT, the conventional choice is

$$\frac{hH}{\lambda} = \Phi\left[\left(\frac{\rho^2 \beta \Delta T g H^3}{\mu^2}\right), \left(\frac{\mu c_p}{\lambda}\right)\right] \tag{5.5a}$$

or

$$\mathrm{Nu} = \Phi[\quad \mathrm{Gr}, \quad \mathrm{Pr} \quad] \tag{5.5b}$$

If it is caused by an imposed flux, ΔT is replaced by Hq/λ so

$$\frac{hH}{\lambda} = \Phi\left[\left(\frac{\rho^2 \beta g q H^4}{\mu^2 \lambda}\right), \left(\frac{\mu c_p}{\lambda}\right)\right]. \tag{5.5c}$$

Thus, for constant wall temperature, the Nusselt number $\mathrm{Nu} = hH/\lambda$ can be expressed as some function of the Grashof number $\mathrm{Gr} = \rho^2 \beta \Delta T g H^3/\mu^2$ and the Prandtl number $\mathrm{Pr} = \mu c_p/\lambda$; all are non-dimensional.[6] In an interesting article on early developments in fluid mechanics and convective heat transfer (Reynolds, Nusselt, Lewis, Schmidt and

[4] Eckert (1981) says that, in his book, McAdams collected, screened, and correlated the available information on heat transfer processes, supplemented it by his own research, and presented through three editions an up-to-date, concise and unified picture of the state of the art. He created a standard text which served the heat transfer community as a reference book for many years.

[5] To state the Π theorem (Buckingham 1915), suppose that we have a relation involving n physical variables in the form $F(Q_1, Q_2, \ldots, Q_n) = 0$. The theorem states this can be reduced to the form $f(P_1, P_2, \ldots, P_{n-k}) = 0$, where the P_i values are dimensionless combinations of form $P_i = Q_1^a Q_2^b \ldots Q_n^m$ and k is the number of fundamental units needed to specify the Q_i values. Thus the time of swing T of a simple pendulum depends on its length L, acceleration due to gravity g and possibly its mass M. So $F(T, L, g, M) = 0$, which involves four quantities. There are three fundamental mechanical units. According to the theorem, this can be expressed in terms of 4 - 3, that is, a single dimensionless variable, $f(T\sqrt{g/L}) = 0$; it does not involve the mass. In fact, a dynamical analysis demonstrates that $T = 2\pi\sqrt{L/g}$ and an approximation to 2π can be found from observational data but dimensional arguments themselves give no information about the 2π.

[6] The Pr grouping was recognised at least by 1921; Pohlhausen listed values for gases and water. The product $\mathrm{Gr} \cdot \mathrm{Pr}$ is defined as the Rayleigh number Ra. $\mathrm{Gr} \cdot \mathrm{Pr}^2$ is the Boussinesq number Bo.

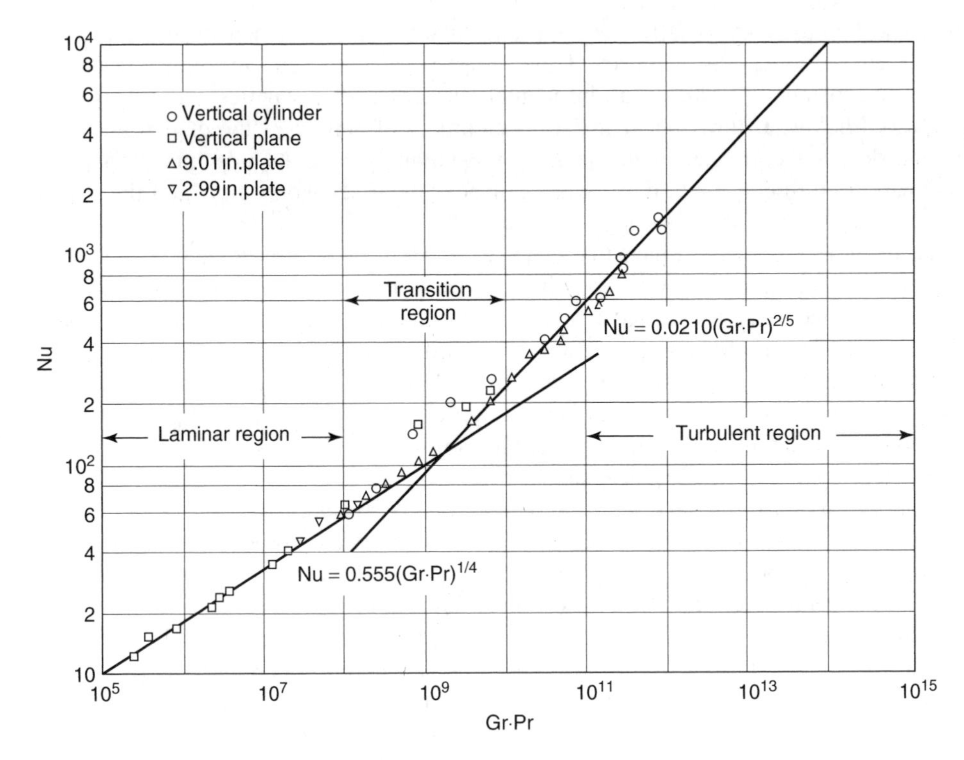

Figure 5.1 Natural convection at a vertical surface: correlation of Nu with Gr · Pr (Eckert, E.R.G and Jackson, T.E., NASA Report 1015, Analysis of turbulent free-convection boundary layer on flat plate, ©1950. Reproduced with permission of National Advisory Committee for Aeronautics)

Prandtl), Eckert (1981) states that this disposition of variables into three non-dimensional groups was first made by Nusselt in 1915.

The dimensional argument gives no further information about how these variables may be related.[7] Much work has been reported with Nu expressed as the product of Gr and Pr, that is, Ra. See for example Figure 5.1 from Eckert and Jackson (1950). Countless examples have appeared since. (The values of the fluid properties, notably the viscosities

[7]The dimensional argument gives no indication of how Nu should depend on Gr and Pr and admits a form as inappropriate as $\mathrm{Nu} = C_1\mathrm{Gr}^{\alpha} + \mathrm{Pr}^{\beta}$, where C_1, C_2, α and β are dimensionless constants. The analysis can be pushed a little further using physical arguments or common sense:

- The heat flow, expressed by h, comes about because of a temperature difference ΔT. Thus $\mathrm{Nu} \propto \mathrm{Gr}^{\alpha}$, where α must be a positive index.
- The value of h must similarly increase with c_p, so β too must be a positive index.
- The product $\lambda\rho c_p$ forms an important group in determining heat flow, and this can only be formed by some product of the Grashof and Prandtl numbers, that is, $\mathrm{Nu} = C\mathrm{Gr}^{\alpha} \cdot \mathrm{Pr}^{\beta}$.
- When the flow becomes turbulent, as it does for large values of ΔT and/or H, the viscosity μ has little controlling effect, so in this regime β must be approximately equal to 2α.

The actual values of C, α and β have to be found experimentally or by a mechanistic analysis.

of some liquids, vary with temperature and in forming non-dimensional groups values are taken at the arithmetic mean of the temperatures bounding the film.) The flow of air at unconstrained vertical and horizontal surfaces and in cavities and rooms will be discussed below. It turns out that the development of certain of the qualitative features of the flow are correlated with the Rayleigh number; these include transition between laminar and turbulent flow, development of pockets of secondary flow and the extent of the concentration of the flow near a surface.

Gr includes the length dimension H as H^3 or H^4. The numerical value of H depends on situation. In building applications it has a value of several metres whereas in natural cooling of electronic equipment it is a few millimetres. Thus typical Grashof numbers are much larger in building applications than in the cooling of electronic equipment; the effect of this however, is reduced since in finding convective coefficients, Gr appears as Gr^{α} where α is around 0.3. Further, H appears in the Nusselt number as H^1, which largely cancels the net effect of H in the Grashof number. It turns out that the convective coefficient h_c is only weakly dependent on H.

Both Nu and Gr · Pr have large numerical values and are expressed logarithmically. Figures of this kind can summarise information found from physically very different situations: they include work on liquids and gases (mainly air), for surface heights of up to 4 m, temperature differences of more than 550 K and pressures up to 65 atm (Fishenden and Saunders 1950: 92). Figure 5.1 shows a laminar flow regime for Gr · Pr values up to about 10^8 followed by a transitional region where the boundary layer assumes a wave-like form, with turbulent flow established after about 10^{10}. The laminar and transitional regimes are shown in the beautiful Mach–Zehnder interferograms of Eckert and Soehnghen (1951)[8] and reproduced for example in McAdams (1954: 169).[9] A description of the process is given by Jannot and Kunc (1998): 'The first stage for transition from laminar to turbulent flow is the boundary layer oscillations with a wavelength in relation to the boundary layer thickness. Such oscillations are always present from time to time ... because of small disturbances coming from the outer flow. A predominantly periodic phenomenon can be observed during the transition onset. Sometimes the oscillations are dampened but more often the wave becomes irregular in shape and ends up by breaking down, transforming into vortices and finally fragmenting into smaller scale turbulence. These oscillations are basically of the same nature as the waves which can

[8]Air has a refractive index $n = 1.000293$ at 0°C and standard atmospheric pressure; $n - 1$ is proportional to density hence inversely proportional to temperature at constant pressure. Thus if temperature varies in the y direction normal to a heated vertical wall, the optical path length in the z direction varies with y. This can be used in various ways to examine the flow field. In the Mach–Zehnder method, light is split into two beams, one of which passes through the flow field. The two beams are subsequently recombined to form interference fringes, from which it is possible to find the variation in density ($\Delta\rho$) and temperature in the whole of the xy plane (integrated in the z direction). Unlike insertion of a probe, the Mach–Zehnder technique does not disturb the flow. Non-steady conditions can be examined by a succession of flashes. The interferometer is comparatively laborious to set up. A further optical technique, the Schlieren method, visually indicates the flow as the gradient of density ($\partial\rho/\partial s$) or temperature in a direction fixed by a slit perpendicular to s, through which the light passes. This is simpler to set up and is qualitatively informative but is not as easy to analyse. Finally, simple projection of a beam of light through the flow onto a screen, the shadowgraph technique, indicates the z-integrated form of ($\partial^2\rho/\partial s^2$). This is the pattern seen in a bathtub when a stream of cold water is poured into a body of hot water; it is very easy to observe but it is difficult to draw quantitative conclusions about the flow. See Hauf and Grigull (1970).

[9]A similar transition can be seen when a lighted cigarette is held in a draught-free room: a steady plume rises some centimetres, then becomes unstable and disperses.

be observed in forced convection along a flat surface.' The authors go on to speculate about the cause of the instability. Kitamura *et al.* (1985) have measured the frequency of the large-scale eddy structure of natural convection at a uniformly heated plate (heat flux q) using air and water and find that the frequency of the oscillations is around $0.03 \times \sqrt{\beta g q/\lambda}$.

Roughened, ribbed or stepped surfaces may exchange more heat than smooth surfaces; see Bhavnani and Bergles (1990). A number of comments may be made.

The correlation between Nu and Gr · Pr for laminar flow is given as

$$\mathrm{Nu} = 0.555(\mathrm{Gr} \cdot \mathrm{Pr})^{1/4}. \tag{5.6}$$

To find a value for h for use in a room at 20°C,

$$h \equiv \frac{\lambda}{H}\frac{hH}{\lambda} = \frac{\lambda}{H}\mathrm{Nu} = \frac{\lambda}{H}0.555(\mathrm{Gr} \cdot \mathrm{Pr})^{1/4} = \frac{\lambda}{H}0.555\left(\frac{\rho^2 \beta g H^3\ \Delta T}{\mu^2}\mathrm{Pr}\right)^{1/4} \tag{5.7a}$$

$$= \lambda\ 0.555\ (\rho^2 \beta g \mathrm{Pr}/\mu^2)^{1/4}\ (\Delta T/H)^{1/4}$$

$$= 0.0257 \times 0.555 \times [1.2^2 \times (1/293) \times 9.81 \times 0.71/(1.8 \times 10^{-5})^2]^{1/4} \times (\Delta T/H)^{1/4}$$

$$= 1.45\ (\Delta T/H)^{1/4}. \tag{5.7b}$$

The factor of 1.45 has units and has a different value when the other quantities are measured in imperial units.

Dascalaki *et al.* (1994) have provided a comprehensive summary of experimental correlations, 58 in all and most deriving from observations on air, between Nu and Gr, evaluated for use in building applications with a value of $\mathrm{Pr} = 0.72$ (the value for air at room temperature). The cases noted relate to vertical plates, horizontal plates facing up and down, and values found in enclosures. Also noted are the range of Grashof number and whether they relate to laminar or turbulent flow. A few additional results are listed in Section 5.5.

To find working values of h, say, appropriate to some specific situation, we take the observational data on natural convection at vertical surfaces, obtained from physically diverse situations and reduced it to dimensionless groupings; it is immaterial to what groupings the original data had been reduced. Workers in the 1930s appear to have non-dimensionalised it as $\mathrm{Nu} = h/(\lambda/H)$ without further discussion and the Nu, Gr, Pr grouping works well enough.[10] In many studies Pr is taken as an explicit parameter. But there may be merit in non-dimensionalising h to give weight to those factors on which it most depends. In forming the Nusselt number, h is divided by λ, which plays a significant role in the wall-to-air transfer. However is multiplied by H, whose role is weak while ρ and c_p, equally as important as λ, do not appear at all, nor does the group expressing buoyancy, $\beta \Delta T g$, the original cause of the fluid motion.

[10] It is sometimes argued that the Nusselt number hH/λ is the ratio of two temperature gradients but this is not so. In the notation of Figure 5.3, heat continuity over the height H of a plate requires h/λ to equal $(\partial\theta/\partial y)_{surface}/\theta_p$. This can be non-dimensionalised as $\mathrm{Nu} = hH/\lambda = (\partial\theta/\partial y)_{surface}/(\theta_p/H)$. Whereas $(\partial\theta/\partial y)_{surface}$ is a gradient, θ_p/H is not; it is the ratio of two independent quantities. The form $h\delta/\lambda$ is the ratio of two gradients.

Parameter h has units W/m^{2}K and might be made dimensionless through division by (i) λ/H, (ii) $\rho c_p(H\beta\Delta T g)^{1/2}$ or (iii) $\mu c_p/H$. The first two combinations contain quantities that facilitate the transfer of heat, so either might prove useful. The last group contains μ, whose effect is to restrain the flow and reduce heat transfer. Thus h might be made non-dimensional as

$$J = \frac{h}{[\lambda/H]^{\alpha}[\rho c_p(H\beta\Delta T g)^{1/2}]^{1-\alpha}} = \frac{h}{(\lambda\rho c_p)^{1/2}(\beta\Delta T g/H)^{1/4}} \quad \text{if} \quad \alpha = \frac{1}{2}. \tag{5.8}$$

This form separates the fluid properties expressed as the effusivity, $(\lambda\rho c_p)^{1/2}$, and the combination of factors associated with the driving force $\beta\Delta T g/H$; if any one of them is zero, there is no fluid motion. J can be plotted as a function of Gr. Note that the quantities $\rho, \beta, \Delta T$ and g appear in the denominator of J and the numerator of Gr; this demonstrates that h depends strongly on these factors. Had Gr · Pr been used as the independent variable, λ and c_p would similarly have been included. By contrast, H appears in the numerators of both J and Gr, so they partly cancel; the dependence of h on H is weak. In his account of experiments on convective loss from a plate, Saunders (1936) provided all the data needed to plot h non-dimensionalised in this way and it is shown in Figure 5.2 as a function of $\log_{10}\sqrt{\text{Gr}}$.

Between $\sqrt{\text{Gr}}$ values of 10^3 and about $10^{4.7}$, J is approximately constant and equal to 0.6. According to equation (5.6), J should equal $0.555/\text{Pr}^{1/4}$ or 0.605. The average heat flow q from the plate (height H, width w) is

$$q = h\Delta T = 0.6(\lambda\rho c_p/t)^{1/2}\Delta T, \tag{5.9a}$$

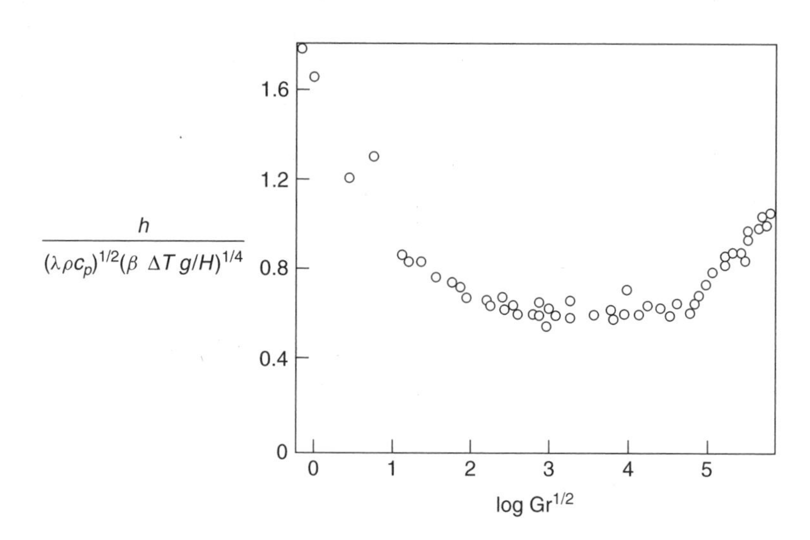

Figure 5.2 Natural convection at a vertical surface: correlation of $h/[(\lambda\rho c_p)^{1/2}(\beta\,\Delta T\,g/H)]^{1/4}$ with $\text{Gr}^{1/2}$ (Reprinted from *International Journal of Heat and Mass Transfer*, vol. 28, M.G. Davies, Similarity between unsteady conduction and natural convection, 2385–2388, ©1985, with permission from Elsevier Science)

where

$$t^2 = H/(\beta \Delta T g). \tag{5.9b}$$

Thus the heat transfer in this steady-state laminar convective process has the same form as conductive transfer in non-steady-state conditions, as is seen by comparing this equation with (12.21).

When $\sqrt{\text{Gr}}$ is greater than about $10^{4.7}$, turbulence sets in at the top of the plate. As $\sqrt{\text{Gr}}$ falls below 10^3, J apparently increases again. Low Gr numbers correspond to low pressures – pressure was a major variable in this study – and at low pressures the conductive losses from the copper connections to the platinum plates together with the radiant losses may have become comparable with the convective loss. If these were not adequately compensated, anomalously high J values would result. The increase is probably an artefact.

The group J presents h values as a scalar with values of order unity while the Nu values are large and must be plotted in log form. The scalar form makes visually clear the range of observed values that h might take at a given Gr value. This information is somewhat suppressed in the plot of log Nu: on a plot of log Nu against log(Gr · Pr) a typical deviation of $\delta(\log \text{Nu})$ of 0.1 for some experimental value from the general correlation corresponds to a deviation of some 25%, but since the full diagram includes a large range of Nu, the deviation makes little visual impact. Figure 5.2 makes clear that Saunders' values of h_c at lower Gr numbers are probably anomalously high.

Kang and Jaluria (1990) and Tewari and Jaluria (1990), describe an experimental study of the heat flow from a strip fixed horizontally on a vertical wall, either flush with the wall or protruding from it. They show the extent of the plume and provide information on the Nusselt number.

5.2.2 *Inclined Surface*

Chen and Tzuoo (1982) review earlier work on convective flow from inclined heated plates and draw attention to vortex and waveform instabilities whose appearance depends on the angle of inclination. It has been noted that at a vertical plate, the heat loss non-dimensionalised as Nu can be expressed uniquely in terms of the independent variables arranged as Gr. Their own analytical study led to a presentation of the heat loss from an inclined plate similarly expressed by a unique function: the vertical axis included a function of Gr and the angle of inclination as well as Nu itself; the horizontal axis included the angle of inclination as well as Gr. But this presentation leads to different curves for air (Pr = 0.7) and for water (Pr = 7). King and Reible (1991) give an expression for natural laminar convection at a surface inclined at θ to the vertical:

$$\text{Nu}_x = 0.587\ (\text{Ra}_x^* \cos\theta)^{1/5}, \tag{5.10a}$$

where Nu_x is the local Nusselt number and Ra_x^* is the modified local Rayleigh number, $\text{Gr}_x^* \cdot \text{Pr}$ and $\text{Gr}_x^* = \text{Gr}_x \cdot \text{Nu}_x$. Thus

$$\text{Nu}_x = 0.514(\text{Gr}_x \cdot \text{Pr} \cos\theta)^{1/4}. \tag{5.10b}$$

5.2.3 Horizontal Surface

The upward heat flow by convection from an unenclosed horizontal surface has received less attention than at a vertical surface. It appears to be inherently unstable (Pera and Gebhart 1973). Kang and Jaluria (1990) also describe an experimental study of the heat flow from a strip fixed on a horizontal surface, either flush with it or protruding, as they did for the wall study. Again, they show the form of the plume and provide Nusselt number information. Horizontal movement of fluid is generated by an upward buoyancy force, and the development of flow with increasing buoyancy is shown by the interferograms of Bahl and Liburdy (1991). They used a disc of 50.8 mm diameter set in a surround of 203.2 mm diameter. The disc was taken to temperatures of 120°C and above with ambient temperature of 25°C. At low temperature the photographs indicated an almost steady narrow plume. With increasing temperature, the plume sways more rapidly and eventually 'the entire structure begins to display dynamic character. The fringes near the surface rise and fall in unison, there is a wave-like motion across the disk, the fringes near the centre and sometimes near the edge of the disk periodically rise and burst open indicating a rising packet of hot fluid.' The authors processed their optical data to give values for the local heat transfer coefficient over the disc.

Results can be expressed as a Rayleigh number $\mathrm{Ra} = (\rho^2 g\beta\Delta T L^3/\mu^2)(\mu c_p/\lambda)$ where the characteristic length L was the disc surface area divided by its perimeter, i.e. diameter/4. They reported results with Ra between about 7×10^3 and 1×10^4. Thus these phenomena would be observed above a disc of 1 m diameter at a temperature difference greater than $(50.8 \times 10^{-3}/1.0)^3(120 - 25)$ or about 0.01 K; this has implications for room heat transfer.

Yousef *et al.* (1982) have summarised previously reported values. Laminar flow results have the form

$$\mathrm{Nu} = C(\mathrm{Gr}\cdot\mathrm{Pr})^{1/4}, \tag{5.11}$$

where Nu and Gr are each based on the length of plate. Values of C ranged between 0.54 and 0.71. The authors' values, found using a Mach–Zehnder interferometer, lay within $\pm 20\%$ of $C = 0.622$ for the range $3 \times 10^6 < \mathrm{Gr}\cdot\mathrm{Pr} < 4 \times 10^7$.

The corresponding forms for turbulent flow were

$$\mathrm{Nu} = C(\mathrm{Gr}\cdot\mathrm{Pr})^{1/3}, \tag{5.12}$$

where reported values for C lay between 0.12 and 0.20. The authors' values lay within $\pm 10.5\%$ of $C = 0.162$ for $\mathrm{Gr}\cdot\mathrm{Pr} > 4 \times 10^7$. Clausing and Berton (1989) improved on this correlation by including the ratio of the plate temperature T_w to ambient temperature T_∞. Using a cryogenic facility they were able to vary T_w/T_∞ between unity (when the viscosity and other properties were known) and 3.1; with low values of T_∞ the radiant correction was small. They expressed their correlation as

$$\mathrm{Nu}_r = 0.140\ \mathrm{Ra}_r^{1/2}\{1.212 - 0.254(T_w/T_\infty) + 0.0405(T_w/T_\infty)^2\}, \tag{5.13}$$

where subscript r denotes reference conditions for air and for nitrogen, $\mathrm{Ra} = (\rho^2\beta g\Delta T L^3/\mu^2)(\mu c_p/\lambda)$, and L the side of the square plate; ($L = 0.6$ m but in fact a 0.6×0.3 horizontal plate was used adjacent to a vertical adiabatic plane, whose viscous shear had negligible influence on the heat transfer). The Rayleigh number range is $3 \times 10^8 < \mathrm{Ra} < 1.5 \times 10^{11}$.

We would expect that if a horizontal plate were enclosed between vertical surfaces as in a room, the position of the plumes would stabilise, leading to less variation in the value of C. Further information about the disposition of the plumes, observed in water, and their modelling is given by Lewandowski *et al.* (2000). On a plate of width a and length $4a$, they suggest three centrally placed plumes each associated with an area $a \times a$, and two end plumes associated with areas $a \times \frac{1}{2}a$.

Hatfield and Edwards (1981) have presented interferograms of the *downward* flow of heat from a heated horizontal plate. Their own correlation of Nu_L with Ra_L (L is the shorter edge of the plate) is somewhat complicated. It approximates to

$$\mathrm{Nu}_L = 1.0\ \mathrm{Ra}_L^{0.191}, \tag{5.14a}$$

but log Nu is not quite linear with log Ra. They quote various earlier results:

$$\mathrm{Nu}_L = 0.068\ \mathrm{Ra}_L^{1/3}, \tag{5.14b}$$

$$\mathrm{Nu}_L = 0.31\ \mathrm{Ra}_L^{1/4}, \tag{5.14c}$$

$$\mathrm{Nu}_L = C\ \mathrm{Ra}_L^{1/5}, \tag{5.14d}$$

with values of C between 0.58 and 0.94.

5.3 *NATURAL CONVECTION AT A VERTICAL SURFACE: ANALYTICAL APPROACH*

Heat transfer by natural convection from one surface of a heated vertical plate in air is associated with a buoyancy force. We may recall the responses to a force in elementary mechanics: (i) when the motion of a body is entirely constrained by viscous action, its velocity is proportional to the force (the only response assumed by Aristotle and expressed in Stokes' relation for the force $6\pi r\mu v$ associated with a sphere, radius r, moving with velocity v in a fluid of viscosity μ); (ii) when there is no restraint, force is balanced by a rate of change of momentum (Newton's second law); (iii) when a body is supported on a plane surface, there may be a restraining frictional force, proportional to the force normal to the surface but independent of velocity. The following analysis is comparatively simple[11] and calls on the first two mechanisms with few empirical assumptions. This is particularly the case for laminar flow. Martin (1984) has sketched the evolution of its theory.

5.3.1 *Heat Transfer through a Laminar Boundary Layer*

The approach to the problem of heat transfer by natural convection at a heated vertical plate through use of dimensional analysis is useful in that it permits observational information found in various physical situations and suitably correlated to be adapted to estimate heat exchange in another situation. A mechanistic analysis, however, is needed to

[11] Simple compared with heat and fluid flow within and between rooms driven by buoyancy. Turbulence itself is anything but simple.

provide other quantities: variation of transfer with height, flow rate in the boundary layer and its thickness. In this problem, we can simplify equations (5.1) to (5.3): since steady state is assumed, terms in $\partial/\partial t$ can be omitted; no heat source is present, so Q is zero; there is virtually no change in pressure; the viscous dissipation term $\mu[\ ]$ in equation (5.3) is negligible; the temperature difference between plate and fluid is assumed to be small, so constant values of λ and μ can be assumed; ρ too is assumed to have a constant inertial effect but the equation for vertical motion includes its variation as the buoyancy effect (the Boussinesq approximation[12]). There is no variation in the z direction.

An important study along these lines was reported by Schmidt and Beckmann as early as 1930. They showed that the independent variables x and y can be replaced by the quotient $y/x^{1/4}$ and two ordinary differential equations result. Their equation 34 can be written as $\mathrm{Nu} = 0.48\mathrm{Gr}^{1/4}$, which is in satisfactory agreement with the correlation $\mathrm{Nu} = 0.555(\mathrm{Gr} \cdot \mathrm{Pr})^{1/4}$ and this is equal to $0.51\mathrm{Gr}^{1/4}$ using Schmidt and Beckmann's value of $\mathrm{Pr} = 0.733$. (They did not explicitly identify these dimensionless groupings.) Note that if $\mathrm{Nu} \propto \mathrm{Gr}^{1/4}$, $h \propto 1/\mathrm{H}^{1/4}$. The two equations were later solved by computer with Pr as parameter (Ostrach 1953 and reported by Burmeister 1983: Section 12.1). This exact approach leads to the relation $\mathrm{Nu} = 0.516(\mathrm{Gr} \cdot \mathrm{Pr})^{1/4}$ for $\mathrm{Pr} = 0.72$.

A simpler, analytical and physically attractive approach, apparently first due to Squire (1953), will be presented here.[13] Rather than considering conditions in the elementary volume of section $\mathrm{d}x\,\mathrm{d}y$ as sketched above (there is no variation in the z direction), the conservation equations are applied to a volume consisting of the full thickness δ of the boundary layer, i.e. $\mathrm{d}x\,\delta l_z$ (x is the vertical distance from the bottom of the plate, δ is in the y direction normal to the plate and l_z denotes unit distance in the z direction, horizontal and along the plate). Forms for the temperature and velocity profiles are imposed, rather than deduced, in the layer.

A plate of height H is maintained at a temperature θ_p above the surrounding bulk air (taken to be at zero). Air in immediate contact with the heated plate is heated by conduction, becomes less dense, rises under the action of buoyancy and draws virtually static air into a boundary layer. Conventionally a vertical x axis is chosen with its origin at the lower edge of the plate; the y direction accordingly denotes the horizontal distance from the plate w denotes the depth of the plate. We consider the situation in the control volume of height δx, thickness δ and depth w. The simplification over the exact solution is to treat the control volume in an integral manner, that is, to consider continuity and momentum in the complete thickness δ. It is assumed that the temperature and flow boundary layers coincide (with thickness δ); this is valid if the Prandtl number is of order unity, (Figure 5.3a).

The upward buoyancy force is balanced by the downward viscous drag at the plate together with the force equivalent to the creation of upward momentum in the control volume. The momentum into the horizontal elementary area $w\,\mathrm{d}y$ is $\rho u_y w\,\mathrm{d}y$ and its rate of flow is accordingly $(\rho u_y w\,\mathrm{d}y)u_y$. Into the control volume as a whole, the flow is $\int_{y=0}^{\delta} \rho u_y{}^2\, w\,\mathrm{d}y$ and the change of this quantity with height δx is $(\mathrm{d}/\mathrm{d}x)[\int_{y=0}^{\delta} \rho u_y{}^2\, w\,\mathrm{d}y]\delta x$.

[12]If a parcel of fluid of density $\rho + \Delta\rho$ is surrounded by fluid of density ρ and $\Delta\rho \ll \rho$, the acceleration it experiences is $(\Delta\rho/\rho)g$.

[13]It relates to convection at a free vertical surface and its application to natural convection at the vertical surfaces of a room is qualitative only. It is given here since it demonstrates in a simple way a number of features of room convective exchange.

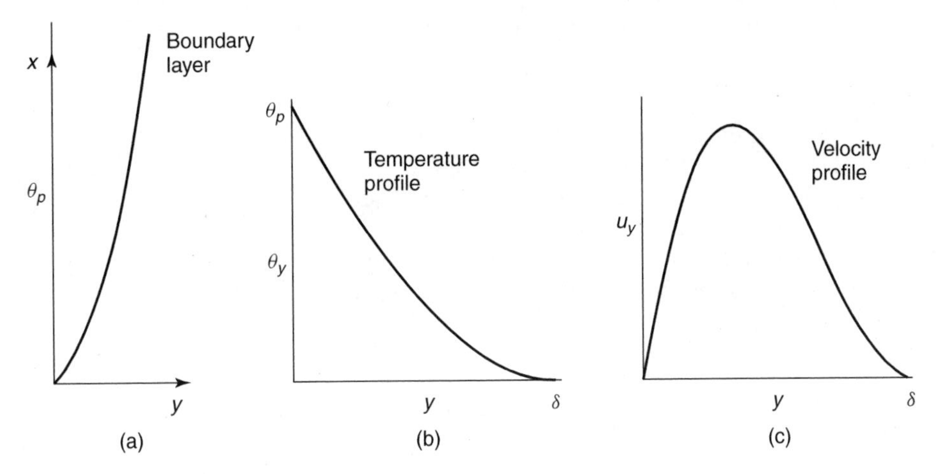

Figure 5.3 Natural convection at a heated vertical surface, model for analytical estimation: (a) co-ordinates and boundary layer, (b) assumed quadratic temperature distribution at height x, (c) assumed cubic velocity distribution at height x

The buoyancy force on the elementary volume is $w\delta x\,\mathrm{d}y\beta\theta_y g\rho$ and for the whole volume it is $\int_{y=0}^{\delta} w\delta x\,\mathrm{d}y\beta\theta_y g\rho$. Finally, there is the viscous drag $\mu w\delta x(\mathrm{d}u_y/\mathrm{d}y)_{y=0}$ at the plate. There is no drag at the edge of the boundary layer. The momentum equation follows by equating these quantities and cancelling $w\,\delta x$:

$$\frac{\mathrm{d}}{\mathrm{d}x}\int_{y=0}^{\delta} \rho {u_y}^2\,\mathrm{d}y = \beta g\rho\int_0^{\delta} \theta_y\,\mathrm{d}y - \mu\left(\frac{\mathrm{d}u_y}{\mathrm{d}y}\right)_0. \tag{5.15}$$

The increased energy in the control volume must result from the conduction of heat into it at the plate, and by a similar argument we obtain

$$\frac{\mathrm{d}}{\mathrm{d}x}\int_{y=0}^{\delta} \rho c_p u_y\theta_y\,\mathrm{d}y = -\lambda\left(\frac{\mathrm{d}\theta_y}{\mathrm{d}y}\right)_0. \tag{5.16}$$

The temperature profile in the control volume must have the form in Figure 5.3b, a gradient corresponding to heat flow into the layer at the plate itself and no further flow into the bulk air at $y = \delta$. Similarly, the upward flow velocity must be as in Figure 5.3c, zero velocity at the plate then rising to a maximum and falling to zero at $y = \delta$. Experimental confirmation of these forms is provided by Schmidt and Beckmann (1930). They made observations of temperature and local velocity in the boundary layer of electrically heated vertical plates 12 and 50 cm high with θ_p values of about 50 K. The boundary layer was about 1 cm thick at the top of the smaller plate. Velocity measurements were made by observing the deflection of a fine quartz fibre. The fibre was motionless in the region of laminar flow, but periodic movement indicated the onset of turbulence toward the top of the plate. The form of their temperature and velocity profiles provided an empirical justification for their expression later as θ_y and u_y. The authors reduced their results to single characteristics plotting θ_x/θ_p against $y/x^{1/4}$ and $u/x^{1/2}$ against $y/x^{1/4}$ but did not suggest expressing them as polynomials. We impose the following forms:

Variation with y $(0 < y < \delta)$	Variation with x $(0 < x < H)$	
$\theta_y = \theta_p(1 - y/\delta)^2$	$\overline{\theta} = C_0 x^0$ (i.e. no variation)	(5.17a,b)
$u_y = 12\overline{u}(1 - y/\delta)^2(y/\delta)$	$\overline{u} = \frac{1}{12} C_1 x^m$	(5.18a,b)
	$\delta = C_2 x^n$	(5.19)

(With a choice of the factor of 12, $\overline{u}$ proves to be the mean velocity at height x.)

Thus we have six useful equations and six unknowns ($\delta, \overline{u}, C_1, C_2, m$ and n). The unknowns δ and $\overline{u}$ vary with height. First the four integrals will be found.

By definition

$$\overline{\theta} = \int_0^\delta \theta_y \, dy \bigg/ \int_0^\delta dy = \frac{1}{3}\theta_p. \tag{5.20a}$$

Similarly,

$$\overline{u} = \int_0^\delta u_y \, dy \bigg/ \int_0^\delta dy = \overline{u}, \tag{5.20b}$$

as follows from the choice of factor. The maximum and minimum values of u_y follow from $du_y/dy = 0$. The maximum value is $\frac{16}{9}\overline{u}$ at $y = \frac{1}{3}\delta$ and u_y has its minimum of zero at $y = \delta$; equation (5.18a) was chosen to achieve this. The integral

$$\int_0^\delta {u_y}^2 \, dy = 12^2 \overline{u}^2 \int [(y/\delta) - 2(y/\delta)^2 + (y/\delta)^3]^2 \, dy = \frac{144}{105}\overline{u}^2\delta, \tag{5.20c}$$

and similarly,

$$\int_0^\delta u_y \theta_y \, dy = \frac{12}{30}\overline{u}\theta_p\delta. \tag{5.20d}$$

These values, based on the assumptions of how u and θ vary with y, i.e. horizontally, are to be used in equations (5.15) and (5.16):

$$\frac{d}{dx}\left(\frac{144}{105}\rho\overline{u}^2\delta\right) = \frac{1}{3}\rho\beta g\theta_p\delta - 12\mu\overline{u}/\delta, \tag{5.21}$$

$$\frac{d}{dx}\left(\frac{12}{30}\rho c_p\overline{u}\theta_p\delta\right) = 2\lambda\theta_p/\delta. \tag{5.22}$$

We now use the assumptions about how u and δ vary with x, i.e. vertically. Inserting equations (5.18b) and (5.19) into (5.21), we obtain

$$\frac{d}{dx}\left(\frac{144}{105}\rho\frac{1}{144}C_1^2 x^{2m} C_2 x^n\right) = \frac{1}{3}\rho\beta g C_0 x^0 C_2 x^n - 12\mu\frac{1}{12}C_1 x^m/(C_2 x^n). \tag{5.23}$$

Each term in this equation must very with x in the same way, so

$$2m + n = 0 + n = m - n. \tag{5.24}$$

So $m = \frac{1}{2}$ and $n = \frac{1}{4}$.

Finally the two constants C_1 and C_2 must be found ($C_0 = \frac{1}{3}\theta_p$). When equations (5.18b) and (5.19) are inserted once more into equations (5.21) and (5.22), together with explicit values for m and n, then x cancels throughout and after some manipulation,

$$C_1 = \left(\frac{\frac{80}{3}\beta g \theta_p}{\frac{20}{21} + \mu c_p/\lambda} \right)^{1/2} [\mathrm{m}^{1/2}\,\mathrm{s}^{-1}] \quad \text{and} \quad C_2 = \left(\frac{240\left(\frac{20}{21} + \mu c_p/\lambda\right)}{(\mu c_p/\lambda)^2(\rho^2 \beta \theta_p g/\mu^2)} \right)^{1/4} [\mathrm{m}^{3/4}]. \tag{5.25}$$

Then δ and $\overline{u}$ together with the volume flow $\delta\overline{u}$ follow from equations (5.18b) and (5.19). The local heat flow from the plate into the flow is

$$q_x = -\lambda(\mathrm{d}\theta_y/\mathrm{d}y) = 2\lambda\theta_p/\delta = 2\lambda\theta_p/(C_2 x^{1/4}). \tag{5.26}$$

The average heat flow over the height x is

$$\overline{q} = \int_{x=0}^{x} q_x\,\mathrm{d}x \bigg/ \int_{x=0}^{x} \mathrm{d}x = \frac{4}{3} q_x. \tag{5.27}$$

The average at height x is greater than the local value at x since the boundary layer is thinner below x and transmits more heat. The average heat transfer coefficient $h = \overline{q}/\theta_p$. All quantities can now be expressed in dimensionless form, with $\mathrm{Pr} = \mu c_p/\lambda$ and $\mathrm{Gr} = \rho^2 \beta \theta_p g H^3/\mu^2$ at height $x = H$, as before:

mean velocity
$$\overline{u}\rho H/\mu = \left(\frac{5}{27}\right)^{1/2} \mathrm{Pr}^0 \left(\mathrm{Pr} + \frac{20}{21}\right)^{-1/2} \mathrm{Gr}^{1/2} = 0.333\mathrm{Gr}^{1/2}, \tag{5.28a}$$

boundary layer thickness
$$\delta/H = (240)^{1/4} \mathrm{Pr}^{-1/2} \left(\mathrm{Pr} + \frac{20}{21}\right)^{1/4} \mathrm{Gr}^{-1/4} = 5.292\mathrm{Gr}^{-1/4}, \tag{5.28b}$$

volume flow
$$\overline{u}\rho\delta/\mu = \left(\frac{2000}{243}\right)^{1/4} \mathrm{Pr}^{-1/2} \left(\mathrm{Pr} + \frac{20}{21}\right)^{-1/4} \mathrm{Gr}^{1/4} = 1.764\mathrm{Gr}^{1/4}, \tag{5.28c}$$

mean heat transfer coefficient
$$\overline{h}H/\lambda = \left(\frac{512}{2430}\right)^{1/4} \mathrm{Pr}^{1/2} \left(\mathrm{Pr} + \frac{20}{21}\right)^{-1/4} \mathrm{Gr}^{1/4} = 0.504\mathrm{Gr}^{1/4}. \tag{5.28d}$$

Equation (5.28c) is the product of (5.28a) and (5.28b). The values in the final column are for air with $\mathrm{Pr} = 0.714$. For use at room temperature of 20°C when $\rho^2 \beta g/\mu^2 = 1.47 \times 10^8\,\mathrm{m}^{-3}\mathrm{K}^{-1}$, we have

$$\overline{u} = 0.061(H \cdot \theta_p)^{1/2}\ [\mathrm{m/s}], \tag{5.29a}$$

$$\delta = 0.048(H/\theta_p)^{1/4}\ [\mathrm{m}], \tag{5.29b}$$

$$\overline{u}\delta = 0.00293(H^3\theta_p)^{1/4}\ [\mathrm{m}^2/\mathrm{s}, \text{ i.e. } \mathrm{m}^3 \text{ per metre horizontal run per second}], \tag{5.29c}$$

$$h = 1.43(\theta_p/H)^{1/4}\ [\mathrm{W/m^2K}]. \tag{5.29d}$$

Equation (5.29d) is in agreement with the form of (5.7b).

5.3.2 Discussion of the Laminar Flow Solution

(i) This approach assumes empirical forms for the temperature distribution θ_y (quadratic) and the velocity distribution u_y (cubic) in the boundary layer. Since they are imposed, rather than having emerged from more fundamental reasoning, they might be in error to some extent. The quantities deriving from their integrals do not in fact depend strongly on the assumed forms so to this extent the conclusions may be expected to be reliable. However, to find the heat flow from the wall at height x, the temperature profile must be differentiated (5.26) and the value obtained depends strongly on the form assumed for θ_y; this is the least reliable step in the argument. With a value of $\text{Pr} = 0.72$, (5.28d) leads to the relation $\text{Nu} = 0.549(\text{Gr} \cdot \text{Pr})^{1/4}$, which should be compared with $\text{Nu} = 0.516(\text{Gr} \cdot \text{Pr})^{1/4}$ from the exact solution – adequate rather than satisfactory agreement.

(ii) As θ_p becomes small, the buoyancy forces become small, δ becomes large and h tends to zero. Clearly the argument breaks down when δ becomes of order H. The analysis of heat flow from a heated plate serves as an approximation to the convective exchange at a vertical wall in a room, so this breakdown is significant when estimates are made of the swings in temperature in a room during a succession of sunny days. Air temperature tends to be above wall temperature during the day and below it by night, so the h value varies through the day, remaining close to zero. Swing estimates are often made by assuming a constant value of around 3 W/m^{2}K.

(iii) The variation of $\overline{u}$, δ and h with plate height H and temperature difference θ_p is given in (5.29) and the relations are physically reasonable. Clearly $\overline{u}$ and δ must increase with H. h decreases with H since any additional section of plate is associated with a boundary layer that is thicker and more resistive than that below it. Parameter $\overline{u}$ must increase with θ_p and so must h since some finite θ_p is needed to establish motion in the first place. Also δ decreases with increasing thermomotive force, much as the width of the wake behind a rod drawn through water decreases with speed.

(iv) The volume flow at height H, given in (5.28c) and (5.29c), consists of the air entrained into the boundary layer up to that point and its value at H is given by differentiation; the rate is greatest at the tip of the plate.

(v) Consider a small volume at a temperature $\frac{1}{3}\theta_p$, the mean temperature of the boundary layer. If unrestrained, it experiences an acceleration $\beta g \frac{1}{3}\theta_p$ and has a velocity $\left(2\beta g \frac{1}{3}\theta_p H\right)^{1/2} = 0.15(\theta_p H)^{1/2}$ at 20°C. This is more than twice the value given in (5.29a) where motion was restrained by viscous action at the plate.

(vi) An unrestrained pocket of air at a temperature ΔT warmer than the bulk air moves through a height L in time t given as

$$t^2 = 2H/(\beta \Delta T g). \tag{5.30}$$

Since the flow is in fact restrained by viscous drag at the plate, the time taken for a particle of air entrained at the bottom of the plate to reach the top is larger than this, but the quantity $\sqrt{H/(\beta \Delta T g)}$ indicates the timescale. Now from equation (5.28d) we have

$$h = \left(\frac{512/2430}{\mathrm{Pr} + 20/21}\right)^{1/4} \left(\frac{\lambda\rho c_p}{\sqrt{H/(\beta\theta g)}}\right)^{1/2} = 0.60 \left(\frac{\lambda\rho c_p}{\sqrt{H/(\beta\theta g)}}\right)^{1/2}. \tag{5.31}$$

The factor 0.60 is the value of J in (5.8); h^2 is proportional to the product of λ and ρc_p for the heat-transmitting medium and inversely proportional to the timescale. A similar form, the factor $\sqrt{\lambda\rho c_p/\mathrm{time}}$, appears for parameters describing time-varying conduction through solid layers, shown in (12.21) and (12.32).

(vii) With increasing height and temperature difference the flow becomes turbulent; the layer becomes turbulent when $\mathrm{Gr}_x \cdot \mathrm{Pr} > 10^9$. For air at 40°C the value of $(\mathrm{Gr}_L \cdot \mathrm{Pr})/(H^3\theta_p)$ is 0.78×10^8. For a panel radiator at 60°C in a room at 20°C, so that $\theta_p = 60 - 20 = 40\,\mathrm{K}$, flow remains laminar up to a height $H = (10^9/(0.78 \times 10^8 \times 40))^{1/3} = 0.68\,\mathrm{m}$. Thus the convective loss from most room panel radiators is by laminar flow. It is not possible to define a leading edge for windows but with lower temperature differences, heights for onset of turbulence are somewhat larger. Bejan and Lage (1990) and Vitharana and Lykoudis (1994) have argued that the transition to turbulence correlates with Gr rather than Gr · Pr but this makes little difference for air.

(viii) An analysis for turbulent flow, similar to the one for laminar flow, is given by Eckert and Jackson (1950) and is summarised below. See also Burmeister (1983: 543). It leads to the expression

$$\mathrm{Nu}_x = 0.0212(\mathrm{Gr}_x \cdot \mathrm{Pr})^{2/5} \tag{5.32a}$$

or

$$\begin{aligned} h &= 0.0212\,\lambda(\mathrm{Gr}_x \cdot \mathrm{Pr}/(H^3\theta_p))^{2/5}(H\theta_p{}^2)^{1/5} \\ &= 0.0212 \times 0.0272 \times (0.78 \times 10^8)^{2/5} \times (H\theta_p{}^2)^{1/5} \\ &= 0.83 \times (H\theta_p{}^2)^{1/5}, \end{aligned} \tag{5.32b}$$

which leaves a slight dependence of h on H. Some work has indicated that when fully turbulent, the boundary layer thickness is constant and that h should be independent of H. In this case $\mathrm{Nu}_x \sim \mathrm{Gr}_x{}^{1/3} f(\mathrm{Pr})$; see Ede (1967b: 124).

(ix) The significance of the Grashof number can be made a little clearer by writing it as

$$\mathrm{Gr} = \rho^2 H^2 (Hg\beta\theta)/\mu^2. \tag{5.33}$$

An unrestrained particle of air at a temperature θ warmer than the surrounding air experiences an acceleration of $g\beta\theta$. If this acts over the distance H, the particle attains a velocity u given by

$$u^2 = 2Hg\beta\theta. \tag{5.34}$$

Thus Gr can be expressed as $(\rho HU/\mu)^2$ where U is some typical velocity. But $\rho HU/\mu$ constitutes a Reynolds number Re where the typical dimension is taken in the direction of the flow. Thus $\sqrt{\mathrm{Gr}}$ is a form of Reynolds number when buoyancy forms the driving force.[14] This will be clear in (5.28a): the left-hand side has the form of a Reynolds number, the right-hand side includes $\mathrm{Gr}^{1/2}$. When Re is based on the dimension *perpendicular* to the flow, it can be taken as a measure of the ratio of the momentum to viscous forces acting over a section of the flow, but this interpretation cannot be extended to the Grashof number since H is the direction of the flow.

(x) From equation 14d,

$$\text{if } \mathrm{Pr} \ll 1, \quad \mathrm{Nu} \propto (\mathrm{Pr}^2 \cdot \mathrm{Gr})^{1/4}, \tag{5.35a}$$

$$\text{if } \mathrm{Pr} \gg 1, \quad \mathrm{Nu} \propto (\mathrm{Pr} \cdot \mathrm{Gr})^{1/4}. \tag{5.35b}$$

$\mathrm{Pr} \sim 0.7$ for air; 0.7 raised to the power $\frac{1}{2}$ or $\frac{1}{4}$ is almost unity and, numerically speaking, can be omitted from an expression for Nu. Nu values are sometimes expressed as a function of Gr alone (see below) but in principle this is inappropriate. If Pr is omitted in forming an explicit expression for h, such as equation (5.31), the effects of λ and of μ are over-represented and the effect of c_p – an essential factor to transfer heat convectively – is omitted altogether.

(xi) The heat transfer to air from a heated plate area A follows upon entrainment of air through a surface of area A outside the boundary layer where the mean temperature is $T_{a,entrainment}$. This is a two-dimensional construct and may differ from the three-dimensionally averaged enclosure air temperature given by (6.55). This point is taken up later.

(xii) It has been assumed that the bulk air outside the boundary layer is stationary but it may be moving vertically so as to assist or oppose the flow in the boundary layer. Kobus and Wedekind (1996) report studies on its effect.

(xiii) The complementary problem is where a constant heat flow $q\,\mathrm{W/m^2}$ is imposed on a vertical plate in a fluid (at zero, say) and its solution is summarised by Burmeister (1983: Section 12.3). If the plate temperature has the value T_H at height H, its value T_x at a distance x from the bottom is $T_H(x/H)^{1/5}$, so initially it increases rapidly with x but this neglects conduction within the plate itself, which would lessen the temperature gradient.

5.3.3 Heat Transfer through a Vertical Turbulent Boundary Layer

A largely similar argument can be advanced when the boundary layer is turbulent and the following presentation is based on Eckert and Jackson (1950). Empirical equations are chosen for the temperature and velocity distributions through the layer based on observational values:

$$\theta_y = \theta_p(1 - (y/\delta)^{1/7}), \tag{5.36}$$

$$u_y = 6.833\overline{u}(y/\delta)^{1/7}(1 - y/\delta)^4, \tag{5.37}$$

where $\overline{u}$ is the mean velocity (the constant is chosen as before to make this so). These expressions reflect the fact that although flow takes place within a certain distance δ from the plate, θ_y and u_y vary much more rapidly near the plate than for laminar flow.

[14] The Richardson number is $\mathrm{Ri}_x = \mathrm{Gr}_x/\mathrm{Re}_x{}^2$. Gryzagoridis (1975) uses it as an independent variable to express Nu for free and forced convection from a vertical plate. Other possibilities for the independent variable are noted by Churchill and Chu (1975).

u_y has its maximum value at $y/\delta = \frac{1}{29} = 0.0345$, which is around 1/10 the laminar flow value. The index of $\frac{1}{7}$ was chosen from the known velocity profile on a flat plate with forced convection. A momentum equation similar to (5.15) is used.

However, instead of the heat continuity equation (5.16), where the assumed form for u_y yields an infinite flux at $y = 0$, we use Reynolds, analogy. See for example Ede (1967b: Ch. 5). Consider the loss of heat from a heated plate θ_p above ambient due to forced convection. At the tip of the plate, the boundary layer is laminar in which temperature and velocity vary steadily throughout its thickness. In due course the boundary layer becomes turbulent and thickens; however, most of the variation in temperature and velocity now takes place in a thin sublayer, with slow variation through the remainder of the turbulent layer. Turbulence replaces the separate actions of viscosity and conductivity and the processes of heat and momentum transfer become identical. Suppose that a packet of fluid of mass m with velocity V in the main stream enters the boundary layer and is brought to rest at the wall. It gains a heat content $mc_p\theta_p$ and loses momentum mV. Then it is argued that

$$\frac{\text{heat flow into fluid at wall } q_w}{\text{sheer stress at the wall } \tau_w} = \frac{\text{heat content } mc_p\theta_p}{\text{change in momentum } mV}. \tag{5.38}$$

The sheer stress is

$$\tau_w = 0.0225\rho V^2(\rho V\delta/\mu)^{-1/4}. \tag{5.39}$$

The calculation of natural turbulent convection at a vertical plate proceeded mainly along the lines of the laminar flow argument, involving the momentum equation, Reynolds analogy and assumed power relations for variation of $\overline{u}$ and δ with height x. Eckert and Jackson's final expressions in non-dimensional form are:

mean velocity $$\overline{u}\rho H/\mu = 0.1734\ \mathrm{Pr}^0\ (0.494\mathrm{Pr}^{2/3}+1)^{-1/2}\ \mathrm{Gr}^{1/2} = 0.147\mathrm{Gr}^{1/2}, \tag{5.40a}$$

boundary layer thickness $$\delta/H = 0.566\ \mathrm{Pr}^{-8/15}\ (0.494\mathrm{Pr}^{2/3}+1)^{1/10}\ \mathrm{Gr}^{-1/10} = 0.700\mathrm{Gr}^{-1/10}, \tag{5.40b}$$

mean heat transfer coefficient $$\overline{h}H/\lambda = 0.0246\ \mathrm{Pr}^{7/15}\ (0.494\mathrm{Pr}^{2/3}+1)^{-2/5}\ \mathrm{Gr}^{2/5} = 0.0184\mathrm{Gr}^{2/5}. \tag{5.40c}$$

The final coefficients are based on a value of $\mathrm{Pr} = 0.714$, as previously.

These relations can be compared with the corresponding values for laminar flow. The powers of the Pr term are nearly the same. With $\mathrm{Pr} = 0.714$, $(0.494\mathrm{Pr}^{2/3}+1) = 1.395$ and $\left(\mathrm{Pr}+\frac{20}{21}\right) = 1.667$ and since they are raised to powers less than unity, they differ little in their effect. Values differ most in their dependence on Gr. In turbulent flow, the boundary layer thickness varies as $\mathrm{Gr}^{-1/10}$, i.e. very little, in agreement with observation. In laminar flow, $\overline{h}$ varies as ${\theta_p}^{0.25}$, or $(\Delta T)^{0.25}$, and in turbulent flow as ${\theta_p}^{0.4}$. These computed values accord with the observed values in Figure 5.1.

Ede (1967b: 86) lists the corresponding values of mean heat transfer coefficient for forced convection at a heated plate. For flow over a length L, the mean heat transfer coefficient is

$$\underset{\text{laminar flow}}{\mathrm{Nu}_L = 0.664\,\mathrm{Pr}^{1/3} \cdot \mathrm{Re}_x^{1/2}} \qquad \underset{\text{turbulent flow}}{\mathrm{Nu}_x = 0.037\,\mathrm{Pr}^{1/3} \cdot \mathrm{Re}_x^{4/5}}. \tag{5.41}$$

Re is the appropriate variable in connection with forced flow and Gr for buoyancy-driven flow; as noted above, Gr is a form of Re^2. Replacing $\mathrm{Gr}^{-1/4}$ by $\mathrm{Re}^{-1/2}$ for example, the forced and natural convective forms are similar. Equation (5.41) differs from equations (5.28) and (5.40) in their form of dependence on Pr but for air this makes little difference.

Eckert and Jackson's analysis assumes that the flow is turbulent from the start of the plate but it is in fact initially laminar and only becomes turbulent later, at around a value of $\mathrm{Gr} \cdot \mathrm{Pr} = 10^9$. They do not note this, but in the case of forced convection across a heated plate, the flow becomes turbulent at $\mathrm{Re}(L_1) = 5 \times 10^5$ and Ede (page 87) gives a modified expression for the mean Nusselt number averaged over the larger distance L_2 which takes account of the initial laminar section:

$$\begin{aligned}\mathrm{Nu}(L_2) &= 0.037\ \mathrm{Pr}^{1/3}\ \{[\mathrm{Re}(L_2)]^{4/5} - (5 \times 10^5)^{4/5} + 0.664 \times (5 \times 10^5)^{1/2}/0.037\} \\ &= 0.037\ \mathrm{Pr}^{1/3}\ \{[\mathrm{Re}(L_2)]^{4/5} - 23\,600\}\ \text{approximately.}\end{aligned} \tag{5.42}$$

Hetsroni *et al.* (1996) have argued that the above analysis does not 'reveal the true mechanism of heat removal in a turbulent boundary layer ... The investigations done during the last decades [from 1967] showed that the near wall flow possesses a rather complicated structure which results from a strong interaction between large-scale vortices emerging in a turbulent boundary layer, and low-speed streaks existing in its sublayer. This process is accompanied by burst formation leading to enhancement of heat removal from the wall.' They state that the model they develop is in fair agreement with experimental data on heat removal in a turbulent boundary layer.

5.4 NATURAL CONVECTION BETWEEN PARALLEL SURFACES

A large number of studies have been reported on fluid flow in an enclosure (height H, separation L between the vertical walls.) There is an important difference between heat and fluid flow at a free surface as discussed above and flow in an enclosure. On the plate, the region outside the boundary layer of the free surface is virtually undisturbed by the flow in the boundary layer; there is only a hint of periodicity in a transition from laminar to turbulent flow. In an enclosure the boundary layers on the surfaces enclose a core region which interacts with it and the form of flow depends strongly on factors whose combination as a Rayleigh number serves to describe the form. Embedded within the flow may be a series of cells, a kind of periodic structure. Attention should be drawn to Ostrach's review article on natural convective flow in enclosures (Ostrach 1972) and its very informative successor (Ostrach 1988).

If $H/L \ll 1$ or $H/L \gg 1$ then the results can be interpreted in a building context as heat flow across a horizontal or vertical cavity; if $H/L \sim 1$ then they can be interpreted as heat flow in a room. Results for cavities are presented in this section, results for rooms are presented in the next section, but in fact they form a continuum, as has been shown by Bejan (1980). His Figure 1 plots the heat flow between vertical walls, expressed as a Nusselt number Nu against the aspect ratio H/L from 0.01 to 100, with the Rayleigh number (values 10^4 to 10^8) as parameter. It appears that $\mathrm{Nu} \sim 1$ for H/L much less than 1, rises to a maximum, dependent on Ra, at some value of $H/L < 1$ and slowly decreases again as H/L increases.

The transfer of heat across enclosures in walls and sloping roofs, and between layers of glass in windows is conveniently summarised as a transfer between parallel surfaces. Many manufactured windows now consist of a sealed, insulated glazing unit of two panes of glass separated by an edge seal and containing gas fill, dry air or argon. The total transfer consists of radiant and convective components. The radiant heat transfer coefficient is $\varepsilon_1\varepsilon_2 \cdot h_r/(\varepsilon_1 + \varepsilon_2 - \varepsilon_1\varepsilon_2)$, where ε_1 and ε_2 are the long-wave emissivities of the two surfaces and h_r is the linearised radiant heat transfer coefficient. It has a value of about $5\,\mathrm{W/m^2K}$ at around room temperatures but is lower for glazing units with low-emissivity coatings. This is independent of the orientation and the separation of the surfaces, ignoring edge corrections. Convective transfer, which has the same order of magnitude as radiant transfer, depends on both separation and orientation. (In observational work, the radiant component must be subtracted from the total observed flux. Sparrow and Bahrami (1980) describe a method of inferring the convective transfer through a mass transfer technique, sublimation of naphthalene, and this automatically excludes radiative transfer and some other complicating factors.)

If the surfaces are close together, less than 30 mm, there is little fluid movement and the convective coefficient h_c tends to λ/δ, where λ and δ are the air conductivity and separation respectively. In this case $\mathrm{Nu}_\delta = h_c\delta/\lambda = 1$, so Nu_δ is the appropriate non-dimensional form for h_c. This remains broadly true for a horizontal enclosure when the upper surface is the hotter. For a horizontal enclosure with upward heat flow, convective patterns develop. Their form was studied in some detail by Bénard (1901) using a liquid layer about 1 mm deep.[15] Figure 5.4a reproduces one of his figures and shows the cellular structure from above. The form is sometimes described as hexagonal, though this is evidently not the case. Bénard's photographs show irregular four-, five- and six-sided cells. Figure 5.4b shows his indication of the fluid motion: upward at the cell centres, downward at the edges, with a ratio of cell spacing to liquid depth between 3 and 4.

Interest in heat flow across cavities dates from the 1920s and Pratt (1966), sketches the 1922 study by Griffiths and Davis. de Graaf and van der Held (1953) reported studies using air as the convective medium. They noted that up to a value of $\mathrm{Gr} \cdot \mathrm{Pr}$ of 1700, no convection occurs. Above this, hexagonal or tetragonal cells form, but with a *falling* air current in the cell centre (unlike the case for a liquid, where it rises.) When $\mathrm{Gr}_\delta \cdot \mathrm{Pr} >$ 45 000 the cell pattern transforms to turbulence.

The marked effect of orientation can be illustrated by results of Wilkes and Peterson (1937). For a specific case of surfaces with a 92 mm separation and 11.1 K temperature difference, they found the following values of $h_c(\mathrm{W/m^2K})$: 0.7 for downward, 2.1 for horizontal and 3.1 for upward flows.

[15] It was known in prehistoric times that if a stream of air impinges on a cylindrical or similar obstacle, it sheds eddies and generates a tone which can be stabilised by a pipe. Eddy shedding, a time-varying periodic phenomenon, was studied by Strouhal, who published an account of it in 1878. Bénard's investigation is mentioned here for historical interest since it appears to be the first objective study of *spatial* periodicity in convective fluid flow. Strictly speaking, it is not relevant to the present discussion since the liquid upper surface is free and is therefore not constrained by the viscous forces that act on a fluid between two parallel surfaces. Very similar patterns are seen for free convection in horizontal air layers (Eckert and Drake 1972: 537) and possible forms of fluid movement are illustrated in Burmeister (1983: 556). The stability of flows of this kind is discussed by Neumann (1990).

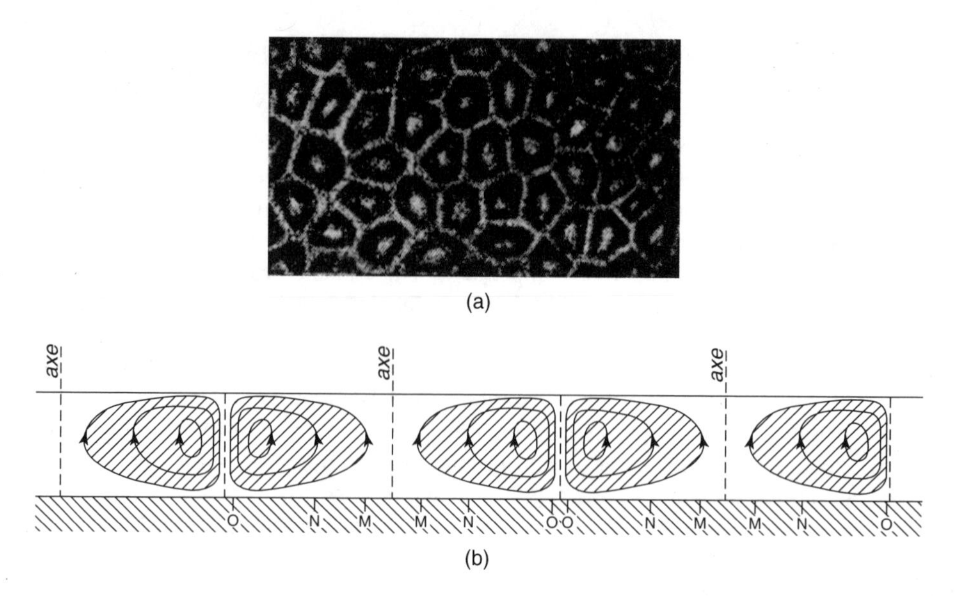

Figure 5.4 (a) Cell formation in a thin layer of liquid, heated from below. (b) Flow lines (Bénard 1901)

A more general form of variation for the convective coefficient with orientation is shown by the curves of de Graaf and van der Held (1953), who included earlier findings by Mull and Reiher (1930). Graaf and Held conducted their experiments using a closed cavity of area 43 cm × 43 cm and gap widths of 6.9, and 12.6 and 22.9 mm. The results are shown in Figure 5.5. Their empirical relations are given in Table 5.1, which lists the values of Nu between the values of Gr_δ. (Thus for a horizontal cavity, Nu_δ is unity when $Gr_\delta < 2000$ and $0.0507 Gr_\delta^{0.40}$ between $Gr_\delta = 2000$ and 50 000. The authors left open the value of Nu_δ in some Gr_δ ranges.) An example was given earlier to show how h values can be found from Nu/Gr values.

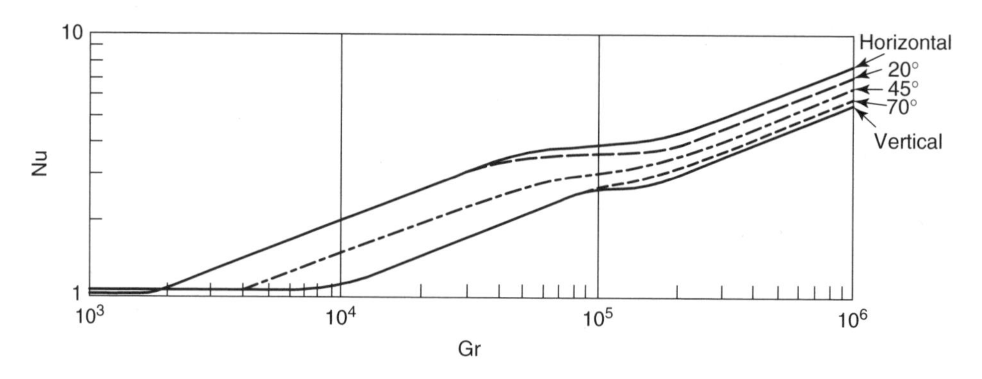

Figure 5.5 Variation of Nu_δ with Gr_δ for convective transfer in an inclined enclosure (*Applied Science Research*, 1953, 393–409, The relation between the heat transfer and the convection phenomena in enclosed plane air layers, J.G.A. de Graaf and E.F.M. van der Held, f. 17, Martinus Nijhoff, The Hague, A3 with kind permission from Kluwer Academic Publishers)

Table 5.1 Values of Nusselt numbers in ranges of Grashof numbers for free convection in an inclined enclosure, after Graaf and Held (1953)

	Gr_δ	Nu_δ	Gr_δ	Nu_δ	Gr_δ	Nu_δ	Gr_δ	Nu_δ	Gr_δ	Nu_δ	Gr_δ
Horizontal	0	1	2×10^3			$0.0507 \times Gr^{0.40}$	5×10^4	≈ 3.8	2×10^5	$0.0426 \times Gr^{0.37}$	large
20°	0	1	2×10^3			$0.0507 \times Gr^{0.40}$	3×10^4	≈ 3.6	2×10^5	$0.0402 \times Gr^{0.37}$	large
30°	0	1	3×10^3			$0.0588 \times Gr^{0.37}$	5×10^4	*	2×10^5	$0.0390 \times Gr^{0.37}$	large
45°	0	1	4×10^3			$0.0503 \times Gr^{0.37}$	5×10^4	*	2×10^5	$0.0372 \times Gr^{0.37}$	large
60°	0	1	5×10^3			$0.0431 \times Gr^{0.37}$	5×10^4	*	2×10^5	$0.0354 \times Gr^{0.37}$	large
70°	0	1	6×10^3	*	10^4	$0.0384 \times Gr^{0.37}$	8×10^4	*	2×10^5	$0.0342 \times Gr^{0.37}$	large
Vertical	0	1	7×10^3	*	10^4	$0.0384 \times Gr^{0.37}$	8×10^4	*	2×10^5	$0.0317 \times Gr^{0.37}$	large

*Values not given for Nu_δ between the adjacent Gr_δ numbers.
Source: de Graaf and van der Held (1953).

Robinson *et al.* (1954) reported similar results, plotting $h_c H$ against $\Delta T H^3$ for a cavity placed horizontally with heat flow up, at 45° with heat flow up, vertical (heat flow horizontal), 45° with heat flow down, and horizontal with heat flow down. This work was undertaken specifically in connection with building heat transfer.

Most of the work on heat flow in an enclosure of interest in a building context has been done since the early 1960s and has involved comparing flow patterns predicted numerically with flow patterns obtained by observation, sometimes optically.

In a vertical cavity the broad pattern of flow is upward near the warm surfaces and downward near the cold surfaces. At a low Rayleigh number $Ra = Gr_\delta \cdot Pr$, there is a single circulation loop with a linear temperature gradient between the surfaces, a cubic velocity profile and a Nusselt number Nu_δ of unity, a regime of pure conduction. It was known in the 1960s, however, that with increase of Ra the flow became more complicated. First, the boundary layers become separate, leaving a central area with no gradient in temperature. Then a series of cells or vortices is formed in the central space with opposite circulation. Finally a regime of turbulent flow fills the space. The development is shown very clearly in Figure 5.6, taken from Zhao *et al.* (1997), who conducted a numerical study for a series of aspect ratios H/δ. Figure 5.6 is for $H/\delta = 15$. For $Ra_\delta = 5000$ there is uniform flow at the two walls; when $Ra = 10\,650$ three feeble cells develop in the central third of the cavity and they strengthen by $Ra = 17\,750$. The pattern changes to two by $Ra = 19\,900$; they persist up to $Ra = 28\,500$ but the flow changes back to a single circulation at $Ra = 28\,750$. The number of cells increases with H/δ, as their further figure shows. With a constant Ra value of about 10 000,

H/δ	15	20	25	30	40
No. of cells	3	5	6	9	11

The authors argue that for $H/\delta > 80$ the flow changes from laminar to turbulent without formation of cells.

(Through comparing earlier studies in air, Dixon and Probert (1975) had placed the conduction/transitional change at $Gr = \rho^2 \beta g \Delta T \delta^3 / \mu^2 = 2 \times 10^3$ and a transitional/boundary layer change at $Gr = 2.8 \times 10^4$. The changes were independent of the height/width ratio for values larger than 1.)

Le Quéré (1990) provides an informative introduction to this interesting phenomenon. He describes the cells as 'cat's eyes' and notes that above Prandtl numbers of about 12,

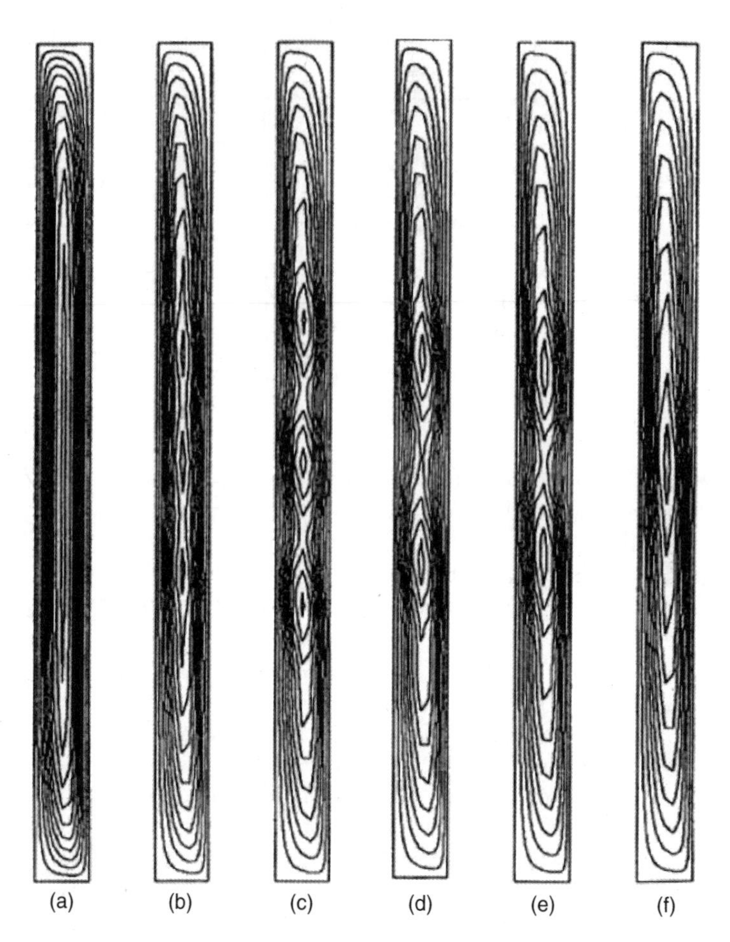

Figure 5.6 Streamline contour plots of airflow in a cavity of aspect ratio $H/\delta = 15$: (a) Ra = 5000, (b) Ra = 10 650, (c) Ra = 17 750, (d) Ra = 19 900, (e) Ra = 28 500, (f) Ra = 28 750. Zhao *et al.* (1997) (© ASHRAE, Inc., www.ashrae.org. Reprinted by permission from *ASHRAE Transactions* 1997, Volume 103, Part 1)

the cells display a travelling wave instability, though this is not of concern in building applications since Pr for air is 0.71.

At the start of this development, there is some weak countercirculation at the top and bottom of an enclosure, as variously shown by the figures of Gilly *et al.* (1981), the visualisations of Linthorst *et al.* (1981) and the interferogram of Ho and Chang (1994). The patterns of flow lines and isothermals found by Korpela *et al.* (1982) and Lee and Korpela (1983) show that for a narrow cavity ($H/\delta = 20$) there is a uniform temperature gradient between the plates other than near the top and bottom of the enclosure with particles circulating in a single loop; this is the regime of (near) pure conduction. When $H/\delta = 10$ there is a large temperature gradient near the bottom of the warm surface and this weakens with height with particle movement again in a single loop. This is not unlike the case of two separate plates but with some interaction of the boundary layers and with replacement of entrainment at the edge by flow circulation in a closed cavity.

At two intermediate spacings however, $H/\delta = 17.5$ and 15, it appears that particles near the centreline of the enclosure circulate around one or other of several smaller cells rather than around the enclosure as a whole and in addition to the major maximum values of the temperature gradient at the bottom and top of the cavity, there are further weak maxima successively located at the cold, warm, cold, warm, cold and warm surfaces.

Schinkel *et al.* (1983) examine stratification aspects. The complicated form of the flow is illustrated by the elegant isosurfaces shown in Figure 7 of Phillips (1996). In an investigation on window cavities, Wright and Sullivan (1994) illustrate the same phenomenon and it has been further studied numerically by Wakitani (1997) and numerically and observationally (using incense smoke) by Lartigue *et al.* (2000). With $H/\delta = 40$ and $\text{Ra} = \text{Gr} \cdot \text{Pr} = 3550$ they found the value of the local Nusselt number Nu to be unity except near the top and bottom of the cavity. At $\text{Ra} = 6800$ some 14 small oscillations in the value of Nu set in; with an increase in Ra, they increase in amplitude, decrease in number and appear to become unstable. Daniels and Wang (1994) state that the multicellular structure sets in at $\text{Ra} = 7880\text{Pr}$ or about 5600 for air. Its explanation is based on an eigenvalue solution. The photographs of Chen and Wu (1993), made with a glycerine-water solution, suggest that some tertiary circulation may occur before turbulence sets in. They give a full description and explanation of the phenomenon. Novak and Nowak (1993) have summarised work on cell formation and from their own calculations plot the local Nusselt number against height in a vertical cavity ($H/L = 30$). The figures (showing seven cells) make clear that while the cell formation is an interesting visual phenomenon, it has little significance for the loss of heat by convection across a window cavity. They remark however that for a window of height 0.9 m and with $\Delta T = 20\,\text{K}$, a change in cavity width from 10 mm to 14 mm reduces the convective loss by about 30%.

A wall cavity is often filled with insulating material and Lock and Zhao (1992) have considered the effect of using a honeycomb fill of square section $H \times H$ where $\delta/H = 5$. It can lead to an effective conductivity less than that of air.

Zhao *et al.* (1999) have reviewed work on convective flow in double-glazed units. Early studies reported results for Nu_δ in terms of Gr_δ and the aspect ratio H/δ using a simple power law relation:

$$\text{Nu}_\delta = C(\text{Gr}_\delta)^a (H/\delta)^b. \tag{5.43}$$

These authors have tabulated values for these constants found from work before about 1980, noting values of C between 0.065 and 0.210, a between 0.250 and 0.333, b between -0.100 and -0.265. They suggest that probably the most accurate and reliable results are those of ElSherbiny *et al.* (1982a), most of whose expressions were expressed (page 102) as

$$\text{Nu}_\delta = [1 + (a(\text{Ra}_\delta)^b)^c]^{1/c}, \tag{5.44}$$

where $\text{Ra}_\delta = \text{Gr}_\delta \cdot \text{Pr}$. Values of the constants are given for values of H/δ up to 110: a is of order 0.03, b is around 0.4 and extremes of c are 6.5 and 18. In a further article, ElSherbiny *et al.* (1982b) examine how the thickness B of the end walls separating the two convecting surfaces may affect the heat transfer. If there is a linear change in temperature in the walls in the δ direction, heat transfer is largely independent of B when B is greater than about 0.7δ; this is not true for small Nusselt numbers.

As noted earlier, insulated glazing units may consist of two plates of glass, sealed, filled with dry air or argon, and coated with a low-emissivity coating. In normal circumstances the plates will be almost parallel, but in conditions of extreme temperatures or barometric pressure, they may bend sufficiently inward to affect the total resistance significantly. A study by Bernier and Bourret (1997) indicated that the unit U value might vary up to 5% above and 10% below the yearly average. Winter conditions down to $-18°C$ were considered.

Heat transfer in an inclined enclosure is of interest particularly for solar collectors. Yang and Leu (1993) have studied this situation numerically and conclude that at a low angle of inclination, a stationary set of cells is set up with their axis in the upward direction. At a larger angle the cells have a horizontal axis and move.

Of less interest for rooms is convective transfer in shallow enclosures driven by temperature differences between opposite walls. A numerical study by Prasad and Kulacki (1984) shows there may be some cells of secondary circulation embedded in the primary flow. The structure of the flow from a localised heated patch on the floor of an enclosure has been studied numerically by Sezai and Mohamad (2000).

The patterns of flow in a cavity which is vented above and below are very different from those in the closed cavity discussed here. Wirtz and Stutzman (1982) describe experiments on natural convection of air between vertical heated parallel plates and offer a series of expressions to relate the Nusselt and Rayleigh numbers. For a discussion of vented enclosures, see Sefcik *et al.* (1991).

5.5 CONVECTIVE EXCHANGE AT ROOM SURFACES

A large literature has built up on fluid flow and heat transfer in a near-square enclosure and some of it is sketched below.[16] Attention may be drawn however to studies on convection in non-rectangular enclosures which may be relevant to heat flow in roof spaces. Poulikakos and Bejan (1983) describe experiments in a small triangular space of height H (152 mm), width L (737 mm) and depth W (559 mm); they use the term 'width' to describe the depth. They found for air:

$$\mathrm{Nu}_H = 0.345\,\mathrm{Ra}_H{}^{0.3}, \quad 10^6 < \mathrm{Ra}_H < 10^7. \tag{5.45}$$

See also Lam *et al.* (1989).

Setting aside the effects of an internal convective source and radiant transfer, room convective exchange supposes a convective exchange of heat between the inside surface of an outer wall of a room via the air to the opposite surface or ceiling/floor (taken to be the warmer). In principle this situation is similar to that discussed above and a crude approximation to the process is provided by supposing the two surfaces to act as two plates, one cold and one hot, on which boundary layers form. This is inadequate, however, since the air velocity forms a closed loop passing over the floor and the ceiling. For a heated plate, the highest entrainment velocity occurs at the leading edge of the plate, which would be at the floor level of the inside wall if the plate model were assumed. Since this is incompatible with flow in a closed loop, convective flow in a room has to be studied as a coupled system.

[16]The results to be discussed relate to situations where the enclosure surfaces are either isothermal or adiabatic since these have most relevance to heat transfer in rooms. A study where one of the enclosure walls is subjected to a constant heat flux is reported by Kimura and Bejan (1984). Their results indicate that the core may be motionless with stratified temperature. They took the Prandtl number to be 7, appropriate for water, but suggested that the results would serve approximately for air.

Broadly in agreement with the earlier numerical studies of Gilly *et al.* (1981) and Markatos and Pericleous (1984), Boehrer (1997) provides a simple model for the flow pattern and disposition of isotherms in an enclosure (height H) with hot and cold opposed walls (L apart) with low values of H/L, (a situation examined earlier by Shiralkar *et al.* (1981) for high Rayleigh numbers.) They fall into three regimes. (i) In the conductive regime, the isotherms are almost parallel to the vertical walls with low horizontal fluid velocity across the central vertical plane of the enclosure. (ii) In the transitional regime, the isotherms are somewhat inclined to the horizontal and there are larger velocities which vary over the height of the enclosure. (iii) In the convective regime, the isotherms are horizontal and the cold-to-hot flow is confined to a thin layer near the floor and a thin layer near the ceiling. An analysis of previously reported work suggests that the boundary layers on floor and ceiling start to form at $\mathrm{Gr}_H \cdot \mathrm{Pr}(H/L)^2 = 10^4$. With increasing Gr_H, the boundary layer ('intrusion') thickness δ decreases as $\delta/H = 3.2[\mathrm{Gr}_H \cdot \mathrm{Pr}(H/L)^2]^{1/5}$ (i.e. δ becomes independent of H). Heat transfer as such is not discussed.

Both Elder (1965) and de Vahl Davis (1968) report that at the warm surface there may be a weak *downward* flow in the outer reaches of the boundary layer and a corresponding upward movement near the cold surface. (Elder showed this using flow visualisation methods with oil of Prandtl number around 1000, so it might not be appropriate for air. However, de Vahl Davis (1968) demonstrated the phenomenon numerically and showed that the flow pattern appeared to depend only slightly upon Pr.)

Khalifa and Marshall (1990) in their Figure 11b indicate that in addition to the main circulation in a room with a heated vertical wall there are small pockets of countercirculation at floor and ceiling levels, adjacent to the heated wall and to the opposite wall. More detail on the form of circulation in a square enclosure is given by Henkes and Hoogendoorn (1993). The left wall (at T_h) was heated, the right wall cooled (T_c), and the floor and ceiling were adiabatic. At $\mathrm{Ra} = \mathrm{Gr}_H \cdot \mathrm{Pr} < 5 \times 10^4$ the flow forms just one cell; the streamlines are circular at the centre and become squarish toward the bounding surface; the isotherms are almost vertical with a somewhat larger gradient bottom left and top right than top left and bottom right. For $5 \times 10^4 < \mathrm{Ra} < 5 \times 10^5$ there are larger flows and gradients near the vertical walls; two centres of circulation have developed in the interior and at the central vertical plane the isotherms are roughly horizontal over the greater part of the height. This pattern consolidates itself for $5 \times 10^5 < \mathrm{Ra} < 5 \times 10^6$: upward and downward flow is concentrated nearer the walls, the two centres of circulation move nearer the walls and a third is formed in the centre. In the figures for $\mathrm{Ra} = 10^8$ the two circulation centres have moved into the top left and bottom right corners and the isotherms are almost horizontal and evenly spaced in most of the enclosure; the upward flow is confined to a thin boundary layer and on approaching the ceiling no longer turns to move gently downward away from the wall but falls sharply and rises sharply again before reaching the central plane. The pattern in the bottom right corner is the same. In a room 3 m high and 3 m between the walls, the value of $\mathrm{Ra} = (\rho^2 \beta g/\mu^2)(\Delta T H^3)\mathrm{Pr} = (1.47 \times 10^8)(\Delta T \times 3^3)(0.71) = 10^8$ implies a value of $\Delta T = T_h - T_c = 0.035$ K, so it is only the last flow pattern that is relevant in room applications.

It is interesting to note that at a lower Rayleigh number the flow pattern may be unstable with a periodically changing pattern. This has been observed by Briggs and Jones (1985) and examined numerically by Winters (1987). His Figure 4 shows nine stages in the moving pattern of streamlines and isotherms round a square enclosure. His value of Ra for onset of oscillations was 2.109×10^6, in reasonable agreement with the measured value of 3×10^6. In an air-filled cubic enclosure of side 0.15 m the estimated period was 3.4 s per cycle.

Hamady *et al.* (1989) show interferograms of the isotherms in a square enclosure with opposed hot and cold surfaces and all other surfaces insulated. $\mathrm{Ra} = 1.1 \times 10^5$. The first shows the cold surface as floor and the hot as ceiling. The enclosure is then rotated through 30°, 60° and 90° (so the isothermal surfaces are now vertical) and through further increments of 30° until the cold surface becomes the ceiling. The seven fringe patterns show the development of the temperature field from one limit to the other. The $\mathrm{Ra} = 10^6$ flow pattern driven by a difference between the vertical walls is shown strikingly in Figure 5 of Fusegi *et al.* (1991), a beautiful colour print. See also Fusegi and Farouk (1990).

From de Vahl Davis (1968: Table 1) the numerical results for $\mathrm{Gr}_H \cdot \mathrm{Pr} \geq 10^4$ and where the height H and length L (the direction of horizontal flow) were equal, we have

$$\mathrm{Nu}_H \approx 0.019(\mathrm{Gr}_H \cdot \mathrm{Pr})^{0.49} \tag{5.46a}$$

This is based on the temperature difference $T_h - T_c$ between the hot and cold vertical surfaces and the corresponding convective coefficient. Based on the difference $T_h - T_i$ $(= T_i - T_c)$, where T_i is the room air temperature, we have

$$\mathrm{Nu}_H = 0.053(\mathrm{Gr}_H \cdot \mathrm{Pr})^{0.49} \tag{5.46b}$$

For $\mathrm{Gr} \cdot \mathrm{Pr}$ in the range 10^3 to 10^4, the corresponding expression is

$$\mathrm{Nu}_H = 0.48(\mathrm{Gr}_H \cdot \mathrm{Pr})^{0.235}. \tag{5.47}$$

In a later account de Vahl Davis (1983) presented results of an accurate analysis of natural convection in air, ($\mathrm{Pr} = 0.71$) in a cavity of square section (and infinite length) so as to provide a 'benchmark' solution. The vertical surfaces were taken to differ by unit temperature and Nu was based on the difference; the horizontal surfaces were adiabatic.

$\mathrm{Gr} \cdot \mathrm{Pr}$	10^3	10^4	10^5	10^6
Nu	1.118	2.243	4.519	8.800

This led to the following approximate relation:

$$\mathrm{Nu} = 0.142(\mathrm{Gr} \cdot \mathrm{Pr})^{0.30}. \tag{5.48}$$

(The results are sufficiently accurate to suggest a slight cubic dependence of log Nu upon $\log(\mathrm{Gr} \cdot \mathrm{Pr})$). The study formed the focal point of an exercise to which 30 workers contributed solutions (de Vahl Davis and Jones 1983). More recently, Leong *et al.* (1999), urging that benchmark solutions should be physically realisable, have provided further such solutions for a cubic enclosure with one pair of opposing faces kept at T_h and T_c, with a linear variation between these values at the other surfaces. Their values of Nu as a function of $\mathrm{Gr} \cdot \mathrm{Pr}$ are a little lower than those listed above.

Bauman *et al.* (1983) describe results on convective exchange between the end walls of a water-filled tank, height $H = 15.2$ cm and length $L = 30.5$ cm, of large breadth and with an insulated base and upper surface. This is more representative of room proportions than the $H = L$ enclosures described above. The flow pattern did not indicate any weak downward flow near the warm end as the earlier investigations suggested; the boundary layers appeared parallel to their respective surfaces and were thinner on the vertical runs than on the horizontal runs. Their graphed results can be expressed as the following two correlations:

$$10^8 < (\mathrm{Gr}_H \cdot \mathrm{Pr}) < 4 \times 10^9, \quad \mathrm{Nu}_H \approx 0.44(\mathrm{Gr}_H \cdot \mathrm{Pr})^{0.27}, \tag{5.49a}$$

$$1.6 \times 10^9 < (\mathrm{Gr}_H \cdot \mathrm{Pr}) < 7 \times 10^9, \quad \mathrm{Nu}_H \approx 1.50(\mathrm{Gr}_H \cdot \mathrm{Pr})^{0.22}. \tag{5.49b}$$

In the latter case, the surface-air convective coefficient h_{sa} (W/m^2 K) would be given as

$$h_c = 2.03((T_h - T_i)/H)^{0.22} \tag{5.50}$$

and this typically leads to values of h_c less than the value of 3.0 W/m^{2}K that is often used. It will be recalled that the flow at a heated plate tended to become turbulent at values of $\mathrm{Gr}_H \cdot \mathrm{Pr}$ greater than 10^9. In the present case, values considerably larger than this are reached without the onset of turbulence.

Most of the numerical studies on flow in an enclosure have used the Navier–Stokes equations in two dimensions and without a model of turbulence, but the three-dimensional study of Ozoe *et al.* (1986) includes the turbulent kinetic energy and its rate of dissipation ε. They considered an air-filled cubic enclosure with a heated floor, one cooled wall, and the other surfaces adiabatic. Their values for heat transfer from the floor were $\mathrm{Nu} = 6.04$ at $\mathrm{Ra} = 10^6$ and $\mathrm{Nu} = 13.27$ at $\mathrm{Ra} = 10^7$, consistent with $\mathrm{Nu} \propto \mathrm{Ra}^{1/3}$.

Delaforce *et al.* (1993) made observations of convective heat transfer in a test cell and expressed h_c values as a continuous function of time over 24 h. Most of the figures show a fairly steady-state regime with short-term fluctuations of order 15 mins. Thus, at a wall with continuous heating, h_c averaged 1.7 W/m^{2}K over much of the day, with most of the extremes at 1.4 and 1.7 but with maximum excursions to 1.1 and 2.4 W/m^{2}K. With no heating, the mean value was around 1.4 W/m^{2}K but with bigger excursions. The authors identified three clusters of values associated with no heating (lowest values but largest spread), continuous heating, and the initial warm-up period during intermittent heating when values clustered around 2.6 W/m^{2}K.

Awbi and Hatton (1999) conducted an investigation of convective coefficients at the heated walls, floor and ceiling of a full-sized enclosure. They plotted their results as a function of the generating temperature difference ΔT and compared their results with some other results and with the correlations suggested in the 1997 *ASHRAE Handbook of Fundamentals* and Book A of the 1986 *CIBSE Guide*. They are summarised in Table 5.2.

Table 5.2 Range of values of heat transfer coefficients (W/m^{2}K) at heated surfaces in a full-sized enclosure. The values are taken from Awbi and Hatton (1999)

ΔK	Walls	Floor	Ceiling
5	2.0–3.4	2.8–3.3 (4.2*)	0.25–0.8
15	2.9–4.5	4.1–4.7 (5.9*)	0.35–1.05

*These high results are those of Min *et al.* (1956). The other five sets of data lie in the comparatively small range indicated.

Apart from the Min *et al.* (1956) results for a floor, observed values lie fairly evenly between the lowest and highest values quoted. Awbi and Hatton expressed their results in terms of a characteristic length, the hydraulic diameter D:

$$D = \frac{4 \times \text{area}}{\text{perimeter}}.$$

$$\text{for a wall, } h_c = (1.823/D^{0.121})(\Delta T)^{0.293}, \tag{5.51a}$$

$$\text{for a floor, } h_c = (2.175/D^{0.076})(\Delta T)^{0.308}, \tag{5.51b}$$

$$\text{for a ceiling, } h_c = (0.704/D^{0.601})(\Delta T)^{0.133}. \tag{5.51c}$$

These values lie within the range indicated in Table 5.2.

A later study (Awbi and Hatton 2000) considered partly forced, partly natural convection and expressed the total coefficient h_c in terms of its natural and forced components h_{cn} and h_{cf}. For a heated wall, floor and ceiling h_c could be expressed as

$$h_c^{3.2} = h_{cn}^{3.2} + h_{cf}^{3.2} \tag{5.52}$$

Values for h_c due to buoyancy and mechanically driven flows are listed in Table 7.17 of Clarke (2001).

In design calculations the convective exchange represented by h_c is usually combined with the radiative exchange Eh_r, which is around $5\,\text{W/m}^2\text{K}$, and a mischoice of h_c may not seriously affect the value of $h_c + Eh_r$ and so a wall U value. But Awbi and Hatton (1999) cite work by Lomas (1996) showing that 'the estimated annual heating energy demand varied by around 27% depending on the value of the convective heat transfer coefficient used', a matter discussed in detail by Beausoleil-Morrison and Strachan (1999), who suggest that characterisation of internal convection may be more significant than characterisation of the building fabric and infiltration rate. In modelling the heat flow across a room surface, the time-varying flux is more sensitive to choice of the local h_c value than is the steady-state flow, since h_c forms the link to readily accessible thermal capacity in the wall. A further problem is that most models assume the room air temperature is uniform, that is, it is fully mixed or 'well-stirred'. It is a matter of common observation that there may be several degrees difference between air temperatures near the floor and ceiling; when a radiator is sited below a window, the plume of hot air may lead to high convective transfer at the glass.

The local value of h_c may vary with position, especially at a window. This is clear from the observed values given in Figure 6 of Wallenten (2001). The scatter of values shows clearly that when room air temperature was higher than the window temperature, with consequent downward flow, the highest h_c values (around $5\,\text{W/m}^2\text{K}$) were found at the top of the window, and low values ($1\,\text{W/m}^2\text{K}$) were found at the bottom. The direction was reversed when the window was warmer than the room air. Larsson and Moshfegh (2002) have recently reported results on the downdraft from a window as it is affected by the width of the sill below it.

Most of the studies on natural convection in enclosures have been concerned with two-dimensional flow–flow in the xy plane when the z direction is large – so the effect of the end walls is unimportant. Bohn and Anderson (1986), however, include three-dimensional considerations at the high Rayleigh numbers which are relevant in building applications.

Three studies report values for convective coefficients associated with high volumetric flow rates where a throughput of air amounts to gentle forced convection and increases the total convective loss. These are examples of three-dimensional flow. Using water as the fluid, Neiswanger *et al.* (1987) demonstrated the enhanced values. Secondly, Spitler *et al.* (1991) give values at the room surfaces as shown for two inlet positions. They are correlated with the jet momentum number $J = V_{flow}U_0/gV_{room}$, where V_{flow}(m^3/s) is the volume flow rate, U_0(m/s) is the velocity of the supply air, and V_{room}(m^3) is the room volume; (g(m/s^2) is the acceleration due to gravity and appears to have been introduced so that J should be dimensionless.)

	h at ceiling	h at vertical walls	h at floor
Inlet in ceiling	$h = 11.4 + 209.7\sqrt{J} \pm 1.6$	$h = 4.2 + 81.3\sqrt{J} \pm 0.6$	$h = 3.5 + 46.8\sqrt{J} \pm 0.3$
	$0.001 < J < 0.03$	$0.001 < J < 0.03$	$0.001 < J < 0.03$
Inlet at side wall	$h = 0.6 + 59.4\sqrt{J} \pm 0.6$	$h = 1.6 + 92.7\sqrt{J} \pm 0.8$	$h = 3.2 + 44.0\sqrt{J} \pm 0.4$
	$0.002 < J < 0.011$	$0.002 < J < 0.011$	$0.002 < J < 0.011$
	Ar < 0.3		Ar < 0.3

Ar is the Archimedes number, $\beta g L_c \Delta T/U_0{}^2$, where β is the coefficient of thermal expansion, L_c is a characteristic length, taken here as 4.6 m, the maximum throw of the jet, and ΔT is the temperature difference between the room inlet and outlet. Thirdly, Fisher and Pedersen (1997) present results on the convective coefficient for the room as a whole, based on the hourly air change rate (ACH) with values for walls $h = 0.19\text{ACH}^{0.8}$, floor $h = 0.13\text{ACH}^{0.8}$ and ceiling $h = 0.49\text{ACH}^{0.8}$, arguing that the physics of the control volume requires this coefficient to be proportional to the ventilative flow rate at a constant inlet air temperature, which they take to be the best reference temperature.

Ganzarolli and Milanez (1995) have made a numerical study of the case of a rectangular room, long in the z direction, with a height-to-width ratio up to 14. Cooling took place at the walls which were maintained at a fixed temperature. Heat was supplied at the floor, either at conditions of constant temperature or of constant heat flow. For $W/H = 14$, and at the high Rayleigh number Ra = Gr · Pr of 10^6, both forms of drive resulted in a strong upflow concentrated near the room centre and a not dissimilar disposition of isotherms. For a floor at constant temperature and the lower value Ra = 10^4, the flow lines and isotherms are located near the walls, leaving almost uniform conditions in the central regions; with conditions of uniform flux and Ra = 10^3, air movement and variation of temperature were found throughout the space. Aydin *et al.* (1999) report a numerical study of natural convection in an enclosure where one wall is heated and the ceiling is cooled.

In modelling convection in mechanically ventilated rooms, convective coefficients are needed for a wide range of circumstances: buoyant flow adjacent to walls, buoyant plumes rising from radiators, fan-driven flows (Beausoleil-Morrison 2001). Beausoleil-Morrison (2002) has set out a scheme classifying internal convective exchange through an examination of the literature. He distinguishes five classes of regime:

(A) The driving force is buoyancy at surfaces and includes heat flow originating through external wall conduction, direct incidence of sunlight on floors and wall, floor and wall heating, and chilled ceiling panels.

(B) The driving force is again buoyancy but due to a localised heat source, e.g. a stove.

(C) A heated or cooled airflow is delivered through suitably mounted diffusers and extracted mechanically.

(D) Air, heated or cooled, is circulated internally without supply and extraction.

(E) Air movement is due to buoyancy and mechanical forces.

An *adaptive convection algorithm* based on this scheme has been incorporated into the Glasgow-based ESP-r simulation program: 'As the simulation progresses, a controller monitors critical simulation variables to assess the flow regime. Based on this assessment, the controller dynamically assigns (for each surface) an appropriate h_c algorithm from amongst the set [of 28 h_c correlation equations] attributed at the problem definition stage.'

A study by Yewell *et al.* (1982) using a long shallow water tank (length L, height H) gives some indication of the time needed to reach steady-state conditions after some change ΔT is made to the temperature of the end walls. Expressed in terms of tabulated constants for water or air, the time can be written as

$$t \sim \frac{H^{1/4}L}{\Delta T^{1/4}}\kappa^{-1}\left(\frac{\rho^2\beta g}{\mu^2}\right)^{-1/4}\mathrm{Pr}^{-1/4}. \tag{5.53}$$

The values for air at 20°C are thermal diffusivity $\kappa = 2.1 \times 10^{-5}\,\mathrm{m^2/s}$, $\rho^2\beta g/\mu^2 = 1.47 \times 10^8\,\mathrm{m^{-3}K^{-1}}$ and the Prandtl number $\mathrm{Pr} = 0.71$. Thus for a room 5 m long and 3 m high with $\Delta T = 1$ K, t is a little more than an hour. Lakhal *et al.* (1999) have reviewed work on unsteady convection. They made a numerical study of airflow in a square enclosure with a cooled ceiling and periodically heated wall.

Several workers, listed by Chadwick *et al.* (1991), have examined the convective heat loss from localised heat sources in rectangular enclosures. They confirm the widely held view that localised heat sources are most effectively located near the floor; this is also the best location for radiant loss.

Finally, Bohn *et al.* (1984) and Kirkpatrick and Bohn (1986) considered various combinations of heated and cooled surfaces. Their enclosure was a cube of side 30.5 cm with heated and cooled surfaces at 45° and 15°C, respectively. The fluid used was ionised water, $\mathrm{Pr} = 7$, rather than air. Experimental results were reported for Rayleigh numbers between about 5×10^9 and 6×10^{10}. The following Nu(Ra) relations were found.

Base	Left wall	Top	Right wall	Nusselt number	
heated	heated	cooled	cooled	at top and bottom,	$\mathrm{Nu} = 1.10\ \mathrm{Ra}^{0.236}$
				at the sides,	$\mathrm{Nu} = 0.141\ \mathrm{Ra}^{0.313}$
heated	cooled	cooled	cooled	at the top, bottom and sides,	$\mathrm{Nu} = 0.346\ \mathrm{Ra}^{0.285}$
heated	heated	heated	cooled	at the bottom,	$\mathrm{Nu} = 2.54\ \mathrm{Ra}^{0.212}$
				at the sides,	$\mathrm{Nu} = 0.233\ \mathrm{Ra}^{0.286}$
				at the top,	$\mathrm{Nu} = 0.223\ \mathrm{Ra}^{0.207}$

The authors also made shadowgraph observations of the flow from which they were able to see the development of the thermals.

5.6 CONVECTIVE EXCHANGE THROUGH AN APERTURE BETWEEN ROOMS

If two adjoining rooms are at different temperatures, there will be an airflow through any aperture, cool air flowing into the warm room at low level with an outward stream of warm air above it. The flow provides a major mechanism to transfer heat from for example a conservatory to an adjoining space. In a room of length L comparable to or larger than the height H, with opposed vertical walls assumed here to be maintained at hot and cold temperatures T_h and T_c and with an insulated floor and ceiling, the rise or fall in air temperature is confined to relatively thin boundary layers with a core at a virtually uniform temperature. Suppose now that a vertical partition is placed in the room, parallel to the hot/cold surfaces: the core becomes divided into warm and cool spaces. Suppose further that there is an opening in the partition, in the form of an unglazed window or a doorway. An airflow develops through the opening from cold to warm in the lower half and warm to cold in the upper, leading to an exchange of heat Q equal to $v\rho c_p \Delta T_{aa}$, where v is the volume flow in the lower or upper half. Boyer *et al.* (1999) report a result due to Passard, who noted that the airflow exchanged between two different zones, separated by a standard doorway with a temperature difference of 0.1 K, is approximately $120\,\text{m}^3/\text{h}$. We can suppose that the exchange is driven by the relatively small difference in air temperature ΔT_{aa} on either side of the partition or by the larger difference in surface temperature $\Delta T_{ss} = (T_h - T_c)/2$, the surface-air temperature when no partition is present. It is possible to derive correlations for the heat flow in dimensional form or non-dimensionalised as a Nusselt number, but they take different forms according to which form of ΔT is adopted.

The earliest analysis is based on ΔT_{aa}. If two adjoining spaces 1 and 2 have air temperatures T_{1a} and T_{2a} and there is a rectangular aperture between them of width w and height H_a, there is a transfer of heat due to the counterflows of air between them. Brown and Solvason (1962) gave an expression for the heat transfer coefficient h:

$$\begin{aligned}\text{Nu}_H &= \frac{hH_a}{\lambda} = \frac{C}{3}\left(\frac{g\Delta\rho H_a^3}{\mu^2/\rho}\right)^{1/2+\varepsilon}\left(\frac{c_p\mu}{\lambda}\right)^{1-\xi} = \frac{C}{3}\left(\frac{g\rho^2\beta\Delta T H_a^3}{\mu^2}\right)^{1/2+\varepsilon}\left(\frac{c_p\mu}{\lambda}\right)^{1-\xi}\\ &= \frac{C}{3}\text{Gr}_H{}^{1/2+\varepsilon}\cdot\text{Pr}^{1+\xi},\end{aligned}\tag{5.54}$$

since $\Delta\rho/\rho = \Delta T/T$ and $1/T = \beta$. Here C is a coefficient of discharge, between 0.6 and 1, and $\Delta\rho$ is the difference in density between the two spaces. If ε and ξ are zero, the expression follows from a simple hydrodynamical analysis in which the air velocity has its maximum value into the cold space at the top of the orifice and into the warm space at the bottom. The model ignores viscosity (whose effect cancels out in this equation). The expression correspondingly ignores the thickness t of the partition (i.e. the length of the opening) and is independent of the height of the room. They argue that the exponents ε and ξ are small compared to $\frac{1}{2}$ and 1. Thus a plot of log(Nu/Pr) against log Gr would be a straight line with a slope of $\frac{1}{2}$.

In fact, the effect of the finite opening length t – one of their experimental variables – is to reduce the volume flow at top and bottom and therefore the heat transfer. They describe measurements in air, leading to the following approximate values of Nu_H/Pr.

Gr_H	Nu_H/Pr	
	$t/H = 0.38$	$t/H = 0.75$
1.5×10^6	230	100
10^7	700	470
4×10^7	1600	1150

The gradients of these correlations are somewhat greater than $\frac{1}{2}$ and increase with decreasing Gr_H (unlike the situation shown in Figure 5.1 for heat exchange at a vertical plate). The lines lie below the relation for log Nu_H when the coefficient of discharge C is 1.

Further investigations include Weber and Kearney (1980). Bauman *et al.* (1983) summarise studies which include observations in which a 'partition' is suspended from the top of a water-filled tank and runs the full width W of the tank, height H, to form an opening below it of height H_a. This effectively blocks movement in the upper space on the warm side: flow rises at the warm wall to a height of around H_a, it then moves horizontally to the partition, rises to the ceiling on the cold side, falls at the cold wall and completes the loop at floor level. It gave an interzone heat transfer coefficient for air of

$$h_{iz} = Q/(WH_a\Delta T_{ss}) = 2.03(H_a/H)^{0.47}(\Delta T_{ss}/H)^{0.22}, \tag{5.55}$$

which is consistent with (5.50). They point out that the difference in air temperature ΔT_{aa} is very small and more difficult to measure than ΔT_{ss}. There are problems too in knowing just what is an appropriate mean air temperature in either space. However, it is the ΔT_{aa} that really drives this flow; setting up a ΔT_{ss} is simply a convenient experimental means of establishing ΔT_{aa} and in practice ΔT_{aa} comes about through other mechanisms, such as solar gains.

The authors also investigated the flow through a model doorway. In their water tank ($H = 15.2$ cm, $L = 30.5$ cm, $W = 83.3$ cm) they found the same Nu numbers for a doorway ($H_a = 11.4$ cm and width $W = 7.6$ cm) and for the full width ($H_a = 11.4$ cm and $W = 83.3$ cm), a surprising result which they put down to an increased airflow through the doorway.

A visual realisation of the resulting temperature (or density) distribution is shown by the Mach–Zehnder photographs of Bajorek and Lloyd (1982). They made observations using air and carbon dioxide under pressure in a long square-section enclosure (63.5 mm × 63.5 mm) with heated and cooled opposing walls and with an aperture of half cross-sectional area in a central partition. The heat flows were described as $Nu = 0.111Gr^{0.30}$ without the partition, and $Nu = 0.063Gr^{0.33}$ with the partition. A sequence of interferograms shows the progress of instability at the top of the warm wall and the base of the cool wall for a Grashof number greater than about 5×10^5.

Haghighat *et al.* (1989) report a detailed study of airflow in a partitioned enclosure of x, y and z dimensions, $L = 10$ m, $W = 4$ m, $H = 3$ m, using a $k - \varepsilon$ turbulence model. A temperature difference was established by taking the end walls at different temperatures and assuming the remaining surfaces to be insulated. The partition was assigned various positions on the x dimension and a doorway (width one-quarter of the enclosure width) was assigned various positions and heights in the partition. They present a comprehensive series of velocity vector fields in the xy and xz planes. They report that the airflow pattern is sensitive to changes in door height and location in the partition, and to the location of the partition. The heat transfer rate too is sensitive to door height and location, but not to the partition location itself. They found good agreement between their computed values of the Nusselt number $\mathrm{Nu}_L = (\text{heat flux}) \times L/(T_H - T_C)\lambda$ and experimental values. Nu_L appears to lie between 150 and 400. Ciofalo and Karayiannis (1991) made a numerical study of a similarly partitioned enclosure and illustrated the streamlines and isotherms, together with the height-averaged Nusselt number as a function of Ra and aperture size, comparing some of their results with those of earlier workers.

A further study of flow in a partitioned square enclosure is described by Acharya and Jetli (1990). They considered a room with hot and cold walls $L(= H)$ apart, assuming perfectly conducting end walls, and on the floor placed a divider of height $h(0 < h < H)$ at $x = \frac{1}{3}L, \frac{1}{2}L$ and $\frac{2}{3}L$. With a centrally placed divider and $h = 0$ (i.e. no partition), the pattern of streamlines and isotherms was skew-symmetric and with $h = H$ the patterns in the two cells were identical, as must be the case. However, for intermediate values of h there appeared to be markedly more circulation in the space near the hot wall than near the cold, with penetration of warm air at high level toward the cold wall. The partition clearly blocks a drift of cold air at floor level. This was broadly true at the spacings $\frac{1}{3}L$ and $\frac{2}{3}L$ The results of this study show the existence of three flow regimes:

- For Ra $\sim 10^4$ there is a weak symmetrical flow.
- For Ra $\sim 10^6$ there is an asymmetric flow, stronger nearer the hot wall.
- For Ra $\sim 10^9$ there is a distinct circulation in the space between the cold wall and the divider.

Now, in a room of height 2.5 m and a temperature difference of 5 K between opposite walls, $\mathrm{Ra} = (\rho^2 \beta g/\mu^2)(\Delta T H^3)\mathrm{Pr} = (1.47 \times 10^8)(5 \times 2.5^3)(0.71) = 8 \times 10^9$. Only the third flow regime appears to be relevant in building studies.

Riffat and Kohal (1994) report a study using a tracer gas (SF_6) of temperature-driven flow through an aperture between two enclosures and provide a figure, again comparing it with earlier work. Experimental values and correlations are plotted as log(Nu/Pr) against log Gr; see also Riffat (1991). Santamouris *et al.* (1995) have compiled a table summarising formulae for interzonal convective heat transfer.

Brown (1962) also examined the flow through a horizontal partition between vertically aligned rooms. Buoyancy-driven air movement also takes place vertically in stairwells; Feustel *et al.* (1985) examined the combined effects of wind forces and the stack effect in a 30 m building and Zohrabian *et al.* (1989, 1990) sketch the circulating flow patterns and report velocity measurements.[17] Klobut and Siren (1994) also describe an observational study. If cool air falls through a horizontal aperture, area A, in a floor of thickness H, into a warm space, the volume flow (m^3/s) is $AC\sqrt{\Delta T g H/T}$, where ΔT is the difference in temperature and T (K) is the mean air temperature. The coefficient of discharge C is given by Peppes *et al.* (2001) as $0.1469(\Delta T/T)^{0.2}$. In this case there will be a similar upflow elsewhere. However, flow in stairwells, etc., is too complicated to be

[17]Suppose that two rooms or adjacent spaces are at different temperatures but have the same pressure at the height of a common aperture. At some point h above the aperture, the pressure difference Δp is $(M p_{atmos} g h/R)(1/T_1 - 1/T_2)$; see equation (2.7). This serves as the driving pressure through any small cracks; see (Shao and Howarth 1992; Andersen 2003).

described by simple expressions. It has been studied through observation using tracer gas and by computational fluid dynamics.

Buoyancy-driven flow takes place through an open window. Some indication of its pattern is given by a study using water performed by Hess and Henze (1984). They made observations using a square cavity of side 1 m with one heated vertical surface and with the plane opposite it fully open or half open. The cavity was immersed in a large tank. The results are presented in terms of a local Nusselt number against the local Rayleigh number in the form $\mathrm{Nu} \propto \mathrm{Ra}^n$, where n varied from 0.25 at the bottom of the heated surface (where $\mathrm{Ra} \sim 10^8$) to 1.1 at the top ($\mathrm{Ra} \sim 10^{11}$) for the fully open condition; similar results were obtained for the half-open condition.

5.7 HEAT EXCHANGE AT AN EXTERNAL SURFACE

In the situations discussed so far, fluid flow has been driven by buoyancy but the heat lost at building exteriors is largely due to forced convection. The effect is expressed by the value of the outer film coefficient h_e, which forms one of the elements in the expression for the wall U value (1.12). For design purposes, CIBSE suggests the following values of film resistance ($\mathrm{m^2K/W}$) for high-emissivity surfaces; the corresponding values in transmittance form ($\mathrm{W/m^2K}$) have been added.

	Direction of heat flow	Sheltered	Normal	Severe
Wall	Horizontal	0.08 (12.5)	0.06 (17)	0.03 (33)
Roof	Upward	0.07 (14)	0.04 (25)	0.02 (50)
Floor	Downward	0.07 (14)	0.04 (25)	0.02 (50)

(CIBSE Guide Book A 1999 Table 3.10)

For a well-insulated wall or roof, the value of h_e is unimportant; it is of most significance at single-glazed windows situated near the edge and at the top of high-rise buildings where the largest local wind speeds occur. There is a large literature on the subject of forced convection at plane surfaces but the variable nature of wind speed and direction, coupled with its complicated interaction with an exterior surface and its surroundings, make detailed findings difficult to use.[18] A pragmatic approach is to adopt a simple correlation between the outside film coefficient h_e and wind speed V. McAdams (1954: 249) suggests the following values (converted to units of $\mathrm{W/m^2K}$):

$$\begin{array}{lll} & V < 5\ \mathrm{m/s} & V > 5\ \mathrm{m/s} \\ \text{smooth surface} & h_e = 3.9V + 5.6 & h_e = 7.2V^{0.78}, \\ \text{rough surface} & h_e = 4.3V + 6.2 & h_e = 7.6V^{0.78}. \end{array} \quad \begin{array}{r} \\ (5.56a) \\ (5.56b) \end{array}$$

More recently, Jayamaha *et al.* (1996) have suggested lower values:

$$h_c = 1.444V + 4.955 \quad (5.57)$$

[18] Observation of snowflakes by night, illuminated by wall-mounted fluorescent lighting near the top of a high-rise building, shows the very turbulent nature of the flow in windy conditions: quiet periods (fractions of a minute) during which flakes may be seen to rise or fall slowly near the building are followed by violent bursts when the airflow nearby is horizontal.

Wind speed in this connection is an ill-defined quantity since the air speed near a surface which drives the surface heat loss is only loosely related to wind speed above the building. Measurements by Ito *et al.* (1972) showed that with a wind speed value of 7 m/s measured 8 m above the roof of a several-storied building, local air speed 30 cm from a vertical surface varied between about 0.5 and 2.5 m/s according to wind direction. They found values for the convective coefficient at the centre of the sixth floor. When the site was windward with the roof value $V > 3$ m/s,

$$h_e = 2.9V + 5.8. \tag{5.58}$$

When leeward and $V > 4$m/s, h_c was about 13 W/m^{2}K, independent of wind speed. The correlation with V measured near the wall was

$$h_e = 9.4V + 8.7, \tag{5.59}$$

independent of wind direction. See also Watmuff *et al.* (1977).

Sharples (1984) has further clarified the question of what constitutes the velocity V and specifies three values: V_s the local air speed, measured 1 m from a surface at which a direct measurement of h_e was made, V_R the speed 6 m above the roof of the building, and V_{10} the velocity at a height of 10 m at a weather station 400 m away. It is clear that values of h_e can be best expressed in terms of V_s but the only information available to a designer comes from some convenient station. Sharples reports results found at the centre and near the edge of the wall on the eighteenth storey and at the centres of the fourteenth and sixth storeys of a high-rise building and which provide 'windward' or 'leeward' locations, according to the wind direction. They are summarised in Table 5.3.

Table 5.3 Regression equations and associated correlation coefficients between h_e, V_s and V_{10} Sharples (1984)

N (observations)		h_e and V_s	h_e and V_{10}	V_s and V_{10}
Floor 18 centre, windward	139	$h_e = 1.3V_s + 4.7$ (0.783)	$h_e = 1.4V_{10} + 6.5$ (0.670)	$V_s = 0.6V_{10} + 4.1$ (0.445)
Floor 18 centre, leeward	114	$h_e = 2.2V_s + 2.4$ (0.830)	$h_e = 1.4V_{10} + 4.4$ (0.829)	$V_s = 0.6V_{10} + 1.0$ (0.936)
Floor 18 edge, windward	304	$h_e = 1.7V_s + 4.9$ (0.750)	$h_e = 2.9V_{10} + 5.3$ (0.592)	$V_s = 1.8V_{10} + 0.2$ (0.793)
Floor 18 edge, leeward	242	$h_e = 1.7V_s + 5.3$ (0.416)	$h_e = 1.5V_{10} + 4.1$ (0.599)	$V_s = 0.2V_{10} + 1.7$ (0.305)
Floor 14 centre, windward	158	$h_e = 0.99V_s + 3.4$ (0.716)	$h_e = 1.6V_{10} + 3.3$ (0.834)	$V_s = 1.2V_{10} + 1.8$ (0.860)
Floor 14 centre, leeward	202	$h_e = 1.7V_s + 0.1$ (0.687)	$h_e = 1.5V_{10} + 1.0$ (0.657)	$V_s = 0.7V_{10} + 0.9$ (0.805)
Floor 6 centre, windward	187	$h_e = 0.65V_s + 1.9$ (0.264)	$h_e = 0.5V_{10} + 3.8$ (0.163)*	$V_s = 0.5V_{10} + 3.3$ (0.472)
Floor 6 centre, leeward	201	$h_e = 2.1V_s - 0.6$ (0.726)	$h_e = 1.4V_{10} + 1.7$ (0.654)	$V_s = 0.4V_{10} + 1.7$ (0.510)

*Apart from the value 0.163, all correlation coefficients are very significant (probability level $P < 0.001$).

We can make the following comments:

- In most cases, $r(h_e, V_s) > r(h_e, V_{10})$, as we might expect.
- The value of h_e (windward, edge) is larger than h_e(windward, centre) at higher wind speeds; there is little difference between h_e(leeward, edge) and h_e(leeward, centre).
- At higher wind speeds, h_e(windward, centre) increases with height; the variation for the leeward values is less marked.
- In leeward conditions, surface wind speeds were less than the weather station speeds. On the windward side they were markedly larger, presumably in part because the undisturbed air speed at the height of the eighteenth storey is larger than the 10 m value.
- A set of $r(V_s, V_{10})$ values must depend on the distance between the sites, quite close in the case reported.

For heat loss from a window we may add to this list the depth D of recess of the glass in its frame and the spacing of the vertical and horizontal members of the frame, the mullions and transoms. This has been addressed by Loveday *et al.* (1994) and Taki and Loveday (1996) who provide a multiple regression equation for h_e as a function of both D and some measure of wind speed. h_e correlates better with air speed nearby than with roof-top values.

Loveday and Taki (1996) further suggest values for h_e found from observations on a panel 0.8 m × 0.5 m in the middle of a block about 8 m wide and located some $3\frac{1}{2}$ m below the roof. V_r is the wind speed at roof level and V_s is the component of the wind velocity in a horizontal plane 1 m from the vertical surface.

$$\begin{aligned}
&\text{windward conditions: } & h_e &= 2.0V_r + 8.91, & r &= 0.63, & &0 < V_r < 15\,\text{m/s},\\
& & h_e &= 16.15V_s^{0.397}, & r &= 0.76, & &0 < V_s < 9\,\text{m/s},\\
& & V_s &= 0.68V_r - 0.5, & & & &\text{incidence between } 20^\circ \text{ and } 160^\circ,\\
& & V_s &= 0.2V_r - 0.1 & & & &\text{incidence less than } 20^\circ,\\
&\text{leeward conditions: } & h_e &= 1.772V_r + 4.93, & r &= 0.82, & &0 < V_r < 16\,\text{m/s},\\
& & h_e &= 16.25V_s^{0.503}, & r &= 0.82, & &0 < V_r < 3.5\,\text{m/s},\\
& & V_s &= 0.157V_r - 0.027. & & & &
\end{aligned}$$

Hagishima and Tanimoto (2003) report further work on this.

It is clear that while the convective coefficient at some point A on a building exterior is correlated with air speed at roof level or at some remote measuring station B, it also depends on a series of other factors: location of A on the building, buildings near A, wind direction, separation of A and B and the nature of the intermediate terrain. This makes it difficult to develop a more comprehensive correlation for h_e. McClellan and Pedersen (1997) summarise some progress in this direction made by earlier authors. The total convective coefficient is taken to be composed of the sum of the forced and natural components[19]:

$$h_e = h_f + h_n \tag{5.60}$$

and each of these is computed from expressions which take account of topography, etc.:

$$h_f = 2.537\, W_f R_f (P V_{az}/A)^{1/2}, \tag{5.61}$$

where P [m] is the perimeter of the surface and A [m^2] its area. The other variables have further parameters. W_f, the wind-modifier, is set to 1 for windward surfaces and $\frac{1}{2}$ for leeward surfaces. The term 'leeward' is defined as a wind direction more than 100° from normal incidence. R_f is the surface roughness multiplier, having values 2.17 for a stucco surface, 1.67 (brick), 1.52 (concrete), 1.13 (clear pine), 1.11 (smooth plaster) and 1.00 (glass). V_{az} [m/s] is the wind speed, modified for height z above ground:

$$V_{az} = V_0.(z/z_0)^{1/\alpha}, \tag{5.62}$$

where z_0 is the height at which standard wind speed measurements are taken, 9.14 m, and α is a terrain-dependent coefficient having values of 7 for flat, open country, 3.5 for rough, wooded country and 2.5 for towns and cities. (An expression of this kind for V_{az} must clearly be somewhat approximate and the authors give an alternative form.) Some values for α given by Davenport (1965) are: open sea 10, flat open country 6.25, woodland forest 3.6, urban area 2.5.

The natural convection component depends on the direction of heat flow:

for upward flow, $$h_n = 9.482(|T_{so} - T_o|)^{1/3}/(7.238 - |\cos\phi|), \tag{5.63a}$$

for downward flow, $$h_n = 1.810(|T_{so} - T_o|)^{1/3}/(1.382 - |\cos\phi|), \tag{5.63b}$$

where T_{so} and T_o are the surface and dry-bulb temperatures respectively and ϕ is the tilt of the surface from the horizontal. The expressions are almost equivalent when the surface is vertical. Wickern (1991) points out that the natural convective loss from an inclined surface consists of two components: one is associated with the component of g along the plate, the other is the loss from a horizontal or near-horizontal plate. In either case the value of h_n is likely to be small compared with the h_f component, which itself varies strongly in conditions of rough weather.

Loveday and Taki (1998) review work generally on convective loss from external surfaces. They argue that both ASHRAE and CIBSE have taken a value of $h_e = 14$ W/m^{2}K for 'normal' exposure (corresponding to a surface windspeed of 2 m/s), but that a value of 23.6 W/m^{2}K would be more representative of actual conditions. To this should be added the radiant transfer coefficient εh_r of about 4 W/m^{2}K.

[19]The article by Chen *et al.* (1986) summarises correlations for mixed convection flows at inclined surfaces.

6

Heat Transfer by Radiation

When a room is heated, the output of the appliance is distributed throughout the room by convection and radiation and is lost to ambient by conduction through the walls and by ventilation or infiltration. Although convective and radiative transfer perform similar functions, they are physically very different mechanisms. In discussing the convective behaviour of an element of fluid, only the conditions immediately adjacent to the element are relevant, as is the case for conduction; radiant exchange with some surface element requires us to consider the disposition of all surfaces surrounding it, together with their emissivities. Further, while conductive exchange is strictly proportional to temperature difference and convective exchange is roughly proportional, radiative exchange is proportional to a difference in fourth powers of temperature.

Since temperature differences in occupied spaces are normally small compared with absolute temperature, the fourth-power problem can be resolved by linearising the exchange; in any case it is not difficult to tackle it exactly. The effect of emissivity can be handled using network theory. Radiant exchange in rooms thus resolves itself into handling the disposition of surfaces which enclose the space – it is largely a geometrical problem. Several methods to tackle it have been put forward; Stefanizzi *et al.* (1990) describe eight methods and compare their predictions with a 'truth' model. In some cases a solution has been developed so that from the simplified viewpoint usually adopted for design purposes, radiant exchange should be transformed to 'look like' convective exchange, and so be merged with it as shown later.

6.1 THE FOURTH-POWER LAW

It is evident from the perception of warmth from a hot surface that all surfaces must be emitting radiant energy and that correspondingly they must be absorbing it. It was first discovered by observation that the total energy emitted was proportional to the fourth power of the absolute temperature T, that is, σT^4 (Stefan 1879). This was required too by classical thermodynamics as shown by Boltzmann (Boltzmann 1884; Eckert and Drake 1972: 590). At the same time, observation indicated the energy density $e_{b\lambda}$ (W/m^2 per unit wavelength λ) from the surface varied with wavelength. The total radiation followed from the assumptions of classical physics, but these assumptions were inadequate to explain its distribution and Planck eventually derived it in 1900 by making the famous assumption that an oscillator in the surface of a cavity vibrating at frequency ν and

Building Heat Transfer Morris G. Davies
© 2004 John Wiley & Sons, Ltd ISBN: 0-470-84731-X

emitting electromagnetic radiation did so in discrete units of $h\nu$, where h became known as Planck's constant. The assumption has negligible consequences for everyday systems.[1] The energy accordingly emitted from unit area of a black-body surface at T (K) between wavelengths λ and $\lambda + \delta\lambda$ is

$$e_{b\lambda} = \frac{2\pi hc^2/\lambda^5}{\exp(hc/k\lambda T) - 1} \text{ W per m}^2 \text{of surface per metre of wavelength,} \tag{6.1}$$

where $h = 6.6256 \times 10^{-34}$ J s (Planck's constant, chosen at the time simply to accord with observation),
$c = 2.997925 \times 10^8$ m/s (velocity of light),
$k = 1.38054 \times 10^{-23}$ J/K (Boltzmann's constant).

Boltzmann's constant is equal to R/N_0, where R is the universal gas constant, 8314 J/kmol K, and N_0 is Avogadro's number, 6.023×10^{26} molecules/kmol.

For a given value of temperature T, $e_{b\lambda}$ increases with increasing λ to a maximum (at λ') and falls again; $e_{b\lambda}$ increases rapidly with temperature and the value of λ' falls. Equation (6.1) can be expressed as

$$\frac{e_{b\lambda}}{T^5} = \frac{2\pi hc^2/(\lambda T)^5}{\exp(hc/k\lambda T) - 1}. \tag{6.2}$$

Since h, c and k are constants, the right-hand side is a function of λT (μ K) and the family of $e_{b\lambda}(T)$ curves fall onto a single curve (McAdams 1954: Figure 4.1). It shows that 2% of the total energy emitted lies below a value of 1600 μ K and 98% below 17 000 μ K. Here we are using the notation $1\,\mu = 1\,\mu\text{m} = 10^{-6}$ m.

A surface first emits perceptible radiation in the visible region above about 700°C. The wavelength λ' at which the intensity of emission has its maximum value is given by

$$\partial(e_{b\lambda})/\partial\lambda = 0, \tag{6.3}$$

leading to the relation

$$\lambda' T = ch/4.965k = 2.898 \times 10^{-3} \text{ mK}. \tag{6.4}$$

(The factor of 4.965 is the solution of the equation $e^{-x} = 1 - x/5$, which is found in evaluating equation (6.3).) Equation (6.4) is Wien's displacement law, also previously deduced from thermodynamics in 1896. For the sun, $\lambda' = 0.52\,\mu$ and our eyes have evolved to be most sensitive to radiation of this wavelength. For room surfaces at 20°C, $\lambda' = 10\,\mu$. The radiation emitted from such relatively low temperatures is often known as thermal radiation and 96% lies between 5.5 μ and 60 μ.

For most building applications it is sufficient to know the total energy emitted from a surface at T, rather than its distribution with λ. The total emittance is found by integration:

$$e_b = \int_{\lambda=0}^{\infty} e_{b\lambda}\, d\lambda = \frac{2\pi^5 k^4}{15c^2h^3} T^4 \equiv \sigma T^4. \tag{6.5}$$

[1]The amplitude of vibration of a pendulum of period 2 s and set to swing in air will slowly decay. According to Planck's postulate, its energy will not decay continuously but in discrete jumps of $(6.6 \times 10^{-34}\text{ J s})(0.5\text{ s}^{-1})$ or 3.3×10^{-34} J. Since its initial energy is typically of order 1 J, these decay steps are unobservable.

Stefan's constant σ, so defined, has a value of 5.6697×10^{-8} W/m^{2}K^4. The net radiative gain or loss of heat for a surface at T_1 depends on the temperatures and emissivities of surrounding surfaces and their mutual disposition, but in the simple case where T_1 is the temperature of a large plane black-body surface which can only exchange radiation with a second similar surface parallel to it, the net transfer is $\sigma(T_1^4 - T_2^4)$ W/m^2. This is the fourth-power law.

6.2 EMISSIVITY, ABSORPTIVITY AND REFLECTIVITY

The black-body surface mentioned above is a fictional surface. Its effect is realised by a small orifice in some cavity whose temperature is known and uniform (Figure 6.1). Radiation emitted by some small area in the cavity wall falls largely on other areas in the cavity where it is partly absorbed and partly reflected. The absorbed fraction becomes re-emitted, and after some reflection the radiation within the cavity becomes typical of the cavity temperature and independent of the reflection and absorption properties of the cavity wall. The small fraction of radiation escaping through the orifice is then typical of the cavity temperature alone and is described as black-body radiation, whose properties are given by equations (6.1), (6.2) and (6.4).

Real surfaces emit less radiation than this and the surface emissivity is defined as

$$\varepsilon = \frac{\text{radiation emitted by the real surface}}{\text{radiation emitted by a black-body surface at the same temperature}}. \tag{6.6}$$

Pure crystalline substances may emit more radiation in one direction than another, depending on the disposition of the atoms in their structure, and two parameters may be needed to describe ε. However, the surfaces of building materials are normally isomorphous and no direction parameters are needed.

The value of surface emissivity generally varies with wavelength, so ε requires a parameter λ. Most building materials have a value of $\varepsilon_\lambda = 0.9$ in the range of thermal radiation and vary little with λ; they are described as thermally grey. Examples are brick and tiles, concrete, glass, white paper, most paints, slate. Metal surfaces, however, may have a lower emissivity: bright aluminium 0.05, brass and copper 0.02 when polished and 0.2 when dull, aluminium paint 0.55, galvanised steel (new) 0.25, (weathered) 0.89,

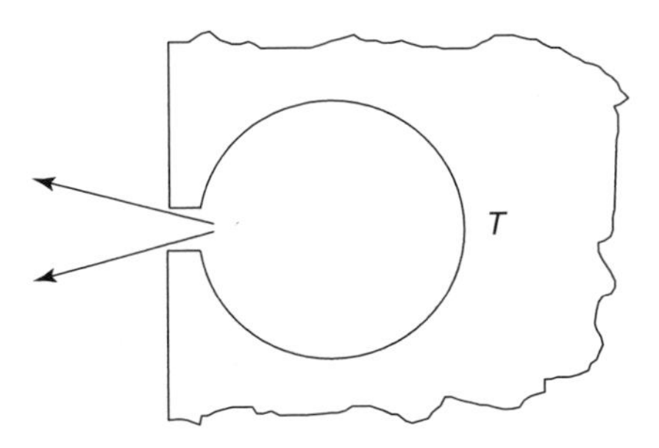

Figure 6.1 The radiation emitted through the small orifice of a cavity at temperature T is characteristic of the temperature alone and not the emissivity of the cavity walls

(Van Straaten 1967; ASHRAE 1993: 36-3; CIBSE 1999: 3-53). See Irvine (1963) for an account of how radiation properties are measured.

Radiation falling on a surface is partly absorbed and partly reflected; we can define the monochromatic absorptivity α_λ as

$$\alpha_\lambda = \frac{\text{radiation absorbed by the real surface in the range } \lambda \text{ to } \lambda + \delta\lambda}{\text{radiation absorbed by a black-body surface in the same range}}. \tag{6.7}$$

The reflectivity ρ_λ is the complement of this:

$$\alpha_\lambda + \rho_\lambda = 1. \tag{6.8}$$

For thermal equilibrium, $\alpha_\lambda = \varepsilon_\lambda$ (Kirchhoff's law, see for example, Eckert and Drake 1972:573).

These quantities are properties of the material surface. The integral $\int_{\lambda=0}^{\infty} \varepsilon_\lambda \, d\lambda$ is also a surface property. However, the corresponding integral for absorptivity, $\int_{\lambda=0}^{\infty} \alpha_\lambda \, d\lambda$, is not a pure surface property of the material since its value also depends on the distribution of radiation falling on the surface, and that depends on the emission characteristics (temperature and emissivity) of surrounding surfaces.

The colour of surfaces is determined, physically speaking, by the reflected (i.e. the non-absorbed) fraction of the visible radiation that falls upon it. Thus, in the visible range, a white surface has a high ρ_λ value and a correspondingly low ε_λ value; the converse is true for dark surfaces. Average values for ε_λ for solar radiation are brick 0.6–0.7, concrete 0.65, aluminium paint 0.5 (similar to its thermal radiation value), white paint 0.3, black paint 0.9; brown, red and green paint all have an average value 0.7.

In the 1970s it became possible to manufacture materials with some specific surface emissivity so as to achieve a 'selective surface'. Thus by coating aluminium with a thin layer of lead sulphide, the normal low emissivity and hence absorptivity in the visible range and up to about 2 μ becomes changed to a value of more than 0.8. Between 2 and 4 μ, ε_λ decreases to about 0.2 and thereafter varies little. The effect of using such a surface as a solar collector panel is therefore to make it effective as an absorber in the wavelength range where solar radiation is intense. However, the radiation losses from the panel occur in the thermal radiation range where the emissivity for this particular surface is low.

6.3 RADIATION VIEW FACTORS

The net exchange of radiation between two surfaces in a rectangular room depends on their temperatures and emissivities; it further depends on their mutual disposition and on all the room's other surfaces. To find the effect of room geometry, we first have to discuss what fraction of radiation leaving one surface falls directly on another and to find its value in some cases. This is the greatest mental effort. Expressions for the direct and total exchange of radiation then follow readily and they can be drafted in terms of a thermal circuit. Finally, the resulting network will be simplified to a star form so that it can eventually be merged with the circuit for convective exchange.

6.3.1 Basic Expression for View Factors

The view factor F_{12} from surface 1 to surface 2 is defined as

$$F_{12} = \frac{\text{the fraction of radiation from surface 1 which falls on surface 2}}{\text{total radiation emitted by surface 1}}. \tag{6.9}$$

This quantity has also been called the angle factor, area factor, geometrical factor, shape factor and configuration factor. Since the definition involves the direction 1 to 2, F_{12} is sometimes written as $F_{1\to 2}$. $F_{21} \neq F_{12}$. We have to find an expression for F_{12} between two small surfaces δA_1 and δA_2, their distance r apart and the angles θ_1 and θ_2 that their normals make with r. According to Paschkis (1936), the construct goes back to Gröber in 1926 and values for a number of geometries were found by Hottel in the early 1930s (McAdams 1954: Ch. 4).

Consider first the small area δA_1. It is assumed to emit diffusely as a black-body surface, so that it appears equally bright when viewed almost normally and at glancing incidence. I_1 denotes the intensity of radiation emitted per unit area and into a unit solid angle (1 steradian). The total flux leaving the surface is $\pi \delta A_1 I$.

To show this result, we note that radiation $\delta A_1 \cos\theta \, d\omega$ is emitted from surface 1 into the solid angle $d\omega$ which makes an angle θ with the normal to δA_1. Part of the emission falls on δA_2. Suppose that δA_2 has the form of a strip on a hemisphere, radius r, with δA_1 at its centre, the strip being symmetrically located about the normal to δA_1. The angle between δA_2 and the normal is θ, the width of δA_2 is $r\, d\theta$, its radius is $r \sin\theta$ and so the area $\delta A_2 = 2\pi r \sin\theta r\, d\theta$. The solid angle that δA_2 subtends at δA_1 is

$$d\omega = \frac{\text{area of strip}}{r^2} = 2\pi \ \sin\theta \, d\theta. \tag{6.10}$$

Thus the total flux emitted by δA_1 is

$$\delta Q = \int_{\theta=0}^{\pi/2} (\delta A_1 \cos\theta)(2\pi \sin\theta \, d\theta \ I) = \pi \delta A_1 I. \tag{6.11}$$

Of the total emission from δA_1, the part that falls on δA_2 is proportional to $\delta A_1 \cos\theta_1$, proportional to $\delta A_2 \cos\theta_2$, and proportional to $1/r^2$, so

$$\delta Q_{12} = I \ \delta A_1 \ \cos \ \theta_1 \ \delta A_2 \ \cos \ \theta_2/r^2 \tag{6.12}$$

and

$$F_{12} = \frac{\delta Q_{12}}{\pi \delta A_1 I} = \frac{I \delta A_1 \cos\theta_1 \delta A_2 \cos\theta_2}{\pi \delta A_1 I r^2} = \frac{\delta A_2 \cos\theta_1 \cos\theta_2}{\pi r^2}. \tag{6.13}$$

(A result originally due to Nusselt may be noted. Suppose that δA_1 is infinitesimal and that δA_2 is finite. Consider a hemisphere of unit radius centred on δA_1 with its axis aligned with the normal to δA_1. A vector from δA_1 taken round the contour of δA_2 defines an area $\delta A_2'$ on the hemisphere, which in turn can be projected onto the plane of δA_1 to define a further area $\delta A_2''$. Then $F(\delta A_1 \to \delta A_2) = \delta A_2''/\pi$. Mavroulakis and Trombe (1998) have recently extended the use of this expression.)

Both surfaces are normally finite and so the view factor follows from integration of the expression

$$F_{12} = \frac{1}{\pi A_1} \iiiint \frac{dA_1\, dA_2 \cos\theta_1 \cos\theta_2}{r^2}. \tag{6.14}$$

We can make the following comments:

- F_{12} (or F_{jk} generally) is dimensionless and less than unity.
- If A_1 radiates to a series of surfaces which completely surround it,

$$F_{11} + F_{12} + F_{13} + \cdots = 1. \tag{6.15}$$

  If A_1 is plane or convex, no portion of it is visible from another portion and $F_{11} = 0$. If A_1 is concave, it can radiate to itself and $F_{11} > 0$. It is implicit in this theory that A_1 is at a uniform temperature T_1, or that any variation in temperature is small enough to be ignored. If the variation is not small, the surface should be subdivided. If surface A_1 is divided into A_a and A_b say, F_{ab} is non-zero only if A_1 is concave.

- It is clear from equation (6.14) that

$$A_1 F_{12} = A F_{21} (= G_{12} \text{ or } G_{21} \text{ say}). \tag{6.16}$$

  It is this product, rather than F_{12}, that is of substantive interest. It has units of m^2 and when multiplied by the linearised radiant heat transfer coefficient $h_r(\mathrm{W/m^2\,K})$ it becomes a physical conductance (W/K) expressing radiant transfer, just as $A_1 h_{c1}$(W/K) expresses convective transfer. F_{12} merits consideration however because it is complex to evaluate.

6.3.2 Examples of View Factors

(a) Suppose that A_1 and A_2 form part of a spherical cavity. Consider first two small areas δA_1 and δA_2 separated by a distance r, patches on a hollow sphere of radius R. Suppose that the angular separation between them subtended at the centre of the sphere is 2ϕ and that the angle between the normal to δA_1 and r is θ_1. Then $\theta_1 = \pi/2 - \phi$ and $\frac{1}{2}r = R \sin\phi$. Then

$$F_{12} = \frac{\cos\theta_1 \cos\theta_2 \delta A_2}{\pi r^2} = \frac{\sin\phi \sin\phi \delta A_2}{\pi 4 R^2 \sin^2\phi} = \frac{\delta A_2}{\text{area of the sphere}}. \tag{6.17}$$

This must apply to a finite area A_2, so

$$F_{12} = A_2/\text{area of sphere}, \tag{6.18}$$

a very simple result. Thus a location on the surface of a sphere has the same view to some area A_2 on the sphere, regardless of the position or shape of A_2. (Hottel and Sarofim 1967: equation 2-37a).

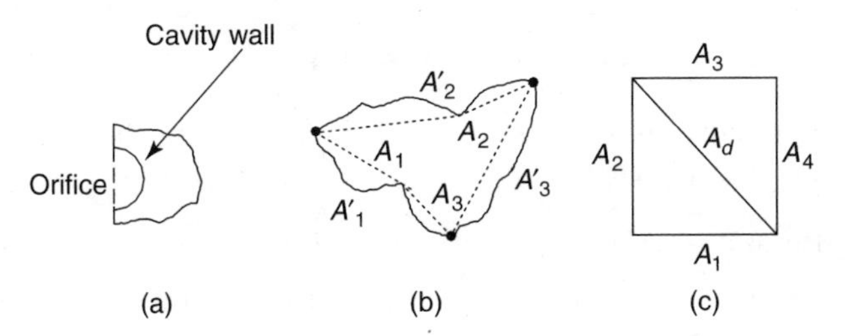

Figure 6.2 (a) The radiation from the black-body walls of a cavity is proportional to the area of the orifice. (b) A long enclosure of uniform section but with irregular walls. The radiant exchange between them is the same as between membranes at the same temperature stretched across them. (c) A long rectangular enclosure

(b) Suppose that A_1 and A_2 are surfaces in a long corridor. Consider first a three-sided long enclosure whose walls A'_1, A'_2 and A'_3 are irregular with high and low portions; the enclosure has everywhere the same section (Figure 6.2b).

We note first that the radiant flux from a cavity (Figure 6.2a) with a black-body surface is not proportional to the area of the cavity walls but to the area of a membrane stretched across its orifice. Accordingly, the exchange between the surfaces in Figure 6.2b is between membranes stretched from edge to edge composed of segments of straight section and forming the rather smaller areas A_1, A_2 and A_3. This ensures that

$$F_{11} = F_{22} = F_{33} = 0 \tag{6.19}$$

for the membrane surfaces. Now

$$\begin{aligned} A_1F_{11} + A_1F_{12} + A_1F_{13} &= A_1, \\ A_2F_{21} + A_2F_{22} + A_2F_{23} &= A_2, \\ A_3F_{31} + A_2F_{32} + A_3F_{33} &= A_3. \end{aligned} \tag{6.20}$$

Also

$$A_1F_{12} = A_2F_{21}, \quad A_1F_{13} = A_3F_{31}, \quad A_2F_{23} = A_3F_{32}, \tag{6.21}$$

so

$$A_1F_{12} = \tfrac{1}{2}(A_1 + A_2 - A_3). \tag{6.22}$$

This is Hottel's 'crossed strings' construction, described in Chapter 4 of McAdams (1954). It can be applied to the case of the rectangular corridor (Figure 6.2c) where $A_3 = A_1$ and $A_4 = A_2$. From equation (6.22), we have

$$A_1F_{12} = A_1F_{14} = \tfrac{1}{2}(A_1 + A_2 - A_d) \tag{6.23a}$$

and

$$A_1F_{12} + A_1F_{13} + A_1F_{14} = A_1, \tag{6.23b}$$

so

$$A_1 F_{13} = \surd(A_1^2 + A_2^2) - A_2. \tag{6.24}$$

(c) A_1 and A_2 are adjacent surfaces in a rectangular room. The surfaces are taken to be $b \times$ unity and $c \times$ unity. Then

$$\begin{aligned} F_{bc} = (1/\pi b) \left\{ \tfrac{1}{4} \right. & [(b^2 + c^2 - 1) \ln(b^2 + c^2 + 1) - (b^2 - 1) \ln(b^2 + 1) - (c^2 - 1) \ln(c^2 + 1) \\ & - (b^2 + c^2 - 0) \ln(b^2 + c^2 + 0) + (b^2 - 0) \ln(b^2 + 0) + (c^2 - 0) \ln(c^2 + 0)] \\ & \left. - (b^2 + c^2)^{1/2} \tan^{-1}(1/(b^2 + c^2)^{1/2}) + b \tan^{-1}(1/b) + c \tan^{-1}(1/c) \right\}. \end{aligned} \tag{6.25}$$

The zero is included to show the structure. Note that $bF_{bc} = cF_{cb}$. In a rectangular room $h \times w \times d$, the conductance G(wall, floor) in m^2 from the wall of area $h \times d$ to the floor of area $w \times d$ is $hd \times F_{bc}$ where $b = h/d$ and $c = w/d$.

(d) A_1 and A_2 are opposite surfaces in a rectangular room. Each is $b \times c$ and they are separated by unit distance. Then

$$\begin{aligned} F_{12} = (1/\pi) \{ & (1/bc) \ln[(b^2 + 1)(c^2 + 1)/(b^2 + c^2 + 1)] \\ & + (2/b)(b^2 + 1)^{1/2} \tan^{-1}(c/(b^2 + 1)^{1/2}) - (2/b) \tan^{-1}(c) \\ & + (2/c)(c^2 + 1)^{1/2} \tan^{-1}(b/(c^2 + 1)^{1/2}) - (2/c) \tan^{-1}(b) \}. \end{aligned} \tag{6.26}$$

In this case $F_{12} = F_{21}$. The conductance of the wall $h \times w$ to the opposite wall, G(wall, wall), is $hw \times F_{12}$; $b = h/d$ and $c = w/d$. These expressions relate to full room surfaces but they can be used to find the conductance between rectangular patches on the surfaces (Section 6.10).

(e) In a rectangular enclosure there are nine numerically distinct view factors. It can be readily checked that application of (6.25) and (6.26) to, say, the floor of a rectangular enclosure leads to a value of unity for the sum of the view factors to the other five surfaces.

For a cubic enclosure, $F_{bc} = 0.200044$ and $F_{12} = 0.199825$. These values are very close but not exactly equal to $\frac{1}{5}$, the value that is usually used.

Values for view factors are given in several texts, including Kollmar (1950) and Sparrow and Cess (1978), who quote the value for inclined adjacent rectangular surfaces; see also Siegel and Howell (1992). Howell (1998) has given an informative account of the Monte Carlo method to determine view factors. In effect, particles are fired from an elementary area δA_1 at random angles into the hemispherical space above it and the frequency is determined with which particles hit some other finite surface A_2.[2] Though inefficient, the method is physically attractive since it mimics the fundamental process of radiant exchange and relates to analytical expressions for F_{12} in the same way as the finite difference method for heat conduction problems relates to analytical expressions for $T(x, t)$. Ways of finding view factors between the human body and room surfaces are given by Cannistraro *et al.* (1992).

[2]This is a statistical method, similar in kind to the statistical method of finding π by dropping matchsticks so that they fall randomly onto parallel lines with spacing equal to the length of the matchsticks. It was first noted in 1777 that an approximate value for π is given by noting the proportion P of matches that fall across a join: $\pi \sim 2/P$. A large number of trials is needed to achieve three-figure accuracy; Howell (1968) notes the same feature when F_{12} is found in this way. The Monte Carlo method is useful when the medium between the surfaces is absorptive but this is not important for radiant exchange in small rooms.

6.3.3 View Factors by Contour Integration

Suppose that two surfaces A_1 and A_2 are not plane, that is, they have hills and valleys, and suppose they have irregular contours. In implementing the basic equation for view factors (6.14), we must take account of the internal topography of each surface. Yet it will be clear that the actual fraction of radiation leaving A_1 and intercepted by A_2, the definition of F_{12}, does not depend on topography; it only depends on the contours of the surfaces. Accordingly, F_{12} can be written in terms of the contour elements of the surfaces, involving a double line integration rather than the double area or fourfold line integration of (6.14). The vector line element $\mathbf{ds}_1$ of the contour of surface A_1 can be resolved into the three components dx_1, dy_1 and dz_1. If r is the separation between $\mathbf{ds}_1$ and $\mathbf{ds}_2$, then

$$F_{12} = (1/2\pi A_1) \int_1 \int_2 \ln(r)\, \mathbf{ds}_1\, \mathbf{ds}_2 \tag{6.27}$$

where A_1 is the area of membrane stretched over the contour of A_1; it is not necessarily plane. Integration is conducted over the closed contours of surfaces 1 and 2. We cannot strictly evaluate the logarithm of a quantity with dimensions, length in (6.27). However, as we will see, differences appear in $\ln r$, so in effect it is the logarithm of a ratio of lengths that is evaluated. The expression is due to Moon (1936). For a proof of the equivalence of the equations see Özisik (1973: 128) or Sparrow and Cess (1978: 128). Shapiro (1985) comments on implementation. It is instructive to see how, in a simple case, the form for F_{12} given by (6.27) can be transformed to that of (6.14).

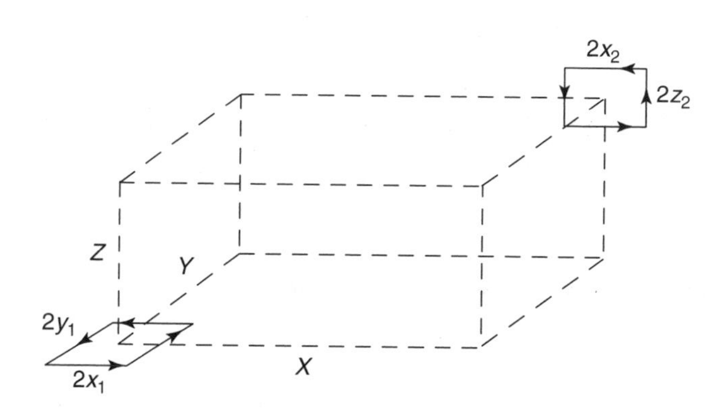

Figure 6.3 Contour summation for two small areas

Consider the two small surfaces 1 and 2 in Figure 6.3 of areas $2x_1 \times 2y_1$ and $2x_2 \times 2z_2$ situated on opposite corners of rectangular prism $X \times Y \times Z$. Since the contour of surface 1 has no component in z and the contour of surface 2 has none in y, only terms of type x_1x_2 enter the analysis. From (6.27), linking each edge of each surface in turn, we have

$$
\begin{aligned}
2\pi A_1 F_{12} &= \sum \ln(r)(2x_1)(2x_2) \\
&= \quad \ln \sqrt{X^2 + (Y + y_1)^2 + (Z - z_2)^2}(+2x_1)(+2x_2) \\
&\quad + \ln \sqrt{X^2 + (Y + y_1)^2 + (Z + z_2)^2}(+2x_1)(-2x_2) \\
&\quad + \ln \sqrt{X^2 + (Y - y_1)^2 + (Z - z_2)^2}(-2x_1)(+2x_2) \\
&\quad + \ln \sqrt{X^2 + (Y - y_1)^2 + (Z + z_2)^2}(-2x_1)(-2x_2).
\end{aligned}
\tag{6.28}
$$

This sum simplifies considerably by assuming (i) that each surface dimension such as $2x_1$ is very small compared with their separation $R = (X^2 + Y^2 + Z^2)^{1/2}$ and (ii) that when a dimensionless quantity $u \ll 1$, $1/(1-u) \approx 1 + u$, and that $\ln(1+u) \approx u$.

Furthermore, R makes an angle θ_1 with the normal to A_1 where $\cos\theta_1 = Z/R$; similarly, $\cos\theta_2 = Y/R$. Then

$$
2\pi A_1 F_{12} = 2A_1 A_2 \cos\theta_1 \cos\theta_2 / R^2. \tag{6.29}
$$

But this is identical with the definition of F_{12} for small areas (6.13). The surface and contour expressions for F_{12} are transforms one of the other.[3]

To see the significance of the contour form, suppose that A_1 and A_2 are large surfaces (rectangular for the sake of argument), and that F_{12} has to be found numerically. Suppose that each dimension is divided into n small lengths. Thus A_1 and A_2 are each composed of n^2 elementary areas. In evaluating (6.14), each of the n^2 elements of A_1 has to be associated with each of the n^2 elements of A_2, so F_{12} is composed of n^4 elements. In the case of (6.27), each of the $4n$ edge elements of A_1 is associated with each of the $4n$ similar elements of A_2, so F_{12} is composed of $16n^2$ elements. Thus, for all but coarse subdivision of the surfaces, numerical evaluation of view factors is more efficiently performed by contour summation. See Rao and Sastri (1996) for further discussion on the merits of finding view factors by contour integral methods. The method cannot be used if the two surfaces have a side in common.

6.4 DIRECT RADIANT EXCHANGE BETWEEN SURFACES

The evaluation of the room view factors, especially those given by (6.25) and (6.26), leads immediately to calculation of radiant exchange in a rectangular room.

6.4.1 Assumptions for Radiant Exchange

The following assumptions are normally made when discussing radiant exchange between surfaces:

- Each surface in the enclosure is supposed to be isothermal. If the temperature varies significantly over say the floor of a room, it can be subdivided into smaller isothermal areas. (The assumption is logically flawed since an edge common to two surfaces cannot have two different temperatures.)

[3] A vector may be generated as the gradient of a scalar (the case of heat flow) or from some form of rotation as in a vortex. In the latter case, Stokes' theorem states that the integral of the vector **V** round a contour is equal to the integral of curl **V** over the area, where curl **V** is found from space differentiation of **V**. The transform between the contour and area integral forms for F_{12} amounts to a double application of Stokes' theorem. Note that $\partial^2(\ln r)/\partial r^2 = -1/r^2$; $\ln r$ forms part of the one expression for F_{12} and $1/r^2$ forms part of the other.

- Each surface is supposed to be thermally grey, that is, there is negligible variation of emissivity ε with wavelength λ in the wavelength range concerned. This is a reasonable assumption in the infrared range of the spectrum but it could not be made in the visible region as our perception of colour variations between surfaces depends on their different values of visible-range absorptivity.
- The value of ε is taken as equal to the absorptivity α, and each is equal to $1 - \rho$ where ρ is the reflectivity.
- It is assumed that when radiation is incident on a surface, the reflected fraction has a uniform angular distribution; that is, the surface reflects diffusely.
- It is similarly assumed that the radiation emitted by the surface has a uniform angular distribution.
- It is assumed that the radiation incident on a surface is distributed uniformly over it, so there is no focusing.
- It is assumed that no radiation is absorbed in the air of the enclosure.

6.4.2 The Thermal Circuit Formulation

A full discussion of radiant exchange in an enclosure is presented in Hottel's chapter in McAdams (1954). Sparrow (1963) presents a series of alternative analyses. The exchange and other issues presented in this chapter are discussed by Liesen and Pedersen (1997). An elegant approach is provided in terms of a thermal circuit, variously shown by Paschkis (1936), Buchberg (1955) and Oppenheim (1956), to whom it is usually credited. We consider an enclosure consisting of several surfaces that exchange radiation one with another. The heat flow to or from a particular surface, e.g. surface 1, may be due to conduction from behind the surface, by convection from the air within the enclosure, or from local heating, perhaps electrical heating. The absorbed fraction of radiation incident on the surface is handled separately and is excluded from this component. Suppose that the net sum of such sources is denoted by Q_1 and suppose it is to be further transferred by radiation (Figure 6.4). The circuit follows from the equality of three expressions for the radiant flux:

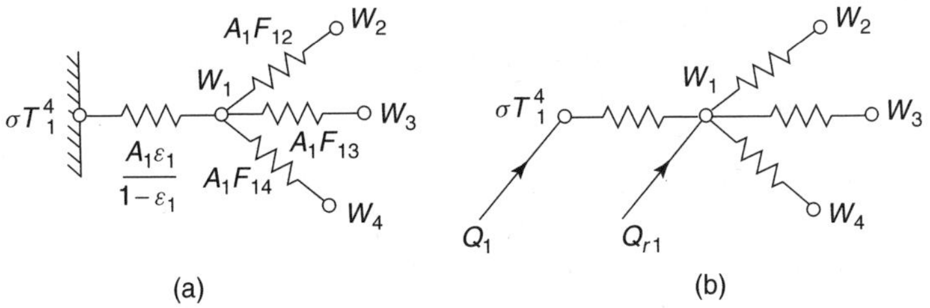

Figure 6.4 (a) Oppenheim's thermal circuit formulation for radiant exchange at surface 1 from other enclosure surfaces. (b) Inclusion in the circuit of radiation Q_{r1} from an internal radiant source acting at W_1

- Suppose that the radiant flux leaving surface j is W_j (the radiosity) with units W/m^2. The radiant flow from surface j falling on surface 1 is $W_j A_j F_{j1}$, which equals $W_j A_1 F_{1j}$. The total energy arriving at surface 1 is accordingly the sum of all such radiant terms $\sum W_j A_1 F_{1j}$, together with Q_1.
- The flux leaving surface 1 is similarly $A_1 W_1$ or $\sum W_1 A_1 F_{1j}$.
- The radiation leaving surface 1 consists of the flow emitted from the surface due to its temperature, $A_1 \varepsilon_1 \sigma T_1^4$, together with the fraction $1 - \varepsilon_1$ of the radiation that is incident on the surface and reflected from it.

Heat balance at the surface therefore requires that

$$Q_1 + \sum W_j A_1 F_{1j} = \sum W_1 A_1 F_{1j} = A_1 \varepsilon_1 \sigma T_1^4 + (1 - \varepsilon_1) \sum W_j A_1 F_{1j}. \quad (6.30)$$

After some manipulation, two expressions follow:

$$Q_1 = \frac{A_1 \varepsilon_1}{1 - \varepsilon_1} (\sigma T_1^4 - W_1) \quad (6.31a)$$

and

$$Q_1 = \sum A_1 F_{1j} (W_1 - W_j). \quad (6.31b)$$

The results can be put into circuit notation as shown in Figure 6.4a. The driving potentials are the surface emittances σT_1^4 here, or σT_j^4 generally, together with the radiosities W_j, which might be considered as having the status of temperatures like σT^4, despite their units of W/m^2. There is a geometrical conductance $A_1 \varepsilon_1 / (1 - \varepsilon_1)$ associated with the potential difference $\sigma T_1^4 - W_1$ in (6.31a) and the conductance $A_1 F_{1j}$ associated with the potential difference $W_1 - W_j$. For a black-body surface, ε_1 is unity, the associated conductance is infinite, and the σT_1^4 and W_1 nodes coincide.

If an enclosure having several surfaces, 1, 2, 3, ... is heated by a heat source acting at surface 1, its total heat loss can be resolved into components proportional to $\sigma(T_1^4 - T_2^4)$, $\sigma(T_1^4 - T_3^4)$, etc. However, the heat flow to surface 2 cannot in general be expressed as $Q_{1,2} = (W_1 - W_2) A_1 F_{12}$, as has been pointed out by Tao and Sparrow (1985). Hottel's presentation of radiant exchange leads to a some what similarly drafted quantity, $Q_{1-2} = (E_{b1} - E_{b2}) A_1 \mathscr{F}_{12}$, but $Q_{1,2} \neq Q_{1-2}$.

As far as surface 1 is concerned, the radiation Q_{r1} falling on it from some internal source of long-wave radiation, such as the component from an electric lamp, is similar in its effect to the radiation from the room bounding surfaces. It can be handled as an input at the radiosity node W_1 (Figure 6.4b). This is simpler than supposing that fractions α and ρ are respectively absorbed and reflected at the surface node itself. This is shown for a simple case in Appendix A of Davies (1992b). The same is true for the diffuse component of short-wave radiation, although α and ρ values will normally differ from their long-wave values and they will vary with wavelength. A specularly reflected component requires separate consideration.

The direct geometrical conductance between nodes W_j and W_k is given as

$$G_{jk} = G_{kj} = A_j F_{jk} = A_k F_{kj}. \quad (6.32)$$

In a rectangular enclosure there are 15 such links and of them, for example, we can write

$$G_{north,floor} = G_{north,ceiling} = G_{south,floor} = G_{south,ceiling} \tag{6.33}$$

and there are two further sets of four. On the other hand, $G_{north,south}$ is unique and there are two further values. Thus the total of 15 conductances is made up as $3 \times 4 + 3 \times 1$, and the set is made up of $3 + 3 = 6$ numerically distinct values.

6.5 RADIANT EXCHANGE IN AN ENCLOSURE

The pattern of 15 radiant exchanges in a six-surface enclosure with black body surfaces is shown in Figure 6.5a. Nine direct links are shown explicitly; the other six are simply indicated so as to avoid non-connections at intersections.

There is no difficulty in using the above formulation to handle exchange between the small areas into which the principal room surfaces can be subdivided (Malalasekera and James 1993; Tuomaala and Piira 2000) and so test the assumption that radiation incident upon a large surface is uniformly distributed over it. However, at the design level, subdivision requires that the convective and conductive processes too relate to subdivided areas and this clearly requires far larger computational resources. The main aim of the following analysis is to simplify a surface-surface pattern of radiant exchange to one where each surface exchanges radiation with a fictitious star node; the star-based network can then be merged with the necessarily star-based network for convective exchange, leading to the usual design method to handle room internal heat exchange. The six-surface enclosure of Figure 6.5 is sufficient for this purpose.

The presence of furnishings and other obstacles is usually ignored in room heat exchange. Methods to tackle it are discussed by Coelho *et al.* (1998) although they relate mainly to combustion chambers.

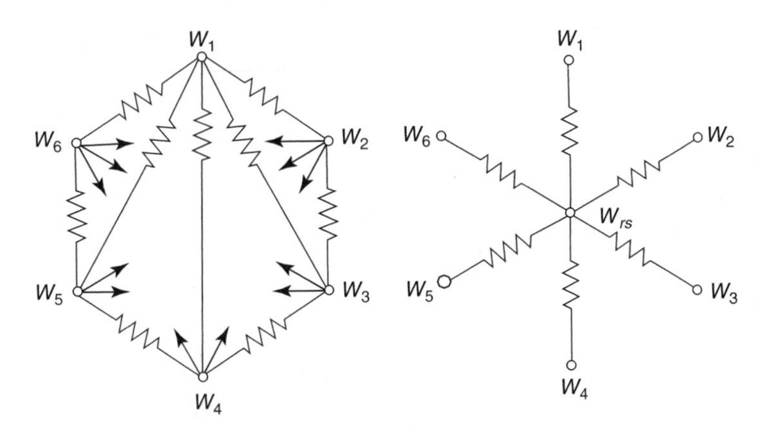

Figure 6.5 (a) Surface-to-surface radiant links in a black-body rectangular enclosure: the delta pattern. (b) Surface to star point links

6.5.1 Net Conductance G^{Δ}_{jk} between Two Nodes

We wish to simplify the full pattern of radiant exchange to the star-based pattern in Figure 6.5b on the basis that the net star-based conductance or resistance between any two nodes W_j and W_k should be approximately equal to the full, or delta-based, pattern of Figure 6.5a.[4] In Figure 6.5a the net conductance between W_j and W_k is greater than the direct value G_{jk} because of the effect of the other 14 paths in parallel. To evaluate the net value, suppose that a heat flow Q_1 due to conduction or surface heating arrives at W_1. Continuity requires that

$$G_{12}(W_1 - W_2) + G_{13}(W_1 - W_3) + G_{14}(W_1 - W_4) \\ + G_{15}(W_1 - W_5) + G_{16}(W_1 - W_6) = Q_1.$$

There are similar relations at the other five nodes, so

$$\begin{bmatrix} G_{11} & -G_{12} & -G_{13} & -G_{14} & -G_{15} & -G_{16} \\ -G_{21} & G_{22} & -G_{23} & -G_{24} & -G_{25} & -G_{26} \\ -G_{31} & -G_{32} & G_{33} & -G_{34} & -G_{35} & -G_{36} \\ -G_{41} & -G_{42} & -G_{43} & G_{44} & -G_{45} & -G_{46} \\ -G_{51} & -G_{52} & -G_{53} & -G_{54} & G_{55} & -G_{56} \\ -G_{61} & -G_{62} & -G_{63} & -G_{64} & -G_{65} & G_{66} \end{bmatrix} \begin{bmatrix} W_1 \\ W_2 \\ W_3 \\ W_4 \\ W_5 \\ W_6 \end{bmatrix} = \begin{bmatrix} Q_1 \\ Q_2 \\ Q_3 \\ Q_4 \\ Q_5 \\ Q_6 \end{bmatrix}. \quad (6.34)$$

G_{11} denotes here the sum of the conductances linked to node 1:

$$G_{11} = G_{12} + G_{13} + G_{14} + G_{15} + G_{16}. \quad (6.35)$$

(This use of the symbol G_{11} is simply convenient. Strictly, G_{11} should denote A_1F_{11}, but this is zero since surface 1 and other surfaces are plane.) Equations (6.34) result from the nodal approach to circuit analysis described in Section 4.2.2.

The equations are not linearly independent since if Q_1 to Q_5 were independently assigned, as they can be, Q_6 must be equal to $-(Q_1 + Q_2 + Q_3 + Q_4 + Q_5)$. Accordingly, we can without loss of generality make $W_6 = 0$ and write

$$\begin{bmatrix} W_1 \\ W_2 \\ W_3 \\ W_4 \\ W_5 \end{bmatrix} = \begin{bmatrix} H_{11} & H_{12} & H_{13} & H_{14} & H_{15} \\ H_{21} & H_{22} & H_{23} & H_{24} & H_{25} \\ H_{31} & H_{32} & H_{33} & H_{34} & H_{35} \\ H_{41} & H_{42} & H_{43} & H_{44} & H_{45} \\ H_{51} & H_{52} & H_{53} & H_{54} & H_{55} \end{bmatrix} \begin{bmatrix} Q_1 \\ Q_2 \\ Q_3 \\ Q_4 \\ Q_5 \end{bmatrix}, \quad (6.36)$$

where the H matrix is the inverse of the G matrix omitting row 6 and column 6.

Now the total resistance between nodes 2 and 4, say, is by definition the difference in radiosity between them when $Q_{r2} = +1$, $Q_{r4} = -1$ and $Q_{r1} = Q_{r3} = Q_{r5} = 0$. This

[4] The term 'delta' used here comes as a generalisation of the procedure described in Section 4.4.2, which showed how a real delta configuration of conductances could be transformed exactly to a star pattern.

resistance will be written as R_{24}^{Δ}. The superscript Δ indicates that the term derives from the generalised delta pattern of links. Thus

$$R_{24}^{\Delta} = (W_2 - W_4)/1 = H_{22} - H_{24} - H_{42} + H_{44}. \tag{6.37}$$

Since W_6 has been set to zero, the resistance between W_1 and W_6 is found by setting Q_1 equal to unity and setting Q_2 to Q_5 equal to zero. So

$$R_{16}^{\Delta} = (W_1 - W_6)/1 = H_{11}. \tag{6.38}$$

The other resistances are found by using one or other of these expressions. There are six numerically distinct values for the resistances R_{jk}^{Δ}.

6.5.2 Star Conductance G_{jk}^{*} or Resistance R_{jk}^{*}

There is a long-standing idea that the exchange of radiation between two surfaces, say surfaces 2 and 4 of a rectangular room, can be estimated by supposing that all the radiation is exchanged via a central node, the radiant star node. This fictitious network is shown in Figure 6.5b, where the star temperature is represented by the node W_{rs}, (which has no meaning as a descriptor of radiosity). How can the star links be chosen so that the star network as a whole best mimics the external effect of the parent delta network?

Now the flow of energy by radiation from a black-body surface A_1 at T_1 to a second black-body surface at T_2 which envelops it is

$$Q = A_1\sigma(T_1^4 - T_2^4) = A_1(W_1 - W_2), \tag{6.39a}$$

so the resistance between these nodes is

$$R_1 = (W_1 - W_2)/Q = 1/A_1. \tag{6.39b}$$

It follows that the resistance between W_1 and the radiant star node W_{rs} must be less than this and will be written as β_1/A_1 where β_1 is less than 1. The total resistance between nodes 2 and 4 of a rectangular room is then $\beta_2/A_2 + \beta_4/A_4$, so in general

$$R_{jk}^{*} = \beta_j/A_j + \beta_k/A_k. \tag{6.40}$$

Thus the net star-based conductance between nodes j and k is $G_{jk}^{*} = 1/R_{jk}^{*}$, to be compared with the delta-based value G_{jk}^{Δ} found above.

6.5.3 Optimal Star Links

The β values have to be adjusted so that the external effect of the star circuit resembles that of the delta circuit as closely as possible. For a three-surface enclosure, such as a very long ridge tent, it is possible to obtain an exact transform. Using the delta-star transform of Section 4.4.2,

$$\beta_1 = \frac{A_1 G_{23}}{G_{12}G_{23} + G_{23}G_{31} + G_{31}G_{12}}, \text{ etc.} \tag{6.41}$$

If the enclosure is a sphere, so that the areas $A_1, A_2 \ldots$ form patches on its surface, an exact delta-star transform is possible, regardless of how many areas into which the spherical surface is divided (Davies 1990). This follows from the simple and exact form for F_{12} for patches on a sphere.

For a rectangular room, however, the transform cannot be exact: no star configuration can achieve the same external effect as the parent delta circuit. We require some measure of the difference between their responses and for this we choose the quantity

$$S_p = \text{Sums of [(deviations as conductances)(deviations as resistances)]} \tag{6.42a}$$

$$= \sum_{j=1}^{6} \sum_{k=j+1}^{6} (G_{jk}^{\Delta} - G_{jk}^{*})(R_{jk}^{*} - R_{jk}^{\Delta}), \tag{6.42b}$$

which has a total of 15 terms and has to be minimised through variation of the β values.

To justify this expression, we note first that to minimise the deviation of $R_{jk}^{*} - R_{jk}^{\Delta}$ it must be non-dimensionalised as $(R_{jk}^{*} - R_{jk}^{\Delta})/R_{jk}^{\Delta}$ since R_{jk} values may be very different in non-cubic enclosures. To avoid spurious cancellation of positive and negative values, the squared value $[(R_{jk}^{*} - R_{jk}^{\Delta})/R_{jk}^{\Delta}]^2$ must be summed. But this favours minimisation of a difference in *resistance* form, an arbitrary choice since the *conductance* form $[(G_{jk}^{*} - G_{jk}^{\Delta})/G_{jk}^{\Delta}]^2$ has equal status. The quantity $[(G_{jk}^{*} - G_{jk}^{\Delta})/G_{jk}^{\Delta}][(R_{jk}^{*} - R_{jk}^{\Delta})/R_{jk}^{\Delta}]$ however is unbiased. Since $G_{jk}^{\Delta} \times R_{jk}^{\Delta} \equiv 1$, it reduces to the right-hand term in (6.42b). A sign reversal ensures the terms are positive.

Thus S_p represents a sum involving all 15 possible combinations of links between nodes and provides a single positive measure to express the difference in response between the delta and star configurations. The quantity $\sqrt{S_p/15}$ is the dimensionless root mean square deviation between the two.

S_p depends on the values chosen for the βs and by systematically varying them, a minimum value for S_p can be found. The resulting star circuit can then be described as the optimal star circuit to express the radiant exchange between the surfaces. There are six β values for a six-sided enclosure but if it is rectangular, there are only three distinct values. Thus we have to find the 3 values of β_j such that $\partial S_p/\partial \beta_j = 0$.

Davies (1983c, 1992b) reports an analysis on enclosures having fixed height $H = 1$ and lengths $L = 10^{0.0}, 10^{0.1}, 10^{0.2}, \ldots, 10^{1.0}$ (a series of 11 values) and depths $D = 10^{-0.0}, 10^{-0.1}, 10^{-0.2}, \ldots, 10^{-1.0}$ (11 values). When $H = L = D$ the enclosure is cubic. With increase of L only, the enclosure assumes the shape of a square-sectioned corridor; with decrease of D only, it has the shape of a square tile; with increase of L and decrease of D it looks like a plank on edge. See Table 6.4.

It turns out that the β_j values generated by computations on the 121 enclosures are quite closely determined by the fractional area $b_j = A_j$/(total enclosure area); the optimal relation between them was

$$\beta_j = 1 - b_j - 3.54(b_j^2 - \tfrac{1}{2}b_j) + 5.03(b_j^3 - \tfrac{1}{4}b_j). \tag{6.43}$$

The standard deviation between β_j values found using the optimisation procedure and β_j values estimated from the regression equation was 0.0068.

Elementary considerations show that the regression line must pass through the points ($b = 0, \beta = 1$) and ($b = \frac{1}{2}, \beta = \frac{1}{2}$). To a first approximation, $\beta_j \approx 1 - b_j$ but the distribution shows a clear cubic dependence on b_j.[5] For a cubic enclosure, $\beta = \frac{5}{6} = 0.833$ from symmetry and this value should be used rather than the relatively poor value of 0.844 from the regression equation.

6.5.4 How Good Is the Delta-Star Transformation?

Although the above procedure yields the 'best' star configuration, that does not necessarily mean that the star circuit closely mimics the delta circuit. Various checks are available.

The two-temperature enclosure

Consider a rectangular enclosure with black-body surfaces, one of which, the floor say, is at temperature T_1; the other five surfaces are at temperature T_2.[6] What error does the optimal star network lead to in estimating the conductance (or resistance) between the floor and the remainder of the enclosure? It can be expressed using a quantity like one of the terms in (6.42b). To set this up, first we note that the total conductance between the radiant star node W_{rs} and the enclosure surface, denoted by G_{rs}, is $2(A_1/\beta_1 + A_2/\beta_2 + A_3/\beta_3)$ and so the resistance between nodes 1 and 2 according to the star circuit is $(\beta_1/A_1 + 1/(G_{rs} - A_1/\beta_1))$. The exact resistance is $1/A_1$ and so the deviation in resistance is $[(\beta_1/A_1 + 1/(G_{rs} - A_1/\beta_1)) - 1/A_1]$. Thus we form the product

$$\begin{aligned} p_1 &= \text{(deviation as a resistance)(deviation as a conductance)} \\ &= [(\beta_1/A_1 + 1/(G_{rs} - A_1/\beta_1)) - 1/A_1] \times [A_1 - 1/(\beta_1/A_1 + 1/(G_{rs} - A_1/\beta_1))] \\ &= \beta_1 G_{rs}/(G_{rs} - A_1/\beta_1) + (G_{rs} - A_1/\beta_1)/\beta_1 G_{rs} - 2. \end{aligned} \quad (6.44)$$

Here p_1 provides a non-dimensional positive measure of this particular difference between the exact and approximate networks.

There are three distinct values p_j when $L \neq D \neq H$, so the quantity

$$\delta_2 = \sqrt{\tfrac{1}{3}(p_1 + p_2 + p_3)} \quad (6.45)$$

is the root mean product deviation between the two networks. The subscript 2 to δ indicates that the deviation refers to a two-surface enclosure, that is, where five of the surfaces have been lumped. For a given enclosure, the procedure to minimise S_p returns three β_j values but these will not be available to the user. Instead they must use the β_j values estimated from the regression equation (6.43), which are marginally less reliable.

[5]The expression has been quoted elsewhere as $\beta_j = 1 - b_j(1 - 3.54(b_j - \frac{1}{2}) + 5.03(b_j^2 - \frac{1}{4}))$. This is algebraically correct but misleading.

[6]This model was adopted in the 1986 and earlier versions of the *CIBSE Guide*, Book A. The inner surface of an outer wall had the value T_1, the 5 surfaces visible from it the value T_2.

Accordingly, two versions of δ_2 are available for an enclosure, δ_{2o} using original values from the minimisation procedure, and δ_{2e} from their estimates. Based on a set of enclosures where $L/H = 1.00, 1.58, 2.51, 3.98, 6.31, 10.0$ and $D/H = 1.00, 0.63, 0.40, 0.25, 0.16, 0.10$, mean values of δ_{2o} and δ_{2e} were 0.3% and 0.6%, respectively. The star circuit provides a satisfactory simplification for radiant exchange when five sides are combined.

The six-temperature enclosure

A more severe test is based on the difference between the 15 pairs of nodes as expressed by S_p above. A further δ can be defined as

$$\delta_6 = \sqrt{S_p/15}, \tag{6.46}$$

which is the root mean product deviation between the exact version and the star version of the net link between nodes j and k. The subscript 6 indicates that the links are considered between all six nodes individually, and again it can be based on original and estimated β values.

For a cube, $\beta = \frac{5}{6}$ for all surfaces. However, F(adjacent) and F(opposite) are close to but not exactly equal to $\frac{1}{5}$. As a result, $S_p = 8.0 \times 10^{-8}$ so δ_{6o} becomes 0.01% as in Table 6.1. A small deviation from the cubic form leads to larger values but most are less than 2%. There is little penalty in using regression rather than optimal β values.

It is useful too to know what may be the largest fractional error in the links. Δ is defined as $(R^{\Delta}_{jk} - R^{*}_{jk})/R^{\Delta}_{jk}$ and may be positive or negative. Table 6.2 lists the extreme values found.

6.5.5 Discussion

The star circuit satisfactorily represents the full delta pattern of exchanges when one surface radiates to the other five at a common temperature. An alternative method of effecting a delta-star transform of radiant conductances based on this model was suggested by Carroll (1980, 1981). In the present notation,

$$G_{rs} = 2(A_1/\beta_1 + A_2/\beta_2 + A_3/\beta_3) \tag{6.47}$$

Table 6.1 The accuracy with which a star circuit can represent radiant exchange between individual surfaces in a rectangular room $L \times D \times H$: values of $\sqrt{\frac{1}{15}S_p}$ (%)

D/H	$L/H = 1.00$		$L/H = 1.58$		$L/H = 2.51$		$L/H = 3.98$		$L/H = 6.31$		$L/H = 10.0$	
	δ_{6o}	δ_{6e}	δ_{6o}	δ_{6e}	δ_{6o}	δ_{6e}	δ_{6o}	δ_{6e}	δ_{6o}	δ_{6e}	δ_{6o}	δ_{6e}
1.000	0.01	0.01	0.97	1.42	1.64	1.75	2.08	2.13	2.37	2.48	2.56	2.76
0.631	1.05	1.49	1.53	1.72	1.86	1.95	2.08	2.20	2.24	2.42	2.35	2.61
0.398	1.72	1.87	1.83	1.92	1.80	1.91	1.76	1.92	1.77	1.97	1.81	2.04
0.251	1.86	1.92	1.74	1.79	1.53	1.61	1.36	1.47	1.28	1.41	1.27	1.40
0.158	1.66	1.70	1.46	1.52	1.21	1.29	1.01	1.10	0.90	0.98	0.86	0.94
0.100	1.36	1.43	1.16	1.25	0.92	1.03	0.73	0.84	0.62	0.72	0.57	0.66

Table 6.2 Extreme fractional errors Δ (%) in the star network for radiant exchange

D/H	$L/H = 1.00$		$L/H = 1.58$		$L/H = 2.51$		$L/H = 3.98$		$L/H = 6.31$		$L/H = 10.0$	
1.00	−0.01	0.00	−0.94	2.50	−1.90	3.44	−2.64	3.93	−3.62	4.12	−4.43	4.16
0.63	−0.00	3.52	−1.42	4.08	−2.00	4.13	−2.96	3.92	−3.68	3.97	−4.18	4.04
0.40	−0.59	4.15	−1.42	3.73	−2.11	3.13	−2.63	2.74	−3.01	3.25	−3.26	3.57
0.25	−0.99	3.15	−1.38	2.91	−1.67	2.34	−1.87	2.20	−2.00	2.66	−2.10	2.95
0.16	−0.88	3.04	−0.99	2.70	−1.06	2.14	−1.10	1.71	−1.13	2.06	−1.15	2.99
0.10	−0.58	2.67	−0.58	2.31	−0.77	1.80	−0.89	1.30	−0.95	1.52	−0.99	1.67

then the total resistance between one surface A_j in the enclosure via W_{rs} to the remainder at a common temperature is $\beta_j/A_j + 1/(G_{rs} - A_j/\beta_j)$. This Carroll took to be the exact value $1/A_j$. Then

$$\beta_j = 1 - A_j/(\beta_j\ G_{rs}) \tag{6.48}$$

and is to be found by iteration. Since $A_j/(\beta_j G_{rs}) \approx A_j/\sum A_j = b_j$, equation (6.48) is similar to the first two terms in (6.43). As the very small mean value of 0.006 for δ_{2e} shows, the β values given by (6.43) very nearly satisfy the condition that the resistance should be $1/A_1$ between one black-body surface and a set of surfaces that envelop it. Thus the Carroll method and the Davies method are similar in their effect. For an enclosure of dimensions $1 \times 2 \times 3$ the values for the three sets of β values are (0.924:0.930), (0.880:0.880), (0.696:0.705) expressed in the form (Carroll: Davies).

For the links between individual nodes whose fractional root mean square values are reported in Table 6.1, values of up to 3% are found. The δ_6 values are higher than the δ_2 values since positive and negative deviations tend to cancel in forming δ_2 but not δ_6. Although δ_{6e} (based on regression values for β) is necessarily higher than δ_{6o} (from the minimisation procedure), the deterioration is not very marked. The extreme values for the deviations (Table 6.2) are necessarily greater than the root mean square values (Table 6.1). Taken as a whole, the deviations are quite modest in relation to the uncertainties that appear in building heat transfer, and the delta-star transformation can be described as satisfactory.

6.5.6 Linearisation of the Driving Potentials

According to conventional theory, heat flow through solids is strictly proportional to temperature difference and convective exchange is approximately proportional. Long-wave transfer, however, is proportional to differences in the fourth power of absolute temperature and in order that mixed-mode transfer calculations can be made without complication in room heat transfer, it is usual to suppose that radiant exchange too is proportional to difference in temperature; this is satisfactory provided the differences are not large. Two steps are needed to linearise the difference:

(i) From each driving potential σT_j^4 or W_j, subtract some representative room potential, W_{rs} say, if we are handling a star circuit. This leaves the heat flow in all links unaltered.

(ii) We can express W_{rs} as σT_{rs}^4 and W_j as σT_{jb}^4 where T_{jb} is the black-body equivalent temperature of surface j.

A heat flow Q_j can then be expressed as

$$Q_j = A_j(W_j - W_{rs}) = A_j\sigma(T_{jb}^4 - T_{rs}^4) = A_j h_r(T_j - T_{rs}), \tag{6.49a}$$

where

$$h_{rj} = \sigma(T_{jb} + T_{rs})(T_{jb}^2 + T_{rs}^2). \tag{6.49b}$$

No approximation has been made and the heat flow is now proportional to a temperature difference $T_j - T_{rs}$. However, h_{rj} depends on T_{rs}, a global variable, and on some special temperature T_{jb}. Some T_{jb} values will lie above T_{rs} and some below T_{rs} but this variation is small compared with their absolute values; if it is ignored, we can assume a global value of $h_r = 4\sigma T_{rs}^3$. With a typical room temperature of 20°C, or 293 K, h_r is about 5.7 W/m^{2}K.

Division of the driving potential differences by $4\sigma T_{rs}^3$ implies that the conductances, so far expressed in m^2, should be multiplied by $4\sigma T_{rs}^3$ to express them in the physical units of W/K, Thus, for example, the geometrical conductance G_{12} of the delta network becomes $G_{12}h_r$ and A_1/β_1 of the star network becomes A_1h_r/β_1, as shown in Figure 6.6

The enclosure wall temperatures derive from observational values averaged over the surfaces and are 2D constructs. T_{rs} is a linear combination of these values so, although fictitious, it is also a 2D construct.

6.5.7 Inclusion of the Emissivity Conductance

Finally, the emissivity conductance $\varepsilon_1 A_1/(1 - \varepsilon_1)$ becomes $\varepsilon_1 A_1 h_r/(1 - \varepsilon_1)$. The total physical conductance between the surface node T_j and the radiant star node T_{rs} will be

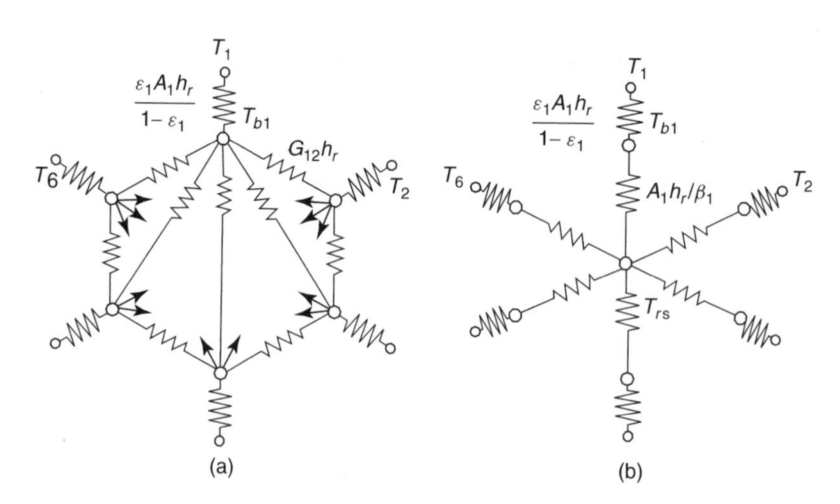

Figure 6.6 (a) The delta network, linearised to indicate surface and black-body equivalent temperature and the emissivity and geometrical conductances in physical form. (b) The star network similarly linearised with emissivity and geometrical conductances

denoted by S_j (W/K):

$$\frac{1}{S_j} = \frac{1-\varepsilon_j}{\varepsilon_j A_j h_r} + \frac{\beta_j}{A_j h_r} \tag{6.50}$$

or

$$S_j = E_j A_j h_r \quad \text{where} \quad 1/E_j = (1-\varepsilon_j)/\varepsilon_j + \beta_j. \tag{6.51}$$

The value of the radiant star temperature T_{rs} is

$$T_{rs} = \sum T_j S_j / \sum S_j. \tag{6.52}$$

This leads readily to the expression given earlier for the radiant heat transfer across a cavity. Consider a cavity wall of area A where the superficial dimensions are large compared to the separation of the surfaces across which radiation is exchanged. In terms of Figure 6.6a, this amounts to an enclosure where surfaces 2, 3, 5 and 6 are negligibly small and all exchange takes place between surfaces 1 and 4. The view factor between the surfaces is unity since each can only see the other. The total resistance R between the surfaces (surfaces 1 and 2, say) is composed of an emissivity/view factor/emissivity chain of resistances:

$$R = \frac{1-\varepsilon_1}{\varepsilon_1 A h_r} + \frac{1}{A h_r} + \frac{1-\varepsilon_2}{\varepsilon_2 A h_r} = \frac{\varepsilon_1 + \varepsilon_2 - \varepsilon_1\varepsilon_2}{A h_r\, \varepsilon_1\varepsilon_2}, \tag{6.53}$$

so that the cavity transmittance $1/AR$ is $\varepsilon_1\varepsilon_2 h_r/(\varepsilon_1 + \varepsilon_2 - \varepsilon_1\varepsilon_2)$. If ε_1 and ε_2 are each 0.9, as is common, the radiant transmittance of the cavity is about 4.7 W/m^2K. To this we must add the convective contribution, say 2.1 W/m^2K for a vertical cavity (Wilkes and Peterson 1938), making 6.8 W/m^2K. The net resistance of the cavity is then $1/6.8 = 0.15$ m^2K/W. A working value of 0.18 m^2K/W is often assumed.

6.6 SPACE-AVERAGED OBSERVABLE RADIANT TEMPERATURE

The quantity W_{rs} and the radiant star temperature T_{rs} derived from it are fictitious constructs and have no physical significance. Yet T_{rs} must clearly be some function of the radiant field and we now consider what significance, if any, might be attached to it. It will be demonstrated that T_{rs} may serve as an estimate of the average observable radiant temperature in an enclosure, although it is not a good estimate.

First a remark about 'large' and 'small' conductances. For a floor of area of 20 m^2 with $h_r = 5.7$ W/m^2K and $\varepsilon = 0.9$, the conductance S_{floor} between the floor and T_{rs} is of order 100 W/K. The convective conductance depends on the direction of heat flow but may be of order 60 W/K. These conductances can be regarded as 'large'. The temperature nodes associated with these conductances–air and surface temperatures–are thus high-conductance or low-impedance nodes. On the other hand, the radiant conductance between the room surfaces and a temperature-sensing element such as the bulb of a thermometer 2 mm in diameter is of order 0.0001 W/K, with a convective link of the same order. These will be regarded as 'small'. The links between occupants or furnishings and the room, although larger than this, can still be classified as small. Perceptions of thermal comfort must therefore be associated with links which are very much smaller than those associated with the bulk movement of heat (of order kilowatts) within and from the room. The temperature node associated with comfort must therefore be a low-conductance or high-impedance node. Although studies of human thermal comfort have been performed since the 1920s, attention does not appear to have been drawn to this distinction.

For consider dry resultant temperature T_c, which is taken as a measure of thermal comfort and is usually expressed as

$$T_c = \tfrac{1}{2}T_a + \tfrac{1}{2}T_r, \tag{6.54}$$

where T_c is typically observed by a small device such as a thermometer bulb and T_a and T_r are respectively the mean air temperature and the radiant temperature in the room[7] without any further speculation about how T_c is linked to these quantities. T_a and T_r are clearly high-conductance nodes. If (6.54) is expressed using a thermal circuit diagram, the links might be represented as equal low conductances (Figure 6.7a) or high conductances (Figure 6.7b). But the high-conductance linkage can be rejected on two grounds:

- Heat is exchanged at large room surfaces; the presence of occupants and small items of furnishing do not greatly affect the exchange. A large conductance via T_c would provide an invalid link between T_a and T_r.
- A heat input of order kilowatts applied convectively at T_a or radiantly at T_r may be expected to produce a modest rise in T_a or T_r. But 1-kW applied to a thermometer bulb will generate a very high temperature there.

Thus T_c, operationally the temperature of a sensing device, must be represented as a low-conductance node (Figure 6.7a), other nodes in the circuit are high-conductance nodes. The value of T_c is due to its convective exchange with the room air, the radiant flux it absorbs for any local source of long-wave radiation, and the radiant flux it absorbs from the room bounding surfaces. We have to examine the latter two processes to see how reliably T_c, a major factor in heating design, can be estimated by assuming that the room radiant temperature can be represented by the fictitious construct T_{rs}, the radiant star temperature. In an enclosure otherwise at the reference temperature of zero, a positive value of T_{rs} may be due to the radiation from an internal source taken to act at T_{rs} or due to a surface of the enclosure that is above zero. These causes are independent and will be considered in turn.

6.6.1 Space-averaged Observable Temperature due to an Internal Radiant Source

If a sensor at some location is shielded from radiation, it records the local or point air temperature T_{ap}. The volume-averaged or mean air temperature T_{av} is the mean of such

T_a T_c T_r

(a)

T_a T_c T_r

(b)

Figure 6.7 (a) T_c linked to T_a and T_r by low conductances. (b) T_c linked by high conductances

[7] T_r is often called the mean radiant temperature (MRT), but this can be ambiguous and will not be used here.

values found at N evenly spaced grid points in the room:

$$T_{av} = \sum T_{ap}/N \quad \text{or} \quad \frac{\iiint T_{ap}\,\mathrm{d}x\,\mathrm{d}y\,\mathrm{d}z}{\iiint \mathrm{d}x\,\mathrm{d}y\,\mathrm{d}z}. \tag{6.55}$$

The local or point radiant temperature T_{rp} is similarly the observed local sensor value in the absence of air. Suppose that the sensor is placed in an air-free room with black-body surfaces at a reference temperature of zero. It is supposed that a long-wave radiant source Q_r is placed at the room centre. The sensor will record a high value when near the source and an almost zero value when near the walls. Suppose the sensor is spherical of radius r so it intercepts radiation on a cross-section πr^2 and absorbs a fraction α. It re-radiates this energy to the walls from its full area $4\pi r^2$. Now the intensity of radiation passing through a spherical surface of radius R centred on the source is $Q_r/4\pi R^2$. Thus the point radiant temperature T_{rp} at a distance R from the source is given by

$$(Q_r/4\pi R^2)\pi r^2 \alpha = \varepsilon 4\pi r^2 h_r (T_{rp} - 0). \tag{6.56}$$

The volume-averaged observable radiant temperature T_{rv} follows by summing these values over a grid, or more generally as

$$T_{rv} = \frac{\iiint T_{rp}\,\mathrm{d}x\,\mathrm{d}y\,\mathrm{d}z}{\iiint \mathrm{d}x\,\mathrm{d}y\,\mathrm{d}z} \tag{6.57}$$

To remove the effect of enclosure size, we express T_{rv} in dimensionless form:

$$\beta_{rv} = T_{rv} h_r \sum A / Q_r,$$

where $\sum A$ denotes the total area of the six surfaces of the enclosure. Since $R^2 = x^2 + y^2 + z^2$, measured from the source, and since $\alpha = \varepsilon$, we have

$$\beta_{rv} = \sum A \iiint \frac{\mathrm{d}x\,\mathrm{d}y\,\mathrm{d}z}{x^2 + y^2 + z^2} \div 16\pi \iiint \mathrm{d}x\,\mathrm{d}y\,\mathrm{d}z. \tag{6.58}$$

Integration takes place over the volume of the enclosure. If the source is placed in the centre, the limits are $-\frac{1}{2}L$ to $+\frac{1}{2}L$, $-\frac{1}{2}D$ to $+\frac{1}{2}D$, $-\frac{1}{2}H$ to $+\frac{1}{2}H$. If the source is placed in the middle of say the $D \times H$ wall, the limits for L become 0 to L. β_{rv} is a purely geometrical quantity. Message has derived an analytical expression for β_{rv} (Davies and Message 1992). On the first row of each panel in Table 6.3, values of β_{rv} are given, found when the source was placed centrally: the enclosure shape is indicated by L/H and D/H. As the source is moved progressively toward one wall, an increasingly large fraction of

Table 6.3 Values for the observable radiant temperature T_{rv}, non-dimensionalised as β_{rv}, in an enclosure due to a source in three positions, and the non-dimensionalised radiant star temperature β_{rs}
First row: β_{rv} with the source at the centre;
Second row: β_{rv} with the source at the centre of the $L \times H$ wall;
Third row: β_{rv} with the source at the centre of the $D \times H$ wall;
Fourth row: β_{rs} found when Q_r is taken to act at T_{rs}.

L/H					
1.00	1.58	2.51	3.98	6.30	10.00
$D/H = 1.00$					
0.916	0.915	0.910	0.903	0.897	0.892
0.548	0.566	0.580	0.590	0.596	0.598
0.548	0.515	0.491	0.474	0.462	0.454
0.843	0.832	0.810	0.789	0.771	0.759
$D/H = 0.630$					
0.916	0.915	0.911	0.906	0.902	0.898
0.588	0.607	0.621	0.629	0.633	0.635
0.531	0.504	0.484	0.471	0.462	0.456
0.827	0.805	0.780	0.758	0.743	0.732
$D/H = 0.398$					
0.921	0.925	0.925	0.924	0.923	0.922
0.635	0.656	0.670	0.678	0.682	0.685
0.520	0.501	0.487	0.477	0.471	0.466
0.777	0.746	0.719	0.699	0.686	0.677
$D/H = 0.251$					
0.939	0.950	0.956	0.960	0.962	0.962
0.688	0.712	0.727	0.737	0.743	0.746
0.520	0.508	0.499	0.493	0.489	0.486
0.708	0.676	0.652	0.636	0.626	0.620
$D/H = 0.158$					
0.973	0.992	1.004	1.012	1.017	1.020
0.748	0.776	0.795	0.806	0.814	0.818
0.531	0.525	0.521	0.518	0.516	0.515
0.640	0.615	0.598	0.587	0.580	0.576
$D/H = 0.100$					
1.024	1.050	1.068	1.080	1.087	1.091
0.818	0.849	0.871	0.885	0.894	0.900
0.553	0.552	0.551	0.551	0.550	0.550
0.589	0.572	0.561	0.554	0.549	0.547

the enclosure is located further from the source, so the average temperature falls and so does β_{rv}. Lines 2 and 3 of each panel show its value when the source is placed in one or other of the walls.

Figure 6.8a shows the thermal circuit for a sensor placed somewhere within a room, intercepting and absorbing the radiation from an arbitrarily placed radiation source Q_r.

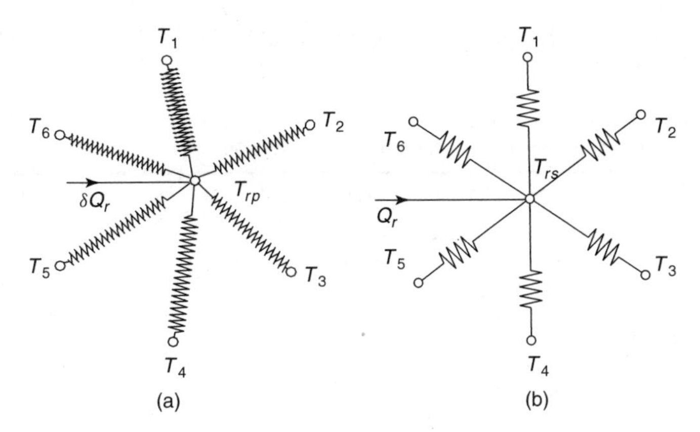

Figure 6.8 (a) Circuit of a small sensor arbitrarily placed in an enclosure, intercepting and absorbing a flux δQ_r from an internal radiant source. (b) The radiant source output Q_r taken to act at the radiant star node T_{rs}

The small conductances linking the sensor to a wall are proportional to the solid angle subtended by the sensor at the wall. The values for T_{rp} when averaged over the space within the enclosure (T_{rv} values) are given as the three β_{rv} values in Table 6.3. Now the volume-averaged radiant temperature is a physically significant quantity, on the same footing as T_{av}. It is too complicated however to evaluate during the design of a heating/cooling system and the question addressed here is how reliably it can be estimated by the value of the fictional radiant star node T_{rs} if we suppose that the radiant input Q_r acts at T_{rs}, as suggested by Figure 6.8b. Although the conductances in Figure 6.8b are large and those in Figure 6.8a are small, T_{rs} must be roughly proportional to T_{rv}. Since T_{rs} in a black-body enclosure with walls at zero is

$$T_{rs} = Q_r \Big/ \sum (A_w h_r / \beta_w), \tag{6.59a}$$

its non-dimensional form is

$$\beta_{rs} \equiv T_{rs} h_r \sum A / Q_r = \sum A_w \Big/ \left(\sum A_w / \beta_w \right). \tag{6.59b}$$

This again is a purely geometrical quantity. Its values are listed in row 4 of each panel in Table 6.3. Notice that in all but two cases, the value of β_{rs} lies within the spread of values of β_{rv}. Thus if we disregard the location of a radiant source in an enclosure, the effect of the source can be modelled adequately by assuming that the radiant fraction acts at the star point of a suitable star-based representation of the room radiant exchange.

6.6.2 Space-averaged Observable Radiant Temperature due to Bounding Surfaces

Section 6.6.1 assumed that the bounding surfaces were at zero and that a sensor responded to an internal source of radiation. Here we examine the complementary problem. Suppose that no source is present, five of the six surfaces are at zero and the sixth has some

Table 6.4 Values of temperature within a black-body rectangular enclosure in which the floor is maintained at unit temperature T and the remaining surfaces are at zero
Upper value $\gamma_{rv} = T_{rv}/T$;
Lower value $\gamma_{rs} = T_{rs}/T$.
The annotations indicate the shapes of the adjacent enclosures.

D/H \ L/H	0.100	0.158	0.251	0.398	0.631	1.00	1.58	2.51	3.98	6.31	10.00	
	telephone kiosk					vertical square tile					plank on edge	
0.100	0.023 0.018	0.028 0.022	0.034 0.024	0.039 0.025	0.044 0.025	0.049 0.025	0.052 0.025	0.055 0.025	0.057 0.025	0.058 0.025	0.059 0.025	
0.158		0.035 0.029	0.043 0.035	0.052 0.038	0.059 0.039	0.066 0.040	0.072 0.041	0.076 0.041	0.079 0.041	0.081 0.041	0.082 0.041	
0.251			0.054 0.046	0.066 0.055	0.077 0.060	0.088 0.063	0.096 0.065	0.102 0.065	0.107 0.066	0.110 0.066	0.112 0.066	
0.398				0.082 0.072	0.098 0.086	0.113 0.095	0.125 0.100	0.134 0.103	0.141 0.105	0.146 0.106	0.149 0.107	
0.630					0.119 0.111	0.140 0.132	0.157 0.146	0.171 0.155	0.181 0.161	0.187 0.165	0.192 0.167	
1.00					cube	0.167 0.167	0.190 0.194	0.209 0.214	0.223 0.227	0.232 0.235	0.239 0.241	square-section corridor
1.58							0.220 0.237	0.245 0.268	0.263 0.290	0.276 0.304	0.284 0.314	
2.51								0.275 0.310	0.298 0.340	0.314 0.359	0.325 0.371	
3.98									0.325 0.374	0.345 0.396	0.359 0.410	
6.31										0.368 0.420	0.384 0.434	
10.00											0.403 0.450	horizontal square title

non-zero value T. Clearly T_{rp} will vary with distance from the warm surface. How well may its volume-averaged value T_{rv} be estimated by the corresponding value of T_{rs}? Values of the ratios $\gamma_{rv} = T_{rv}/T$ and $\gamma_{rs} = T_{rs}/T$ are compared in Table 6.4, taken from Davies and Message (1992).

Table 6.4 demonstrates broad agreement between γ_{rv} and γ_{rs}, that is, between T_{rv} and T_{rs}, for enclosures where *D/H* is not much less than 1, but when *D/H* is small then T_{rv} is around double T_{rs}. It is easy to verify that this must be so by considering an enclosure having the shape of a long and squat ridge tent where three surfaces exchange radiation: the base and the two sloping sides, whose combined area is little greater than the area of the base. Suppose that the base is at unit temperature and that the sloping surfaces are at zero. The average observable radiant temperature T_{rv} must be around $\frac{1}{2}$. However, for such a three-sided enclosure, an exact delta-star transform is possible (Section 4.4.2) and it turns out that for quite straightforward reasons, the radiant star temperature T_{rs} tends to that of the base, i.e. unity.

6.7 STAR-BASED MODEL FOR RADIANT EXCHANGE IN A ROOM

The previous section examined the extent to which radiant transfer in a room can be simplified to a star pattern. If the attempt is successful, radiant and convective transfers can be merged, leading to a thermal model which is simple enough for design purposes. Of the two measures for the global radiant temperature in an enclosure, T_{rv}, a 3D construct, is physically significant but tedious to evaluate, whereas T_{rs}, a 2D construct, is physically meaningless but easy to calculate.

The analysis has demonstrated that for representing radiant exchange between surfaces – transfers of order kilowatts – a suitably sized star-based network is satisfactory. When the network is used to estimate the average observable radiant temperature, it is reasonably successful at estimating the effect of an internal source, since the effect of a source depends on where it is placed, a factor usually omitted in initial design calculations. Provided the enclosure does not differ greatly from cubic dimensions, the effect of surface temperatures can be estimated adequately by a star network; T_{rs} proves to be an adequate, if not a good, proxy for T_{rv}.

These results have been demonstrated for an enclosure with black-body surfaces, that is, the T_w and T_{wb} nodes coincide. When a surface is thermally grey rather than black, there is the conductance $\varepsilon_w A_w h_r/(1 - \varepsilon_w)$ between T_w and T_{wb}. Strictly speaking, the fraction of the radiant output Q_r that falls on surface w should be taken to act at T_{wb}, but if we suppose the full Q_r to act at T_{rs}, we largely take account of the additional warming effect produced by the source when its radiation is intercepted by occupants and furnishings. Two caveats may be noted:

- Of the radiant input to an enclosure, only that part which traverses the occupiable space and can be intercepted by sensors or occupants should be treated as input at T_{rs}. The radiation from the back of a wall-mounted radiator should be input at the wall T_{wb} node.
- The local observable radiant temperature is not a unique quantity but depends on the shape of the sensor. If the sensor is flat (not spherical as assumed above) and the flat surface is presented to the source, it will record a higher temperature than when it is placed edge on to the source. If the two sensor surfaces have different emissivities, we should have a further variation in observed temperature with the orientation of the sensor.

Radiant studies can be conducted to include much of the fine detail in a room, but they may not be needed for the design of heating and cooling appliances in most applications. A radiant star model somewhat along the lines of the above model has been tacitly assumed in the past and the foregoing argument clarifies its validity. It will be used in the next chapter to set up a room thermal model.

6.8 REPRESENTATION OF RADIANT EXCHANGE BY SURFACE-SURFACE LINKS

Radiant exchange between black-body surfaces (Figure 6.5a) is necessarily a direct exchange between surfaces. When the emissivity conductances are introduced as in Figure 6.6a, the network represents it as an exchange via the black-body equivalent nodes.

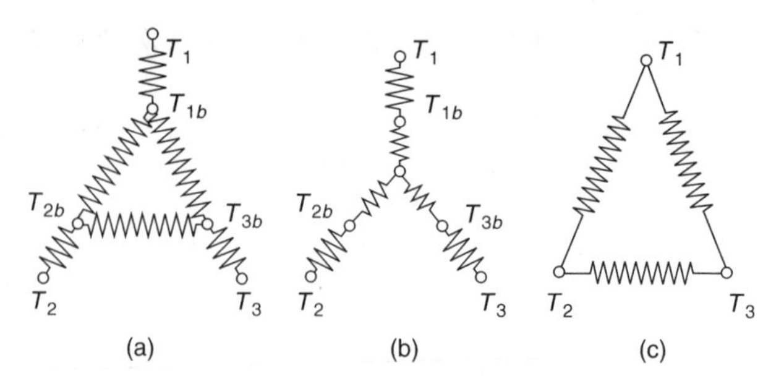

Figure 6.9 Representation of radiant exchange by surface-to-surface links

The pattern can, however, be transformed to one of direct surface-surface links. Figure 6.9 shows this for the simple case of a three-surface enclosure.

Figure 6.9a shows the emissivity and geometrical conductances linked to the black-body equivalent nodes. The delta form of geometrical links can be transformed exactly and unconditionally to the star form in Figure 6.9b (Section 4.4.2). If no heat is input at any of the black-body equivalent nodes, this in turn can be transformed exactly to Figure 6.9c, which provides direct links between the surface nodes. Each direct link is a function of all the six links in Figure 6.9a. This form cannot include the warming effect produced by an internal radiant source on room furnishings and occupants due to the direct radiation they intercept.

The same transform can be performed using matrix algebra for an enclosure with four or more distinct surfaces. It will not be developed here since there seems no application for it, but the transform provides the justification for the quantity g_{ij} in the 1993 *ASHRAE Handbook of Fundamentals* (page 26.2) where g_{ij} is defined as the radiation transfer factor between interior surface i and interior surface j.

6.9 LONG-WAVE RADIANT EXCHANGE AT BUILDING EXTERIOR SURFACES

Building walls and roofs gain heat by solar radiation during the day and will lose heat by long-wave radiation to the sky during day and night.[8] There may be a long-wave exchange of either sign with the ground and surrounding buildings or other features. There is a further exchange, usually a loss, by forced convection. The long-wave exchange can in principle be found using the view factors from the exterior surface to all surrounding surfaces including the ground, together with the surface temperatures, but this is difficult to implement since it involves computing the view factors for detailed and specific configurations, together with assumptions for the temperatures. When exchange takes place only between the surface, sky and ground, it can be modelled as shown in Figure 6.10.

T_{wb} and T_{gb} are the black-body equivalent temperatures of the wall surface and the ground respectively, with associated conductances of form $E = \varepsilon A/(1 - \varepsilon)$. The sky will

[8]Cole (1976: Figure 6), quoting Dines and Dines (1927), suggests radiant losses of some 10–20 W/m^2 to an overcast sky and 100 W/m^2 to a clear sky for a latitude of 53° in the UK. For comparison, the mainly convective loss from a wall of U value 1 W/m^2 K with internal and external temperatures of 20°C and 0°C is some 20 W/m^2.

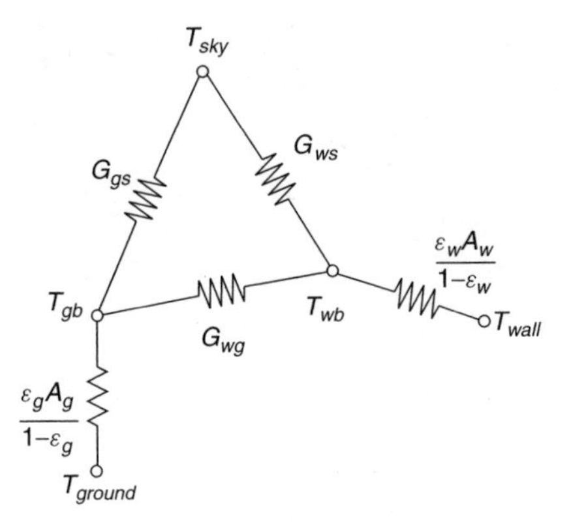

Figure 6.10 Thermal circuit for radiant exchange between a surface, the ground and the sky

be simply assigned the temperature given for example by equation (4.2). The geometrical conductance (m^2) between T_{wb} and T_{gb} is

$$G_{wg} = A_{wall} F_{wall,ground} \tag{6.60a}$$

and since A_{wall} only sees the sky and ground,

$$G_{ws} = A_{wall}(1 - F_{wall,ground}). \tag{6.60b}$$

Since the radiant exchange between the ground and the sky is unaffected by A_{wall},

$$G_{gs} = A_g \tag{6.60c}$$

and

$$T_{gb} = \varepsilon_g T_{ground} + (1 - \varepsilon_g) T_{sky}. \tag{6.61}$$

Assuming that the radiant exchanges can be taken as proportional to temperature differences, the heat lost Q_r from A_{wall} to the ground and sky is

$$Q_r = h_r(T_{wall} - T_{wb})\varepsilon_w A_{wall}/(1 - \varepsilon_{wall}) = (T_{wb} - T_{gb})G_{wg} + (T_{wb} - T_{sky})G_{ws}. \tag{6.62a}$$

On eliminating T_{wb} and noting that $G_{wg} + G_{ws} = A_{wall}$, we have

$$Q_r = A_{wall}\varepsilon_w h_r(T_{wall} - [F_{wall,ground} T_{gb} + (1 - F_{wall,ground})T_{sky}]). \tag{6.62b}$$

An elementary expression for $F_{wall,ground}$ is available if A_{wall} has the form of a very long plane sloping surface in contact with the ground (Figure 6.11). The sky and ground are represented as plane surfaces A_s and A_g.

According to Hottel's crossed strings construction, the view factor geometrical conductances between these areas are

$$G_{ws} = \tfrac{1}{2}(A_{wall} + A_s - A_g) \Rightarrow \tfrac{1}{2}A_{wall}(1 + \cos\beta)$$
$$\text{since } A_g + A_w \cos\beta = A_s, \tag{6.63a}$$

$$G_{wg} = \tfrac{1}{2}(A_{wall} + A_g - A_s) \Rightarrow \tfrac{1}{2}A_{wall}(1 - \cos\beta), \tag{6.63b}$$

$$G_{gs} = \tfrac{1}{2}(A_g + A_s - A_{wall}) \Rightarrow A_g = A_s, \tag{6.63c}$$

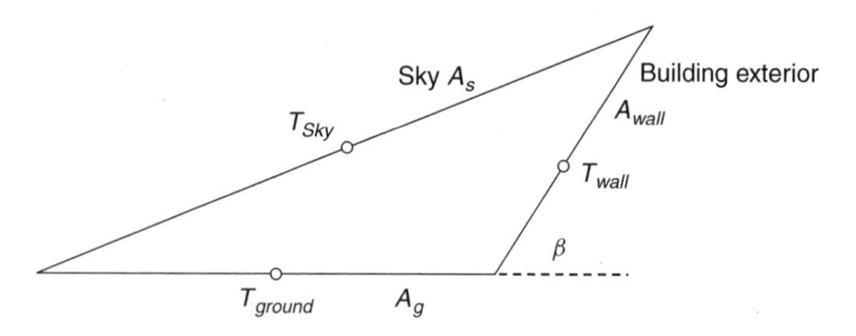

Figure 6.11 Long-wave radiant exchange between a very long building exterior surface, ground and sky

so

$$F_{wall,ground} = \tfrac{1}{2}(1 - \cos\beta). \tag{6.64}$$

The outward loss of heat by radiation from T_{wall} to the sky and the ground is then

$$Q_r = A_{wall}\varepsilon_w h_r\{T_{wall} - [\tfrac{1}{2}(1 - \cos\beta)T_{ground} + (1 - \tfrac{1}{2}(1 - \cos\beta))T_{sky}]\}. \tag{6.65}$$

Equation (6.65) is qualitative; it is only correct for an infinitely long strip. For a sloping element of finite size, the view factor to the ground can be found by contour integration. Suppose that the sides of A_{wall} have projections δx_1, δy_1 and δz_1 on the planes of Cartesian coordinates, as shown in Figure 6.12, so that the slope $\beta = \arctan(\delta z_1/\delta x_1)$. The line forming the intersection of the ground and the plane containing A_{wall} forms the y axis. A_{wall} is situated at $y = 0$, $x = -x'$ and $z = +z'$. The ground visible from A_{wall} is the area $0 < x < \infty$ and $-\infty < y < \infty$ and $F_{wall,ground}$ can be found by summing around a suitably large ground contour using equation (6.27). Only a single stage of summation is needed since A_{wall} is very small compared with the ground area. $F_{wall,ground}$ decreases with increasing height z' of A_{wall} above the ground.

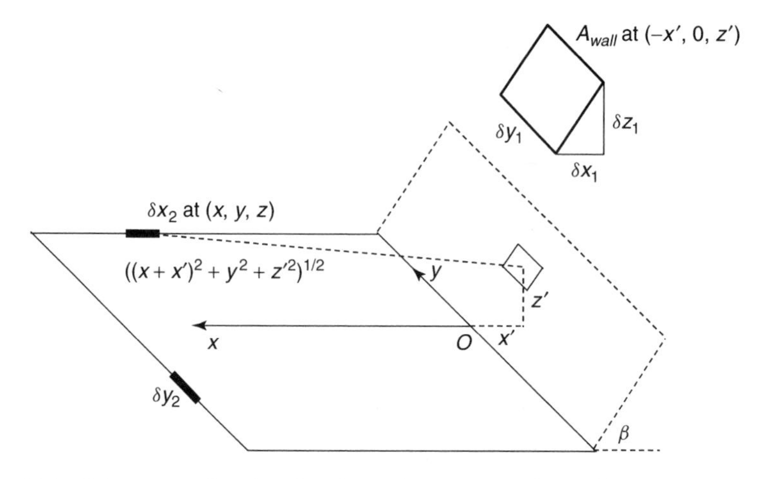

Figure 6.12 Diagram to find the view factor from an exterior wall surface to the ground

Values for T_{sky} are given in equations (4.2); $h_r = \sigma(T_{wall} + T_{sky})(T_{wall}^2 + T_{sky}^2)$ and is around $4\,\mathrm{W/m^2K}$. Values for T_{ground} must be assumed; they depend on the nature of the terrain (tarmac, forest, etc.) and on the previous weather history. To the radiant loss must be added the convective loss of $A_w h_{ce}(T_{wall} - T_e)$ and the two together form part of the overall heat balance relations which include wall conduction and room internal exchange. The tacit replacement here of differences in fourth powers of temperature by their linear difference is less valid than linearisation of internal temperatures since the external differences are larger than internal differences. This can be remedied by restoring the fourth-power differences and iterating, but the uncertainty or difficulty of quantifying other factors affecting external loss generally make this of dubious worth. Sky models are discussed by McClellan and Pedersen (1997).

6.10 APPENDIX: CONDUCTANCE BETWEEN RECTANGLES ON PERPENDICULAR AND PARALLEL SURFACES

The geometrical conductances between surfaces in a rectangular room were given in (6.25) and (6.26). They can be used to find the conductances between rectangular patches on these surfaces. Consider the situation in Figure 6.13. (The rectangular space $X \times Y \times Z$ depends on the disposition of the patches, not the room they are in.) We wish to find the conductance between the area indicated as $21 + 22$ on the wall to $22 + 23$ on the floor. Their areas are $A_1 = (Z - c)b_2$ and $A_2 = (X - a)(Y - b_1)$. It is clear that

$$G(21 + 22, 22 + 23) = G(21, 22) + G(21, 23) + G(22, 22) + G(22, 23)$$

and we require expressions to evaluate these terms. They can be found by continued use of two principles:

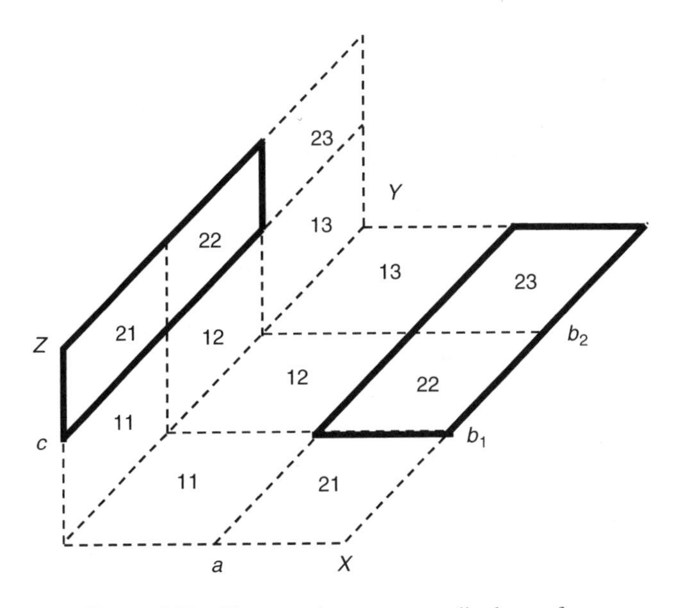

Figure 6.13 Elements in two perpendicular surfaces

- Conductances can be formed as sums. The conductance $G(11, 21)$ between areas 11 on the wall and 21 on the floor can be found as $G(11, 11 + 21) - G(11, 11)$, each of which can be found from (6.25).

- There is a further reciprocity relationship. Consider the conductance between wall area 21 (area A_1) and floor area 23 (area A_2). From equation (6.14) this is given by

$$G(21, 23) = A_1 F_{12} = \int_0^{b1} \int_c^Z \int_a^X \int_{b2}^Y \frac{(\mathrm{d}y_1\, \mathrm{d}z_1)(\mathrm{d}x_2\, \mathrm{d}y_2) \cos\theta_1 \cos\theta_2}{\pi[(x_1 - x_2)^2 + (y_1 - y_2)^2 + (z_1 - z_2)^2]}, \tag{6.66}$$

where $\mathrm{d}A_1$ is situated at x_1, y_1, z_1.

If instead areas A_1 and A_2 are chosen to be wall area 23 and floor area 21, then

$$G(23, 21) = A_1 F_{12} = \int_{b2}^{Y} \int_c^Z \int_a^X \int_0^{b1} \frac{(\mathrm{d}y_1\, \mathrm{d}z_1)(\mathrm{d}x_2\, \mathrm{d}y_2) \cos\theta_1 \cos\theta_2}{\pi[(x_1 - x_2)^2 + (y_1 - y_2)^2 + (z_1 - z_2)^2]}. \tag{6.67}$$

The limits of the integrals are the same, so

$$G(21, 23) = G(23, 21)$$

The result is true for all such crossed pairs. The result leads to a factor of $\frac{1}{2}$ in values for conductances between elements in perpendicular surfaces and $\frac{1}{4}$ for parallel surfaces.

The conductances between the several separated areas are given in Table 6.5; 21/13 denotes the conductance between area 21 on the wall to 13 on the floor (so with this notation, $13/21 \neq 21/13$). The arrays of areas in right-hand quantities above and below their horizontal lines indicate graphically the rectangular areas on the wall and the floor whose

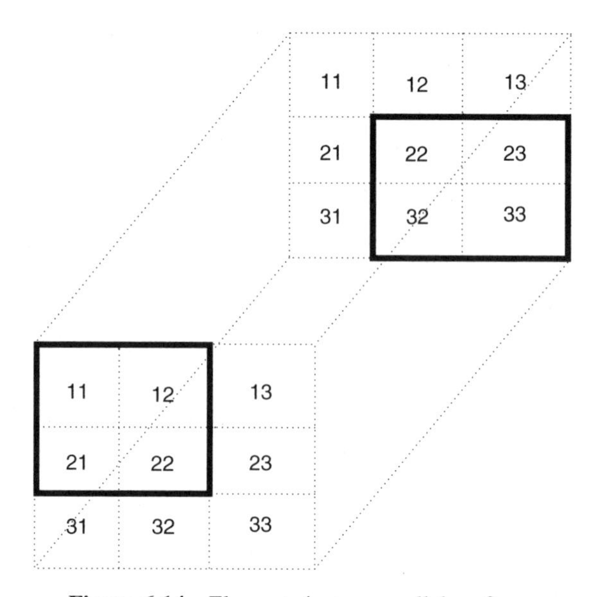

Figure 6.14 Elements in two parallel surfaces

Table 6.5 Geometrical conductances between elements on the perpendicular surfaces in Figure 6.13

$$\frac{11}{21} = \left[\frac{11}{\substack{11\\21}} - \frac{11}{11}\right]$$

$$\frac{21}{11} = \left[\frac{\substack{21\\11}}{11} - \frac{11}{11}\right]$$

$$\frac{21}{21} = \left[\frac{\substack{21\\11}}{\substack{11\\21}} - \frac{\substack{21\\11}}{11} - \frac{11}{\substack{11\\21}} + \frac{11}{11}\right]$$

$$\frac{11}{12} = \frac{12}{11} = \frac{1}{2}\left[\frac{11\ 12}{11\ 12} - \frac{11}{11} - \frac{12}{12}\right]$$

$$\frac{11}{13} = \frac{13}{11} = \frac{1}{2}\left[\frac{11\ 12\ 13}{11\ 12\ 13} - \frac{11\ 12}{11\ 12} - \frac{12\ 13}{12\ 13} + \frac{12}{12}\right]$$

$$\frac{11}{22} = \frac{12}{21} = \frac{1}{2}\left[\frac{11\ 12}{\substack{11\ 12\\21\ 22}} - \frac{11\ 12}{11\ 12} - \frac{11}{\substack{11\\21}} - \frac{12}{\substack{12\\22}} + \frac{11}{11} + \frac{12}{12}\right]$$

$$\frac{21}{12} = \frac{22}{11} = \frac{1}{2}\left[\frac{\substack{21\ 22\\11\ 12}}{11\ 12} - \frac{11\ 12}{11\ 12} - \frac{\substack{21\\11}}{11} - \frac{\substack{22\\12}}{12} + \frac{11}{11} + \frac{12}{12}\right]$$

$$\frac{21}{22} = \frac{22}{21} = \frac{1}{2}\left[\frac{\substack{21\ 22\\11\ 12}}{\substack{11\ 12\\21\ 22}} - \frac{\substack{21\ 22\\11\ 12}}{11\ 12} - \frac{11\ 12}{\substack{11\ 12\\21\ 22}} - \frac{\substack{21\\11}}{\substack{11\\21}} - \frac{\substack{22\\12}}{\substack{12\\22}} + \frac{11\ 12}{11\ 12} + \frac{\substack{21\\11}}{11} + \frac{\substack{22\\12}}{12} + \frac{11}{\substack{11\\12}} + \frac{12}{\substack{12\\22}} - \frac{11}{11} - \frac{12}{12}\right]$$

$$\frac{11}{23} = \frac{13}{21} = \frac{1}{2}\left[\frac{11\ 12\ 13}{\substack{11\ 12\ 13\\21\ 22\ 23}} - \frac{11\ 12\ 13}{11\ 12\ 13} - \frac{11\ 12}{\substack{11\ 12\\21\ 22}} - \frac{1213}{\substack{12\ 13\\22\ 23}} + \frac{11\ 12}{11\ 12} + \frac{12\ 13}{12\ 13} + \frac{12}{\substack{12\\22}} - \frac{12}{12}\right]$$

$$\frac{21}{13} = \frac{23}{11} = \frac{1}{2}\left[\frac{\substack{21\ 22\ 23\\11\ 12\ 13}}{11\ 12\ 13} - \frac{11\ 12\ 13}{11\ 12\ 13} - \frac{\substack{21\ 22\\11\ 12}}{11\ 12} - \frac{\substack{22\ 23\\12\ 13}}{12\ 13} + \frac{11\ 12}{11\ 12} + \frac{12\ 13}{12\ 13} + \frac{\substack{22\\12}}{12} - \frac{12}{12}\right]$$

$$\frac{21}{23} = \frac{23}{21} = \frac{1}{2}\left[\frac{\substack{21\ 22\ 23\\11\ 12\ 13}}{\substack{11\ 12\ 13\\21\ 22\ 23}} - \frac{\substack{21\ 22\ 23\\11\ 12\ 13}}{11\ 12\ 13} - \frac{11\ 12\ 13}{\substack{11\ 12\ 13\\21\ 22\ 23}} - \frac{\substack{21\ 22\\11\ 12}}{\substack{11\ 12\\21\ 22}} - \frac{\substack{22\ 23\\12\ 13}}{\substack{12\ 13\\22\ 23}} + \frac{\substack{21\ 22\\11\ 12}}{11\ 12} + \frac{\substack{22\ 23\\12\ 13}}{12\ 13}\right.$$
$$\left. + \frac{11\ 12}{\substack{11\ 12\\21\ 22}} + \frac{12\ 13}{\substack{12\ 13\\22\ 23}} + \frac{11\ 12\ 13}{11\ 12\ 13} + \frac{\substack{22\\12}}{\substack{12\\22}} - \frac{11\ 12}{11\ 13} - \frac{12\ 13}{12\ 13} - \frac{\substack{22\\12}}{12} - \frac{12}{\substack{12\\22}} + \frac{12}{12}\right]$$

Table 6.6 Geometrical conductances between elements on the parallel surfaces in Figure 6.14

$$\frac{11}{12} = \frac{1}{2}[11\ 12 - 11 - 12]$$

$$\frac{11}{13} = \frac{1}{2}[11\ 12\ 13 - 11\ 12 - 12\ 13 + 12]$$

$$\frac{11}{22} = \frac{1}{4}\left[\begin{matrix}11\ 12\\ 21\ 22\end{matrix} - 11\ 12 - 21\ 22 - \begin{matrix}11\\ 21\end{matrix} - \begin{matrix}12\\ 22\end{matrix} + 11 + 12 + 21 + 22\right]$$

$$\frac{11}{23} = \frac{1}{4}\left[\begin{matrix}11\ 12\ 13\\ 21\ 22\ 23\end{matrix} - 11\ 12\ 13 - 21\ 22\ 23 - \begin{matrix}11\ 12\\ 21\ 22\end{matrix} - \begin{matrix}12\ 13\\ 22\ 23\end{matrix}\right.$$
$$\left. + 11\ 12 + 12\ 13 + 21\ 22 + 22\ 23 + \begin{matrix}12\\ 22\end{matrix} - 12 - 22\right]$$

$$\frac{11}{33} = \frac{1}{4}\left[\begin{matrix}11\ 12\ 13\\ 21\ 22\ 23\\ 31\ 32\ 33\end{matrix} - \begin{matrix}11\ 12\ 13\\ 21\ 22\ 23\end{matrix} - \begin{matrix}21\ 22\ 23\\ 31\ 32\ 33\end{matrix} - \begin{matrix}11\ 12\\ 21\ 22\\ 31\ 32\end{matrix} - \begin{matrix}12\ 13\\ 22\ 23\\ 32\ 33\end{matrix} + \begin{matrix}11\ 12\\ 21\ 22\end{matrix}\right.$$
$$\left. + \begin{matrix}12\ 13\\ 22\ 23\end{matrix} + \begin{matrix}21\ 22\\ 31\ 32\end{matrix} + \begin{matrix}22\ 23\\ 32\ 33\end{matrix} + 21\ 22\ 23 + \begin{matrix}12\\ 22\\ 32\end{matrix} - 21\ 22 - 22\ 23 - \begin{matrix}12\\ 22\end{matrix} - \begin{matrix}22\\ 32\end{matrix} + 22\right]$$

conductance values can be found using (6.25). These expressions are simply identities. Consider for example the conductance between patch 11 on the wall and patch 13 on the floor, $G(11, 13)$, notated as 11/13. When the sequence of terms on the right is expanded, $G(11, 13)$ and its reciprocal $G(13, 11)$ occur only once, but all other pairings occur twice with opposite signs and cancel.

Table 6.6 gives the conductances between elements of the two parallel surfaces shown in Figure 6.14. The quantities on the right-hand side denote rectangular areas which are directly opposed on the two surfaces and whose conductances can be found using (6.26). (Suppose that the opposed surfaces were so far apart that all the pairs $G(ij, mn)$ were near enough equal, having the small value ε. Thus the quantity notated as 11 12, with 2^2 or four components, $G(11, 11)$, $G(11, 12)$, $G(12, 11)$, $G(12, 12)$, has the value 4ε and 11/12 has the value $\frac{1}{2}[2^2\varepsilon - \varepsilon - \varepsilon] = \varepsilon$, as it must. Similarly for example, $11/12 = \frac{1}{4}[4^2\varepsilon - 2^2\varepsilon - 2^2\varepsilon - 2^2\varepsilon - 2^2\varepsilon + \varepsilon + \varepsilon + \varepsilon + \varepsilon] = \varepsilon$.) The conductance between the two framed areas is the sum of 16 components, the first being $G(11, 22)$. This is a cumbersome procedure and contour integration may well provide a more efficient method of computation.

Clarke (1985, 2001) provides an alternative formulation for computing the conductances between elements on such surfaces.

7 Design Model for Steady-State Room Heat Exchange

Section 1.3 described a simple thermal model for a room: to establish an internal temperature T_i with ambient temperature T_e, combined conduction losses of $\sum AU$ and a ventilation loss of V, the heat need Q is

$$Q = \left(\sum AU + V\right)(T_i - T_e) = (L + V)(T_i - T_e), \tag{7.1}$$

where $L \equiv \sum AU$ and denotes the total loss conductance. Its thermal circuit is shown in Figure 4.2b (with T_a replacing T_i). In this expression the ventilation rate V is exactly defined (even though the numerical value to be attached to it may not be known reliably). A, a surface area, may be a little ambiguous, depending on whether it relates to external or internal areas, and these may differ since flow through corners is usually ignored. U, the wall transmittance, is composed of conduction resistances of type thickness/conductivity (X/λ), which are defined exactly in principle, although λ in particular may not be known closely; U also includes film resistances in which fixed values of convective resistance are combined with some measure of radiant resistance. T_e, the external or ambient temperature, is largely the air temperature, although some allowance for the sky radiant temperature may be included.

The definition of the remaining quantities Q and T_i raises more serious issues. Setting aside solar gains, the heat input to a room is partly convective and partly radiative and the mechanisms follow different physical laws. The simple expression ignores this. The internal temperature T_i is an index temperature, several levels removed from observational data, and has air and radiative components. From the 1960s onward there have been attempts to model room internal exchange in connection with heating/cooling design that take better account of room internal exchange than does equation (7.1). The problem is more linked to radiative exchange than convective exchange and the analyses in Chapter 6 lead to a somewhat better model, discussed below. The model is useful in a simple design context but has no role in present building simulation work.

Building Heat Transfer Morris G. Davies
© 2004 John Wiley & Sons, Ltd ISBN: 0-470-84731-X

7.1 A MODEL ENCLOSURE

To illustrate the evolution of a room internal model, consider the enclosure of Figure 7.1, an indefinitely long room with heat losses by ventilation, to the exterior through a window, the exterior via an external wall or roof, to the ground through the floor and on to an adjoining room through a partition wall. Flanking walls are omitted since they simply reproduce one or other of these links. (For steady-state analyses, the window and external links are similar. The walls and floor, however, may provide thermal storage which affects the time-varying response while the window provides virtually no storage.)

Figure 7.2 shows the room heat transfer in several alternative forms. Parts (a) to (c) show different forms of modelling the internal exchange of heat by radiation and convection, and parts (i) to (iv) show forms of modelling the conductive and infiltration losses to ambient. The steady-state loss network is shown in Figure 7.2(i). The losses by ventilation, through the window and the outer wall are to temperature sources – ambient temperature for losses by ventilation and through the window, and sol-air temperature for losses through opaque surfaces, when the effect of radiation absorbed at external surfaces is added to ambient; see equation (4.17). The loss to ground is in reality a three-dimensional loss, partly to ambient and partly to the constant deep ground temperature. There is the further exchange through the partition wall. (Parts (ii) to (iv) illustrate the means of modelling these losses in time-varying conditions; see later).

Figure 7.2a shows separate convective and radiative links with the surfaces. Assuming black-body surfaces, the radiative links follow from the view factors, as discussed in Section 6.4; if the effect of emissivity is included, direct links can formally be set up as indicated in Section 6.8. If the delta pattern of radiant links is replaced by a star pattern, it can be shown as in Figure 7.2b; there are now two star nodes, the air node T_a (earlier called T_{av}) and the radiant star node T_{rs}. We wish to reduce the pattern still further so as to have a single star node, the rad-air node T_{ra}, an index with radiant and air components, with links to the surfaces which express convection and radiation. This is shown in Figure 7.2c, which includes a fictitious conductance X.

Of the models representing conductive loss to the exterior, we are concerned here only with the steady-state loss, shown in Figure 7.2(i), where thermal capacity has no effect. (If the enclosure is driven sinusoidally, the thermal characteristics of a wall can be represented exactly by six lumped elements, as shown in Figure 7.2(ii) and discussed in Section 15.6. It can be represented more robustly as shown in Figure 7.3(iii), the basis

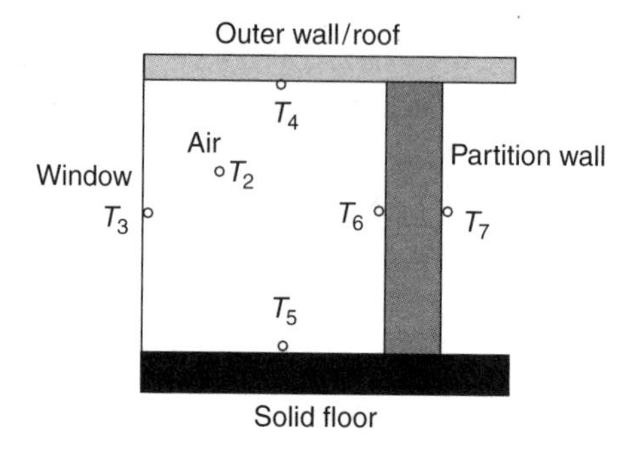

Figure 7.1 A simple enclosure to illustrate internal heat exchange and external loss

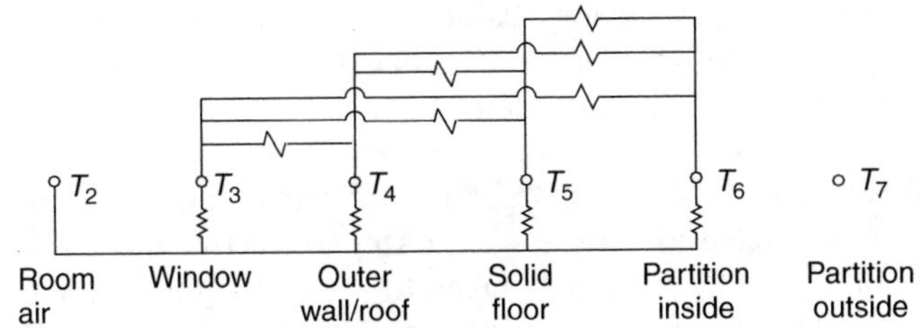

(a) Convective links between each surface and air. Separate radiant links between surfaces, here 6 links; $\frac{1}{2}n(n-1)$ links in a room of n surfaces.

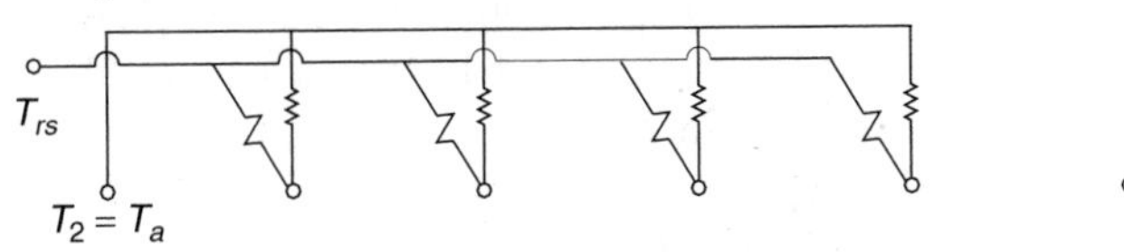

(b) Convective links as above. Radiant links between each surface and the radiant star node T_{rs}.

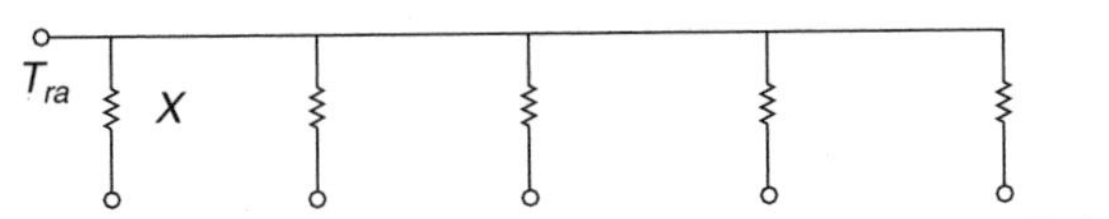

(c) Radiant and convective links merged; T_{rs} replaced by the rad-air node T_{ra} and air temperature T_2 linked by the conductance X.

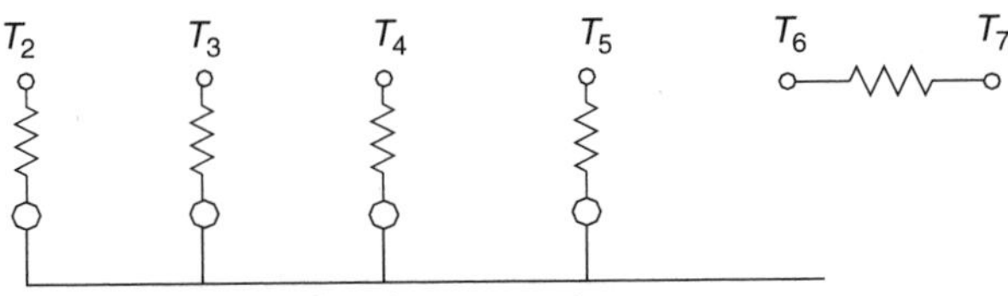

(i) Steady-state conductances driven by mean values.

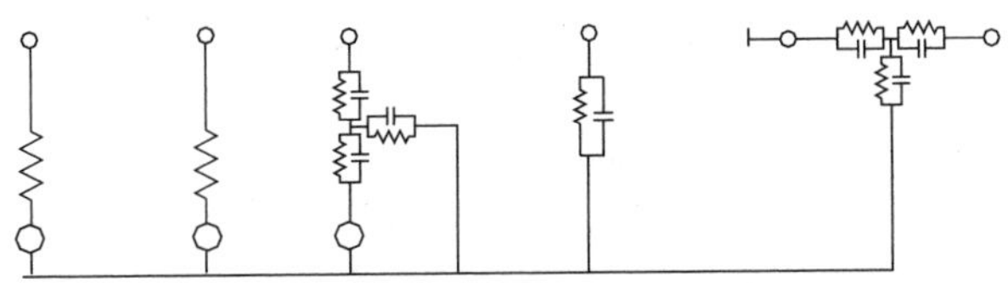

(ii) Model for periodic excitation driven by Fourier components.

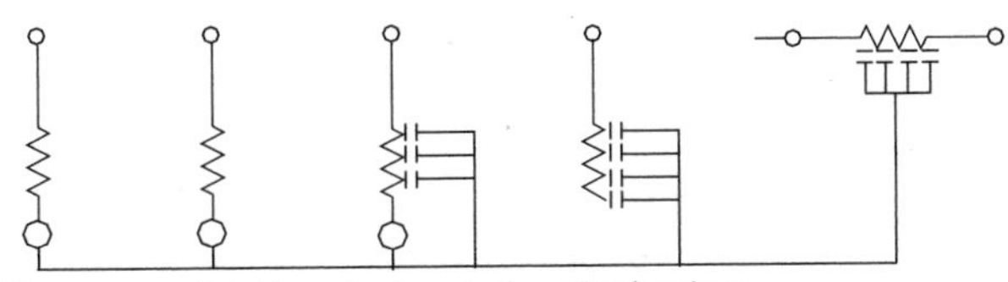

(iii) Finite difference model driven by hourly (or other) values.

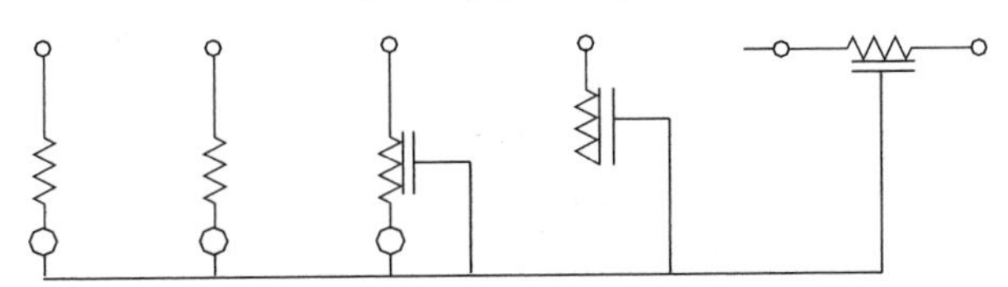

(iv) Transfer coefficient model driven by hourly (or other) values.

Figure 7.2 Modelling radiant and convective heat transfer in a room and loss from it. Any internal model can be linked to any loss model. Heat can be input at any node

for the finite difference method in Section 10.4. For an example, see Figure 1 of Mathews *et al.* (1997). The representation in Figure 7.2(iv) may be useful when the room response is estimated using transfer coefficients (Chapter 19).)

The basis for the step to Figure 7.2c with the rad-air node T_{ra} is the series-parallel transformation (Section 4.4.3), where the separate convective link C_1 and radiant link S_1 to a surface are formally added to give $S_1 + C_1$. Any convective input Q_c continues to act at T_a. The radiant input Q_r, previously taken to act at T_{rs}, must now be handled as an augmented input $Q_r(1 + C_1/S_1)$ at T_{ra} with withdrawal of the excess $Q_r C_1/S_1$ from T_a. The transformation is exact for a single surface. It continues to be exact for two or more surfaces if $C_1/S_1 = C_2/S_2 = C_3/S_3$, etc., a relation that may be expected to remain approximately true since radiative and convective conductances are roughly proportional to area. The rad-air temperature (Davies 1983c, 1992b) is a weighted mean of the radiant and air temperatures:

$$T_{ra} = (ST_{rs} + CT_a)/(S + C), \tag{7.2a}$$

where

$$S = \sum S_j \quad \text{and} \quad C = \sum C_j. \tag{7.2b}$$

The conductance X between T_{ra} and T_a is

$$X = (S + C)C/S \tag{7.3}$$

equation (4.30) and proves to be large compared with the links S_j and C_j to individual surfaces.

It should be noted that the series-parallel transform has been conducted unsymmetrically: T_{rs} has disappeared but T_a is retained. The transform could in principle have been made so as to retain T_{rs} instead. The choice is determined by the need to retain the T_a node explicitly since the ventilation loss acts there. In the model where both T_a and T_{rs} appear (Figure 7.2b), a convection-like loss to the exterior from T_a is made explicit and a radiant flow to the exterior through an open window could also have been made. But in choosing to form T_{ra} to include T_a and to eliminate T_{rs}, it is not possible to show such a radiant loss to the exterior.

It may also be noted that T_{ra} is far removed from local or point temperature measurements. The logic of the argument leading to T_{ra} is shown in Figure 7.3. The real (i.e. independent) determinants of room temperature are the ambient temperature, building thermal construction and ventilation rate, together with the components Q_r and Q_a of the heat source. As a result of these, a distribution of local measurable surface temperature is set up, together with a volume distribution of measurable local air temperatures. These form the starting point for the logic and replace ambient temperature and details of construction. Q_a is not needed since its effect is represented by the T_{ap} distribution. But Q_r is needed since it is the effect of Q_r together with values of surface temperatures (and their emissivities) that leads to the space distribution of T_{rp} values, a further measurable quantity; Q_r is also a component of T_{rs}. These feature in Figure 7.3.

Two- and three-dimensional means of observational values must first be formed. Physical weighting factors – areas, emissivities, h_r and h_{cj} values – are then needed, together

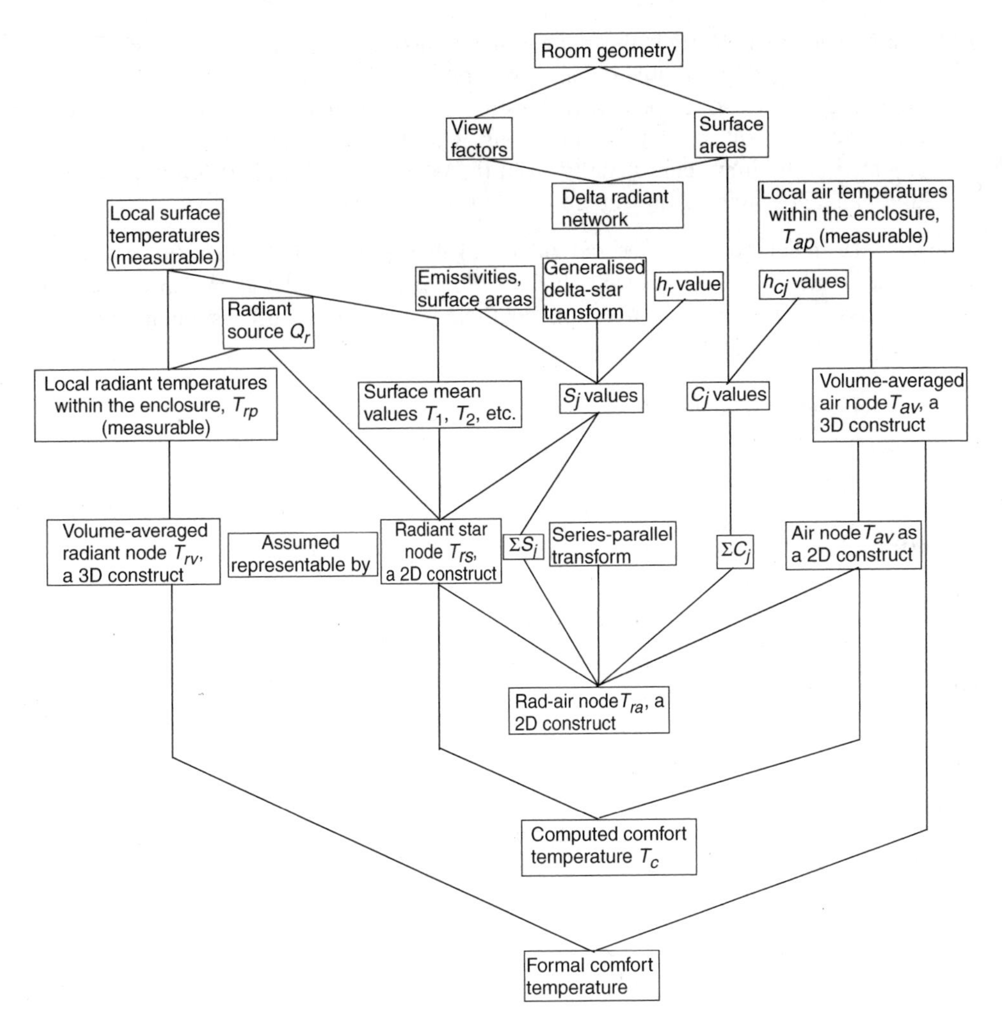

Figure 7.3 How to take the global measures, rad-air temperature and the computed and formal values of comfort temperature and relate them to measurable temperatures

with the delta-star and series-parallel transforms. It will be noted that the mean value of the perceived comfort temperature, the formal comfort temperature, derives from the 3D averages of local air and radiant temperatures. But the rad-air temperature and the computed comfort temperature derive from 2D averaged forms of radiant and air temperatures; see Sections 6.56 and 5.3.1.

Various approximations have had to be made:

- The star pattern of radiant exchange does not exactly reproduce the original delta pattern, and the regression β_j values differ from the original β_j values for some given enclosure. Further deviations may be involved for non-rectangular rooms.

- The argument assumes that the radiant star temperature T_{rs} may serve as proxy for the space-averaged observable radiant temperature T_{rv}, for the effect of internal radiant sources and of surface temperatures. This assumption is not very satisfactory since T_{rv} varies with the position of the source, but against this, the local radiant temperature T_{rp}, which is the important quantity from the occupant's viewpoint, itself varies from a high value near the radiant source to a low value remove from it.
- Lastly, the development of the rad-air model involved assuming that the ratio C_j/S_j was the same for each surface. This may be near enough true in most cases, but low emissivities or high convection coefficients on large surfaces might make the assumption inappropriate.

It was noted earlier (equation 6.52) that

$$T_{rs}(\text{unheated}) = \sum T_j S_j \Big/ \sum S_j. \tag{7.4}$$

The effect of a radiant input Q_r at T_{rs} is to raise T_{rs} by an amount Q_r/S, so

$$T_{rs} = T_{rs}(\text{unheated}) + Q_r/S \tag{7.5}$$

and it follows that

$$T_{ra} = \frac{CT_a}{S+C} + \frac{ST_{rs}}{S+C} = \frac{CT_a}{S+C} + \frac{ST_{rs}(\text{unheated})}{S+C} + \frac{Q_r}{S+C}. \tag{7.6}$$

This makes clear that T_{ra} is a mix of air and two radiant components of temperature. T_{ra} is structurally similar to the generalised sol-air temperature (Section 4.4.1), the index set up to handle gain at the outside of a wall or roof, which has the form

$$T_{sa} = \frac{h_{ce}T_e}{\varepsilon h_r + h_{ce}} + \frac{\varepsilon h_r T_{sky}}{\varepsilon h_r + h_{ce}} + \frac{\alpha I}{\varepsilon h_r + h_{ce}}. \tag{7.7}$$

The following correspondences are apparent:

- CT_a and $h_{ce}T_e$ correspond; ST_{rs} and $\varepsilon h_r T_{sky}$ correspond; Q_r and αI correspond.
- $S + C$, the total exchange between the surface and the index temperature, corresponds to $\varepsilon h_r + h_{ce}$.

T_{ra} is associated with large conductances. Like T_{ra}, the comfort temperature T_c also derives from T_{rs} (or T_{rv}) and T_a, but is associated with small conductances. Note too that since T_c is linked to T_{rs}, and T_{rs} is removed in forming T_{ra}, the T_c and T_{ra} nodes cannot appear in the same thermal circuit. Like T_{ra}, T_c has components due to the air, the unheated wall temperature and the effect of radiation intercepted from an internal source.

It would be attractive if we could dispense with the global quantities T_a, T_{rv}, T_{rs}, T_{ra} and T_c and allow the computationally simple T_i to serve for all of them. This is effectively what was done in the UK before 1970 and the practice is still current. T_a, T_{rv}, T_{rs}

and T_c do not normally differ much and many practising engineers regard the distinction between them as hair-splitting and academic, unnecessary in practical applications. However, a distinction of this kind is too widely recognised and is of too long-standing to be swept aside. The next section presents the model to include them and considers its consequences.

7.2 THE RAD-AIR MODEL FOR ENCLOSURE HEAT FLOWS

The case of the enclosure internal exchange represented by a single index temperature in steady-state conditions is given by combining Figure 7.2c (the rad-air approximation) and Figure 7.2(i). The values at all the temperature nodes are mean temperatures associated with a surface and they are therefore two-dimensional constructs. This is obvious for node 3 onward. Now T_{rs} is a 2D construct (Section 6.5.6), each $T_{a,surface}$ for convective exchange is a 2D construct (Section 5.3.1) and so T_{ra} which is formed from them is also a 2D construct. The ventilation exchange too is associated with an area (Section 2.3.1). It is conventionally assumed that the air is fully mixed so that each $T_{a,convection} = T_{a,ventilation} = T_{av}$ but this may not be so. Heat can be input at any of the nodes.

The circuit can be analysed by the nodal method (Section 4.2.2) and to illustrate its use to calculate heat need, two elementary situations will be outlined: (1) heat is supplied by an internal hot body source, emitting a fraction p by long-wave radiation and $1 - p$ by convection; (2) heat is input at the floor.

Temperatures due to an uncontrolled heat source

The loss conductance L_j between T_{ra} and T_e through the bounding surface j, area A_j, is found as

$$1/L_j = 1/(C_j + S_j) + \sum(X_j/A_j\lambda_j) + 1/C_e \tag{7.8}$$

and the sum of these conductances will be denoted by L. The rad-air expression for the heat need Q to establish a specified value T_c of the comfort temperature follows from analysing the circuit in Figure 7.4, which replaces Figure 4.2b as the basic room model. The radiant component pQ has to be treated as an augmented input $pQ(1 + C/S)$ at T_{ra} with withdrawal of the excess pQC/S from T_a. The relations between the net inputs and temperatures at these nodes is given in equation (4.21). T_c is given as

$$T_c = \tfrac{1}{2}T_{rs} + \tfrac{1}{2}T_a \tag{7.9a}$$

and T_{ra} follows from the relation

$$T_{ra} = (T_{rs} + \alpha T_a)/(1 + \alpha), \tag{7.9b}$$

where

$$\alpha = C/S = \sum C_j / \sum S_j. \tag{7.9c}$$

Section 6.6.1 showed that by taking a radiant input Q_r from an internal source to be input at the fictitious node T_{rs}, account is taken of the heating effect on internal

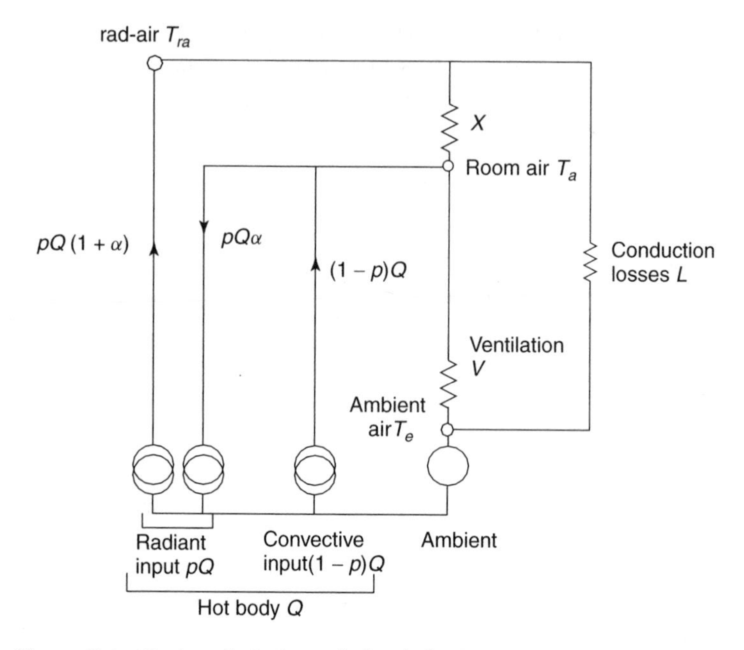

Figure 7.4 Basic rad-air thermal circuit for heat transfer in an enclosure

furnishings produced by the radiation they intercept. Since T_{ra} is partly composed of T_{rs}, it too includes the heating effect of intercepted radiation, and so does T_c. After manipulation, the heat input Q to attain some specified comfort temperature is

$$Q = \frac{VX + XL + LV}{X + \frac{1}{2}L(1-\alpha) + p\left[\frac{1}{2}V(1+\alpha)^2 - \frac{1}{2}L(1-\alpha^2)\right]}(T_c - T_e), \tag{7.10a}$$

or alternatively the comfort temperature resulting from some uncontrolled input is

$$T_c - T_e = \frac{Q\left\{X + \frac{1}{2}L(1-\alpha) + p\left[\frac{1}{2}V(1+\alpha)^2 - \frac{1}{2}L(1-\alpha^2)\right]\right\}}{VX + XL + LV}. \tag{7.10b}$$

A number of functions of temperature follow. The average air temperature is

$$\frac{T_a - T_e}{T_c - T_e} = \frac{X + L - pL(1+\alpha)}{X + \frac{1}{2}L(1-\alpha) + p\left[\frac{1}{2}V(1+\alpha)^2 - \frac{1}{2}L(1-\alpha^2)\right]} = F_2. \tag{7.11}$$

The rad-air temperature is

$$\frac{T_{ra} - T_e}{T_c - T_e} = \frac{X + pV(1+\alpha)}{X + \frac{1}{2}L(1-\alpha) + p\left[\frac{1}{2}V(1+\alpha)^2 - \frac{1}{2}L(1-\alpha^2)\right]} = F_1. \tag{7.12}$$

F_1 and F_2 are used in connection with the environmental temperature model discussed below, where environmental temperature T_{ei} is similar in status to T_{ra}. They are in effect equations (A5.68) and (A5.69) of CIBSE (1986) and also in their more rationally drafted forms, equations (5.169) and (5.170) of CIBSE (1999), where the numerical factors (18.0, etc.) result from a choice of $h_c = 3.0\,\mathrm{W/m^2K}$ and the coefficients $\frac{1}{3}$ and $\frac{2}{3}$ in the definition of T_{ei}.

The radiant star temperature is

$$\frac{T_{rs} - T_e}{T_c - T_e} = \frac{X - L\alpha + p[V(1+\alpha)^2 + L\alpha(1+\alpha)]}{X + \frac{1}{2}L(1-\alpha) + p\left[\frac{1}{2}V(1+\alpha)^2 - \frac{1}{2}L(1-\alpha^2)\right]}. \tag{7.13}$$

The difference between the air temperature and the radiant temperature is

$$\frac{T_a - T_{rs}}{T_c - T_e} = \frac{L(1+\alpha) - p[(L+V)(1+\alpha)^2]}{X + \frac{1}{2}L(1-\alpha) + p\left[\frac{1}{2}V(1+\alpha)^2 - \frac{1}{2}L(1-\alpha^2)\right]}. \tag{7.14}$$

A number of comments may be made:

- The total heat need Q decreases as we move to a more radiative system (i.e. as p increases) if the ventilation loss V exceeds $L(1-\alpha)/(1+\alpha)$, as expected on physical grounds.
- T_a must decrease and T_{rs} must increase as p increases.
- The equations are interrelated. Thus the difference $T_a - T_{rs}$ can be formed from two of the other equations. The total need can be written as

$$Q = V(T_a - T_e) + L(T_{ra} - T_e) \tag{7.15}$$

  and this leads to the value above.
- X tends to be numerically large. If X is very large, T_{ra} and T_a tend to the same value, T_i say, and T_{rs} and T_c also tend to T_i. This is the traditional room thermal model. The rad-air expression for Q (7.10a) tends to the simple form

$$Q = (L+V)(T_i - T_e) \text{ or } \left(\sum AU + V\right)(T_i - T_e), \tag{7.16}$$

  as in equation (7.1).
- Suppose instead that the enclosure contains internal sources of heat totalling Q (fractions p and $1-p$ radiantly and convectively) and that a further source Q' acts at the air node T_a controlling the comfort temperature to some specified value. The value of Q' (positive for heating, negative for cooling) will then be

$$Q' = \frac{(VX+XL+LV)(T_c-T_e) - Q\left\{X+\frac{1}{2}L(1-\alpha)+p\left[\frac{1}{2}V(1+\alpha)^2-\frac{1}{2}L(1-\alpha^2)\right]\right\}}{X+\frac{1}{2}L(1-\alpha)}. \tag{7.17}$$

  If X is large, this reduces to the elementary form

$$Q' = (L+V)(T_i - T_e) - Q. \tag{7.18}$$

Heat input at a surface

Suppose that a floor heating system is used. The floor node will be denoted by T_5, as in Figures 7.1 and 7.2, and will be linked to T_e by the fabric loss conductance F_5; L_5 denotes the conductance between T_{ra} and T_e via T_5; the total conductance L includes L_5. It can be shown that to maintain a comfort temperature T_c, the heat input Q_5 acting at T_5 must be

$$Q_5 = \frac{(VX + XL + LV)}{(L_5/F_5)\left(X + \frac{1}{2}V(1+\alpha)\right)}(T_c - T_e). \tag{7.19}$$

The difference between the air temperature and the radiant star temperature is

$$T_a - T_{rs} = \frac{(1+\alpha)(-V)}{X + \frac{1}{2}V(1+\alpha)}(T_c - T_e). \tag{7.20}$$

A heated floor must lead to a condition where the air is cooler than the average radiant temperature.

This star-based model differs somewhat from the star-based model proposed by Seem *et al.* (1989a) to combine radiant and convective transfers. To develop the rad-air model, the delta pattern of view factor radiant exchanges was transformed through a minimisation process to an approximate star pattern; the emissivity links were then added in. These radiant links were then merged with the convective links using the series-parallel transform. The links are independent of the values of convective and long-wave energy input to the enclosure and so can be found in advance. In Seem's model the minimisation process was effected upon a system of links which included the view factor, emissivity and convective components and the inputs acting at some instant; it appears to require evaluation at each time step. Since the rad-air model includes the two internal nodes, T_a and T_{ra}, an estimate of comfort temperature can be found (19.20). In Seem's model, t_{star} provides comfort temperature.

7.3 PROBLEMS IN MODELLING ROOM HEAT EXCHANGE

Figure 7.3 shows the steps to arrive at a room index temperature and it also lists the required approximations. Another method, apparently simpler and exact, has been proposed but its logic is flawed. This will be demonstrated below to illustrate the care required when modelling heat transfer in a room where a distinction is to be made between air and radiant temperatures.

7.3.1 The Environmental Temperature Model

The environmental temperature model for heat transfer in an enclosure was a 1960s development of the traditional building model, where convective flow and radiant flow to a surface were simply lumped as $h_c + \varepsilon h_r$. It was intended as a means of distinguishing between the separate processes of radiant and convective exchange (Loudon 1970; Danter 1974) and formed the basis for the design method advanced by IHVE (1970) and CIBSE

(1986, 1999) for the design of heating and cooling systems. Its status has been discussed by Billington (1987).

The argument leading to environmental temperature (CIBSE 1986: A5-11, A5-12) is based on a cubic enclosure, one of whose sides, area A, is at T_1 and the remaining surfaces, area $5A$, have a common temperature T_2. The mean air temperature is T_a. There are convective links of Ah_c and $5Ah_c$, together with a radiative link simply denoted AEh_r, all shown in Figure 7.5a. The analysis sought to find an index temperature consisting of suitably weighted values of T_1, T_2 and T_a which drove the same heat flow Q_1 to T_1 (and beyond). It begins with the continuity equation

$$Q_1 = Ah_c(T_a - T_1) + AEh_r(T_2 - T_1). \tag{7.21}$$

The mean surface temperature T_m is

$$T_m = \tfrac{1}{6}T_1 + \tfrac{5}{6}T_2. \tag{7.22}$$

After manipulation, Q_1 can be written as

$$Q_1 = A\left(\tfrac{6}{5}Eh_r + h_c\right)(T_{ei} - T_1), \tag{7.23a}$$

where

$$T_{ei} = \frac{\tfrac{6}{5}Eh_rT_m + h_cT_a}{\tfrac{6}{5}Eh_r + h_c}. \tag{7.23b}$$

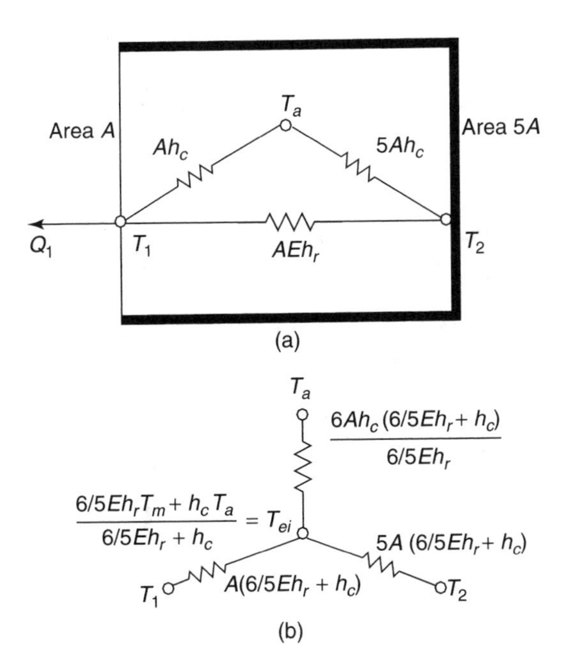

Figure 7.5 (a) The cubic enclosure for the environmental temperature model and the physical (delta) pattern of links. (b) The derived (star) pattern of links

This establishes T_{ei} as a weighted mean temperature, associated with a link of $A\left(\frac{6}{5}Eh_r + h_c\right)$ to T_1 (combining radiant and convective transfer). By analogy there is a link of $5A\left(\frac{6}{5}Eh_r + h_c\right)$ between T_{ei} and T_2 (Figure 7.5b). The heat flow from walls to air is

$$Q_a = Ah_c(T_1 - T_a) + 5Ah_c(T_2 - T_a) \tag{7.24a}$$

$$= \frac{6Ah_c\left(\frac{6}{5}Eh_r + h_c\right)}{\frac{6}{5}Eh_r}(T_{ei} - T_a). \tag{7.24b}$$

The star network results from a delta-star transform (Section 4.4.2) by setting $C_1 \equiv AEh_r$, $C_2 \equiv 5Ah_c$ and $C_3 \equiv Ah_c$. The corresponding K values are shown in Figure 7.5b. The CIBSE analysis was conducted partly through use of working values: $h_c = 3\,\mathrm{W/m^2K}$, $h_r = 5.7\,\mathrm{W/m^2K}$ and $E = 0.9$. With these values,

$$T_{ei} \approx \tfrac{2}{3}T_m + \tfrac{1}{3}T_a. \tag{7.25}$$

The transmittance $\frac{6}{5}Eh_r + h_c = 9.2\,\mathrm{W/m^2K}$ served as the inner film transmittance to find U values. Finally, the $T_{ei} - T_a$ link (7.24b) was written as $h_a \sum A$ or $4.5 \sum A(\mathrm{W/K})$.

T_{ei} was taken to be a general measure of the room temperature, having convective and radiant components, and suitable to estimate heat loss from the room.

7.3.2 The Invalidity of Environmental Temperature

If the argument is applied to an enclosure with heat loss to a surface A_1, emissivity ε_1 and convective coefficient h_{c1}, where A_1 is small compared with the total enclosure area (Figure 7.6), it is found that

$$T_{ei} = \frac{\varepsilon_1 h_r T_2 + h_{c1} T_a}{\varepsilon_1 h_r + h_{c1}}. \tag{7.26}$$

Now this is not a valid expression since a property of surface 1, its emissivity, has become associated with surface 2. Furthermore, T_{ei} preserves this form as A_1 decreases to zero, as of course it can. But if A_1 is zero, T_{ei}, the intended index for the enclosure as a whole, is based on the ε and h_c values of a non-existent surface and takes no account of the ε

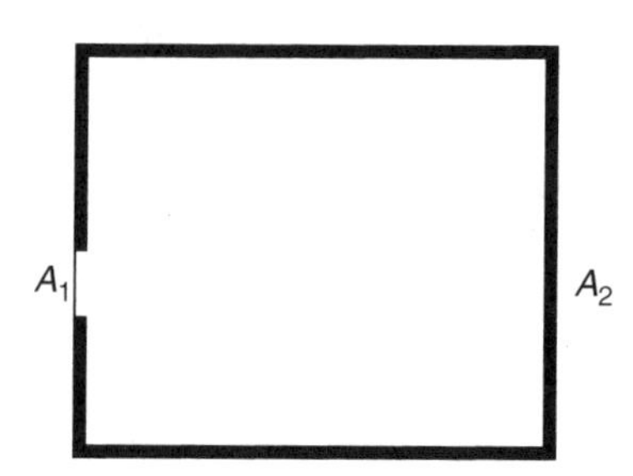

Figure 7.6 An enclosure with a small area A_1 at T_1 and the remainder at T_2

and h_c values of the enclosure as it is. It cannot be an enclosure index.[1] From a pragmatic standpoint this may not matter since the ε and h_c values are not known sufficiently reliably and $\varepsilon_1 \approx \varepsilon_2$ but it makes clear that the principle used to find an index which combines air and radiant temperatures is flawed. The argument in fact entails further inconsistencies:

- The analysis assumes a single value h_c for both surfaces. If separate values h_{c1} and h_{c2} are assumed, the link between T_{ei} and T_1 will contain a term in h_{c2}. This is clearly incorrect: convection at one surface is independent of convection at another surface.
- If the quantity E is disaggregated into its emissivity and view factor components, for the cubic enclosure

$$1/E = (1 - \varepsilon_1)/\varepsilon_1 + 1 + (1 - \varepsilon_2)/5\varepsilon_2, \tag{7.27}$$

  and inclusion of this relation similarly leads to a term in ε_2 in the expression for the link between T_{ei} and T_1. This conflicts with the Oppenheim–Buchberg formulation for radiant exchange: the link between T_1 and the rest of the enclosure via its black-body equivalent node depends on the emissivity of surface 1 and not on any other surface emissivity.

Thus we have a paradoxical situation: the enclosure model of Figure 7.5a is an acceptable model, the equation for heat continuity is valid, the algebraic manipulation correct. Yet the concept arrived at, the index temperature, is absurd. What has gone wrong?

7.3.3 *Flaws in the Argument*

Brief though the argument is, it contains a number of shortcomings (Davies 1992a, 1996a).

(a) The enclosure of Figure 7.5a, while valid, is too simple to conduct the merging of radiant and convective exchanges. The table shows the number of convective and black-body radiative links in an enclosure with n surfaces at distinct and specified temperatures.

Number of surfaces n	1	2	3	4	5	6
Number of convective links	1	2	3	4	5	6
Number of radiative links	0	1	3	6	10	15

In order to merge the radiation and convection processes, the radiation network, consisting of $\frac{1}{2}n(n-1)$ links, must first be transformed to one of n links. For $n = 3$, an exact delta-star transform is possible. For $n > 3$ the transform can only be achieved in

[1]The Oxford mathematician Charles L. Dodgson, better known as Lewis Carroll, author of *Alice in Wonderland*, may have had in mind a model along the lines of this reductio ad absurdum argument in his invention of the Cheshire Cat which, sitting up in a tree, 'vanished quite slowly, beginning with the end of the tail, and ending with the grin, which remained some time after the rest of it had gone'. Its form remained but without substance, just as T_{ei} does here as A_1 vanishes.

some approximate manner, as shown earlier. For $n = 2$, the environmental temperature enclosure, a radiant star node can only be placed arbitrarily on the link between T_1 and T_2.[2]

(b) The value of temperature experienced by a sensor in a room is a weighted average of the temperature of the air contacting it and the long-wave radiation that it intercepts. A representative room index temperature T_i must be an average of the air and radiant temperatures. For the air this is simply an average over local values in the space, T_{av}. For radiation, the local radiant value at some location depends not only on temperatures of the surfaces visible from it, but also on their emissivities and relative disposition, together with the radiation intercepted from any internal source; the room average value T_{rv} is the space-averaged value of local values. Then

$$T_i = aT_{rv} + (1 - a)T_{av}; \tag{7.28}$$

a depends on the relative effectiveness of convective and radiative transfer. In all the source literature on environmental temperature, however, the average (or mean) *radiant* temperature T_{rv} and the average *surface* temperature T_m (7.22) were taken to be the same. This introduced some semantic confusion since T_{rv} depends on surface emissivity, the configuration of room surfaces and the presence of an internal radiant source whereas T_m depends on none of these quantities. (It will be noted that both terms on the right-hand side of (7.28) are averages over three dimensions, whereas those defining T_{ei} in (7.23b) are two- and three- dimensional averages, respectively.)

(c) In deriving T_{ei}, the model in fact made no mention of any internal heat source and so failed to take account of the radiant fraction of its emission which fell on the surface at T_1. This transfer would have provided a further term in the definition of T_{ei}. Thus the effect of a radiant source Q_r had to be argued into the procedure later (see below).

(d) A source of heat in a room delivers part of its output convectively and the environmental temperature model took this to be input at T_a. However the model network of Figure 7.5a did not include the black-body equivalent nodes on the radiant conductance AEh_r at which the radiant output Q_r, should be input in due proportion. Instead the argument was further developed so that, by implication, Q_r was delivered to a form of star point temperature T_{rs}' located on AEh_r. Q_r was split into an augmented component at T_{ei} with withdrawal of the excess from T_a, in accordance with the series-parallel transformation. T_{rs}' might be described as a form of mean radiant temperature, depending on the radiant input, surface emissivities and the relative disposition of the surfaces.

(e) Why is the *mean* of any quantity constructed? Consider the loss of heat by conduction to ambient from the enclosure in Figure 7.5a; u_1 and u_2 denote the transmittances from the two surfaces to ambient, so the conduction loss is

$$Q_{conduction} = Au_1(T_1 - T_e) + 5Au_2(T_2 - T_e), \tag{7.29}$$

which can be written

$$Q_{conduction} = A(u_1 + 5u_2)(T_{m,conduction} - T_e), \tag{7.30a}$$

[2]If T_{rs} is located on the view factor component Ah_r of AEh_r so as to define segments of $\frac{6}{5}Ah_r$ and $6Ah_r$, then T_{rs} and T_a could be logically merged, as shown in Section 7.2. If $\varepsilon_1 = \varepsilon_2$ and $h_{c1} = h_{c2}$, the series-parallel transform is exact; otherwise it is only approximate.

where

$$T_{m,conduction} = (u_1 T_1 + 5u_2 T_2)/(u_1 + 5u_2). \quad (7.30b)$$

This is the *mean surface* temperature for the case of conduction loss and it depends on the u values of the wall. $T_{m,conduction}$ and T_m are only equal if u_1 and u_2 happened to be equal, which is rarely the case. Similarly, for convective exchange,

$$T_{m,convection} = (h_{c1} T_1 + 5h_{c2}\, T_2)/(h_{c1} + 5h_{c2}) \quad (7.31)$$

is the mean surface temperature for internal convective exchange. In general, $T_m \neq T_{m,conduction} \neq T_{m,convection}$.

Although T_1 and T_2 represent nodes at which some local heat source might act, T_m cannot be represented in a thermal circuit and sources cannot be taken to act at T_m. Thus the construct T_m has no relevance to any form of room heat exchange and should not appear as an *input* to any formulation of room transfer. (There is no difficulty in using it as an *output* variable in comfort studies.)

(f) Even though T_m had been a valid quantity in itself, it is invalid to combine it with T_a since T_m results from *two-dimensional* averaging whereas T_a results from *three-dimensional* averaging. It is not sufficient that each is temperature; one might enquire whether any useful quantity could be devised from the mean of the temperature of a window frame, a *one-dimensional* construct, with the mean of the glass within it, a *two-dimensional* construct.[3]

(g) A further problem with the environmental temperature procedure is its placing of comfort temperature T_c on the $h_a \sum A$ conductance between T_a and T_{ei} (Figure 7.7). This is invalid for three reasons:

- Though differing in derivation, the rad-air and environmental temperature models are similar in appearance and T_{ra} and T_{ei} are comparable at a pragmatic level. It was shown earlier that T_c and T_{ra} cannot appear simultaneously in a thermal circuit. Similarly, T_c should not be placed in a circuit that includes T_{ei}.
- Comfort temperature is a mean value associated with measurement of temperature by a sensor or some object which is small compared with room dimensions. It is a low-conductance (high-impedance) link. But $h_a \sum A$ is a large conductance, so T_c should not be placed on it. Put another way, a large heat input, kilowatts, at any of the nodes

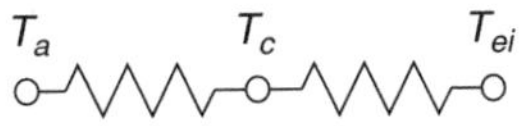

Figure 7.7 Comfort temperature T_c placed on the $T_a - T_{ei}$ link $h_a \sum A$

[3]There is no problem in combining together quantities X_1 and X_2 where X_1 and X_2 have different units in the form $Y = a_1 X_1 + a_2 X_2$ since a_1 will have units such that $a_1 X_1$ has the units of Y; the same goes for $a_2 X_2$. The problem comes when we attempt to devise a mean value in the form $(a_1 X_1 + a_2 X_2)/(a_1 + a_2)$. The denominator here may be invalid: a_1 and a_2 have different units unless X_1 and X_2 are of the same kind and have the same units. Suppose we purchase 1 kg of bananas and 2 kg of potatoes. There is a total cost but we have no useful mean *weight* for the purchase, formed similarly to T_{ei}, nor is it valid to add together the prices per kilogram of bananas and potatoes, the equivalent of $\frac{6}{5} E h_r + h_c$.

in Figure 7.7 is associated with a modest rise in temperature since the heat is lost through the ventilation and conduction conductances linked to T_a and T_{ei}, respectively. An input of kilowatts to a sensor or small object within the room, however, would lead to gross temperatures. A T_c node placed on $h_a \sum A$ will not represent this response.

- With the working values for h_c, etc., together with T_c defined as

$$T_c = \tfrac{1}{2}T_a + \tfrac{1}{2}T_m, \tag{7.32}$$

it can be written (CIBSE 1986: equation A5.122) as

$$T_c = \tfrac{1}{4}T_a + \tfrac{3}{4}T_{ei}. \tag{7.33}$$

In this case a location can be found on $h_a \sum A$ where the temperature has the same numerical value as T_c. If the calculation is conducted algebraically, we find

$$T_c = \tfrac{1}{2}\left[T_a\left(1 - h_c/\tfrac{6}{5}Eh_r\right)\right] + \tfrac{1}{2}\left[T_{ei}\left(1 + h_c/\tfrac{6}{5}Eh_r\right)\right]. \tag{7.34}$$

For an enclosure with somewhat lower emissivities, E will be less than 0.9. As E decreases, T_c, if located on $h_a \sum A$, moves toward T_{ei} and falls on it when $h_c = \tfrac{6}{5}Eh_r$. With further decrease in E, T_c lies beyond the range between T_a and T_{ei} and so cannot be placed on $h_a \sum A$. Its placement is therefore opportunistic; if T_c cannot in general be placed on $h_a \sum A$, its placement is invalid.

The concept of environmental temperature was evolved in the 1960s in response to a very urgent need to have some means, realisable using a hand-held calculator, to estimate the large swings and excessively high temperatures that were being experienced in the lightweight, glass-clad buildings that had proved so popular in the post-war period. In addition to the long-established U value, dynamic parameters were needed. In the UK these took the form of wall admittances Y and cyclic transmittances fU, measures of wall response to sinusoidal excitation of period 24 hours (discussed in Chapter 15). They proved to be immensely valuable. Some logical basis was needed for the form for inside film transmittance, one element in these parameters (the others were the thickness, thermal conductivity and capacity of the wall layers). The values of Y and fU were not sensitive to the choice of form for the inside film and the pragmatic means of finding it as sketched above was quite adequate. Analysis of the steady state however was on a different level and it is unfortunate that these logical flaws were not recognised and resolved before the procedure, originally a useful expedient, was elevated to the status of a design procedure based on an apparently comprehensive thermal model. Its shortcomings were to do with radiant exchange, identifying mean radiant temperature with mean surface temperature, a confusion which should not have occurred in view of what was known by the early 1960s about radiant transfer. It should be said though that the procedure drew on and sharpened views that were widely accepted at the time in Europe (Uyttenbroeck 1990).

7.4 WHAT IS MEAN RADIANT TEMPERATURE?

One of the problems that continues to plague discussion of room internal heat transfer is the interpretation of the term 'mean radiant temperature', which is used to denote several different constructs. The definition of mean air temperature (MAT) is quite clear: it is the arithmetic mean $\sum T_{ap}/N$ of N observations of local or point air temperature at uniformly spaced locations in the room. It is volume-averaged and therefore a global construct. Since further indices are to be formed from MAT and mean radiant temperature (MRT), we should expect them to have equal status as global variables, but this may not be so.

MRT may be seen as the local observable temperature T_{rp} in an air-free room, having surfaces of differing temperatures and emissivities. 'Mean' here implies that the sensor is supposed to be spherical and black-body, so it has no directional properties. This is a local value since its value varies with position. (Kamal and Novak (1991) define a local mean radiant temperature as LMRT $= \sum \varepsilon_i T_i F_{L,i}$, summing over all the surfaces (emissivity ε_i temperature T_i) in an enclosure; $F_{L,i}$ is the local view factor from a small black sphere wherever placed, to surface i. They assumed this to be a measurable quantity. This does not appear to be a well-based construct however. Consider its value at the centre of a cubic enclosure, five of whose surfaces are black and at temperature unity, and the sixth with zero emissivity (and arbitrary temperature). LMRT $= \frac{5}{6}$. The sphere picks up radiation coming from the direction of the sixth surface, which consists of the reflected components from the other surfaces; the *measured* temperature is unity.)

MRT might be taken as a local quantity as above, but to include the effect of radiation the sensor intercepts from any internal radiant source. See for example CIBSE (1986: A5-4) and CIBSE (1999: A5-3). It is noted there that in special circumstances (centre of a cubic enclosure with emissivities of 1 and no internal radiant source) MRT and mean surface temperature are the same. Although this is correct, it is misleading, as noted above.

MRT might be the volume-averaged value T_{rv} of local observations, including intercepted radiation. T_{rv} is on the same footing as MAT and is a global value. This is the sense in which many workers appear to view MRT.

When the effects of internal radiant and convective exchange are lumped, it is implicit that the radiant component of flow to some surface at T_j is proportional to the difference $T_{rs} - T_j$. The radiant star node T_{rs} is a convenient fictional node when radiant exchange is simplified from a delta to a star configuration. MRT might be viewed as such a star temperature (Carroll 1981).

It is also implicit in such a simplified radiant exchange model that the radiant flux Q_r from an internal radiant source acts at T_{rs}. Thus MRT might denote this higher value of T_{rs}. Both this and previous constructs are global values.

The term 'mean radiant' has been used to denote the black-body equivalent temperature of a surface T_{jb}. Each room surface now has its own MRT. In this sense $(\mathrm{MRT})_1$ or T_{1b} will be greater than T_1 itself if surface 1 is surrounded by warmer surfaces.

The 'MRT of surface 1 of an enclosure' has been defined by Walton (1981) as the area and emissivity weighted mean surface temperature of all the surfaces seen from surface 1. $(\mathrm{MRT})_1 = \sum A_j\varepsilon_j T_j / \sum A_j\varepsilon_j$, where summations exclude surface 1.

CEN and various authors define MRT as the uniform surface temperature of an enclosure in which an occupant would exchange the same amount of radiant heat as in the actual non-uniform enclosure. This definition does not refer to any observable temperature. Indeed statements of this kind are not definitions but rather spell out the consequences of some foregoing provisions. Being warm, the occupant will lose more heat by radiation when near a cool surface than a warm surface, so with this definition MRT varies with the position of the occupant. It appears compatible with the CIBSE definitions.

8 Moisture Movement in Rooms

In a ventilated enclosure with no moisture-absorbing materials and no sources of water vapour, the inside vapour pressure p_i must be equal to the ambient pressure p_e but if any source is present $p_i > p_e$. An adult may be expected to input more than a kilogram of moisture in a 24 h period by metabolic action and there may be further inputs from activities such as having a bath, cooking or drying clothes indoors; Oreszczyn and Pretlove (1999: Table 1) provide a detailed list for domestic sources. These sources raise the inside vapour pressure. Most of the load will normally be lost by ventilation, some small fraction may be lost by diffusion through solid surfaces. Some vapour may condense on cool surfaces and in soft furnishings, evaporating later.

Buildings may be affected when the relative humidity in the internal space reaches high levels for long periods, leading to unsightly mould growth, especially in corners of outer walls. It is a long-recognised problem (Section 8.10). Erhorn (1990) has summarised its principal determinants.

- For their germination and growth, moulds require a relative humidity above 80–85% with optimal conditions of 90–95%.
- Moulds prefer a slightly acid environment with pH values of 4.5 to 6.5. A few grow with values around 2 and around 8.
- Growth proceeds at temperatures above 0°C and is greatest around 30–45°C.
- Moulds have minimal nutrient requirements and growth is largely sustained by air-borne particles.
- Moulds require less oxygen than humans.
- Moulds do not require light for growth.

Clarke (2001), also Clarke *et al.* (1999), report that most moulds grow at values of relative humidity greater than 88%; none was observed below 74% (see his Table 7.23). The time for growth in laboratory conditions varied from 78 to 103 days. This is a long time compared with room thermal response times so that the steady-state analysis presented below should prove adequate for humidity calculations. Grant *et al.* (1989) conclude that

Building Heat Transfer Morris G. Davies
© 2004 John Wiley & Sons, Ltd ISBN: 0-470-84731-X

'otherwise susceptible surfaces can be kept free of mould if relative humidities ... are maintained below 80% relative humidity'. High humidity may further lead to the rotting of timber,[1] interstitial condensation and the spalling of masonry. The classical advice to avoid mould growth has always been 'heat, ventilate, insulate'.

People are generally unaware of moisture levels over a large range of moisture content expressed as a relative humidity, but become conscious of it at extremes of dryness and dampness. Sustained damp conditions may affect the health of the occupants. Paton (1993) lists a number of symptoms reported to be associated with damp and mould-infested housing. Respiratory symptoms include persistent cough, wheeze, asthma, sore throat, earache; non-respiratory symptoms include diarrhoea and vomiting, aches and pains, tiredness, headaches, fevers, rashes and eye irritation. There appears to be no clear mechanism to establish a causal connection between humidity per se and health; they are probably concomitant aspects of underlying factors such as poverty and education as well as mould growth. Baughman and Arens provide a detailed account of the relation between relative humidity and the presence of house mites (a problem in homes with thick carpeting and evident at lower humidities than those for mould growth), of moulds themselves, bacteria and viruses, and the non-biological pollutants formaldehyde and ozone (Baughman and Arens 1996; Arens and Baughman 1996). The relation between ventilation rate, humidity, dust mite allergens and asthma in the UK is discussed by Howieson *et al.* (2003). They remark that air change rates of 0.5 air changes an hour are likely to produce vapour pressures which regularly rise above the critical equilibrium humidity (60%) for growth of house dust mites. Use of mechanical heat recovery, however, will allow a ventilation rate of 1.3 ac/h and this removes moisture and pollutants without energy penalty.

Closer control of humidity than is needed for human occupation may be necessary for the preservation of ancient manuscripts and this is to be achieved through air conditioning. This chapter is mainly concerned with the complementary mechanism – transport of moisture through the walls (and flat roofs and subfloor cavities) in so far as it is similar to heat flow. Other aspects have been widely reported; see for example the series of

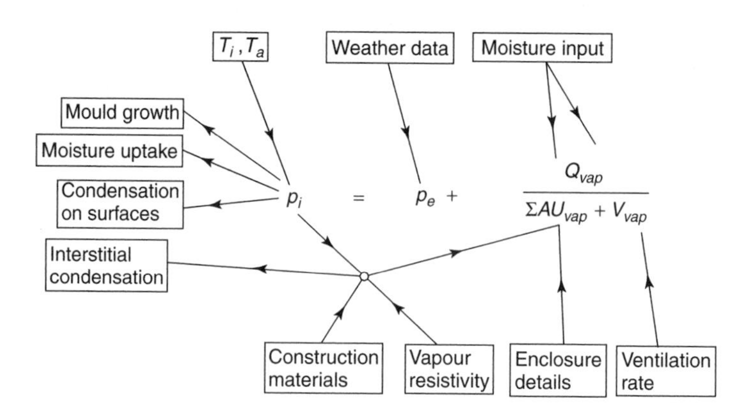

Figure 8.1 Steady-state vapour balance in a room

[1]Cases have been reported where inadequate ventilation has led to damp conditions in the softwood beams below old timber floors, making them susceptible to attack by the larvae of death-watch beetles. Harris (1996) points out that a build-up of moisture beneath a suspended timber floor can lead to wet or dry rot.

studies by Hall and colleagues, (1977–1995). A form of steady-state moisture balance equation is shown in Figure 8.1 and relates to the case where room air is replaced by the infiltration of ambient air (Becker 1984: equation 1; Letherman 1988: equation 6). It is similar in structure to equation (1.5) for sensible heat flow in the same circumstances, though while V and $\sum AU$ are of the same order of magnitude, $\sum AU_{vap}$ is very much smaller than V_{vap}. Further, it omits consideration of the substantial quantity of moisture that can condense on room surfaces and be absorbed in furnishings. Jones (1993, 1995) has shown how, in principle, account may be taken of such adsorption[2] and subsequent desorption of water vapour; see also El Diasty *et al.* (1993a). Water vapour cannot be used as a tracer gas to investigate interzonal air movement but the simultaneous release of tracer gas and moisture can be used to study the absorption of moisture (Plathner and Woloszyn 2002).[3] Vapour pressure has the units of pascals: 1 Pa = 1 N/m^2.

8.1 VAPOUR LOSS BY VENTILATION

The quantity e can be defined as the increase δm in water vapour present per unit volume per unit rise in water vapour pressure[4]:

$$e = \delta m/(V_r\ \delta p_i), \tag{8.1}$$

where V_r is the room volume. Now, according to the gas law,

$$p_i V_r = (m/M)RT, \tag{8.2}$$

where m is the mass of water vapour present in the room and M is the molecular weight of water, i.e. 18 kg. So at 20°C,

$$e = M/RT = 18/(8314 \times 293) = 7.4 \times 10^{-6}\ \text{kg/m}^3\ \text{Pa}. \tag{8.3}$$

If v is the volume flow into the room (m^3/s), the difference in moisture content of the incoming and outgoing airflows is $ve(p_i - p_e)$, so the vapour conductance V_{vap} in Figure 8.1 is ve or $(NV_r/3600)e$, where N is the number of air changes per hour.

Example 8.1

Room inside conditions are 21°C and 60% relative humidity or 1490 Pa, outside conditions 10°C and 80% relative humidity or 981 Pa. Find the loss of water vapour from a room of 40 m^3 with 2 volume air changes per hour. The loss is

$$(2/3600) \times 40 \times (7.4 \times 10^{-6}) \times (1490 - 981) = 8.4 \times 10^{-5}\ \text{kg/s or } 7.2\ \text{kg per day}.$$

[2]Adsorption denotes the take-up of moisture by a material when its volume remains unchanged; absorption is a similar process but the volume increases (Wong and Wang 1990); absorption occurs in wood.

[3]Heidt *et al.* (1991) make a comparison of tracer gas methods to determine airflows in buildings. Laporthe *et al.* (2001) present a comparison of the commonly used tracer gases SF_6 and N_2O; SF_6 is preferred. CO_2 may be used.

[4]The quantity e with units kg/m^3 Pa is analogous to the volumetric specific heat s, units J/m^3 K (Section 2.3.1).

8.2 VAPOUR RESISTIVITY

The term 'moisture' denotes water in liquid and vapour phases. A flow in liquid form takes place as a result of a difference in hydraulic pressures. A flow in vapour form follows mainly from a difference in vapour pressure. In solid materials both mechanisms may operate but the transport is complicated by the interaction between water and material molecules, by the presence of salts, electric potentials, capillary flow and internal evaporative or condensation behaviour. Some condensation in a vertical wall may be removed by gravity. Liesen and Pedersen (1999) remark that 'there is not a single known theory that can be used to cover all situations for moisture transfer in a porous solid'. The elementary theory to be described here is based on the ideas set out qualitatively by Rowley (1939) and derives from procedures described by Cammerer (1952) and Glaser (1958a, 1958b, 1958c, 1959) and assumes that vapour transport alone contributes to the loss of moisture; see also Billington (1967: Ch. 5).

It is convenient first to state Fick's law for the rate of diffusion of water vapour in air due to a concentration gradient such as exists between the moist air in a room and a bounding surface of a wall on which it condenses or through which it diffuses[5]:

$$m_{vap} = -D\,\mathrm{d}c/\mathrm{d}x, \tag{8.4}$$

where m_{vap} is the mass transfer of vapour (kg/m^{2}s), c is the concentration of water vapour (kg/m^3, i.e. its density), x is the direction of the concentration gradient and D is the mass diffusion coefficient,[6] 2.5×10^{-5} m^2/s in air at 20°C and lower in solid materials (Ede 1967b: 259).[7] de Vries (1987) gives D as $21.7 \times 10^{-6}(101325/\mathrm{p})(\mathrm{T}/273.16)^{1.88}$, where p is total pressure (Pa) and T is absolute temperature. It is more conveniently expressed as a gradient of vapour pressure:

$$m_{vap} = -(DM/RT)\,\mathrm{d}p/\mathrm{d}x. \tag{8.5}$$

DM/RT is the permeability or vapour conductivity of air and equal to about 1.85×10^{-10} (kg/s)/(m Pa) or as a resistivity, 5.4×10^9 Pa/(kg/s m). The significance of the units is made clearer when written as (Pa/m$_x$)/(kg/(s m$_y$m$_z$)). When the resistivity is multiplied by a layer thickness in the x direction, the layer resistance has units Pa/(kg/(s m$_y$m$_z$)). When a thermal flow in watts is expressed as J/s, the formal similarity between thermal and moisture flows becomes clear:

thermal conductivity (J/s m K)	vapour conductivity or permeability (kg/s m Pa)
thermal transmittance (J/s m^2 K)	vapour transmittance or permeance (kg/s m^2 Pa)
thermal conductance (J/s K)	vapour conductance (kg/s Pa)

[5]The corresponding law for liquid diffusion is known as Darcy's law.

[6]There are two dimensionless groups including D. The Schmidt number $\mathrm{Sc} = \mu/\rho D$; the Sherwood number $\mathrm{Sh} = h_D L/D$ and is the ratio of mass transfer to that under pure diffusion. Dimensionless groups are listed by Ede (1967b: 278).

[7]At normal barometric pressure and 15°C the mean free path of an air molecule is 6.4×10^{-8} m and 4.2×10^{-8} m for a molecule of water vapour. It increases in proportion to absolute temperature. Capillary-porous bodies have pores of different sizes and Fick's law holds for diffusion in pores of characteristic sizes greater than about 10×10^{-8} m (of order or greater than the mean free path). In smaller pores the flow becomes molecular rather than macroscopic and the diffusion law is not obeyed. The same restriction applies to laminar flow. See Luikov (1966: 220).

Solid materials have thermal resistivities that are less than that of air, but their vapour resistivities are greater than that of air. McLean and Galbraith (1988) discuss fundamentals and give an account of how they may be measured; see also Galbraith and McLean (1990). They point out that while the resistivity of plasterboard is almost independent of the relative humidity (RH) values across the sample, the resistivity for plywood with a 0–60% difference in RH is some 10 times larger than the 100% to 80% difference in RH. A few values for resistivity are listed in Table 8.1, taken from Tables 3.48 to 3.51 in Book A of the 1999 *CIBSE Guide*. The difficulties in making the measurements cast doubt on their reliability. A further article (Galbraith *et al.* 1993a) describes how measurements of permeability of identical samples of polystyrene and particle board were made in 13 European laboratories, with detailed instructions on their conduct. For the polystyrene, values ranged from 1.36 to 3.30×10^{-12} (kg/s)/(m Pa) and for the particle board 0.828 to 2.90×10^{-11} (kg/s)/(m Pa), an unacceptably wide spread. Spooner (1980) reported a similar spread of values for measurement of thermal conductivity. In any case, values of transport quantities vary strongly with moisture content. Kallel *et al.* (1993) cite values for the vapour mass transfer coefficient D_{mv} for mortar: $10^{-9}\ m^2/s$ with 1% moisture and $7 \times 10^{-11}\ m^2/s$ with 6%.[8] Daian and Saliba (1991) present graphs of D against the degree of saturation for samples of mortar. Values vary from around $10^{-7}\ m^2/s$ at low saturation (less than 20%) and increasing sharply toward $10^{-1}\ m^2/s$ and in some cases much higher above 80% saturation, an enormous fractional variation.

The mass transfer (kg/s m^2) through some layer due a 1 Pa pressure difference is very small. Numerical values of quantities associated with moisture transfer differ by many orders of magnitude from heat transfer values. Diffusion, however, is physically a slow process and 1 Pa is a very small pressure difference. (1 Pa corresponds to a change of height of about 85 mm. The stagnation pressure associated with an air speed of 30 m/s (108 kph) is about 540 Pa.)

Since the values of diffusion coefficients in general depend strongly on the moisture content of the material, measurements that fully characterise them have to be made at a series of values for the amount of moisture stored and these measurements are discussed in Section 8.6.2.

Table 8.1 Selected values of vapour resistivity

	Density (kg/m^3)	Resistivity (Pa/(kg/s m) or s^{-1})
Glass		Infinite
Linoleum	1200	9000×10^9
Glass wool		$5–7 \times 10^9$
Lightweight brick	<1000	$25–50 \times 10^9$
Cast concrete	>1000	$30–80 \times 10^9$
Sandstone	–	$75–450 \times 10^9$
Pine	–	$45–1850 \times 10^9$

[8]Defining D_{ml} as the liquid mass transfer coefficient and W as the fractional moisture content by weight, they note:

$W > 3\%$, D_{ml} is of order 100 D_{mv}; liquid flow dominates in the early drying process.

$W < 1\%$, D_{ml} is of order 0.1 D_{mv}; vapour diffusion is more important in the late stage of drying.

8.3 VAPOUR LOSS BY DIFFUSION THROUGH POROUS WALLS

As an example of the loss of vapour by conduction through a wall, expressed by the $\sum AU_{vap}$ term in Figure 8.1, we consider the wall described in Table 8.2, where a layer of insulation is enclosed between load-bearing elements. It is taken from CIBSE (1999: 7-7) and the example follows Glaser's procedure. The thermal resistance of each element is d/λ and their sum totals 3.145 m^{2}K/W. Suppose first that there is a relatively small mean temperature difference, 21°C inside and 10°C outside. To find the moisture transfer, suppose that the relative humidities inside and outside are 60% and 80%, respectively, so the vapour pressure inside is $0.6 \times \text{svp}(21°\text{C}) = 0.6 \times 2484 = 1490$ Pa. The vapour resistance of each element is $d \times$ vapour resistivity and their total is 9.3×10^9(m^2 Pa)/(kg/s). The reciprocal of the vapour resistance is defined as the permeance with units (kg/s)/(m^2 Pa) and is analogous to the thermal transmittance with units W/m^{2}K. The vapour resistances of the films are not zero but with values of $\rho c_p RT/(h_c M)$, around 0.05×10^9 Pa/(kg/s m^2), they are negligibly small compared with those of the wall elements. Thus the mass transfer is $(1490 - 981)/9.3 \times 10^9 = 55 \times 10^{-9}$ kg/m^2 s; multiplying by the latent heat of 2.47×10^6 J/kg, this amounts to a latent heat flow of 0.14 W/m^2, which is small compared with the sensible heat loss of $(21 - 10)/3.145$ or 3.50 W/m^2. The mass flow also amounts to about 0.05 kg a day for a 10 m^2 wall area and this is negligible compared with the loss of 7.2 kg a day due to ventilation, noted above for the same values of p_i and p_e. The vapour transmittance $U_{vap} = 1/(9.3 \times 10^9)$ and the total vapour conductance $\sum AU_{vap}$ follows by summing over all areas.

The temperature profile through the wall implies a saturated vapour pressure (svp) profile. Since the actual vapour pressure cannot be greater than the saturated vapour pressure at some location, a test must be made to see whether the vapour pressure distribution is valid. Accordingly, the temperatures must be found at the interfaces of the construction, nodes 1 to 5. The temperature at node 5 is $10 + (21 - 10) \times 0.06/3.145 = 10.2$°C. See row 1 of Table 8.3. The nodal values of vapour pressure can be found similarly, row 3.

Table 8.2 Thermal and moisture values for an example wall

Nodes	0 Inside	1	2	3	4	5	6 Outside
	Inside film	Light-weight plaster	Aerated concrete	Mineral fibre insulation	Outer brick	Outside film	
Thickness (mm)	–	12	100	75	105	–	
Thermal conductivity (W/m K)	–	0.16	0.16	0.035	0.84	–	
Thermal resistance (m^{2}K/W)	0.12	0.075	0.625	2.14	0.125	0.06	total = 3.145
Vapour resistivity (GPa s m/kg)	0	50	30	6	50	0	
Vapour resistance (GPa s m^2/kg)	0	0.6	3.0	0.45	5.25	0	total = 9.3×10^9

Source: CIBSE (1999: Table 7.7).

Table 8.3 Temperature and vapour pressure profiles at interfaces through the wall

	0 Inside	1	2	3	4	5	6 Outside	
1 Temperatures, small difference	21.0	20.6	20.3	18.1	10.6	10.2	10.0	
2 Corresponding sat vap pressure	2484	2421	2382	2079	1281	1244	1226	
3 Vapour pressure	1490	1490	1458	1293	1269	981	981	all values below svp
4 Temperature, large difference	21.0	20.4	20.0	16.8	5.9	5.3	5.0	
5 Corresponding sat vap pressure	2484	2393	2337	1915	930	890	871	
6 Initial vapour pressure	1490	1490	1444	1217	**1183**	784	784	one value above svp
7 Corrected vapour pressure	1490	1490	1407	992	930	784	784	one value at svp

Source: CIBSE (1999: Tables 7.8 to 7.10).

From the nodal values of temperature, the corresponding svp values can be found, line 2. Since vapour pressure everywhere lies below the svp value, no condensation takes place within the wall.

Moisture is stored in internal walls and furnishings as well as outer walls. However, the transport of moisture by diffusion is largely associated with outer walls. Dutt (1979) notes that moisture movement from living space into attics is almost entirely by air movement, not by diffusion.

With a larger difference in temperature across the wall, condensation may take place. Before considering that, we discuss the simpler situation where condensation takes place on an exposed surface.

8.4 CONDENSATION ON A SURFACE

It is a matter of simple observation that in a warm humid room, water vapour may condense in sizeable quantities on cool surfaces, notably window panes.[9] Its rate of deposition is found through the simultaneous solution of the continuity equations for heat and mass.

If no condensation takes place, continuity of heat flow requires that

$$(T_{ra} - T_c')h_i = (T_c' - T_e)h_e, \tag{8.6}$$

[9]This is normally the interior surface but Bong *et al.* (1998) describe a study of condensation on the external surface of a conditioned room in the very humid conditions of Singapore.

where T_{ra}, T_c' and T_e are respectively the rad-air, glass and ambient temperatures. It will be assumed for convenience that the rad-air and room air temperatures are equal. Parameters h_i and h_e are the inside and outside film transmittances, respectively. The glass transmittance is large and can be neglected. Thus, in the absence of condensation, the glass temperature is

$$T_c' = (h_i T_{ra} + h_e T_e)/(h_i + h_e). \tag{8.7}$$

Suppose that condensation takes place at the rate of m kg/m^2 s. The latent heat yield must now be included:

$$(T_{ra} - T_c)h_i + mL = (T_c - T_e)h_e, \tag{8.8}$$

where L (J/kg) is the latent heat of water vapour. The mass transfer equation relates m to the concentration difference or the pressure difference that causes it:

$$m = h_D \Delta c = h_D(M/RT)(p_i - p_c), \tag{8.9}$$

where h_D (m/s) is the mass transfer coefficient and p_i and p_c are respectively the vapour pressures in air and at the glass surface. It will be assumed that p_c is the saturated vapour pressure at the glass temperature T_c, so

$$p_c = \text{svp}(T_c). \tag{8.10}$$

The mass transfer coefficient h_D (m/s) is usually taken to be

$$h_D = h_c/\rho c_p. \tag{8.11}$$

This is Lewis' relation (Ede 1967b: 215), where h_c is the convective coefficient at the glass surface and ρ and c_p are the density and specific heat of air.

Example 8.2

Find the rate of condensation on vertical single glazing when $T_{ra} = 21°\text{C}$, $T_e = 5°\text{C}$, inside relative humidity 60% (values in Table 8.3). Take $h_i = 1/0.12$ and $h_e = 1/0.06\,\text{W/m}^2\text{K}$ (as in Table 8.2) so $h_i + h_e = 25\,\text{W/m}^2\text{K}$. It is convenient to present the equations in the form

$$(T_c - T_c')(h_i + h_e) = mL = \frac{h_c ML}{\rho c_p RT}(p_i - \text{svp}(T_c)). \tag{8.12}$$

The value of T_c' is 10.33°C. The values of svp(10.0) and svp(11.0) are 1228.0 and 1312.7 Pa, respectively (see page 6.4 in the 1993 *ASHRAE Handbook of Fundamentals*, where latent heat is also listed) and by linear interpolation (sufficient for this purpose), $\text{svp}(T_c') = 1256\,\text{Pa}$. This is less than the room vapour pressure of 1490 Pa, so some condensation will take place and lead to a slight increase in glass temperature. The constant multiplier on the right of (8.12) is taken as

$$(3.0 \times 18 \times 2.47 \times 10^6)/(1.24 \times 1006 \times 8314 \times 283) = 0.0454.$$

Again by linear interpolation, equation (8.12) is found to be satisfied by a value of $T_c = 10.7°C$. For this value, $mL = 9.2\,\text{W/m}^2$ and so the mass flow is $9.2/2.47 \times 10^6 = 3.7 \times 10^{-6}\,\text{kg/m}^2\text{s}$ or about 0.6 kg per day for $2\,\text{m}^2$ of glazing. This is less than likely ventilation losses but not negligibly so.[10]

The value of m depends on the values selected for h_i and h_e. Probably the least secure factor is the choice of the convective coefficient h_c which varies over a window surface because of frames, etc. Simple observation shows that condensation does not take place uniformly over the surface.

The associated condensation heat flow is

$$q = 0.0454(p_i - \text{svp}(T_c)). \tag{8.13}$$

This can be compared with expressions for the heat loss due to evaporation from a swimming pool surface at T_c due to forced convection across its surface into air where the vapour pressure is p_i. When in active use,

$$q = (0.089 + 0.078V)(p_i - \text{svp}(T_c)), \tag{8.14a}$$

where V is the air velocity (m/s) at the water surface.[11] Smith *et al.* (1998) give an expression for a quiet pool, with lower coefficients:

$$q = (0.064 + 0.064V)(p_i - \text{svp}(T_c)). \tag{8.14b}$$

Evaporation rates from a pool surface into low-speed air currents found by several workers are listed by Pauker *et al.* (1995). Their own study led to the expression

$$J = a(p_w - p_i)^b \pm 16\%, \tag{8.15}$$

where J is the evaporation rate ($\text{g/m}^2\,\text{h}$), p_w (kPa) is the vapour pressure at the water surface, p_i (kPa) is the vapour pressure in ambient air,

$$a = 74.0 + 97.97V + 24.91V^2 \tag{8.16a}$$

and

$$b = 1.22 - 0.19V + 0.038V^2. \tag{8.16b}$$

V is air speed (m/s). The associated heat flow in SI units is

$$q = (2.47 \times 10^6/3.6 \times 10^6) \times 10^{-3b} a(p_w - p_i)^b \pm 16\%. \tag{8.17}$$

With $V = 0$ we have $q = 0.011(p_w - p_i)^b$, which roughly accords with the above values since the pressure difference is raised to a positive index. These values relate to fresh water. They point out that vapour pressure is lowered by the presence of salts and minerals. Tang and Etzion (2004) report a study on evaporation rates at a free water surface and find a value of $b = 0.82$.

[10] With these values, the coefficient for mass transfer in (8.9) is $1.8 \times 10^{-8}\,\text{kg/(s m}^2\,\text{Pa)}$. Quoted values for the rate of evaporation at a water surface at low temperatures are around $2 \times 10^{-8}\,\text{kg/(s m}^2\,\text{Pa)}$.

[11] Conversion from imperial units is based on $1\,\text{Btu/h ft}^2 \equiv 3.155\,\text{W/m}^2$, $1\,\text{ft/min} \equiv 5.08 \times 10^{-3}\,\text{m/s}$ and 1 in Hg $\equiv$ 3386 Pa.

A comparable problem is the evaporation of moisture from the exposed earth of a crawl space beneath a ground floor. Kurnitski (2000) and Kurnitski and Matilainen (2000) report rates of 2.4–4.9 g/m^2 h, depending on the rate of ventilation (0.5 to 3 air changes per hour) taking place due to pressure differences of around 1 Pa. The rate could be reduced to almost zero using a suitable cover.

Sparrow *et al.* (1983) report a study of the evaporative loss of water from small circular pans into still air. They note that since the temperature of the water surface is lower than ambient, the temperature gradient is directed downward while the vapour concentration gradient is upward. The height of the side wall had a significant effect in restraining the slow flow of cooled air radially outward.

The basic process under discussion is that of liquid water evaporating into a space with vapour pressure p_v in the absence of air. For this situation, Eames *et al.* (1997: equation 4) cite an expression for the rate of evaporation of water due to Knudsen (1950), and also in Roberts (1940: 178), from which the associated heat flux is given as

$$q = \frac{2\varepsilon}{2-\varepsilon}\left(\frac{ML^2}{2\pi RT_s}\right)^{1/2}(\text{svp}(T_s) - p_v). \tag{8.18}$$

The evaporation coefficient ε = (experimental rate of evaporation)/(theoretical rate of evaporation). They survey results of 15 previous experiments which suggest that ε should lie between 0.01 and 1. With $\varepsilon = 0.1$, $T_s = 283$ K and $M = 18$ kg for water, the value of q is around $300(\text{svp}(T_s) - p_v)$W/m^2, which is much larger than the values noted above. The actual rate depends on the boundary layer at the surface and probably on a thin boundary layer with a temperature gradient in the surface of the water.

8.5 CONDENSATION IN A WALL: SIMPLE MODEL

The effect of placing insulation within a wall is to bring the temperature of the outer layer to a low value in cold conditions and this may lead to condensation within the wall. The situation is illustrated in the continuation of Table 8.3, where we take interfacial temperatures for the lower ambient temperature of 5°C (row 4), together with a relative humidity of 90%. The corresponding svp values are given in row 5, and proceeding as previously, the notional vapour pressure distribution is given in row 6. We see that at node 4, the interface between the insulation and the outer brick, the vapour pressure exceeds the svp value there. This cannot be; condensation must take place there and the vapour pressure will have the svp value of 930 Pa. In the case of the glass, the vapour flow was sufficient to raise the temperature a little at the point of condensation, but in the current case the vapour flow and latent heat yields are too small to affect the temperature profile significantly.

The vapour flow now falls into two independent paths. Up to the point of condensation

$$m = (1490 - 930)/((0.6 + 3.0 + 0.45) \times 10^9) = 138 \times 10^{-9}\,\text{kg/m}^2\text{s},$$

(resistance values in Table 8.2). Beyond the point of condensation

$$m = (930 - 784)/(5.25 \times 10^9) = 28 \times 10^{-9}\,\text{kg/m}^2\text{s}.$$

The difference between these values, $110 \times 10^{-9}\,kg/m^2s$, represents the rate of accumulation of water, amounting to a little less than 0.01 mm per day. Hens remarks (Hens and Fatin 1995) that one day's driving rain might introduce more moisture into the outer leaf of a cavity wall than condensation by diffusion over a whole winter season. The condensation process leads to a heat gain within the wall of less than $0.3\,W/m^2$, which is small compared with sensible loss in these conditions of about $5\,W/m^2$. Condensation could be avoided using a vapour barrier near the interior surface. In this example the change imposed to bring about condensation was a decrease in outdoor temperature. (Figure 6 of Simpson *et al.* (1991) however shows a marked change in the distribution of relative humidity through a wall due to an increase in inside relative humidity.)

Even though the values for vapour resistivity may not be known reliably, the above procedure has proved popular with designers since it is easy to conduct. (If the wall is drawn so that the layers are represented proportional to their vapour resistivities, the vapour pressure (vp) gradient in absence of condensation becomes a straight line lying below the saturated vapour pressure (svp) profile. If vp intersects svp, it must be redrawn with segments touching the svp curve; the points of contact indicates the locations of condensation and evaporation.) Letherman (1989) has described how the procedure may be used to synthesise a construction that avoids condensation in given conditions of temperature and humidity.

Merrill and TenWolde (1989) describe a study of severe damage due to condensation in walls and roofs which took place in Wisconsin in the mid 1980s. The walls consisted of interior gypsum board, 90-mm of rock wool bat insulation and an outer construction of plywood sheathing, heavy asphalt-coated building paper and hardboard lapped siding. It appeared that 'condensation and decay of the sheathing ... were primarily caused by an unfortunate combination of high indoor humidity and cold weather'. The homes were exceptionally airtight and the greater the number of occupants per unit area of floor, the greater the moisture-related problems. The residents suffered more often from respiratory problems.

Since the vapour pressure of any water within the wall has the value of the local saturated pressure (which follows from the temperature profile) and will normally be higher than the actual vapour pressure of the room or of ambient, it might be assumed that moisture which condensed and was accumulated in winter might evaporate and be lost again during the summer. Pedersen (1992) has presented a sample calculation to examine the possibility, based on a flat roof consisting from top to bottom of roofing felt, a thin layer of wood fibreboard, expanded polystyrene and concrete. A calculation using monthly means for outdoor temperature and humidity suggested that the downward flux of moisture in summer was insufficient to remove the winter condensation so that the roof would accumulate moisture over a year. If however an hour-by-hour analysis was made, taking account of large possible solar gains during the summer, the winter accumulation would be lost. The hour-by-hour calculation leads to a surprisingly high value of vapour pressure in the fibreboard during July. Bellia and Minchiello (2003) have recently reported a similar steady-state study based on a current European Standard. Künzel (1998) has described a 'smart vapor retarder' which is more permeable in summer than winter.

The situation suggested above is typical of the north European climate with cool winters and temperate summers. The converse problem is encountered in the hot humid conditions of parts of the southern United States, particularly with concrete masonry construction. Interior cooling, together with the effectively impermeable vinyl wall coverings used in hotels, leads to high moisture levels and possible mould growth. Fungal spores may enter the room, leading to a musty odour. This situation has been described by Shakun (1992). Hosni *et al.* (1999a) report and amplify recommendations of Burch (1993) to the effect that in these circumstances the material used as an interior wall finish should have a permeance of at least 5 perms[12] (2.8×10^{-10} kg/s m^2 Pa), an exterior vapour retarder should be used, infiltration of outside air into the wall construction should be minimised, construction using moist building materials should be avoided, and the indoor air temperature should not be lowered below the design set-point temperature. (See too Burch and TenWolde 1993).

Though easy to use, the procedure is oversimplified. It might be taken to suggest that the region of condensation was confined to a small thickness of the wall which becomes fully wetted, thus blocking the flow of vapour, but this is not so: the wetted thickness may occupy a sizeable fraction of the wall thickness which varies in time; an outward vapour transfer continues through it. Taylor (1993) has included the effect of a finite-thickness film. The model can include a cavity but does not take account of any cavity ventilation; this is treated by Cunningham (1983, 1984). Vapour diffusion does in fact appear to be the principal mechanism for moisture transport in walls but there are others which are discussed below.

Even then, theory assumes tacitly that condensation only occurs when the relative humidity becomes 100% but this has been questioned. Galbraith *et al.* (1993b) report that in hygroscopic materials the transport of moisture in liquid form in fact takes place at humidities of 60–70%. They assume a form of flow as

$$\begin{aligned} q_{\text{moisture}} &= \text{vapour flow} + \text{liquid flow} \\ &= -(D_1{}^*\tau + D_2{}^*\phi^m)\Delta p_1. \end{aligned} \tag{8.19}$$

$D_1{}^*(\sim 10^{-12}\text{ s})$ and $D_2{}^*(\sim 10^{-13}\text{ s})$ are diffusion coefficients, found by experiment, for water in vapour and liquid forms; $\tau = V_1/(V_1 + V_2)$ where V_1 is the void space within the material free of liquid and V_2 is the volume of liquid; ϕ is the relative humidity; m is a coefficient of order 9 for particle board, plywood and plasterboard and 1 for expanded polystyrene; p_1 is the vapour pressure. Values for vapour resistivity are available for a range of materials, but the information needed to implement this procedure is limited.

It may be noted that the simple one-dimensional analyses take no account of the possible removal of condensate by gravity, of rain on outside surfaces or of rising damp.

8.6 CONDENSATION IN A WALL: MORE DETAILED MODELS

Various engineering endeavours – drying and humidification in chemical and food processing, moisture migration in soils as well as building climate control – lead to a need to analyse the processes of combined heat and mass transfer with phase change in a porous medium. The complex of factors involved has been set out usefully by Eckert and Faghri (1980):

> The structure of the solid matrix varies widely in shape. It may, for instance, be composed of cells, fibres or grains. There is, in general, a distribution of void sizes

[12] 1 perm ≡ 1 grain/(hr ft^2 in Hg) = 5.72×10^{-11} kg/(s m^2 Pa); 1 grain = 1/7000 lb.

and the structure may also be locally irregular. Energy transport in such a medium occurs by conduction in all of the phases as well as by convection with those phases which are able to move. Mass transfer occurs within the voids of the medium. In an unsaturated state these voids are partially filled with a liquid, whereas the rest of the voids contains some gas [most frequently water and air]. Evaporation or condensation occurs at the interface between the water and the air so that the air is mixed with water vapor. A flow of the mixture of air and vapor may be caused by external forces, for instance, by an imposed pressure difference. The vapor will also move relative to the gas by diffusion from regions where the partial pressure of the vapor is higher to those where it is lower. The partial pressure of vapor at the interface to the liquid is determined by the sorption isotherm, which makes it dependent on moisture content as well as on temperature. The saturation pressure is also different from that on a plane one and is influenced by the presence of air. The flow of liquid is caused by external forces, like imposed pressure difference, gravity, and internal forces, like capillary, intermolecular and osmotic forces.

Rossen and Hayakawa (1977) similarly summarise the mechanisms which might in principle be taken account of: (i) liquid movement due to capillary forces; (ii) liquid diffusion due to concentration gradients; (iii) liquid diffusion at pore surfaces due to gradients at the surface; (iv) gravity-induced liquid moisture flow; (v) vapour diffusion within the partly air-filled pores due to vapour pressure gradients, or capillary condensation mechanism; (vi) vapour flow due to differences in total pressure as noted above. Further information on the basics of moisture movement is given by Wong and Wang (1990), who distinguish between capillary-porous, hygroscopic-porous and non-porous media.

A general formulation to describe the simultaneous transport in a material of heat, liquid and vapour is provided by Onsager (1931a: equation 2.3)[13]:

$$\text{heat [W/m}^2\text{]} \qquad J_q = L_{qq}X_q + L_{ql}X_l + L_{qv}X_v; \tag{8.20a}$$

$$\text{liquid [kg/m}^2\text{s]} \qquad J_l = L_{lq}X_q + L_{ll}X_l + L_{lv}X_v; \tag{8.20b}$$

$$\text{vapour [kg/m}^2\text{s]} \qquad J_v = L_{vq}X_q + L_{vl}X_l + L_{vv}X_v. \tag{8.20c}$$

The quantities L_{ij} are called kinetic coefficients and there exists a reciprocity relationship between the interactive coefficients such that $L_{lq} = L_{ql}$, $L_{vq} = L_{qv}$ and $L_{vl} = L_{lv}$. Luikov (1966: 16) remarks that from the macroscopic viewpoint, the relation is an axiom. The coefficients would be zero if the processes proceeded independently. The rate of increase of entropy (ibid.: equation 1.56) is

$$\mathrm{d}S/\mathrm{d}t = J_q X_q + J_l X_l + J_v X_v \tag{8.21}$$

and is always positive. According to the theory outlined by Onsager, it is the *rate of increase of the entropy* that plays the role of the thermal potential so that the factor $1/T_{abs}$

[13]Mitrovic (1997) remarks that Onsager is considered to be the founder of a theory of transport processes occurring in systems undergoing irreversible transformations but points out that it might date from 1911. He draws attention to the relationship between expressions for diffusion processes given in the mid nineteenth century by Maxwell, Stefan and Fick.

is included with the temperature gradient. The thermal driving force X_q in one dimension is $-(1/T^2)(\partial T/\partial x)$ (ibid.: equation 1.59) or $-(1/T)(\partial T/\partial x)$ (ibid.: equation 1.64). The moisture driving forces X_l and X_v (taken below to be the same) are $-(\partial/\partial x)(\mu/T)$ (ibid.: equation 1.60) or $-T(\partial/\partial x)(\mu/T)$ (ibid.: equation 1.64), where the isobaric-isothermal potential of mass transfer μ (ibid.: equation 1.53) is

$$\mu = A(T) + RT \ \ln(p_v/(p_a + p_v)). \tag{8.22}$$

$A(T)$ is a coefficient, p_a and p_v are the partial pressures of dry air and vapour, respectively.

The coefficients are described as phenomenological since they make no reference to the various physical mechanisms (vapour diffusion, etc.) which determine the details of the process. They can in principle be determined as such, and Pierce and Benner (1986) describe their determination for moisture movement driven by a temperature difference (11 to 30 K) in a sample of fibreglass 5 cm thick, finding for example $L_{vq}/T^2 \approx 2 \times 10^{-9}$ kg/m s K and $(L_{vl} + L_{vv})R \approx 2.5 \times 10^{-7}$ kg/m s. (R is described as the universal gas constant but in effect has units J/kg K.) They provide expressions relating the L_{ij} values to various physical quantities (relative humidity, liquid and vapour densities) and to the liquid and vapour mass transport conductivities k_l and k_v, reporting values of $J_v \approx 10^{-5}$ kg/m^2 s, $k_l \approx 10^{-15}$ s and k_v close to 2.5×10^{-5} m^2/s. Although the formal structure set up by Onsager is simple, the various physical mechanisms involved in heat and mass transfer in solids lead to considerable complication in implementing them, as equations in Luikov (1975) show. The individual studies sketched below are concerned with situations where simplifications can be made.

Henry (1939) established the time-varying equations for simultaneous heat and mass transfer with phase change in a material and other formulations are given by Luikov (1966, 1975). They may be difficult to implement realistically because of the possibly complicated nature of the pore structure of the materials; see for example Khan (2002) on the properties of concrete. A large literature has built up, particularly since about 1980, on observational and numerical studies of moisture movement in walls; Wijeysundera and Wilson (1994) cite a selection. A short account will be given here of moisture behaviour in two elements present in wall constructions: a layer of a porous material such as fibreglass, a non-storage element whose thermal and vapour transfer properties are near to those of air, and in sandstone which provides storage and where the void volume is small compared with the total. They deal with quasi-steady state conditions when condensation or evaporation takes place but when the thermal storage term $\rho c_p \partial T/\partial t$ (12.1) can be neglected. Its inclusion, together with a corresponding term for the moisture, considerably complicates the analysis; some account is given in Section 17.11.

The condensation process is associated with a yield of some 2.4 MJ/kg at wall temperatures but when moisture is first absorbed at relative humidities less 100% on dry glass-fibre threads, typically about 10 μm thick, to form a single-molecule layer (Ozaki *et al.* 2001), the yield is about four times this value (Tao *et al.* 1992b). With further deposition, the yield quickly falls to the normal value, reaching it at a moisture density of about 0.0015 kg moisture/kg dry material, corresponding to a relative humidity of 25%. The capacity of fibreglass to store moisture in this way is small, with a volume fraction less than 0.0009 at 90% relative humidity ϕ, but the adsorption and desorption curves are very different: in the diagram cited by Simonson *et al.* (1993), the desorption value is more then five times larger than the adsorption value.

8.6.1 Condensation in Glass Fibre

Glass fibre is widely used as an insulating material. In the dry state its conductivity is basically that of air but it depends on its density in a non-linear manner (Batty *et al.* 1981). Consider the heat flow across a thickness X limited by solid surfaces. When its density is very low, the heat transfer is near that across a cavity, and due to radiation and convection. With increase in density the radiant component decreases sharply due to absorption of radiation in the fibres and there is a slight fall in convective transfer. For the effect of fibre orientation on thermal radiation in fibrous media, see Lee (1989). With further increase in density, solid material progressively replaces air, increasing the conductive transfer. Thus the net conductivity has a minimum which the results of Batty *et al.* Figure 4 suggest to be about 0.03 W/m K at a bulk density of around 60–80 kg/m^3.

The problems with use of fibreglass have been set out clearly by Mitchell *et al.* (1995). Thick layers of the material are used to conserve energy in buildings and refrigerated spaces but if the convection of air (of order 1 mm/s) and moisture diffusion through a roof insulated with fibreglass are not prevented, the accumulation of moisture may halve the thermal resistance of the layer and lead to mould growth, degradation of wood and metallic corrosion. Vapour which condenses in the insulation in walls may evaporate later but condensation in the insulation around chilled water pipes is likely to be permanent; a vapour barrier is essential.

Some facts about moisture in fibrous insulation are given by Wijeysundera and Hawlader (1992): 1 m^3 of material can contain up to 300 kg of moisture and the rates of condensation they illustrate are around 1 kg/m^3 h. The net thermal conductivity $\lambda(\theta)$ depends on the mass of liquid per unit volume of the dry slab θ (kg/m^3) and on the thermal conductivities of the dry slab λ_d and of water λ_w. When the fibres tend to lie in planes and have the same direction, $\lambda(\theta)$ depends on the direction of the thermal gradient. They cite expressions due to Batty *et al.* (1981). For heat flow perpendicular to the plane of the fibres,

$$\lambda(\theta) = \frac{\lambda_d \rho(\lambda_w - \lambda_w \varepsilon)}{\rho(\lambda_w - \lambda_w \varepsilon) - (\lambda_w - \lambda_d)\theta}, \tag{8.23a}$$

where ε is the fibre volume fraction in the dry slab and ρ is the density of water. For heat flow in the plane of the bead-covered fibres,

$$\lambda(\theta) = \frac{\lambda_d[\rho(\lambda_w + 2\lambda_d) + 2(\lambda_w - \lambda_d)\theta]}{\rho(\lambda_w + 2\lambda_d) - (\lambda_w - \lambda_d)\theta}. \tag{8.23b}$$

It is of interest to present a qualitative account of the model studied by Wijeysundera *et al.* (1989, 1996). They considered the vertical transfer of moisture from a warm, humid region upward into a horizontal layer of insulation whose upper surface was impermeable to moisture. The temperature of the upper surface was taken to fall suddenly to below the dew point of the space below. A cooling wave is propagated through the slab. Condensation takes place at the cold surface, lowering the vapour pressure to the saturated vapour pressure for that temperature. The authors supposed that the transport process fell into four stages. (i) There is a brief initial stage of 30 min, during which temperature and vapour concentration gradients are established in the insulation. The vapour pressure profile reaches its steady-state condition faster than temperature. (ii) These fields remain constant; vapour diffuses and condenses to form a wet region near the upper surface but its density is insufficient to affect the transport significantly and the wet/dry interface remains stationary. (iii) When the liquid accumulation at the cold surface exceeds some critical value, a liquid/wet interface is formed which starts to move down due to the generated liquid pressure toward the wet/dry interface; the insulation now has areas where it is liquid-filled, wet, and dry. (iv) After about 80 h the liquid front moves into the dry region and eventually reaches the exposed surface. The moisture content of the insulation affects the thermal and moisture conductivity to varying degrees.

Using transport equations similar to those in Section 17.11, the authors develop this model and compare its predictions with experimental values for a slab thickness of 6–7 cm. For the most part, good agreement was found for temperature distribution, heat flux upward from the cold surface (around 38 W/m^2 for a 26 K temperature difference), and total moisture gain with values up to 300 kg/m^3. (The heat flow was partly due to sensible flow, which increased a little with time, and partly due to the latent heat from condensation, which decreased a little; the total remained roughly constant.) Good agreement was found too for the moisture distribution up to 70 h; after that, the observed liquid content at the exposed surface was larger than its calculated value. The study provides a useful test of the capabilities of the transport equations underlying the model. It does not offer practical guidance however since walls and roofs are not usually intended to be impermeable to moisture. A similar situation has been studied by Murata (1995) but in which the cool surface is vertical. Temperature and vapour concentration gradients establish themselves; condensation starts and continues until the pendular drops coalesce but then they move due to surface tension and gravity. Murata's analysis is based on Whitaker's theory (1977)

It was well known by the 1970s that if a layer of porous insulation separated a warm and a cool environment, neither of which is saturated, a wet zone might develop in the insulation, flanked by two dry zones; transport of vapour might be by diffusion or by slow convection, and its effect was to reduce the thermal resistance of the insulation. Early work has been listed by Ogniewicz and Tien (1981). The problems of heat and mass transfer in fibreglass has been studied in detail by Motakef and El-Masri (1986) and Shapiro and Motakef (1990) and their analysis is summarised here, including the transport equations. They consider a slab of fibreglass, one side of which is exposed to typical room temperature and humidity and the other to corresponding ambient quantities. No account is taken of any sorption characteristics the fibreglass may have. Figure 8.2a shows temperature and dew point profiles for the situation where no condensation takes place.[14] Suppose now that the inside humidity is sufficient to lead to condensation within the slab. If the slab is initially at room conditions throughout and its one surface is then exposed to ambient, the authors' model supposes a sequence of events consisting of an initial transient reaction, a first steady-state regime with a central region of condensation flanked by dry zones, a further transient state, leading to a further and final steady regime of condensation.

In each of the steady-state regimes, the heat and moisture flows in the dry regions are determined by the Fourier and Fick laws respectively and are independent. In the wet region, condensation takes place, denoted by Γ_x kg/s per unit cross-sectional area, per unit distance in the direction of flow (i.e. kg/s m^3) and varies with position. There is a corresponding heat input of $L\Gamma_x$ which, as Henry (1939) pointed out, will diffuse through

[14]The temperature and vapour pressure profiles can be reduced to a common basis by expressing the vapour profile as a dew point temperature as in Figure 8.2a, or by expressing the temperature profile as a saturated vapour pressure distribution as in row 5 of Table 8.3.

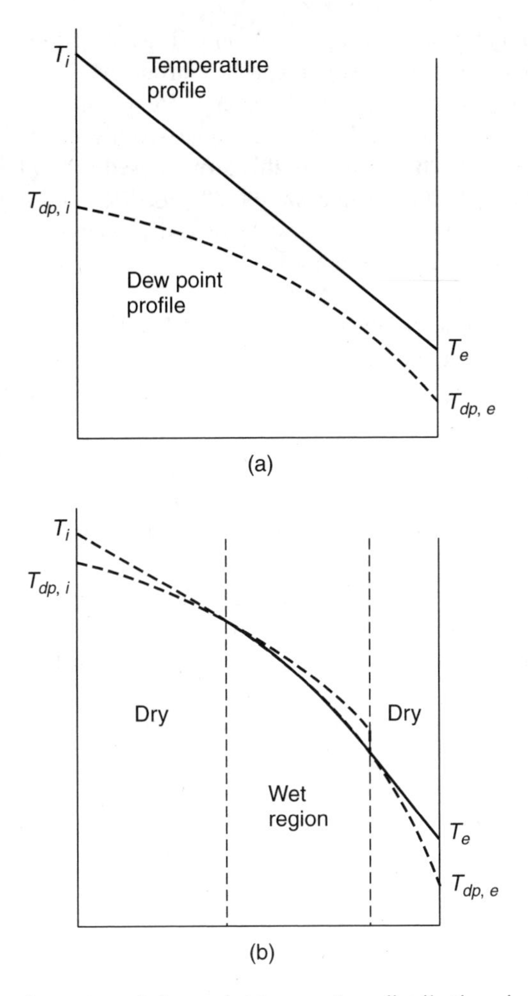

Figure 8.2 (a) Thermodynamic and dew point temperature distributions in a porous slab without condensation. (b) Temperature distribution with higher humidity inside, leading to a first steady-state region of condensation

the medium and will affect the extent to which the medium can absorb vapour. The heat flow equation is

$$\lambda \frac{\mathrm{d}^2 T_x}{\mathrm{d}x^2} + L\Gamma_x = 0 \tag{8.24}$$

and the moisture flow equation is

$$\frac{M\tau D}{RT}\frac{\mathrm{d}^2 p_x}{\mathrm{d}x^2} - \Gamma_x = 0. \tag{8.25}$$

In the initial and further transient states, a term of type $\partial^*/\partial t$ is needed on the right-hand side of these equations. The non-steady state is considered in Section 17.11.

For very porous materials, the values of λ and D are near those for air.[15] For denser materials, the dimensionless tortuosity τ can be introduced. If Γ_x is non-zero, the temperature profile becomes curved. Since the water content is increasing, the regime cannot strictly be described as steady and the term 'quasi-steady state' is usually used. It is sometimes described too as 'transient' but this term is used here when the right-hand side of (8.24) is non-zero, having the value $\rho c_p \partial T_x/\partial t$, as discussed in Chapter 12. Thermal 'transient' changes in walls have a timescale of order hours, and for an enclosure the timescale is up to several days (Figure 14.4). The timescale for the drying out of walls (the quasi-steady state) may amount to months.

In Motakef's first regime, when once established, the boundaries of the wet region do not move. Vapour is supposed to condense into small stationary separate globules (the 'pendular' state[16]) and to accumulate steadily so that the continuity equation for the liquid is

$$\rho_w \, \mathrm{d}w_x/\mathrm{d}t = \Gamma_x, \tag{8.26}$$

where ρ_w is the density of water and w_x is the volume of water per volume of the matrix.

In this state the rate of condensation Γ_x increases from zero at the inner edge of the wet zone to the outer edge, where it falls abruptly to zero (Figure 8.2b). The condensation rate varies as the cube of the temperature difference across the wet zone. Wyrwal and Marynowicz (2002) have performed a similar calculation for a wall of 75 mm glass wool inside and 150 mm of light concrete outside, with surface finishes and suitable film coefficients. Values of temperature and humidity assumed were 22°C and 60% relative humidity inside and −3°C and 90% RH outside. They found a wet zone of 35 mm in the glass wool, with bounding temperatures of 9.8°C and 0.5°C, terminating at the interface with the concrete where the condensation rate was 3.5×10^{-5} kg/m^{3}s (from their equation, their numerical value is wrong).

Eventually the droplets increase in size to such an extent that they coalesce. Values quoted by Wyrwal and Marynowicz indicate that for glass wool the critical value of w_x is 0.95. Surface tension forces now draw water into the dry regions, so the wet region gets bigger. After a further transient stage, a new steady state is set up. The distribution of Γ_x is now roughly parabolic, rising at the inner edge and falling steadily toward the outer edge. The water distribution is similarly parabolic with its maximum rather nearer the cold edge. It has become mobile, diffusing inwards and outwards and evaporating again at the two boundaries of the wet zone. In this state there is no further accumulation of condensate: the heat flow leaving the outer dry region is equal to the flow into the inner region; similarly, the inward and outward vapour flows are equal. Thus the heat flow across the slab is independent of the condensation process within it.

If a sudden change in temperature is imposed at the surfaces of a (dry) slab of thickness X, the effects of the change are virtually complete after a time lapse of about $5\rho c_p X^2/(\pi^2\lambda)$ (Section 12.21), about 0.1 h for a 40 mm thickness of fibreglass. As values

[15] The theory omits a temperature gradient, another factor which affects moisture flow. There is a temperature gradient (the Soret effect) in addition to a vapour pressure gradient, but it is small (Eckert and Drake 1972: 716).

[16] This is in contrast to the 'funicular' state, where there are continuous threads of moisture in the pores. Liesen and Pedersen (1999) quote Harmathy (1969): 'The liquid phase flow is never completely extinct, but there is ample evidence that in the pendular state, the mobility of moisture in the liquid phase is so small that in most practical problems it need not be considered.' Most building walls are in the pendular state.

given by Shapiro and Motakef show, when phase change is involved, timescales of tens of hours are found.

Moisture movement in fibreglass has been studied by a number of workers; see for example Tao *et al.* (1991, 1992a). Later Simonson *et al.* (1996) examined condensation in fibreglass insulation, performing simulations and finding satisfactory agreement with observed values in a set of five slabs, each 2 cm thick, cooled at the lower surface to −20°C.

The possibility of slow bulk air movement through a wall was mentioned above: see (3.11). It may result from wind pressure, stack effect or air circulation fans within the building and its effect has been studied by Vafai and Sarker (1986). The ratio of the convected flow $\rho c_p V(T_i - T_o)$ to the conducted flow $(\lambda/L)(T_i - T_o)$ through a thickness of material L is defined as the Peclet number Pe and is a major variable in this study. According to Tien and Vafai (1990), infiltration through breaks in otherwise impermeable boundaries may indeed be the dominant mechanism for heat and mass transfer. Exfiltration leads to the accumulation of condensation and frost, infiltration to its removal and possible deposition elsewhere. Mitchell *et al.* (1995) have examined both flow conditions through a 90 mm slab of fibreglass with inside and outside temperatures of +20 and −20°C, respectively. To explain their temperature–time observations at positions through the slab, a distinction had to be made between the heat yield to adsorption and the heat yield to condensation. A high exfiltration rate at high relative humidity led to a large accumulation of condensation on the cold side, as expected. A low infiltration rate however led to a relatively high rate of drying; a high infiltration rate of cold air maintains the cold region colder than does a low rate. The authors conducted a parallel model study and demonstrated broad agreement.

8.6.2 The Sorption Characteristic for Capillary-Porous Materials

The above analyses can be regarded as an extension to the classical vapour flow solution, now taking into account the adsorption/condensation process. However, it does not consider the ability of most dense building materials to absorb and store moisture. A material transmits a heat flow imposed by a temperature gradient together with a flow of moisture due to a gradient in vapour pressure; there are parallel quantities for thermal and moisture storage. The heat stored is described by $\rho c_p \Delta T$ (J/m^3), where T is measured on a temperature scale of arbitrary zero, and the associated change in heat flow is described by the local value of $\rho c_p \partial T/\partial T$. The value of ρc_p for dense materials varies little with moisture content or temperature (unlike the λ value), but somewhat nominal values normally have to be accepted in design calculations. The role of thermal storage is studied later in the book. The moisture content of a material can similarly be expressed in kilograms of moisture per unit volume (w say) or per unit mass of dry material. However, moisture content is not proportional to its associated vapour pressure; it depends on *relative humidity* ϕ, which has values between 0 and 1, often expressed as a percentage. The moisture content may be almost proportional to ϕ over part of its range but it increases sharply as ϕ approaches 1. Moisture content also decreases a little with temperature. While the heat content (J/m^3) of a material cannot change in steady-state conditions, its moisture content may increase in virtually steady conditions due to condensation of vapour and this will be discussed later in this chapter.

Sasaki *et al.* (1987) present some information about the uptake of water in sands, noting that it may be contained as bound, capillary and free water; $p\%$ denotes the proportion of space between the dry grains of sand. Their investigation led to a number of regimes:

- At $p = 0$ sand consists of dry sand and air alone.
- Up to $p = 9.6$ water is bound to the sand particles and at this value small menisci appear (wedge water) and capillary action begins. The apparent thermal diffusivity and apparent conductivity increase rapidly.
- At $p \sim 20$ the diffusivity has reached about three times its $p = 0$ value and now declines slowly. A number of results for apparent conductivity are given. All have increased by a factor of 4 or more and continue to rise, but more slowly.
- At $p = 19.5$ the capillary water forms continuous layers throughout the sand.
- At $p = 40.9$ free water starts to accumulate.
- At $p = 93.9$ the air is almost entirely replaced by water.

These regimes are brought about by the externally imposed relative humidity, as indicated in Figure 2.2. The relation between moisture absorbed (in units say of kg moisture/kg dry material) and relative humidity ϕ for some material is described as its sorption isotherm (Figure 8.3). Roughly speaking, the water content is proportional to ϕ up $\phi = 0.6$ approximately and then increases until it is capillary saturated; more water may be contained under pressure. For this situation, Pel *et al.* (1996) explain three regimes that can be distinguished: 'At high moisture contents the moisture transport is dominated by liquid transport. ... With decreasing moisture content the large pores will be drained and will therefore no longer contribute to liquid transport. [Consequently,] the moisture diffusivity will decrease. Below a so-called critical moisture content, the water in the sample no longer forms a continuous phase. Hence the moisture has to be transported by vapour and this vapour will therefore be governed by the vapour pressure. For small moisture contents the moisture diffusivity starts to increase again.' Their Figure 5 shows this by plotting observed diffusivity against moisture content.

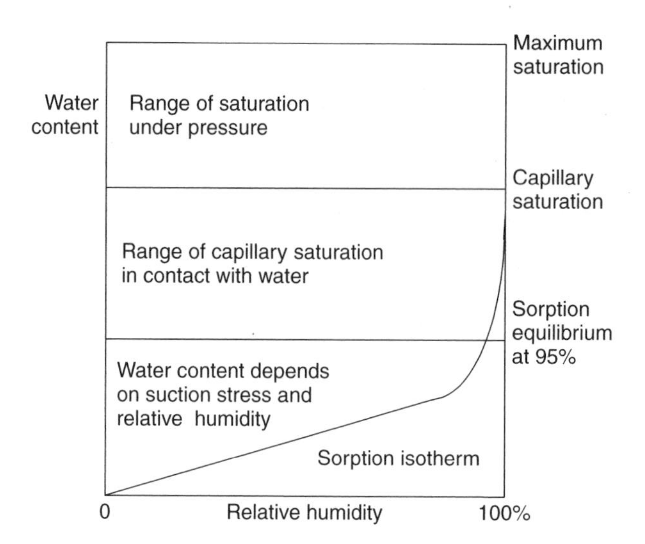

Figure 8.3 Representation of the water retention function of a sandstone specimen, a hygroscopic, capillary-active material

A series of assumptions that may be made in setting up a mathematical model for water vapour sorption at building surfaces are listed by Thomas and Burch (1990). These include one-dimensional heat and mass transfer, diffusion transfer proportional to concentration gradient, with neglect of thermal diffusion, and absence of free liquid, neglect of any hysteresis between adsorption and desorption data,[17] the heat and mass transfer analogy (8.11), constant density and specific heat of the dry porous material, constant specific heat of water as liquid and as vapour, the vapour being an ideal gas, paint coatings to have constant vapour resistance, negligible vapour storage and negligible thermal resistance. Figure 4 of Pel *et al.* (1996) shows that as relative humidity increases from 0 to 100%, the moisture content of a material (gypsum) can be considerably less than is found during decrease. (These authors used nuclear magnetic resonance methods to determine the water content. The radio frequency signal is tuned to a frequency appropriate to hydrogen.)

Richards *et al.* (1992) describe measurements of γ (kg/kg) against ϕ for a number of materials, mostly wood-based for which saturation values of γ are around 0.25 kg/kg. They were found by taking small samples (weighing about 1 g) and desiccating them above dry calcium chloride to 1.4% relative humidity, then placing them successively above a series of saturated salt solutions: the relative humidity above a saturated potassium carbonate solution at 25°C for example is $43.2 \pm 0.3\%$. There were eight such solutions, starting with lithium chloride at 11.3%. Above pure water it is 100%. The diffusion process is very slow: specimens took up to 6 weeks to reach equilibrium above the salt solutions. Taking samples to successively higher humidities is an *adsorption* process and the reverse is a *desorption* process. Values of γ(desorption) tended to be somewhat higher than values of γ(adsorption). An empirical equation for γ can be written in the form $\gamma = a_1\phi/[(1 + a_2\phi)(1 - a_3\phi)]$, which ensures the initial linearity of γ with ϕ. They write the relation for moisture diffusion in some layer in the wall as

$$\frac{\partial}{\partial x}\left\{\mu(T,\phi)\frac{\partial}{\partial x}[\phi P_g(T)]\right\} = \rho G(T,\phi)\frac{\partial \phi}{\partial t}, \tag{8.27}$$

where μ (kg/s m Pa) is the layer permeability, $P_g(T)$ (Pa) is the saturated vapour pressure at temperature T, ρ(kg/m^3) is the layer density and G (the specific moisture capacity) is the slope of the slope of the sorption isotherm function, strongly dependent on ϕ but only weakly on T. This form omits mention of any liquid water diffusion.

The same authors (Burch *et al.* 1992b) also report measurements of permeability. The specimen under test was sealed in the top of a cup containing one of the solutions. The cup was placed in a glass vessel containing a quantity of another solution, so that the mass transfer taking place due to the difference in vapour pressure in some range of relative humidity could be found by weighing the cup at weekly intervals. In processing the data, the authors took account of the two surface (or air layer) resistances. A single surface resistance was around 0.25×10^9 Pa/(kg/sm^2)), five times larger than the value found using the Lewis relationship. These tests however were carried out in isothermal conditions, so the h_c value entering the Lewis relation would be near zero. Permeabilities of between 3×10^{-13} and 10^{-10} kg/Pa s m were found. Most increase markedly with relative humidity,

[17]Suppose r_m is the radius of the part-spherical water meniscus in a narrow cylinder of radius r_c; then $r_m \geq r_c$ and its tangent at the wall makes an angle θ with the wall, $\theta \geq 0$. Ozaki *et al.* (2001) attribute the hysteresis effect to θ being larger during adsorption than desorption. The significance of the angle of contact is discussed by Wayner (1982).

which the authors attribute to water-bound diffusion acting in parallel to the diffusion of water vapour through the air-filled pore spaces of the material. Temperature has little effect. Permeability can be summarised as

$$\mu = \exp(A_0 + A_1\phi + A_2\phi^2), \tag{8.28}$$

where A_0 is between about -21 and -28.[18]

(Galbraith *et al.* (1998) noted that five equations had been set to express moisture permeability μ as a function of relative humidity ϕ using empirical constants A, B, C and D:

(1) $\mu = A + B\exp(C\phi)$, (2) $\mu = \exp(A + B\phi)$ (see footnote 18), (3) $\mu = A + B\phi^C$,

(4) $\mu = \exp(A + B\phi + C\phi^2)$, (5) $\mu = A + B\phi + C\exp(D\phi)$.

They cite appropriate values for A to D for each equation and for various materials: plasterboard, concrete, plywood and expanded polystyrene. All give values for the coefficient of determination r^2 (≤ 1) greater than 0.88. Forms 4 and 5 can show an initial decrease in μ (down to a minimum value around 40% relative humidity). Form 2 can only express an increase and seems the least successful.)

Since it may take up to several weeks to reach steady state at a given level of relative humidity, a comprehensive determination of the material's diffusion coefficients may be very time-consuming and Arfvidsson and Cunningham (2000) present a method by which the timescale can be much reduced using a transient method. The weights are noted of specimens subjected to steps of relative humidity (5 hours for sandstone and 24 hours for lightweight concrete) so this stage can be completed relatively quickly. The method requires knowledge however of the sorption curve, which is found from steady-state methods. For sandstone they find that for the transport of liquid water, D_w falls from about $1.5 \times 10^{-8}\,m^2/s$ at a moisture content w of about $3.7\,kg/m^3$ to about $0.1 \times 10^{-8}\,m^2/s$ at $w = 10\,kg/m^3$. For the transport of vapour, D_p falls from 3.2×10^{-11} s at $3.6\,kg/m^3$ to 1×10^{-11} s beyond $10\,kg/m^3$. Somewhat similar results are reported for the lightweight concrete though the vapour diffusion coefficient showed a small fall followed by a marked rise with increase in w.

8.6.3 Moisture Movement in Capillary-Porous Materials

Masmoudi and Prat (1991) provide two useful and simple figures to illustrate the grains of a porous material, initially saturated with water between the grains, and later in a drying process when only a thin film of water remains bound to the grains. Eckert and Faghri (1980) comment on an early study of moisture movement during a demonstration of such a drying process carried out by Krischer. He examined the behaviour of moist sand contained in a tube 50 cm long, initially at a uniform temperature of 20°C. Both ends were impervious to moisture and at the beginning of the experiment, the temperature of one end was raised to 70°C and maintained there for 5 months. (Presumably lateral

[18] Since μ has dimensions and $\exp(x)$ is dimensionless, the relation might be better expressed as $\mu = a_0\exp(A_1\phi + A_2\phi^2)$.

heat losses were small). The first decay time for this arrangement as a thermal system is $\rho c_p X^2/\pi^2\lambda$ – around 6 hours assuming typical values for sand – and a thermal steady state is reached at five times this value. But moisture diffusion is very much slower. With an initial (uniform) moisture content of 0.025 kg/kg sand, Krischer found that, after the 5 months, 0.325 m of the length of sand had dried out and the moisture was concentrated into the remaining 0.175 m with an almost linear concentration gradient having a value of 0.17 kg/kg at the cold end. With the higher initial concentration of 0.1 kg/kg, the final distribution was strictly linear between the hot and cold ends, with values of 0.005 and 0.20 kg/kg. Following an earlier study by Dinulescu and Eckert (1980), Eckert and Faghri (1980) performed moisture flow calculations relating to this study, finding that a period of 4 months was needed to approach the steady state. Their figures suggest a value for the moisture diffusion coefficient D of $1.08 \times 10^{-9}\,\text{m}^2/\text{s}$; the assumed values for sand give a value for κ of about $1 \times 10^{-6}\,\text{m}^2/\text{s}$, so D/κ, the Luikov number, is around 10^{-3}. The Luikov number describes the relative speed of the moisture and thermal diffusion processes. This study does not have direct bearing on moisture flow through walls since the tube was impervious to moisture flow at each end, but it shows very clearly the relative slowness of moisture movement. Kallel *et al.* (1993) report somewhat similar results in the drying of a mortar slab 20 cm thick: the time required to achieve equilibrium moisture content was around 46 days whereas thermal equilibrium was reached in 5–6 hours.

Some further insight into the movement of moisture in a dense material is gained from the model described by Chen and Pei (1989). Suppose that a layer of material with one impervious surface and one exposed surface is initially wet throughout. A stream of hot air with some known humidity starts a process of drying at the exposed surface. The initial rate of drying is constant. According to a two-zone model originally due to Luikov, the process can eventually be described by a model indicated in Figure 8.4,

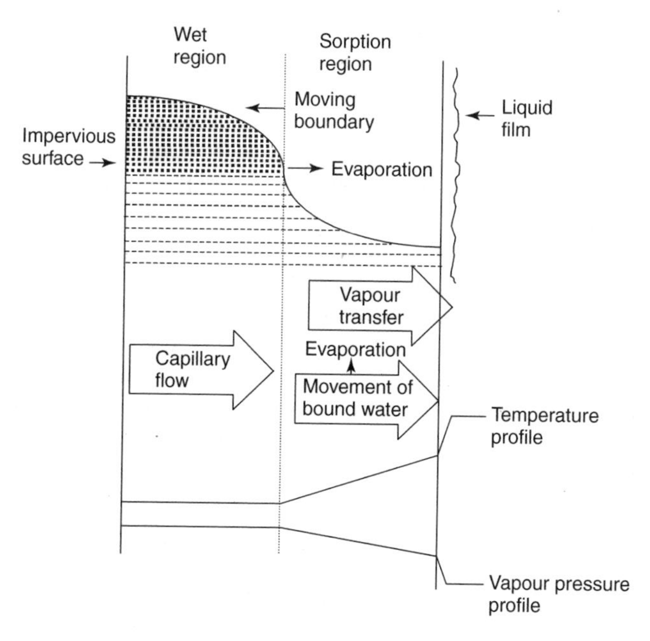

Figure 8.4 Later stage in the drying out of a dense material

in which various transport mechanisms are shown: capillary flow in the area still wet, evaporation at a receding front and vapour flow in the sorption region in which there is further evaporation of bound water. The authors state:

> [When] the surface moisture content reaches its maximum sorptive value, no free water exists. The surface temperature will rise rapidly, signaling the start of the second falling rate period, [for some materials there is an intervening period when there are wet and dry patches on the surface], during which a receding evaporation front often appears, dividing the system into two regions, the wet region and the sorption region. Inside the evaporation front, the material is wet, i.e., the voids contain free water and the main mechanism of moisture transfer is capillary flow. Outside the front, no free water exists. All water is in the sorptive or bound state and the main mechanisms of moisture transfer are movement of bound water and vapor transfer. Evaporation takes place at the front as well as in the whole sorption region, while vapor flows through the sorption region to the surface.

Dayan and Gluekler (1982) describe the retreat of the dry/wet interface in wet concrete due to intense heating at its surface. Drying of porous bodies has been studied too by Schadler and Kast (1987). Kaviany and Mittal (1987) describe ancillary studies on the drying of the wet region and Paláncz (1987) describes the progress of the evaporation front.

The authors set out the governing equations and solve them numerically for various situations where experimental values were available for comparison. The water content of a sample of brick, 5 cm thick, was reduced to about a quarter of its initial value after 16 h exposure to air at 80°C.

Moisture movement in building walls is more complicated than the situation this analysis addresses: temperature and humidity have to be specified at both surfaces and the variation outside may lead to condensation in winter and evaporation in summer. Wall transfer can be illustrated by the work of Künzel and Kiessl (1996), who describe a study on sandstone, a dense porous building material.[19] Its moisture characteristic is shown schematically in Figure 8.3. The relation between moisture content and relative humidity is suggested in Figure 2.2 and is shown as the sorption isotherm in Figure 8.3. A value of about 95% relative humidity leaves much of the void volume empty; it can only be fully saturated under pressure.[20]

These authors' model for vapour transfer is shown in Figure 8.5, where one of the micropores is idealised as a fine cylinder aligned in the direction of heat flow, left to right. It is assumed that the vapour pressure is higher inside than outside. Starting from dry conditions, vapour diffusing to the right moves into a sub-dew-point region and condenses into globules where it builds up. As in Motakef's model, the drops coalesce, leading to an inward transport of liquid. But in Motakef's model, condensation takes place on fibres of sparse density; here the condensation is within pores. Pores of many shapes and sizes are present and the net inward capillary transport may exceed the outward diffusion transport.

[19] Beck *et al.* (2003) investigated the moisture properties of limestone used for restoring historic buildings in France.

[20] The capillary tension p_c (having a negative value) and the relative humidity ϕ are related by Kelvin's law, $p_c = \rho_L(RT/M)\ln(\phi)$, where ρ_L is liquid water density and M its molecular weight. Häupl *et al.* (1997): Figure 1, illustrate numerical values.

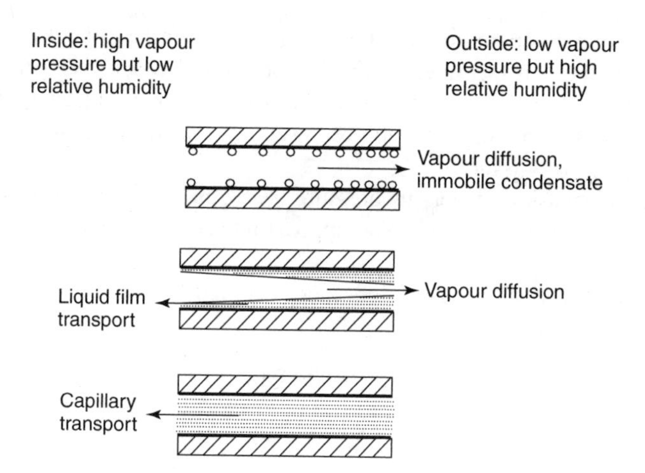

Figure 8.5 Idealised representation of the condensation process for moisture migration through a fine pore in heavy porous wall material

The authors tested these ideas by comparing computed and observed profiles of water content (kg/m^3) in a sandstone prism 25 cm long. The outer surface was exposed to various weather patterns: rainy, bright, etc. Equations similar to (8.20) and (8.21), but including storage terms of type $\partial^*/\partial t$ on the right-hand sides, were implemented using a finite difference model. The results show that, starting with the dry sample, the increase in moisture content shows no signs of levelling toward an equilibrium state after a period of 80 days. This timescale is much larger than the one found for fibreglass by Motakef since it takes account of the sorption isotherm of the material under investigation. The study demonstrates very good agreement between observed and calculated values for the total moisture content in the sample, together with profiles of the moisture density that follow four patterns of weather. The authors caution that a model cannot replace a field test since the model may omit some mechanism of major importance.

Although not directly relevant to moisture movement in walls, an earlier study by Udell (1985) examined the transport of heat, liquid water and its vapour through a column of compacted sand grains, 25 cm long and 5 cm in diameter, using distilled air-free water under the action of heat flux of around 1700 W/m^2. Three zones were established: a vapour zone in which there was a fall in temperature, a more or less isothermal two-phase zone 11–23 cm in length, and a liquid zone in which there was a further fall in temperature. The behaviour of the central zone was similar to that of a wick in a heat tube. He remarked that 'for thermodynamic equilibrium to exist between phases in such liquid-wetting media, the vapor must be slightly superheated as predicted by the Kelvin equation and the liquid will be in a stable superheated state due to interfacial tension effects'.

8.7 APPENDIX: THE SATURATED VAPOUR PRESSURE RELATION

The gradient of the saturated vapour pressure curve is given by the Clausius-Clapyron equation:

$$\frac{\Delta p}{\Delta T} = \frac{L}{T(v_v - v_l)}, \tag{8.29}$$

where T is absolute temperature, v_v and v_l are the volumes of 1 kg of water in the vapour and liquid phases respectively at that temperature and L is the corresponding latent heat.[21] Most of the heat required to vaporise the liquid is taken in freeing molecules from intermolecular forces; a small proportion represents the work done by the vapour against the externally imposed pressure.

For a detailed discussion of vapour pressure, see Chi *et al.* (1991). A simple form for the svp(T) relation can be found if some approximations are made:

- At the low vapour pressures associated with occupied conditions, v_l is small compared with v_v and can be neglected.
- In these conditions, the vapour obeys Boyle's law, so

$$v_v = 1/\rho_v = RT/Mp; \tag{8.30}$$

  see equation (2.7) with M referring to water.
- L decreases with temperature and if the specific heats $c_{p,l}$ and $c_{p,v}$ of water in the liquid and vapour phases, respectively, are taken to be constant, L_T at temperature T is related to L_0 at the reference temperature T_0 as

$$L_T = L_0 - (c_{p,l} - c_{p,v})(T - T_0) \quad \text{or} \quad L_0 - \Delta c_p(T - T_0), \tag{8.31}$$

  where $\Delta c_p = (c_{p,l} - c_{p,v})$.

So

$$\frac{\mathrm{d}p}{p} = \frac{ML\mathrm{d}T}{RT^2} = \frac{M}{R}\left(\frac{L_0 + \Delta c_p}{T^2} - \frac{\Delta c_p}{T}\right)\mathrm{d}T. \tag{8.32}$$

This is to be integrated between the reference state 0 and current state 1:

$$\ln\frac{p_1}{p_0} = \frac{M}{R}\left((L_0 + \Delta c_p T_0)\left(\frac{1}{T_0} - \frac{1}{T_1}\right) - \Delta c_p \ln\left(\frac{T_1}{T_0}\right)\right). \tag{8.33}$$

Example 8.3

Given data for 100°C, find the saturated vapour pressure at 20°C. The data are taken from Tables C2.1 and C2.2 of the 1986 *CIBSE Guide*.

At 100°C, $c_{p,l} = 4219$ J/kg K, $c_{p,v} = 2010$ J/kg K, so $\Delta c_p = 2209$ J/kg K and $L_{100} = 2.2257 \times 10^6$ J/kg. $(L_0 + \Delta c_p T_0) = 3.082 \times 10^6$ J/kg

So

$$\ln\frac{p_{20}}{p_{100}} = \frac{18}{8314}\left(3.082 \times 10^6\left(\frac{1}{373} - \frac{1}{293}\right) - 3.082 \times 10^6 \ln\left(\frac{293}{373}\right)\right) = -3.729.$$

[21] The relation is exact and is one of the earliest applications of the second law of thermodynamics. It is not derived here since the book makes no other application of the second law.

The saturated vapour pressure $p_{100} = 1.0133 \times 10^5$ Pa. So $p_{20} = 1.0133 \times 10^5 \times \exp(-3.729) = 2434$ Pa. This agrees with the tabled value. But the good agreement may be spurious. In the temperature range selected, $c_{p,l}$ falls from 4219 to 4178 J/kg K at 35°C before rising a little, and $c_{p,v}$ falls from 2010 to 1870 J/kg K at 20°C. The assumption of a constant value of Δc_p is not close.

The following expression is quoted by Hill and MacMillan (1988) for the saturated vapour pressure p_s at temperature T between the triple point and the critical point temperature T_c:

$$\ln(p_s/p_c) = (T_c/T)(a_1\tau + a_2\tau^{1.5} + a_3\tau^3 + a_4\tau^{3.5} + a_5\tau^4 + a_6\tau^{7.5}), \tag{8.34}$$

where $\tau = 1 - T/T_c$, $T_c = 647.14$ K, the critical pressure $p_c = 22.064$ MPa and

$$a_1 = -7.85823,\ a_2 = 1.83991,\ a_3 = -11.7811,\ a_4 = 22.6705,\ a_5 = -15.9393,$$
$$a_6 = 1.77516.$$

An accuracy of 20–40 ppm is claimed below 100°C and 200 ppm above it.

Values for saturated vapour pressure are given in Chapter 6 of the 1993 *ASHRAE Handbook of Fundamentals* and Book C of the *CIBSE Guide*. A simple expression, correct to a few percent, is given by Dayan and Gluekler (1982):

$$p(\text{kPa}) = 107 \times 10^{21} T_a^{-5} \exp(-7000/T_a) \qquad (273 < T_a < 450\ \text{K}). \tag{8.35}$$

Equation (8.33) gives the value of the saturated vapour pressure p_s (of order 10^3 Pa at room temperatures) above a liquid surface. Thus the total pressure $p_a + p_i$ (of order 10^5 Pa) is largely due to the dry air component p_a. The ratio of the partial pressure of water vapour p_i to its saturated value p_s at that temperature is the relative humidity ϕ, often expressed as a percentage. Humidity can be expressed as a dew point, the temperature at which vapour starts to condense to form a mist or fog, either on a surface or in the space itself. For use in air-conditioning calculations, information on moist air can be presented on a psychrometric chart. The ASHRAE charts take temperature as the horizontal axis but replace pressure by the humidity ratio W, where

$$W = \text{mass of moisture per mass of dry air} = (M_w/M_a)(p_i/p_a). \tag{8.36}$$

M_w and M_a are the molecular weights of water, 18.0 and 29.0 kg, respectively. The chart includes wet bulb and enthalpy characteristics.

The science of psychrometrics, the behaviour of water vapour in air, is fundamental to the design of air-conditioning plant. It is based on the relation between saturated vapour pressure and temperature, together with atmospheric air. Extensive tables of derived quantities – enthalpy, entropy, etc. – as a function of temperature are given in various ASHRAE and CIBSE handbooks. Their use is described for example by Jones (1994).

8.8 APPENDIX: SATURATED VAPOUR PRESSURE OVER A CURVED SURFACE

The saturated vapour pressure above a liquid surface results from molecules of liquid which approach the surface with a sufficiently high velocity to escape the forces of attraction between the molecules; molecules return to the liquid and a state of dynamic equilibrium is established. The liquid surface acts as a potential barrier and has an associated energy called surface tension or vapour/liquid interfacial tension σ. For water at 20°C, σ has a value of 0.073 J/m^2 (or N/m). Compared with a plane surface, there is a smaller volume of liquid to restrain a molecule leaving a small droplet; it escapes more readily, so the vapour pressure needed to prevent net evaporation outside a cloud of droplets is a little larger than above a plane surface. Conversely, the value above a concave surface to prevent condensation is a little below the plane surface value; put another way, the saturated vapour pressure outside a concave meniscus is lower than the plane surface value. Condensation takes place in micro-crevices at lower vapour pressures than on plane surfaces, which explains the partial filling of interstices in porous materials (Section 2.3).

The difference in pressure across a spherical surface with radius of curvature r is $2\sigma/r$.[22] The difference between the vapour pressure and the liquid pressures is defined as the capillary pressure, so

$$p_0 - p = (2\sigma/r). \tag{8.37}$$

See for example equation 4 of Konev and Mitrovic (1986). The difference is significant only when r is equal to molecular dimensions. Vapour pressure and saturated vapour pressure are related by Kelvin's equation:

$$p_v/p_{svp} = \exp(-Mp_c/(\rho_l RT)), \tag{8.38}$$

where p_c is the capillary pressure $p_0 - p$, sometimes expressed as $\rho_l g \psi$ where ψ (m) is called the capillary potential. A derivation is given in the latter equations of Udell (1983). See also Udell (1985: equation 17), Galbraith and McLean (1990: equation 2) and Masmoudi and Prat (1991: after equation 11). The role of surface tension on condensation in a porous medium is discussed in detail by Majumdar and Tien (1990).

8.9 APPENDIX: MEASURES OF THE DRIVING POTENTIAL FOR WATER VAPOUR TRANSPORT

A statement of the temperature difference provides the only measure in general use for the potential difference to describe the heat flow across a boundary layer but there are several measures which may be used to describe the transfer of moisture. Two have already been noted: the concentration or density (kg/m^3) of the water vapour and its pressure (Pa). Webb (1991) notes four more: the mass fraction of water vapour (dimensionless), its molar fraction (dimensionless), the molar density (kg/mol) and the specific humidity or humidity ratio W (kg vapour/kg dry air). They are simply related using the gas equation $p_i V = (m_i/M_i)RT$, where subscript i refers either to vapour or dry air, m denotes its mass in the volume V and $M = 18$ kg for water and 29 kg for air. The total pressure

[22]Consider the excess pressure δp inside a spherical vapour bubble of diameter $2r$ in the liquid. Suppose that the pressure does work by increasing the bubble size to $2(r + \delta r)$. Equating the work done to the increase in surface energy, we have $\delta p\, \delta r\, 4\pi r^2 = \sigma 4\pi[(r + \delta r)^2 - r^2]$. The result follows, neglecting the term in $(\delta r)^2$. Water rises to a height $2\sigma \cos\alpha/(r\rho g)$ in a cylindrical tube of diameter $2r$. This is the height measured to the base of the meniscus; α is the angle of contact, often taken as zero.

$p_t = p_w + p_a$ but at room temperatures p_w is of order 10^3 Pa and is small compared with atmospheric pressure of about 10^5 Pa.

Denoting the driving potential by β, we can write

$$m_{vap} = K_\beta \Delta\beta,$$

where m_{vap} has units kg/m^{2}s, but the units of the mass transfer coefficient K_β depend on the choice of β. The coefficient is often denoted by h_D. If concentration is chosen as β, the units of h_D are m/s (Ede 1967b: 215); if humidity ratio is chosen, h_D has units kg/m^{2}s (Threlkeld 1962: 192).

The relation that $h_D = h_c/\rho c_p$ (8.11) was first noted by Lewis (1922). Threlkeld (1962: Figure 10.2) later showed, on the basis of earlier work, that h_D depended too on the mean temperature T between the wetted surface and the air stream. It follows from his figure that

$$h_D = \frac{h_c}{\rho c_p} \frac{1}{0.868 + 0.000416T},$$

where T is in °C. Thus h_D may be a little greater than $h_c/\rho c_p$ but h_c may not be known reliably.

Webb goes on to point out that various authors speak of a Lewis number Le, but there was no agreed definition of Le. Of the 23 sources he examined, 13 authors defined it as $\lambda/\rho c_p D$ (the values for air, D is the diffusion coefficient), two as $\rho c_p D/\lambda$, five as $h_c/h_D \rho c_p$ and two left it undefined.[23] With the first of these definitions, Le = Sc/Pr, where Sc is the Schmidt number $\mu/\rho D$ and Pr the Prandtl number $\mu c_p/\lambda$. The Luikov number Lu is defined similarly to one of these: Lu $= \rho_s c_{ps} D_s/\lambda_s$, where D_s is the moisture diffusion coefficient and subscript s denotes values for the solid material (Eckert and Faghri 1980).

Luikov (1966: 249; 1975: equation 13) has provided a further driving potential, the *moisture transfer potential* or *moistness*, with symbol θ and units °M. This potential is used in connection with solid materials as well as boundary layers. It is some function of the moisture content and the external temperature and has the same value in all parts of, say, a multilayer wall when it is in thermodynamic equilibrium. The moisture content of some layer can be expressed by the ratio u (kg of moisture/kg dry material). When $u = 0, \theta = 0$°M; u increases with relative humidity (RH) and at RH = 100% (the maximum sorptional moisture content) it has the value u_{mes} (say). θ is then defined to be 100°M, so that generally

$$\theta = (u/u_{mes}) \times 100.$$

Porous materials can contain more moisture than this, so θ can take values greater than 100°M. Thus, following his example, for filter paper at 25°C, $u_{mes} = 0.277$ kg/kg. If in fact its moisture content is 0.5 kg/kg,

$$\theta = (0.5/0.277) \times 100 = 180° \text{ M}.$$

[23] With the definition $\lambda/\rho c_p D$, Le could be, and sometimes is, written as κ/D, where κ is the diffusivity of air. Although this is perfectly correct, it could be misleading since the κ value for a dense material is associated with the non-steady state, whereas D is valid in the steady state. Le appears to be comparing unlike quantities. Defined as $\lambda/\rho c_p D$, it clearly compares like quantities: conduction and diffusion through air.

With this content, filter paper is in equilibrium with peat having a moisture content of 2.1 kg/kg. Presumably, u_{mes} for peat is 1.17 kg/kg. In a further example, sand having $u = 0.1$ kg/kg and $\theta = 600°$M is in contact with peat having $u = 3.0$ and $\theta = 350°$M, so moisture actually passes from the seemingly dry sand to the wet peat.[24] The inferred value of u_{mes} for peat is now 0.857 kg/kg, which is inconsistent with the above value of 1.17; curiously, $1/0.857 = 1.17$.

The equations for heat and mass transfer in terms of the mass transfer potential are given by Luikov (1966: 259). Ozaki *et al.* (2001) have recently suggested a further potential as a driving force in the gaseous phase.

8.10 APPENDIX: MOULD GROWTH IN ANTIQUITY

'The Lord gave Moses and Aaron the following regulations about houses affected by spread of mould. ... Anyone who finds mould in his house must go and tell the priest about it. The priest shall order everything to be moved out of the house before he goes to examine the mould, otherwise everything in the house will be declared unclean. Then he shall go to the house and examine the mould. If there are greenish or reddish spots that appear to be eating into the wall, he shall leave the house and lock it up for seven days. On the seventh day he shall return and examine it again. If the mould has spread, he shall order the stones on which the mould is found to be removed and thrown into some unclean place outside the city. After that he must have all the interior wall scraped and the plaster dumped in an unclean place outside the city. Then other stones are to be used to replace the stones that were removed, and new plaster will be used to cover the walls. If the mould breaks out again ... (the house) must be torn down and its stones, its wood, and all its plaster must be carried out of the city to an unclean place.' The text goes on to specify what ritual the priest should conduct if the remedial action should prove successful. (Leviticus 14: 33–45). Parts of Leviticus may date from about 1300–1200 BC, but this passage was probably added nearer the fifth century BC. The Hebrew *tsara'at* apparently denotes a disfigurement of some kind due to an organic cause (not an injury); it is somewhat abstract and may be applied to the skin, to fabrics and to building walls. It was translated into Greek and later into Latin as *lepra* and into earlier English versions as *leprosy* (Wenham 1979). It gets rendered more specifically as 'mould' or 'mildew' in connection with buildings in modern translation.

[24]Luikov (1966: 256) assumes that the density of moisture flow is directly proportional to the gradient of the mass transfer potential θ. The flow of water from the sand to the peat can probably be expressed in terms of some potential gradient, similar to the relation for heat flow, $\lambda_1 \partial T_1/\partial x = \lambda_2 \partial T_2/\partial x$, at the boundary between layers 1 and 2 of a wall, but he does not justify his assumption for the flow from sand to peat. The value of θ is based on the value of u_{mes} but it is not clear what significance u_{mes} may have for transfer when u_1 and u_2 are significantly larger than their respective u_{mes} values.

9
Solar Heating

The temperature in an enclosure is determined to a great extent by ambient temperature, solar incidence and mechanical heating or cooling. Sky temperature, rain and wind speed normally have less effect. Casual gains may or may not be significant. In a cold climate, solar gains may serve to reduce the heat needed to achieve thermal comfort, though solar energy conversion systems differ from other energy conversion systems in that the amount of energy available is variable and is not easily controlled. In warm conditions, solar gain may lead to unacceptably high temperatures in an uncontrolled building while in a controlled building, cooling may be needed to combat such gains. This chapter summarises the main factors which enable an estimate to be made of the heat input to building surfaces, external and internal. Figure 9.1 provides an overview. The effect they have on the internal environment will be considered in Chapter 19. A full account of solar radiation is given by Iqbal (1983).

9.1 FACTORS AFFECTING RADIATION REACHING THE EARTH

If the atmosphere consisted of nitrogen and oxygen alone, without water vapour and carbon dioxide, the radiant temperature of the sky by night would be the nearly absolute zero value of outer space, since nitrogen and oxygen do not absorb in the infrared. However, the atmosphere contains about 0.03% of carbon dioxide by volume, uniformly distributed, and some 0.2% to 2% water vapour near ground level. The presence of these two gases leads to an effective radiant temperature for the night sky in cloudless conditions which is cooler than the air temperature at ground level by some 10 K in a hot moist climate and 25 K in a cold dry climate (Bliss 1961); see Section 4.1.3. By day the effect of solar radiation is superposed on this exchange and is the dominant driver.

Any account of the fascinating details of the generation of energy in the sun lies outside the scope of this book. It must be sufficient to note that the sun's diameter is some 1.393×10^6 km and its surface has a black-body equivalent temperature of about 5762 K. The resulting radiation has a nearly black-body spectrum with absorption bands mainly at the short-wave end. The energy reaching the earth's orbit is the solar constant and is defined as the energy received per unit area perpendicular to the direction of propagation at the earth's mean distance from the sun and outside the atmosphere. It has a value of about 1370 W/m^2 and of this, 7% is in the ultraviolet (wavelength $\lambda < 0.38\,\mu$m), 47% in the visible range ($0.38 < \lambda < 0.78\,\mu$m) and 46% in the infrared ($\lambda > 0.78\,\mu$m). See Duffie and Beckman (1980: 5).

Building Heat Transfer Morris G. Davies
© 2004 John Wiley & Sons, Ltd ISBN: 0-470-84731-X

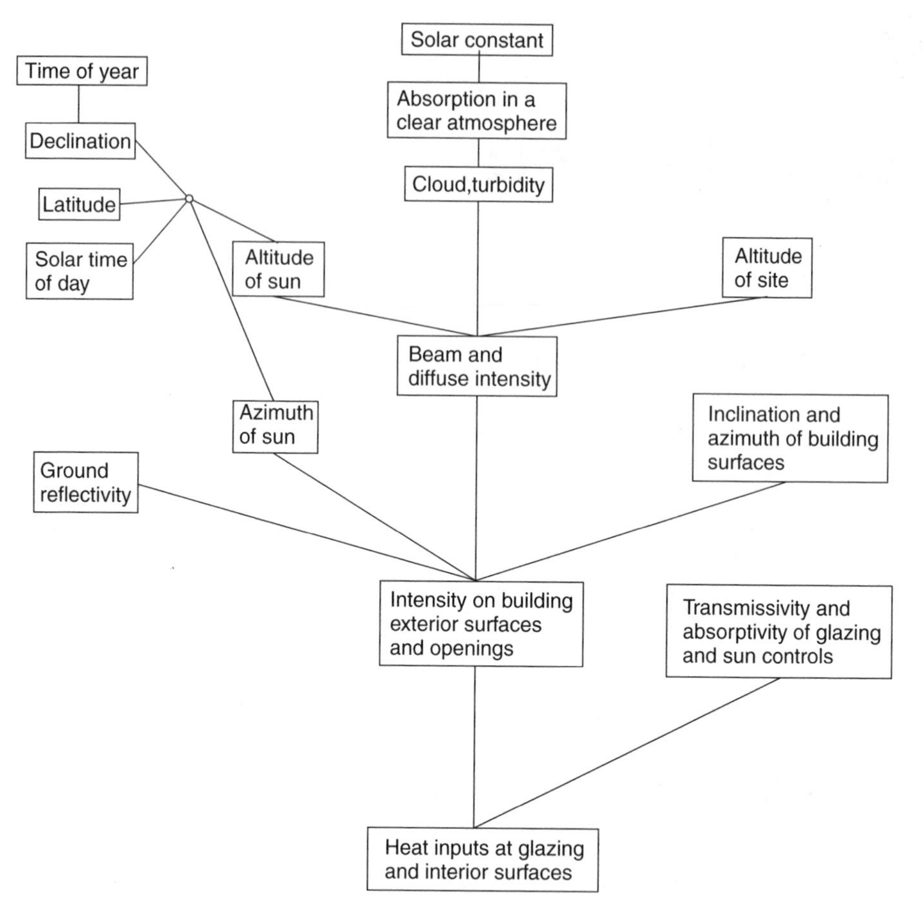

Figure 9.1 Factors affecting solar gain on building surfaces

This flux becomes reduced during passage through the atmosphere. Its interaction with air and water vapour molecules and with dust particles results in its being scattered. Air molecules are small compared with typical radiation wavelengths and here the scattering is approximately proportional to λ^{-4}; that is, short-wave radiation is scattered more than long-wave radiation, so that the light from a clear sky vault is blue while the depleted beam radiation from the low-lying sun is mainly red. The scattering by water vapour depends on the vapour content and varies as λ^{-2}. The earth's surface receives radiation in direct and diffuse forms.

Some radiation is absorbed in the upper atmosphere (around 25 km) where O_2 molecules become dissociated and form O_3 molecules, which in turn become dissociated. This removes radiation of wavelength less than $0.32\,\mu m$ (fortunately, since this ultraviolet radiation is very harmful). Certain bands in the infrared are strongly absorbed by water vapour and little radiation longer than $2.2\,\mu m$ is received at ground level.

The optical thickness of the atmosphere through which beam radiation passes can be described by the air mass m, which is the ratio of the thickness at an oblique solar altitude (β°) to its value at zenith ($\beta = 90^\circ$). For values of β greater than 20°, $m = 1/\sin\beta$ but

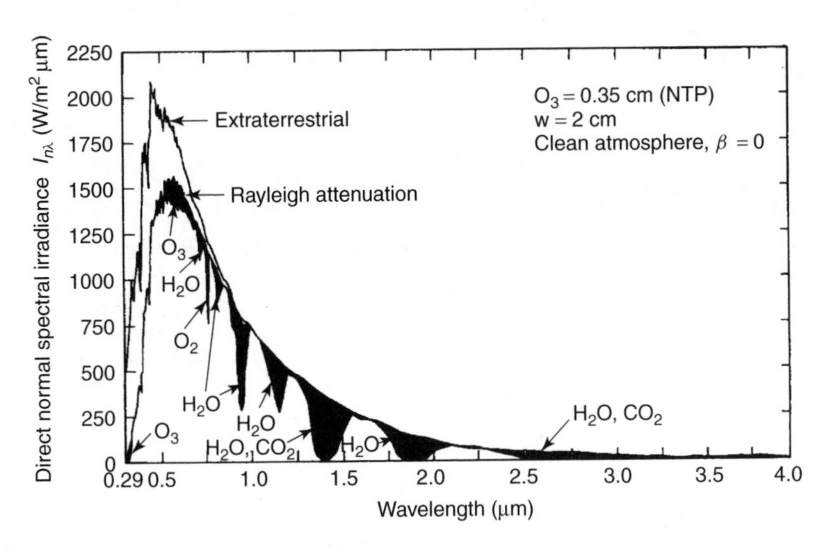

Figure 9.2 The effect of various molecular absorbers on the spectral distribution of solar radiation (Iqbal, M., 1983, *An Introduction to Solar Radiation*, Academic Press, Toronto, reproduced by the permission of Academic Press)

for lower values the curvature of the earth must be taken into account (Kondratyev 1969). Figure 9.2, taken from Iqbal (1983), shows the extraterrestrial distribution of radiation and a number of absorption bands. ('O_3 = 0.35 cm (NTP)' indicates that the results are based on an assumed ozone component amounting to a height of 0.35 cm if all the ozone in a vertical column in the atmosphere were brought to normal temperature and pressure, the total height being around 8000 m. In 1966 it varied between about 0.2 cm at the equator and 0.4 cm at the poles but it has risen considerably since. '$w = 2$ cm' indicates that 2 cm of precipitable moisture was assumed.) Similar information is given in Thekaekara (1974).

9.2 EARTH'S ORBIT AND ROTATION

The period of the earth's revolution around the sun is by definition 1 year and is in a counterclockwise direction as viewed from above the north pole. The plane containing its orbit is the ecliptic plane. The orbit is in fact an almost circular ellipse with semimajor and semiminor axes $a = 1.4968 \times 10^{11}$ m and $b = 1.4966 \times 10^{11}$ m, respectively; the sun is at one of the foci. This second-order difference in length of axes leads to a first-order value for the separation of the foci, $2c = 2\sqrt{a^2 - b^2}$, giving an eccentricity $e = c/a$ of 0.017. The least and greatest separations between earth and sun are shown in Figure 9.3a The fluxes falling on the upper atmosphere are 1418 W/m^2 on 2 January and 1325 W/m^2 on 2 July.[1]

[1]The energy flux from the sun is $\pi d^2 \sigma T^4 = \pi \times (1.393 \times 10^9)^2 \times (5.67 \times 10^{-8}) \times 5762^4 = 3.81 \times 10^{26}$ W. Thus the intensity over the sphere containing the earth's orbit is this value divided by $4\pi r^2$, where $r = 1.4967 \times 10^{11}$ m, and amounts to 1353 W. This is the value adopted by Duffie and Beckman, but it may be a little higher. Using the expression $e = mc^2$, where $c = 2.9979 \times 10^8$ m/s, the sun's flux corresponds to loss of mass equal to 4.2 million tonnes per second.

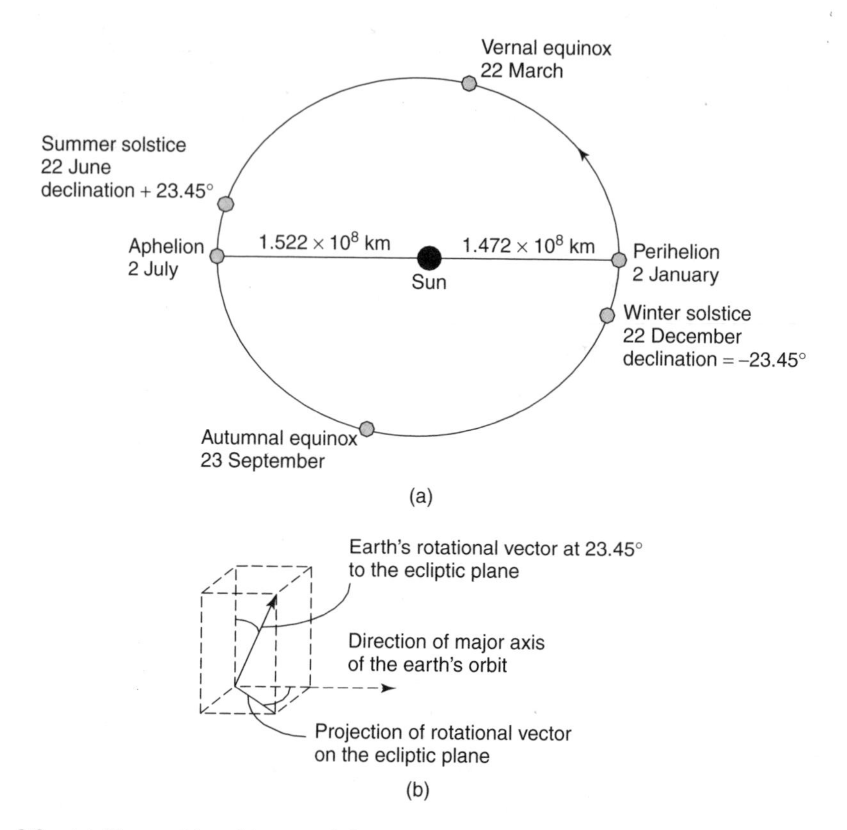

Figure 9.3 (a) The earth's orbit around the sun. (b) Direction of the earth's axis in relation to its orbit around the sun; seen from above the north pole, the earth rotates anticlockwise

The solar day is associated with the earth's rotation in space, also anticlockwise as seen from above the north pole, and is defined to be the interval between the time the sun crosses the local meridian and the time it crosses it the following day. Its value varies a little throughout the year and its average value is by definition 24 h. The year consists of 365.242 days.[2]

The earth's axis of rotation is inclined at an angle of 23.45° to the normal to the ecliptic plane; (or alternatively, the angle between the earth's equatorial plane and the ecliptic plane is 23.45°). The earth's axis is fixed in space.[3] In broad terms, the northern hemisphere is tilted away from the sun during the months of October to February so that solar incidence is low, leading to winter conditions; there is correspondingly high incidence in the southern hemisphere, leading to summer conditions. From April to August the situation is reversed. The projection of the axis on the ecliptic plane makes a small angle with the major axis of the ecliptic plane (Figure 9.3b). The angle between the earth–sun vector and the earth's equatorial plane, by definition the angle of declination, has its maximum (negative) value of 23.45° on 22 December, a little before the time of

[2]The period of rotation of the earth relative to the fixed stars is less than the solar day by $24 \times 60/365.242$ or 3.94 min.

[3]More exactly, the axis describes a conical motion of period 26 000 years.

minimum separation (2 January). The angle of declination is associated with the division of 24 hours into periods of daylight and darkness. On the longest day of the year, 22 June, the angle of declination has its maximum positive value. On 22 March and 22 September, when periods of light and darkness are equal, the angle of declination is zero; that is, the earth–sun vector lies in the equatorial plane.[4] The angle of declination d in degrees is given with sufficient accuracy for building geometry as

$$d = 23.45 \sin(360(284 + N)/365) \quad \text{or} \quad 23.45 \sin(360(N - 22\ \text{March})/365), \qquad (9.1)$$

where N is the day of the year ($N = 1$ on 1 January).

9.3 THE SUN'S ALTITUDE AND AZIMUTH

We wish to find the altitude and azimuth of the sun at some site P given the time of year, the time of day and the latitude. The altitude β is the angle between the sun's rays and the horizontal, and the azimuth is the angle between south (or north, according to convention) and the projection of the rays on the horizontal. The time of year determines the declination, the angle d between the sun's rays and the equatorial plane. The time of day is represented as the hour angle h, the angular displacement of the sun east or west of south due to the earth's rotation; east negative, west positive. The latitude L is the angular displacement of P from the equatorial plane.[5]

These angles can be related through a coordinate system (Figure 9.4a). O is the centre of the earth which rotates about the Oz axis, the xy plane lies in the equatorial plane and the sun's rays lie in the xz plane. It is convenient to define vector directions at P: **S** is south, **V** is vertical, **B** is to the sun, and **A** is the projection of **B** on the horizontal.

The vector **B** makes an angle d with the equatorial plane and so has direction cosines $\cos d$, 0, $\sin d$ with the x, y, z directions, respectively. The Oz axis makes angles $-L$

[4] The period of rotation of the earth around the sun involves just one degree of freedom since the ratio (mean separation between planet and earth)^3/(time of rotation of planet)^2 has the same value for all the planets (Kepler's third law). The earth, however, has three degrees of freedom: two associated with the direction of its axis in space and the third with its rate of rotation.

[5] The expressions for β and γ to be given are frequently quoted and sometimes derived in modern texts but no indication is given of their origin. They have obvious bearing on the design and calibration of sundials which were current in antiquity. The physical basis to find the angles would have been known to Ptolemy (c. 100–170 AD) but Greek trigonometry was too primitive to provide expressions in closed form. Copernicus used the geometrical and arithmetical methods of the Greeks. Trigonometry was developed in detail through Arabian work in the tenth century but only became widely known in the west in the sixteenth century, so expressions relating β, γ, L, d and h might have appeared thereafter. I do not know whether Galileo, Tycho Brahe or Kepler knew them, but one might expect that they did. The expression for β in terms of L, d and h was certainly known however by the mid eighteenth century and applied for navigation purposes. The texts by Heath (1760) and Rios (1805) give worked examples of this type: given a ship's latitude, the date and the sun's elevation, find the local time. The examples involve a series of unexplained arithmetical operations with logarithms of trigonometric functions of L, d and β and they use extensive tabulated information of log sin x; the method of evaluation rests on the relation of L, d and β to the hour angle h but no explicit statement of the relation is given. (In fact, knowing the exact local time would probably not have been particularly useful to find longitude until the corresponding time at Greenwich was known reliably. That only became possible in principle with the advent of the reliable chronometer developed by John Harrison (1693–1776), which allowed one to determine longitude at sea accurate to some 15 km. Only when such intricate chronometers could be mass-produced could this method supersede the alternative lunar-distance method.)

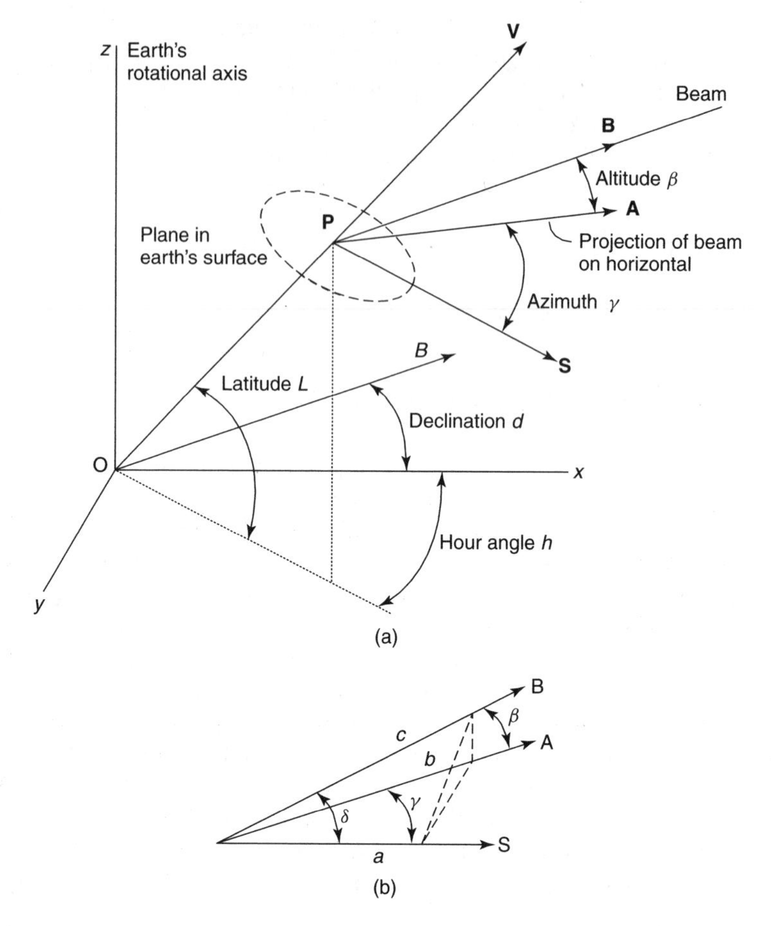

Figure 9.4 (a) Independent variables are declination (time of year), hour angle (time of day) and latitude; auxiliary vectors are **S, B, A, V**; dependent variables are sun's altitude and azimuth. The sun's rays lie in the plane Oxz. (b) How to find the azimuth angle γ

with **S** and $90 - L$ with **V**, hence its direction cosines. The other direction cosines involve the hour angle h. The projection of OP on the xy plane is $OP \cos L$ and makes angles h and $90 - h$ with the x and y axes, respectively, from which the direction cosines follow. Finally, the vector **S** makes an angle $-(90 - L)$ with the xy plane; it makes angles h and $90 - h$ with the x and y axes, respectively, again leading to direction cosines. They are given in Table 9.1.

The angle θ between two vectors having direction cosines a_1, b_1, c_1 and a_2, b_2, c_2 is given by

$$\cos\theta = a_1 a_2 + b_1 b_2 + c_1 c_2. \tag{9.2}$$

Now the angle of altitude $\beta = 90 - \angle\mathbf{BV}$, so

$$\begin{aligned}\sin\beta &= \cos(\angle\mathbf{BV}) = \cos d \cos L \cos h + 0 \times \cos L \sin h + \sin d \sin L \\ &= \cos L \cos d \cos h + \sin L \sin d.\end{aligned} \tag{9.3}$$

Table 9.1 Direction cosines between position directions and coordinate axes

	B	V	S
Ox	cos *d*	cos *L* cos *h*	sin *L* cos *h*
Oy	0	cos *L* sin *h*	sin *L* sin *h*
Oz	sin *d*	sin *L*	− cos *L*

A value of $h = 0$ denotes solar noon and β is symmetrical about noon, as it must be. At solar noon, $\beta = 90 + d - L$ and is largest at the summer solstice. The sun is then vertically overhead at latitudes $L = \pm 23.45°$ (Tropics of Cancer and Capricorn.)

For sunrise and sunset $\beta = 0$, so

$$\cos h = -\tan d \ \tan L. \tag{9.4}$$

In winter in the northern hemisphere, d is negative, $\cos h$ is positive and $-90 < h < +90$. In summer $-180 < h < -90$ denotes sunrise and $90 < h < 180$ denotes sunset. In fact, due to atmospheric refraction, the sun remains visible a little longer than the times implied by this expression. According to Walraven (1978), the actual elevation of the sun at sunrise or sunset in degrees is $-0.833 - 0.0388\,h^{1/2}$, where h is the local height above sea level (m).

To find the azimuth angle γ we note that in Figure 9.4b

$$\frac{a}{b} = \cos\gamma, \quad \frac{b}{c} = \cos\beta, \quad \frac{a}{c} = \cos\delta. \tag{9.5}$$

But

$$\frac{a}{b}\frac{b}{c}\frac{c}{a} \equiv 1 \quad \text{and} \quad \delta = \angle(\mathbf{SB}).$$

So

$$\cos\gamma = \frac{\cos\delta}{\cos\beta} = \frac{\sin L \cos d \cos h - \cos L \sin d}{\cos\beta}. \tag{9.6}$$

(These expressions for angles are sufficient to find the incidence of the sun's rays on building surfaces. By contrast, the mirrors of a solar tower generator have to be positioned with an accuracy of 0.05°. As Walraven points out, accurate values must take account of the perturbations of the earth's orbit by the moon and planets, luni-solar precession, nutation of the earth's orbit as well as refraction by the atmosphere.)

The angle θ between the sun's rays and the normal to a surface is given by

$$\begin{aligned}\cos\theta = &+ \sin L \sin d \cos s \\ &- \cos L \sin d \sin s \cos a \\ &+ \cos L \cos d \cos s \cos h \\ &+ \sin L \cos d \cos s \cos a \cos h \\ &+ \cos d \sin s \sin a \sin h,\end{aligned} \tag{9.7}$$

where s is the slope, the angle between the surface and the horizontal ($s = 90^\circ$ for a vertical surface; if $s > 90$ then the surface faces downward) and a, the azimuth, is the deviation between the local meridian and the normal to the surface; $a = 0$ denotes south, east negative, west positive.

If the earth's orbit were strictly circular with the sun at the centre of the circle and the earth's rotational axis were normal to the ecliptic plane, each solar day would be strictly 24 h. As a result of the deviations from this situation, solar time and civil (or clock) time at some location differ a little:

$$\text{Solar time} = \text{local standard time} + \text{longitude adjustment} + \text{equation of time}. \quad (9.8)$$

The local standard time depends on the time zone for the location, with allowance for summer time or daylight saving time. In the UK, Ireland and Portugal the time zone is based on the longitude of Greenwich, 0°. The standard meridian for western Europe is 15°E, for the eastern US it is 75°W, and so on. An adjustment for longitude must be made for the location of the site within the zone: a complete revolution of the earth, 360°, corresponds to 24 h, so at a site 15° west of the standard meridian, solar noon falls one hour after 1200. Finally, according to Woolf (1968),

$$\text{Equation of time (EOT)} = -(0.1236 \sin x - 0.0043 \cos x + 0.1538 \sin 2x + 0.0608 \cos 2x) \quad \text{[h]}, \quad (9.9)$$

where $x = 360(N - 1)/365.242$ and N denotes the day of the year; $N = 1$ denotes 1 January. It has four zero values: 15 April, 15 June, 1 September and 24 December.

The EOT correction is scarcely of significance when estimating the intensity of beam radiation (which depends on the sun's altitude β and therefore on the time of day). However, its effect on the times of sunrise and sunset is quite perceptible. EOT has its maximum value of rather more than 16 min (or −16 min, according to the sign convention) around 1 November, so the sun sets some 16 min earlier than the value given by (9.4), which depends on latitude and time of year alone. At this time of year the relatively rapid shortening of afternoon daylight is quite noticeable. Furthermore, while 22 December is indeed the shortest day, the time of earliest sunset is 12 December but sunrise continues to fall later until 29 December. A lengthening of the afternoon is perceptible in early January.

9.4 INTENSITY OF RADIATION

We have considered the intensity of radiation falling on the atmosphere, its attenuation in passing through the atmosphere and how to find the path length through the atmosphere. Figure 9.5 shows a range of radiation intensities normal to the beam through cloudless skies.

Hottel (1976) has described a procedure to estimate the intensity of beam radiation which includes considering climate type and the altitude of the site. First, define certain climate-dependent ratios:

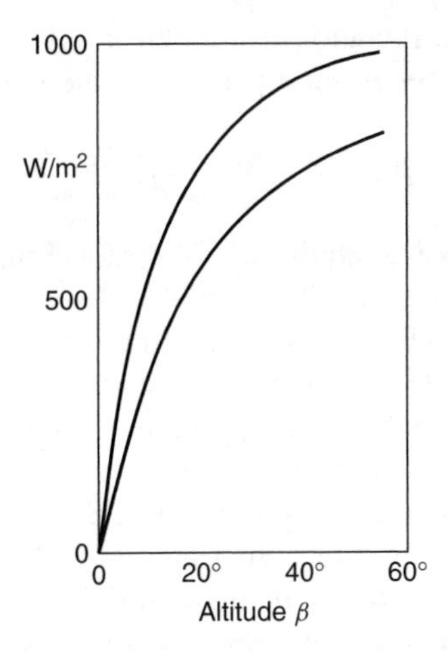

Figure 9.5 Direct normal intensity of solar radiation from cloudless skies. A range of values from several authors. Adapted from Loudon (1967)

Climate type	$a_0/a_0{}^*$	$a_1/a_1{}^*$	k/k^*
Tropical	0.95	0.98	1.02
Mid-latitude summer	0.97	0.99	1.02
Subarctic summer	0.99	0.99	1.01
Mid-latitude winter	1.03	1.01	1.00

Two visibility haze models are presented:

- 23 km visibility haze model:

$$a_0{}^* = 0.4237 - 0.00821(6 - A)^2,$$
$$a_1{}^* = 0.5055 + 0.00595(6.5 - A)^2,$$
$$k^* = 0.2711 + 0.01858(2.5 - A)^2.$$

- 5 km visibility haze model:

$$a_0{}^* = 0.2538 - 0.0063(6 - A)^2,$$
$$a_1{}^* = 0.7678 + 0.0010(6.5 - A)^2,$$
$$k^* = 0.249 + 0.081(2.5 - A)^2.$$

Here A is the altitude of the site (km). The clear sky beam normal radiation (W/m^2) is then

$$G_{cnb} = (\text{solar constant})(1 + 0.033\cos(360N/365))(a_0 + a_1\exp(-k/\sin\beta)),$$

where $N = 1$ denotes 1 January and β is the sun's altitude.

The intensity of the diffuse radiation from a clear sky is less than the intensity of the direct radiation. Reported values of the intensity on the horizontal suggest

$$\beta = 20^\circ,\ 50\text{–}100\,\mathrm{W/m^2} \qquad \beta = 40^\circ,\ 85\text{–}170\,\mathrm{W/m^2} \qquad \beta = 60^\circ,\ 110\text{–}200\,\mathrm{W/m^2}.$$

For fuller information on the availability of solar radiation, including its means of measurement, see Duffie and Beckman (1980: Ch. 2).

The annual mean global irradiance on a horizontal plane on the earth's surface ($\mathrm{W/m^2}$) averaged over 24 h (Budyko 1958) amounts to between 100 and 150 over Canada, northern Europe and northern Asia and the southerly parts of South America, and between 150 and 200 for the northern US, southern Europe, central and south-east Asia. Areas with irradiance greater than 250 include the south-west US, the Caribbean, all of north Africa to the Gulf (with the Red Sea area receiving around 300), southern Africa and northern Australia. Although the annual totals in high-latitude areas are as much as one-third those of equatorial regions, the high-latitude summer/winter variation is much larger than low-latitude variation.

9.5 SOLAR INCIDENCE ON GLAZING

When beam radiation of intensity I_B falls on an opaque surface of absorptivity α, the absorbed radiation is $\alpha I_B \cos\theta$, where θ is the angle between the beam and the normal to the surface. For glass or some other transparent material, some of the radiation is reflected, some absorbed and the rest transmitted. These fractions depend on the angle of incidence and the absorption characteristics of the medium (Figure 9.6).

The angles of incidence θ_1 and refraction θ_2 are related as

$$\sin\theta_1/\sin\theta_2 = \mu, \text{ the refractive index.} \tag{9.10}$$

The reflected beam consists of components polarised in the plane and perpendicular to the plane containing the incident and reflected rays together with the normal. The parallel and perpendicular reflected fractions of radiation $r_\parallel$ and $r_\perp$ respectively are given as

$$r_\parallel = \frac{\tan^2(\theta_1 - \theta_2)}{\tan^2(\theta_1 + \theta_2)} \qquad r_\perp = \frac{\sin^2(\theta_1 - \theta_2)}{\sin^2(\theta_1 + \theta_2)}. \tag{9.11}$$

Suppose that a beam of unit intensity falls on the surface. A fraction r is reflected and $1 - r$ refracted, as Figure 9.6 illustrates; r denotes either $r_\parallel$ or $r_\perp$. If L denotes the thickness of the glass, the path length of the refracted ray is $L/\cos\theta_2$. Some of the radiation is absorbed in traversing this distance and the relative intensity reaching the far surface is

$$a = \exp(-kL/\cos\theta_2). \tag{9.12}$$

where k is the extinction coefficient, having a value of around $4\,\mathrm{m^{-1}}$ for clear glass and larger values (e.g. $100\,\mathrm{m^{-1}}$) for heat-absorbing glasses. The absorption coefficient a tends to unity for thin, low-absorbing glass. For coloured glasses, k depends on wavelength since the colour transmitted is evidently the radiation which has not been absorbed. Further,

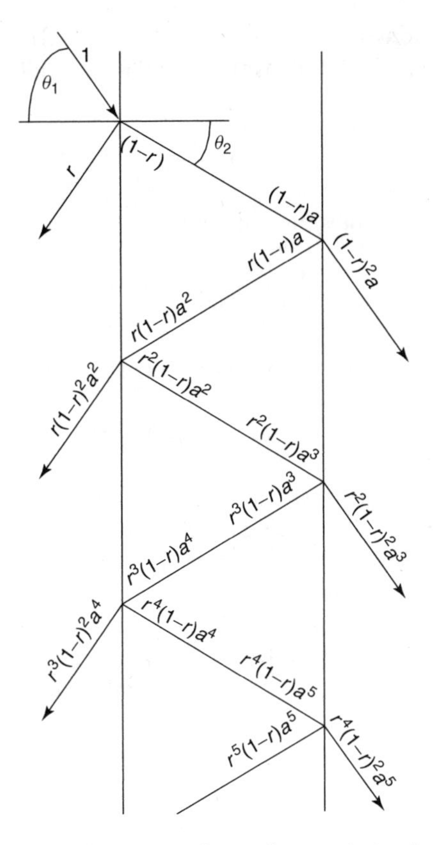

Figure 9.6 Reflection, absorption and transmission in a glass slab

the refractive index μ depends a little on wavelength. Both a and r should strictly have subscript λ. Thus a fraction $(1-r)a$ reaches the far surface, where again a fraction r is reflected and $1-r$ refracted, leading to the actual reflected and refracted components $r(1-r)a$ and $(1-r)^2a$ as shown. The pattern is repeated indefinitely. The total reflected component, the reflectivity, is

$$\rho = r + r(1-r)^2a^2[1 + r^2a^2 + r^4a^4 + \ldots] = r + r(1-r)^2a^2/(1-r^2a^2), \quad (9.13)$$

(noting that $1 + x + x^2 + x^3 + \ldots = 1/(1-x)$.) (9.14)

The component absorbed in a single traverse is $(1-r)(1-a)$ and the sum, the absorptivity, is

$$\alpha = (1-r)(1-a)[1 + ra + r^2a^2 + r^3a^3 + \ldots] = (1-r)(1-a)/(1-ra). \quad (9.15)$$

The total transmitted fraction, the transmissivity,[6] is

$$\tau = (1-r)^2a[1 + r^2a^2 + r^4a^4 + \ldots] = (1-r)^2a/(1-r^2a^2). \quad (9.16)$$

[6]These quantities are also called reflectance, absorptance and transmittance. The above forms are preferred here because they avoid confusion with transmittance used in a thermal context and having dimensions W/m^2K.

The sum of these fractions is $\rho + \alpha + \tau = 1$. Each should be subscripted as $\|$ or $\perp$. Now the incident radiation of unit intensity is unpolarised and can be written as

$$1 \equiv \frac{1}{2_{\|}} + \frac{1}{2_{\perp}}. \tag{9.17}$$

Thus the net absorptivity and transmissivity are

$$\alpha = \frac{1}{2}\alpha_{\|} + \frac{1}{2}\alpha_{\perp} \quad \text{and} \quad \tau = \frac{1}{2}\tau_{\|} + \frac{1}{2}\tau_{\perp}. \tag{9.18}$$

At near-normal incidence, $\theta_2 = \theta_1/\mu$ and $\sin\theta_2 = \tan\theta_2 = \theta_2$. Accordingly, the reflectivity r at a single surface is

$$r = ((\mu - 1)/(\mu + 1))^2 \tag{9.19}$$

and has a value of 0.0434 for glass with a refractive index $\mu = 1.526$. It varies little up to a value of $\theta_1 = 50^\circ$ and from 70° increases rapidly to unity at 90°. Since thin clear glass absorbs little radiation, so that a is effectively unity, the reflectivity is $2r/(1+r)$ or 0.0832 at normal incidence and the transmissivity is $1 - 0.0832 = 0.917$. The transmissivity correspondingly decreases to zero at $\theta_1 = 90^\circ$.

The optics of double glazing is shown in Figure 9.7. Sheet 1 reflects, absorbs and transmits beam radiation as described above. The figure illustrates the subsequent action of the transmitted beam.

The total reflected component is

$$\rho_{12} = \rho_1 + \rho_2\tau_1^2(1 + \rho_1\rho_2 + \rho_1^2\rho_2^2 + \ldots) = \rho_1 + \rho_2\tau_1^2/(1 - \rho_1\rho_2). \tag{9.20}$$

The radiation absorbed in the first pane is

$$\alpha_{1\text{ of }2} = \alpha_1 + \alpha_1\rho_2\tau_1(1 + \rho_1\rho_2 + \rho_1^2\rho_2^2 + \ldots) = \alpha_1 + \alpha_1\rho_2\tau_1/(1 - \rho_1\rho_2). \tag{9.21}$$

The radiation absorbed in the second pane is

$$\alpha_{2\text{ of }2} = \alpha_2\tau_1(1 + \rho_1\rho_2 + \rho_1^2\rho_2^2 + \ldots) = \alpha_2\tau_1/(1 - \rho_1\rho_2). \tag{9.22}$$

The transmitted component is

$$\tau_{12} = \tau_1\tau_2(1 + \rho_1\rho_2 + \rho_1^2\rho_2^2 + \ldots) = \tau_1\tau_2/(1 - \rho_1\rho_2). \tag{9.23}$$

Expressions of this kind were found by Parmelee (1945). Jones (1980) quotes these equations but also gives expressions for transmission through vertical venetian slats and uses a thermal circuit to analyse the effect of absorption.

The presence of a second pane leads to more radiation being reflected back to the exterior than is the case for single glazing, and more radiation is absorbed at the outer pane than at a single pane, as might be expected. The transmissivity, however, is the same in both directions. The normal transmissivity neglecting absorption is now 0.85 compared with the single-glazing value of 0.917. It will be recalled that double glazing has around half the thermal transmittance (U value) of single glazing: a change from single glazing

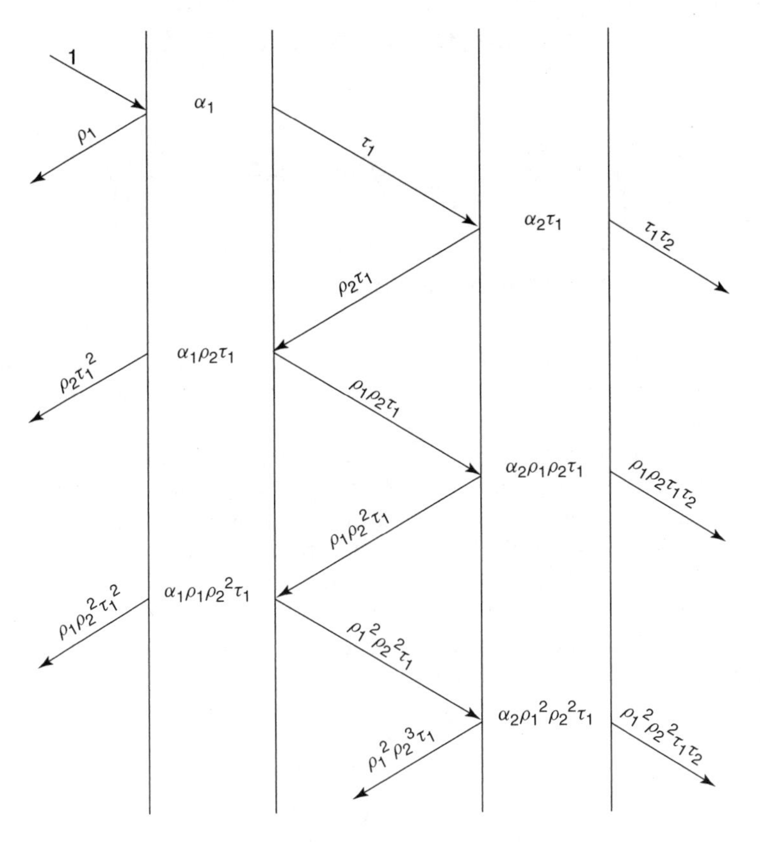

Figure 9.7 Reflection, absorption and transmission in double glazing

to double glazing has more effect on the aperture's thermal characteristics than its optical characteristics. The transmissivity, etc., for glazing with optical thin films and/or tinted glass has been studied by Pfrommer *et al.* (1994).

In general a temperature node is needed at each location where heat is input to an enclosure, but for double glazing it is not necessary to model explicitly the radiation absorbed in the outer pane. Suppose that the outer and inner panes are denoted as 1 and 2, respectively. If the beam intensity of radiation falling on a glazed area A_g at time t is I_t, the inputs at the two nodes are $Q_1 = I_t A_g \alpha_{1 \text{ of } 2}$ and $Q_2 = I_t A_g \alpha_{2 \text{ of } 2}$ as shown in Figure 9.8, where T_e and T_{ra} have the reference values of zero. R_e, R_c and R_i are respectively the outside, cavity and internal film resistances (around $0.06/A_g$, $0.18/A_g$ and $0.12/A_g$ K/W, respectively). For now values of the ambient and room temperatures T_e and T_{ra} are taken as the reference temperature value of zero.

It can readily be shown that the heat flow from T_2 into the enclosure at T_{ra} is

$$\frac{T_2}{R_i} = \frac{R_e + R_c}{R_e + R_c + R_i}\left(\frac{R_e}{R_e + R_c}Q_1 + Q_2\right). \tag{9.24}$$

This is another form of equations (4.21). As far as the enclosure is concerned, the real inputs Q_1 and Q_2 can be replaced by the reduced absorbed input

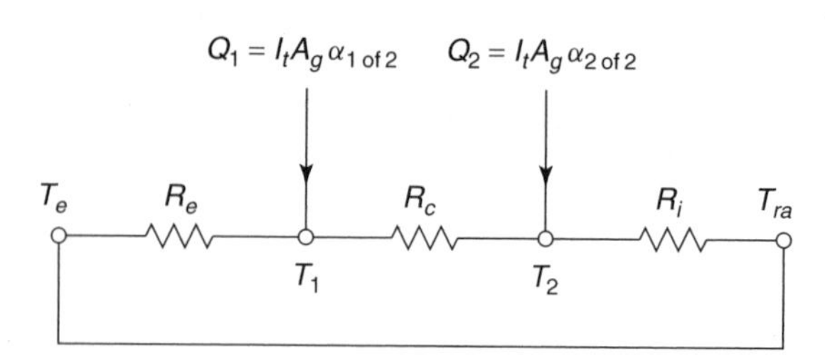

Figure 9.8 Thermal circuit for double glazing

$Q_{abs} = R_e Q_1/(R_e + R_c) + Q_2$ acting at the T_2 node (independently of the value of R_i) without explicit inclusion of T_1. This procedure would lead to a lower value for T_1 than its correct value.

T_e and T_{ra} have finite values and in the absence of solar radiation, the temperature at node 2 has some intermediate value. The above value of T_2 is the incremental temperature due to solar gains. In addition to the equivalent flux of $R_e Q_1/(R_e + R_c) + Q_2$ acting at T_2 (the window interior surface), the transmitted fraction $I_t A_g \tau_{12}$ is received mainly on the floor and furnishings, where it is partly absorbed and partly reflected. The reflected component, after further absorption and reflection, is largely absorbed at the room surfaces although some small fraction – the component that permits us to view a room interior from outside – is retransmitted to the exterior.

Values for the reflected, absorbed and transmitted components for triple glazing are given in the CIBSE (1999: A5-80). The effect of the three absorbed fractions can be found by an expression similar to (9.24). CIBSE (1999) also discusses the transmission by slatted blinds.

9.6 THE STEADY-STATE SOLAR GAIN FACTOR

It is convenient here to discuss the basis for the solar gain factor S, introduced in the 1960s, which forms part of the scheme to implement the CIBSE 'admittance procedure' and is used to estimate the daily mean and peak values of temperature in a room which may build up during a succession of warm and sunny days. It is a measure of the time-varying response of glazing, averaged over a day, to the solar radiation falling on the glazing. S is defined as the fraction of the solar irradiance on a window which acts to raise the internal temperature and is a function of the window-absorbed and transmitted fractions. It is based on the principle shown in Figure 9.9. The heat flow Q_j applied at the node T_j generates the same temperature at T_{ra} as the reduced input $(H/(H + F))Q_j$ acting at T_{ra}. Q here denotes a steady-state or time-averaged value. The value of the reduced input is independent of L.

Consider now an enclosure with solar gains and with losses by ventilation, through a window, through an outer wall and into other surfaces such as the floor and internal walls which can be treated as adiabatic in steady-state conditions. H denotes the film conductance between a surface and T_{ra} and F denotes the conductance of the fabric and the outer film. The gain due to absorption in the glass of a single-glazed window is straightforward. For double glazing the input is the full absorbed component for the inner leaf and a scaled value for the outer, the quantity Q_{abs} as explained above. The transmitted flux is absorbed at the other surfaces as shown in Figure 9.10a.

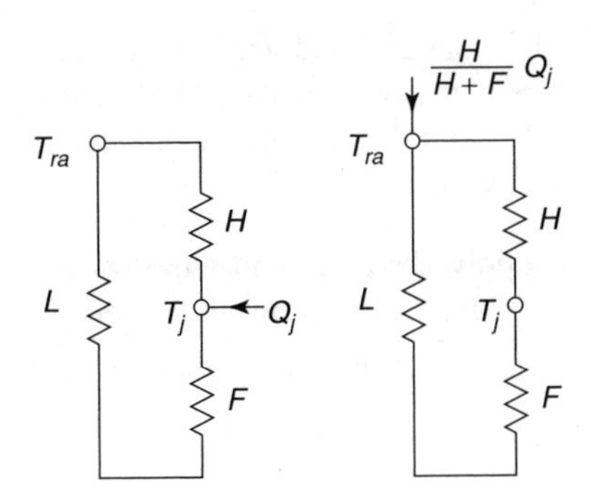

Figure 9.9 The value of T_{ra} is the same for both inputs

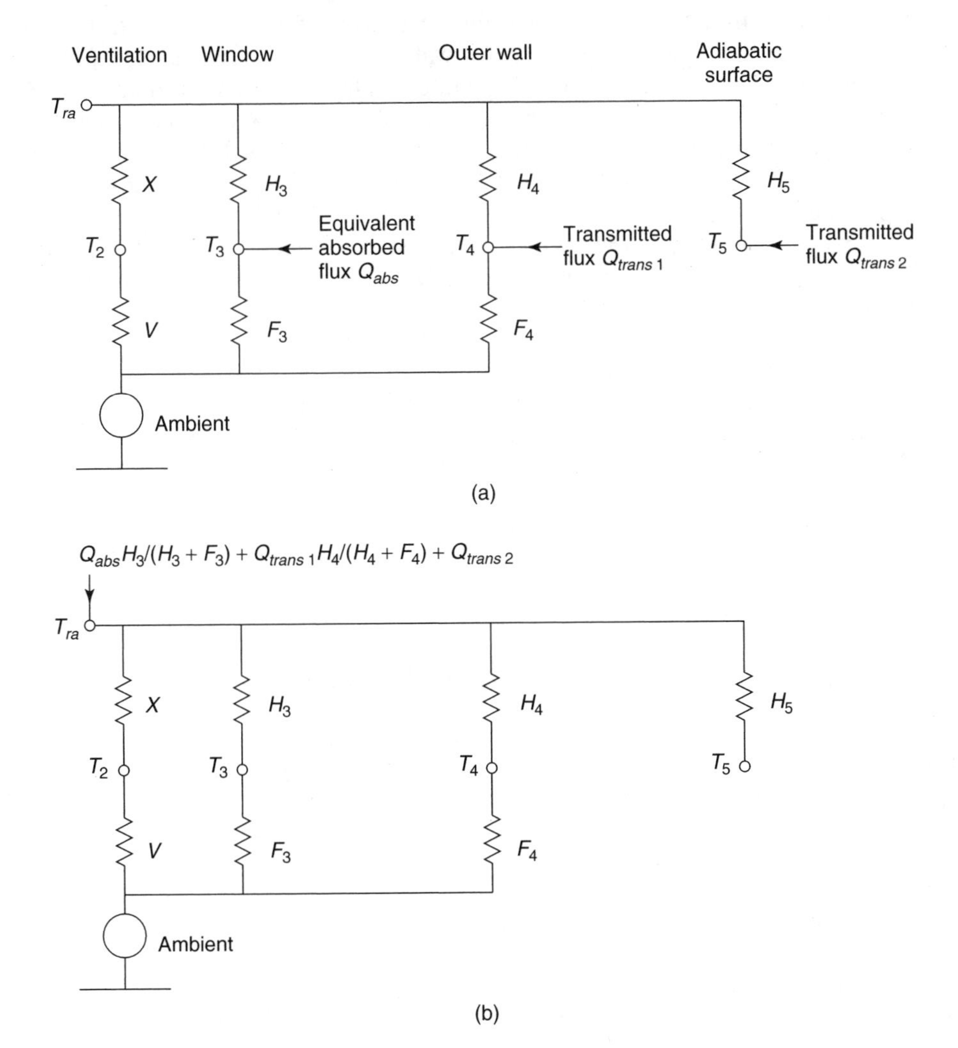

Figure 9.10 (a) Solar inputs at locations within an enclosure. (b) Equivalent solar inputs acting at the room index temperature T_{ra}

According to Figure 9.9, their net effect at T_{ra} can be found by inputting scaled values of each at T_{ra} itself, as shown in Figure 9.10b, since the contribution of $Q_{trans\,1}$, for example, is independent of the details of thermal elements in branches 2, 3 and 5.

The steady-state solar gain factor is defined as

$$S = \frac{\text{daily total of window-absorbed and window-transmitted solar gains as acting at the room index temperature}}{\text{daily total solar radiation incident upon the window}}$$

$$= \frac{\Sigma_{abs} H_3/(H_3 + F_3) + \Sigma_{trans\,1} H_4/(H_4 + F_4) + \Sigma_{trans\,2}}{\sum I_t A_g \cos\theta\delta t}.$$

Summation is carried out over daylight hours. Σ_{abs} is formed from the window-absorbed flux Q_{abs},[7] $\Sigma_{trans\,1}$ from the window-transmitted flux which falls on surfaces from which there is a loss to the exterior (e.g. an outer wall or the floor), and $\Sigma_{trans\,2}$ from the flux which falls on internal surfaces (e.g. partition walls) which are in effect adiabatic in steady-state conditions. I_t (noted in Figure 9.8) consists partly of beam radiation for which absorptivity and transmissivity values vary with angle of incidence θ, and a diffuse component which is usually taken to act at 60° incidence. On a south-facing window in winter the sun rises south of the E–W direction and summation takes place for values of $|\theta| < 90^\circ$ when beam radiation immediately starts to contribute to S. In summer it rises north of this direction; diffuse radiation then makes some small contribution but beam radiation only contributes when $|\theta| < 90^\circ$.

CIBSE (1999: A5-15) suggests $S = 0.75$ for clear single glazing and $S = 0.55$ for absorbing glass. A selection of values are given in Table 16.1. S values are used together with information on solar incidence during warm and sunny days (such as in Figure 9.5) to estimate the response of an enclosure to steady excitation. In Chapter 15 measures for wall response similar to the wall U value are derived, but for excitation which varies sinusoidally in time. 'Alternating' solar gain factors S_a (Table 16.1) enter the calculation from which estimates of the daily variation of room temperature about its mean value can be found, so leading to the peak temperature and its time (Chapter 16), although the required approximations and simplifications mean that only rough values can be found.

9.7 SOLAR GAIN CONTRIBUTION TO HEAT NEED

The theory of solar irradiation has a range of applications in the building field. They include the use of solar thermal collectors to heat water or air (active systems), solar ponds and photovoltaic devices, topics which are outside the scope of this book; see Sayigh (1979), Hastings (1994), International Energy Agency (1997), Peuser *et al.* (2002). In addition there are passive systems, which function without needing external energy to operate pumps or fans,[8] in which glazing with a south-facing aspect (i.e. between south-east and south-west) admits solar radiation to a space; this energy flow may reduce the

[7] Notice that the flux absorbed in the outer pane of a double-glazed window (Q_1 in Figure 9.8) is subjected to two stages of scaling down in forming its contribution to S. The first is to reduce it to its equivalent value at T_2, where it forms part of Q_{abs}, the second is to form the reduced equivalent of Q_{abs} acting at the room index node. It is immaterial whether the scaling factors are expressed as ratios of resistances or conductances.

[8] Figure 1.5 of Clarke (2001) lists several passive solar elements used architecturally.

heat load provided by conventional fuel appliances. Thermal capacity is required to reduce fluctuations of temperature and, in combination with thermal insulation, to maintain warm conditions by night.

Normal glazing with a southerly aspect readily admits solar radiation, especially during sunny periods, but it affords poor insulation compared with insulated walls and allows large and continuous thermal losses. There are good reasons to construct sunspaces and other large glazed areas in which energy considerations are not dominant. From the energy conservation viewpoint, a central question is whether an increase in glazed area during some season or the whole year is likely to increase or decrease the conventionally provided heat load. A detailed discussion of passive design lies outside this book. Zalewski *et al.* (2002) report a study of the energy-gathering potential of various types of solar wall, including the non-ventilated wall discussed below and the Trombe wall, in which apertures at the bottom and top of the glazed area of wall allow a circulation of air between the space between wall and glazing and the room.[9] Some comments may be made in this regard on steady-state aspects. It shows that the usefulness of solar construction depends not only on the details of the solar collecting device but also on the temperature to be achieved in the occupied space and the ventilation and conduction losses other than through the solar wall. (The most cost-effective use of solar energy appears to be to extend the period of use of open-air swimming pools; the collector temperature remains low so that most of the incident solar energy raises the pool temperature.)

Consider the simple steady-state model of an enclosure shown in Figure 9.11, having a glazed area A_g in an elevation of area A. The steady-state loss conductances are $A_g U_g$ through the glazing, $(A - A_g)U_w$ through the remainder of the elevation (U value U_w), L through other enclosure surfaces and V by ventilation. Solar radiation having a daily mean value of I (W/m^2) incident on the window area leads to an input of SIA_g to the enclosure. This gain, together with the output Q, wholly or partly supplied by a heater, maintains the internal design temperature T_d against an ambient temperature T_e. The steady-state or daily mean heat balance equation at the room index node T_{ra} (having the value T_d) is

$$Q = (A_g U_g + (A - A_g)U_w + L + V)(T_d - T_e) - SIA_g. \tag{9.25}$$

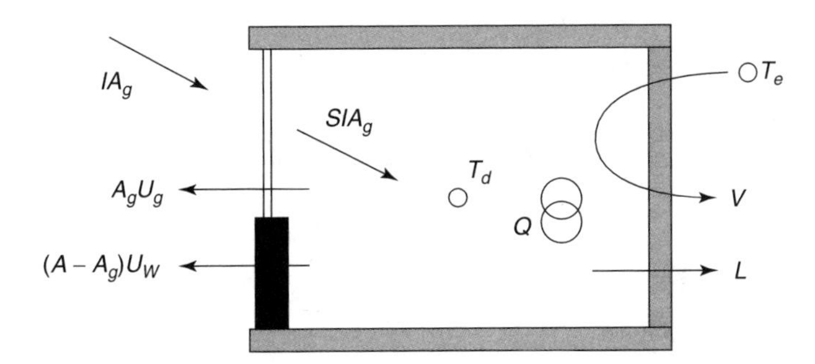

Figure 9.11 Steady-state model of an enclosure partly heated by solar gain

[9]The energy they report collected per square metre of wall during a heating season appears to vary strongly with latitude; for example, 134 kWh/m^2 at 44° and 59 kWh/m^2 at 49° for a non-ventilated solar wall.

To see in general terms how the heat need Q varies as the window area varies, it is convenient to redraft the equation as

$$T_h \equiv \frac{Q}{AU_w + (L+V)} = \underbrace{(T_d - T_e)}_{\text{degree-day value}} + \frac{AU_g}{AU_w + (L+V)}\left(T_d - \underbrace{\left(T_e + \frac{SI}{U_g}\right)}_{\text{sol-air temperature}} - (T_d - T_e)\frac{U_w}{U_g}\right) \underbrace{(A_g/A)}_{\text{window area, design variable}}. \quad (9.26)$$

In this form, through division by the total loss coefficient, the heat need Q is expressed in units of temperature as the quantity T_h. When ambient temperature T_e is lower than the design or comfort temperature T_d, some heating is needed and only periods when T_d is greater than T_e need be considered. T_h partly depends on the degree-day value $T_d - T_e$ for the site. Of the area-dependent terms, $(T_d - T_e)U_w/U_g$ can be neglected since U_w/U_g is likely to be small. The dependence of T_h on $T_d - (T_e + SI/U_g)$ is shown in Figure 9.12.

For zero glazing, the heat needs are simply large and small on cool and warm days, respectively. On a cool day, the change in heat need with increasing window area depends upon the insolation. In dull conditions, I is small and $T_d - (T_e + SI/U_g)$ is positive, so T_h increases with A_g/A and the glazing is disadvantageous. In sufficiently bright conditions, $T_d - (T_e + SI/U_g)$ is negative and T_h decreases, so the glazing is advantageous. Without air conditioning, T_h or Q cannot be negative. A similar situation pertains on a bright and warm day. The mean heat need over the three days depends on all three days until Q(bright and warm) becomes zero. It then depends on the needs for the two cool days until the benefit of the sunny day is lost. Thereafter the mean need depends on the dull day alone. See Figure 9.12.

The mean need averaged over a number of days has a smooth form, but it must have the curvature in Figure 9.12. Possible forms in relation to climate type are suggested in Figure 9.13. There may or may not be a minimum value, that is, an optimum area of glazing. The daily mean characteristics of the glazing are expressed by the ratio S/U_g. As

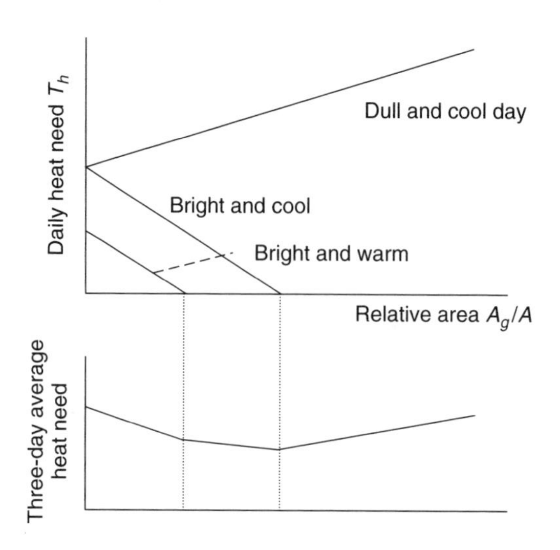

Figure 9.12 Heat need as a function of window area: three weather types and their average

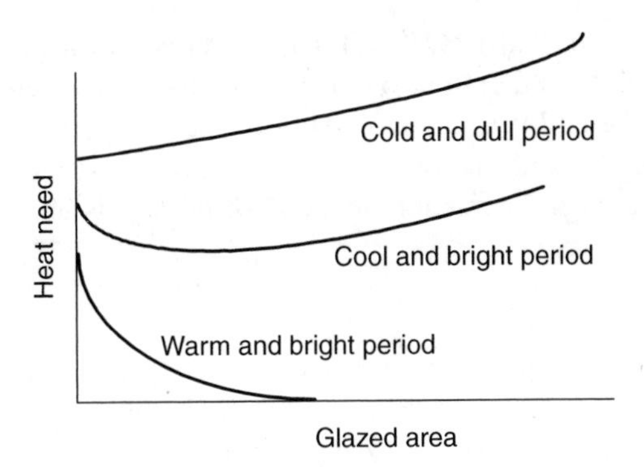

Figure 9.13 Heat need as affected by the interaction of climate with glazing

noted earlier, $(S/U_g)_{double\ glazing} > (S/U_g)_{single\ glazing}$ and performance can be improved further by providing additional insulation by night.

In the indirect system, a mass wall is placed behind glazing as shown in Figure 9.14. The mass wall, thickness d and resistance $d/A\lambda$, is flanked by an internal resistance R_i and an external resistance R_e. R_e includes the resistance of the cavity between the wall and the glazing, the glazing cavity if double glazed, and the outside film resistance.

The steady-state heat balance equation at T_d is now

$$Q = (L + V + 1/(R_e + d/A\lambda + R_i))(T_d - T_e) - SIA\ R_e/(R_e + d/A\lambda + R_i), \quad (9.27)$$

which can be written as

$$T_h \equiv \frac{Q}{L+V} = \underbrace{(T_d - T_e)}_{\text{degree-day value}} + \frac{AU'}{(L+V)}\underbrace{(T_d - (T_e + SIAR_e))}_{\text{sol-air temperature}} \underbrace{\frac{R_e + R_i}{R_e + d/A\lambda + R_i}}_{\text{wall thickness, design variable}}. \quad (9.28)$$

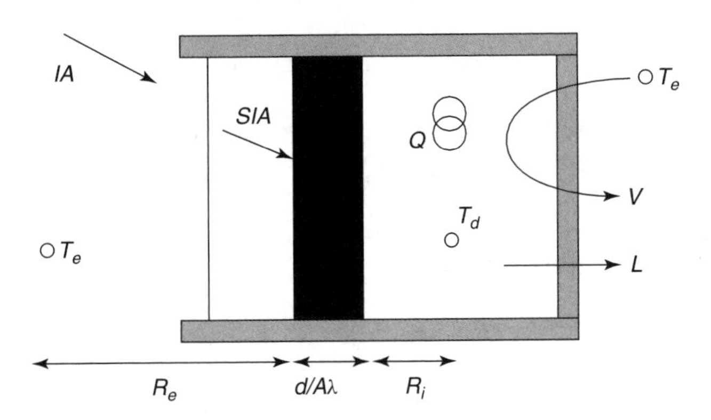

Figure 9.14 Steady-state model of an enclosure partly heated by indirect solar gain

This is formally similar to (9.26). $AU' = 1/(R_e + R_i)$ is introduced to preserve the similarity. The case of wall thickness $d = 0$ corresponds to $A_g = A$ and $d = \infty$ corresponds to $A_g = 0$. Figures 9.12 and 9.13 apply equally to indirect systems.

Sol-air temperature T_{sa} is normally associated with the incidence of radiation I (W/m^2) on an external surface where a fraction α, the absorptivity, is absorbed:

$$T_{sa} = T_e + (\alpha I/h_e \text{ or } \alpha I r_e); \tag{9.29}$$

h_e and r_e denote the outside film link in transmittance (W/m^{2}K) and resistance (m^{2}K/W) forms, respectively, between the node where the radiation is absorbed and the ambient node. Thus the forms $T_e + SI/U_g$ in (9.26) and $T_e + SIAR_e \equiv T_e + SIr_e$ in (9.28) are forms of sol-air temperature.

It will be clear that solar gain can be utilised when lower rather than higher values are set for the design temperature T_d. Further, for a direct-gain system, if there is an optimum area of glazing, its value increases with the size of the fabric and ventilation losses L and V, and an increase of these losses leads to an increase in the heat need Q.

Davies (1980, 1982a, 1986, 1987) conducted a study of possible savings in back-up heating for a site in the UK at 53.4°N and with a design temperature of 18°C. It was based on a passive solar building with a large solar wall designed in the late 1950s (Morgan 1966) and it used 50 years of daily values for ambient temperature and estimates of the total incident radiation recorded nearby at Bidston Observatory. The study suggested substantial savings for a double-glazed window with night insulation, significant savings for a double-glazed window and a double-glazed mass wall and minor savings for a single-glazed window and wall. For direct-gain systems, windows larger than their thermally 'optimal' values may be appropriate as a result of the extra daylight admitted and consequent saving in electric lighting.

Following the energy crisis of 1973 there was an upsurge of interest in passive solar construction and Balcomb in particular has developed methods to estimate the value of solar gains in passive solar buildings. See SEIS (1981), Chapter 16 by Balcomb in Kreider and Krieth (1981) and articles by J.D.Balcomb (1984) and S. Balcomb (1984). Szokolay (1980) illustrates a large number of solar-heated buildings. However, the use of time-averaged measures such as solar gain factors to estimate seasonal energy needs is hardly satisfactory; solar gains may lead to large daily variations in measures of room temperature which depend on the room thermal capacity, a major parameter so far largely ignored in this connection. More reliable estimates of response will be given using hour-by-hour values for excitation variables (casual gains, solar incidence and ambient temperature, together with the effect of ventilation) and measures of the fabric response. Following a study of time-varying conduction in walls, an approach to handling hourly response is presented in the final chapter.

10 The Wall with Lumped Elements

Thermal mass in a building can reduce peak cooling loads and the size of swings in indoor temperature. Consider a wall with exterior and interior temperatures T_e and T_i, initially equal. Suppose that radiation (solar or long-wave) is incident on the interior surface. If the wall has no thermal mass (i.e. capacity), T_i quickly rises to an equilibrium value, heat is lost to the air and may have to be removed immediately by the cooling plant if room temperature is not to exceed some specified (comfort) temperature. If however the wall has mass – it is brick or concrete, say – T_i rises slowly, with response times of order hours; either the cooling load needed is now less or the peak temperature reached is less. If the radiation is received at the wall exterior, similar considerations apply, although excitation at T_e has less effect on inside conditions than the same excitation acting at T_i. In addition to reducing cooling loads and peak temperatures, thermal mass delays its effects by a few hours. Ruud *et al.* (1990) reported cooling energy savings of 18% during the daytime through use of mass, although there was no reduction in peak demand. How these effects can be estimated for a given space depends on a range of factors: size and resistive/storage characteristics of its bounding elements, forms of thermal excitation, patterns of occupancy, ventilation and control. A useful introduction to the problem is given by Balaras (1996).

This chapter is concerned with the storage of sensible heat in a discretised wall with conduction as the dominant mechanism. Heat can be stored in many forms however, as shown in Figure 1 of Tahat *et al.* (1993). Winwood *et al.* (1997) report a detailed study of sensible energy storage in the fabric of the building, notably massive floors, where heat transfer is effected by air movement through embedded pipes. Latent heat storage through the use of phase change materials is discussed by Hawes *et al.* (1993).

It is convenient to start a study of the dynamic behaviour of walls by confining attention to a wall composed of a succession of purely resistive and purely capacitative elements. This allows us to present certain ideas which are simpler as applied to discretised elements than to real wall elements in which resistance and capacity are distributed. Further, the finite difference technique to estimate wall heat flow is based on expression of the wall in discretised form.

10.1 MODELLING CAPACITY

A pure capacitative element is one whose conductivity is infinite, so that all points are at the same temperature. No real material has infinite conductivity, but a thin sheet

Building Heat Transfer Morris G. Davies
© 2004 John Wiley & Sons, Ltd ISBN: 0-470-84731-X

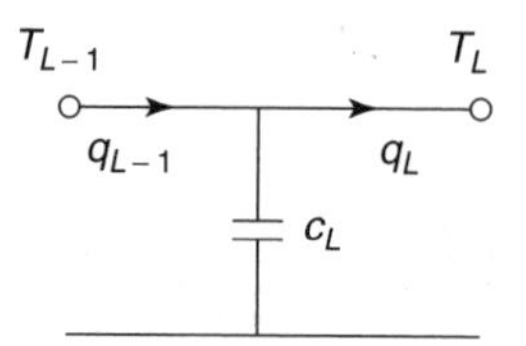

Figure 10.1 The basic capacitative element

of metal in a wall is effectively isothermal and will be assigned a capacity of $c =$ (thickness X)(density ρ)(mass specific heat c_p) with units J/m^{2}K. Room internal partitions may sometimes be treated as though isothermal.

A heat flow q into the capacity produces an increase in temperature: $q = c\,\mathrm{d}T/\mathrm{d}t = c\,\mathrm{d}(T - T_{ref})/\mathrm{d}t$, where T_{ref} is some convenient reference temperature, e.g. 0°C. If the capacity is element L in a succession of n layers and denoted c_L, it can be modelled as shown in Figure 10.1. The connection to the reference temperature is shown here, but to avoid unnecessary detail the connection will normally be omitted.

Since $T_{L-1} = T_L$ and $q_{L-1} = c_L\,\mathrm{d}T_\mathrm{L}/\mathrm{d}t + q_L$, its effect can be represented by the transmission matrix

$$\begin{bmatrix} T_{L-1} \\ q_{L-1} \end{bmatrix} = \begin{bmatrix} 1 & 0 \\ c_L D & 1 \end{bmatrix} \begin{bmatrix} T_L \\ q_L \end{bmatrix}, \qquad (10.1)$$

where D denotes $\mathrm{d}/\mathrm{d}t$. We shall be concerned with three types of time variation:

- Temperatures in the wall are increasing linearly, so that the time-varying value T_{Lt} consists of a $t = 0$ value $T_L{}'$ and a time-varying component θt, that is, $T_{Lt} = T_L{}' + \theta_L t$. A matrix formulation is given later. Equation (17.29) refers to layer 2 of distributed resistance r_2 and capacity c_2 placed between two similar layers. If layer 2 has no resistance, the terms including r_2 are zero.

- Temperatures in the wall are varying sinusoidally with amplitude $T_L{}'$, so that $T_{Lt} = T_L{}' \exp(\mathrm{j}2\pi t/P) = T_L{}' \exp(\mathrm{j}\omega t)$ where $\mathrm{j} = \sqrt{-1}$, P is the period and $\omega = 2\pi/P$ is the frequency in rad/s:

$$\begin{bmatrix} T_{L-1} \\ q_{L-1} \end{bmatrix} = \begin{bmatrix} 1 & 0 \\ \mathrm{j}\omega c_L & 1 \end{bmatrix} \begin{bmatrix} T_L \\ q_L \end{bmatrix}; \qquad (10.2)$$

j enters the formulation because the heat flow into a capacity is a quarter of a cycle or $\pi/2$ radians ahead of the associated temperature in sinusoidally varying conditions. Also $\mathrm{j} \equiv \exp(\mathrm{j}\pi/2)$.

- Temperatures in the wall are decaying exponentially from an initial value $T_L{}'$, so that $T_{Lt} = T_L{}' \exp(-t/z)$, where z is a decay time:

$$\begin{bmatrix} T_{L-1} \\ q_{L-1} \end{bmatrix} = \begin{bmatrix} 1 & 0 \\ -c_L/z & 1 \end{bmatrix} \begin{bmatrix} T_L \\ q_L \end{bmatrix}. \qquad (10.3)$$

The first two cases apply when the wall of which the capacity is part is excited by external means, either by a steadily increasing or a sinusoidally varying temperature. The third case applies when the wall has been excited in some way; from time $t = 0$ it is no longer excited but, with specified boundary conditions (bounding nodes being either isothermal or adiabatic), the temperature decays to its final value, usually zero. In this case the decay is composed of a series of modes, one for a one-capacity system, two for a two-capacity system, etc. The temperature components are of form $A_{ij}\exp(-t/z_j)$, where j denotes the number of the mode, z_j the corresponding decay time and A_{ij} (K) depends on position in the wall and its initial ($t = 0$) state. One purpose of the following discussion is to show how to find the response of a thermal system, initially at zero temperature when subjected to ramp excitation for $t > 0$. This is an essential step in deriving wall transfer coefficients (see later). It will be shown by elementary means for a single-capacity wall and by formal means for a two-capacity wall.

10.2 FORMS OF RESPONSE FOR A SINGLE-CAPACITY CIRCUIT

10.2.1 The r-c Circuit

The simplest circuit to model dynamic behaviour is shown in Figure 10.2a with elements r and c. The heat flow q_2 to T_2 is zero and $T_2 = T_1$. In general, to preserve heat continuity,

$$(T_0 - T_1)/r_1 = c\,\mathrm{d}T_2/\mathrm{d}t. \tag{10.4}$$

Its response to a ramp excitation is composed of independent transient and slope components.

Transient response

Suppose that T_1 has the value T_I up to time zero and that from $t = 0$ onward, $T_0 = 0$. Then

$$\mathrm{d}T_2/T_2 = -\mathrm{d}t/r_1c = -\mathrm{d}t/z, \tag{10.5}$$

where z is the response time or decay time r_1c. The value of T_2 is given subsequently as

$$T_{2t} = T_I \exp(-t/z). \tag{10.6}$$

T_2 falls to $1/\mathrm{e} = 0.368$ of its initial value after the time z. If $T_0 = T_1 = T_2 = 0$ initially but at $t = 0$, T_0 steps to a constant value T_I, then

$$T_{2t} = T_I(1 - \exp(-t/z)). \tag{10.7}$$

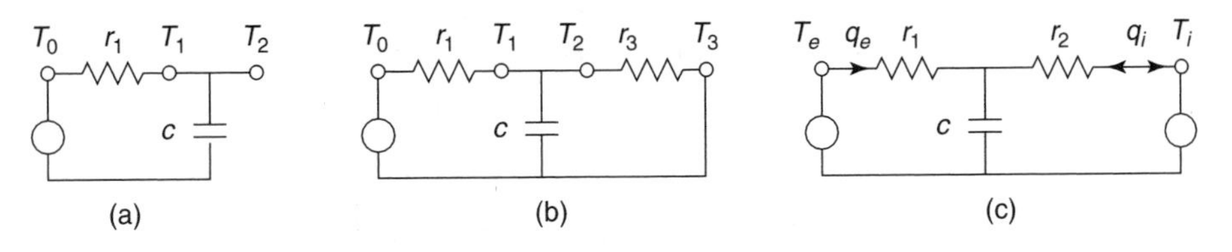

Figure 10.2 Three circuits: (a) r-c, (b) r-c-r, (c) r-c-r

Slope response

Suppose instead that T_0 has been increasing for an indefinite period at a rate θ (K/s) and that the time origin is chosen so that $T_0 = \theta t$. At $t = 0$, T_1 is negative but increases at rate θ. Then

$$\begin{aligned} T_1 = T_2 &= T_2{}' + \theta t \\ &= -\theta z + \theta t. \end{aligned} \qquad (10.8)$$

from the continuity equation.

Ramp response

Since the continuity equation is linear in T, solutions can be added. To evaluate the ramp response, we add the slope solution to the first transient solution with an initial value of $T_1 = T_2 = +\theta z$. So

$$T_{2t} = (-\theta z + \theta t) + \theta z \exp(-t/z) = \theta(t - z(1 - \exp(-t/z))). \qquad (10.9)$$

This is the solution with the initial condition that $T_2 = 0$, and that from $t = 0$ onward, $T_0 = \theta t$. Shortly after $t = 0$, T_{2t} is substantially unaltered. As time progresses, the transient component dies out and for large values of time ($t > 5z$ say), T_{2t} approximates to $\theta(t - z)$.

10.2.2 The r-c-r Circuit: Ramp Solution

Figure 10.2a shows an adiabatic right-hand condition and Figure 10.2b an isothermal condition. Continuity then requires that

$$(T_0 - T_1)/r_1 = c\,\mathrm{d}T_2/\mathrm{d}t + (T_2 - T_3)/r_3, \qquad (10.10)$$

where $T_3 = 0$. For the slope solution, we take

$$T_0 = \theta t \quad \text{and} \quad T_{2t} = T_2{}' + \theta_2 t. \qquad (10.11)$$

When these quantities are substituted into the continuity equation, some terms include t and some do not. Since the equation must be true at all times, it provides values for θ_2 and $T_2{}'$ and it is found that

$$T_{2t} = \frac{r_3\theta}{r_1 + r_3}(t - z), \qquad (10.12)$$

where z is the decay time $cr_1r_3/(r_1 + r_3)$ of the transient solution. If to this we add a transient solution with $T_0 = 0$ and an initial value for T_2 of $+r_3\theta z/(r_1 + r_3)$,

$$T_{2t} = \frac{r_3\theta}{r_1 + r_3}(t - z(1 - \exp(-t/z))). \qquad (10.13)$$

This is the solution of the problem where c is initially at zero temperature, $T_3 = 0$, and $T_0 = \theta t$ for $t > 0$. But our principal interest in developing transient solutions is not to find temperatures within a wall; rather it is to find the heat flow into or out of the wall due to some series of imposed temperatures. In reverse order, these are the required stages:

(i) Express the present flux into or out of the wall as a function of present and previous temperatures and also a function of previous fluxes, as shown in equation (11.1). The functions are the *wall transfer coefficients*, to be determined in Chapters 11 and 17.

(ii) The transfer coefficients follow from the fluxes due to an imposed triangular pulse of temperature. The series of these fluxes are the *wall response factors.*

(iii) The triangular pulse is formed by superposing three ramps in temperature and the fluxes are formed as the sum of three individual ramp responses.

For stage (iii) we want the flux due to a single ramp in temperature and this has a simple form for the r-c-r circuit. The heat flow q_{00} into T_0 due to the ramp increase θt at T_0 is

$$q_{00,t} = \frac{\theta t - T_{2t}}{r_1} = \frac{\theta t}{r} + \frac{\theta c r_3^2}{r^2} - \frac{\theta c r_3^2}{r^2} \times \exp\left(-\frac{t}{z}\right). \quad (10.14a)$$

This can be written more generally as

$$q_{00,t} = \underset{\substack{\text{slope solution}\\ \text{time variation}}}{\theta t/r} + \underset{\substack{\text{slope solution}\\ \text{constant}}}{q_{00,0}} + \underset{\substack{\text{from slope and}\\ \text{transient solutions}}}{\sum q_{00,j}} \times \underset{\substack{\text{from transient}\\ \text{solution only}}}{\exp(-t/z_j)}. \quad (10.14b)$$

- The first subscript to q on the left denotes that it is the flux at node 0 that is intended, the second that it is driven by a temperature source also acting at node 0. The third subscript denotes time.
- The first two subscripts to the qs on the right-hand side have the same meaning as above. These qs, however, have the status of coefficients (but proportional to θ): $q_{00,0}$ is the constant term associated with the slope component of the solution and $q_{00,j}$ the coefficient of the jth component of the transient solution. $j = 1$ only for the r-c-r circuit.
- The term $q_{00,t}$ consists of three components. The first, $\theta t/r$, always has this simple form; $r = r_1 + r_3$ here and $r = \sum r_L$ generally.
- The slope constant term $q_{00,0}$ can be found generally by multiplication of the respective slope transmission matrices in (17.29) together with (17.34). For the current r-c-r model the matrices correspond to $[r_1, c_1 = 0][c, r = 0][r_3, c_3 = 0]$.
- The third term, $q_{00,j}$, has a simple form here, the negative of $q_{00,0}$ but in general it is a complicated function of the various r_L, c_L values (Section 17.5.1).

These components are illustrated for the r-c-r circuit in Figure 10.3. The transient term here is a simple exponential decay. More generally it consists of the sum of exponential terms with decreasing values of z_j, whose effect is to lead to a faster initial response.

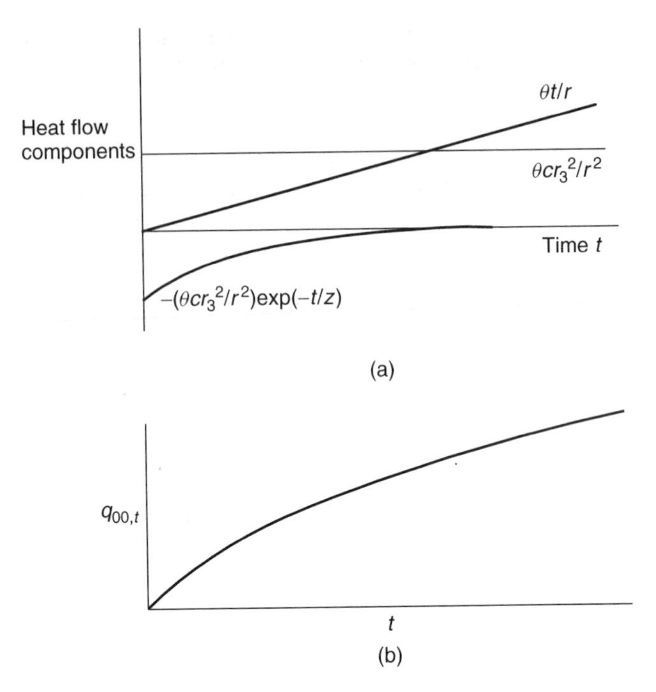

Figure 10.3 Components of heat flow into an *r*-*c*-*r* circuit due to a ramp excitation $T_0 = \theta t$. (b) Net heat flow

The term $q_{33,t}$ is found by replacing r_3 by r_1 in (10.14a). Clearly $q_{33,t} \neq q_{11,t}$ unless the wall is symmetrical. Also $q_{30,t}$, found as T_{2t}/r_3 is symmetrical in r_1 and r_3 and $q_{30,t} = q_{03,t}$.

10.2.3 The r-c-r Circuit: Periodic Solution

It is convenient here to present the elementary forms for the wall parameters when it is undergoing sinusoidal excitation, that is, the temperatures at the external and internal nodes T_e and T_i are varying sinusoidally as $T \sin \omega t$ or $T \cos \omega t$, or generally as $T \exp(j\omega t)$. The corresponding flows are q_e and q_i, shown in Figure 10.2c. They are related as

$$\begin{bmatrix} \mathbf{T}_e \\ \mathbf{q}_e \end{bmatrix} = \begin{bmatrix} 1 & r_1 \\ 0 & 1 \end{bmatrix} \begin{bmatrix} 1 & 0 \\ j\omega c & 1 \end{bmatrix} \begin{bmatrix} 1 & r_2 \\ 0 & 1 \end{bmatrix} \begin{bmatrix} \mathbf{T}_i \\ \mathbf{q}_i \end{bmatrix} = \begin{bmatrix} 1 + j\omega r_1 c & r_1 + r_2 + j\omega r_1 c r_2 \\ j\omega c & 1 + j\omega c r_2 \end{bmatrix} \begin{bmatrix} \mathbf{T}_i \\ \mathbf{q}_i \end{bmatrix}$$

$$= \begin{bmatrix} e_{11} & e_{12} \\ e_{21} & e_{22} \end{bmatrix} \begin{bmatrix} \mathbf{T}_i \\ \mathbf{q}_i \end{bmatrix}. \tag{10.15}$$

Now the cyclic transmittance $\mathbf{u}$ is defined as the heat flow $\mathbf{q}_i$ *into* the enclosure due to unit variation in $\mathbf{T}_e$ when $\mathbf{T}_i$ is held at zero, (i.e. $\mathbf{T}_i$ is isothermal). Inserting $\mathbf{T}_i = 0$ in the equation

$$\mathbf{T}_e = e_{11} T_i + e_{12} \mathbf{q}_i, \tag{10.16}$$

we have

$$\mathbf{u} = \left(\frac{\mathbf{q}_i}{\mathbf{T}_e}\right)_{T_i=0} = \frac{1}{e_{12}} = \frac{1}{r_1 + r_2 + j\omega r_1 c r_2} = \frac{r_1 + r_2 - j\omega r_1 c r_2}{(r_1 + r_2)^2 + (\omega r_1 c r_2)^2}. \tag{10.17}$$

Now the steady-state U value is $1/(r_1 + r_2)$ so the equations show that

- the magnitude of $\mathbf{u}$, $((r_1 + r_2)^2 + (\omega r_1 c r_2)^2)^{-1/2}$, is less than U;
- since $\omega r_1 c r_2$ is positive, $\mathbf{q}_i$ has its maximum value after $\mathbf{T}_e$ has its maximum.

Noting too that

$$\mathbf{q}_e = e_{21}\mathbf{T}_i + e_{22}\mathbf{q}_i \tag{10.18}$$

and that

$$e_{11}e_{22} - e_{21}e_{12} = 1, \tag{10.19}$$

(a result which is true for the product of transmission matrices representing both periodic and transient flow), the reciprocal relation follows:

$$\left(\frac{\mathbf{q}_e}{\mathbf{T}_i}\right)_{\mathbf{T}_e=0} = \frac{-1}{e_{12}}. \tag{10.20}$$

In using a U value, the room temperature is normally taken to be higher than ambient and so there is a steady flow of heat *out of* the room driven by $\mathbf{T}_i$. In using this dynamic $\mathbf{u}$ value, there is a fluctuating flow of heat *into* the room, driven by $\mathbf{T}_e$ or the corresponding sol-air temperature. The implication of the negative sign should be kept in mind.

The fluctuating component of heat flow $\mathbf{q}_i$ *out of* the room driven by a fluctuating indoor temperature is the admittance. Equation (10.16) refers to the value of $\mathbf{q}_i$ *into* the room. Accordingly, the value of $\mathbf{q}_i$ *out of* the room is

$$\mathbf{y}_\mathbf{i} = \left(\frac{\mathbf{q}_i}{\mathbf{T}_i}\right)_{T_e=0} = +\frac{e_{11}}{e_{12}} = \frac{1 + j\omega r_1 c}{r_1 + r_2 + j\omega r_1 c r_2}. \tag{10.21}$$

It readily follows that the magnitude of $\mathbf{y}_\mathbf{i}$ is greater than $U = 1/(r_1 + r_2)$ and, using the expression for the phase of $\mathbf{y}$ (Section 10.4.5), $\mathbf{q}_i$ has its maximum value *before* $\mathbf{T}_i$ has its maximum. Thus

$$\begin{array}{ccccc} \text{magnitude}(\mathbf{u}) & < & U \text{ value} & < & \text{magnitude}(\mathbf{y}), \\ \text{phase}(\mathbf{u}) & < & 0 & < & \text{phase}(\mathbf{y}). \end{array}$$

These results are true for walls of any construction. The derivation of $\mathbf{u}$ and $\mathbf{y}$ values for walls is presented in Chapter 15.

10.3 THE TWO-CAPACITY WALL

We wish to determine temperatures in a system composed of the sequence r_1, c_2, r_3, c_4, r_5 (Figure 10.4). It is quite easy to use the method developed in Section 10.2.3 to find

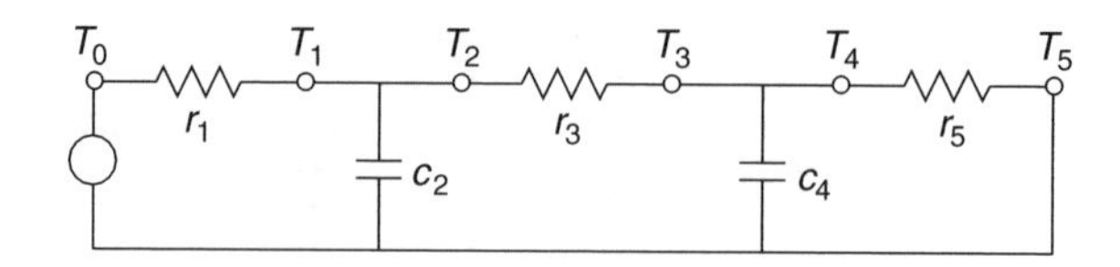

Figure 10.4 The two-capacity circuit

the ratio $\mathbf{q}_i/\mathbf{T}_e = \mathbf{u}$, which applied when $\mathbf{T}_e$ varies sinusoidally, with period 24 hours in building applications. We shall in passing derive a solution for the case when the temperatures at c_2 and c_4 have specified initial values and from $t = 0$, T_0 and T_5 are zero; it provides a simple example of a class of step-excitation problems. Our main interest however is in the case when the system is initially at zero temperature, and for $t > 0$, T_0 rises at the rate θ(K/s) while T_5 remains at zero. It leads to a ramp solution similar to (10.14). From a combination of ramp solutions we can derive the set of response factors, and use them to find the wall conduction transfer coefficients (Chapter 11). These coefficients provide the means to find the heat flow into a room (q_i) due to values of ambient temperature (T_e) specified at hourly values. They allow a much more detailed examination of wall dynamic behaviour than is possible using the values of $\mathbf{u}$ and $\mathbf{y}$, which are only valid when T_e has been varying sinusoidally for a long period.

The logic for the problem of ramp response is shown in Figure 10.5. The full solution is formed by adding the slope and transient solutions. The decay times of the transient solution depend on which one of four possible sets of boundary conditions is selected and they are discussed in Section 10.3.1. In fact, only one boundary condition proves to be of interest; this leads to two independent temperature distributions which slump from some initial condition with decay times z_1 and z_2.

10.3.1 Wall Decay Times

To ensure continuity of temperature and flow between the elements, the layer transmission matrices are multiplied.

$$\begin{bmatrix} T_0 \\ q_0 \end{bmatrix} = \begin{bmatrix} 1 & r_1 \\ 0 & 1 \end{bmatrix} \begin{bmatrix} 1 & 0 \\ -c_2/z & 1 \end{bmatrix} \begin{bmatrix} 1 & r_3 \\ 0 & 1 \end{bmatrix} \begin{bmatrix} 1 & 0 \\ -c_4/z & 1 \end{bmatrix} \begin{bmatrix} 1 & r_5 \\ 0 & 1 \end{bmatrix} \begin{bmatrix} T_5 \\ c_5 \end{bmatrix}. \tag{10.22}$$

This is of form

$$\begin{bmatrix} T_0 \\ q_0 \end{bmatrix} = \begin{bmatrix} e_{11} & e_{12} \\ e_{21} & e_{22} \end{bmatrix} \begin{bmatrix} T_5 \\ q_5 \end{bmatrix} \quad \text{or} \quad \begin{aligned} T_0 &= e_{11}T_5 + e_{12}q_5, \\ q_0 &= e_{21}T_5 + e_{22}q_5, \end{aligned} \tag{10.23}$$

where

$$\begin{aligned} e_{11} &= 1 - (r_1c_2 + r_1c_4 + r_3c_4)/z + r_1c_2r_3c_4/z^2, \\ e_{12} &= (r_1 + r_3 + r_5) - (r_1c_2r_3 + r_1c_2r_5 + r_1c_4r_5 + r_3c_4r_5)/z + r_1c_2r_3c_4r_5/z^2, \\ e_{21} &= -(c_2 + c_4)/z + c_2r_3c_4/z^2, \\ e_{22} &= 1 - (c_2r_3 + c_2r_5 + c_4r_5)/z + c_2r_3c_4r_5/z^2. \end{aligned} \tag{10.24}$$

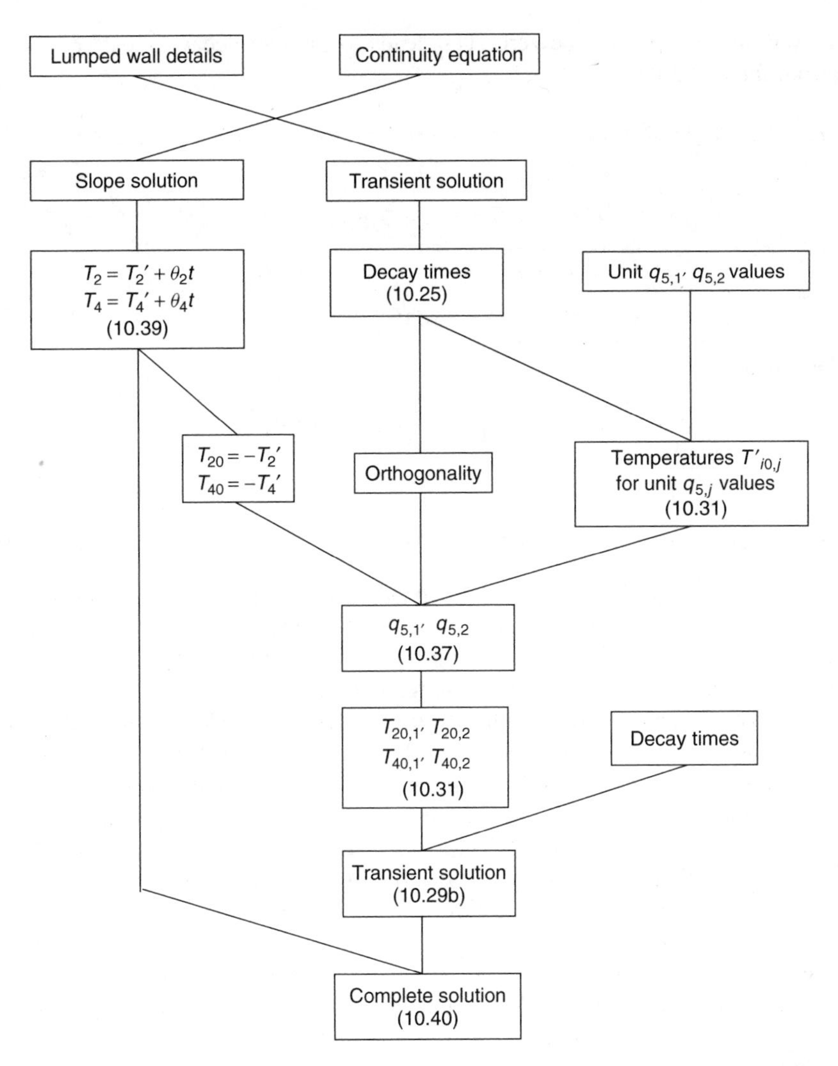

Figure 10.5 Calculation of the ramp response of a two-capacity wall. Equation numbers are in parentheses

The decay times of the system depend on what boundary conditions are imposed.[1] If it is assumed, as is most usual, that the wall is isothermal at its bounding nodes, then

[1]These conditions are sometimes referred to by name. The Dirichlet condition (or 'first kind' condition) denotes that a surface temperature is prescribed, the Neumann condition (or 'second kind' condition) denotes that a surface gradient or flux is prescribed and the Robin condition (the 'third kind') denotes that the heat exchange is proportional to a temperature difference, that is, Newton's law of cooling applies. It is argued on the cover of ASME's *Journal of Heat Transfer* **110**(1) that the concept of the heat transfer coefficient should be attributed to Fourier, not Newton. Kolesnikov (1987) sketches the use of these boundary conditions in classical problems tackled in the eighteenth and nineteenth centuries. These include the oscillations of rods and membranes as well as thermal applications by Fourier. The name of Dirichlet was first attached to the relevant boundary condition by Riemann in 1857. Kolesnikov states the boundary condition of the third kind as the assignment of a linear combination of the unknown function with its normal derivative at the interface, so that $\alpha T + \beta q = \zeta$. With the

$T_0 = T_5 = 0$ and so e_{12} must be zero. This leads to the two decay times, $z_{12,1}$ and $z_{12,2}$, found from the equation

$$(r_1 + r_3 + r_5)z_{12}^2 - (r_1c_2r_3 + r_1c_2r_5 + r_1c_4r_5 + r_3c_4r_5)z_{12} + r_1c_2r_3c_4r_5 = 0, \quad (10.25a)$$

so

$$z_{12} = \{(r_1c_2r_3 + r_1c_2r_5 + r_1c_4r_5 + r_3c_4r_5) \pm \sqrt{\begin{array}{c}(r_1c_2r_3 + r_1c_2r_5 + r_1c_4r_5 + r_3c_4r_5)^2 \\ -4(r_1 + r_3 + r_5)r_1c_2r_3c_4r_5\end{array}}\}/2(r_1 + r_3 + r_5), \quad (10.25b)$$

with the relations that

$$z_{12,1} + z_{12,2} = (r_1c_2r_3 + r_1c_2r_5 + r_1c_4r_5 + r_3c_4r_5)/(r_1 + r_3 + r_5) \quad (10.25c)$$

and

$$z_{12,1} \times z_{12,2} = r_1c_2r_3c_4r_5/(r_1 + r_3 + r_5). \quad (10.25d)$$

(If $r_1c_2 = c_4r_5 = z'$ say, then $z_{12,1} = z'$ independently of r_3 (as is physically obvious) and $z_{12,2} = z'r_3/(r_1 + r_3 + r_5)$. If r_3 is much larger than r_1 and r_5, the z_{12} values tend to r_1c_2 and c_4r_5.) To make use of this transient solution, we may add a further solution based on *imposed temperatures*, steady or time-varying, at T_0 and T_5.

We could however choose the condition corresponding to node 0 being isothermal and node 5 being adiabatic and in this case $e_{11} = 0$, leading to values for $z_{11,1}$ and $z_{11,2}$. To the transient solution could then be added a *temperature* at node 0 and a *flow* at node 5. If the system is adiabatic at node 5, its transient behaviour should be independent of the value of r_5 and r_5 does not appear in e_{11}. The case of $e_{22} = 0$ reverses these considerations. Finally, if $e_{21} = 0$, then there is only one decay time, $z_{21,1}$, since it depends on neither r_1 nor r_5. But e_{21} is zero in steady-state conditions and $z_{21,0} = \infty$ provides a formal solution.

Thus there is a total of $2 + 2 + 2 + 1 = 7$ decay times but the system has only 5 degrees of freedom; there must be internal relations between the various decay times. First we note that the quadratic function $f = x^2 + bx + c$ can be expressed as $f = x^2 - (x_1 + x_2)x + x_1x_2$ where x_1 and x_2 are the roots when $f = 0$. Thus the e_{ij} can be written as

$$\begin{aligned} e_{11} &= 1 - (z_{11,1} + z_{11,2})/z + z_{11,1} \cdot z_{11,2}/z^2, \\ e_{12} &= (r_1 + r_3 + r_5)[1 - (z_{12,1} + z_{12,2})/z + z_{12,1} \cdot z_{12,2}/z^2], \\ e_{21} &= (-(c_2 + c_4)/z)[1 - z_{21}/z], \\ e_{22} &= 1 - (z_{22,1} + z_{22,2})/z + z_{22,1} \cdot z_{22,2}/z^2. \end{aligned} \quad (10.26)$$

Now the determinant of any element transmission matrix is unity and the determinant of the overall matrix is also unity:

$$e_{11} \cdot e_{22} - e_{21} \cdot e_{12} = 1 \quad (10.27)$$

Chapter 4 definitions of a pure temperature source T and a pure heat source q, this is not a realisable condition at an external boundary: a pure temperature source has zero impedance and is taken to act at some node T_e say; if a pure heat source is then also taken to act at T_e, the entire flux q flows to the reference temperature and has no effect on the physical system which has T_e as one of its bounding nodes. Beck and Litkouhi (1988) list two further boundary conditions.

and this is true for all values of z. Equating the terms in z^{-1} we find

$$z_{11,1} + z_{11,2} + z_{22,1} + z_{22,2} = (r_1 + r_3 + r_5)(c_2 + c_4)$$
$$= (\text{total resistance } r)(\text{total capacity } c). \quad (10.28a)$$

This is true for a wall with n capacities:

$$\sum z_{11,j} + \sum z_{22,j} = \sum r_i \sum c_i. \quad (10.28b)$$

It is also true in principle for a wall with layers of distributed resistance and capacitance, but this cannot be demonstrated exactly since the wall has an infinite series of $z_{11,j}$ and $z_{22,j}$ values and their sum converges only slowly.

In the above five-element wall, there are three further relations between the $z_{ij,k}$ values, found by equating terms in z^{-2}, z^{-3} and z^{-4}. For example, equating terms in z^{-2} and using (10.28a),

$$(z_{11,1} + z_{11,2})(z_{22,1} + z_{22,2}) + z_{11,1} \cdot z_{11,2} + z_{22,1} \cdot z_{22,2}$$
$$= (z_{11,1} + z_{11,2} + z_{22,1} + z_{22,2})(z_{12,1} + z_{12,2} + z_{21}). \quad (10.28c)$$

Thus for this chain of five elements, there are seven decay times and three internal relations between them. In general,

$$\begin{pmatrix}\text{number of}\\ \text{decay times}\end{pmatrix} + 1 - \begin{pmatrix}\text{number of internal}\\ \text{relations between them}\end{pmatrix} = \begin{pmatrix}\text{number of elements}\\ \text{in the chain}\end{pmatrix}.$$

The value $+1$ corresponds to the steady-state value of $z_{21,0} = \infty$.

The structure of the set of decay times can be illustrated by a three-capacity ($n = 3$) wall. The number of elements is $2n + 1 = 7$. Take the nominal values, $r_1 = 1$, $c_2 = 2$, $r_3 = 3$, $c_4 = 4$, $r_5 = 5$, $c_6 = 6$ and $r_7 = 7$. Thus the term 123, one of those formed by multiplying the seven matrices, may be interpreted as $r_1 c_2 r_3$, and with these values has the numerical value 6. Multiplication demonstrates that the element e_{12} for example is given as

$$e_{12} = (1+3+5+7) - (123+125+127+145+147+167+345+347+367+567)/z$$
$$+ (12345 + 12347 + 12367 + 12567 + 14567 + 34567)/z^2 - 1234567/z^3.$$

Since e_{12} is to be zero, we have a cubic equation and the three values of $z_{12,k}$ follow readily using a standard computer routine. Similarly for e_{11}, e_{21} and e_{22}. There are three solutions to $e_{11} = 0$, $e_{12} = 0$ and $e_{22} = 0$ but only two finite values for $e_{21} = 0$, a total of 11 in all. If r_L and c_L have units of m^2 W/K and J/m^{2}K respectively, the units of $z_{ij,k}$ are seconds.

j	z_{21j}	z_{22j}	z_{11j}	z_{12j}
1	∞	102.999214	62.598906	28.376181
2	16.324555	13.329913	7.955289	7.678022
3	3.675445	3.670872	1.445805	1.445797
		119.999999	72.000000	

It may be noted that $\sum z_{11} + \sum z_{22} = 72 + 120 = 192$, and that $\sum r \cdot \sum c = 16 \times 12 = 192$, in agreement with (10.28b). The largest value is $z_{22,1} = 103$ s, broadly because the condition $e_{22} = 0$ corresponds to all the heat draining away externally from capacities, all at positive temperatures, through the largest resistance (7 m^{2}K/W). In higher modes of decay, heat flows internally between adjacent capacities. The decay times form a sequence of decreasing values:

$$(z_{21,1} > z_{22,1} > z_{11,1} > z_{12,1}) > (z_{21,2} > z_{22,2} > z_{11,2} > z_{12,2}) > (z_{21,3} > z_{22,3} > z_{11,3} > z_{12,3}).$$

10.3.2 Unit Flux Temperatures

We return to the two-capacity wall with the boundary condition that T_0 and T_5 should be zero (Figure 10.4). The decay times are given by the two solutions to $e_{12} = 0$. Suppose that the system has been excited in some way and that at $t = 0$ the temperature of c_2 is T_{20}. The first subscript indicates location, the second indicates the $t = 0$ value. This total is associated with two eigenvalue temperature distributions, so that

$$T_{20} = T_{20,1} + T_{20,2} \tag{10.29a}$$

and at a later time t, the transient component of the solution is

$$T_{2t} = T_{20,1}\exp(-t/z_1) + T_{20,2}\exp(-t/z_2) \tag{10.29b}$$

with similar relations for T_4. $T_{20,1}$ and $T_{20,2}$ are as yet independent. $T_{20,1}$ and $T_{40,1}$, however, are related; since $T_5 = 0$, $T_{40,1}$ can be found by back multiplication:

$$\begin{bmatrix} T_4 \\ q_4 \end{bmatrix} = \begin{bmatrix} 1 & r_5 \\ 0 & 1 \end{bmatrix} \begin{bmatrix} T_5 \\ q_{5,1} \end{bmatrix} = \begin{bmatrix} 1 & r_5 \\ 0 & 1 \end{bmatrix} \begin{bmatrix} 0 \\ q_{5,1} \end{bmatrix}, \tag{10.30}$$

where $q_{5,1}$ is the heat flow at node 5 (as yet unknown) associated with the first eigenfunction. Thus

$$T_{40,1} = r_5 q_{5,1}. \tag{10.31a}$$

Two further stages of back multiplication give

$$T_{20,1} = (r_3 + r_5 - r_3 c_4 r_5 / z_1) q_{5,1}. \tag{10.31b}$$

Similarly for eigenfunction 2.[2]

To find the response of the wall for some given initial condition, i.e. given the values of T_{20} and T_{40}, we have to know the values of $T_{20,1}$, $T_{20,2}$ and $T_{40,1}$, $T_{40,2}$; that is, we have

[2]These equations can be written as

$$\frac{T_{20,i}}{T_{40,i}} = \frac{r_3}{r_5} + 1 - \frac{r_3 c_4}{z_i} \quad (= 1/\alpha_i.)$$

Since

$$\frac{T_{20,i}}{T_{40,i}} \times \frac{T_{40,i}}{T_{20,i}} \equiv 1,$$

to find $q_{5,1}$ and $q_{5,2}$. To do this, we use the orthogonality property of the eigenfunction distributions.

10.3.3 The Orthogonality Theorem and the Transient Solution

Consider the quantity

$$
\begin{aligned}
F_{12} &= \text{Sum over the two capacities of} \\
&\quad \text{(temperature distribution for eigenfunction 1)(energy distribution for eigenfunction 2)} \\
&= T_{20,1}(c_2 T_{20,2}) + T_{40,1}(c_4 T_{40,2}) \\
&= [(r_3 + r_5 - r_3 c_4 r_5/z_1)q_{5,1}][c_2(r_3 + r_5 - r_3 c_4 r_5/z_2)q_{5,2}] + [r_5 q_{5,1}][c_4 r_5 q_{5,2}]. \qquad (10.32)
\end{aligned}
$$

From the solution of $e_{12} = 0$ in (10.24) we have

$$z_1 + z_2 = (r_1 c_2 r_3 + r_1 c_2 r_5 + r_1 c_4 r_5 + r_3 c_4 r_5)/(r_1 + r_3 + r_5), \qquad (10.33a)$$

$$z_1 \cdot z_2 = (r_1 c_2 r_3 c_4 r_5)/(r_1 + r_3 + r_5) \qquad (10.33b)$$

and we find that F_{12} is zero. The result can be generalised and expressed in three forms. In the notation of Figure 10.4, where elements $1, 3, 5, \ldots$ are resistances and elements $2, 4, 6, \ldots$ are capacities and j and k denote eigennumbers, we have

$$F_{jk} = \sum T_{i,j} T_{i,k} \cdot c_i = 0 \ (i = 2, 4, 6, \ldots), \qquad (10.34a)$$

$$F_{jk} = \sum q_{i,j} q_{i,k} \cdot r_i = 0 \ (i = 1, 3, 5, \ldots), \qquad (10.34b)$$

$$F_{jk} = \sum \Delta q_{i,j} \Delta q_{i,k}/c_i = 0 \ (i = 2, 4, 6, \ldots \text{ and } \Delta q_{i,j} = q_{i-1,j} - q_{i,j}). \qquad (10.34c)$$

(It is easy and illustrative to check these results for the seven-element chain where all resistances and all capacities are given the value unity. This is in effect a dimensionless approximation to the finite difference model for a uniform layer with some measure of surface films. It will be found using the expression for the solution of a cubic equation but without needing a calculator that the decay times are $z_1 = 1/(2 - \sqrt{2})$, $z_2 = 1/2$, $z_3 = 1/(2 + \sqrt{2})$ and that the first, second and third eigenfunction distributions are approximations to one, two and three half-waves, respectively.)

the product of the right-hand term with its complement formed by interchange of r_1 and r_5, c_2 and c_4, must lead back to (10.24) with $e_{12} = 0$. It can readily be shown that

$$\frac{T_{40,1}}{T_{20,1}} \times \frac{T_{40,2}}{T_{20,2}} = -\frac{c_2}{c_4}$$

and that

$$\frac{T_{40,1}}{T_{20,1}} + \frac{T_{40,2}}{T_{20,2}} = 1 + \frac{r_3}{r_1} - \frac{c_2}{c_4}\left(\frac{r_3}{r_5} + 1\right).$$

A result of this kind is to be expected, since any two distributions are independent of each other; as eigenfunction distribution j decays in time, no heat is transferred to eigenfunction k. A proof of this result for two layers with distributed resistance and capacity will be given later and it is true for any number of layers and any pair of eigenvalues: $F_{jk} = 0$ except when $j = k$. The result allows us to calculate the heat flows $q_{5,1}$ and $q_{5,2}$ out of the wall in terms of T_{20} and T_{40}. Summing the two eigenfunction components at T_2 and T_4, equations (10.31a) and (10.31b), we have

$$(r_3 + r_5 - r_3c_4r_5/z_1)q_{5,1} + (r_3 + r_5 - r_3c_4r_5/z_2)q_{5,2} = T_{20,1} + T_{20,2} = T_{20}, \quad (10.35a)$$

$$r_5q_{5,1} + r_5q_{5,2} = T_{40,1} + T_{40,2} = T_{40}. \quad (10.35b)$$

T_{20} and T_{40} are determined by the initial condition, which is not yet specified. Multiply the first equation by $c_2(r_3 + r_5 - r_3c_4r_5/z_1)q_{5,1}$ and the second by $c_4r_5q_{5,1}$ and add:

$$\begin{aligned}&[(r_3 + r_5 - r_3c_4r_5/z_1)q_{5,1}][c_2(r_3 + r_5 - r_3c_4r_5/z_1)q_{5,1}] + \{[(r_3 + r_5 - r_3c_4r_5/z_2)q_{5,2}]\\ &\quad \times [c_2(r_3 + r_5 - r_3c_4r_5/z_1)q_{5,1}] + [r_5q_{5,1}][c_4r_5q_{5,1}] + [r_5q_{5,2}][c_4r_5q_{5,1}]\}\\ &= T_{20}c_2(r_3 + r_5 - r_3c_4r_5/z_1)q_{5,1} + T_{40}c_4r_5q_{5,1}. \end{aligned} \quad (10.36)$$

But according to the orthogonality theorem, certain terms in the curly brackets {} equal zero:

$$[(r_3 + r_5 - r_3c_4r_5/z_2)q_{5,2}][c_2(r_3 + r_5 - r_3c_4r_5/z_1)q_{5,1}] + [r_5q_{5,2}][c_4r_5q_{5,1}] = 0.$$

Thus we have an expression for $q_{5,1}$:

$$q_{5,1} = \frac{T_{20}c_2(r_3 + r_5 - r_3c_4r_5/z_1) + T_{40}c_4r_5}{c_2(r_3 + r_5 - r_3c_4r_5/z_1)^2 + c_4r_5{}^2} = \frac{T_{20} \cdot c_2T'_{20,1} + T_{40} \cdot c_4T'_{40,1}}{c_2(T'_{20,1})^2 + c_4(T'_{40,1})^2}. \quad (10.37a)$$

Similarly,

$$q_{5,2} = \frac{T_{20}c_2(r_3 + r_5 - r_3c_4r_5/z_2) + T_{40}c_4r_5}{c_2(r_3 + r_5 - r_3c_4r_5/z_2)^2 + c_4r_5{}^2} = \frac{T_{20} \cdot c_2T'_{20,2} + T_{40} \cdot c_4T'_{40,2}}{c_2(T'_{20,2})^2 + c_4(T'_{40,2})^2}, \quad (10.37b)$$

where $T'_{20,1}$ indicates the $t = 0$ value of temperature at c_2 associated with unit value of $q_{5,1}$ from (10.31). It has units m^2K/W.

In this case, and for any number of lumped elements, $q_{5,1}$ and $q_{5,2}$ can equally well be found by direct solution of the simultaneous equations. This is not possible for real wall layers; the method using the orthogonality theorem will be used later.

From (10.31) we have $T_{20j} = (r_3 + r_5 - r_3c_4r_5/z_j)q_{5,j}$ and $T_{40,j} = c_4r_5q_{5,j}$, where $j = 1, 2$. Equation (10.29b) then gives the transient solution to the two-capacity circuit of Figure 10.4 when T_0 and T_5 are zero and T_2 and T_4 have the initial values T_{20} and T_{40}

10.3.4 Step and Steady-Slope Solutions

The values of T_{20} and T_{40} for use in (10.37) depend on the initial conditions. If the wall is initially at the uniform temperature T_I, then $T_{20} = T_{40} = T_I$. If $T_0 = T_5 = 0$ from $t = 0$

onward, the wall is subjected to a step excitation and the values of T_{2t} and T_{4t} can be found immediately.

If however the wall is initially at zero temperature, but subjected to a ramp excitation at T_0, T_{2t} and T_{4t} will be found by superposition of suitable steady-slope and transient solutions. Suppose that $T_5 = 0$ always but that T_0 has been increasing for an indefinite time at rate θ (K/s) and that the time origin is chosen so that $T_0 = \theta t$. T_2 and T_4 similarly increase at uniform rates and at $t = 0$ they have negative values. So $T_2 = T_2' + \theta_2 t$ and $T_4 = T_4' + \theta_4 t$. Continuity at T_2 and T_4 requires that

$$\frac{T_0 - T_1}{r_1} = c_2\theta_2 + \frac{T_2 - T_3}{r_3} \quad \text{and} \quad \frac{T_2 - T_3}{r_3} = c_4\theta_4 + \frac{T_4 - T_5}{r_5}. \tag{10.38}$$

These equations include time-dependent and time-independent terms. It is found that

$$T_2' = \frac{-r_1[c_2(r_3 + r_5)^2 + c_4 r_5{}^2]\theta}{(r_1 + r_3 + r_5)^2}, \quad (10.39a) \qquad T_2 = T_2' + \frac{(r_3 + r_5)\theta t}{(r_1 + r_3 + r_5)}, \quad (10.39b)$$

$$T_4' = \frac{-r_5[c_2 r_1(r_3 + r_5) + c_4(r_1 + r_3)r_5]\theta}{(r_1 + r_3 + r_5)^2}, \quad (10.39c) \quad T_4 = T_4' + \frac{r_5\theta t}{(r_1 + r_3 + r_5)}. \quad (10.39d)$$

Evaluation of these temperatures involves solution of simultaneous equations and the method can be used for a sequence of n lumped capacities. This cannot readily be done however for layers with distributed resistance and capacity, and the T_L' values will be found by matrix multiplication (Chapter 17). Indeed, these results can equally be found by multiplication of five of the matrices shown in (17.29); multiplication is easy here since $c_1 = c_3 = c_5 = 0$ and $r_2 = r_4 = 0$.

10.3.5 Ramp Solution

We now set up the transient solution whose temperature distribution at $t = 0$ is equal and opposite to the slope solution at $t = 0$, that is, $T_{20} = -T_2'$ and $T_{40} = -T_4'$ (positive values, since T_2' and T_4' are negative). Then $q_{5,1}$ and $q_{5,2}$ follow from (10.37) and $T_{20,1}$ and $T_{20,2}$ from (10.31). The final solution for the system of two capacities, initially at zero temperature, $T_5 = 0$ but T_0 having the value θt from $t = 0$ onward is then

$$T_{2t} = \left(\frac{-r_1[c_2(r_3 + r_5)^2 + c_4 r_5^2]}{(r_1 + r_3 + r_5)^2} + \frac{(r_3 + r_5)t}{r_1 + r_3 + r_5}\right)\theta \quad + T_{20,1}\exp(-t/z_1) + T_{20,2}\exp(-t/z_2), \tag{10.40a}$$

$$T_{4t} = \left(\frac{-r_5[c_2 r_1(r_3 + r_5) + c_4 r_5(r_1 + r_3)]}{(r_1 + r_3 + r_5)^2} + \frac{r_5 t}{r_1 + r_3 + r_5}\right)\theta + T_{40,1}\exp(-t/z_1) + T_{40,2}\exp(-t/z_2). \tag{10.40b}$$

The values of $T_{20,1}$, etc., are somewhat complicated functions of the wall parameters, consisting of the product of slope and transient combinations of the resistances and capacities. They are proportional however to θ since T_2' and T_4' (10.39) are proportional to θ.

The heat flows associated with the ramp excitation follow immediately as $q_{00,t} = (\theta t - T_{2t})/r_1$ and $q_{50,t} = T_{4t}/r_5$. These q values consist of the three terms explained earlier, see equation 10.14. It can readily be checked that the slope constant terms $q_{00,0}$ and $q_{50,0}$ can be found from the matrix form (17.29) and (17.34).

10.3.6 Examples

Consider a system with the simple values $r_1 = 1, c_2 = 2, r_3 = 3, c_4 = 4, r_5 = 5$ (r in m^2K/W and c in J/m^2K) whose transient solution is based on T_0 and T_5 being isothermal, so that $e_{12} = 0$ in (10.24). Then $z_{12,1}$ and $z_{12,2}$, (or simply z_1 and z_2) are given by the quadratic equation

$$e_{12} = (1+3+5) - (1\times 2\times 3 + 1\times 2\times 5 + 1\times 4\times 5 + 3\times 4\times 5)/z$$
$$+ (1\times 2\times 3\times 4\times 5)/z^2 = 9 - 96/z + 120/z^2 = 0,$$

so $z_1 = (16+\sqrt{136})/3 = 9.2206\,\text{s}$ and $z_2 = (16-\sqrt{136})/3 = 1.4460\,\text{s}$.

When $q_{5,1}$ and $q_{5,2}$ are each taken to be unity, the temperatures per unit heat flow are

$$T'_{20,1} = 1.4929,\ T'_{40,1} = 5,$$
$$T'_{20,2} = -33.4929,\ T'_{40,2} = 5.$$

Thus the temperature profile for eigenfunction 1 has everywhere the same sign and as it decays, all heat must be lost to the exterior. For eigenfunction 2 the capacity temperatures have opposite signs and they exchange heat during decay. $T'_{20,2}$ is numerically large here compared to $T'_{40,2}$ since c_2 is a comparatively small capacity linked to a small resistance whereas c_4 is large and linked to a large resistance.

We wish to determine the values of T_2 and T_4 corresponding to some specified initial conditions and forms of excitation.

Example 10.1

Two-sided step change: the system has an initial temperature of T_I. From $t = 0$ onward, $T_0 = T_5 = 0$. From equation (10.37) we have

$$q_{5,1} = \frac{T_I\times 2\times 1.4929 + T_I\times 4\times 5}{2\times 1.4929^2 + 4\times 5^2} = 0.2200T_I,$$

$$q_{5,2} = \frac{T_I\times 2\times(-33.4929) + T_I\times 4\times 5}{2\times(-33.4929)^2 + 4\times 5^2} = -0.0200T_I.$$

So

$$T_{20,1} = T'_{20,1}q_{5,1} = 1.4929\times 0.2200\ T_I = 0.3285\ T_I,$$
$$T_{40,1} = T'_{40,1}q_{5,1} = 5\times 0.2200\ T_I = 1.1002\ T_I,$$
$$T_{20,2} = T'_{20,2}q_{5,2} = -33.4929\times(-0.0200)T_I = 0.6715\ T_I,$$
$$T_{40,2} = T'_{40,2}q_{5,2} = 5\times(-0.0200)T_I = -0.1002\ T_I,$$

and

$$T_{2t}/T_I = 0.3285\exp(-t/9.2206) + 0.6715\exp(-t/1.4460),$$

$$T_{4t}/T_I = 1.1002\exp(-t/9.2206) - 0.1002\exp(-t/1.4460).$$

Example 10.2

The system has an initial temperature of zero. From $t = 0$ onward, $T_0 = T_5 = T_I$. This follows from the above solution by replacing T_I by $-T_I$ and adding $+T_I$.

Example 10.3

One-sided step change: the system has an initial temperature of zero. From $t = 0$ onward, $T_0 = T_I$ and $T_5 = 0$. In the final state, $T_2 = T_I \times (3+5)/(1+3+5) = (8/9)T_I$ and $T_4 = T_I \times 5/(1+3+5) = (5/9)T_I$. The transient solution is based on the negative of these values, so from (10.37) we have

$$q_{5,1} = \frac{-(8/9)T_I \times 2 \times 1.4929 - (5/9)T_I \times 4 \times 5}{2 \times 1.4929^2 + 4 \times 5^2} = -0.1318T_I \text{ [W/m}^2\text{]},$$

and

$$q_{5,2} = \frac{-(8/9)T_I \times 2 \times (-33.4929) - (5/9)T_I \times 4 \times 5}{2 \times (-33.4929)^2 + 4 \times 5^2} = +0.02067T_I \text{ [W/m}^2\text{]}.$$

$T_{20,1}$, etc., follow as above. To the transient solution (which decays to zero) we must add the final steady-state solution, so

$$T_{2t}/T_I = 8/9 - 0.1967\exp(-t/9.2206) - 0.6922\exp(-t/1.4460),$$

$$T_{4t}/T_I = 5/9 - 0.6589\exp(-t/9.2206) + 0.1033\exp(-t/1.4460).$$

It is easily checked that at $t = 0$ each temperature is zero. When t is so small that $\exp(-t/z) = 1 - t/z$, it is found that $T_{2t}/T_I = 0.5t$ and that T_{4t}/T_I is zero. Immediately after $t = 0$, a sizeable flow of heat to T_2 causes a first-order increase of temperature there but only a second-order change at T_4.

Example 10.4

The system has an initial temperature of zero. After $t = 0$, $T_0 = \theta t$ and $T_5 = 0$. As in Example 10.3, the final solution consists of a transient solution decaying to zero, upon which we superpose a steady-slope solution chosen so the net effect at $t = 0$ is $T_{2t} = T_{4t} = 0$. The $t = 0$ values of the slope solution, found using (10.39) are

$$T_2' = (-228/81)\theta = -2.8148\theta \quad \text{and} \quad T_4' = -(480/81)\theta = -5.9529\theta;$$

The relations $T_{20} = -T_2'$ and $T_{40} = -T_4'$ provide the values for $q_{5,1}$ and $q_{5,2}$ and so the values of $T_{20,1}$ to $T_{40,2}$. The complete solution is

$$T_{2t}/\theta = 8t/9 - 2.8148 + 1.8139\exp(-t/9.2206) + 1.0009\exp(-t/1.4460),$$

$$T_{4t}/\theta = 5t/9 - 5.9259 + 6.0753\exp(-t/9.2206) - 0.1494\exp(-t/1.4460).$$

Again, the $t = 0$ values are zero. When t is small, T_{2t} is zero (as is T_{4t}) since with ramp excitation the heat flow to T_2 is very small. With the approximation $\exp(-t/z) = 1 - t/z + \frac{1}{2}(t/z)^2$, $T_{2t}/\theta = \frac{1}{4}t^2$, but T_{4t} continues to be zero.

$$q_{00,t} = (\theta t - T_{2t})/1 = \theta[t/9 + 2.8148 - 1.8139\exp(-t/9.2206) - 1.0009\exp(-t/1.4460)] \quad [\mathrm{W/m^2}],$$

$$q_{50,t} = \quad T_{5t}/5 \quad = \theta[t/9 - 1.1852 + 1.2151\exp(-t/9.2206) - 0.0299\exp(-t/1.4460)] \quad [\mathrm{W/m^2}].$$

When t is small,

$$q_{00,t} = 1.0000t - 0.2500t^2 \qquad \text{that is, an immediate but falling response}$$

and

$$q_{50,t} = 0.0000t - 0.0000t^2 \qquad \text{so that there is no initial response.}$$

Although the response of a multilayer wall (Chapter 17) requires a more complicated analysis than this, the underlying approach is the same: evaluate the slope response of the wall, evaluate its transient response and combine them to find the wall response to ramp excitation. The thermal behaviour of a layered wall can be largely explained qualitatively in terms of this simple *r-c-r-c-r* model. Section 10.7 gives a numerical example using parameters typical of building applications. Chapter 11 shows how to evaluate response factors and wall transfer coefficients for a lumped model.

10.4 FINITE DIFFERENCE METHOD

A sequence of resistors and capacitors provides the basis for the well-known finite difference method to compute heat flows through walls. Suppose that a layer of some wall material such as concrete or insulation of thickness X (resistance $R = X/A\lambda$ and capacity $C = A\rho c_p X$) is divided into N slices of uniform thickness $\delta x = X/N$. The thermal behaviour of the layer can be modelled approximately by supposing that the thermal capacity of a slice can be represented as a localised capacity C/N at the centre of the resistance R/N of the slice, that is, a resistance of $R/2N$ on either side. Thus, within the slab, each capacity is flanked by two resistances R/N. As N increases, this simplified model tends to the parent distributed layer. The model is consistent with the Fourier continuity equation (see later) and can be derived from it using Taylor's theorem.

10.4.1 Subdivision of the Wall

A suitable value of N can be found by comparing the sinusoidal transmission matrix of the model wall with that of the slab, based on some suitable period P. Suppose that values of the driving data are known at hourly intervals. Variation between these values is normally taken to be linear. Suppose that hourly values of ambient temperature T_e were 0, 1, 0, 1, 0, ... This is a sawtooth form with period $P = 2$ h. Fourier analysis represents it as a series of terms with periods $2P/1, 2P/3, 2P/5, \ldots$ and amplitudes $+1/1^2, -1/3^2, +1/5^2, \ldots$. If T_e were daily-periodic but non-sinusoidal, a component of $P = 2$ h would provide only a small part of the total daily variation, so the $2P/3$ and higher components in the $2P$ Fourier component would provide a small part of a small part of the total variation and could be ignored. Thus the model will be designed to cope with periodic excitation of period 2 h. If short-term response is of interest, a smaller value of P must be chosen.

The transmission matrix of the slice is

$$\begin{bmatrix} 1 & R/2N \\ 0 & 1 \end{bmatrix} \begin{bmatrix} 1 & 0 \\ j2\pi C/PN & 1 \end{bmatrix} \begin{bmatrix} 1 & R/2N \\ 0 & 1 \end{bmatrix} = \begin{bmatrix} 1 + j2\pi RC/2PN^2 & (R/N)[1 + j2\pi RC/4PN^2] \\ j2\pi C/PN & 1 + j2\pi RC/2PN^2 \end{bmatrix}. \tag{10.41}$$

The transmission matrix of the model wall itself consists of the product of N of these units:

$$\begin{bmatrix} f_{111} + jf_{112} & f_{121} + jf_{122} \\ f_{211} + jf_{212} & f_{221} + jf_{212} \end{bmatrix} = \begin{bmatrix} 1 + j2\pi RC/2PN^2 & (R/N)[1 + j2\pi RC/4PN^2] \\ j2\pi C/PN & 1 + j2\pi RC/2PN^2 \end{bmatrix}^N. \tag{10.42}$$

Now it will be shown later that, in the current notation, the corresponding transmission matrix of a homogeneous slab has elements e_{ijk} given by

$$\begin{bmatrix} e_{111} + je_{112} & e_{121} + je_{122} \\ e_{211} + je_{212} & e_{221} + je_{222} \end{bmatrix} = \begin{bmatrix} \cosh\{(\pi RC/P)^{1/2}(1+j)\} & \sinh\{(\pi RC/P)^{1/2}(1+j)\}/[(\pi C/PR)^{1/2}(1+j)] \\ \sinh\{(\pi RC/P)^{1/2}(1+j)\} \times [(\pi C/PR)^{1/2}(1+j)] & \cosh\{(\pi RC/P)^{1/2}(1+j)\} \end{bmatrix} \tag{10.43}$$

The values of the elements f_{ijk} of the model tend to the elements e_{ijk} of the slab as N tends to infinity.

Two convenient measures of the thermal characteristics of a wall are the cyclic transmittance **u** and the cyclic admittance **y** noted earlier. They relate the heat flow at the wall inner surface due to unit variation in a sinusoidal driving temperature of period P applied either at the wall outer surface, or the wall inner surface, the other node being isothermal. The 'wall' may be defined to include either or both of the surface films, but is considered here to be a homogeneous slab alone. Since the drive is sinusoidal, the flow is complex and can be expressed in terms of Cartesian components, e_{ij1} and e_{ij2}, or as a magnitude (W/m^{2}K) and phase ϕ (rad) or a time lead/lag $(P/2\pi)\phi$:

$$\mathbf{u} = \frac{\text{inward heat flow at inner surface}}{\text{temperature at outer surface}} = \frac{1}{e_{121} + je_{122}}, \tag{10.44}$$

$$\mathbf{y} = \frac{\text{outward heat flow at inner surface}}{\text{temperature at inner surface}} = \frac{e_{111} + je_{112}}{e_{121} + je_{122}}. \tag{10.45}$$

The model estimates are found with fs replacing es.

As an example, consider a standard double-thickness brick construction ($X = 0.22$ m, $\lambda = 0.8$ W/m K, $\rho = 1700$ kg/m^3, $c_p = 800$ J/kg K, diffusivity 5.9×10^{-7} m^2/s). Table 10.1 shows the variation of the 24 h and 2 h transmittance and admittance with the number of slices N as it progresses toward the exact values in the last row.

The number of slices to model the wall is chosen so as to provide values of these parameters which are sufficiently close to the exact values. The 95% accuracy can be achieved with fewer slices for $P = 24$ h than $P = 2$ h. $N = 3$ may be satisfactory for daily excitation. But if the $N = 3$ value of $y(2\,\text{h})$ – 20.6 W/m^{2}K – were used instead of a value nearer its exact value of 30.8 W/m^{2}K to estimate temperature due to an imposed heat flow of period 2 h, (as $T = q/y$), such a temperature would be overestimated by some 50%. This is significant since indoor heat sources can change by steps, and in the frequency domain, the low-period components of their representation are important. On the other hand, the effect of an externally imposed excitation of low period is so damped that it is unimportant. The 95% criterion is only met for the 2 h transmittance with $N = 19$ but adequate $P = 2$ h response is obtainable with $N = 7$.

Ceylan and Myers (1980) represented a concrete wall ($X = 0.203$ m, diffusivity 1.27×10^{-6} m^2/s) by three nodes and compared the heat flow from it, driven by hourly values of air temperature, with values from a series of other, more accurate means. Fair agreement was found, although it was substantially improved by a 9-node representation. Clarke (1985) reports results for a similar wall ($X = 0.2$ m, diffusivity 4.8×10^{-7} m^2/s) excited externally and shows (Figure 3.1) that a 15-node representation is little better than one with 3 nodes. Waters and Wright (1985) examined nodal placing in a series of multi-layer walls. For a construction of 40 mm polyurethane and 220 mm brickwork, and using a total of 7 nodes, they found that errors were least when 1 node was placed in the polyurethane and 6 in the brickwork, when the polyurethane surface and the brickwork surface were subjected to a step in temperature. Thus X/N for the brickwork was 0.037 m. They have argued that the quantity $\lambda\delta t/\rho c_p(\delta x)^2$ should have the same value for all layers through the wall in order to make truncation errors similar everywhere. A suggestion

Table 10.1 Transmittance and admittance of an N-capacity model for a homogeneous slab driven at period P. Values in bold are within 95% of the exact values

	$P = 24$ h				$P = 2$ h			
	Transmittance		Admittance		Transmittance		Admittance	
N	Mag (W/m^{2}K)	Phase (h)	Mag (W/m^{2}K)	Phase (h)	Mag (W/m^{2}K)	Phase* (h)	Mag (W/m^{2}K)	Phase (h)
1	2.021	**−3.75**	6.375	1.02	0.202	−0.48	7.264	0.01
2	2.727	−3.82	**8.166**	2.33	0.088	−0.89	14.371	0.03
3	**2.916**	−3.70	8.351	2.73	0.071	−1.20	20.629	0.07
4	2.982	−3.65	8.378	2.88	0.077	−1.39	25.135	0.12
5	3.012	−3.62	8.383	**2.95**	0.089	−1.49	27.771	0.15
6	3.028	−3.61	8.383	2.99	0.101	−1.55	29.149	0.18
7	3.037	−3.60	8.383	3.01	0.111	**−1.58**	**29.858**	0.19
8	3.043	−3.59	8.382	3.03	0.119	−1.60	30.236	0.21
exact	3.063	−3.57	8.379	3.08	0.154	−1.66	30.814	0.25

*See Section 10.4.5 for the evaluation of the phase of the transmission parameter.

has been advanced that if $\lambda < 0.5\,\mathrm{W/m\,K}$, X/N should not exceed 0.02 m and if $\lambda > 0.5\,\mathrm{W/m\,K}$, X/N should not exceed 0.03 m. With this criterion, the above 0.22 m wall should be divided into eight slices.

Section 17.8 describes a technique which reduces the required number of nodes by allowing the capacities and resistances to be unequal.

10.4.2 Computational Formulae

Models of the kind discussed here are set up to estimate temperatures or heat flows at inside and outside nodes which result from heat flows or temperatures acting there. Successive temperature nodes will be labelled T_{m-1}, T_m and T_{m+1} (Figure 10.6). From continuity at node T_m,

$$c\,\mathrm{d}T_m/\mathrm{d}t = (T_{m-1} - T_m)/r + (T_{m+1} - T_m)/r \tag{10.46a}$$

or

$$(\rho c_p(\delta x)^2/\lambda)\,\mathrm{d}T_m/\mathrm{d}t = T_{m-1} - 2T_m + T_{m+1}. \tag{10.46b}$$

Time is now taken to be discretised, so that temperatures are supposed known at time level i, and are to be determined at time level $i+1$, an interval δt later. For use with hourly meteorological data, $\delta t = 1$ hour. Also $\mathrm{d}T_m/\mathrm{d}t$ is replaced by $(T_{m,i+1} - T_{m,i})/\delta t$ and the constants are combined non-dimensionally as

$$p = \lambda\delta t/\rho c_p(\delta x)^2. \tag{10.47}$$

During the interval δt, T_m moves toward a value between the flanking nodes T_{m-1} and T_{m+1} and if these values are themselves fixed, the variation is exponential, quickly at first and slower later. Thus we can write

$$T_{m,i+1} - T_{m,i} = p(T_{m-1,i} - 2T_{m,i} + T_{m+1,i}). \tag{10.48}$$

Since this estimates the change in terms of values at the beginning of the interval, it overestimates the change in T_m. Alternatively, at the end of the interval

$$T_{m,i+1} - T_{m,i} = p(T_{m-1,i+1} - 2T_{m,i+1} + T_{m+1,i+1}) \tag{10.49}$$

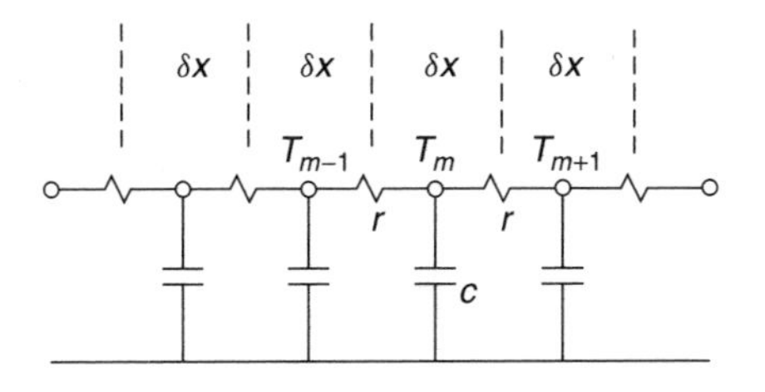

Figure 10.6 Section through a discretised wall

and the change is underestimated. A value based on the average

$$T_{m,i+1} - T_{m,i} = \tfrac{1}{2}p(T_{m-1,i} - 2T_{m,i} + T_{m+1,i}) + \tfrac{1}{2}p(T_{m-1,i+1} - 2T_{m,i+1} + T_{m+1,i+1}) \tag{10.50}$$

should provide a better approximation.

According to the first possibility, the Euler method, $T_{m,i+1}$ is given explicitly as

$$T_{m,i+1} = pT_{m-1,i} + (1 - 2p)T_{m,i} + pT_{m+1,i}, \tag{10.51a}$$

where the right-hand terms are known. Values can be determined sequentially through the wall. In the second case,

$$pT_{m-1,i+1} - (1 + 2p)T_{m,i+1} + pT_{m+1,i+1} = -T_{m,i}. \tag{10.51b}$$

This is one of a set of equations; there is a similar equation relating to values at $m - 2, m - 1$ and m, etc. The temperatures at the layer or wall flanking nodes are known from assumed boundary conditions. Thus by solution of the set, values can be found simultaneously through the wall. This is the pure implicit method. Similarly, in the third case, the Crank–Nicolson method (Crank and Nicolson 1947) uses

$$\tfrac{1}{2}pT_{m-1,i+1} - (1 + p)T_{m,i+1} + \tfrac{1}{2}pT_{m+1,i+1} = -\tfrac{1}{2}pT_{m-1,i} + (1 - p)T_{m,i} + \tfrac{1}{2}pT_{m+1,i}, \tag{10.52}$$

which again requires solution of a set of equations. Waters (1981) notes some of the users of these equations.

Bhattacharya (1985, 1993) has argued that all three forms are special cases of the explicit form

$$\frac{T_{m,i+1}}{T_{m,i}} = \exp\left[-\frac{2T_{m,i} - T_{m-1,i} - T_{m+1,i}}{T_{m,i}}p\right]. \tag{10.53}$$

The expression can be evaluated by subdividing the time step δt without the need to store intermediate temperatures. It provides better accuracy than the Crank–Nicolson form, but with some small time penalty for $p > 1$.

To see the significance of the choice of p, it is convenient to consider the simple case where T_{m-1} and T_{m+1} are zero at all times and T_m has some initial value T_I. We wish to determine the value of T_m for $t > 0$. The exact analytical solution is

$$T_{m,t} = T_I \exp(-t/z), \tag{10.54}$$

where

$$z = \tfrac{1}{2}rc = \tfrac{1}{2}\rho c_p(\delta x)^2/\lambda, \tag{10.55a}$$

so

$$p = \lambda\delta t/\rho c_p(\delta x)^2 = \delta t/2z. \tag{10.55b}$$

The finite difference values are as follows:

- Euler

$$T_{m,i+1} = (1 - 2p)T_{m,i} = (1 - \delta t/z)T_{m,i}; \tag{10.56a}$$

- Crank–Nicolson

$$T_{m,i+1} = \frac{1-p}{1+p}T_{m,i} = \frac{1-\frac{1}{2}\delta t/z}{1+\frac{1}{2}\delta t/z}T_{m,i}; \tag{10.56b}$$

- Pure implicit

$$T_{m,i+1} = \frac{1}{1+2p}T_{m,i} = \frac{1}{1+\delta t/z}T_{m,i}. \tag{10.56c}$$

The Bhattacharya form in this case is exact.

Now in exponential cooling, T_m falls to $1/e = 0.368$ of its initial value between $t = 0$ and $t = z$, that is, 63.2% of its total fall. If the time interval δt were to equal z, it would represent a very coarse analysis of the cooling process. But if $\delta t = z$ then $p = \frac{1}{2}$, so p should be smaller than this for accuracy.

To see the implications of these expressions, we evaluate T_m/T_I at times δt, $2\delta t$ and $3\delta t$ (Table 10.2). Thus, with $p = 0.1$ the Euler and pure implicit methods, though respectively overestimating and underestimating the change in T_m, provide possibly acceptable values. The Crank–Nicolson values are acceptable. With $p = 0.3$ the Euler and pure implicit methods are no longer acceptable, although the Crank–Nicolson values may be.

With a value of $0.5 < p < 1.0$ the Euler expression gives a first time step value for T_m that is negative, a physical impossibility. Values for $2\delta t, 3\delta t, \ldots$ alternate in sign but diminish in magnitude. If $p > 1$ then successive values increase in magnitude. The Crank–Nicolson method provides stable values up to $p = 1$, and thereafter values oscillate. The pure implicit method gives stable results for all values of p; this is its virtue. The time step δt can be made indefinitely large without leading to instability. For a more detailed discussion, see Myers (1971) or Özisik (1994).

10.4.3 Discussion

Finite difference methods are intuitively attractive since the basic computational formula expresses approximately the physical cause for propagation of a thermal signal. All parameters, such as conductivity, can be reassigned at each time step. The physics is totally

Table 10.2 Values of T_m/T_I at 0, δt, $2\delta t$ and $3\delta t$

	(T_m/T_I)(exact)	(T_m/T_I)(Euler)	(T_m/T_I)(Crank–Nicolson)	(T_m/T_I)(pure implicit)
With $p = 0.1$				
0	1.0000	1.0000	1.0000	1.0000
δt	0.8187	0.8000	0.8182	0.8333
$2\delta t$	0.6703	0.6400	0.6694	0.6944
$3\delta t$	0.5488	0.5120	0.5477	0.5787
With $p = 0.3$				
0	1.0000	1.0000	1.0000	1.0000
δt	0.5488	0.4000	0.5385	0.6250
$2\delta t$	0.3012	0.1600	0.2899	0.3906
$3\delta t$	0.1653	0.0640	0.1561	0.2441

obscured by analytical methods. There are some negative aspects, however. The ratio

$$p = \frac{\lambda \delta t N^2}{\rho c_p X^2}. \tag{10.57}$$

Now accuracy increases as p decreases and N increases. If N is doubled and p is to be kept constant, δt must be reduced to $\delta t/4$ and the computational labour in the Euler method increases by a factor of $2 \times 4 = 8$. The implicit methods require solution of a set of simultaneous equations at each time step and the number of operations varies as N^3. This has to be done for each massive wall of an enclosure. The method is computationally laborious.

The method provides information about all nodes in the wall but little of this is normally needed. All that is required are the temperatures and heat flows at the bounding nodes.

Methods to compute values at the bounding nodes alone, based on analytical solutions to the wall continuity equation, will be presented in later chapters.

10.4.4 Evaluation of Complex Quantities

The magnitudes of $\mathbf{u}$ and of $\mathbf{y}$ are found routinely as $1/\sqrt{f_{121}^2 + f_{122}^2}$ and $\sqrt{(f_{111}^2 + f_{112}^2)/(f_{121}^2 + f_{122}^2)}$, respectively. The phase of $\mathbf{y}$ is simply $\arctan\{(f_{112}f_{121} - f_{111}f_{122})/(f_{111}f_{121} + f_{112}f_{122})\}$ and lies between 0 and $\pi/2$.

The phase of $\mathbf{u}$, however, depends on the thermal thickness of the wall and the phase of the flux may lag behind the temperature which drives it by a complete cycle or more. In conductance form,

$$\mathbf{q}_i = \mathbf{u}\mathbf{T}_o; \tag{10.58a}$$

in resistance form,

$$\mathbf{T}_o = \mathbf{r}\mathbf{q}_i = (f_{121} + \mathrm{j} f_{122})\, \mathbf{q}_i. \tag{10.58b}$$

Now the operation $b = \arctan(a)$ returns a value of b between $-\pi/2$ and $+\pi/2$ and having the same sign as a. When $N = 1$ in Table 10.1, f_{121} and f_{122} are positive so the phase of $\mathbf{r}$, $\phi(\mathbf{r}) = \arctan(f_{122}/f_{121})$, is between 0 and $\pi/2$, that is, a phase shift of less than one quarter-cycle or less than half an hour when $P = 2\,\text{h}$, as must be the case for a single capacity. As N increases, either or both f_{12k} may become negative and this has implications for the real phase shift between $\mathbf{T}$ and $\mathbf{q}$:

$$\begin{aligned}
&\text{if } f_{121} > 0 \text{ and } f_{122} > 0, \phi(\mathbf{r}) = \arctan(f_{122}/f_{121});\\
&\text{if } f_{121} < 0 \text{ and } f_{122} > 0, \phi(\mathbf{r}) = \arctan(f_{122}/f_{121}) + \pi;\\
&\text{if } f_{121} < 0 \text{ and } f_{122} < 0, \phi(\mathbf{r}) = \arctan(f_{122}/f_{121}) + \pi;\\
&\text{if } f_{121} > 0 \text{ and } f_{122} < 0, \phi(\mathbf{r}) = \arctan(f_{122}/f_{121}) + 2\pi;\\
&\text{if } f_{121} \approx 0 \text{ and } f_{122} > 0, \phi(\mathbf{r}) = +\pi/2;\\
&\text{if } f_{121} \approx 0 \text{ and } f_{122} < 0, \phi(\mathbf{r}) = -\pi/2;\\
&\text{finally, since } \mathbf{u} = 1/\mathbf{r}, \phi(\mathbf{u}) = -\phi(\mathbf{r}).
\end{aligned} \tag{10.59}$$

The first few entries in Table 10.1 for the phase of **u** when $P = 2\,\text{h}$ show these changes. When $N = 1$, both f_{121} and f_{122} are positive, leading to a lag of 0.48 h. When $N = 2$, f_{121} is negative, so leading to a lag of between 0.5 and 1.0 h. With $N = 3, 4, 5$ both terms are negative, so the lag is between 1.0 and 1.5 h. When $N > 5$, f_{121} is positive once more and a lag of between 1.5 and 2 h is found. It gradually converges on the exact value of -1.66 h.

10.5 THE ELECTRICAL ANALOGUE

The flow of an electric current through a resistance is proportional to the potential difference across it; the steady heat flow is similarly proportional to the temperature difference across it. Further, the current into a capacitor is proportional to $\partial V/\partial t$, while the heat flow into a thermal capacity is proportional to $\partial T/\partial t$. Early investigations of unsteady heat flow used the analogy between electrical and thermal quantities. It was employed by Paschkis (1936) and Beuken (1936). The scheme adopted by Paschkis and Baker (1942) will serve as an example of the scaling procedure needed, including the conversion between imperial and SI units:

- *Potential*: $1^\circ\text{F} \equiv \frac{5}{9}\,\text{K}$ is modelled as 2 V.
- *Quantity*: 1 btu $\equiv$ 1056 J is modelled as 6.08×10^{-4} C. [1 C = 1 coulomb]
- *Time*: 1 h $\equiv$ 3600 s is modelled as 200 s.

Thus a thermal change which occupies a real time of 1 h is to be modelled in a time of 200 s. This is the principal scaling factor. Similarly, a thermal process occupying only a fraction of a second can be stretched out to some minutes. The analogue components are consequent upon this choice. The derived units are

- *Capacity*: a layer of material having a thermal capacity of 1 btu/°F = 1901 J/K will be modelled by a capacity of 6.08×10^{-4} Cs/2 V or 3.04×10^{-4} F.
- *Flow*: a heat flow of 1 btu/h = 0.293 J/s = 0.293 W will be modelled by a current of 6.08×10^{-4} C/200 s or 3.04×10^{-6} A.
- *Resistance*: a layer of material having a thermal resistance of 1°Fh/btu = 1.894 K s/J will be modelled by a resistance of 2 V × 200 s/6.08×10^{-4} C or $0.658 \times 10^{6}\,\Omega$.

The heat continuity equation in a homogeneous slab,

$$\frac{\partial T}{\partial t} = \frac{\lambda}{\rho c_p}\frac{\partial^2 T}{\partial x^2}, \tag{10.60}$$

has an electrical analogue, the *RC* cable, whose voltage distribution is given by

$$\frac{\partial V}{\partial t} = \frac{1}{R^* C^*} \frac{\partial^2 T}{\partial x^2}, \qquad (10.61)$$

where R^* and C^* are respectively the resistance per unit length and capacity per unit length of the cable. Both $\rho c_p/\lambda$ and R^*C^* have units s/m^2, so the equations show the similarity between thermal and electrical flows for potential and time, although they do not involve quantity.

This provided the basis to take a wall, a distributed thermal system, and represent its properties using a discrete sequence of electrical elements. The fineness of division was a topic of enquiry among the workers using it but that is no longer of interest. It is summarised in Davies (1983b: 204). A classic example of its use applied to an enclosure is that of Nottage and Parmlee (1954, 1955). According to Shavit (1995),

> Nelson (1965) simulated an entire house, including the envelope, roof, furnace, air-conditioning, ductwork [and internal transfer].... The time scale was set so that the fastest responding system (furnace, air-conditioning, or control system) was properly represented. This enabled him to study the time behavior of the control systems and the various control strategies such as night setback, morning warmup, rate of furnace cycling, thermostats with compensator, droops, etc. This was the first detailed work that considered the complete interaction between the house envelope, the equipment, and the control. By merely changing the function generator, Nelson was able to analyze the *performance* of the house in various geographical locations.

An electric analogue model has to be driven in some way by application of a time-varying potential source to model variation in ambient temperature or a current source for heat inputs at one or more nodes in the circuit. There was no difficulty in providing step, ramp or periodic sources. To evaluate seasonal energy needs, however, a sequence of, say, hourly values of ambient temperature, solar incidence and casual gains were needed, for which a direct electric drive was difficult to construct. Burberry *et al.* (1979) have described a hybrid model where the building was represented as an assembly of *RC* elements but where a season's meteorological information was stored electronically and converted to electric signals.

The analogue using real resistances and capacitors was well established in the 1960s and one might have expected that by 1970, when electronic (digital) computing had become generally widespread, it would have been exploited to estimate the thermal behaviour of a room on the basis of a discretised thermal circuit. In one respect it was: wall finite difference calculations, previously conducted manually, could readily be carried out electronically. Surprisingly, comparable calculations for an enclosure were not much reported until about 1980. They will be discussed in Chapter 13.

There are parallels between simple mechanical, electrical and thermal systems. While the parallels between mechanical and electrical systems are close, thermal systems only partly fit into the scheme.

- Mechanical system

force	=	spring rate	×	displacement	+	viscous drag	×	velocity gradient	+	inertia	×	acceleration
F	=	s	×	x	+	$\mu(A)$	×	$(\mathrm{d}x/\mathrm{d}t)/y$	+	m	×	$\mathrm{d}^2x/\mathrm{d}t^2$

- Electrical system

voltage	=	(1/capacity)	×	charge	+	resistance	×	current (I)	+	inductance	×	$\mathrm{d}I/\mathrm{d}t$
V	=	$(1/C)$	×	Q	+	R	×	$\mathrm{d}Q/\mathrm{d}t$	+	L	×	$\mathrm{d}^2Q/\mathrm{d}t^2$

- Thermal system

temp diff.	=	(1/capacity)	×	energy	+	resistance	×	flow
ΔT	=	$(1/C)$	×	H	+	R	×	$\mathrm{d}H/\mathrm{d}t$

The electrical/thermal parallel is sufficient to enable building and similar thermal processes to be examined as indicated above. However it will be noted:

- The products, force × displacement and voltage × charge each have the units of energy (J). The corresponding thermal product, temperature difference × energy, is physically meaningless.
- There is no thermal equivalent of inertia and inductance.[3]
- There is no mechanical or electrical equivalent of entropy, (energy or heat)/(absolute temperature), a measure of the disorder in a system.

10.6 TIME-VARYING ELEMENTS

In the foregoing discussion it has been tacitly assumed that the values of the resistances and capacities do not vary in time. This is not realistic however. The value of the ventilation link may vary widely throughout the day; curtaining and movable insulation may increase the window resistance by night, and the convection coefficient between warm air and a cold ceiling, for example, is larger than with the reverse situation.

There is no difficulty in modelling such varying resistances using finite difference, transfer coefficient or analogue methods. There is a difficulty, however, when the periodic approach is used; this assumes that all temperatures vary sinusoidally in time. However, under the action of suddenly increased ventilation, room temperatures may fall rapidly and in this case the basic assumption that $\mathrm{d}T/\mathrm{d}t$ can be replaced by $\mathrm{j}\omega T$ no longer holds. As will be seen in Chapter 16, the periodic approach, at any rate for the 24 h period, provides a useful initial technique to gain an indication of the response of a room at the early design stage. This section illustrates the consequences of introducing a time-varying ventilation rate in a very simple model into the periodic analysis; the consequences will be compared with those from a finite difference approach.

In Figure 10.7 an external wall is represented as a single lumped capacity c at T_c flanked by outside and inside convective transmittances k_1 and k_2. The room temperature is T_i and the space is supposed limited by a perfectly reflecting, perfectly insulating wall, so that it can be ignored. Heat is introduced by the sinusoidally varying source q, and the space ventilation is represented by k_3; k_3 will be assumed to vary periodically but not

[3]The property noted here as (thermal) capacity is often perfectly reasonably described as thermal mass or inertia.

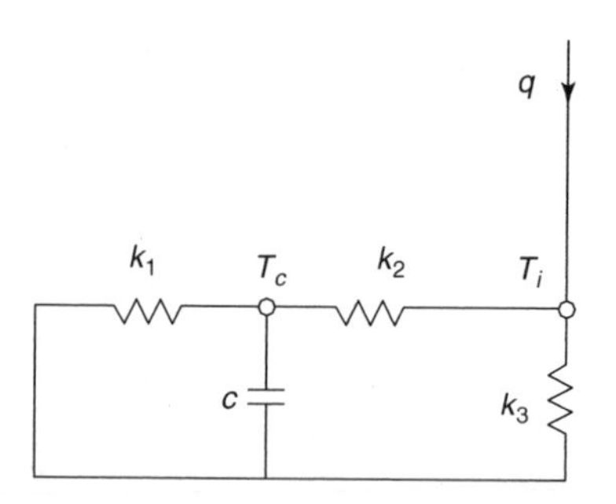

Figure 10.7 A basic circuit with variable ventilation rate

necessarily sinusoidally. The continuity equations at T_c and T_i are

$$T_c k_1 + (T_c - T_i)k_2 = -c\,\mathrm{d}T_c/\mathrm{d}t = -\mathrm{j}\omega c T_c \tag{10.62a}$$

and

$$T_i k_3 + (T_i - T_c)k_2 = q, \tag{10.62b}$$

leading to the relation

$$T_i = \frac{(k_1 + k_2 + \mathrm{j}\omega c)q}{k + \mathrm{j}\omega c(k_2 + k_3)} = m\ \mathrm{e}^{j\phi} q, \tag{10.63}$$

where $k = k_1 k_2 + k_2 k_3 + k_3 k_1$ and the magnitude m and phase ϕ of T_i are given by application of (15.28). Properly speaking, m and ϕ are fixed through the cycle. We shall 'misuse' the expression and at each time t through the cycle use values of m and ϕ based on the current value of k_3.

The consequences of this approach can be tested by evaluating T_i throughout its (daily) cycle using the finite difference method. Consider continuity at two times t_1 and t_2 a short interval δ apart. Then

$$\tfrac{1}{2}(T_{c1} + T_{c2})k_1\delta + \tfrac{1}{2}(T_{c1} - T_{i1} + T_{c2} - T_{i2})k_2\delta = (T_{c1} - T_{c2})c \tag{10.64a}$$

and

$$T_{i1} k_3 + (T_{i1} - T_{c1})k_2 = q_1, \tag{10.64b}$$

with a similar relation at time t_2. T_{c2} is given as

$$T_{c2} = \frac{\tfrac{1}{2}(q_1 + q_2)\delta k_2/(k_2 + k_3) + T_{c1}\left(c - \tfrac{1}{2}k\delta/(k_2 + k_3)\right)}{\left(c + \tfrac{1}{2}k\delta/(k_2 + k_3)\right)} \tag{10.65a}$$

and

$$T_{i2} = (q_2 + T_{c2}k_2)/(k_2 + k_3). \tag{10.65b}$$

To effect a comparison, a concrete wall of thickness 100 mm will be assumed so that $c = 0.1 \times 2100 \times 840\,\mathrm{J/m^2K}$, $k_1 = 18$ and $k_2 = 8\,\mathrm{W/m^2K}$, corresponding to inside and outside lumped radiative and convective coefficients. The ventilated space is taken to be 3 m wide, so $k_3 = 3 \times (N/3600) \times 1200\,\mathrm{W/m^2K}$ where N is the hourly ventilation rate.

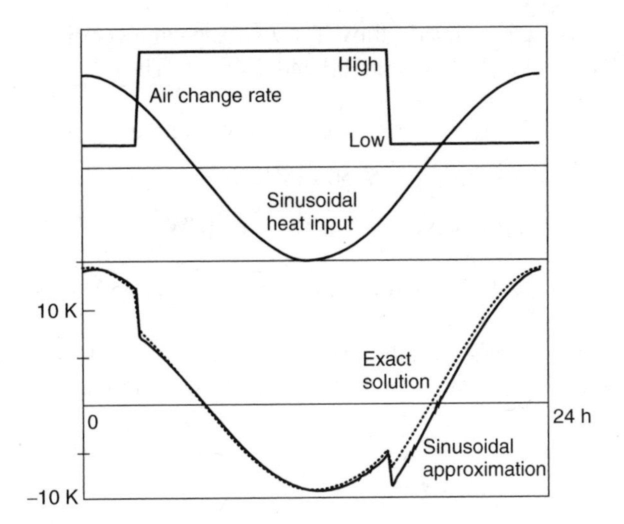

Figure 10.8 (a) Times of high and low ventilation rates in relation to a heat input of $100\cos(2\pi t/24)$ W/m^2 at the room index temperature T_i in Figure 10.7. (b) The exact solution for T_i and the sinusoidal approximation for T_i

We will assume values for N of $N = 1$ before 3 h and after 16 h and $N = 5$ air changes between these times. Also q will be assumed as $100\cos(2\pi t/24)$ W/m^2 and T_i will peak a time $(\phi/2\pi) \times 24$ hours later. An interval of $\delta = 12$ min will be selected to show the detailed response when step changes in ventilation are imposed. If a constant ventilation rate is used, the values of T_i given by (10.63) and (10.65) agree to 4 decimal places.

Figure 10.8 shows the response of the room index temperature. The sinusoidal solution consists of the sinusoidal segments for the periods of low and high ventilation (14.33 K magnitude, 0.65 h lag and 9.14 K, 0.41 h, respectively). It does not differ significantly from the exact solution over much of the 24 h period. The mean difference (approximate value − exact value) is −0.33 K. Since most of the values in calculations of this kind are not usually known reliably, the approximate sinusoidal solution, which is very easily found, will be acceptable.

10.7 APPENDIX

It is instructive to develop the *r-c-r-c-r* model of Figure 10.4 using values for the elements which are typical of wall constructions, although the wall is not a recommended design. They are shown in Figure 10.9. The three resistances are typical of the resistances of an

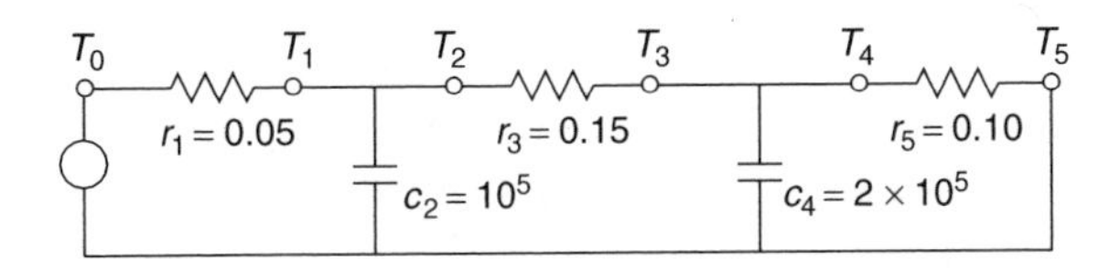

Figure 10.9 Two-capacity model: typical element values for a building wall

outside film, a cavity and an inside film. The two capacities correspond to materials of ρc_p value 10^6 J/m^2 K and of thickness 100 and 200 mm. Thus

$$r_1 + r_3 + r_5 = r = 0.3\,\text{m}^2\text{K/W (so the } U \text{ value is } 3.33\,\text{W/m}^2\text{K)},$$

$$r_1 c_2 r_3 + r_1 c_2 r_5 + r_1 c_4 r_5 + r_3 c_4 r_5 = 5250\,\text{s m}^2\text{K/W},$$

$$r_1 c_2 r_3 c_4 r_5 = 1.5 \times 10^7\,\text{s}^2\,\text{m}^2\text{K/W}.$$

The decay times are $z_1 = 13\,904$ s (or 3.86219 h) and $z_2 = 3596$ s (or 0.99892 h). (Thus $z_1 + z_2 = 5250/0.3$ and $z_1 \times z_2 = 1.5 \times 10^7/0.3$.)

If the wall layers have non-zero temperatures and from $t = 0$ onward T_0 and T_5 are zero, the subsequent decay is described by T_2 and T_4 with initial values given by

$$\frac{T_{20,i}}{T_{40,i}} = \frac{r_3}{r_5} + 1 - \frac{r_3 c_4}{z_i},$$

so $T_{40,1}/T_{20,1} = 2.9211$ and $T_{40,2}/T_{20,2} = -0.1712$, and their product is equal to $-c_2/c_4 = -\frac{1}{2}$ (see footnote in Section 10.3.2).

If, on the other hand, T_0 has been rising for an indefinite period at 1 K per hour, so that $\theta = 1/3600$ K/s, and time zero is chosen so that $T_0 = 0$ at $t = 0$, the $t = 0$ values at T_2 and T_4 are given as

$$T'_{20} = -(\theta/r^2) r_1 [c_2 (r_3 + r_5)^2 + c_4 {r_5}^2] = -1.2731\,\text{K},$$

$$T'_{40} = -(\theta/r^2) r_5 [c_2 r_1 (r_3 + r_5) + c_4 (r_1 + r_3) r_5] = -1.6204\,\text{K}.$$

Values for $T_{20,1}$, $T_{40,1}$, $T_{20,2}$ and $T_{40,2}$ must now be found such that the initial state of the transient solution is equal and opposite to the $t = 0$ state of the slope solution; that is

$$T_{20,1} + T_{20,2} = -T'_{20} \quad \text{or} \quad T_{20,1} + T_{20,2} = +1.2731,$$

$$T_{40,1} + T_{40,2} = -T'_{40} \quad \text{or} \quad 2.9211\ T_{20,1} + (-0.1712)\ T_{20,2} = +1.6204.$$

Solution of these equations gives the values of $T_{20,1}$, etc., and by adding the slope and transient solutions we can write down the complete solution for values at T_2 and T_4:

$$T_{20,t} = \frac{0.25}{0.30}\frac{t}{3600} - 1.2731 + 0.5945 \exp\left(\frac{-t}{13904}\right) + 0.6787 \exp\left(\frac{-t}{3596}\right),$$

$$T_{40,t} = \frac{0.10}{0.30}\frac{t}{3600} - 1.6204 + 1.7365 \exp\left(\frac{-t}{13904}\right) - 0.1162 \exp\left(\frac{-t}{3596}\right).$$

This is the solution for temperatures in the wall, initially at zero, where T_5 is held at zero but from time $t = 0$ onward, T_0 is raised at the rate of 1 K per hour. We can examine the initial response at T_2 and T_4 by expanding the exponential terms up to powers of 2:

- *Terms in t^0:* coefficients total zero, in accordance with the initial condition which was imposed.
- *Terms in t^1:* coefficients for both T_2 and T_4 total zero, in accordance with the boundary conditions imposed.
- *Terms in t^2:* coefficient for T_2 is finite, $O(10^{-8})$, but the coefficient for T_4 remains exactly zero, showing that T_2 shows some response before T_4, as expected.

This solution will be developed in the next chapter to illustrate the evaluation of the wall transfer coefficients.

11

Wall Conduction Transfer Coefficients for a Discretised System

In Chapter 10 the measures **u** and **y** were mentioned as a means to describe the dynamic thermal behaviour of a wall. They are of limited utility, however, and the series of wall conduction transfer coefficients a, b, c and d provide a more flexible procedure. Chapter 17 presents the derivation of coefficients for walls with distributed resistance and capacity. It is a lengthy and complicated calculation but consideration of a wall represented as a chain of resistive and capacitative elements enables us to derive the coefficients in a simple way. Suppose that in the two-capacity circuit of Figure 10.4 the temperature node T_5 is held at zero. Suppose that T_0 is undergoing continuous random variation and that its value at $t = 0$ is known, together with values at previous times $t = -\delta, t = -2\delta, t = -3\delta$, etc. Suppose that δ is chosen to be sufficiently small so that variation between successive values can be taken as linear. It is to be shown that the heat flow density $q(\mathrm{W/m^2})$ to T_5 at time $t = 0$ can be expressed in terms of recent values of the driving temperature, together with recent values of the flow itself:

$$q_{50,0} = \underbrace{b_0 T_{0,0} + b_1 T_{0,-\delta} + b_2 T_{0,-2\delta}}_{\text{temperature terms}} - \underbrace{d_1 q_{50,-\delta} - d_2 q_{50,-2\delta}}_{\text{flux terms}} \tag{11.1}$$

The first subscript to q-(5)-indicates to which node the flow is directed (T_5), the second subscript (0) indicates the node at which the driving temperature acts (T_0) and the third subscript denotes time. Thus the flow now is expressible in terms of the temperature now and its values at times δ and 2δ earlier, together with the two earlier values of the flow. For building applications, the most common choice of δ is 1 h, and it is usual to express the time in integer form. Thus at time level i,

$$q_{50,i} = b_0 T_{0,i} + b_1 T_{0,i-1} + b_2 T_{0,i-2} - d_1 q_{50,i-1} - d_2 q_{50,i-2} \quad \text{with } T_5 = 0. \tag{11.2a}$$

Three similar equations can be written down, corresponding to the locations of the flow and temperature.[1]

[1]The coefficients a, b and c have units $\mathrm{W/m^2K}$, so when they are applied to a wall area, the corresponding capitals can be used; d is dimensionless. For a wall, Q (W) replaces $q(\mathrm{W/m^2})$.

Building Heat Transfer Morris G. Davies
© 2004 John Wiley & Sons, Ltd ISBN: 0-470-84731-X

$$q_{55,i} = c_0 T_{5,i} + c_1 T_{5,i-1} + c_2 T_{5,i-2} - d_1 q_{55,i-1} - d_2 q_{55,i-2} \quad \text{with } T_0 = 0, \qquad (11.2b)$$

$$q_{00,i} = a_0 T_{0,i} + a_1 T_{0,i-1} + a_2 T_{0,i-2} - d_1 q_{00,i-1} - d_2 q_{00,i-2} \quad \text{with } T_5 = 0, \qquad (11.2c)$$

$$q_{05,i} = b_0 T_{5,i} + b_1 T_{5,i-1} + b_2 T_{5,i-2} - d_1 q_{05,i-1} - d_2 q_{05,i-2} \quad \text{with } T_0 = 0. \qquad (11.2d)$$

If both T_0 and T_5 vary, the net flow q_5 out is

$$q_{5,i} = q_{55,i} - q_{50,i}. \qquad (11.3)$$

If the wall is modelled as a three-capacity wall, similar equations apply; the series is extended to include terms corresponding to $t = -3\delta$. These expressions for heat flow are unlike those associated with routine circuit analysis. The coefficients are similar to regression coefficients. The successive values of temperature multiplying the a_k, b_k and c_k values are mathematically independent of each other, although in a real situation they will be closely associated. The successive values of flux multiplying the d_k coefficients however are very highly correlated. The derivation of the transfer coefficients follows. First we find the flux at one or other surface of the wall when it is subjected to a triangular variation in one of its bounding temperatures; the response is expressed as an infinitely long series of 'response factors'. The information they provide can be expressed succinctly through use of two out of four short series of wall 'conduction transfer coefficients'.

11.1 THE RESPONSE FACTORS $\phi_{50,K}$

Our task is to derive as an intermediate quantity, the wall response factors which describe the response of the wall when it is excited by a triangular pulse in temperature (Figure 11.1b). The process will be illustrated by finding the heat flow on one side of a wall due to excitation at the other.

T_0, say, is supposed to have been at zero for an indefinite time. A triangular pulse can be formed by superposing three ramp excitations. At $t = -\delta$ an increase of rate θ is imposed on the exterior of the wall, at $t = 0$ an additional decrease of 2θ and at $t = +\delta$ an increase of θ again. The net result (Figure 11.1b) is a pulse of base 2δ and height $\Theta = \theta\delta$.

The heat flow to T_5 due to a ramp applied at T_0 from $t = 0$ onward is $q_{50,t} = T_{4t}/r_5$, where T_{4t} is given by equation (10.40b). In Example 10.4 of Section 10.3.6,

$$q_{50,t} = \theta[t/9 - 1.1852 + 1.2151\exp(-t/9.2206) - 0.0299\exp(-t/1.4460)][\mathrm{W/m^2}].$$

The pulse is centred at $t = 0$. The net response is therefore the sum of three component responses:

- $q_{50,t}$ with $t - \delta$ replacing t,
- $q_{50,t}$ with t as it stands but -2θ replacing θ,
- $q_{50,t}$ with $t + \delta$ replacing t.

The wall response factors $\phi_{50,k}$ are defined as the net heat flows at times $t = 0$, δ, $2\delta, \ldots, k\delta$ per pulse of unit height $\Theta = \theta\delta = 1$ K; they have units W/m^{2}K. The zeroth

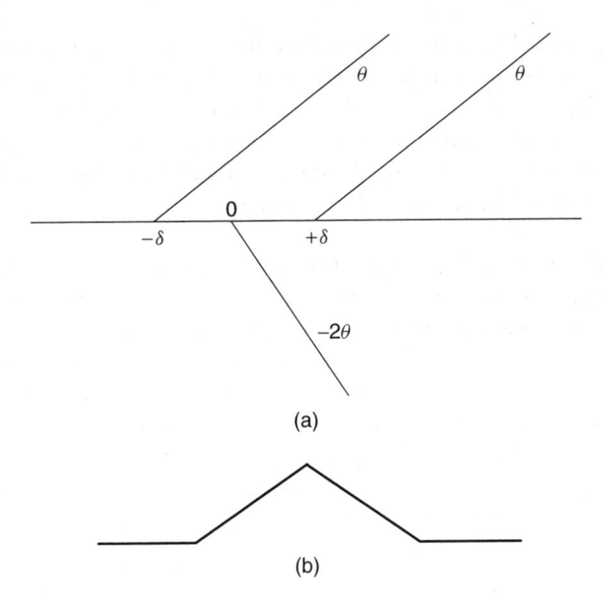

Figure 11.1 (a) Three temperature ramps. Left to right, the temperature gradients are $+\theta$, -2θ and $+\theta$. (b) The resulting triangular pulse of base 2δ and height $\Theta = \theta\delta$

factor, $\phi_{50,0}$, the response at $t = 0$, depends on the first ramp alone and so has the value given by substituting $t = \delta$ in (10.40b); $\phi_{50,1}$ depends on the first two ramps and the contributions of the $1/(r_1 + r_3 + r_5)$ terms cancel. For $\phi_{50,2}$ and higher terms, the contributions of both this and the constant term $T_4{}'$ cancel. Then

$$\begin{aligned}\phi_{50,k} = \{&T_{40,1}[\exp(-(k+1)\delta/z_1) - 2\ \exp(-k\delta/z_1) + \exp(-(k-1)\delta/z_1]\\ &+T_{40,2}[\exp(-(k+1)\delta/z_2) - 2\ \exp(-k\delta/z_2) + \exp(-(k-1)\delta/z_2]\}/(r_5\Theta)\end{aligned} \tag{11.4a}$$

$$\begin{aligned}= \{&T_{40,1}[\exp(-\delta/z_1) - 2 + \exp(\delta/z_1)]\exp(-k\delta/z_1)\\ &+T_{40,2}[\exp(-\delta/z_2) - 2 + \exp(\delta/z_2)]\exp(-k\delta/z_2)\}/(r_5\Theta)\end{aligned} \tag{11.4b}$$

$$= \{T_{40,1}[\beta_1 - 2 + 1/\beta_1]\beta_1{}^k + T_{40,2}[\beta_2 - 2 + 1/\beta_2]\beta_2{}^k\}/(r_5\Theta), \tag{11.4c}$$

where $\beta_1 = \exp(-\delta/z_1)$, similarly β_2.

Since z_2 is normally considerably smaller than z_1, $\exp(-\delta/z_2) \ll \exp(-\delta/z_1)$ and as k increases, the contribution of $T_{40,2}$ fades in comparison with $T_{40,1}$ terms. Eventually the response factors decay exponentially, so

$$\phi_{50,k+1}/\phi_{50,k} = \exp(-\delta/z_1) = \beta_1. \tag{11.5}$$

If a pulse of height $T_{0,0}$ acts at T_0 at time $t = 0$, T_0 being zero before $t = -\delta$ and after $t = +\delta$, the heat flow to T_5 at times $0, \delta, 2\delta, \ldots, k\delta$, will be $\phi_{50,0}T_{0,0}$, $\phi_{50,1}T_{0,0}$, $\phi_{50,2}T_{0,0}, \ldots, \phi_{50,k}T_{0,0}$ and tends to zero.

At the end of Section 10.3 it was pointed out that when a ramp excitation was imposed at T_0, there was virtually no immediate effect at T_4. Thus if t is measured in units of δ and δ is small compared to z_1 and z_2, there will initially be a near-zero response at T_4. Since the triangular pulse consists of three ramps whose net effect is largely to annul each other, the net effect at T_4 will initially be small and will never become large. In physical terms, the thermal signal imposed at T_0 is strongly attenuated as it passes through the wall, as is intuitively obvious.

We wish to find the heat flow to T_5 due to random continuous variation in T_0. The variation is illustrated by the full line in Figure 11.2. It is clear that if the variation is sampled at sufficiently frequent intervals, expressed as δ, the variation between sample values at $t = 0, \delta, 2\delta, \ldots$ can be taken as linear (the dotted line). The linear variation between $T_{0,0}$ and $T_{0,1}$ can be constructed from the superposed effect of the first two triangular pulses in the range $0 < t < \delta$ and the variation between $T_{0,1}$ and $T_{0,2}$ can be similarly constructed using pulses 2 and 3.

Consider the value of q_5 at time $t = 0$, due to variation in T_0 up to that time. The component due to the current value there is $\phi_{50,0}T_{0,0}$. That due to $T_{0,-\delta}$ one interval earlier is $\phi_{50,1}T_{0,-\delta}$, etc. Adding the components,

$$q_{50,0} = \phi_{50,0}T_{0,0} + \phi_{50,1}T_{0,-\delta} + \cdots + \phi_{50,k}T_{0,-k\delta} + \cdots. \tag{11.6a}$$

Using time-level notation for the subscripts, at time level i we have

$$q_{50,i} = \sum_{k=0}^{\infty} \phi_{50,k}T_{0,i-k}; \tag{11.6b}$$

other notation is sometimes used.[2]

In the steady state, T_0 has the constant value $T_{0,ss}$, so

$$q_{50,t} = \left(\sum_{k=0}^{\infty} \phi_{50,k}\right) T_{0,ss}, \tag{11.7a}$$

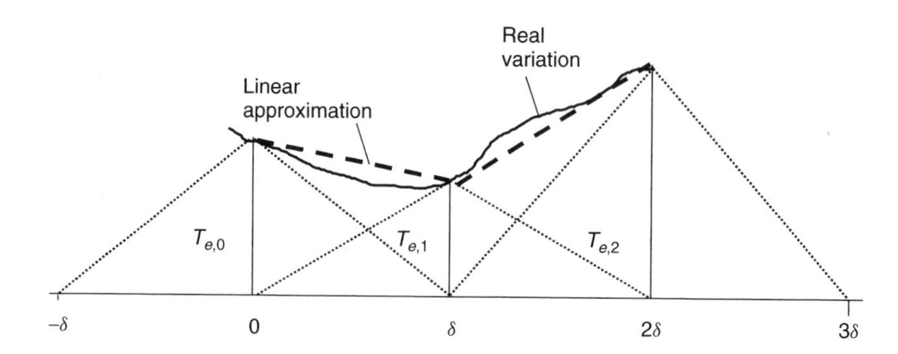

Figure 11.2 Continuous variation in ambient temperature T_e, its sampled values at $t = 0, \delta, 2\delta$, the assumed linear variation between these values, and their representation by triangular pulses

[2]The quantity ϕ is often denoted by capital letters: ϕ_{eej} as X_j, $\phi_{ei,j}$ as Y_j, $\phi_{ii,j}$ as Z_j. Subscripts e and i denote exterior and interior.

but then

$$q_{50,t} = UT_{0,ss}, \tag{11.7b}$$

so

$$\left(\sum_{k=0}^{\infty} \phi_{50,k}\right) = U. \tag{11.7c}$$

This is consistent with the formation of $\sum_{k=0}^{\infty} \phi_{50,k}$. All terms except the value of $\theta\delta/r$ in the first ramp response cancel, so the total consists simply of the single term $\theta\delta/r = \Theta/r = 1/r$ or the wall U value.

Use of response factors implies a long series of values of $\phi_{50,k}$ but the information can be coded in shorter form using the d coefficients, and in this case the b coefficients as shown below.

11.2 THE *d* COEFFICIENTS

The values of the d coefficients depend on the values of the wall resistances and capacities (i.e. the wall decay times) and on the choice of the time step δ, but not on the way the wall is being excited. Suppose that the two-capacity wall of Figure 10.4 has been excited in some way in the past but that from $t = -2\delta$ onward both T_0 and T_5 are zero. Now T_4 is composed of independent terms proportional to $\exp(-t/z_1)$ and $\exp(-t/z_2)$. Furthermore, $q_5 = T_4/r_5$. Thus at times $t = -2\delta$, $t = -\delta$ and $t = 0$ we can write

$$q_{50,i-2} = A\exp(-(i-2)\delta/z_1) + B\exp(-(i-2)\delta/z_2), \tag{11.8a}$$

$$q_{50,i-1} = A\exp(-(i-1)\delta/z_1) + B\exp(-(i-1)\delta/z_2), \tag{11.8b}$$

$$q_{50,i-0} = A\exp(-(i-0)\delta/z_1) + B\exp(-(i-0)\delta/z_2). \tag{11.8c}$$

Elimination of A and B permits us to express q_{50} 'now' in terms of its two previous values. Writing $\beta_1 = \exp(-\delta/z_1)$ and $\beta_2 = \exp(-\delta/z_2)$, we find

$$q_{50,i} = (\beta_1 + \beta_2)q_{50,i-1} - \beta_1\beta_2 q_{50,i-2}. \tag{11.9}$$

But from (11.2a) with driving temperatures T_0 of zero,

$$q_{50,i} = \quad - d_1 q_{50,i-1} - \quad d_2 q_{50,i-2}, \tag{11.10}$$

so by identification,

$$d_1 = -(\beta_1 + \beta_2), \tag{11.11a}$$

$$d_2 = +\beta_1\beta_2. \tag{11.11b}$$

Conversely, the β values are the roots of the equation

$$\beta^2 + d_1\beta + d_2 = 0. \tag{11.12}$$

A relation of this type is true for any number of capacities. If $N = 4$, giving values z_1 to z_4, we have

$$d_1 = -(\beta_1 + \beta_2 + \beta_3 + \beta_4), \tag{11.13a}$$

$$d_2 = +(\beta_1\beta_2 + \beta_1\beta_3 + \beta_1\beta_4 + \beta_2\beta_3 + \beta_2\beta_4 + \beta_3\beta_4), \tag{11.13b}$$

$$d_3 = -(\beta_1\beta_2\beta_3 + \beta_1\beta_2\beta_4 + \beta_1\beta_3\beta_4 + \beta_2\beta_3\beta_4), \tag{11.13c}$$

$$d_4 = +(\beta_1\beta_2\beta_3\beta_4) \tag{11.13d}$$

and

$$\beta^4 + d_1\beta^3 + d_2\beta^2 + d_3\beta + d_4 = 0. \tag{11.14}$$

If J is defined as

$$J = (1 - \beta_1)(1 - \beta_2)(1 - \beta_3)(1 - \beta_4), \tag{11.15a}$$

then

$$\begin{aligned} J = 1 &- (\beta_1 + \beta_2 + \beta_3 + \beta_4) + (\beta_1\beta_2 + \beta_1\beta_3 + \beta_1\beta_4 + \beta_2\beta_3 + \beta_2\beta_4 + \beta_3\beta_4) \\ &- (\beta_1\beta_2\beta_3 + \beta_1\beta_2\beta_4 + \beta_1\beta_3\beta_4 + \beta_2\beta_3\beta_4) + (\beta_1\beta_2\beta_3\beta_4) \end{aligned} \tag{11.15b}$$

$$= d_0 + d_1 + d_2 + d_3 + d_4 \tag{11.15c}$$

$$= \sum_{k=0}^{4} d_k; \tag{11.15d}$$

d_0 is defined as unity.

Thus the d_k coefficients are transformations of the decay times z_j for some specified value of δ. If $\delta \gg z_j$, $\beta_j \ll 1$ and that value of z_j can be ignored in considering the response. If $\delta \ll z_j$ then $(1 - \beta_j) \sim \delta/z_j$ and so J, the product of these terms, can become very small. The significance of this will be discussed later.

11.3 THE TRANSFER COEFFICIENTS $b_{50,K}$

In a heavyweight building, the values of an internal surface temperature may be correlated with themselves up to more than a fortnight previously (Figure 14.4). This implies that if (11.6b) is used to handle data at hourly intervals, a series of up to 400 terms may have to be summed for each hourly time level in order to compute a flux such as q_{50} over a period of time. However, for a two-capacity wall, the effect at $t = 0$ of all excitation previous to $t = -2\delta$ can be represented by the values of the flux at $t = -\delta$ and $t = -2\delta$. Thus $q_{50,i}$ can be expressed in terms of the recent temperature excitation $T_{0,i}$, $T_{0,i-1}$ and $T_{0,i-2}$, together with $q_{50,i-1}$ and $q_{50,i-2}$:

$$q_{50,i} = b_0 T_{0,i} + b_1 T_{0,i-1} + b_2 T_{0,i-2} - d_1 q_{50,i-1} - d_2 q_{50,i-2} \tag{11.16a}$$

or

$$d_0 q_{50,i-0} + d_1 q_{50,i-1} + d_2 q_{50,i-2} = b_0 T_{0,i} + b_1 T_{0,i-1} + b_2 T_{0,i-2}. \tag{11.16b}$$

But

$$q_{50,i-0} = \phi_{50,0}T_{0,i-0} + \phi_{50,1}T_{0,i-1} + \phi_{50,2}T_{0,i-2} + \phi_{50,3}T_{0,i-3} + \cdots, \quad (11.17a)$$

$$q_{50,i-1} = \phi_{50,0}T_{0,i-1} + \phi_{50,1}T_{0,i-2} + \phi_{50,2}T_{0,i-3} + \cdots, \quad (11.17b)$$

$$q_{50,i-2} = \phi_{50,0}T_{0,i-2} + \phi_{50,1}T_{0,i-3} + \cdots. \quad (11.17c)$$

Substituting these values into (11.16b), we have

$$\begin{aligned} & d_0[\phi_{50,0}T_{0,i-0} + \phi_{50,1}T_{0,i-1} + \phi_{50,2}T_{0,i-2} + \phi_{50,3}T_{0,i-3} + \cdots] \\ & + d_1[\qquad \phi_{50,0}T_{0,i-1} + \phi_{50,1}T_{0,i-2} + \phi_{50,2}T_{0,i-3} + \cdots] \\ & + d_2[\qquad \phi_{50,0}T_{0,i-2} + \phi_{50,1}T_{0,i-3} + \cdots] \\ & = \quad b_0T_{0,i-0} + b_1T_{0,i-1} + b_2T_{0,i-2}. \end{aligned} \quad (11.18)$$

Now the successive values $T_{0,i-0}$, $T_{0,i-1}$ and $T_{0,i-2}$ are unrelated, so we may equate their coefficients:

$$b_0 = d_0\phi_{50,0}, \quad (11.19a)$$

$$b_1 = d_0\phi_{50,1} + d_1\phi_{50,0}, \quad (11.19b)$$

$$b_2 = d_0\phi_{50,2} + d_1\phi_{50,1} + d_2\phi_{50,0}. \quad (11.19c)$$

Note that b_0 is equal to the zeroth response factor. For a four-capacity wall, the argument of Section 11.1 provides values up to d_4 and the argument of Section 11.2 provides the ϕ_{90} values at node 9, so we have

$$b_0 = d_0\phi_{90,0}, \quad (11.20a)$$

$$b_1 = d_0\phi_{90,1} + d_1\phi_{90,0}, \quad (11.20b)$$

$$b_2 = d_0\phi_{90,2} + d_1\phi_{90,1} + d_2\phi_{90,0}, \quad (11.20c)$$

$$b_3 = d_0\phi_{90,3} + d_1\phi_{90,2} + d_2\phi_{90,1} + d_3\phi_{90,0}, \quad (11.20d)$$

$$b_4 = d_0\phi_{90,4} + d_1\phi_{90,3} + d_2\phi_{90,2} + d_3\phi_{90,1} + d_4\phi_{90,0}. \quad (11.20e)$$

If the interval δ is large compared with z_4, d_4 (11.13d) may be near-zero, but the value of b_4 is not necessarily negligible.

For the two-capacity wall in steady-state conditions,

$$(d_0 + d_1 + d_2)q_{50,ss} = (b_0 + b_1 + b_2)T_{0,ss}, \quad (11.21a)$$

so

$$\sum b / \sum d = q_{50,ss}/T_{0,ss} = \text{wall } U \text{ value}. \quad (11.21b)$$

11.4 THE RESPONSE FACTORS $\phi_{00,k}$, $\phi_{55,k}$ AND TRANSFER COEFFICIENTS a, c

The previous two sections dealt with the response of a two-capacity system due to a triangular temperature pulse imposed at a point (T_0) remote from the point of excitation; it resulted in a delayed and small response. The response q_{00} at the adjacent point (T_0), given by the value of $(T_0 - T_1)/r_1$ (or q_{55} at T_5) due to a triangular pulse can be found similarly. The flow of heat at T_0 is immediate and positive during the first interval δ, goes negative during the second interval, and thereafter remains negative, gradually returning to zero.

The series of transfer coefficients a_k and c_k can be found in the same way as the b_k values. Their values are much larger than the values of b and, like the values of d, they alternate in sign (a_0 positive, a_1 negative) but $\sum a/\sum d$, like $\sum b/\sum d$, has the value of the wall transmittance.

11.5 SIMPLE CASES

It is convenient to note results for some simple walls. In the *r-c* and *r-c-r* circuits of Figure 10.2, with $r_1 = r_2 = r$, we do not need to solve simultaneous equations or use the orthogonality theorem. The coefficients will be found to have the following values.

	r-c circuit	*r-c-r* circuit, where $z = \frac{1}{2}rc$
b_0	$1/r$	$(2r)^{-1}[1 \quad - (z/\delta)(1 - \exp(-\delta/z))]$
b_1	0	$(2r)^{-1}[- \exp(-\delta/z) + (z/\delta)(1 - \exp(-\delta/z))]$
c_0	$1/r + c/\delta$	$(2r)^{-1}[1 \quad + (z/\delta)(1 - \exp(-\delta/z))]$
c_1	$-c/\delta$	$(2r)^{-1}[- \exp(-\delta/z) - (z/\delta)(1 - \exp(-\delta/z))]$
d_0	1(by definition)	1(by definition)
d_1	0	$- \exp(-\delta/z)$

(11.22)

Simple results can be found too for the *r-c-r-c-r* model. The heat flow $q_{00,i}$ at T_0 at time level i due to ramp excitation at T_0 is

$$q_{00,i} = i/r + A_1(1 - \beta_1{}^i) + A_2(1 - \beta_2{}^i), \tag{11.23}$$

where $\beta_1 = \exp(-\delta/z_1)$. The A coefficients result from fitting the two transient solutions to the T_2', T_4' values (10.39) of the slope solution at $t = 0$:

$$A_1 = \frac{\alpha_1(-T_2' + \alpha_2 T_4')/r_5}{\alpha_1 - \alpha_2}, \quad A_2 = \frac{\alpha_2(T_2' - \alpha_1 T_4')/r_5}{\alpha_1 - \alpha_2}; \tag{11.24}$$

α_1 is the ratio $T_{40,1}/T_{20,1}$ of the transient solution (Section 10.3.3) and has the value 2.9211 below, following from Section 10.7; $\alpha_2 = T_{40,2}/T_{20,2} = -0.1712$.

The zeroth response factor is

$$\phi_{00,0} = 1/r + A_1(1 - \beta_1) + A_2(1 - \beta_2). \tag{11.25a}$$

The ith factor is found from the $(i-1)$th, ith and $(i+1)$th fluxes:

$$\phi_{00,i} = -A_1\beta_1{}^{i-1}(1-\beta_1) - A_2\beta_2{}^{i-1}(1-\beta_2). \tag{11.25b}$$

The transfer coefficients follow, noting that $d_1 = -(\beta_1+\beta_2)$ and $d_2 = \beta_1\beta_2$:

$$a_0 = 1/r + A_1(1-\beta_1) + A_2(1-\beta_2)\ (=\phi_{00,0})\ \text{and is positive,} \tag{11.26a}$$

$$a_1 = -(\beta_1+\beta_2)/r - A_1(1-\beta_1)(1+\beta_2) - A_2(1-\beta_2)(1+\beta_1)\ \text{and is negative,} \tag{11.26b}$$

$$a_2 = \beta_1\beta_2/r + A_1(1-\beta_1)\beta_2 + A_2(1-\beta_2)\beta_1\ \text{and is positive.} \tag{11.26c}$$

Recalling that $\sum d_k = (1-\beta_1)(1-\beta_2)$ from (11.15), it is easily checked that these values satisfy (11.36).

For this two-capacity model, there is no z_3 so $d_3 \equiv 0$. Although the response factors continue to infinity, it can be checked that a_3 too is zero. It may be noted however that for a multilayer wall, with a series of z values which only tend to zero slowly, d_n rapidly tends to zero with increasing n and soon becomes effectively zero; a_n may not be zero.

11.6 HEAT STORED IN THE STEADY STATE

To derive a general expression for the heat stored (Davies 2004), we consider a four-capacity model, initially at zero temperature. For external excitation,

$$q_{00,i} = a_0T_{i-0} + a_1T_{i-1} + a_2T_{i-2} + a_3T_{i-3} + a_4T_{i-4} - d_1q_{00,i-1} - d_2q_{00,i-2} - d_3q_{00,i-3} - d_4q_{00,i-4}. \tag{11.27}$$

From time level $i = 0$ onward, $T_0 = 1$ K and $T_9 = 0$. From equation (11.2c) the heat flows into the circuit are

$$\begin{aligned}
q_{00,0} &= a_0, && \text{i.e.} \quad a_0,\\
q_{00,1} &= a_0 + a_1 - d_1[a_0], && \text{or} \quad a_0(1-d_1) + a_1,\\
q_{00,2} &= a_0 + a_1 + a_2 - d_1[a_0 + a_1 - d_1[a_0]] - d_2\{a_0\}, && \text{or} \quad a_0(1-d_1+d_1{}^2-d_2) + a_1(1-d_1) + a_2, \quad \text{etc.}
\end{aligned} \tag{11.28}$$

We define $D_1 \equiv -d_1$, $D_2 \equiv d_1{}^2 - d_2$ with the further definitions $D_3 \equiv -d_1{}^3 + 2d_1d_2 - d_3$, $D_4 \equiv d_1{}^4 - 3d_1{}^2d_2 + 2d_1d_3 + d_2{}^2 - d_4$, etc. So we have

$$\begin{aligned}
q_{00,0} &= a_0,\\
q_{00,1} &= a_0(1+D_1) + a_1,\\
q_{00,2} &= a_0(1+D_1+D_2) + a_1(1+D_1) + a_2,\\
q_{00,3} &= a_0(1+D_1+D_2+D_3) + a_1(1+D_1+D_2) + a_2(1+D_1) + a_3,\\
q_{00,4} &= a_0(1+D_1+D_2+D_3+D_4) + a_1(1+D_1+D_2+D_3) + a_2(1+D_1+D_2) + a_3(1+D_1) + a_4,\\
q_{00,5} &= a_0(1+D_1+D_2+D_3+D_4+D_5) + a_1(1+D_1+D_2+D_3+D_4) + a_2(1+D_1+D_2+D_3) + a_3(1+D_1+D_2) + a_4(1+D_1).
\end{aligned} \tag{11.29}$$

There is no further term a_5 to be added, but at each time level we add an extra coefficient to the series which form the multipliers of the as. On rearrangement, we obtain

$$
\begin{aligned}
q_{00,0} &= (a_0), \\
q_{00,1} &= (a_0 + a_1) \quad + D_1(a_0), \\
q_{00,2} &= (a_0 + a_1 + a_2) \quad + D_1(a_0 + a_1) \quad + D_2(a_0), \\
q_{00,3} &= (a_0 + a_1 + a_2 + a_3) \quad + D_1(a_0 + a_1 + a_2) \quad + D_2(a_0 + a_1) \quad + D_3(a_0), \\
q_{00,4} &= (a_0 + a_1 + a_2 + a_3 + a_4) + D_1(a_0 + a_1 + a_2 + a_3) \quad + D_2(a_0 + a_1 + a_2) \quad + D_3(a_0 + a_1) \quad + D_4(a_0), \\
q_{00,5} &= (a_0 + a_1 + a_2 + a_3 + a_4) + D_1(a_0 + a_1 + a_2 + a_3 + a_4) + D_2(a_0 + a_1 + a_2 + a_3) + D_3(a_0 + a_1 + a_2) + D_4(a_0 + a_1) + D_5(a_0).
\end{aligned}
\tag{11.30}
$$

The total heat flowing into the circuit is proportional to the sum to infinity of all elements in this array. The flow out at T_9 is composed of a similar array with bs replacing as. However, $b_0 + b_1 + b_2 + b_3 + b_4 = a_0 + a_1 + a_2 + a_3 + a_4$ so the charge S' stored in the circuit is proportional only to the sum of the terms above the line:

$$
\begin{aligned}
S' \propto {} & [(a_0 + (a_0 + a_1) + (a_0 + a_1 + a_2) + (a_0 + a_1 + a_2 + a_3)) \\
& - (b_0 + (b_0 + b_1) + (b_0 + b_1 + b_2) + (b_0 + b_1 + b_2 + b_3))] \times (1 + D_1 + D_2 + D_3 + D_4 + \cdots).
\end{aligned}
\tag{11.31}
$$

Now as the steady state is approached with 1 K temperature difference,

$$
q_{00,i} \text{ tends to } (a_0 + a_1 + a_2 + a_3 + a_4)(1 + D_1 + D_2 + D_3 + D_4 + D_5 + \cdots). \tag{11.32a}
$$

However, the steady-state heat flow is

$$
(a_0 + a_1 + a_2 + a_3 + a_4) \Big/ \sum d, \tag{11.32b}
$$

so

$$
(1 + D_1 + D_2 + D_3 + D_4 + D_5 + \cdots) = 1 \Big/ \sum d. \tag{11.32c}
$$

The heat flows evaluated are the average values over the sampling period δ. Thus the value of the heat stored in one-sided excitation is

$$
\begin{aligned}
S' &= [4(a_0 - b_0) + 3(a_1 - b_1) + 2(a_2 - b_2) + (a_3 - b_3)]\delta \Big/ \sum d \qquad (11.33a) \\
&= -[(a_1 - b_1) + 2(a_2 - b_2) + 3(a_3 - b_3) + 4(a_4 - b_4)]\delta \Big/ \sum d,
\end{aligned}
$$

so

$$
S' = -\sum k(a_k - b_k)\delta \Big/ \sum d \tag{11.33b}
$$

since

$$
4(a_0 - b_0) + 4(a_1 - b_1) + 4(a_2 - b_2) + 4(a_3 - b_3) + 4(a_4 - b_4) = 0. \tag{11.34}
$$

The sum of this together with the quantity S'' for inside excitation, when terms $c_k - b_k$ replace $a_k - b_k$, provides the expression for the total wall capacity.

For the *r-c-r-c-r* circuit, the quantity $2a_0 + a_1$ is associated with the inflow of heat at T_0 to finally establish steady conditions:

$$2a_0 + a_1 = (2 - (\beta_1 + \beta_2))/r + (A_1 + A_2)(1 - \beta_1)(1 - \beta_2); \tag{11.35}$$

$2b_0 + b_1$ similarly describes the outflow at T_5. It can be checked that the difference between these quantities leads back to the simple expression for the heat stored.

11.7 DISCUSSION

The foregoing arguments lead to sets of response factors ϕ_{00}, ϕ_{55} and ϕ_{50} and transfer coefficients **a, c** and **b**. It turns out that $\phi_{05,k} = \phi_{50,k}$ exactly (and this is true generally unless the wall contains a ventilated cavity); thus the b values derived above can also be used to determine q_{05}. The argument can clearly be extended to any number of discrete resistance/capacity units. It is sufficient at this point to note some general properties of the coefficients for a wall consisting of N capacities. The bounding nodes are written as 0 and $M = 2N + 1$, respectively.

$d_0 = 1$ by definition; d_1 is negative, d_2 positive, d_3 negative, etc., since the sequence of flux values are highly correlated. All b values are positive (on physical grounds) but are numerically small; a_0 and c_0 are positive, a_1 and c_1 are negative, etc., and may become of order unity or larger. Values for real walls are given in Chapters 17 and 18. We can write

$$\begin{aligned}\sum \phi_{00,k} &= \sum \phi_{0M,k} = \sum \phi_{M0,k} = \sum \phi_{MM,k}\\ &= \frac{\sum a_k}{\sum d_k} = \frac{\sum b_k}{\sum d_k} = \frac{\sum c_k}{\sum d_k} = \text{wall transmittance } 1/\sum r_j.\end{aligned} \tag{11.36}$$

Furthermore, if coefficients up to $n = 4$ are needed (for a 4-capacity/5-resistance 'wall') then

$$4(a_0 + c_0 - 2b_0) + 3(a_1 + c_1 - 2b_1) + 2(a_2 + c_2 - 2b_2) + (a_3 + c_3 - 2b_3)\delta \Big/ \sum d_k = \text{wall capacity } \sum c_i, \tag{11.37}$$

or

$$-\sum^{N} k((a_k + c_k - 2b_k)\delta / \sum d_k = \sum c_j. \tag{11.38}$$

Further,

$$rc \equiv \sum r_i \sum c_i = \frac{-\sum^{N} k(a_k + c_k - 2b_k)\delta}{\sum^{N} a_k \left[\text{or } \sum^{N} b_k \text{ or } \sum^{N} c_k\right]}. \tag{11.39}$$

The extension of this theory from a two-capacity to a multiple-capacity circuit entails no more ideas. Further methods have to be developed however to derive transfer coefficients for a real wall consisting of several homogeneous resistive/capacitative layers. The basic solutions needed for use in such materials are given in the next chapter and their application is discussed in Chapter 17.

11.8 APPENDIX

We continue the example at the end of Chapter 10 to show how the transfer coefficients of the wall can be found. We have to find first the heat flows caused by a ramp in temperature imposed at T_0. With a value of $\theta = 1/3600\,\mathrm{K/s}$, the flux into the wall exterior is

$$q_{00,t} = \frac{\theta t - T_{2,t}}{r_1} = \frac{\theta t}{r} + \frac{\theta[c_2(r_3+r_5)^2 + c_4 r_5{}^2]}{r^2} - \frac{T_{20,1}}{r_1}\exp\left(\frac{-t}{z_1}\right) - \frac{T_{20,2}}{r_1}\exp\left(\frac{-t}{z_2}\right)$$
$$= \frac{1}{3600}\frac{t}{0.3} + 25.463 - 11.889\exp\left(\frac{-t}{13904}\right) - 13.574\exp\left(\frac{-t}{3596}\right). \quad (11.40)$$

The flux out of the wall interior is

$$q_{50,t} = \frac{T_{4,t}}{r_5} = \frac{\theta t}{r} + \frac{\theta[c_2 r_1(r_3+r_5) + c_4(r_1+r_3)r_5]}{r_2} - \frac{T_{40,1}}{r_5}\exp\left(\frac{-t}{z_1}\right) - \frac{T_{40,2}}{r_5}\exp\left(\frac{-t}{z_2}\right)$$
$$= \frac{1}{3600}\frac{t}{0.3} - 16.204 + 17.365\exp\left(\frac{-t}{13904}\right) - 1.1617\exp\left(\frac{-t}{3596}\right). \quad (11.41)$$

The flux into the wall interior due to a ramp increase at the interior, the $q_{55,t}$ flux, can be found by direct application of the boundary conditions $T_0 = 0$ and $T_5 = \theta t$, or by taking the $T_0 = \theta t$ and $T_5 = 0$ condition to apply to a construction with the elements in Figure 10.9 reversed. We find

$$q_{55,t} = \frac{1}{3600}\frac{t}{0.3} + 25.463 - 25.364\exp\left(\frac{-t}{13904}\right) - 0.0994\exp\left(\frac{-t}{3596}\right).$$

In general, the constant terms in the expressions for $q_{00,t}$ and $q_{55,t}$ are not equal. At times $t = 0, 1, 2$ and 3 h the heat fluxes (W/m^2) are as follows.

t	$q_{00,t}$	$q_{50,t}$	$q_{55,t}$
0	0.0	0.0	0.0
3600	14.63103	0.10676	9.18204
7200	23.21286	0.65243	17.00446
10 800	29.32153	1.72482	23.79351

Thus the ramp leads to a large flux at the immediate surface and a small flux at the remote surface. Most of the heat input is stored.

The response factors are the result of the combined effect of ramps of $+\theta$ at $t = -1$ h, -2θ at 0 and $+\theta$ at $+1$ h, so that

$$\phi_{00,0} = 14.63103 = 14.63103,$$

$$\phi_{00,1} = 23.21286 - 2 \times 14.63103 = -6.04919,$$

$$\phi_{00,2} = 29.32153 - 2 \times 23.21286 + 14.63103 = -2.47317,$$

$$\phi_{50,0} = 0.10676 = 0.10676,$$

$$\phi_{50,1} = 0.65243 - 2 \times 0.10676 = 0.43892,$$

$$\phi_{50,2} = 1.72482 - 2 \times 0.65243 + 0.10676 = 0.52671,$$

$$\phi_{55,0} = 9.18204 = 9.18204,$$

$$\phi_{55,1} = 17.00446 - 2 \times 9.18204 = -1.35961,$$

$$\phi_{55,2} = 23.79351 - 2 \times 17.00446 + 9.18204 = -1.03338.$$

The contribution of the z_2 component to $\phi_{50,t}$, say, decreases more rapidly with time than does the z_1 contribution and is about 1% of the z_1 contribution at $t = 6$ h. Thereafter $\phi_{50,t}$ depends on the z_1 term alone and $\phi_{50,n+1}/\phi_{50,n} = \exp(-\delta/z_1) = 0.77$. The values of ϕ themselves become negligible after about $5z_1$, 20 h in this case.

Only the first three ϕ values are needed here to transform this information into series of transfer coefficients. We first find the d coefficients:

$$\beta_1 = \exp(-\delta/z_1) = \exp(-3600/13904) = 0.77189,$$

$$\beta_2 = \exp(-\delta/z_2) = \exp(-3600/3596) = 0.36748,$$

so

$$d_1 = -(\beta_1 + \beta_2) = -1.13967,$$

and

$$d_2 = \beta_1 \beta_2 = +0.28365,$$

$$d_0 = 1 \text{ by definition.}$$

so

$$\sum d = (1 - \beta_1)(1 - \beta_2) = 0.14429.$$

The a coefficients are found from the ϕ_{00} values and the d values:

$$a_0 = 1 \times (14.63103) = 14.63103,$$

$$a_1 = 1 \times (-6.04919) + (-1.13967)(14.63103) = -22.71930,$$

$$a_2 = 1 \times (-2.47317) + (-1.13967)(-6.04919) + (+0.28365)(14.63103) = 8.56923,$$

$$\sum a = 0.48096.$$

Similarly for the b and c coefficients:

$$b_0 = 1 \times (0.10676) = 0.10676,$$

$$b_1 = 1 \times (0.43892) + (-1.13967)(0.10676) = 0.31728,$$

$$b_2 = 1 \times (0.52671) + (-1.13967)(0.43892) + (+0.28365)(0.10676) = 0.05691,$$

$$\sum b = 0.48095.$$

$$c_0 = 1 \times (9.18204) = 9.18204,$$

$$c_1 = 1 \times (-1.35961) + (-1.13967)(9.18204) = -11.82132,$$

$$c_2 = 1 \times (-1.03338) + (-1.13967)(-1.35961) + (+0.28365)(9.18204) = 3.12024,$$

$$\sum c = 0.48096.$$

The example illustrates the method presented in Chapter 17 to find the transfer coefficients for a multilayer wall. As remarked above, all b values are positive. We note too that

$$\sum a \Big/ \sum d = \sum b \Big/ \sum d = \sum c \Big/ \sum d$$
$$= 0.48096/0.14429 = 3.333 = 1/0.3 = \text{wall } U \text{ value.}$$

There is the further relationship that the heat stored S' in the steady state with $T_0 = 1$ and $T_5 = 0$ is

$$S' = c_2(r_3 + r_5)/r + c_4 r_5/r,$$

which here is

$$10^5 \times (0.15 + 0.10)/0.3 + 2 \times 10^5 \times 0.1/0.3 = 1.5 \times 10^5\ \text{J/m}^2.$$

It was shown above that S' is also

$$S' = [2(a_0 - b_0) + (a_1 - b_1)]\delta \Big/ \sum d$$
$$= [2(14.63103 - 0.10676) + (-22.71930 - 0.31728)] \times 3600/0.14429 = 1.5 \times 10^5\ \text{J/m}^2.$$

The value S'' with $T_0 = 0$ and $T_5 = 1$ is given similarly (and happens to equal S' for this example). $S' + S''$ is the total wall capacity, 3×10^5 J/m^2. This is the value of $c_2 + c_4$ in Figure 10.9.

12

The Fourier Continuity Equation in One Dimension

Real walls consist of layers of material with distributed resistance and capacity. In one-dimensional heat flow in steady-state conditions, a non-zero value of the term $\lambda\, \mathrm{d}^2T(x)/\mathrm{d}x^2$ indicates that more heat flows into some element of thickness δx than out and Chapter 3 discussed the consequences of the balance of the difference, either with a temperature-driven convective/radiative exchange with the surroundings, or with some externally imposed source of heat. Two boundary conditions are needed to determine $T(x)$. In unsteady flow, the difference balances the rate of change of internal energy, $\mathrm{d}(\rho c_p\, \delta x\, T(t))/\mathrm{d}t$. A single initial condition is needed to determine $T(t)$ in the element δx.

When the element forms part of a one-dimensional construction, flow in the element is subject to the Fourier continuity equation (Fourier 1822: Section 128)[1]:

$$\lambda \frac{\partial^2 T(x,t)}{\partial x^2} = \rho c_p \frac{\partial T(x,t)}{\partial t}. \tag{12.1}$$

The small element δx is now replaced by the two-dimensional element $\delta x\, \delta t$ and this forms part of any solution to (12.1). The solution may be visualised as a surface in three-dimensional space with independent x and t axes, the third representing temperature. A solution requires boundary and initial conditions to be specified, but since this is a partial differential equation, a number of different forms of solution will be found to satisfy it and they are presented in this chapter. The treatment is based upon Davies (1995). The useful ones among them will be used later in connection with building response. The form of the equation and its handling when the conductivity varies with temperature are given in the appendix to Brown and Stephenson (1993).

An important non-dimensional parameter, the Fourier number F_0 is implicit in the above equation. Consider a slab of thickness X; at time $t = 0$ it is subjected to an excitation such as a pulse of heat or a step or ramp in surface temperature. At a later time t',

[1] Alternatively, this equation can be regarded as a combination of the Fourier conduction equation $q(x,t) = -\lambda\, \partial T(x,t)/\partial x$ with (2.19) and $\partial q(x,t)/\partial x = -\rho c_p\, \partial T(x,t)/\partial t$. If more (positively directed) heat flows into δx than out, $\partial q(x,t)/\partial x$ is negative but is associated with an increase in $T(x,t)$, so a negative sign is needed.

Building Heat Transfer Morris G. Davies
© 2004 John Wiley & Sons, Ltd ISBN: 0-470-84731-X

equation (12.1) can be redrafted as

$$\frac{\lambda t'}{\rho c_p X^2}\frac{\partial T(x,t)}{\partial (x/X)^2} = \frac{T(x,t)}{\partial (t/t')}. \tag{12.2}$$

The grouping $\lambda t'/\rho c_p X^2$ is F_0 and can be interpreted as a non-dimensional measure of time. If the slab surfaces are taken to be adiabatic or isothermal, the solution $T(x, t)$ can be expressed in terms of F_0 alone. If instead one or both of the slab surfaces is flanked by a convective or radiative film, two further non-dimensional measures of time, F_1 and F_2 can be defined (Section 13.2.7).[2]

Equation (12.1), the diffusion equation, is superficially similar to the wave equation,

$$\frac{\partial^2 y(x,t)}{\partial x^2} = c^2\frac{\partial^2 y(x,t)}{\partial t^2}, \tag{12.3}$$

where c is the velocity of propagation of a signal, the displacement y, in a longitudinal or transverse direction. (For a string under tension T and having mass m per unit length, $c^2 = T/m$.) It describes a mechanical disturbance carried through some medium – a stretched string or an elastic solid – with constant velocity and no attenuation: the effect of some excitation is only perceived at a distant point after a precisely definable lapse of time. Solutions serve equally for negative values of time. The two equations, however, describe very different phenomena. According to (12.1), if a heat-conducting solid is subjected to a disturbance at some point, no such finite time lapse exists. If a quantity of heat w(J/m^2) is injected into the surface of a semi-infinite slab, the resulting temperature $T(x, t)$ (12.36) implies an infinite rate of propagation, although the change in temperature at distant points immediately after the disturbance is imposed is vanishingly small. (If 1 J/m^2 is injected into the exposed surface of thick concrete, the disturbance 10 cm inside and 1 min later is around 1×10^{-30} K)[3]. If two materials at different temperatures are brought into contact, then according to classical theory which assumes a continuous distribution of material, there may be an infinite heat flow, but sustained for an infinitesimal time. The infinite rate of propagation of a thermal signal is often taken to be unreasonable, but it is a necessary consequence of the Fourier law; if one end of a rod is suddenly displaced longitudinally, the other end will be calculated as displaced by the same amount instantaneously if we ignore the elasticity and density of the rod. The heat flow will depend on the molecular structure of the materials and Tzou (1997) discusses in depth the modifications that have to be made to (12.1) and their basis. The treatment extends to such diverse topics as temperature pulses in liquid helium and laser heating of metal films. These developments, however, have no bearing on the processes to be discussed here, where time delays for perceptible changes through building walls are of order hours, not milliseconds.

[2]In common with other non-dimensional groupings such as the Reynolds number Re and Prandtl number Pr, the Fourier number is usually written as a pair of letters, Fo. The notation is modified here to indicate the power of h, the film coefficient [W/m^{2}K], in the definitions of F_0, F_1 and F_2; F_0 is independent of h.

[3]The analysis in Section 13.1.2 does indicate an approximate time of arrival of a perceptible signal, a kind of surge, following some step change.

In expressions for temperature, λ and ρc_p appear as the diffusivity, $\kappa = \lambda/\rho c_p$. The solutions can be classified as follows:

(a) Those where x and t may appear as a product, e.g. $\kappa x^3 t$. A solution of this kind where t appears as t^1 will be described as a *slope* solution.

(b) Those where x and t appear in independent exponential functions. This group divides into b(i) and b(ii): b(i) where the response consists of *standing waves* in space that *decay exponentially* in time, and b(ii) where the response has the form of *progressive waves decaying exponentially* in space and are *periodic* in time.

(c) Those where x and t appear as a quotient, $x^2/\kappa t$.

The solutions most used in a building context are the first two terms in type (a), to give wall U values, and b(ii) to find the parameters for excitation of daily frequency. Type b(ii) can be extended through use of transformations to find wall parameters for excitation at hourly intervals; this is the usual method. The parameters, however, can be found instead through use of the slope solution, type (a), together with the time domain solution, type b(i): this approach is physically simpler and will be presented later. Some of the solutions of type (c) appear in heat transfer texts generally; occasionally they may serve as benchmark solutions for building heat flow, but are not of use for design purposes.

In discussing the solutions, three factors will be considered: whether x may be infinite, whether negative values of t are possible, and whether a transmission matrix formulation is possible.

12.1 PROGRESSIVE SOLUTIONS

The following forms satisfy the Fourier equation (Carslaw and Jaeger 1959: 52):

$$T(x,t) = A, \tag{12.4a}$$

$$T(x,t) = Bx, \tag{12.4b}$$

$$T(x,t) = C(x^2 + 2\kappa t), \tag{12.4c}$$

$$T(x,t) = D(x^3 + 6\kappa x t), \tag{12.4d}$$

$$T(x,t) = E(x^4 + 12\kappa x^2 t + 12\kappa^2 t^2), \tag{12.4e}$$

$$T(x,t) = F(x^5 + 20\kappa x^3 t + 60\kappa^2 x t^2). \tag{12.4f}$$

Further solutions, involving higher powers of t, can be written down. The first two solutions are trivial, representing constant temperature and constant gradient. The wall U value derives from repeated application of (12.4b). The third describes the case of a slab, adiabatic at $x = 0$, and driven at $x = X$, either by a temperature which increases at a uniform rate or by injection of a constant heat flux. The fourth relates to the situation where the slab is at zero at $x = 0$; it is excited at $x = X$, either by a linearly increasing temperature source or by a linearly increasing heat source. In either case the heat flow from $x = 0$ increases linearly with time.

The further solutions apply when excitation includes a component proportional to t^2 and they are not of interest in a building context. The others however can be combined:

$$T(x,t) = A + Bx + C(x^2 + 2\kappa t) + D(x^3 + 6\kappa xt). \tag{12.5}$$

The constants can be found from initial and boundary conditions. Suppose that a slab of thickness X is being driven at $x = 0$ by a temperature source increasing at the rate θ_0(K/s), with θ_1 similarly acting at $x = X$, and further that at $t = 0$, both $T(0,0)$ and $T(X,0)$ are zero. Then

$$\begin{aligned} T(x,t) = \; & \theta_0\{(\rho c_p X^2/6\lambda)(-(1 - x/X) + (1 - (x/X)^3) + t(1 - x/X)\} \\ & + \theta_1\{(\rho c_p X^2/6\lambda)(-x/X \quad + (x/X)^3) \quad + \quad t(x/X)\}\,. \end{aligned} \tag{12.6}$$

The temperature profile at $t = 0$ is wholly negative. The minimum value at $t = 0$ for excitation at the left surface only (so that $\theta_1 = 0$) is at $x/X = 0.423$, which is a little left of centre, as might be expected. The solution is not useful as it stands, since the initial condition cannot be realised; a transient component too is needed and will be included shortly.

Solutions of this kind are not suitable for semi-infinite solids since two boundary conditions are needed. There is no restriction on t. As will be shown in Section 17.2, a transmission matrix can be set up to handle excitation of this kind in a multilayer wall.

12.2 SPACE/TIME-INDEPENDENT SOLUTIONS

Families of solutions are possible in which x and t appear in separate functions and are of form

$$T(x,t) = A\exp(x/\xi)\exp(t/\zeta), \tag{12.7}$$

where ξ and ζ have units of distance and time respectively. This satisfies the Fourier equation if

$$\xi^2 = \kappa\zeta. \tag{12.8}$$

Either of two assumptions can be made for ζ and they lead to quite different solutions. If ζ is taken to equal $-z$ where z is real and positive, then with imposition of suitable boundary conditions on a finite-thickness slab, we obtain a series of z values which correspond to decay times and have a structure similar to the series of periodic times associated with the vibrations of a stretched string or an organ pipe. They are properties of the system and it does not matter how the system is excited. The modes of decay or vibration are described as eigenfunctions.[4] If instead ζ is taken to be the imaginary number $\zeta = P/(\mathrm{j}2\pi)$ where $\mathrm{j} = \sqrt{-1}$, solutions can be found which describe the response when the slab is excited sinusoidally with period P. Thus P is an independent variable while the series of z values are characteristics of the system.

[4]From the German *eigen* 'own', indicating that the modes are characteristics of the system. The French equivalent is *propre*, but the English look-alike *proper*, giving the term 'properfunction', is a mistranslation.

12.2.1 The Transient Solution

We assume that $\zeta = -z$, where z is a positive value of time. Then

$$\frac{x}{\xi} = \frac{x}{\pm\sqrt{-\kappa z}} = \pm \mathrm{j}\left(\frac{\rho c_p x^2}{\lambda z}\right)^{1/2} \tag{12.9}$$

so, noting that $\exp(\mathrm{j}u) = \cos u + \mathrm{j}\sin u$, we obtain

$$T(x,t) = A\left(\cos\left(\frac{\rho c_p x^2}{\lambda z}\right)^{1/2} \pm \mathrm{j}\sin\left(\frac{\rho c_p x^2}{\lambda z}\right)^{1/2}\right)\exp\left(-\frac{t}{z}\right). \tag{12.10}$$

This represents a wave train standing in space and decaying exponentially in time. From $t = 0$ to $t = z$ its amplitude falls from A to $A/\mathrm{e} = 0.368A$.

Writing $\alpha = \sqrt{\rho c_p/\lambda z}$, it is easily checked that the forms

$$T(x,t) = [A' \exp(+\mathrm{j}\alpha x) + B' \exp(-\mathrm{j}\alpha x)]\exp(-t/z) \tag{12.11a}$$

$$= [A\sin(\alpha x) + B\cos(\alpha x)]\exp(-t/z). \tag{12.11b}$$

provide a solution. They are valid for an infinite medium and for $t < 0$ but we will apply them to a slab $(0 < x < X)$ and the sum of the resulting eigenfunction temperature distributions is valid only for $t > 0$.

In handling wall conduction, we are normally concerned only with the temperatures and heat flows at the bounding surfaces and not with internal values. The values of the temperatures T_0 and T_1 and the heat flows q_0 and q_1 at $x = 0$ and $x = X$, respectively, follow from (12.11b). Noting that $q(x,t) = -\lambda T(x,t)$, we find that

$$T_0 = T_1 \cos u + q_1(\sin u)/v, \tag{12.12a}$$

$$q_0 = -T_1(\sin u) \times v + q_1 \cos u, \tag{12.12b}$$

or

$$\begin{bmatrix} T_0 \\ q_0 \end{bmatrix} = \begin{bmatrix} \cos u & (\sin u)/v \\ -(\sin u) \times v & \cos u \end{bmatrix} \begin{bmatrix} T_1 \\ q_1 \end{bmatrix}, \tag{12.12c}$$

where

$$u^2 = \rho c_p X^2/\lambda z = cr/z \quad \text{and} \quad v^2 = \lambda\rho c_p/z = c/rz. \tag{12.13}$$

Both the $\lambda \times \rho c_p$ and the $\lambda/\rho c_p$ groupings appear here.

To obtain useful results for an isolated slab, conditions at $x = 0$ and $x = X$ must be assumed. If each node is taken always to be at zero, the most useful assumption, the coefficient B in (12.11b) must be zero and $\alpha X = \pi, 2\pi, \ldots, n\pi$, so

$$z_n = \frac{\rho c_p X^2}{\pi^2 \lambda}\frac{1}{n^2} \text{ or as a Fourier number, } (F_0)_n = \frac{\lambda z_n}{\rho c_p X^2} = \frac{1}{n^2\pi^2}, \tag{12.14}$$

and equation (12.11b) becomes

$$T(x,t,n) = A_n \sin(\alpha_n x)\exp(-t/z_n) = A_n \sin(\pi n x/X)\exp(-t/z_n). \qquad (12.15)$$

The form of $T(x)$ when $n = 1$ is a half-sine wave. If $n = 3$ then $T(x)$ consists of three half-waves, etc., and the sum of a series of odd-n eigenfunctions is symmetrical about the centreline of the slab. If $n = 2$ then $T(x)$ consists of two half-waves, i.e. a full sine wave, etc., and the sum of even-n eigenfunctions is skew-symmetric.

The temperature distribution has an important property. Consider the product

$$P_{nm}(x) = \sin(\alpha_n x)\sin(\alpha_m x). \qquad (12.16)$$

Its integral,

$$\int_0^X P_{nm}(x)\,\mathrm{d}x = \begin{cases} X/2 & \text{if } m = n, \\ 0 & \text{if } m \neq n. \end{cases} \qquad (12.17)$$

This is easily checked. Since $0 < x < X$, $P_{nm}(x)$ has a range of values, positive and negative, and in simple statistical terms, the size of the sum of such products, $\sum P_{nm}(x)$ at regular values of x measures the extent of the correlation between $\sin(\alpha_n x)$ and $\sin(\alpha_m x)$. Equation (12.17) shows that the correlation is zero. That is, temperature distributions based on n and m are independent of each other. A similar result was seen in Chapter 10 and it occurs again in Chapter 17.

The result permits us to calculate the response of a slab when it is subjected to a step or other change at $t = 0$.

Step excitation

Suppose that the slab is initially at a uniform temperature T_I. From time $t = 0$ the temperatures at the two surfaces $x = 0$ and $x = X$ are held at zero. We wish to know the subsequent temperature distribution, $T(x,t)$. Following standard Fourier analysis, it is assumed that $T(x,t)$ can be found from an infinite series:

$$T(x,t) = \sum A_n \sin(\alpha_n x)\exp(-t/z_n), \qquad (12.18a)$$

so at $t = 0$ we have

$$\sum A_n \sin(\alpha_n X) = T_I. \qquad (12.18b)$$

To evaluate the series of values of A_n, both sides of (12.18b) are to be multiplied by $\sin(\alpha_n x)$ and integrated from 0 to X. It is found that $A_n = 4T_I n\pi$, where $n = 1, 3, 5, \ldots$. The initial condition and the excitation are symmetrical, so only odd values of n are needed. Then

$$\frac{T(x,t)}{T_I} = \frac{4}{\pi}\sum \frac{1}{n}\sin\left(\frac{n\pi x}{X}\right)\exp\left(-\frac{t}{z_n}\right) \qquad (n = 1, 3, 5, \ldots); \qquad (12.19)$$

See Carslaw and Jaeger (1959: 96). A similar result is given in the footnote.[5] The heat flow at the surface $x = 0$ is

$$\frac{q(0,t)}{T_I} = -\frac{4\lambda}{X}\sum \exp\left(-\frac{t}{z_n}\right). \tag{12.20}$$

In the early stages of cooling, the cooling imposed at $x = X$ has no effect on the heat flow at $x = 0$, and during this time, when the heat flow depends on the step at $x = 0$ alone, we have

$$q(0,t) = -T_I(\lambda\rho c_p/\pi t)^{1/2}; \tag{12.21}$$

see Carslaw and Jaeger (1959: 61, equation 11).

Thus a heat flow generated by a temperature in unsteady conditions is proportional to $(\lambda\rho c_p/t)^{1/2}$. This form appears repeatedly. Although q is proportional to λ alone in steady conditions, on physical grounds, it must also depend on ρc_p in time-varying conditions, and on dimensional grounds, we should expect it to vary as $\sqrt{\lambda}$ and as $\sqrt{\rho c_p}$. But since capacity is only effective in time-varying conditions, ρc_p is associated with time as $\rho c_p/t$.

Ramp excitation

We can now use the steady-slope solution (12.6) and a suitable transient solution to find the response of a slab to ramp excitation. If the initial distribution for the transient solution is the negative of the slope distribution at $t = 0$ and with θ_1 equal to zero, the subsequent transient distribution is given by

$$\frac{T(x,t)}{\theta_0} = \frac{2\rho c_p X^2}{\lambda}\sum \frac{1}{n^3\pi^3}\sin\left(\frac{n\pi x}{X}\right)\exp\left(-\frac{t}{z_n}\right); \tag{12.22}$$

here n takes values 1, 2, 3, ..., since the initial distribution is neither symmetric nor skew-symmetric and at $t = 0$ has its maximum at $x/X = 0.423$, the value of the minimum noted above.[6]

If this is added to the solution of (12.6) with $\theta_1 = 0$, we have a solution to the problem of a slab of thickness X at zero temperature up to $t = 0$. After $t = 0$, $T_0 = \theta_0 t$, while T_1 continues to be zero (Churchill 1958: 214, equation 2. Carslaw and Jaeger do not appear

[5]Consider a slab initially at zero. At $t = 0$ the temperature at $x = 0$ is raised to T_I while the temperature at $x = X$ remains at zero. Then

$$\frac{T(x,t)}{T_I} = \frac{X-x}{X} - \frac{2}{\pi}\sum \frac{1}{n}\sin\left(\frac{n\pi x}{X}\right)\exp\left(-\frac{t}{z_n}\right) \qquad (n = 1, 2, 3, \ldots).$$

Carslaw and Jaeger's expression has as its initial condition $T(x) = T_I(x/X)$ with $T(0) = T(X) = 0$ for $t > 0$. To transform their expression to meet the present conditions, add to it $-T_I(x/X)$, multiply by -1 and replace x by $X - x$.

[6]Evaluation involves the integrals $\int_0^{n\pi} u^r \sin u \, du$ where $r = 1, 2, 3$. Most terms are either zero or cancel.

to have addressed this problem.) Later we shall be concerned only with the heat flows at the bounding surfaces $x = 0$ and $x = X$ that this excitation generates. At the surface subjected to ramp excitation,

$$\frac{q(0,t)}{\theta_0} = \frac{\lambda t}{X} + \frac{1}{3}\rho c_p X - 2\rho c_p X \sum \frac{1}{n^2\pi^2} \exp\left(-\frac{t}{z_n}\right)$$

$$= \frac{t}{r} + \frac{1}{3}c - 2c \sum \frac{1}{n^2\pi^2} \exp\left(-\frac{t}{z_n}\right). \qquad (12.23a)$$

At the far, isothermal surface,

$$\frac{q(X,t)}{\theta_0} = \frac{\lambda t}{X} - \frac{1}{6}\rho c_p X - 2\rho c_p X \sum \frac{(-1)^n}{n^2\pi^2} \exp\left(-\frac{t}{z_n}\right)$$

$$= \frac{t}{r} - \frac{1}{6}c - 2c \sum \frac{(-1)^n}{n^2\pi^2} \exp\left(-\frac{t}{z_n}\right), \qquad (12.23b)$$

where $n = 1, 2, 3, \ldots$. These equations have the form of (10.14), where n was 1 only.[7]

The higher decay modes fade rapidly with time and the $n = 1$ mode is the longest-lived. Since $\exp(-4.5) = 0.01$, or 1%, the transient components of both flows die out after a time of about $4.5z_1$. Then the difference $q(0,t) - q(X,t) = \theta_0[\frac{1}{3} - (-\frac{1}{6})]\rho c_p X$ or $\frac{1}{2}\theta_0 \rho c_p X$; more heat flows in than out and the mean temperature increases. Since $\sum n^{-2} = \pi^2/6$ and $\sum(-1)^{n-1}n^{-2} = \pi^2/12$, it can be checked that both flows are zero at time $t = 0$, as must be. When t is small (Carslaw and Jaeger 1959: 63, equation 4) then

$$q(0,t) = \theta_0 t(\lambda\rho c_p/\pi t)^{1/2} \qquad (12.24a)$$

and

$$q(X,t) = 0. \qquad (12.24b)$$

These limits can be conveniently checked by writing (12.23) in dimensionless form. For example, equation (12.23a) becomes

$$\frac{q(0,t)}{\theta_0 \rho c_p X} = F_0 + \frac{1}{3} - 2\sum \frac{1}{n^2\pi^2} \exp(-n^2\pi^2 F_0), \qquad (12.25)$$

where F_0 is the Fourier number $\lambda t/\rho c_p X^2$.[8] The right-hand side of the equation is a function of the single variable F_0.

[7] A similar expression holds for the problem of a slab initially at zero, insulated at its rear surface, which is subjected to a constant heat flux q at its front face from $t = 0$ onward. The rear surface temperature is then given (Carslaw and Jaeger 1959: 112) as

$$\frac{T(X,t)}{q} = \frac{t}{c} - \frac{1}{6}r - 2r \sum \frac{(-1)^n}{n^2\pi^2} \exp\left(-\frac{t}{z_n}\right) \quad \text{where} \quad z_n = cr/(n^2\pi^2) \text{ with } n = 1, 2, 3, \ldots.$$

Vozar and Sramkova (1997) have used this expression as a starting point for an experimental determination of thermal diffusivity.

[8] When $t = 0$, $q(0,0)$ must be zero. It can be readily checked that the right-hand side of (12.25) is zero. $F_0 = 0$ and $\exp(-n^2\pi^2 F_0) = 1$. Thus the right-hand side $= \frac{1}{3} - (2/\pi^2)\sum n^{-2}$, which is zero since $\sum n^{-2} = \pi^2/6$.

The right-hand sides of (12.23) consist of three terms: one proportional to t, one constant and one consisting of decaying components, all of them having simple analytical expression. Equation (10.14) for the r-c-r circuit has this form. It will be shown later that the form holds for multilayer walls. The t-proportional term continues to be simple, but the others result from complicated functions of the resistances and capacities of the layers and have to be expressed numerically. An early example of using this kind of solution is reported by Dufton (1934), who considered the case of a slab, initially at zero, one surface being held at zero and the other having imposed on it a constant flux. He found an expression for the exposed surface temperature and used the solution to examine the merits of intermittent room heating.

This form of solution is used in Chapter 17.

12.2.2 The Periodic Solution

The other basic assumption that can be made for ζ in (12.8) is that $\zeta = P/(\mathrm{j}2\pi)$ where P is a positive value of time. Then

$$\frac{x}{\xi} = \frac{x}{\pm(\kappa P/\mathrm{j}2\pi)^{1/2}} = \pm(1+\mathrm{j})\left(\frac{\pi\rho c_p x^2}{\lambda P}\right)^{1/2} = \pm(1+\mathrm{j})\alpha x, \tag{12.26}$$

where

$$\alpha = \sqrt{\pi\rho c_p/\lambda P}. \tag{12.27}$$

The solution to the Fourier equation is then

$$\begin{aligned} \mathbf{T}(x,t) &= [A'\exp(\alpha x + \mathrm{j}\alpha x) + B'\exp(-\alpha x - \mathrm{j}\alpha x)]\exp(\mathrm{j}2\pi t/P) \qquad (12.28\mathrm{a}) \\ &= [A\sinh(\alpha x + \mathrm{j}\alpha x) + B\cosh(\alpha x + \mathrm{j}\alpha x)]\exp(\mathrm{j}2\pi t/P). \qquad (12.28\mathrm{b}) \end{aligned}$$

Vector quantities are in bold. $\mathbf{T}(x,t)$ is periodic in time with period P; t may be negative. $\mathbf{T}(x,t)$ is also periodic in x but is strongly attenuated. Consider the semi-infinite solid with an exposed surface at $x = 0$ where a temperature variation of

$$\mathbf{T}(0,t) = T_0\exp(\mathrm{j}2\pi t/P) \tag{12.29}$$

is imposed. Then

$$\mathbf{T}(x,t) = T_0\exp(-\alpha x - \mathrm{j}\alpha x)\exp(\mathrm{j}2\pi t/P). \tag{12.30}$$

(A' must be zero to prevent T tending to infinity as x becomes large.) Then

$$\mathbf{T}(x,t) = \underbrace{T_0}_{\substack{\text{surface}\\ \text{value}}}\ \underbrace{\exp(-\alpha x)}_{\substack{\text{space}\\ \text{attenuation}}}\ \underbrace{\exp\left(\mathrm{j}2\pi\left(\frac{t}{P} - \frac{x}{\Lambda}\right)\right)}_{\substack{\text{sinusoidal wave}\\ \text{moving to the right}}}. \tag{12.31}$$

This has a space periodicity of wavelength $\Lambda = 2\pi/\alpha = (4\pi\lambda P/\rho c_p)^{1/2}$. The amplitude of the wave however is strongly attenuated as it progresses. When $x = \Lambda$, $T(\Lambda, t)$ has amplitude $T_0 \exp(-2\pi) = 0.00019T_0$ (attenuation to 0.19%), so it has become virtually extinct. For brickwork excited at the daily period,

$$\Lambda \approx (4\pi \times 0.8 \times (24 \times 3600)/(1700 \times 800))^{1/2} = 0.8\,\text{m}.$$

Thus with typical wall thicknesses of 20–30 cm, the effect of sunshine absorbed on the outer surface is quite perceptible, although much reduced, on the inner surface some hours later. The variation below the surface of the earth over a yearly period is apparent at greater depths. With the values for common earth given in Table 2.1,

$$\Lambda \approx (4\pi \times 1.28 \times (365 \times 24 \times 3600)/(1460 \times 880))^{1/2} = 20\,\text{m}.$$

It is typically reported that virtually no variation is perceptible below a depth of about 10 m; the temperature there is the yearly average surface temperature.

In a semi-infinite medium, the depth at which the amplitude of temperature variation has fallen to 1/e of its value at the surface is the *periodic penetration depth*, p_p, found when $\alpha p_p = 1$ (12.31) so $p_p = \sqrt{\lambda P/\pi \rho c_p}$. For common earth, values of (P, p_p) are (1 min, 4.4 mm), (1 hr, 34 mm), (1 day, 166 mm), (1 month, 0.91 m), (1 year, 3.2 m).

The amplitude of variation of the heat flow in relation to the temperature variation at the exposed surface of a thick slab defines the characteristic admittance **a** of the material:

$$\mathbf{a} = \frac{\mathbf{q}(0, t)}{\mathbf{T}(0, t)} = \left(\frac{2\pi \lambda \rho c_p}{P}\right)^{1/2} \exp(\mathrm{j}\pi/4), \tag{12.32}$$

where **a** has units W/m^{2}K (those of a wall U value). Its magnitude is proportional to $(\lambda \rho c_p/P)^{1/2}$, a form noted earlier. The phase of the flow leads the phase of the temperature by $\pi/4$, or 45°; this is 1/8 of a cycle, or 3 h in a 24 h period. On physical grounds, the lead is expected since heat must have been accumulated in storage before it can exhibit a temperature, and it is a familiar experience that although solar gain into a south-facing room may peak at solar noon, the maximum temperature it causes comes some hours later. Heat flow through a resistive element is in phase with the temperature difference causing it, and the flow into a capacity element is a quarter of a cycle ahead of temperature; the flow into a semi-infinite solid is midway between these limits.

If the conducting medium is a slab of finite thickness X, temperatures and flows at the two surfaces follow from (12.28b) and are related as follows:

$$\mathbf{T}_0 = \mathbf{T}_1 \cosh(\tau + \mathrm{j}\tau) \qquad + \mathbf{q}_1(\sinh(\tau + \mathrm{j}\tau))/\mathbf{a}, \tag{12.33a}$$

$$\mathbf{q}_0 = \mathbf{T}_1(\sinh(\tau + \mathrm{j}\tau)) \times \mathbf{a} + \mathbf{q}_1 \cosh(\tau + \mathrm{j}\tau), \tag{12.33b}$$

or

$$\begin{bmatrix} \mathbf{T}_0 \\ \mathbf{q}_0 \end{bmatrix} = \begin{bmatrix} \cosh(\tau + \mathrm{j}\tau) & (\sinh(\tau + \mathrm{j}\tau))/\mathbf{a} \\ (\sinh(\tau + \mathrm{j}\tau)) \times \mathbf{a} & \cosh(\tau + \mathrm{j}\tau) \end{bmatrix} \begin{bmatrix} \mathbf{T}_1 \\ \mathbf{q}_1 \end{bmatrix}, \tag{12.33c}$$

where

$$\tau^2 = \pi \rho c_p X^2/\lambda P = \pi c r/P \quad \text{and} \quad \mathbf{a}^2 = \mathrm{j}2\pi \lambda \rho c_p/P = \mathrm{j}2\pi c/rP, \tag{12.34}$$

drawing a parallel with the definitions of u and v in (12.13).

These equations are similar to those for transient change (12.12). Here a, the magnitude of $\mathbf{a}$, is equivalent to v and τ is equivalent to u. Parameter τ is the cyclic thickness of the slab, dimensionless but effectively in radians. If τ is small, less than 0.05 say, the wall is thermally thin in the sense that there will only be a small temperature difference between the surfaces. If $\tau > 3$, it is thick in that any excitation experienced at one surface is practically extinct at the other. (If X were to equal Λ, then $\tau = 2\pi$.) For a single thickness of brick ($X = 110\,\text{m}$) $\tau = 0.86$. Note that λ and ρc_p are combined in quotient form $\lambda/\rho c_p$ in τ; diffusivities ($\kappa = \lambda/\rho c_p$) of materials of very different densities may not differ much: 0.6×10^{-6} and $1.4 \times 10^{-6}\,\text{m}^2/\text{s}$ for brickwork and glass-fibre slab, respectively. By contrast λ and ρc_p are in product form $(\lambda \rho c_p)^{1/2}$ in $\mathbf{a}$ and the characteristic admittance of brickwork is 30 times that of glass fibre.

This form of solution is used in Chapter 15.

12.3 THE SOURCE SOLUTION AND ITS FAMILY

There is a further family of solutions to the Fourier continuity equation in which x and t appear as the quotient x^2/t. It is easy to check that the function

$$T(x, t) = At^{-1/2} \exp(-\rho c_p x^2/4\lambda t) \tag{12.35}$$

satisfies the equation. The heat contained in the element δx is $\rho c_p \delta x\, T(x, t)$. Noting that the definite integral $\int_0^\infty \exp(-a^2u^2)\,\text{d}u = \sqrt{\pi}/2a$, the total heat content between $x = 0$ and $x = \infty$ is found to be $\sqrt{\pi \lambda \rho c_p} A$, which is independent of time. Thus it must correspond to the pulse of heat $w(\text{J/m}^2)$ which was input at $x = 0$ and $t = 0$ and flowed in the positive x direction. The temperature following such a pulse of heat is given as

$$\frac{T(x, t)}{w} = \frac{1}{\sqrt{\pi \lambda \rho c_p}} \frac{1}{t^{1/2}} \exp\left(\frac{\rho c_p x^2}{4\lambda t}\right); \tag{12.36}$$

see Carslaw and Jaeger (1959: 51). The solution has the form

$$T(x, t) = At^m f(u), \tag{12.37a}$$

where

$$u = \sqrt{\rho c_p x^2/4\lambda t}, \tag{12.37b}$$

and substituting this into the Fourier equation leads to the differential equation

$$\frac{\text{d}^2 f(u)}{\text{d}u^2} + 2u\frac{\text{d}f(u)}{\text{d}u} - 4mf(u) = 0. \tag{12.38}$$

This suggests there may be solutions to the Fourier equation involving x and t as x^2/t in addition to (12.36), but it does not clearly indicate what they might be. They are however simple to construct.

According to the Fourier conduction equation,

$$q(x, t) = -\lambda \partial T(x, t)/\partial x, \tag{12.39a}$$

so

$$T(x,t) = (-1/\lambda)\int q(x,t)\,\mathrm{d}x. \tag{12.39b}$$

The Fourier continuity equation can be written in terms of the heat flux:

$$\lambda\frac{\partial^2 q(x,t)}{\partial x^2} = \rho c_p\frac{\partial q(x,t)}{\partial t}. \tag{12.40}$$

It follows that if we have a solution for $T(x,t)$ corresponding to some form of physical excitation, the same solution will serve for the heat flow $q(x,t)$ and the corresponding temperature solution can be found from it using (12.39b). It is then a matter for further enquiry to find what physical excitation it is that generates this form of $T(x,t)$. Further solutions can be generated in this way and are summarised in Table 12.1.

The function $\exp(-u^2)$, where $u^2 = \rho c_p x^2/4\lambda t$, in equation (12.35) has a bell-shaped form, value 1 and zero gradient at $x = 0$ and it tends to zero at large negative and positive

Table 12.1 Sequence of source-derivative solutions for a semi-infinite solid. The solid is initially at zero temperature. Excitation is applied at $t = 0$; $u^2 = \rho c_p x^2/4\lambda t = x^2/4\kappa t$. (Davies, 1995, with permission from Elsevier Science)

Form of excitation applied at $x = 0$	Value at $x = 0$	Equation in Carslaw and Jaeger (1959)
Heat pulse w *(J/m²)*		
$\dfrac{q(x,t)}{w} = \dfrac{1}{\sqrt{\pi}t}u\exp(-u^2)$	$q = 0$	
$\dfrac{T(x,t)}{w} = \dfrac{1}{\sqrt{\pi\lambda\rho c_p}}\dfrac{1}{\sqrt{t}}\exp(-u^2)$	$T \propto \dfrac{1}{\sqrt{t}}$	Eqn (2) on p. 50
Step in temperature T_s *(K)*		
$\dfrac{q(x,t)}{T_s} = \dfrac{\sqrt{\lambda\rho c_p}}{\sqrt{\pi}}\dfrac{1}{\sqrt{t}}\exp(-u^2)$	$q \propto \dfrac{1}{\sqrt{t}}$	
$\dfrac{T(x,t)}{T_s} = 1 - \dfrac{2}{\sqrt{\pi}}\displaystyle\int_0^u \exp(-u^2)\,\mathrm{d}u = \mathrm{erfc}(u)$	$T = T_s$	Eqn (10) on p. 60
Step in heat input q_s *(W/m²)*		
$\dfrac{q(x,t)}{q_s} = \mathrm{erfc}(u)$	$q = q_s$	
$\dfrac{T(x,t)}{q_s} = \dfrac{2\sqrt{t}}{\sqrt{\lambda\rho c_p}}\left(\dfrac{1}{\sqrt{\pi}}\exp(-u^2) - u\,\mathrm{erfc}(u)\right)$	$T \propto \sqrt{t}$	Eqn (7) on p. 75
Ramp in temperature θ *(K/s)*		
$\dfrac{q(x,t)}{\theta t} = \dfrac{2\sqrt{\lambda\rho c_p}}{\sqrt{t}}\left(\dfrac{1}{\sqrt{\pi}}\exp(-u^2) - u\,\mathrm{erfc}(u)\right)$	$q \propto \sqrt{t}$	
$\dfrac{T(x,t)}{\theta t} = (1 + 2u^2)\mathrm{erfc}(u) - \dfrac{2}{\sqrt{\pi}}u\exp(-u^2)$	$T = \theta t$	Eqn (4) on p. 63

values of u or x. Its integral $\int \exp(-u^2)\,du$ between $u = 0$ and $u = \infty$ is $\sqrt{\pi}/2$. The normalised integral between 0 and u' is defined as the error function $\mathrm{erf}(u')$:

$$\mathrm{erf}(u') = (2/\sqrt{\pi}) \int_0^{u'} \exp(-u^2)\,du. \tag{12.41a}$$

Computationally,

$$\mathrm{erf}(u) = (2/\sqrt{\pi}) \sum_{n=0}^{\infty} (-1)^n u^{2n+1}/((2n+1)n!); \tag{12.41b}$$

see Carslaw and Jaeger (1959: 482). The function $\mathrm{erf}(u)$ has the value 0 at $u = 0$ and tends to unity as u tends to infinity. It is rather more convenient to use the complementary error function

$$\mathrm{erfc}(u) \equiv 1 - \mathrm{erf}(u). \tag{12.42}$$

A number of points may be noted regarding these solutions.

(i) They do not apply to a finite-thickness slab. Further, they are valid for $t > 0$ only; the $t^{1/2}$ term in (12.36) cannot be evaluated for $t < 0$ and the exponential term tends to infinity for $t < 0$.

(ii) The instantaneous input of a pulse of heat (J/m^2 in one-dimensional flow) constitutes the most violent form of thermal excitation possible. The sequence of solutions based upon it – the step in temperature T_s, the step in flux q_s and the ramp in temperature θ – represent successively gentler forms of excitation. The next members in the sequence would be a ramp in heat flux and an imposed temperature change proportional to t^2.

(iii) In principle, solutions can be found too by differentiating (12.35). The resulting expression will be found to satisfy the continuity equation but the solution corresponds to imposition of a positive step in temperature at $x = +0$ and a negative step at $x = -0$ (a temperature dipole source). Clearly this cannot be realised; there is no point in constructing solutions by further differentiation.

(iv) The case of the step in temperature is presented in many heat transfer texts. $T(x, t)$ is a function of $\rho c_p x^2/4\lambda t$ alone (there is no term in t) and the temperature profiles through the slab at different times are geometrically similar. It is not of direct importance in building heat transfer since building temperatures do not undergo step changes. However, it forms the basis for one of the many methods to find thermal diffusivity: two samples of the material at uniform and different temperatures are brought into contact. Greenwood (1991) discusses the short-term and long-term solutions for the heat flow on contact.

(v) The relationships between T_0, q_0 at $x = 0$ and T_1, q_1 at $x = X$ in a solid can be expressed by a transmission matrix for the slope solution and for the time-transient and time-periodic solutions. Then by multiplication of the layer matrices we can easily relate the T_0, q_0 values with T_n, q_n values through a wall of n layers. However, it does not appear possible to express these quantities by a transmission matrix for any of the family of source solutions, and since the solutions only apply to semi-infinite bodies, matrix multiplication would not be appropriate.

(vi) The last of these relations, the value of $T(x, t)/\theta t$, can be found using Duhamel's theorem, dating from 1833. Consider (in this application) a semi-infinite solid, initially at zero temperature. From time $t = 0$ onward, the temperature of the surface at $x = 0$ is held at T_s and the temperature can written as $T_s F(x, t)$ (Table 12.1). Suppose instead that the surface is subjected to a succession of n steps of size T_s/n applied at $t = 0, t = \lambda, t = 2\lambda$, etc. The response at x can be expressed as $(T_s/n)(F(x,t) + F(x,t-\lambda) + F(x,t-2\lambda) + \ldots)$. As n increases, this staircase excitation tends to the form of a ramp. According to Duhamel's theorem, the response is then given as

$$T(x,t) = \int_0^t \phi(\lambda)\partial F(x,t-\lambda)/\partial t \, \mathrm{d}\lambda \tag{12.43a}$$

where $\phi(\lambda)$ is the form of variation in surface temperature imposed at $x = 0$ and here is θt. The theorem is stated generally on page 31 of Carslaw and Jaeger (1959) and used in the current context on page 63. Blackwell (1983) cites the expression in one dimension as

$$T(x,t) = T_0 + \int_0^t \phi(x,t-\lambda)(\mathrm{d}q/\mathrm{d}\lambda)\,\mathrm{d}\lambda + \sum_{i=0}^{n} \phi(x,t-\lambda_i)\Delta q_i, \tag{12.43b}$$

where $\Delta q_i = q_i - q_{i-1}, q_0 \equiv 0$ and $\phi(x,t)$ is the temperature response of a body initially at zero temperature and subjected to a unit step in heat flux. The integral term allows for continuous variation of heat flux with time; the summation term accounts for discrete steps in heat flux.

(vii) It will be noticed that in the expression for $T(x,t)/w$ (response to a heat pulse), the thermal conductivity λ and the volumetric specific heat ρc_p appear in product form as a constant and quotient form together with x and t in the exponential term. Thus in principle it should be possible to determine both quantities by observing variation of temperature at two locations in a semi-infinite medium due to the source w. Heating curves resulting from assumed values of λ and ρc_p can readily be found. The values can then be varied so as to yield curves acceptably close to those observed.

Alternatively, if the values of λ and ρc_p are known reliably, the heat flux at the surface can be inferred from internal temperature measurement. This is described as an inverse problem and following Beck (1970) and Beck and Arnold (1977), several studies along these lines have been reported. See Weber (1981), an interesting approach; see also Blackwell (1983), Beck *et al.* (1985), Raynaud (1986), Marquardt and Auracher (1990), Alifanov (1994), Beck *et al.* (1996), Chen and Lin (1998) and Martin and Dulikravich (2000). Woodbury (1990) has shown that the effect of the time constant of a thermocouple is to diminish the magnitude of the predicted heat flux and to displace its distribution in time, and as Marquardt and Auracher remarked, 'In contrast to direct problems, where the solution depends continuously on the initial and boundary conditions, inverse problems are unstable in the sense, that small changes in the data (for example, the measured interior temperatures) can produce large or even unbounded deviations in the solution. Therefore, these so-called ill-posed problems are difficult to solve, especially if measurement noise is present.' Despite this warning, the extensive experience in the use of inverse methods might provide a useful basis to determine conduction losses from buildings. In 2000 Howell as editor of the *Journal of Heat Transfer* invited a number of heat transfer experts, mostly in the US, to predict future lines of enquiry and the use of inverse methods was identified by several contributors.

Some further solutions are listed in Table 14.1.

12.3.1 *Further Source-Based Solutions*

Step excitation of a finite slab with no film resistance

We consider again the case of a slab, initially at a uniform temperature, whose surfaces from $t = 0$ onward are kept at zero temperature. Equation (12.19) presents an eigenfunction solution. An error function formulation is also available.

The slab is taken to be bounded by the surfaces $x = -X$ and $x = +X$. Suppose that the surfaces are in fact just two planes in an infinite solid ($-\infty < x < \infty$) initially at zero temperature, and that in the element δx centred on $x = x_1$ between the surfaces a quantity of heat $\upsilon\delta x$ is released at $t = 0$. According to (12.36), the contribution to temperature at some point x_2 is given by

$$\delta T(x_2, t) = \frac{\upsilon\delta x}{2\sqrt{\pi\lambda\rho c_p}}\frac{1}{t^{1/2}}\exp\left(\frac{\rho c_p(x_2 - x_1)^2}{4\lambda t}\right). \tag{12.44}$$

The factor 2 in the denominator appears since the heat is taken here to flow in both directions. If heat is released uniformly between the surfaces $x = \pm X$ the temperature at x_2 is found by integrating this expression between $x_1 = -X$ and $+X$. After some manipulation, this leads to two error function terms. Further, the heat input generates a temperature $T_I = \upsilon/\rho c_p$, between the surfaces and we find

$$T(x_2, t) = \tfrac{1}{2}T_I\{\text{erf}[(\rho c_p/4\lambda t)^{1/2}(X + x_2)] + \text{erf}[(\rho c_p/4\lambda t)^{1/2}(X - x_2)]\}; \tag{12.45}$$

See Crank (1975: equation 2.15). The expression gives the temperature at $t > 0$ anywhere in an infinite solid, initially at zero except for the section between $\pm X$ where the initial temperature is T_I. At the right-hand surface $x_2 = X$, the second term is always zero. When t is small $[(\rho c_p/4\lambda t)^{1/2}(X + x_2)]$ is large and erf $[(\rho c_p/4\lambda t)^{1/2}(X + x_2)]$ is near unity, so $T(X, t) = \frac{1}{2}T_I$. This is midway between the temperatures either side of $x_2 = X$. Eventually the cooling imposed at $x_2 = -X$ makes itself felt at $x_2 = X$ and $T(X, t)$ falls below $\frac{1}{2}T_I$.

Now the infinite solid could consist of layers having various thicknesses and uniform temperatures up to $t = 0$. If from $t = 0$ onward, conduction takes place between them, the resultant temperature at some field point x_2 will be the sum of the component contributions. Suppose in fact that the continuum consists of an infinite succession of layers of thickness $2X$ at alternating temperatures $\pm T_I$, centred on the layer $-X < x < +X$ at $+T_I$. The contributions to temperature at x_2 due to the nearby layers are

$$\begin{aligned}
&\text{layer} -2 \quad \delta T(x_2, t) = +\tfrac{1}{2}T_I\{\text{erf}[(\rho c_p/4\lambda t)^{1/2}(5X + x_2)] + \text{erf}[(\rho c_p/4\lambda t)^{1/2}(-3X - x_2)]\},\\
&\text{layer} -1 \quad \delta T(x_2, t) = -\tfrac{1}{2}T_I\{\text{erf}[(\rho c_p/4\lambda t)^{1/2}(3X + x_2)] + \text{erf}[(\rho c_p/4\lambda t)^{1/2}(-X - x_2)]\},\\
&\text{layer } 0 \quad \delta T(x_2, t) = +\tfrac{1}{2}T_I\{\text{erf}[(\rho c_p/4\lambda t)^{1/2}(X + x_2)] + \text{erf}[(\rho c_p/4\lambda t)^{1/2}(X - x_2)]\},\\
&\text{layer } 1 \quad \delta T(x_2, t) = -\tfrac{1}{2}T_I\{\text{erf}[(\rho c_p/4\lambda t)^{1/2}(-X + x_2)] + \text{erf}[(\rho c_p/4\lambda t)^{1/2}(3X - x_2)]\},\\
&\text{layer } 2 \quad \delta T(x_2, t) = +\tfrac{1}{2}T_I\{\text{erf}[(\rho c_p/4\lambda t)^{1/2}(-3X + x_2)] + \text{erf}[(\rho c_p/4\lambda t)^{1/2}(5X - x_2)]\},\\
&\text{etc.}
\end{aligned} \tag{12.46}$$

This device ensures that $T(\pm X, t) = 0$. Carslaw and Jaeger (1959: 97), give the sum as

$$T(x_2, t)/T_I = 1 - \sum_{n=0}^{\infty} -1^n \{\mathrm{erfc}[(\rho c_p/4\lambda t)^{1/2}((2n+1)X + x_2)] + \mathrm{erfc}[(\rho c_p/4\lambda t)^{1/2}((2n+1)X - x_2)]\}. \tag{12.47}$$

The corresponding eigenfunction expression (for a slab $0 < x < X$) is

$$\frac{T(x,t)}{T_I} = 4\sum \frac{1}{n\pi} \sin\left(\frac{n\pi x}{X}\right) \exp\left(-\frac{t}{z_n}\right) \qquad (n = 1, 3, 5, \ldots), \tag{12.48}$$

where

$$z_n = \frac{\rho c_p X^2}{n^2 \pi^2 \lambda}. \tag{12.49}$$

It is unusual to have two exact and very unlike expressions for the same quantity. For early values of time, one or two terms of the error function expression are sufficient but many terms in the eigenfunction expression are needed. Conversely, for large value of time, the first eigenfunction term is sufficient but several error function terms are necessary. These solutions represent an idealisation that is not often justified in practice: at least a film resistance is needed.

Step excitation of a semi-infinite slab with film resistance

A solution for a semi-infinite slab with a film transmittance h is given by Carslaw and Jaeger (1959: 71, equation 1). The slab is initially at zero throughout and at $t = 0$ the temperature of the fluid wetting its exposed face increases to T_e.[9] Defining $u^2 = \rho c_p x^2/4\lambda t$ as before and defining $v = u + \sqrt{h^2 t/(\lambda \rho c_p)}$, the solution is

$$T(x,t) = \frac{T_e}{\exp(+u^2)}[\exp(+u^2)\mathrm{erfc}(u) - \exp(+v^2)\mathrm{erfc}(v)]. \tag{12.50}$$

For larger values of x,

$$\exp(x^2)\mathrm{erfc}(x) = \frac{1}{\sqrt{\pi}}\left(\frac{1}{x} - \frac{1}{2x^3} + \frac{1.3}{2^2 x^5} - \frac{1.3.5}{2^3 x^7} + \cdots\right); \tag{12.51}$$

see Carslaw and Jaeger (1959: 483, equation 5). Table 12.2 is a short table showing values of useful functions. All decrease from unity. Function $\mathrm{erfc}(x)$ relates to variation of temperature within a semi-infinite solid whose surface has undergone a step in temperature. Function $\exp(x^2)\mathrm{erfc}(x)$ provides the value for the surface temperature of a semi-infinite solid where the temperature of the adjacent fluid undergoes a step in temperature. Function

[9] Alternatively, the fluid temperature remains at zero but a pure heat source of value hT_e acts at the surface from $t = 0$ onward. This is an example of Thévenin's theorem.

Table 12.2 Values of erfc(x), exp(x^2)erfc(x) and exp($-x$)

x	erfc(x)	exp(x^2)erfc(x)	exp($-x$)
0.0	1.0	1.0	1.0
0.1	0.887537	0.8965	0.904837
0.2	0.777297	0.8090	0.818731
0.3	0.671373	0.7346	0.740818
0.4	0.571608	0.6708	0.670320
0.5	0.479500	0.6157	0.606531
0.6	0.396144	0.5678	0.548812
0.7	0.322199	0.5259	0.496585
0.8	0.257899	0.4891	0.449329
0.9	0.203092	0.4565	0.406570
1.0	0.157299	0.4276	0.367879
1.2	0.089686	0.3785	0.301194
1.4	0.047715	0.3387	0.246597
1.6	0.023652	0.3060	0.201897
1.8	0.010909	0.2786	0.165299
2.0	0.004678	0.2554	0.135335
2.2	0.001863	0.2356	0.110803
2.4	0.000689	0.2185	0.090718
2.6	0.000236	0.2036	0.074274
2.8	0.000075	0.1905	0.060810
3.0	0.000022	0.1790	0.049787

exp($-x$) relates to the temperature of a finite-thickness slab of infinite conductivity when the temperature of the adjacent fluid undergoes a step in temperature.

If h is infinite, exp($+v^2$)erfc(v) is zero and the solution reduces to that in Table 12.1. At the surface itself, $x = 0$ so $u = 0$ and with finite h we obtain

$$\begin{aligned} T(0, t) &= T_e[1 - \exp(+v^2)\mathrm{erfc}(v)] \\ &= T_e[1 - \exp((h^2t/(\lambda\rho c_p))\mathrm{erfc}(\surd(h^2t/(\lambda\rho c_p))]; \end{aligned} \tag{12.52}$$

that is, temperature is now to be expressed using time non-dimensionalised as $F_2 = h^2t/(\lambda\rho c_p)$ instead of $F_0 = \lambda t/(\rho c_p x^2)$ as has appeared hitherto. An approximate polynomial solution to this problem is given in Section 12.7.

12.4 SOLUTIONS FOR THE TEMPERATURE PROFILE AND TAYLOR'S SERIES

The foregoing solutions can be coordinated through use of Taylor's series. We suppose as usual that at the point x and time t, T and all its space differentials are known. The temperature at a point distant h from x can be expressed as

$$T(x+h, t) = T(x, t) + \frac{h\partial T(x, t)}{\partial x} + \frac{h^2\partial^2 T(x, t)}{2!\partial x^2} + \frac{h^3\partial^3 T(x, t)}{3!\partial x^3} + \frac{h^4\partial^4 T(x, t)}{4!\partial x^4} + \cdots. \tag{12.53}$$

Now

$$q(x,t) = -\lambda \frac{\partial T(x,t)}{\partial x} \tag{12.54a}$$

and in time-varying heat flow,

$$\frac{\partial}{\partial t} \equiv \kappa \frac{\partial^2}{\partial x^2}. \tag{12.54b}$$

So

$$\frac{\partial q(x,t)}{\partial t} = \kappa \frac{\partial^2 q(x,t)}{\partial x^2} = \kappa \frac{\partial^2}{\partial x^2}\left(-\lambda \frac{\partial T(x,t)}{\partial x}\right) = -\kappa\lambda \frac{\partial^3 T(x,t)}{\partial x^3} \tag{12.55a}$$

and

$$\frac{\partial^2 T(x,t)}{\partial t^2} = \frac{\partial}{\partial t}\frac{\partial T(x,t)}{\partial t} = \kappa \frac{\partial^2}{\partial x^2}\left(\kappa \frac{\partial^2 T(x,t)}{\partial x^2}\right) = \kappa^2 \frac{\partial^4 T(x,t)}{\partial x^4}. \tag{12.55b}$$

Equation (12.53) can now be expressed in terms of $T(x,t)$ and $q(x,t)$:

$$T(x+h,t) = T(x,t) + \frac{h^2 \partial T(x,t)}{2!\kappa \partial t} + \frac{h^4 \partial^2 T(x,t)}{4!\kappa^2 \partial t^2} + \cdots + \frac{h}{-\lambda} q(x,t) + \frac{h^3 \partial q(x,t)}{3!(-\lambda\kappa)\partial t} + \cdots. \tag{12.56}$$

Thus the temperature at some incremental position $x+h$ can be expressed in terms of two independent functions of time, one specifying temperature and the other the heat flow at some point x. If $x = 0$ has some special status, we can rewrite this equation setting $x = 0$ and replacing the displacement h by x itself:

$$T(x,t) = T(0,t) + \frac{x^2 \partial T(0,t)}{2!\kappa \partial t} + \frac{x^4 \partial^2 T(0,t)}{4!\kappa^2 \partial t^2} + \cdots + \frac{x}{-\lambda} q(0,t) + \frac{x^3 \partial q(0,t)}{3!(-\lambda\kappa)\partial t} + \cdots. \tag{12.57}$$

It can readily be checked that all the solutions given earlier are special cases of this equation. Equation (12.4) involves a finite number of terms and the later ones involve an infinite number. But the exercise does not provide useful information and it will not be presented here.

12.5 TRANSFORM METHODS

Our concern with the Fourier continuity equation stems from its use to estimate the heat flow through walls, a topic which has been of general interest since the 1930s. Some of the analytical approaches have come to be associated with the names of Laplace, Green, Sturm-Liouville, discussed for example by Özisik (1980: Section 1-7). Bouzidi (1991) cites some authors whose work might well have proved valuable in a building context but have not in fact featured in discussions of wall heat flow.

The Laplace transformation is a mathematical procedure used extensively in solving differential equations which occur in mathematical physics, such as the diffusion equation (12.1). Chapter 12 of Carslaw and Jaeger (1959) gives a detailed discussion on using Laplace transformations to investigate heat conduction in solids. If $f(x, t)$ is some continuous function of time, $0 < t < \infty$, it can be transformed to the frequency domain as

$$\overline{f}(x, p) = \int_0^\infty f(x, t) \exp(-pt)\, \mathrm{d}t, \tag{12.58}$$

where p is a complex number whose real part is large enough to ensure that the integral converges.[10] The transformation converts (12.1) to an ordinary differential equation

$$\lambda \frac{\mathrm{d}^2\overline{T}(x, p)}{\mathrm{d}x^2} = \rho c_p\{p\overline{T}(x, p) - T(x, 0)\}, \tag{12.59}$$

which is to be solved for the given boundary conditions. The solution is then transformed back to the time domain. Tables of transforms are available. As a simple example, consider a semi-infinite medium, initially at zero, so that $T(x, 0) = 0$. Thus $\overline{T}(x, p)$ is of form $B\exp(+qx)$ or $A\exp(-qx)$, where $q = \sqrt{p\rho c_p/\lambda}$. We reject the first since T must remain finite as x tends to infinity. Thus $\overline{T}(x, p) = A\exp(-qx)$. Suppose further that for $t > 0$ the surface temperature $T(0, t)$ has the constant value T_s, so $\overline{T}(0, p) = \int_0^\infty T_s \exp(-pt)\, \mathrm{d}t = T_s/p$. Therefore

$$\overline{T}(x, p) = T_s[\exp(-qx)/p]. \tag{12.60}$$

The inverse transform of $\exp(-qx)/p$ is $\mathrm{erfc}(x/(2\sqrt{\lambda t/\rho c_p}))$, a standard form (Carslaw and Jaeger 1959: 494), so transforming back to the time domain, we obtain

$$T(x, t) = T_s\ \mathrm{erfc}(x/(2\sqrt{\lambda t/\rho c_p})). \tag{12.61}$$

This is the solution to the problem of a semi-infinite medium initially at zero and whose surface temperature at $x = 0$ is held at the value T_s from $t = 0$ onward. The result was noted in Section 12.3.

The Z transform is closely related and serves when the function of time is known at discrete intervals, δ. Suppose that values of ambient temperature T_e are known at intervals of δ, hourly perhaps, and that T_e drives a heat flow q_{ie} through a wall into a room at zero temperature. The Z transforms of these values are

$$Z\{T_e\} = T_e(0)z^0 + T_e(\delta)z^{-1} + T_e(2\delta)z^{-2} + \cdots, \tag{12.62a}$$

$$Z\{q_{ie}\} = q_{ie}(0)z^0 + q_{ie}(\delta)z^{-1} + q_{ie}(2\delta)z^{-2} + \cdots, \tag{12.62b}$$

where $z = \exp(p\delta)$.

It is shown in Chapter 15 that the overall properties of a series of layers excited sinusoidally can be found as the product of their respective individual transmission matrices. Following Jury (1964), Stephenson and Mitalas (1971) have argued that the ratio of output to input can be expressed as the ratio of two polynomials:

$$\frac{q_{ie}(0)z^0 + q_{ie}(\delta)z^{-1} + q_{ie}(2\delta)z^{-2} + \cdots}{T_e(0)z^0 + T_e(\delta)z^{-1} + T_e(2\delta)z^{-2} + \cdots} = \frac{b_0z^0 + b_1z^{-1} + b_2z^{-2} + \cdots + b_Mz^{-M}}{1 + d_1z^{-1} + d_2z^{-2} + \cdots + d_Nz^{-N}} \quad (M \geq N). \tag{12.63}$$

[10]If $f(t)$ is some sound pressure level recording of sufficient duration, $\overline{f}(\mathrm{j}\omega)$ is its spectral density, a complex function of ω (or $\omega/2\pi$ hertz).

For a wall modelled by just two capacities, $N = 2$; for a thick heavy wall with $\delta = 1$ h, N may be 6. After cross-multiplication and equating terms to the same power of z, we have an equation for q_{ie} 'now' in terms of T_e 'now', and values T_e together with values of q_{ie} at the previous times $t = -\delta, -2\delta$, etc., as shown in (11.1). Using the Z transform, Stephenson and Mitalas go on to illustrate how coefficients such as b and d may be found.

However, such transform methods prove to be unnecessary; it is sufficient to use purely time-domain solutions, similar to those given earlier for discretised systems. This has the advantage that nothing beyond generally familiar mathematics is needed; the additional concepts needed for the transforms are not invoked. Further, a physical interpretation of results at various stages in the analysis is easier in the time domain than in the transform approach. There is little to choose between them computationally: the most time-demanding operation in the time-domain method is searching for the wall decay times and this is essentially identical with the search for poles in the transform approach.

Green's function solutions may be used in transient heat flow problems; they are discussed by Beck (1984). Green's functions do not appear to have contributed much to studies of heat flow through building walls and are of no concern here.

12.6 USE OF THE SOLUTIONS

Figure 12.1 shows how the solutions are used. The thermal details of a wall are supposed to be known. If the wall is to be examined using finite difference computation, the

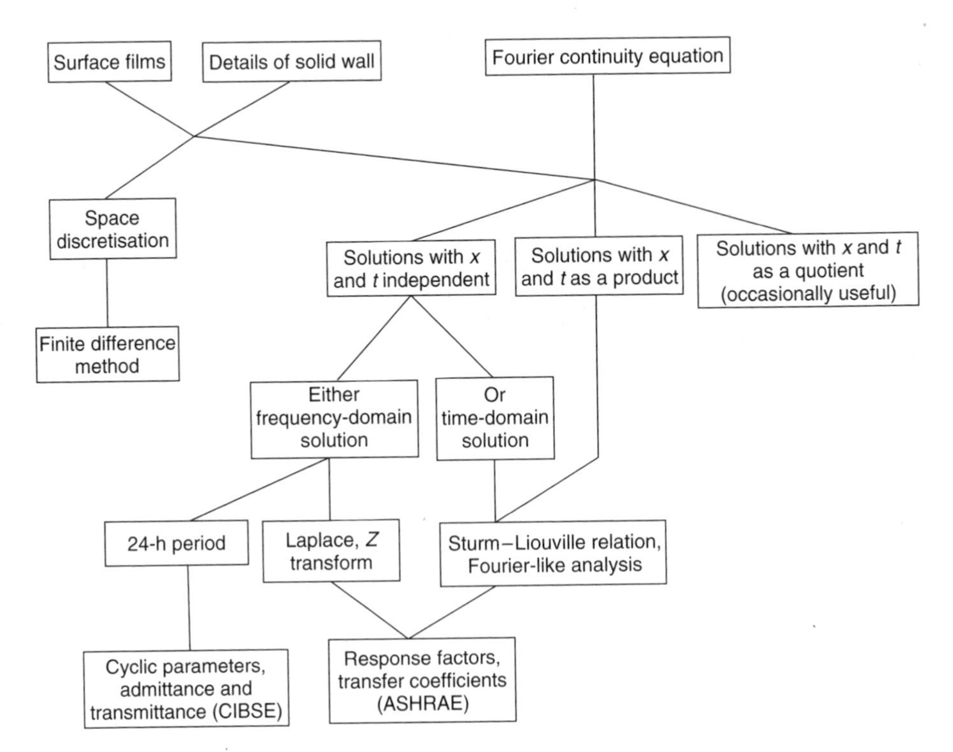

Figure 12.1 Scheme to find the wall heat flow parameters (Reproduced by the permission of *ASHRAE Transactions*)

solutions are not relevant. To find the 24 h cyclic parameters (used for example in the *CIBSE Guide*), we require only the frequency-domain solution (Chapter 15). To find the response factors and transfer coefficients (used in the *ASHRAE Handbook*), either the frequency-domain solution may be used, together with Z transform, or the time-domain solution may be adopted, together with the steady-slope solution; the time-domain solution is given in Chapter 17. One or other of these approaches is used in simulation packages such as HVAC Sim+, TRNSYS, ESP-r and SERIRES. Error function solutions (in which x and t appear as a quotient) cannot readily be used for layered structures but some solutions will be used in Chapter 14.

The task of a thermal model is to permit estimation of room internal temperatures in an uncontrolled building, or estimation of the heating or cooling load needed when room temperature is to be fixed. In either case we need to know the thermal characteristics of the room concerned, together with the time-varying quantities that influence it: internal heat loads of various kinds, solar incidence and ambient temperature, all of which serve as drivers. We may also need the ventilation rate, possible variation of internal convective coefficients with temperature difference, and variation of external convection with wind speed. Chapter 13 discusses some of these matters in simplified models and Chapter 14 discusses some basic analytical models. Figures 12.2 and 14.12 show how to handle them when more detail is supplied. They are developed in Chapters 16 and 19.

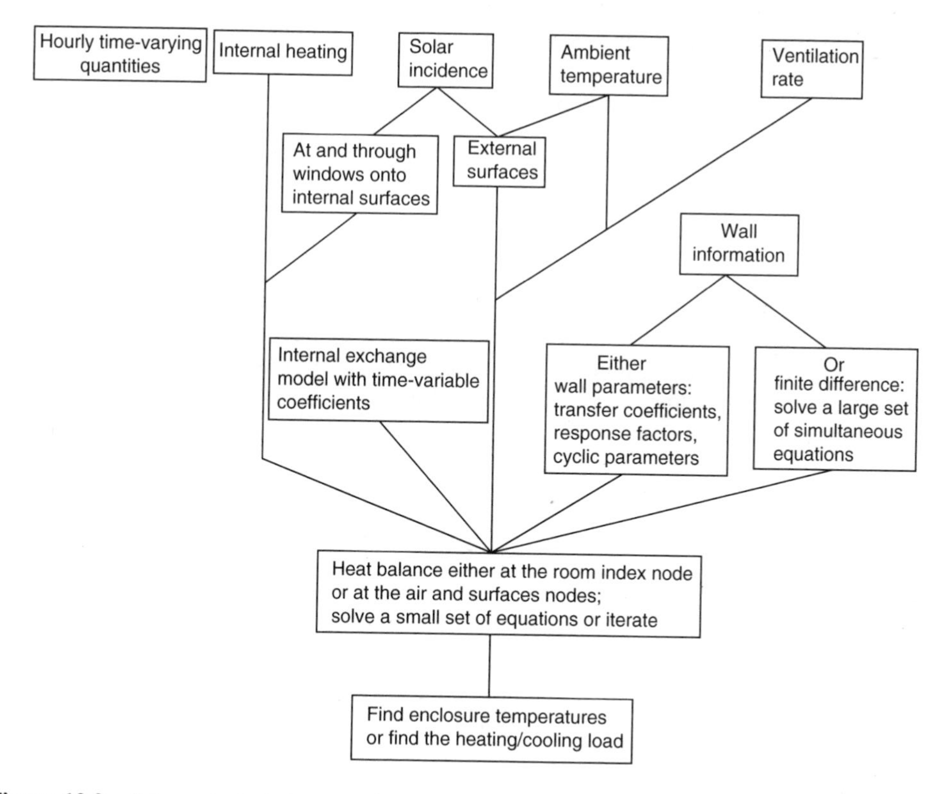

Figure 12.2 Scheme to find the room thermal response parameters (Reproduced by the permission of *ASHRAE Transactions*)

12.7 APPENDIX: PENETRATION OF A SIGNAL INTO AN INFINITE SLAB

It is not possible using classical theory to arrive at an expression saying just how far a thermal signal has penetrated into a semi-infinite solid but in this section we find an approximate value. We consider a semi-infinite slab, initially at the uniform temperature T_I. At $t = 0$ the temperature of the fluid wetting it at its exposed surface $x = 0$ is reduced to zero. The exact expression for temperature (Carslaw and Jaeger 1959: 71, equation 1) is

$$\theta = T(x, t)/T_I = \mathrm{erf}(0.5/\sqrt{F_0}) + \exp(B + F_2)\,\mathrm{erfc}(0.5/\sqrt{F_0} + \sqrt{F_2}), \qquad (12.64)$$

where $F_0 = \lambda t/\rho c_p X^2$, $F_2 = h^2 t/\lambda\rho c_p$ and $B = hx/\lambda$. (This is another form of (12.50): the initial temperature there was zero but here it is T_I.) This excitation leads to a finite but vanishingly small disturbance at remote points immediately after $t = 0$. The temperature profile in the slab at some elapsed time t has a smooth form and the cooling behaviour can be described fairly accurately by representing the profile as a polynomial. This provides simple though of course approximate expressions for the penetration depth of the thermal change and for the surface temperature.

Suppose that at time t the effect of the step change in fluid temperature at the surface has reached a penetration depth p and that y denotes the position in the slab in relation to p: $y = p - x$, where x is measured from the exposed surface (Figure 12.3). If the boundary layer resistance $1/h$ is represented as λ/h, the notional temperature gradient in the layer has the same gradient as that of the slab profile at the surface. The temperature $T(x, t)$ or $T(y, t)$ can then be expressed approximately in polynomial form:

$$T(y, t) = T_I + a_1 y + a_2 y^2 + a_3 y^3 + \cdots \qquad (0 < y < p), \qquad (12.65a)$$

$$T(y, t) = T_I \qquad (y < 0). \qquad (12.65b)$$

Since the cooling rate $\partial T/\partial t$ is proportional to $\partial^2 T/\partial x^2$, inclusion of just the two terms $a_1 y$ and $a_2 y^2$ would imply a uniform rate of cooling between $x = 0$ and p, and zero beyond, which is physically not acceptable. Inclusion of $a_3 y^3$ ensures the cooling rate varies uniformly from a maximum value at the surface to zero at $x = p$ but there is a discontinuity in the cooling rate at $x = p$, again not acceptable. With four terms, the

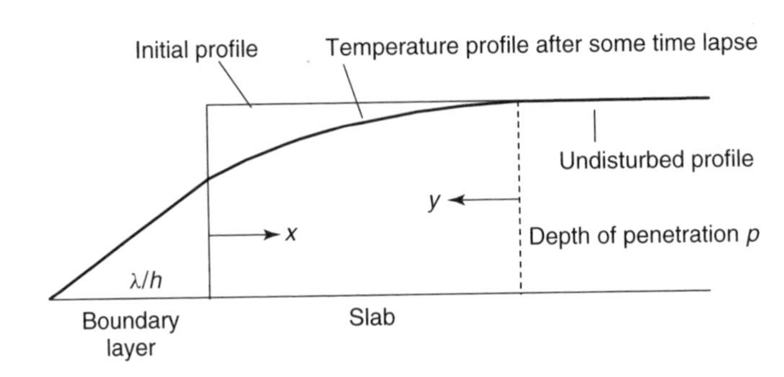

Figure 12.3 Temperature profile in a slab following a step change in the fluid temperature at the exposed surface

cooling rate can vary from a maximum at the surface and can patch smoothly to zero beyond $x = p$, a physically acceptable model. Further terms may be expected to improve accuracy and a sixth-power expression will be used here.[11]

The coefficients are to be found from the boundary equations:

- At $y = 0$ the heat flux must be zero, so $\partial T/\partial y = 0$ and $a_1 = 0$.
- At $y = 0$ the cooling rate too must be zero, so $\partial^2 T/\partial y^2 = 0$ and $a_2 = 0$.
- At $y = 0$, $\partial^3 T/\partial y^3$ will be taken as zero, so $a_3 = 0$.
- At the surface $(y = p)$, heat balance requires that

$$hT(y, t) = -\lambda \, \partial T(y, t)/\partial y. \tag{12.66a}$$

- Heat balance at the surface holds at all times, so by differentiating (12.66a) with respect to time and using the Fourier continuity equation, we have on either side of the surface that

$$h \, \partial^2 T(y, t)/\partial y^2 = -\lambda \, \partial^3 T(y, t)/\partial y^3. \tag{12.66b}$$

- Finally (12.66b) can be differentiated to give

$$h \, \partial^4 T(y, t)/\partial y^4 = -\lambda \, \partial^5 T(y, t)/\partial y^5. \tag{12.66c}$$

Thus the Fourier continuity equation is observed at the surface itself but not within the slab; the boundary conditions imposed at the current depth of penetration $x = p$ are only approximate. The final three conditions allow us to evaluate the non-zero values of a. The solution for the profile is

$$T(y, t) = 1 + \frac{A_4(y/p)^4 + A_5(y/p)^5 + A_6(y/p)^6}{192 + 180B + 69B^2 + 13B^3 + B^4}, \tag{12.67}$$

where p is the momentary depth of penetration of the profile, $y = p - x$, $B = hp/\lambda$. Also

$$A_4 = -(120B + 105B^2 + 33.75B^3 + 3.75B^4), \tag{12.68a}$$

$$A_5 = (72B + 78B^2 + 30B^3 + 3.75B^4) \tag{12.68b}$$

and

$$A_6 = -(12B + 15B^2 + 7B^3 + B^4). \tag{12.68c}$$

At the surface we have $y = p$, so the surface temperature θ_{ip} as a function of the non-dimensional penetration depth B is

$$\theta_{ip} = \theta(0, F_2) = \frac{192 + 120B + 27B^2 + 2.25B^3}{192 + 180B + 69B^2 + 13B^3 + B^4}. \tag{12.69}$$

[11] This is a somewhat cautious choice. Eckert and Drake (1972: 18, 184) consider the case of a step change in temperature at the surface, representing the temperature profile by a polynomial of second degree. They show that the heat flow at the surface, so found, is greater than its exact value by a factor of $\sqrt{\pi/3}$. This choice has recently been discussed by Wood (2001).

The relation between B and time $F_2 = h^2t/\lambda\rho c_p$ can be found from certain properties of the exact solution. At the surface of the slab,

$$\theta_i = \quad \theta(0, F_2) \quad = \exp(F_2)\text{erfc}(\sqrt{}F_2) \tag{12.70a}$$

and

$$\partial^2\theta(0, F_2)/\partial b^2 = \exp(F_2)\text{erfc}(\sqrt{F_2}) - 1/\sqrt{\pi F_2}. \tag{12.70b}$$

Use of (12.70) in the polynomial expression gives an explicit expression for the time $F_0 = F_2/B^2$ associated with B:

$$F_0 = \frac{\lambda t}{\rho c_p p^2} = \frac{(192 + 180B + 69B^2 + 13B^3 + B^4)^2}{\pi(360 + 342B + 135B^2 + 27B^3 + 2.25B^4)^2} \overset{B=\infty}{\Rightarrow} \frac{1}{\pi.2.25^2} = 0.0629. \tag{12.71}$$

Equations (12.69) and (12.71) express surface temperature and time as a function of B, so elimination of B between them gives explicit values of temperature θ_{ip} and time (expressed as F_0 or F_2). Table 12.3 compares them with exact values found from (12.70a). Values of B were chosen so that the exact temperature θ_i has values 0.90, 0.80, etc. θ_{ip} lies a little above θ_i but the largest discrepancy in a range of 1.0 is about 0.013. This may be no greater than the physical uncertainty in F_2. Since the polynomial form gives satisfactory estimates of surface temperature, we may have confidence in its estimate of the depth of penetration (Table 12.4).

The value $B = \infty$ corresponds to $h = \infty$ and therefore denotes a step change in the temperature of the surface itself. When $B = \infty$ in (12.71) then

$$F_0 = 1/(\pi \times 2.25^2) = 0.063. \tag{12.72}$$

Thus the approximate depth of penetration p due to a disturbance imposed on the surface of a slab after a time t is

$$p = (\lambda t/(\rho c_p \times 0.063))^{1/2}. \tag{12.73}$$

The time taken by a disturbance in passing through a slab of finite thickness $2X$ is four times these values. They are consistent with those of $[F_o]_d$, $[F_o]_e$ and $[F_o]_f$, derived from a quite different argument, in Table 13.2.

Table 12.3 Error in surface temperature of a semi-infinite solid during cooling, as estimated by a sixth order polynomial

θ_i	1.0	0.9	0.8	0.7	0.6	0.5	0.4	0.3	0.2	0.1
$\theta_{ip} - \theta_i$	0.0000	0.0068	0.0113	0.0134	0.0133	0.0113	0.0081	0.0045	0.0017	0.0002

Table 12.4 Approximate time t (expressed as the Fourier number $F_0 = \lambda t/\rho c_p p^2$) for a signal to penetrate a distance p into a slab as a function of the film transmittance h (expressed as the Biot number $B = hp/\lambda$)

B	0.001	0.01	0.1	1.0	10	100	1000	∞
F_0	0.0905	0.0905	0.0903	0.0878	0.0726	0.0641	0.0630	0.0629

13

Analytical Transient Models for Step Excitation

In the period before electronic computing became current, a sizeable literature had appeared concerned with the response of a slab of material, finite or infinite, with distributed conductive and capacitative properties, with various initial conditions (uniform temperature, uniform gradient), with or without surface films and subjected to a variety of boundary conditions, fixed or varying. Many of the solutions are presented in Chapters 2 and 3 of Carslaw and Jaeger (1959). They are 'analytical' in the sense that explicit expressions are given for the temperature $T(x,t)$ as a function of position in the slab and subsequent time. Most are eigenfunction solutions and it must be pointed out that except for the simplest cases, explicit expressions for the eigenvalues or decay times are not available. Recourse was had to numerical search and that very much restricted the use of the solutions before electronic computing made searching easy.

This chapter presents four classical models which merit discussion, although they are too restricted to have immediate application in the building field. They are shown in Figure 13.1. A slab having distributed resistance and capacity is shown as a resistance with two parallel lines. A link to the reference temperature is to be understood.

13.1 SLAB WITHOUT FILMS

The case of a slab, initially at the uniform temperature T_I whose surfaces are maintained at zero from $t = 0$ onward was presented in (12.19) and (12.47). It was remarked that for an initial period, cooling took place at one surface independently of cooling at the other. The argument leading to (12.73) provides a measure of the time taken by a signal penetrating some distance into a slab. The models of a slab, with and without a surface film, provide an alternative approach to find this time. It is based on Davies (1999).

It is more convenient to suppose that the origin $x = 0$ is at the centre of the slab, now taken to have a thickness $2X(-X < x < +X)$, so that from $t = 0$ onward, $T(-X,t) = T(+X,t) = 0$. The solution $T(x,t)$ is given in Carslaw and Jaeger (1959: 97, equation 8):

$$\frac{T(x,t)}{T_I} = \frac{4}{\pi}\sum\frac{(-1)^{j-1}}{2j-1}\cos\left(\frac{(2j-1)\pi x}{2X}\right)\exp\left(-\frac{\pi^2(2j-1)^2}{4}F_0\right) \quad (j = 1, 2, \ldots, \infty), \qquad (13.1)$$

Building Heat Transfer Morris G. Davies
© 2004 John Wiley & Sons, Ltd ISBN: 0-470-84731-X

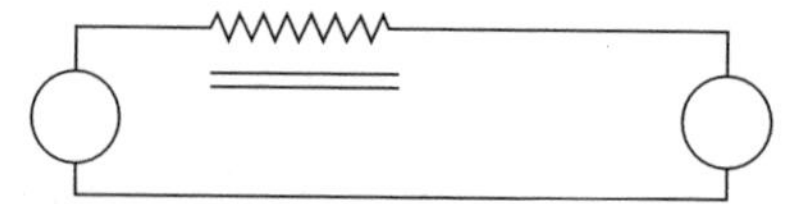

(a) Slab without films, symmetrically excited.

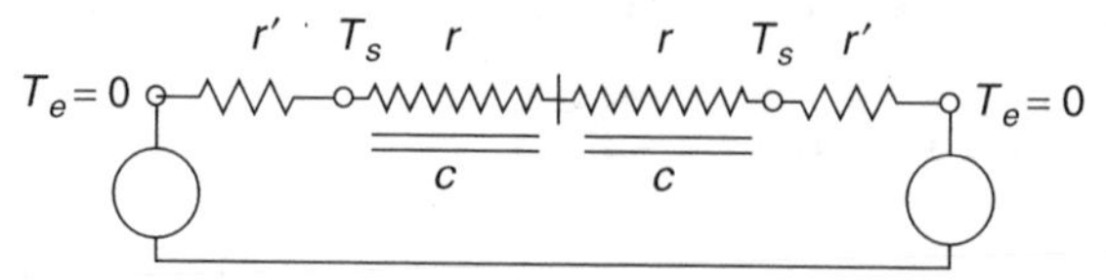

(b) *Either* film r' and slab r, c, adiabatic on rear surface; *or* film r', slab $2r$, $2c$, film r', symmetrically excited (Groeber's model).

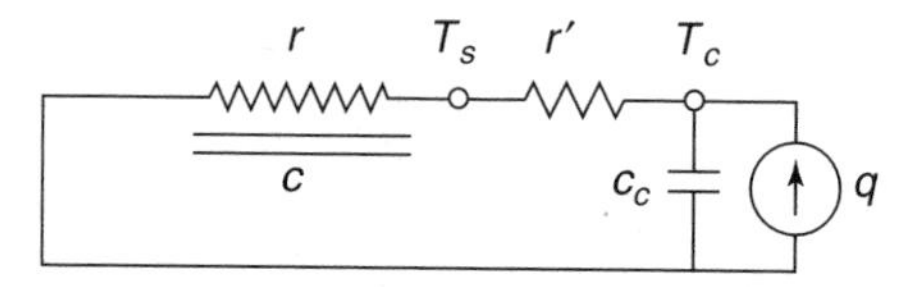

(c) Slab r, c, film r' and core c_c (Jaeger's model, flux driven).

(d) Film r_e, slab r, c, film r_i, ventilated space r_v, film r_c, core c_c; (Pratt's model).

Figure 13.1 Models to examine the transient response of a slab

where the Fourier number $F_0 = \lambda t/\rho c_p X^2$. The expression enables us to recognise stages in the cooling process, at the surfaces and midplane.

13.1.1 Cooling at the Surface

The heat flow q from the surface $x = -X$ is given in dimensionless form as

$$\phi_f = \frac{q(-X,t)X}{T_I\lambda} = \sum 2\exp\left(-\frac{\pi^2(2j-1)^2}{4}F_0\right) = \phi_1 + \phi_2 + \phi_3 + \cdots + \phi_j + \cdots, \tag{13.2}$$

where ϕ_j is the jth eigenfunction contribution to ϕ_f.

If the material is semi-infinite, the non-dimensionalised surface heat flux is

$$\phi_i = (\pi F_0)^{-1/2}; \tag{13.3}$$

see Carslaw and Jaeger (1959: 61, equation 11) and Table 12.1.

It is obvious on physical grounds that in the early stages of cooling, ϕ_f and ϕ_i should be nearly equal. This proves to be the case, although $\phi_i - \phi_f$ is finite but vanishingly small at small values of t.

However, the cooling imposed at the slab surface $x = +X$ leads to an additional loss of heat from the slab, so the flux ϕ_f at $x = -X$ eventually becomes substantially less than ϕ_i.

Figure 13.2 shows in detail the development of $\phi_f(= \phi_1 + \phi_2 + \phi_3 + \cdots)$ from its asymptotic value of ϕ_i during early cooling, through the period when ϕ_j values higher than ϕ_1 are decaying rapidly with time, to its late cooling asymptotic value of ϕ_1 alone. (The contributions of ϕ_2 and higher terms for low values of F_0 are omitted so as not to complicate the graph. They total ϕ_i.)

The quantity $\phi_i - \phi_1$ has a minimum value at $F_0 = 0.2094$, when $\phi_i = 1.233$ and $\phi_1 = 1.193$, a difference of 0.040 or 3.2%. As a first approximation, cooling at $x = -X$ could be said to proceed unaffected by cooling at $x = +X$ for a time given by F_0 of about 0.2, that is, determined by (13.3), and thereafter by the first term in (13.2). The uncertainty in the values of λ, ρ and c_p in building applications means the value of F_0 is uncertain and an error of 3.2% would not be important here.

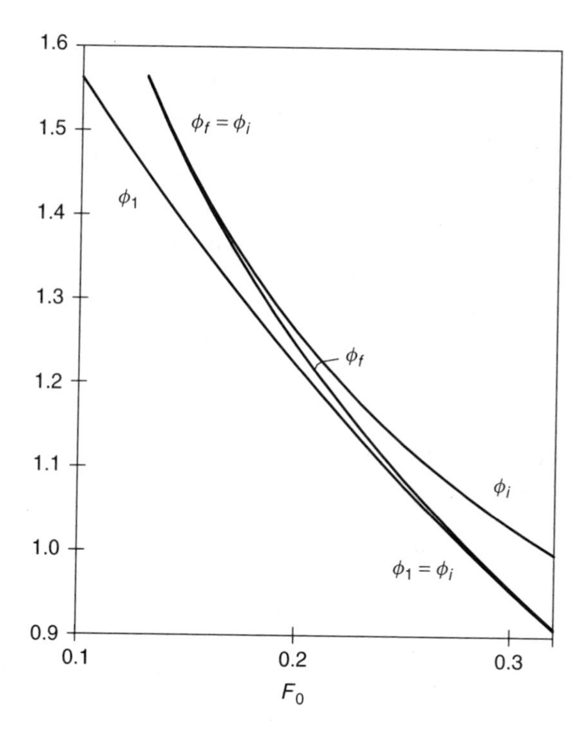

Figure 13.2 Cooling curve for a double-sided slab, initially at a uniform temperature T_I whose surfaces themselves are at zero temperature from $t = 0$. The vertical axis plots the non-dimensionalised heat flow $\phi = qX/T_I\lambda$ and the horizontal axis plots the non-dimensionalised time $F_0 = \lambda t/\rho c_p X^2$. The graph shows the development of the surface flux ϕ_f from the value ϕ_i, for a semi-infinite solid, to the value ϕ_1, the first eigenvalue component. (Reprinted from *Journal of Heat Transfer*, vol. 121, M.G. Davies, The time delay for a perceptible thermal disturbance in a slab, with permission from ASME)

The estimate however is coarse, since by now ϕ_f is quite perceptibly less than ϕ_i. Also $\mathrm{d}\phi_f/\mathrm{d}F_0$ is correspondingly greater than $\mathrm{d}\phi_i/\mathrm{d}F_0$ and the difference in this region changes uniformly with F_0. However, $\mathrm{d}^2(\phi_i - \phi_f)/\mathrm{d}F_0^2$ shows a weak maximum at $F_0 = 0.181$, which gives some qualitative measure of the region in which ϕ_f departs from ϕ_i. A more conservative estimate for the duration of early cooling is given by the time when $\phi_i - (\phi_1 + \phi_2)$ has its minimum. This is found at $F_0 = 0.124$, when $\phi_i = 1.602$ and $\phi_i - (\phi_1 + \phi_2) = 0.0023$, a deviation of only 0.14%.

At $F_0 = 0.2$ the second eigenfunction ϕ_2 still contributes significantly to ϕ_f, as Figure 13.2 shows, but in due course it becomes effectively zero. We cannot assign any non-arbitrary value to F_0 to indicate when this is so, since ϕ_2/ϕ_f never becomes strictly zero.

An estimate can be obtained by noting that in steady-state conditions the heat flow in one dimension is everywhere constant, so $\partial^2 q/\partial t \partial x = 0$. Now $\partial^2 q/\partial t \partial x$ is proportional to $\partial^2 T/\partial t^2$, so $\partial^2 T/\partial t^2$ is a measure of departure from steady conditions, although it is not a sufficient condition for a steady state. Within the slab under discussion, the amplitude of temperature of the second eigenfunction, T_2 say, is at all times less than T_1 but initially $\partial^2 T_2/\partial t^2$ is greater and latterly less than $\partial^2 T_1/\partial t^2$. When they are equal, their respective departures from steady conditions are equal. Hence this condition denotes a time after which T_2 and therefore ϕ_2 are becoming negligible compared with the first eigenvalue quantities.

The condition that $\partial^2 T_2/\partial t^2$ should equal $\partial^2 T_1/\partial t^2$ just within the surface of the slab gives $F_0 = \ln(81)/2\pi^2 = 0.223$, so we expect that shortly afterwards ϕ_2 will be negligible compared with ϕ_1. In the late phase of cooling, the rate of cooling is, fractionally speaking, the same everywhere in the slab and fastest at the centre.

By taking account of the first two eigenfunction contributions to ϕ_f we should expect ϕ_f to depart from its early asymptotic form of ϕ_i some time after F_0 equals about 0.12 and to have assumed its late asymptotic form of ϕ_1 shortly after F_0 equals 0.22. From Figure 13.2 these values, estimated by eye, appear to be about 0.15 and 0.26. In Section 13.2.3 this analysis is applied to the case where cooling takes place through a resistive layer.

13.1.2 Cooling at the Midplane

After the surface temperatures have been reduced to zero, the temperature at the midplane remains unaltered for some while. Since solutions to the Fourier continuity equation suggest that an infinitesimal disturbance is propagated instantaneously, we cannot specify some delay similar to that associated with propagation of a mechanical shock or a sound wave. But according to the argument above, the time when $\partial^2 T(0, t)/\partial t^2$ has its maximum value indicates the maximum deviation from the initial and final steady states. This occurs at $F_0 = 0.0613$, close to the value of 0.063 given by the polynomial approach in (12.72).

The argument provides a partial answer to the philosophical objection to the infinite rate of propagation of thermal signal, as required by some solutions of the Fourier continuity equation. It is a common observation that following thermal excitation, there is a perceptible delay before any substantial change is noted at a point distant from the point of excitation. If the bowls of silver and stainless steel teaspoons are dipped simultaneously into hot water, the effect is noticed first in the handle of the silver spoon. Eckert and Drake (1972: 23) list a number of articles addressing this issue. The above argument is consistent with an infinite rate of propagation of an exceedingly small change but demonstrates that the rate of change must increase perceptibly after some finite delay. After some excitation the solution to the wave equation (12.3) leads to an unchanged signal after a precisely defined delay, but the solution to the diffusion equation leads to an eventual surge. There is no need in this context, at any rate, to introduce a term of type $\tau\partial^2 T(x,t)/\partial t^2$ into the Fourier continuity equation, where τ is a positive scalar known as the thermal relaxation time.[1] See Tang and Arak (1996) and Kronberg *et al.* (1998).

Exponential cooling becomes established after the time when $\partial^2 T(0,t,2)/\partial t^2 = \partial^2 T(0,t,1)/\partial t^2$, so

$$F_0 = \ln(27)/2\pi^2 = 0.167. \tag{13.4}$$

It is also convenient to find the response time at the midplane – the time when the temperature there has fallen to $1/e = 0.368$ of its initial value. This has a value

$$F_0 = (4/\pi^2)(1 + \ln(4/\pi)) = 0.503. \tag{13.5}$$

We cannot write down a similar quantity for the surface since $T(X,t) = 0$ always and the initial heat flux is infinite.

13.2 THE FILM AND SLAB, ADIABATIC AT REAR: GROEBER'S MODEL

We consider the case of a slab of material, $-X < x < 0$, adiabatic on its rear surface $x = 0$ and in contact with a film transmittance h at the left surface; the transmittance represents convective or radiative exchange between the surface and the adjoining external space at T_e. (Figure 13.1b). The initial temperature of the slab is T_I. From $t = 0$ onward, the temperature T_e is zero. We wish to determine the subsequent temperature $T(x,t)$. Alternatively, the slab may be taken to be of thickness $2X$ and in contact with a film h at its right surface representing heat transfer to the internal space at T_i. If the double-thickness slab is symmetrically excited, the heat flow q across its central plane $x = 0$ is always zero.

The solution to this problem was given by Groeber in 1925. It is presented in many heat transfer texts since it is the most comprehensive of the basic solutions, including finite values of slab thickness, conductivity and film transmittance. The above argument on early and late stages of cooling can be applied to show the structure of cooling curves at the surface and the surface response time. Stages of cooling at the adiabatic surface (or midplane) can be identified.

The solution is also appropriate if the fluid temperature remains constant but a flux of heat $q = hT_e$ is withdrawn from the surface, as follows from Thévenin's theorem (Section 12.3.1). It was first reported by Heisler (1947), (although his method unnecessarily involved the temperature distribution within the slab itself.)

[1] The need for the quantity τ was put forward in 1958. It has application in situations such as microwave and laser heating of very short duration for surface melting of metal and sintering of ceramics. Barletta and Zanchini (1997) report a result due to Kaminski that sand may have a τ value of about 20 s. This is too small to affect building response times and τ does not appear to have been included in any studies of heat flow in walls.

13.2.1 Solution

The transient solution given by (12.11b) can be put in the form

$$T(x,t) = [A\sin(ux/X) + C\cos(ux/X)]\exp(-t/z), \qquad (13.6)$$

where

$$ux/X = \alpha x = (\rho c_p/\lambda z)^{1/2} x \qquad (13.7)$$

Since

$$q(x,t) = -\lambda \partial T(x,t) \partial x \quad \text{and} \quad q(0,t) = 0,\ A = 0.$$

At $x = +X$ the positively directed conducted flow is $-\lambda(-C(u/X)\sin u)\exp(-t/z)$ and equals the heat loss to the medium, $hC\cos(u)\exp(-t/z)$, so

$$u\tan u = hX/\lambda = B. \qquad (13.8)$$

B, the Biot number, is a non-dimensional measure of the thickness of the slab. If B is small, <0.1 say, the slab will not sustain a large temperature difference between its surfaces and midplane.

The equation has multiple solutions corresponding to the intersections of the linear function $f_1 = u/B$ with the series of branches of the function $f_2 = \cot u$.

$$0 < u_1 < \pi/2, \quad \pi < u_2 < 3\pi/2, \quad 2\pi < u_3 < 5\pi/2, \ldots.$$

Carslaw and Jaeger (1959: 491) give values of u_1 to u_6 for values of B between 0.001 and 100.

If B is small, $u_1 \approx \sqrt{B}$. The approximation

$$1/u_1^2 = 4/\pi^2 + 1/B \qquad (13.9a)$$

is correct to about 2% and the further approximation, based on a sixth-order polynomial for the temperature profile,

$$u_1^2 = B(720 + 480B + 96B^2)/(720 + 720B + 272B^2 + 38\tfrac{6}{7}B^3) \qquad (13.9b)$$

provides five-digit accuracy for u_1 up to $B \approx 1$. When $B = \infty$ so that $u_1 = \pi/2$, the deviation is 0.00101.

Equation (13.8) gives the complete set of eigenvalues for the film/slab system (thickness X). For the film/slab/film system (thickness $2X$) there is a further set corresponding to the condition that $T(0,t) = 0$ and defined by $-u/\tan u = B$. This is not relevant to the current problem but is used in Section 17.9.

The terms u_1, u_2, etc., are eigenvalues and characterise temperature profiles through the slab. Within the slab, the temperature profiles are sinusoidal and the quantity $2u_j/2\pi$ is the number of cycles of profile within the slab. If the heat transfer coefficient h is infinite, the slab surface temperatures themselves are reduced to zero and $B = \infty$. The solution where $u_1 = \pi/2$ implies that the region $-X < x < +X$ contains half a complete sinusoid, for $u_2, = 3\pi/2$ the slab contains 3/2 sinusoids, etc.

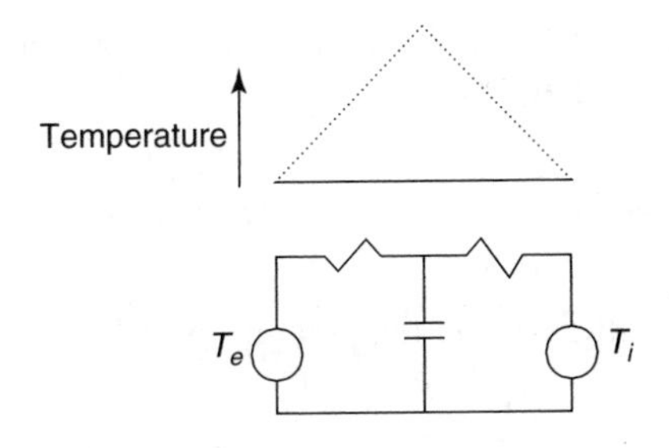

Figure 13.3 The film, slab, film model of Figure 13.1b with $B = 0$ and the notional temperature profile

For finite values of B, the sinusoidal portions of the temperature profile within the slab are less than these values. They are now 'completed' in the boundary layers themselves. This can be demonstrated graphically by plotting the temperature profile from T_e through the slab to T_i as a function of resistance between these nodes. In this way, a notional gradient can be assigned to heat flow in the film and there is no change in gradient immediately either side of the slab surfaces. If this argument is taken to the opposite limit ($B \to 0$), all the temperature variation is within the films and we have the profile of Figure 13.3.

The eigenvalues describe the series of decay times. From equation (13.7) we have

$$z_j = \frac{\rho c_p X^2}{\lambda u_j^2}. \tag{13.10}$$

When $B = \infty$, $u_1 = \pi/2$, etc. The z_j values are now those implied by (13.1).

The eigenvalues have the property that

$$\sum(1/u_j^2) = 1/B + 1/2. \tag{13.11}$$

(Consider the symmetrical system film/slab/film of thickness X. Its decay times $z_{11,j} = z_{22,j}$ in Chapter 10 notation are written here as z_j. Its $\sum r \sum c$ value is $(1/h + X/\lambda + 1/h)\rho c_p X$. The result follows using (10.28b) and (13.10).)

It is a property of the eigenfunctions that

$$\int_0^X \cos(u_j x/X)\cos(u_k x/X)\,\mathrm{d}x = \begin{cases} (X/u_j)\sin u_j & \text{if } u_k = 0, \\ (X/2u_j)(u_j + \sin u_j \cos u_j) & \text{if } u_k = u_j, \\ 0 & \text{if } u_k \neq u_j. \end{cases} \tag{13.12}$$

These relations are similar to those in (12.17).

The values of u_j derive from the slab *boundary conditions* alone. To find the values for the coefficient C in (13.6), we use the *initial condition* that the temperature everywhere in the slab is T_I at $t = 0$; that is,

$$\sum_{j=1}^{\infty} C_j \cos(u_j x/X) = T_I \quad \text{for all values of } x. \tag{13.13}$$

Multiplication of this equation by $\cos(u_k x/X)$, integration between 0 and X, and application of (13.12) gives the value of C_j:

$$C_j = 2T_I \sin u_j/(u_j + \sin u_j \cos u_j) = T_I[2B/((u_j^2 + B^2 + B)\cos u_j)]. \quad (13.14)$$

Finally, using (13.8), the value for the temperature is

$$\theta(x,t) = \frac{T(x,t)}{T_I} = \sum_{j=1}^{\infty} \frac{2B}{u_j{}^2 + B^2 + B} \frac{\cos(u_j x/X)}{\cos u_j} \exp(-t/z_j); \quad (13.15a)$$

see Carslaw and Jaeger (1959: 122). It can be written as

$$\theta(x,t) = \frac{T(x,t)}{T_I} = \sum_{j=1}^{\infty} \frac{2\sin u_j \cos(u_j x/X)}{u_j + \sin u_j \cos u_j} \exp(-t/z_j). \quad (13.15b)$$

The expression can serve as a benchmark solution to test a finite difference algorithm (Bland 1992). The value T_s at either surface ($x = \pm X$) is

$$\theta(\pm X,t) = \frac{T(\pm X,t)}{T_I} = \frac{T_s(t)}{T_I} = \sum_{j=1}^{\infty} \frac{2B}{u_j{}^2 + B^2 + B} \exp(-t/z_j). \quad (13.16)$$

Table 13.1 shows the relation between the physical changes during cooling and the solution given by (13.15).

If a state of 'no perceptible change' shortly after excitation is to be modelled by a series of eigenfunction terms, some must be negative; at an adiabatic surface they may be expected to alternate in sign. (An elementary illustration is provided by Example 10.3 in Section 10.3.6. It is readily checked that when $t = 0$, $\partial(T_{4t}/T_I)/\partial t$ is near zero. It is not exactly zero since, according to elementary thermal conduction theory, a vanishingly small signal is propagated at an infinite rate to the isothermal node T_5.)

Table 13.1 The relation between the physical changes during cooling and the solution to the Groeber problem given by (13.15)

	Response at exposed surface		Response at adiabatic surface or midplane	
	Physical effect	Eigenfunction solution terms have same sign	Physical effect	Eigenfunction solution terms alternate in sign
$t = 0$	No change yet imposed	Sum converges slowly to initial value	No change imposed	Sum converges slowly to initial value
t small	Sizeable change	Sum converges rapidly	No perceptible change	Rapid convergence
t large	Temperature now falling slowly	Response determined by first eigenfunction alone	Temperature falling faster than at surface	Response determined by first eigenfunction alone

13.2.2 Limiting Forms

If the heat transfer coefficient h tends to infinity, u_j tends to $(2j-1)\pi/2$ and (13.15) tends to (13.1).

If the conductivity λ tends to infinity, the temperature becomes equal through the slab, C_1 tends to 1, C_2 and higher coefficients tend to zero. Then

$$\theta(x,t) = \exp(-t/z), \tag{13.17}$$

where $z = \rho c_p X/h$, which is the decay time for the model in Figure 10.2b.

If the thickness X tends to infinity, numerical results from the Groeber model tend toward those given by (12.47). An analytical comparison is not possible here.

If both X and h are infinite, $T(x,t)/T_I = \text{erfc}(u)$ where $u^2 = \rho c_p x^2/4\lambda t$ (Table 12.1). An analytical demonstration is possible here because the finite X, finite λ, infinite h model has both error function (12.47) and eigenfunction (12.19) forms of solution.

13.2.3 Early and Late Stages of Cooling at the Surface

As time increases, the higher eigenfunctions die away quickly. It is to be shown that, as in Section 13.1, when we consider the first two eigenfunctions ($j = 1, 2$) certain stages in the cooling at the surface can be distinguished.

Since the surface temperatures remain finite, it is more convenient to work using temperature than flux. We recall a further version of non-dimensionalised time,

$$F_2 = h^2 t/\lambda \rho c_p. \tag{13.18}$$

The temperature of the finite thickness slab at $x = -X$ is

$$\frac{T(-X,t)}{T_I} = \theta_f = \sum \frac{2B}{B^2 + B + u_j{}^2} \exp\left(-\frac{u_j{}^2 F_2}{B^2}\right) = \theta_1 + \theta_2 + \theta_3 + \cdots, \tag{13.19}$$

where θ_j is the jth eigenfunction contribution. This is to be compared with the temperature at the surface of a semi-infinite medium (Carslaw and Jaeger 1959: 71 equation 2); see also (12.50) but note that here we are considering a change from T_I to zero, and not the converse. The surface temperature is

$$\theta_i = \exp(F_2)\ \text{erfc}(\sqrt{F_2}). \tag{13.20}$$

As in the previous section, at some time during cooling, the value of θ_1 is found to be very little less than θ_i. To demonstrate this, we first note that for a slab where B is small, θ_1 is initially almost unity, but falls rapidly with F_2. For a slab having a larger value of B, θ_1 is initially less than unity but falls away less rapidly. Thus the θ_1, F_2 characteristic for B(moderate) must cross over that for B(small). In general, the θ_1, F_2 characteristics cross each other to form an envelope determined by the condition

$$\partial\theta_1/\partial u_1 = 0. \tag{13.21}$$

This leads to a time F_0 (written as $[F_0]_e$, subscript e for envelope) when the first eigenfunction component θ_1 touches the envelope of θ_1 values, given by

$$[F_0]_e = \frac{1}{2u_1{}^2}\left(\frac{B^2 + B - u_1{}^2}{B^2 + B + u_1{}^2}\right). \tag{13.22}$$

Values are given in Table 13.3. The value of θ_1 at this time is

$$\theta_{1e} = \frac{2B}{B^2 + B + u_1{}^2}\exp(-u_1{}^2[F_0]_e). \tag{13.23}$$

Values of θ_{1e} in relation to θ_i values are shown in Table 13.2. (A value of B was selected and the value of $[F_0]_e$ found from (13.22), from which $F_2 = F_0 B^2$. With this value of F_2, θ_i was found from (13.20). B values were chosen so as to give values of $\theta_i = 0.9, 0.8$, etc.; θ_{1e} follows from (13.23), hence $\theta_i - \theta_{1e}$.)

It will be seen that over the full range of dimensionless slab thicknesses, the envelope of the first eigenfunction characteristics lies only very little below the semi-infinite value θ_i. Since the surface temperature θ_f of the finite-thickness slab for some specific value of B must initially follow θ_i and latterly θ_1, this demonstrates there can be only a small transitional range of θ_f between these asymptotic conditions. Figure 13.4 shows the θ_1 component for $B = 2$. Replacing θ_f by ϕ_f, in Figure 13.2 shows the same feature. Time is presented as $\arctan(\sqrt{F_2}) = \arctan(h\sqrt{(t/\lambda\rho c_p)})$. This transforms an initial infinitely fast rate of fall to a finite rate proportional to h; it also compresses the infinite timescale into a span of 0 to $\pi/2$.

The corresponding time $[F_0]_e$ could be taken as a crude measure of the transition from early to late cooling. Following the arguments of the last section, a more conservative estimate of time is that when $\theta_i - (\theta_1 + \theta_2)$ has its minimum value. Values are listed as $[F_0]_d$ in Table 13.3, where d denotes departure from early cooling at the surface of a finite-thickness slab.

Again, following the argument of Section 13.1, an estimate can be made of the time $[F_0]_f$ after which the surface temperature θ_f is virtually determined as θ_1 alone. It is given by the condition that

$$\partial^2\theta_2/\partial t^2 = \partial^2\theta_1/\partial t^2 \tag{13.24}$$

and leads to a value

$$[F_0]_f = \frac{1}{u_2{}^2 - u_1{}^2}\ln\left(\frac{u_2{}^4(B^2 + B + u_1{}^2)}{u_1{}^4(B^2 + B + u_2{}^2)}\right); \tag{13.25}$$

see Table 13.3. Although the Biot numbers cover a fractional range of 5000, the times describing the stages reached in cooling, expressed as a Fourier number, vary over quite

Table 13.2 Deviation from θ_i of the envelope θ_{1e} of θ_i values

θ_i	1.0	0.9	0.8	0.7	0.6	0.5	0.4	0.3	0.2	0.1
$\theta_i - \theta_{1e}$	0.0000	0.0022	0.0043	0.0060	0.0073	0.0081	0.0083	0.0075	0.0058	0.0032

Source: Reprinted from *Journal of Heat Transfer*, vol. 121, M.G. Davies, The time delay for a perceptible thermal disturbance in a slab, with permission from ASME.

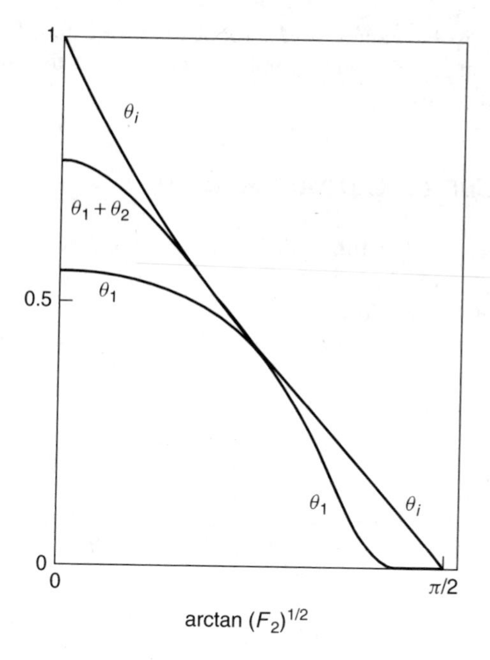

Figure 13.4 Cooling curve for a double-sided slab at whose surfaces the fluid temperature is zero from $t = 0$. The vertical axis is the non-dimensionalised temperature and the horizontal axis the non-dimensionalised time $F_2 = h^2 t/\lambda\rho c_p$ plotted as arctan $\sqrt{F_2}$. The actual surface temperature θ_f initially has the value θ_i of an infinitely thick slab and latterly the value θ_1 of the first eigenvalue component

Table 13.3 Values of $F_0 = \lambda t/\rho c_p X^2$ at three stages of cooling at the surface of a slab*

B	$[F_0]_d$	$[F_0]_e$	$[F_0]_f$
0.01	0.155	0.333	1.002
0.1	0.155	0.330	0.541
1	0.154	0.311	0.356
10	0.140	0.236	0.259
50	0.129	0.210	0.231
∞	0.124	0.203	0.223

*Subscript definitions:

d is the approximate time of departure from early cooling at the slab surface given by the minimum of $\theta_i - (\theta_1 + \theta_2)$;

e is the time at which the θ_1 characteristic for B touches the envelope of all θ_1 characteristics;

f is the time after which cooling is dominated by the first eigenfunction alone.

Source: Reprinted from *Journal of Heat Transfer*, vol. 121, M.G. Davies, The time delay for a perceptible thermal disturbance in a slab, with permission from ASME.

a small range. Note that the values of F_0 for $B = \infty$ are those already found for the surface-cooled slab: a large value of B implies that in effect the surfaces themselves are brought to near zero at $t = 0$.

13.2.4 Cooling Curves: Exposed Surface

Table 13.4 gives values for the time taken for the temperature of the slab, initially at unit temperature to cool to θ_t. It presents information in a tabular form similar to a chart of Bachmann (1938), who plotted θ_f against $\log(h^2t/\lambda\rho c_p)$. (This has the effect of putting the time origin at $-\infty$). Bachmann's chart shows very clearly the departure of the characteristics for $B = 0.1, 0.5$, etc., from the envelope $\theta_i = \exp(F_2)\mathrm{erfc}(\sqrt{F_2})$.

The foregoing theory presents a basis for the time when a B characteristic departs from the envelope and the later time after which cooling is estimated exclusively by the first

Table 13.4 Values of time $F_2 = h^2t/\lambda\rho c_p$ for values of temperature θ_t at the exposed surface of a slab of thickness $B = hX/\lambda^*$. (A selection from Table 4 of Davies (1978))

θ_t	*B*						
	0.1	0.2	0.5	1.0	2	5	10
1.0	0.0	0.0	0.0	0.0	0.0	0.0	0.0
0.95	2.12_{10}^{-3}	2.13_{10}^{-3}	2.13_{10}^{-3}	2.13_{10}^{-3}	2.13_{10}^{-3}	2.13_{10}^{-3}	2.13_{10}^{-3}
0.9	7.46_{10}^{-3}	9.22_{10}^{-3}	9.27_{10}^{-3}	9.27_{10}^{-3}	9.27_{10}^{-3}	9.27_{10}^{-3}	9.27_{10}^{-3}
0.85	1.34_{10}^{-2}	2.06_{10}^{-2}	2.28_{10}^{-2}	2.13_{10}^{-3}	2.13_{10}^{-3}	2.13_{10}^{-3}	2.13_{10}^{-3}
0.8	1.96_{10}^{-2}	3.35_{10}^{-2}	4.46_{10}^{-2}	4.47_{10}^{-2}	4.47_{10}^{-2}	4.47_{10}^{-2}	4.47_{10}^{-2}
0.75	2.63_{10}^{-2}	4.73_{10}^{-2}	7.57_{10}^{-2}	7.73_{10}^{-2}	7.73_{10}^{-2}	7.73_{10}^{-2}	7.73_{10}^{-2}
0.7	3.34_{10}^{-2}	6.20_{10}^{-2}	0.114	0.124	0.124	0.124	0.124
0.65	4.11_{10}^{-2}	7.79_{10}^{-2}	0.157	0.190	0.190	0.190	0.190
0.6	4.94_{10}^{-2}	9.49_{10}^{-2}	0.204	0.277	0.282	0.282	0.282
0.55	5.84_{10}^{-2}	0.114	0.255	0.386	0.411	0.411	0.411
0.5	6.82_{10}^{-2}	0.134	0.311	0.512	0.592	0.592	0.592
0.45	7.91_{10}^{-2}	0.156	0.372	0.654	0.844	0.849	0.849
0.4	9.13_{10}^{-2}	0.182	0.441	0.813	1.19	1.23	1.23
0.35	0.105	0.210	0.520	0.993	1.62	1.79	1.79
0.3	0.121	0.243	0.610	1.20	2.15	2.69	2.69
0.25	0.140	0.282	0.717	1.45	2.77	4.20	4.21
0.2	0.163	0.330	0.848	1.75	3.54	6.78	7.04
0.15	0.193	0.391	1.02	2.14	4.54	10.8	13.2
0.1	0.235	0.478	1.25	2.69	5.93	16.6	28.8
0.05	0.306	0.626	1.66	3.62	8.33	26.7	62.3
1/e	9.99_{10}^{-2}	0.199	0.491	0.926	1.46	1.56	1.56

*Early cooling pertains for values of F_2 above the upper line and late or exponential cooling below the lower line.

term θ_1 of the eigenfunction series (13.18), and thus is exponential. Using values of $[F_0]_d$ and $[F_0]_f$ (Table 13.3), lines are included indicating the duration of early cooling and the establishment of exponential cooling. Notice that most of the cooling falls into one or other of these stages; the transitional stage is comparatively short.

In view of the general importance of Groeber's solution, several authors besides Groeber and Bachmann have presented results for surface and midplane temperatures. They include Schack (1930), reproduced in Fishenden and Saunders (1950), Ede (1945), Heisler (1947) and others, listed by Carslaw and Jaeger (1959: 134).

13.2.5 Surface Response Time

If a simple system such as the r-c circuit of Figure 10.2a is initially at temperature T_I and then allowed to discharge, its subsequent temperature is

$$\theta_t = T_t/T_I = \exp(-t/rc) = \exp(-t/z), \tag{13.26}$$

where z is the decay time, rc. The temperature falls to $1/e = 0.368$ of its initial value in the time z. The time to fall to 1/e, or to accomplish $1 - 1/e = 0.632$ of the total change from one steady state to another steady state, provides a useful measure of the speed of response in more complicated systems.

In the current problem, when B is small, <0.1 say, then $C_1 \approx 1$, C_2, etc., are near zero, $z_1 \approx \rho c_p X/h$, so the surface temperature is

$$\theta_f = 1\exp(-ht/\rho c_p X) \tag{13.27}$$

and is equal to 1/e when $t = \rho c_p X/h$ or $F_2 = h^2 t/\lambda \rho c_p = hX/\lambda = B$.

The value of F_2 when $\theta_f = 1/e$ is given in the final row of Table 13.4. Notice that $F_2 \approx B$ up to $B = 0.5$. With further increase in B, the condition $\theta_f = 1/e$ falls in the transitional stage and finally in the early stage of cooling, after which $F_2 = 1.562$, independent of B.

13.2.6 Cooling at the Midplane

Cooling at the midplane (or adiabatic surface) is shown in Figure 13.5 and four associated times can be specified. $\theta(0, t)$ is given by

$$\theta(0, t) = \sum C_j \exp(-u_j{}^2 F_0), \tag{13.28a}$$

where

$$C_j = \frac{2B}{(u_j{}^2 + B^2 + B)\cos u_j}; \tag{13.28b}$$

see equations (13.10) and (13.15).

- For some while no cooling takes place. A search has to be made for the time $[F_0]_a$ when $\partial^2\theta(0, t)/\partial t^2$ has its maximum value. Values are listed in Table 13.5.

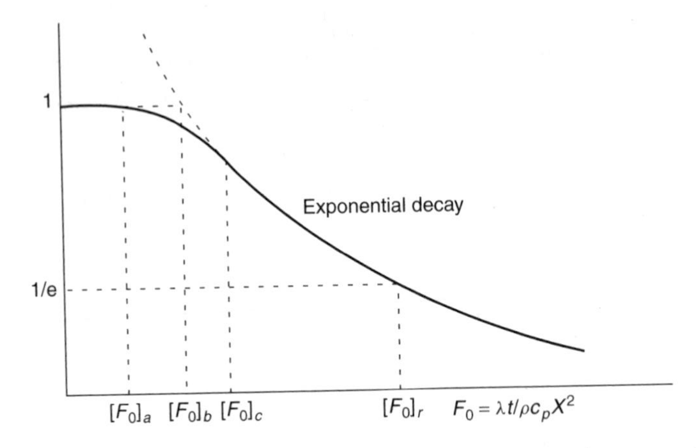

Figure 13.5 Dimensionless temperature against Fourier number showing stages in the cooling process at midplane. Key to subscripts: a is the end of the undisturbed period, b is the delay time of the first eigenfunction, c is the onset of exponential cooling, r is the response time

Table 13.5 Times associated with cooling at the slab midplane ($F_0 = \lambda t/\rho c_p X^2$)*

B	$[F_0]_a$	$[F_0]_b$	$[F_0]_c$	$[F_0]_r$	$[F_0]_{S/L}$
0.001	0.092	0.167	1.002	1000.5	1000.5
0.01	0.092	0.167	0.769	100.5	100.5
0.1	0.091	0.165	0.536	10.5005	10.5000
1.0	0.085	0.152	0.321	1.5031	1.5000
10	0.070	0.114	0.197	0.6038	0.6000
100	0.062	0.100	0.170	0.5132	0.5100
1000	0.061	0.098	0.167	0.5042	0.5010

*$[F_0]_a$ is the approximate start of cooling; $[F_0]_b$ is the time delay of the first eigenfunction; $[F_0]_c$ is the approximate onset of exponential cooling; $[F_0]_r$ is the time for temperature to fall from unity to the response time value 1/e; $[F_0]_{S/L}$ is the storage/loss approximation to $[F_0]_r$.

- At $x = 0$ and $t = 0$ the first eigenfunction $\theta(0, t, 1)$ has a value greater than unity. It can be written as

$$\theta(0, t, 1) = 1.0 \exp(-{u_1}^2(F_0 - [F_0]_b)), \tag{13.29}$$

where

$$[F_0]_b = \ln(C_1)/{u_1}^2; \tag{13.30}$$

that is, with an amplitude of unity but with a time delay $[F_0]_b$. A similar time delay features in Krischer's model (Section 14.1).

- Using an argument similar to the one above, $\theta(0, t, 2)$ becomes negligible compared to $\theta(0, t, 2)$ after a time given by

$$[F_0]_c = \frac{1}{{u_2}^2 - {u_1}^2} \ln\left(\frac{{u_2}^4(B^2 + B + {u_1}^2)\cos u_1}{{u_1}^4(B^2 + B + {u_2}^2)\cos u_2}\right) \tag{13.31}$$

and values are listed in Table 13.5.

- At the *surface* of the slab, the temperature falls to 1/e of its initial value during the phase of late, exponential cooling when B is small and during the phase of early cooling when B is large. At *midplane*, however, a value of $\theta(0, t) = 1/\mathrm{e}$ is always reached during exponential cooling. The time taken is t_r, and when non-dimensionalised as $[Fo]_r = \lambda t_r/\rho c_p X^2$, it is given as

$$[Fo]_r = (1 + \ln C_1)/u_1^2; \tag{13.32}$$

values are listed in Table 13.5.

Table 13.5 also lists values of the dimensionless 'storage/loss' time, S/L:

$$[Fo]_{S/L} = \frac{\lambda(S/L)}{\rho c_p X^2} = \frac{1}{B} + \frac{1}{2}. \tag{13.33}$$

This value provides a surprisingly close approximation to $[Fo]_r$ for all Biot numbers B, as shown by the final two columns of Table 13.5. More simply, $S/L \approx t_r$. To provide a rationale for S/L, suppose that a steady heat flow $2L$ is supplied to the centre plane at T_0 of the model in Figure 13.1b. The flow in the one direction is

$$L = (T_0 - T_S)\lambda/X = T_S/r'. \tag{13.34a}$$

The heat stored in the slab is

$$S = \tfrac{1}{2}(T_0 + T_S)\rho c_p X. \tag{13.34b}$$

Equation (13.33) is found by eliminating T_0 and T_S and noting that $B = X/r'\lambda$. Although there is no apparent physical reason or indeed any algebraic reason at high B numbers, it so happens that the right-hand sides of (13.32) and (13.33) are virtually identical when B is small and differ little when B is large. Noting (13.11) and summarising these various relations, we obtain

$$\sum \frac{1}{u_j^2} = \frac{\lambda(S/L)}{\rho c_p X^2} = \frac{1}{B} + \frac{1}{2} \approx \left(\frac{\lambda t_r}{\rho c_p X^2} = \frac{1}{u_1^2}\left(1 + \ln \frac{2B}{(u_1^2 + B^2 + B)\cos u_1}\right)\right). \tag{13.35}$$

During the initial stage of cooling, the cooling rate is largest at the surface and during the final stage it is largest at the midplane.

13.2.7 Discussion

Attention should be drawn to the three possible ways of non-dimensionalising time. Each can be interpreted as the ratio of two similar quantities. The Fourier number is

$$Fo = \frac{\lambda/X}{\rho c_p X/t} = \frac{\text{ability of the slab to conduct heat from surface to surface}}{\text{ability of the slab to store heat}}; \tag{13.36}$$

it is the appropriate form to describe the cooling process within the slab, as illustrated in Tables 13.3 and 13.5. The quantity

$$\sqrt{F_2} = \frac{h}{\sqrt{\lambda \rho c_p / t}} = \frac{\text{ability of the boundary layer to transmit heat}}{\text{ability of the slab to accept heat in unsteady conditions}} \tag{13.37}$$

is the form of time for use in cooling curves (Table 13.4).

F_0 and F_2 involve the thermal constants as $\lambda/\rho c_p$ and $\lambda \rho c_p$, respectively. (It was noted earlier that since the conductivities of building materials vary roughly in proportion to their densities, the first of these combinations varies little from material to material, whereas the second shows a large variation.)

The third non-dimensional form of time is

$$F_1 = \frac{h}{\rho c_p X / t} = \frac{\text{ability of the slab to lose heat}}{\text{ability of the slab to store heat}}; \tag{13.38}$$

it is the appropriate form when the slab has a high conductivity or is thin, so that B is small and the slab is nearly isothermal. If λ is infinite, neither F_0 nor F_2 can be used. The response time at the surface of a thin slab is given by the value $F_1 = 1$ and, rather surprisingly, it holds approximately up to $B \sim 0.5$.

The Biot number B can be similarly interpreted:

$$B = \frac{h}{\lambda / X} = \frac{\text{ability of the boundary layer to transmit heat}}{\text{ability of the slab to transmit heat from surface to surface}}. \tag{13.39}$$

When B is small, most cooling takes place in the late phase, and vice versa. Table 13.4 gives surface cooling curves for Biot numbers where significant parts of the cooling fall into both phases.

Conventionally, h denotes a single physical process – convection, or radiation in Carslaw and Jaeger (1959) but it can be generalised to comprise several physical processes together with step changes at the surface itself or at any of the intermediate nodes. The author has used it in this way to estimate the response to the switching on and off of the classroom lights in a passive solar-heated school, and in this application h was made up of internal convective and radiative exchanges, together with the effect of ventilation and an external mixed convection/radiation loss, five links in all, variously acting in series and parallel. Furthermore, a step in heat input rather than of temperature can be taken as the input at any of the temperature nodes between that of the slab surface – the real dependent variable – and ambient temperature T_e, which is assumed constant, such as at the node representing the room mean air temperature, where the convective component of the heat input acts. The matrix formulation needed to handle the thermal circuit is presented in Davies (1986 III: Section 3.12).

de Monte (2000) has recently presented an analysis of a system composed of a film resistance, two layers having thermal capacity and resistance, and a further resistance, at a known initial temperature and subjected to step changes in the temperature of the fluids.

13.3 JAEGER'S MODEL

Jaeger (1945) discusses a model which includes a pure capacity c_c, a core, as shown in Figure 13.1c. If the core is interpreted as the internal mass of a building and the film as the combined radiative and convective link between the internal mass and the outside wall, Jaeger's solutions provide exact values for the temperatures of the internal exposed surfaces. He presents four sets of solutions for the core temperature T_c and the temperature within the slab according to whether (i) the slab left surface is adiabatic or isothermal at zero, and (ii) whether the core is initially at a temperature T_I or is initially at zero and driven by a heat flow q. Figure 13.1c illustrates the isothermal/flow-driven case. In all cases the initial temperature of the slab is zero.

To determine the decay times, the respective transmission matrices are multiplied:

$$\begin{bmatrix} \cos u & (\sin u)/v \\ -(\sin u) \times v & \cos u \end{bmatrix} \begin{bmatrix} 1 & r' \\ 0 & 1 \end{bmatrix} \begin{bmatrix} 1 & 0 \\ -c'/z & 1 \end{bmatrix} \quad \text{or} \quad \begin{bmatrix} e_{11} & e_{12} \\ e_{21} & e_{22} \end{bmatrix}, \tag{13.40}$$

where $u^2 = cr/z$ and $v^2 = c/rz$.

The model in Figure 13.1c is isothermal at its left node and adiabatic at its right node, so the eigenvalues are determined by the condition that $e_{11} = 0$. Defining $B = r/r'$, as previously, and defining $B' = c/c_c$, the values of u_j are given by solutions of the equation

$$\cot u - Bu/(BB' - u^2) = 0. \tag{13.41}$$

The search for values of u_j is discussed in Section 17.9. (When the left node too is adiabatic, $e_{21} = 0$ and u_j values are determined by

$$\tan u + Bu/(BB' - u^2) = 0. \qquad (13.42))$$

The final temperature reached by the core is

$$T_f = q(r + r'). \tag{13.43}$$

At time t the core temperature is

$$\frac{T_c(t)}{T_f} = 1 - \frac{2B'^2B^3}{B+1} \sum_{j=1}^{\infty} \frac{\exp(-u_j^2 F_0)}{u_j^2 P_j}, \tag{13.44}$$

where F_0, the Fourier number, is

$$\lambda t/\rho c_p X^2 = t/rc \tag{13.45a}$$

and

$$P_j = u_j^4 + u_j^2 B(B + 1 - 2B') + B^2 B'(1 + B'). \tag{13.45b}$$

The temperature at some point x in the slab is

$$\frac{T(x,t)}{T_f} = \frac{B}{B+1}\frac{x}{X} + \frac{2B'B^2}{B+1} \sum_{j=1}^{\infty} \frac{(u_j^2 - B'B)\sin(u_j x/X)}{u_j^2 P_j \sin u_j} \exp(-u_j^2 F_0). \tag{13.46}$$

By taking the $t = 0$ values for the temperature of the slab surface and the core, and also the $t = 0$ value for the slab surface when the core itself is initially at T_I, we obtain the following relations:

$$2B\sum u_j^2/P_j = 1, \quad 2B^2B'\sum 1/P_j = 1, \quad [2B^3B'^2/(B+1)]\sum(1/u_j^2P_j) = 1, \tag{13.47}$$

summing j from 1 to infinity. These provide useful computational checks.

This solution can be readily adapted to the case where, up to the time $t = 0$, a steady heat flow imposed at the core maintains a uniform temperature gradient from the core at T_I through the film and slab to zero. At $t = 0$ the heat supply is cut off so that the core and slab cool. The solution can be constructed, first by subtracting the initial distribution from the above distribution and then obtaining the negative of the difference.

13.4 PRATT'S MODEL

Pratt and Ball (1963) and Pratt (1981) describe a more detailed model than Jaeger's, providing a film resistance between ambient and the slab, and, more importantly, offering the possibility of ventilating the space between the core, that is, internal partition walls, and the outer wall. The model is shown in Figure 13.1d. The authors intended the circuit to model an enclosure, but being a one-dimensional assembly, it can equally be regarded as a wall, adiabatic at its interior surface. Pratt and Ball in fact supposed that a heat flow acted at all times at the air temperature T_i so that for $t < 0$ there was a uniform gradient through the wall to an external temperature T_e; at $t = 0$ temperature T_e stepped to zero, resulting in an immediate step fall in T_i, followed by a gradual fall everywhere toward a new steady state. It is a little simpler to dispense with the heat source and suppose instead that up to $t = 0$, temperature everywhere has the value T_I and that at $t = 0$ the external temperature steps to zero.

Parameter r_e denotes the wall external film resistance, r_i its internal value and r_c the value between air and core. No account is taken of radiant exchange between core and wall. The ventilation conductance between a room and ambient due to infiltration is nsV, where n is the number of air changes per unit time, s is the volumetric specific heat of air (about 1200 J/m^{3}K) and V is the room volume. In the present study, if A is the wall or core area and w their separation, $V = Aw$, so the ventilation link as a transmittance is nsw, and as a resistance it is $r_v = 1/nsw$. $r(= X/\lambda)$ denotes the wall resistance itself and the four other resistances are non-dimensionalised as Biot numbers: $B_e = r/r_e$, $B_i = r/r_i$, $B_c = r/r_c$ and $B_v = r/r_v$. As in the Jaeger model, B' denotes the ratio c/c_c.

To evaluate the eigenvalues, we multiply the transmission matrices of the six elements in Figure 13.1d:

$$\begin{bmatrix} 1 & r_e \\ 0 & 1 \end{bmatrix}\begin{bmatrix} \cos u & (\sin u)/v \\ -(\sin u)\times v & \cos u \end{bmatrix}\begin{bmatrix} 1 & r_i \\ 0 & 1 \end{bmatrix}\begin{bmatrix} 1 & 0 \\ 1/r_v & 1 \end{bmatrix}\begin{bmatrix} 1 & r_c \\ 0 & 1 \end{bmatrix}\begin{bmatrix} 1 & 0 \\ -c'/z & 1 \end{bmatrix}, \tag{13.48}$$

where $u^2 = cr/z$ and $v^2 = c/rz$. Again the system is isothermal to the left and adiabatic to the right, so the overall element $e_{11} = 0$. It is convenient to define certain groups:

$$
\begin{aligned}
H_1 &= (B_i + B_c + B_v),\\
H_2 &= (B_v + B_c)(B_i + B_e) + B_i B_e,\\
H_3 &= B_i B_e(B_v + B_c) + B' B_c(B_v + B_i),\\
H_4 &= B' B_c B_v(B_i + B_e) + B' B_c B_i B_e,\\
H_5 &= B' B_v B_i B_e B_c.
\end{aligned}
\tag{13.49}
$$

The eigenvalues are given by solutions to the equation

$$\tan u = (H_2 u^3 - H_4 u)/(H_1 u^4 - H_3 u^2 + H_5). \tag{13.50}$$

The temperature of the air is given as

$$\frac{T_i(t)}{T_I} = \sum_{j=1}^{\infty} M_j \exp(-u_j^2 F_0) + M' \exp(-\beta F_0), \tag{13.51}$$

where

$$M_j = 2B_i B_e \frac{B' B_c - u_j^2}{E(u_j)}\left(1 + \frac{B_i B_v}{B_e H_1}\frac{(B' B_c - u_j^2)(\cos u_j + B_e \sin u_j / u_j)}{\beta - u_j^2}\right), \tag{13.52a}$$

$$\beta = B' B_c(B_i + B_v)/H_1, \tag{13.52b}$$

$$F_0 = \lambda t/(\rho c_p X^2) = t/rc \tag{13.52c}$$

and

$$E(u_j) = [-H_1 u_j^4 + (2H_2 + H_3)u_j^2 - H_5]\cos u_j + [-(3H_1 + H_2)u_j^4 + (H_3 + H_4)u_j^2 + H_5]\sin u_j/u_j. \tag{13.52d}$$

The further term $M' \exp(-\beta F_0)$ requires comment. $\exp(-\beta F_0)$ describes the fall in temperature associated with the core capacity c_c through its linked resistances r_c, r_v and r_i, as though the remaining elements r, c and r_e were zero. This is clearly inappropriate. The response of the complete assembly cannot be constructed by including a solution pertaining to part of it. The eigenvalue u_j implies a temperature profile which is a component of the real temperature profile and is defined everywhere in the system; the temperature at T_i is to be determined completely by the sum of terms of type $M_j \exp(-u_j{}^2 F_0)$ alone. Furthermore, equation 19 of Pratt and Ball (1963) appears to contain some errors, noted by Davies and Bhattacharya (1984), and when these are corrected, the coefficient M' reduces identically to zero, as it should. The inclusion of this term appears to be a conceptual blunder.

Note the presence of the term $\beta - u_j^2$ in the denominator. β is fixed while u_j increases with j. There would be evident difficulties should some u_j value happen to equal β exactly, but the argument in Section 13.5 suggests that the u_j values adjust themselves so that this does not occur. (A denominator term of this kind is not peculiar to the current problem. Carslaw and Jaeger (1959: 104) refer to the region $-X < x < X$, initially at zero, whose surfaces are at a temperature $T(1 - \exp(-\alpha t))$ for $t > 0$. The expression for

$T(x, t)$ includes the factor $F = [4\alpha X^2 - (\lambda/\rho c_p)\pi^2(2j+1)^2]$ in the denominator, and they remark without further comment that this must not be zero. However, α here is an *externally imposed* variable, independent of the eigenvalues of the slab concerned and it appears that it could be chosen to make F zero.)

As an example of this model, consider the space enclosed between a double-thickness brick wall and a single-thickness wall taken to be an internal partition wall and adiabatic at its midplane. The space is ventilated. Up to $t = 0$ the temperature is everywhere at unit value. At $t = 0$ the external temperature steps to zero. We wish to find the temperature of the internal air.

The wall is assumed to have X, λ, ρ and c_p values of 0.2 m, 0.84 W/m K, 1700 kg/m^3 and 800 J/kg K, respectively, so $r = X/\lambda = 0.2381$ m^{2}K/W and $c = \rho c_p X =$ 272 000 J/m^{2}K. Further assumed values are $r_e = 0.05$ and $r_i = r_c = 0.4$ m^{2}K/W. These are values for natural convection. Radiant exchange is excluded to avoid complication. The core will be assumed to have a capacity corresponding to $\frac{1}{4}$ that of the outer wall, i.e. $c_c = 68\,000$ J/m^{2}K. The separation w is 2 m and the air change rate is 5 air changes per hour, so the ventilation resistance $r_v = 1/nsw = 1/((5/3600) \times 1200 \times 2.0) =$ 0.3 m^{2}K/W. The corresponding dimensionless parameters are $B_e = 4.7619$, $B_i = B_c =$ 0.5952, $B_v = 0.7936$, $B' = 4.0$, $\beta = 1.6667$.

The first five decay times are $z_1 = 12.48$, $z_2 = 7.12$, $z_3 = 1.07$, $z_4 = 0.37$ and $z_5 =$ 0.18 h. The corresponding fractional temperature amplitudes, $A_j = T_{ij}(0)/T_I$ are 0.6130, 0.0348, −0.0849, 0.0473, −0.0235. They diminish in magnitude and alternate in sign and their sum converges on the $t = 0$ value for T_i/T_I which is equal to $(1/r_i + 1/r_c)/(1/r_i + 1/r_c + 1/r_v) = 0.6$. Three- figure accuracy was achieved in this example when $j = 38$. The variation of air temperature is given as

$$T_i(t)/T_I = \sum [T_{ij}(0)/T_I] \exp(-t/z_j) \tag{13.53}$$

and has the following values.

Time (h)	0	1	2	3	6	12	24
$T_i(t)/T_I$	0.600	0.566	0.536	0.500	0.394	0.241	0.091

The immediate response is due to ventilation; if the ventilation rate were zero, there would be a delay of some hours before any perceptible change in T_i would be noted.

The model ignores the thermal capacity of the air. In this example it is 2400 J/m^{2}K, much less than for either wall. It could be modelled by a capacity in parallel to the ventilation resistance. If the wall temperatures can be assumed to have changed negligibly during the cooling of the air immediately after $t = 0$, T_i would change from its $t < 0$ value of unity to $(1/r_i + 1/r_c)/(1/r_i + 1/r_c + 1/r_v)$ exponentially as $\exp(-nt)$, which amounts to a decay time of $1/n$ or 0.2 h in this example.

In order to draw attention to an unexpected feature of the model, consider the situation where $r_e = r_i = r_c = 0.1$ m^{2}K/W, $r_v = 0.001$ m^{2}K/W, $c = 200\,000$ and $c_c = 2000$ J/m^{2}K. This corresponds to a normal massive outer wall with normal convection coefficients but a very high ventilation rate and a very small core capacity. $B_e = B_i = B_c = 2$, $B_v = 200$ and $B' = 100$. If the ventilation rate were infinite, the Pratt model would become two independent models: the Groeber model for the outer wall and a lumped $r - c$ model for

Table 13.6 The Pratt model with a high ventilation rate and low core capacity

j	1	2	3	4	5	6	7	8	9	10
z_j (s)	13 560	2434	853	414	242	**202**	157	110	81	63
A_j	0.0073	−0.0000	0.0014	−0.0000	0.0005	**0.0097**	−0.0000	0.0002	−0.0000	0.0001
$\sum A_j$	0.0073	0.0073	0.0087	0.0087	0.0092	**0.0189**	0.0189	0.0191	0.0191	0.0196

the core. With the present assumed values, we may expect a series of decay times for the wall, closely equal to those of a Groeber solution, together with a single value due to the core, closely equal to $(0.1 + 0.001) \times 2000 = 202$ s. Values are given in Table 13.6.

The first five decay times and z_7 and further terms are closely equal to Groeber values. They are not exactly equal since, as far as the wall is concerned, its outer film is 0.100 and its inner film is effectively 0.101 m^{2}K/W. Its z_j values are virtually independent of c_c when $c_c \ll c$. However, z_6 is closely equal to the $r - c_c$ decay time; in effect, the core has introduced its own decay time. The amplitudes A_j are listed. Apart from A_6, they show declining values and alternating signs; A_6 however proves to be *larger* than A_1 The consequence of this for the convergence to $(1/r_i + 1/r_c)/(1/r_i + 1/r_c + 1/r_v)$, here 0.0196, will be seen in the values of $\sum A_j$. Only after A_6 has been included does convergence take place. The first five values of $\beta - u_j{}^2$ are positive; for $j \geq 6$ it is negative. Choosing a smaller value of c_c leads to a similar situation at a higher value of j.

This detail is needed to estimate the value of $T_i(t)$. After T_i's instantaneous fall to $(1/r_i + 1/r_c)/(1/r_i + 1/r_c + 1/r_v)T_I$, the temperature of the core must show a further rapid fall, since it has so little thermal capacity, with a corresponding effect on T_i. Hence the need for the $A_6 \exp(-t/z_6)$ contribution – large but short-lived. Once the core temperature is practically equal to T_i, further cooling is virtually determined by wall characteristics alone.

This example is not of immediate practical interest. The ventilation rate chosen is such as to bring T_i down immediately to 2% of its $t < 0$ value; any further time variation is clearly unimportant. Nevertheless, it does show the importance of determining all the z_j values; by suitable choice of c_c, its decay time could be made fairly close to any one of the wall z_j series and so might be missed. A computational root-finding procedure is needed which identifies all solutions to $e_{11} = 0$ or $e_{12} = 0$.

Zmeureanu *et al.* (1987) and Pasqualetto *et al.* (1998) give examples of using Pratt's model. Choudhury and Warsi (1964) developed it further. Instead of the step in ambient temperature which Pratt assumed, they supposed the enclosure to be driven by a unit pulse in ambient temperature. They evaluated the indoor temperature for some hours following the pulse and called the temperature–time relations 'weighting functions'. They illustrated the temperature history in a small room enclosed by heavy masonry walls following a pulse. When the room internal capacity is neglected and the room is unventilated, the room air temperature reaches a low peak after about 4 h; ventilation causes the peak to occur a little earlier and enhances the cooling, as expected. When the capacity of internal walls is included, the peak temperature is lower.

In a companion paper (Warsi and Choudhury 1964) the authors extend the analysis to a room having an outer wall of three layers. The approach leads to some very bulky equations for the response. This, together with the lumping together of all five internal surfaces as a single pure capacity, rather than as individual surfaces with resistive/capacitative characteristics, and the absence of a way to drive the model at any internal node, prevent it from being a practical design technique. This was provided a few years later by Mitalas and Stephenson (1967) through response factors and transfer coefficients.

13.5 A ONE-DIMENSIONAL SYSTEM CANNOT HAVE TWO EQUAL DECAY TIMES

Table 13.6 shows a series of regular decay times associated with the outer massive wall, together with an 'anomalous' decay time, 202 s, whose value largely depends on the capacity of the inner core and its link with the room air. By suitable choice of core values, one might suppose that the anomalous value could coincide *exactly* with one of the regular series. The possibility can be investigated using the simpler model of Figure 13.6. Elements 1 to 5 represent the wall, v_6 represents a ventilation transmittance and elements 7 to 9 represent the internal core structure. (Element 9, not appearing in Pratt's model, is included here to make the usual assumption of an isothermal inner node; the inner node in Pratt's' model was adiabatic.) One might conjecture that if v_6 were very large (i.e. its resistance is very small), the wall would have two decay times largely independent of the core values, and vice versa, and that by suitable choice, the core decay time might coincide with one of the wall decay times. This does not happen.

The analysis takes the usual form. Transfer matrices for the nine elements are written down in order and multiplied. Taking T_0 and T_9 to be zero, the element e_{12} in the product matrix is set to zero, resulting in a cubic equation. The values of z follow.

It is convenient to choose nominal values: $r_1 = r_3 = 1\,\mathrm{m^2K/W}$, $c_2 = c_4 = 1\,\mathrm{J/m^2K}$, also $c_8 = 1$ and $r_9 = 1$. When $r_5 = 1$, v_6 is infinite (i.e. its resistance is zero) and r_7 is infinite; the c_2, c_4 network is not coupled to c_8. By inspection of the circuit, the c_2, c_4 network has decay times of $z = 1$ and $\frac{1}{3}$ s and c_8 has $z = 1$ s.

We now impose some loose coupling: $r_5 = 0.9\,\mathrm{m^2K/W}$, $v_6 = 10\,\mathrm{W/m^2K}$ and $r_7 = 10\,\mathrm{m^2K/W}$. The decay times have the values $z_1 = 1$ (exactly and independent of r_7), $z_2 = 0.90944$, $z_3 = 0.33328$. Thus $z_1 - z_2 = 0.09056$. By assuming a value of $1\,\mathrm{W/m^2}$ through r_9, the temperature distribution through the network can be found for each value of z_j. The temperatures of the capacities are noted in Table 13.7.

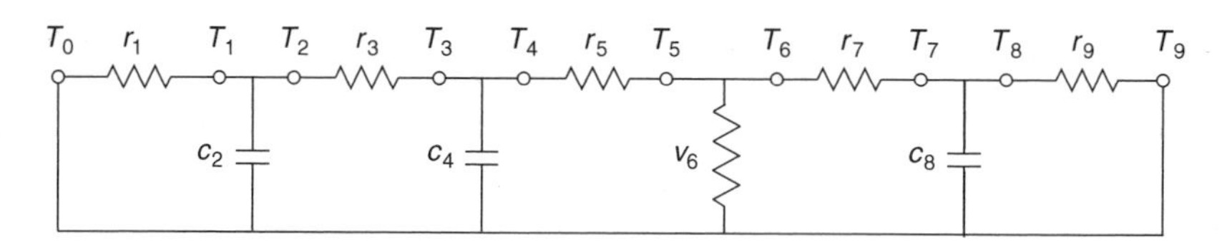

Figure 13.6 A nine-element model to examine the possibility of two exactly equal decay times

Table 13.7 Temperature distribution through the network of Figure 13.6 for its three decay times

j	z_j	T_2	T_4	T_8
1	1.00000	+10.000	+10.000	+1.000
2	0.90944	−0.053	−0.047	+1.000
3	0.33328	+191.75	−191.85	+1.000

With a change of c_8 from 1.0 to 1.09861 and r_9 from 1.0 to 0.99984, $z_1 - z_2$ is reduced from 0.09056 to 0.01336 s, which proves to be a minimum. It increases with any change in c_8 or r_9. The temperatures change very considerably but retain their signs.

	T_2	T_4	T_8
For z_1	positive,	positive,	positive.
For z_2	negative,	negative,	positive.
For z_3	positive,	negative,	positive.

If instead c_8 is chosen to be $\frac{1}{3}$ J/m^{2}K then $z_1 = 0.99957$, $z_2 = 0.30322$ and $z_3 = \frac{1}{3}$ exactly. The difference $z_3 - z_2$ of 0.03011 can be reduced through variation of c_8 and r_9 to a minimum value of 0.00257. In either case the values of T_2, T_4 and T_8 are as follows: for z_1, positive, positive, positive; for z_2, negative, positive, positive; for z_3, positive, negative, positive. The three-capacity linear system has three modes of decay whose temperature distributions are qualitatively distinct. It is not possible for one to 'develop' into another. While two z values may be close (and in the above circuit they may be made closer by choosing larger values of v_6 and r_7), they remain distinct.

(It may be added that with the above choice of elements and $c_8 \approx 1.0$, the z_1 values of T_2 and T_4 prove to be very sensitive to the actual choice of c_8. Although temperatures and heat flows associated with the nine elements follow routinely from the transmission matrices (leading back to a value of $T_0 = 0$ as must be the case), they cannot be readily explained in simple terms. The value of T_6 is not of interest. It could be removed if the star configuration of elements 5, 6 and 7 were replaced by its delta equivalent.)

With the values of temperature in the table as initial values, after $t = 0$, each capacity will heat or cool at the same fractional rate. (For z_3, c_2 and c_4 mainly exchange heat between each other.) If instead the $t = 0$ values are prescribed (e.g. they are all equal), the subsequent temperature at node $i (i = 2, 4, 8)$ is given as

$$T_i(t) = A_{i1} \exp(-t/z_1) + A_{i2} \exp(-t/z_2) + A_{i3} \exp(-t/z_3). \quad (13.54)$$

The values of A_{ij} can be found as shown in Chapter 10. T_6 might have been held at some arbitrary temperature up to $t = 0$. But if it is allowed to float after $t = 0$, it immediately changes to a value determined by the values of T_4 and T_8.

The circuit of Figure 13.6 is sufficient for the purpose in hand – to analyse the transient solution consequent on some imposed initial condition. The circuit may, however, be driven externally in some way: steady, step, ramp or sinusoidally varying temperatures or heat flows might be considered to act at T_0 or T_9; a heat source might act at T_6. In this case the circuit must make clear whether the ventilation exchange takes place with the space at T_0 or the space at T_9 and the element v_6 must be positioned accordingly.

13.6 DISCUSSION

This chapter has examined a sequence of transient conduction models which might have application in the field of building response. The models of the isolated slab and the slab with films are of general importance; they are not particularly complicated and may serve as benchmark solutions to test more sophisticated programs. As the source material makes clear however, the Pratt model is algebraically laborious and it models the response to a step in temperature, though in practice building temperatures do not suffer step changes. While it may serve for benchmarking purposes, numerically formulated rather than analytically formulated solutions prove to be more appropriate for buildings. The next chapter considers how the thermal characteristics of a wall and an enclosure may be analysed approximately using less rigorous models. Later chapters show how the response of a wall of arbitrary one-dimensional construction can be described using a short series of transfer coefficients and how to use them for setting up a room model.

14

Simple Models for Room Response

This chapter presents a selection of room thermal models which have been or are significant in describing the thermal behaviour of a room. It is generally recognised that buildings with thick, massive walls, such as medieval churches and castles, are less affected by short-term variations in ambient temperature than are thin-walled constructions. Extreme examples are a cave on the one hand with walls of effectively infinite thickness, and a tent on the other. The other principal factor determining response is the rate of heat loss, either by conduction through the walls or by ventilation; high losses lead to a rapid response to changing ambient conditions. The heat stored in the walls in steady-state conditions will be denoted by S and the rate of loss by L. The ratio of these quantities, S/L, has the units of time and may be anything from a few minutes to several days. A building with a high value of S/L has a more stable thermal environment than one with a low value. The earliest models to describe the thermal behaviour of buildings were based on this concept. They ultimately proved unsuccessful as will be shown below but they led to a series of models using just a few lumped capacities to represent storage.

14.1 WALL TIME CONSTANT MODELS

One of the earliest dynamic thermal models of a room is that of Krischer (Esser and Krischer 1930), based on two ideas:

- The ratio S/L is significant, where

$$\begin{aligned} &S \text{ is the heat stored in steady state conditions, units J/m}^2\text{ K or J/K or J,}\\ &L \text{ is the heat loss in steady state conditions, units W/m}^2\text{ K or W/K or W.}\end{aligned} \tag{14.1}$$

- After a delay time t_u, heat loss following some step change in excitation should be exponential.

Suppose that a wall interior is maintained at T_I up to $t = 0$, either by a temperature or a heat source, and its exterior temperature is zero. In the case of a lumped RC system

Building Heat Transfer Morris G. Davies
© 2004 John Wiley & Sons, Ltd ISBN: 0-470-84731-X

(Figure 14.1(i)), the heat stored $S = T_I C$ and the heat loss L is T_I/R, so

$$\frac{S}{L} = \frac{T_I C}{T_I/R} = CR, \tag{14.2}$$

which is the decay time z, and the delay time $t_u = 0$ (Section 10.2a).

Suppose instead that the wall is a simple slab of area A with no surface films. If the temperature at $x = 0$ is T_I (maintained by a heat source, say) the heat stored in steady conditions is $S = \frac{1}{2} T_I A \rho c_p X = \frac{1}{2} T_I C$ and the heat flow is $L = Q = T_I \lambda A/X = T_I/R$. Suppose now that at $t = 0$ the heat supply is cut off. The solution is given in Carslaw and Jaeger (1959: 97, equation 14):

$$\frac{T(x,t)}{T_I} = \frac{8}{\pi^2} \sum_{j=1}^{\infty} \frac{1}{(2j-1)^2} \cos\left(\frac{(2j-1)\pi x}{2X}\right) \exp\left(\frac{-\pi^2 (2j-1)^2 \lambda t}{4\rho c_p X^2}\right). \tag{14.3}$$

At $x = 0$ the terms have the same sign, indicating that the temperature there drops sharply after $t = 0$ when the heat flow is zero. The flow from the wall ($x = X$) is

$$\frac{Q(X,t)}{T_I} = \frac{4}{\pi R} \sum \frac{(-1)^{j-I}}{(2j-1)} \exp\left(\frac{-\pi^2 (2j-1)^2 \lambda t}{4\rho c_p X^2}\right). \tag{14.4}$$

The signs alternate, indicating that Q does not change perceptibly immediately after $t = 0$. After a sufficient lapse of time, all terms higher than $j = 1$ are negligible and the flow can be described by a model consisting of two lumped elements (Figure 14.1(iii)):

$$\frac{Q(X,t)}{T_I} = \frac{1}{R'} \exp\left(\frac{-(t - t_{delay})}{C'R'}\right), \tag{14.5}$$

where $R' = \pi R/4$, $C' = 16C/\pi^3$ and $t_{delay} = (4CR/\pi^2)\ln(4/\pi)$.

Krischer (1942: 9, equation 13) argued that a delay time t_u should be taken, where

$$t_u = \frac{S - S_{fr}}{L} \tag{14.6}$$

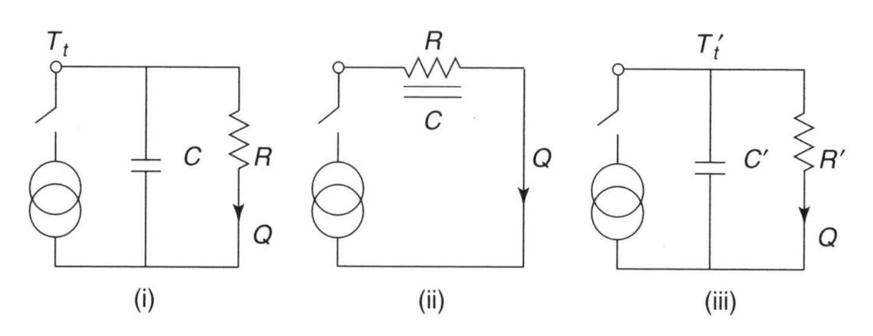

Figure 14.1 (i) The lumped RC circuit, (ii) the single slab and (iii) a lumped circuit equivalent to (ii) after a delay time t_u

and S_{fr} is the heat stored in the first eigenfunction at time t_u when the heat stored in higher terms is effectively zero. Thus, according to (14.3) with $j = 1$ only, we have

$$S_{fr} = \int_0^x AT(x,t)\rho c_p \, dx = \frac{16T_I \rho c_p X}{\pi^3} \exp\left(\frac{-\pi^2 \lambda t_u}{4\rho c_p X^2}\right). \quad (14.7)$$

Krischer wrote W_{st} for S, W_{fr} for S_{fr} and q_{st} for L. Substitution of S_{fr} together with the values $S = \frac{1}{2}TA_I\rho c_p X$ and $L = T_I\lambda/AX$ into (14.6) should give the value of t_u. Writing t_u in dimensionless form as

$$x = \pi^2 \lambda t_u / 4\rho c_p X^2, \quad (14.8)$$

equation (14.6) becomes

$$x = \pi^2/8 - (4/\pi)\exp(-x). \quad (14.9)$$

But this has no solution for any value of x, positive or negative, so (14.6) is defective to find the delay time. Krischer did not in fact cite (14.3). He took the first eigenfunction at $t = 0$ to be of form

$$T(x,t) = A \sin mx + B \cos mx \quad (14.10)$$

and determined the constants A and B from the initial temperature distribution. This led to a means of determining S_{fr}/S, written ψ, and the authors provided an extensive table of ψ values for more detailed walls; t_u was given as $(1-\psi)S/L$ and the heat loss as $L\exp(-(L/\psi S)(t - t_u))$.

The analysis was concerned with the heat loss from the wall exterior rather than the more important question of the temperature within the wall after switching a heat source off or on.

This approach is no longer of practical interest but it dominated German research for a long period. Of particular interest at the time was Bruckmayer's concept of a brick wall having the same storage potential as a real wall. Following Bruckmayer (1940), consider a two-layer wall (brick outside, cork inside) together with outside and inside films, forming the sequence of resistances r_1 to r_4. The capacities of the solid layers are c_2 and c_3. If T_0 is taken as zero, then in steady-state conditions, the heat flow is

$$L = T_4/(r_1 + r_2 + r_3 + r_4). \quad (14.11a)$$

The temperature at the interface between the brick and cork is

$$T_2 = T_4(r_1 + r_2)/(r_1 + r_2 + r_3 + r_4) \quad (14.11b)$$

and the heat stored in the brick and cork is

$$S = c_2 \times \tfrac{1}{2}(T_1 + T_2) + c_3 \times \tfrac{1}{2}(T_2 + T_3). \quad (14.11c)$$

The storage/loss ratio is

$$S/L = c_2\left(r_1 + \tfrac{1}{2}r_2\right) + c_3\left(r_1 + r_2 + \tfrac{1}{2}r_3\right). \quad (14.12)$$

This is independent of the inner film resistance r_4. Then according to Bruckmayer (1951a: equation 5), the thickness of the equivalent brick wall is

$$X_{brick\ wall} = \left(\left(\frac{2\lambda}{\rho c_p} \right)_{brick} \left(\frac{S}{L} \right)_{real\ wall} + \left(\frac{\lambda_{brick}}{h_e} \right)^2 \right)^{1/2} - \left(\frac{\lambda_{brick}}{h_e} \right); \tag{14.13}$$

h_e is the external film coefficient, equal to $1/r_1$. With a Biot number B defined as $h_e X_{brick\ wall}/\lambda_{brick}$, this equation can be rearranged as (14.14). Bruckmayer provided an extensive list of references.

The quantity S/L has the units of time but Esser and Krischer placed no interpretation upon it. Bruckmayer (1940) described it as the 'flow through' time, the time needed for the entire heat content to be lost if passing through at its initial rate and later identified it (Bruckmayer 1951b) as the time constant – the time needed for a temperature to fall to $1/e = 0.368$ of its initial value. He suggested that the S/L concept might be applied to the whole building. The ratio was sometimes called the Q/U ratio.

Pratt (1981: 129) too identified the S/L ratio (heat stored/heat transmitted) as a time constant. The result followed from a special case of his building model, described in Section 13.4. (It will be recalled that the model is that of a room, so simplified that all five internal elements are taken to have a lumped capacity linked to the room air node and where the air node is directly linked to ambient by a ventilation conductance. The outer wall is a distributed capacity/resistance element (Figure 13.1d)). When the internal capacity is zero and the ventilation rate is zero, the model reduces to that of a wall, adiabatic at its inside surface – the Groeber model in Section 13.2. Pratt's result followed from three equations. Firstly, he noted the S/L ratio as

$$\frac{S}{L} = \frac{\rho c_p X^2}{\lambda} \left(\frac{1}{B} + \frac{1}{2} \right). \tag{14.14}$$

Secondly, the inside temperature (less the effect of the constant heat input assumed) is

$$T_i(t)/T_I = \exp(-[u_1^2]\lambda t/\rho c_p X^2) \tag{14.15}$$

and since the internal storage and ventilation rate are both assumed to be zero, this is also the temperature of the inner surface of the outer wall, now adiabatic. The time constant t_r is the value of t when $T_i(t)/T_I$ has fallen to 1/e.

Thirdly, an approximate value $[u_1]$ for the first eigenvalue, u_1, is given as follows[1]:

$$\frac{1}{[u_1^2]} \approx \frac{1}{B} + \frac{1}{2}. \tag{14.16}$$

After rearrangement,

$$\frac{\lambda(S/L)}{\rho c_p X^2} = \frac{1}{B} + \frac{1}{2} = \frac{1}{[u_1^2]} = \frac{\lambda t_r}{\rho c_p X^2}. \tag{14.17}$$

[1]The computation following the original analysis was performed manually and the extraction of eigenvalues was very laborious. The author used an approximate method to evaluate u_1 and u_2, noting that $u_2 \gg u_1$.

If this analysis is correct, it follows that

$$t_r = S/L. \tag{14.18}$$

The conclusion is correct but the above reasoning is defective in two respects:

- It is the sum $\sum u_j^{-2}$ that is equal to $1/B + \frac{1}{2}$, not u_1^{-2} alone; see equation (13.11). If the approximation $\tan u \approx u + \frac{1}{3}u^3$ is taken, $1/u_1^2 = 1/B + \frac{1}{3}$.
- According to (14.15), the effect of a step in ambient temperature is felt immediately at the inner face of the outer wall; this is incorrect. With the assumed ventilation rate of zero, there must be some delay before the step is felt at this surface. To model the response, we must include at least a term in u_2. The author actually provided a form based on two approximately evaluated eigenvalues:

$$\frac{T_i(t)}{T_I} = \frac{[u_2^2]}{[u_2^2]-[u_1^2]}\exp(-[u_1^2]\lambda t/\rho c_p X^2) - \frac{[u_1^2]}{[u_2^2]-[u_1^2]}\exp(-[u_2^2]\lambda t/\rho c_p X^2). \tag{14.19}$$

There were no further terms, and the author neglected the second term.

Pratt, defining the thermal time constant t_r as the time for air temperature T_i (Figure 14.1d) to change by 1 – 1/e of its initial value following a fall in ambient temperature, provided a general approximate expression for t_r (Pratt 1981: 139, equation 2.8) based on his approximation $[u_1]$ to the first eigenvalue, u_1:

$$t_r = \frac{c\left(r_e + \frac{1}{2}r\right) + c_c(r_e + r + r_i + r_c) + c(r/r_v)\left(\frac{1}{2}r_e + \frac{1}{6}r + \frac{1}{2}r_i + r_e r_i/r\right) + c_c(r_c/r_v)(r_e + r + r_i)}{1 + (1/r_v)(r_e + r + r_i)}, \tag{14.20}$$

where $1/r_v$ is the ventilation rate. If the core capacity c_c is zero, T_i is simply a passive node located on the total resistance $r_i + r_v$ between the wall inner surface and ambient, so the room model reduces to a wall model. With this simplification,

$$t_r = \frac{c\left[r_e(r_i + r_v) + \frac{1}{2}r(r_e + r_i + r_v) + \frac{1}{6}r^2\right]}{r_e + r + r_i + r_v}. \tag{14.21}$$

To test the accuracy of this expression, suppose that $c = 2$, $r_e = 2$, $r = 4$, $r_i = r_v = 1$ in SI units (so the wall is symmetrical). Then $t_r = 3.7$ s.

With this choice of values for the elements, the model becomes the Groeber model with $1/r_e = 1/(r_i + r_c) = h$ and $r = 2X/\lambda$. As a passive node linked to the wall inner surface, T_i has the same time constant as the exposed surface itself and values of $h^2 t/\lambda\rho c_p$ for $T_{surface}/T_I = 1/e$ are listed in the bottom row of Table 13.4. In this example, $B = \frac{1}{2}r/r_e = 1$ and for this value,

$$F_2 = h^2 t_r/\lambda\rho c_p = 0.926. \tag{14.22}$$

Thus the time constant is

$$t_r = 0.926\frac{\lambda\rho c_p}{h^2} = 0.926\frac{r_e^2\left(\frac{1}{2}c\right)}{\frac{1}{2}r} = 1.8\ \text{s}. \tag{14.23}$$

This is half Pratt's approximate value. Equation (14.20) appears dubious as a way to estimate the response of a structure. When a step change T_I is imposed on ambient temperature, the temperatures of the solid surfaces remain momentarily unchanged, but air temperature T_i falls instantly to $T_i r_v/(r_i + r_v)$. The time constant refers to the time taken to reach 1/e of this latter value. Thus when $t = t_r$, T_i will be less than T_I/e.

Hoffman and Feldman (1981) used the storage/loss ratio to model a wall with a change imposed at time t_1 leading to a response at the later time t_2 proportional to $[1 - \exp(-(t_2 - t_1)/(S/L))]$. Values of S/L are listed with values between 0.05 h for a window and 93 h for a concrete wall with about the same resistance. Most of the walls studied were fairly thin, 70 mm, and for thin walls it may be satisfactory to represent them using a single lumped capacity, as implied by use of a single time constant. Davies (1983a) however examined how a wall should be discretised when it was intended to model the response to sinusoidal excitation of 24 h period. In a series of 29 wall and roof constructions it was found that in most cases walls required three capacities to express their storage and transmittance characteristics. Indeed, for the thickest walls with a cavity, the three-capacity representation was not very satisfactory. Swaid and Hoffman (1989) give a further example of the thermal time constant concept.

Raychaudhuri (1965: Table 2) gives two examples of the time taken for a wall inside temperature to fall to 1/e of the step change imposed on ambient temperature computed by summing eigenvalues and he compares them with Bruckmayer's S/L value. For a 229 mm brick wall they are 15.5 and 15.2 h, respectively, and for a 114 mm wall, they are 4.9 and 4.8 h. This is a further example of the situation described by equation (13.35).

14.2 ENCLOSURE RESPONSE TIME MODELS

14.2.1 Response Time by Analysis

We have considered the models of Raychaudhuri, Pratt and Ball, Pratt, and Hoffman and Feldman, in connection with wall behaviour, as is valid when the wall inside surface is in effect adiabatic or is not linked to any further storage.

All were envisaged however as *enclosure* models in which the thermal storage of internal walls was included. The consequences of Pratt's expression for t_r (14.20) can be examined readily for the case of zero ventilation ($1/r_v = 0$) by representing the external wall as a lumped capacity with zero resistance, that is $r = 0$. In this case

$$t_r = cr_e + c_c(r_e + (r_i + r_c)), \tag{14.24}$$

where $r_i + r_c$ forms a single resistance with T_i as a passive node located on it (Figure 14.2). T_1 and T_2 are respectively the wall temperature and the core temperature. Suppose that

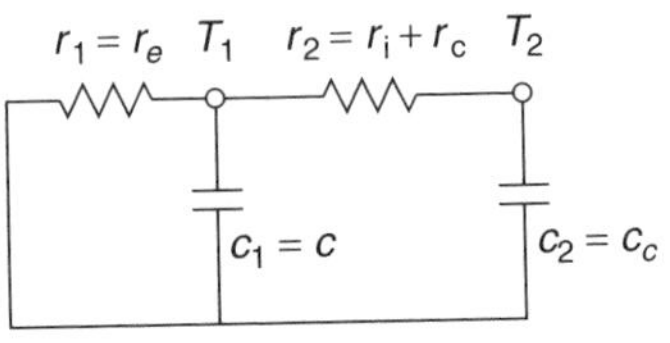

$r_1 = r_e$ T_1 $r_2 = r_i + r_c$ T_2

$c_1 = c$ $c_2 = c_c$

Figure 14.2 Pratt's model with zero ventilation rate and zero slab resistance

$r_1 = 1$, $c_1 = 2$, $r_2 = 3$ and $c_2 = 4$ SI units. These are nominal values but they are proportional to possible building values.

The circuit has decay times $z_1 = 16.54$ and $z_2 = 1.45$ s. Suppose that initially T_1 and T_2 have unit value and that after $t = 0$ ambient temperature is zero; T_1 and T_2 fall. Then

$$T_1 = \frac{z_1 - c_2 r_2}{z_1 - z_2}\exp(-t/z_1) + \frac{c_2 r_2 - z_2}{z_1 - z_2}\exp(-t/z_2)$$
$$= 0.301\exp(-t/16.54) + 0.699\exp(-t/1.45), \qquad (14.25a)$$

$$T_2 = \frac{z_1}{z_1 - z_2}\exp(-t/z_1) + \frac{-z_2}{z_1 - z_2}\exp(-t/z_2)$$
$$= 1.096\exp(-t/16.54) - 0.096\exp(-t/1.45). \qquad (14.25b)$$

T_1 falls rapidly and has the value $1/e = 0.368$ when $t = 2.66$ s. T_2 falls slowly and is $1/e$ when $t = 18.07$ s. Since T_i is a node located between T_1 and T_2, its response time t_r can have any value between these values, according to the relative sizes of r_i and r_c.

According to Pratt's expression, $t_r = 18.0$ s, which is virtually equal to the 18.07 s just noted when the core temperature $T_2 = 1/e$. The expression again somewhat overestimates the time constant.

The right-hand side of (14.24) is the S/L value for the circuit. This is another example where $S/L(= 18.0\,\text{s})$ is a close approximation to the response time at an adiabatic node (18.07 s). The model is intermediate between the simple *r*-*c* circuit of Figure 10.2a and the Groeber model of Figure 14.1b. It may be noted that the right-hand side of (14.24) is also the sum of the decay times, $z_1 + z_2$.

Raychaudhuri applied his method (using the values of u_1 and u_2 for each wall) to find the time constant of several test enclosures and noted the decrease in its value with ventilation rate.

The enclosure storage loss time t_{SL} is easy to evaluate; one of its components is the steady-state heat loss $L = \sum AU$, which is routinely found. The enclosure response time or time constant t_r is easy to interpret. To test the belief, once strongly held, that t_{SL} should provide an estimate of t_r, the author (Davies 1984) undertook a study using a number of simple models for which exact solutions were available in Carslaw and Jaeger (1959). Manipulation is needed in some cases to adapt those solutions which assume an *initial* state of zero temperature to the form for a *final* state of zero temperature. They are noted in Table 14.1. Here r_e denotes the film resistance, and r and c the resistance and capacity of the slab. In case 5, c_c is the capacity of the core. $B = r/r_e$ and $B^* = c/c_c$. The left-hand column of thermal circuits indicates how the storage loss time t_{SL} is found. It is immaterial whether the source is supposed to be a temperature or a heat source; for convenience we may suppose that the temperature generated at the field point is unity. Time t_{SL} depends of course on the wall boundary conditions but it is simply a steady-state concept and may be used in connection with any dynamic situation for the wall concerned.

The central column shows cases where the initial condition is a constant temperature everywhere, imposed by a temperature source. At $t = 0$ the temperature becomes zero. A temperature source has zero resistance. The right-hand column illustrates cases where there is initially a uniform temperature gradient, positive or negative, imposed by a heat source which has an infinite resistance. At $t = 0$ the flow becomes zero.

Once the models are set up, the time t_r is the value of t when $\theta = 1/e$. Non-dimensionalised as $F_0 = \lambda t_r/\rho c_p X^2$, it is a function of B or B^* only. Since the first decay time

Table 14.1 Models to test the hypothesis that the storage-loss time t_{SL} should provide an estimate of the response time t_r. $E_j = \exp(-t/z_j)$

Case 1: Isolated slab, adiabatic surface (p. 97, equations 8 and 14)

$$t_{SL} = \frac{\frac{1}{2}c}{1/r} = \frac{\rho c_p X^2}{\lambda}\frac{1}{2}$$

$$\tan u = \infty$$

$$u_1 = \pi/2,\ u_2 = 3\pi/2,\ u_3 = 5\pi/2$$

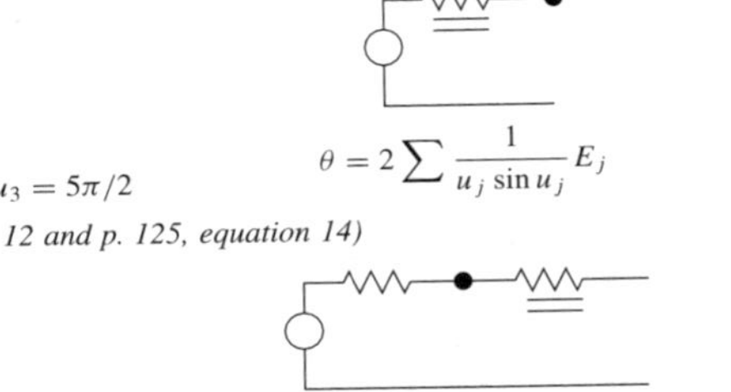

$$\theta = 2\sum \frac{1}{u_j \sin u_j} E_j$$

$$\theta = 2\sum \frac{1}{u_j^2} E_j$$

Case 2: Film and slab, right node adiabatic, exposed surface (p. 122 equation 12 and p. 125, equation 14)

$$t_{SL} = \frac{c}{1/r_e} = \frac{\rho c_p X^2}{\lambda}\frac{1}{B}$$

$$\tan u = \frac{B}{u}$$

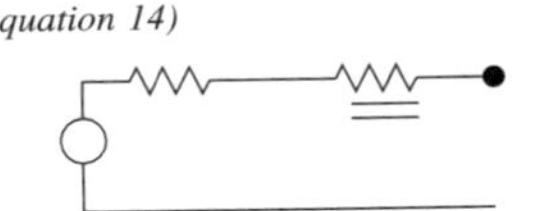

$$\theta = 2\sum \frac{B}{(B + B^2 + u_j^2)} E_j$$

$$\theta = 2\sum \frac{B}{(1 + B)(B + B^2 + u_j^2)\cos u_j} E_j$$

Case 3: Film and slab, right node adiabatic, adiabatic surface (p. 122 equation 12 and p. 125, equation 14)

$$t_{SL} = \frac{c\frac{1}{2}(r_e/(r_e + r) + 1)}{1/(r_e + r)} = \frac{\rho c_p X^2}{\lambda}\left[\frac{1}{B} + \frac{1}{2}\right]$$

$$\tan u = \frac{B}{u}$$

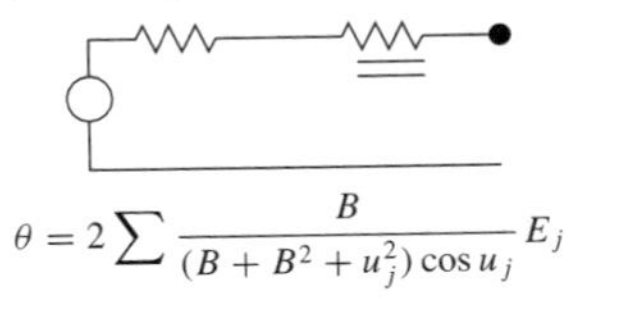

$$\theta = 2\sum \frac{B}{(B + B^2 + u_j^2)\cos u_j} E_j$$

$$\theta = 2\sum \frac{B}{(1 + B)(B + B^2 + u_j^2)\cos^2 u_j} E_j$$

Case 4: Film and slab, right node isothermal, exposed surface (p. 126, equation 17)

$$t_{SL} = \frac{\frac{1}{2}c}{r_e^{-1} + r^{-1}} = \frac{\rho c_p X^2}{\lambda}\frac{1}{2(B + 1)}$$

$$\tan u = \frac{-u}{B}$$

negative gradient

$$\theta = 2\sum \frac{B + 1}{B + B^2 + u_j^2} E_j$$

Case 5: Slab and core, right node adiabatic, interface (p. 129, equation 9 and p. 128, equation 7)

$$t_{SL} = \frac{\frac{1}{2}\rho c_p X + c_c}{1/r} = \frac{\rho c_p X^2}{\lambda}\left[\frac{1}{2} + \frac{1}{B^*}\right]$$

$$\tan u = \frac{B^*}{u}$$

$$\theta = 2\sum \frac{B^*}{(B^{*2} + B^{*2} + u_j^2)\cos u_j} E_j$$

$$\theta = 2\sum \frac{B^{*2}}{(B^* + B^{*2} + u_j^2)u_j^2} E_j$$

z_1 is of order t_{SL}, it is convenient to present t_r/t_{SL} as a function of z_1/t_{SL}, as shown in Figure 14.3. The base case is the simple *r-c* circuit of Figure 10.2a, for which $t_r = t_{SL} = z = cr$, so that $\log_{10}(t_r/t_{SL})$ and $\log_{10}(z/t_{SL})$ are zero. It is represented as the point (0, 0) in Figure 14.3.

For an isolated slab, isothermal at one surface and adiabatic at the other, initially at a uniform temperature, whose isothermal surface is at zero for $t > 0$, the subsequent temperature is given by (13.1). From this it readily follows that

$$z_1/t_{SL} = 8/\pi^2 = 0.81 \qquad \text{so } \log(z_1/t_{SL}) = -0.091, \tag{14.26a}$$

$$t_r/t_{SL} = (8/\pi^2)\ln(4e/\pi) = 1.006 \qquad \text{so } \log(t_r/t_{SL}) = 0.003. \tag{14.26b}$$

This too is denoted by a point in the diagram; t_r is very nearly equal to t_{SL}.

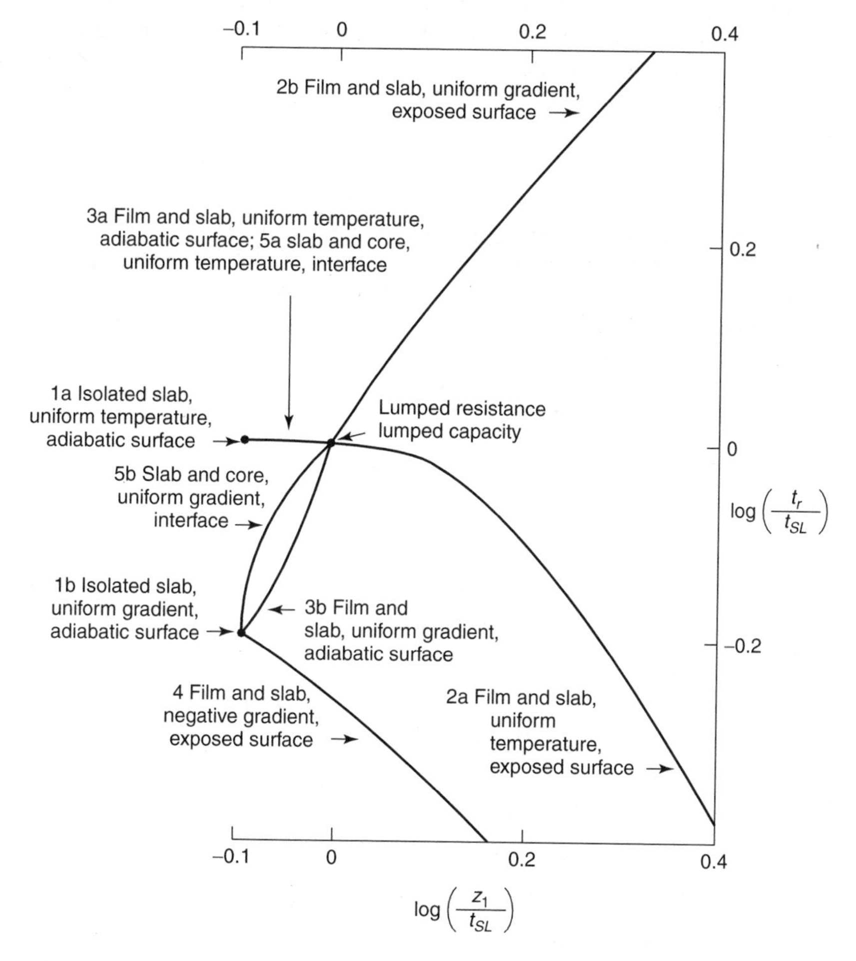

Figure 14.3 The relationship between t_r/t_{SL} and z_1/t_{SL} for some simple models as a function of position and initial condition (Reprinted from *Applied Energy*, vol. 18, M.G. Davies, Heat storage/loss ratio for a building and its response time 179–238, © 1984, with permission from Elsevier Science)

For the Groeber problem (film and slab, uniform initial temperature), t_r is very nearly equal to t_{SL} at the adiabatic surface (case 3a) for all B numbers, as we have already seen, and for small B numbers it is also true at the exposed surface (case 2a). The diagram shows this. It will also be seen that t_r and t_{SL} are near enough equal at the interface between the slab and an adjacent core (case 5a).

For the other cases, however, the ratio t_r/t_{SL} departs significantly from unity and although they may have the same order of magnitude, t_{SL} cannot be treated as a reliable estimate of t_r.

14.2.2 Response Time by Computation

The response time of an enclosure can be readily found while finding its response to some repeated daily excitation (Section 19.4). To compute temperatures in the enclosure of Figure 19.8, all values are initially set to zero and steady-cyclic values are eventually reached after cycling for some days. Thus the mean value of say air temperature T_a over a 24 h period gradually increased. Progress toward the final state can be described by the expression

$$T_{ad} = T_{af}(1 - \exp(-d/z)),$$

where T_{ad} is the mean value of T_a during day d and T_{af} is its final value. Here d has units of days but only takes integral values; z accordingly has units of days. When the enclosure in Section 19.4 was considered as a module, it either had an exposed concrete ceiling or the ceiling was treated acoustically. Two steadily maintained ventilation rates were also computed. The evaluated response times (days) were as follows.

	Exposed ceiling	Acoustic ceiling
Low ventilation	5.8	6.8
High ventilation	1.9	3.6

The response time decreases with increasing ventilation rate; it is a little less with the concrete ceiling than the acoustically treated ceiling because of the higher conductivity of concrete.

14.2.3 Response Time by Observation

An estimate of the enclosure response time can also be found by evaluating the autocorrelation function of a long time series of temperature measurements. The argument will be illustrated here using the author's observations of the response of a school classroom with a floor, ceiling and east and west walls of heavy construction (Morgan 1966; Davies 1986, 1987 IV). The south wall was almost entirely glazed, admitting large solar gains and allowing large thermal losses; the north wall was lightweight.

Consider the ceiling temperature T. If values of T(now) are plotted against T(now), they lie on a straight line of unit slope, with a correlation coefficient r of unity. If T(now) is plotted against T(now - 1 h), the points show a slight scatter about the line and r is a little less than 1. Thus if $T(t)$ is plotted against $T(t\text{-lag})$, r is a function of the lag. If r is significantly different from zero, $T(t)$ and $T(t\text{-lag})$ must be correlated to some extent.

For a lag of a few hours, $T(t)$ and $T(t\text{-lag})$ must be correlated since the ceiling is massive and its temperature can only change slowly. For a lag of several months, they must be uncorrelated as far as the ceiling is concerned but they may be independently correlated with solar gains and seasonal changes in ambient temperature, (so that because of the daily periodicity of these drivers, $T(t)$ will be better correlated with $T(t\text{-}24\,\text{h})$ and $T(t\text{-}48\,\text{h})$ than with $T(t\text{-}12\,\text{h})$ and $T(t\text{-}36\,\text{h})$). For a run of several years' data, r will tend to be positive for lags of about 3 months and 12 months, and negative for lags of about 6 months.

A plot of r against lag constitutes an autocorrelation function, which is of interest here in so far as it provides information on the thermal memory or thermal persistence of the ceiling; r is computed as

$$r(\text{lag}) = \frac{\sum T(i)T(i - \text{lag})}{\left\{\sum[T(i)]^2 \sum[T(i - \text{lag})]^2\right\}^{1/2}}, \tag{14.27}$$

where $T(i)$ are the deviations from the mean value, (so that about half are negative.)

Plots are shown in Figure 14.4 for the winter period (1 November to 28 February) and early summer (1 March to 31 May), demonstrating the persistence of the ceiling slab. It is longer in winter than summer since ventilation rates in winter were low. The building was not heated by a normal heating system but depended on solar gains, the electric lighting system and metabolic heat, which together were barely adequate, so the windows were normally kept closed. In summer the large solar gains through the south window area had to be removed by ventilation; the daily variation in indoor temperature they caused was restrained by the large thermal mass of the walls, floor and ceiling.

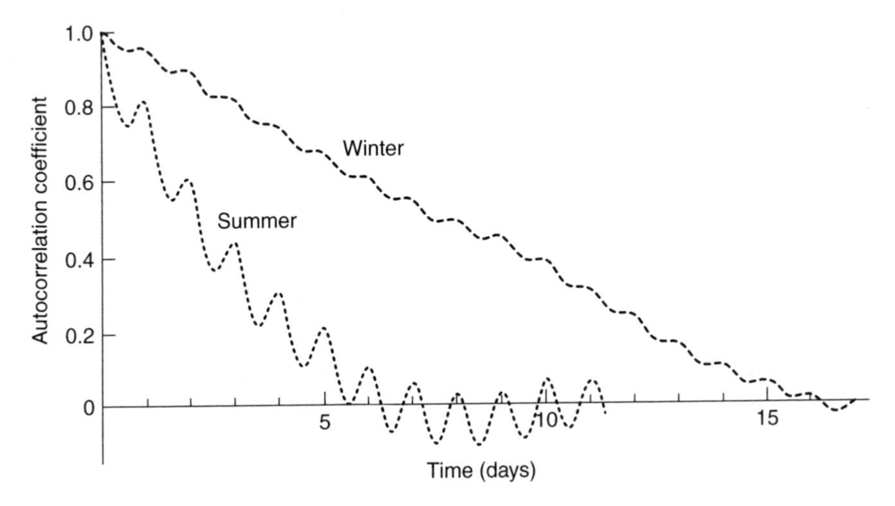

Figure 14.4 Covariation of ceiling temperature with itself at an interval of up to 15 days (Reproduced by the permission of John Wiley)

The autocorrelation function represents a reliable finding since it is based on a large quantity of simple observational data subjected to statistical processing without any dynamic theory. Its use in a building context has been discussed by Loveday and Craggs (1993). To use it to estimate a response time, so far considered as a product of mechanistic analysis, the succession of temperatures will be treated as a first-order Markov process.

First we note that if the enclosure is not excited in any way in the interval δt between times $i-1$ and i, the deviations T_{i-1} and T_i are related as

$$T_i = T_{i-1}\exp(-\delta t/t_r) = T_{i-1}\rho, \tag{14.28a}$$

where

$$\rho = \exp(-\delta t/t_r) \approx 1 - \delta t/t_r, \tag{14.28b}$$

since $\delta t \ll t_r$.

If the enclosure is subjected to some random excitation during the interval δt,

$$T_i = T_{i-1}\rho + \varepsilon_i, \tag{14.29}$$

Now this is the equation of a first-order Markov process in which ε_i is a random variable. It is a property of such a process that the autocorrelation coefficients at intervals of 0, δt, $2\delta t, \ldots, k\delta t, \ldots$ are $1, \rho, \rho^2, \ldots, \rho^k, \ldots$ and the area under the curve is $\delta t \sum \rho^k$ or $\delta t/(1-\rho)$. But $\delta t/(1-\rho) = t_r$. Thus if the ρ values of the Markov process can be identified with the r values of the autocorrelation function, t_r can be estimated as $\delta t \sum r^k$, summing over the positive values of r. From the autocorrelations in Figure 14.4 it was found that $t_r(\text{winter}) = 7.8$ days and $t_r(\text{summer}) = 2.9$ days.

These values were in satisfactory agreement with purely theoretical estimates based on assumptions for the thermophysical characteristics of the enclosure, and also on a partly observational estimate based on values for cooling during a still, cold winter night.

(The agreement undoubtedly came about partly by chance. The theoretical model for the estimates was necessarily simplified and reliable values for the thermal constants were not available. Nor were the changes ε_i random, as is assumed in a first-order Markov process; they result from sustained drives such as solar gains and slowly varying ambient temperature and they are correlated. Further, the r^k values of the autocorrelation function show features which differ from the simple geometrical progression $1, \rho, \rho^2, \ldots$ since (i) when the effect of diurnal variation is removed by 24 h averaging, the autocorrelation function has zero gradient at a lag of zero, while the value for the Markov process is $1-\rho$, (ii) the r^k values do not fall away in simple geometric progression, and (iii) they eventually become slightly negative. Although the autocorrelation function gives a reliable impression of the thermal memory of an element in a building, the way it is used here will not necessarily give a close estimate of a response time.)

14.2.4 Response Time and HVAC Time Delays

The response time of a room may range from an hour to several days. The response time of a hot water radiator amounts to tens of minutes but a heating, ventilating or air-conditioning system responds more quickly. The time constants of heating/cooling coils may be less than a minute, transportation times within ducts and pipes may be less than 10 s and sensor time constants are rarely greater than a minute (Athienitis 1993). The next section discusses how to form building models so simplified as to provide a realistic measure of their response time, and which enable them to be linked to the faster-acting heating or cooling system.

14.3 MODELS WITH FEW CAPACITIES

The use of an electrical analogue computer using real resistances and capacitors to simulate building response was well established in the 1960s (Section 10.5) and one might have expected that by 1970, when electronic computing had become generally widespread, it would have been exploited to estimate the thermal behaviour of a room on the basis of a discretised thermal circuit. In one respect it was: wall finite difference calculations, previously conducted manually, could readily be carried out electronically. Surprisingly, comparable calculations for an enclosure were not much reported until about 1980. They will be discussed in this section. They do not attempt to represent an enclosure in the detail suggested by, for example, the circuit in Figure 19.8 and are therefore less ambitious in their aims; they may thereby gain in robustness and simplicity and prove adequate for certain tasks.

Figure 14.5 shows three conceptual models of increasing detail.[2] The first models the degree-day concept: the heat need over a season, the plant load, is related to the time during which ambient temperature remains below some reference value T_{base} which is less than the comfort temperature T_c that is to be achieved. The enclosure is described by its fabric conductance and an assumed ventilation rate. The heat need corresponding to the difference $T_c - T_{base}$ is supposed to be supplied by casual gains. For many buildings, a significant part of the casual gains will be due to solar incidence, and if we want to include if explicitly in a steady-state model, it could be modelled as in Figure 14.5b. A better estimate, perhaps for air-conditioning use, will be obtained by including the thermal capacity of external and internal walls (Figure 14.5c); some internal surfaces may be irradiated, some not.

It is immaterial whether the sequence of resistors representing a wall is drawn as a horizontal sequence as in Figure 13.1 or a vertical sequence as in Figure 14.5. When the figure is linked to a wall transmission matrix, necessarily horizontal, the horizontal display will be used. When the resistors constitute one wall in the circuit of an enclosure, they will be displayed vertically. In this way, any number of surfaces can be included. A link between a capacity and reference temperature is to be understood if not indicated.

[2]There is no accepted convention in setting out the details of a building thermal circuit, and the same basic disposition of components can be displayed in quite dissimilar forms. The convention adopted here is to represent the reference temperature T_{ref} as a horizontal line at the bottom of the figure, similar to the representation of earth potential in electric circuit diagrams. Pure temperature sources must be connected to T_{ref}. Pure heat sources need not be connected. Ambient temperature is shown as a source immediately above T_{ref}. Room internal temperature (air, rad-air, possibly the radiant star node) will normally be located top left and other room surfaces as nodes to its right; heat can be input at these nodes. Losses by ventilation and conduction are denoted by vertical conductances linked to ambient. Internal radiant and convective exchanges or combined values may be displayed horizontally or vertically as is convenient.

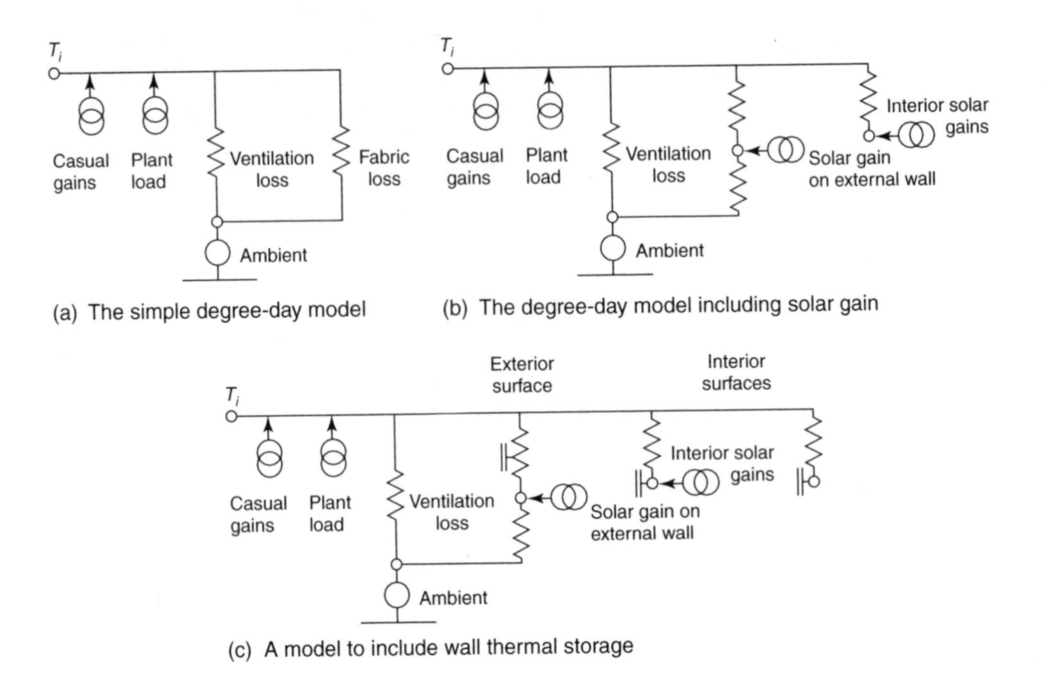

Figure 14.5 Basic room thermal models

14.3.1 One-Capacity Wall Models

If wall capacity is to be included in a model, its first-order representation is as a single capacity. This was adopted by Lorenz and Masey (1982) and has recently been taken up by Gouda *et al.* (2002). They express a multilayer as a T-section consisting of resistances r_{out} and r_{ins} and a single capacity c which is $\sum c_i$. For a two-layer wall, layers 1 and 2,

$$r_{out} = \left[\tfrac{3}{2} r_1 c_1 + \left(r_1 + \tfrac{1}{2} r_2\right) c_2\right] / c. \tag{14.30}$$

This is clearly related to Bruckmayer's approach ((14.12) with some differences in notation):

$$r_{ins} = r_{total} - r_{out}, \tag{14.31}$$

where r_{total} includes the resistance of the inside and outside films (r_i and r_e) together with the solid layer resistances (r_1 and r_2). But this is flawed since r_{out} is formed from solid layer resistances only and does not include the outside film resistance as such.

A rational model taking account of the capacity of an enclosure by a single unit has been developed by Mathews and his colleagues in some detail (Mathews 1986; Mathews and Richards 1989, 1993; Mathews *et al.* 1991, 1994a, 1994b; Lombard and Mathews 1992, 1999).

A wall can be modelled approximately by a single T-section. To illustrate the reduction, consider the two-capacity wall in Figure 14.6 with surface nodes T_0 and T_2. When excited

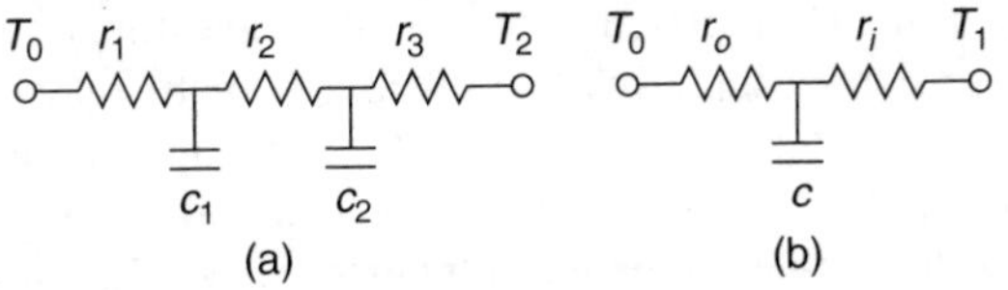

Figure 14.6 Reduction of two capacities in series to an approximate one-capacity model

sinusoidally its transmission matrix is

$$\begin{bmatrix} 1 & r_1 \\ 0 & 1 \end{bmatrix}\begin{bmatrix} 1 & 0 \\ j\omega c_1 & 1 \end{bmatrix}\begin{bmatrix} 1 & r_2 \\ 0 & 1 \end{bmatrix}\begin{bmatrix} 1 & 0 \\ j\omega c_2 & 1 \end{bmatrix}\begin{bmatrix} 1 & r_3 \\ 0 & 1 \end{bmatrix}$$

$$= \begin{bmatrix} 1+j\omega(r_1c_1+r_1c_2+r_2c_2)-\omega^2 r_1c_1r_2c_2 & r_1+r_2+r_3+j\omega(r_1c_1r_2+r_1c_1r_3+r_1c_2r_3+r_2c_2r_3)-\omega^2 r_1c_1r_2c_2r_3 \\ j\omega(c_1+c_2)-\omega^2 c_1r_2c_2 & 1+j\omega(c_1r_2+c_1r_3+c_2r_3)-\omega^2 c_1r_2c_2r_3 \end{bmatrix}. \tag{14.32}$$

Suppose this is to be represented approximately by a one-capacity model with nodes T_0 and T_1 and resistances r_o and r_i. Its transmission matrix is

$$\begin{bmatrix} 1+j\omega r_o c & r_o+r_i+j\omega r_o c r_i \\ j\omega c & 1+j\omega c r_i \end{bmatrix}. \tag{14.33}$$

First, we neglect elements in ω^2 on the right-hand side of (14.32). Then by identification of the real elements in term 12 in the two matrices, and similarly the imaginary element in term 21,

$$r_o + r_i \equiv r_1 + r_2 + r_3, \tag{14.34a}$$

as must obviously be the case. Further,

$$c \equiv c_1 + c_2, \tag{14.34b}$$

although a condition of this kind is not always imposed. Identification of the imaginary element in terms 11 gives

$$r_o \equiv (r_1c_1 + r_1c_2 + r_2c_2)/(c_1 + c_2); \tag{14.34c}$$

(r_i is similarly given by terms 22 but this is not independent of the foregoing results.)

The $r_o c r_i$ element in the one-capacity matrix is then equal to $(r_1c_1r_2 + r_1c_1r_3 + r_1c_2r_3 + r_2c_2r_3) + c_1{r_2}^2c_2/(c_1 + c_2)$, which differs from the linearised two-capacity value by $c_1{r_2}^2c_2/(c_1 + c_2)$.

If the wall index nodes are isothermal, the product of the decay times, $z_1z_2 = r_1c_2r_2c_2r_3/(r_1 + r_2 + r_3)$. Thus neglect of the term $\omega^2 r_1c_2r_2c_2r_3$ in comparison with $r_1 + r_2 + r_3$ is equivalent to assuming that the period of excitation, P, is much larger than $2\pi\sqrt{z_1z_2}$; whether or not this is valid depends on the weight of the wall, and the present argument assumes it is valid.

A wall reduction of this kind may be useful in connection with monthly energy needs but it brings two important limitations. Firstly, it removes wall surface nodes. These may be needed to express the radiant component of temperature in estimating comfort temperature and as locations where radiant input may act; heat is sometimes taken to act at the node of the lumped capacity but this may be physically inappropriate. Secondly, the response to some step in heat input, say, is too slow. For external sinusoidal excitation – daily varying temperature – the phase lag between temperature and the heat flow to inside that it causes will be too small and the phase difference between an internally acting heat source and internal temperature will be too large.

14.3.2 One-Capacity Enclosure Models

A number of models have been proposed where the entire enclosure capacity – walls, floor and ceiling – have been lumped. There is first the problem of combining several such surfaces acting effectively in parallel. Consider the two T-sections in Figure 14.7. We can write

$$\begin{bmatrix} T_1 \\ Q_{11} \end{bmatrix} = \begin{bmatrix} a_1 & b_1 \\ c_1 & d_1 \end{bmatrix} \begin{bmatrix} T_0 \\ Q_{10} \end{bmatrix} \quad \text{and} \quad \begin{bmatrix} T_1 \\ Q_{21} \end{bmatrix} = \begin{bmatrix} a_2 & b_2 \\ c_2 & d_2 \end{bmatrix} \begin{bmatrix} T_0 \\ Q_{20} \end{bmatrix}. \tag{14.35}$$

Since these relate to whole areas of wall, Q has units of W and b and c have units K/W and W/K, respectively. The combined effect can be expressed as the matrix

$$\begin{bmatrix} T_1 \\ Q_{11} + Q_{21} \end{bmatrix} = \begin{bmatrix} a & b \\ c & d \end{bmatrix} \begin{bmatrix} T_0 \\ Q_{10} + Q_{20} \end{bmatrix}, \tag{14.36}$$

where

$$\begin{aligned} a &= (a_1 b_2 + a_2 b_1)/(b_1 + b_2), & b &= b_1 b_2/(b_1 + b_2), \\ c &= (c_1 + c_2) - (a_1 - a_2)(d_1 - d_2)/(b_1 + b_2), & d &= (b_1 d_2 + b_2 d_1)/(b_1 + b_2). \end{aligned} \tag{14.37}$$

In this application,

$$a_1 = 1 + j\omega R_{10} C_1, b_1 = R_{10} + R_{11} + j\omega R_{10} C_1 R_{11}, c_1 = j\omega C_1, d_1 = 1 + j\omega R_{11} C_1, \text{ etc.,} \tag{14.38}$$

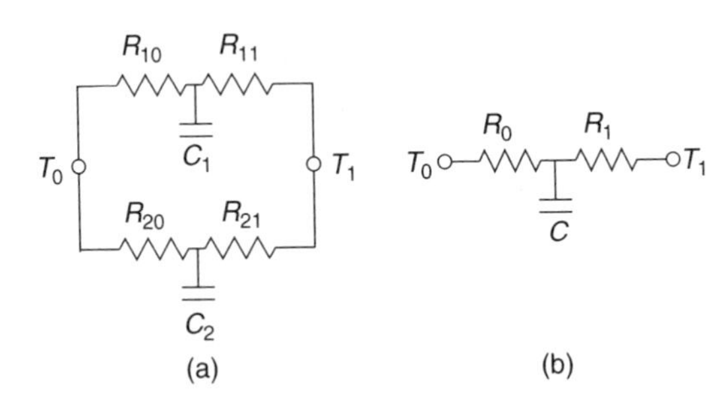

Figure 14.7 Reduction of two capacities in parallel to an approximate one-capacity model

from (14.33); capitals replace lower case letters since we are considering an area. Evaluation reveals that the elements have the following form:

$$a = 1 + j\omega A_1 - \omega^2 A_2 - j\omega^3 A_3 \ldots, \qquad b = B_0 + j\omega B_1 - \omega^2 B_2 - j\omega^3 B_3 \ldots,$$
$$c = j\omega(C_1 + C_2) - \omega^2 \Gamma_2 - j\omega^3 \Gamma_3 \ldots, \qquad d = 1 + j\omega\Delta_1 - \omega^2\Delta_2 - j\omega\Delta_3 \ldots. \quad (14.39)$$

Again, terms in ω^2 and above will be dropped. Then the single equivalent capacity C is simply the sum of the two constituent parts C_1 and C_2, as expected. The total resistance $R_0 + R_1$ (the coefficient B_0 above) is the sum of $R_{10} + R_{11}$ and $R_{20} + R_{21}$ in parallel. $R_0 = A_1/C$. Hence the equivalent wall. It should be noted that the quantities R_{10} and R_{11}, R_{20} and R_{21} in Figure 14.7 are based on wall information alone, that is, they do not include the outside and inside films. The reduction has the effect of lumping the wall surfaces, which is not a valid transformation: mean surface temperature including several separate surfaces is strictly not an allowable construct in thermal modelling. However, in view of the many approximations that reduction of this kind entails, it may be allowable at a pragmatic level.

Mathews and his colleagues argued somewhat along these lines. The thermal storage provided by interior partitions cannot strictly be included as a parallel reduction since no resistance similar to R_0 exists, nor are internal surfaces driven by sol-air temperature. They took empirical account of internal storage by increasing C. A version of their model is shown in Figure 14.8. Lumped inner and outer films are explicit, as is the ventilation conductance. Ambient temperature and solar radiation received at the outer surface are shown separately in the figure but can be combined as a sol-air temperature. Internal convective gains are taken to act at the room index temperature (described as air temperature) and radiant gains are taken to act at the lumped room internal surface.

They report a study to validate their model based on 32 buildings (70 cases) of widely differing type and location. Details of construction and meteorological information were available and ventilation rates were assumed. They measured daily means and swings in temperature and plotted values against estimates based on the model. Predicted indoor air temperature correlated well with measured values (correlation coefficient $r = 0.982$) and predicted and measured daily swings similarly ($r = 0.965$.) Also, 95% of predicted hourly values were within 3 K of measured values. The authors suggest that the model

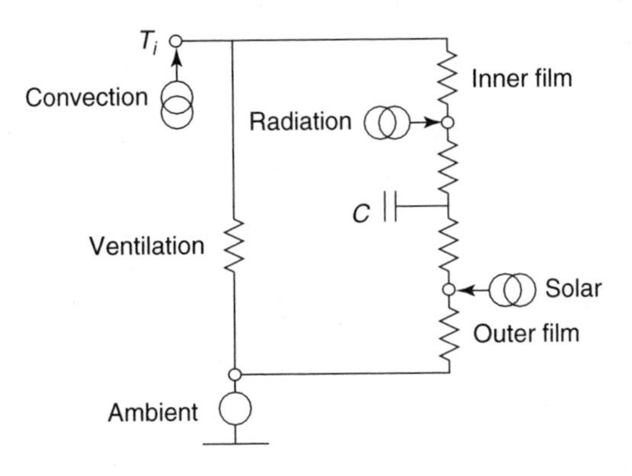

Figure 14.8 First-order thermal model for a whole building zone

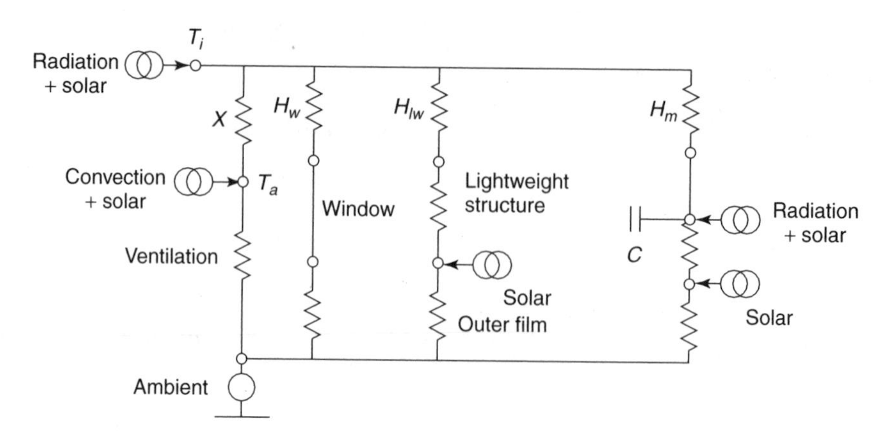

Figure 14.9 Reformulated CEN first-order thermal model for an enclosure

is adequate for passive design purposes but is not sufficiently detailed to simulate energy requirements in a building with active climate control.

A model which has been discussed by the Comité Européen de Normalisation (European Standards Committee) is also a first-order model; a redrafted form is shown in Figure 14.9 (CEN 1998). This model distinguishes between the room air temperature T_a proper and the room index temperature T_i, similar to the rad-air temperature. Thus there is a conductance $X = A_t(h_c + Eh_r)h_c/Eh_r$ between T_i and T_a. A_t is the total room area. The model lumps any lightweight structure with the window (shown separately in Figure 14.9) and there are conductances of type $H = A(h_c + Eh_r)$ between T_i and the respective internal surfaces. The lumped capacity C, however, is based on wall sinusoidal admittances and its link to T_i, $H_m = A_m(h_c + Eh_r)$, is based on an equivalent area A_m, not an actual area. CEN's use of an equivalent outdoor temperature implies the solar incidence and outer films of the figure.

14.3.3 Enclosure Models with Two or More Capacities

Models have been proposed in which the thermal capacity of a room is represented by two or three discrete capacities. Figure 14.10 shows a model by Laret (1980). It has the transmission matrix

$$\begin{bmatrix} 1 & R_1 \\ 0 & 1 \end{bmatrix}\begin{bmatrix} 1 & 0 \\ -C_1/z & 1 \end{bmatrix}\begin{bmatrix} 1 & R_2 \\ 0 & 1 \end{bmatrix}\begin{bmatrix} 1 & 0 \\ K & 1 \end{bmatrix}\begin{bmatrix} 1 & 0 \\ -C_2/z & 1 \end{bmatrix} = \begin{bmatrix} e_{11} & e_{12} \\ e_{21} & e_{22} \end{bmatrix}. \tag{14.40}$$

The model is isothermal at the outer node T_e and adiabatic at the inner node T_i, so the decay times z_1 and z_2 are given by the relation

$$e_{11} = 0. \tag{14.41}$$

The model so reduces the complex pattern of heat transfer in a real enclosure that no account can be taken of where heat is actually input. The single internal node lumps the

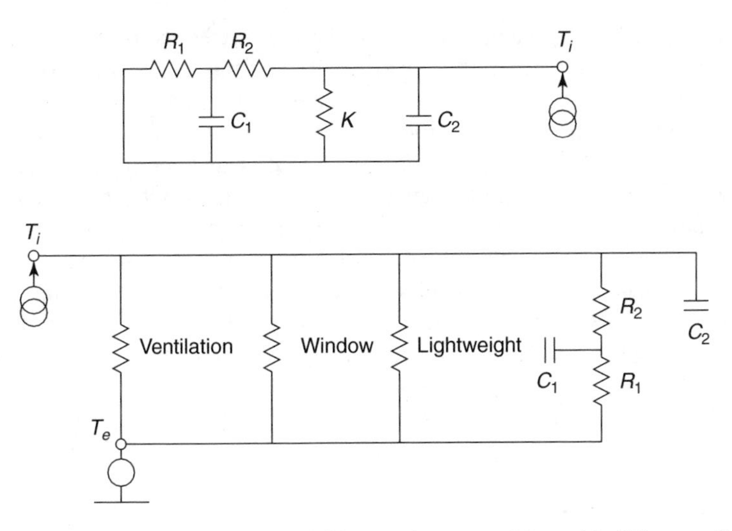

Figure 14.10 Laret's two-capacity model in matrix-compatible and building configurations

air and all surfaces and serves as the point where convective and radiative inputs act. The C_1 node is not accessible. The pattern is set up as follows:

(i) Heat losses due to ventilation and through windows and any lightweight elements providing negligible thermal storage are lumped and assigned the conductance K (units W/K), which is directly calculable.

(ii) The steady-state heat loss through external walls which have thermal storage is assigned a resistance $R_1 + R_2$, (units K/W), also directly calculable.

(iii) The longer and shorter decay times are to be calculated as

$$z_1 = \frac{\text{heat stored in the steady state}}{\text{heat loss in steady state}} \quad \text{with } T_e = 0 \text{ and } T_i = 1\,\text{K}, \tag{14.42a}$$

$$z_2 = 1\,\text{h}. \tag{14.42b}$$

The steady-state heat loss is simply $K + 1/(R_1 + R_2)$. The heat stored is the sum of the energy in external walls in which there is a linear gradient, as expressed in (14.11c), the energy in internal walls at a uniform unit temperature and the heat similarly stored in the air.

Thus, while K and $R_1 + R_2$ are based directly on physical measures of the enclosure, the values of C_1, C_2 and the ratio R_1/R_2 are derived quantities, evaluated to fit certain preconceived measures of the enclosure; they do not follow from solutions of the Fourier continuity equation.

Mathews' simplification and Laret's simplification are complementary. Mathews starts with a potentially exact representation of exterior walls and then reduces it by logical, if drastic, steps to a single capacity; internal storage is then incorporated in an empirical manner. Laret handles the exterior walls empirically from the start but includes internal storage in a more physical manner; C_2 might be interpreted as a form of 'air capacity' but with a numerical value of some 4–5 times the real thermal capacity of the air.

It may be noted too that Laret's model is a formal simplification of Pratt's model (Figure 14.1d and Section 14.4) but while the idea that the building time constant might be estimated as the storage/loss ratio is an *outcome* of Pratt's analysis, it constitutes an *input* to Laret's model (Laret 1980: 267). As Figure 14.3 shows, the idea itself is not well founded.

Further details in implementing the model are given by Lorenz and Masey (1982). Crabb *et al.* (1987) describe an application of Laret's model in relation to a school in winter, assigning values of $1/R_1 = 374$ W/K, $1/R_2 = 12\,873$ W/K; here the conductance K represents the losses by ventilation and conduction through doors, windows and the lightweight roof and is 1329 W/K for one air change per hour. $C_1 = 46 \times 10^7$ J/K and $C_2 = 1 \times 10^7$ J/K, computed as five times the thermal capacity of the air. Decay times are quoted as 101 h and 12 min. They demonstrate satisfactory agreement between prediction and observation. Parameter values were later found by minimising the function

$$J = \sum_{n=1}^{N} (T_{i,n}^G - T_{i,n}^M)^2 \qquad (14.43)$$

through variation of the model parameters. T^M is the measured internal temperature, T^G is the value generated by the model on assumed values of the parameters, and N for a week of half-hourly observations is 336.[3] They used the model in connection with an optimum start controller for the heating system (Penman 1990; Coley and Penman 1996). The authors argue that a two-capacity (second-order) model is 'the simplest model capable of representing the essential features of building thermal response'. Candau and Piar (1993) report the results of observing temperature in a test cell as it responds to ambient temperature and a known heat input. The response to the heat input could be represented by terms which included two time constants, 24.9 h and 2.48 h. Zaheer-uddin (1990) and Dewson *et al.* (1993) discuss the least-squares method to determine model parameters. Hanby and Dil (1995) use the two-capacity model in connection with stochastic modelling of heating and cooling systems. This inverse method of evaluating constants has been used to find thermal conductivity (Alifanov 1994; Yang 1998).

Tindale (1993) has described a development of Laret's model. Figure 14.11 shows the circuit in building form. The model adopts the rad-air formulation so that the room index temperature T_i becomes replaced by T_{ra} and by T_a, with the conductance X between them together with their heat inputs. (They are the augmented radiant input $(1 + \alpha)Q_r$ at T_{ra}, the convective input Q_c at T_a together with removal of the excess radiation, αQ_r.) The model also makes clear the contributions to internal storage from the air and the internal partitions, shows the link to the rad-air node in an adjoining room and makes explicit the wall outer film, so that absorbed solar radiation can be modelled. Tindale's main development, however, was to give a better representation of the distributed mass in an outer wall than a single capacity permits. This was provided by an additional RC pair in parallel with the usual RCR unit for the outer wall. He notes that it would have made better physical sense to have placed an additional capacity in the outer wall proper, so as to represent it as an $RCRCR$ unit, since the RC pair is a non-physical construct. However,

[3] This technique is successful if it leads to an acceptably small value for J. Five parameters might not be sufficient to achieve this.

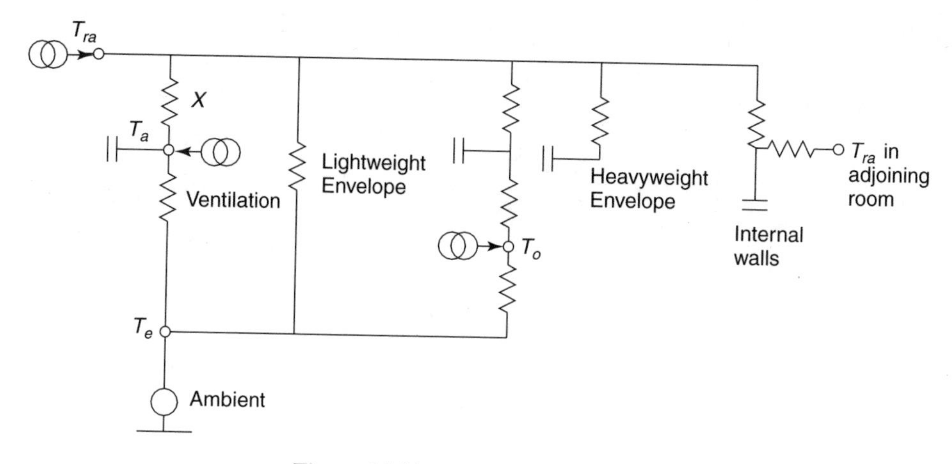

Figure 14.11 Tindale's 3TC model

the main aim of the modification was to give a better representation of the storage of the outer wall due to *inside* changes in temperature. For this purpose, the $RCR + RC$ model as shown might serve well enough. In principle one might seek to minimise the value of Δ in (15.24) but on the basis of the 12 and 22 subscripted elements in the respective transmission matrices, as noted at the end of Section 15.7. In the event, the elements were sized through comparison of Bode diagrams. The model is called the three time constant (3TC) model. With four capacities it must strictly have four time constants. However, since the air capacity – no longer taken to include internal partition capacity as in Laret's model – is very small, the model may be described as one of three time constants.

Chen and Athienitis (1993: Figure 1) discuss a more advanced form of three-capacity model for a room. Four internal nodes are assumed: the room air; a window surface; the ceiling slab, which is treated as a lumped capacity; and the surface of the remaining walls and floor, which are represented as a distributed capacity/resistance element. This model can be expressed as in Figure 14.12.

It follows the Chen–Athienitis circuit in that it shows explicit convective links between the air and surfaces and radiative links between the surfaces (like Figure 7.2a), together with the air and roof slab capacities and losses. No internal heat input is assumed; the driver is sinusoidal variation of ambient temperature. The present formulation differs in two respects. Firstly, when driven sinusoidally, the distributed element can be represented exactly by three complex (or six real) lumped elements, as shown in Section 15.6; the authors model it with two. Secondly, Figure 14.12 models the drive as a pure temperature source, whereas the authors represent it as a pure heat source by using Norton's theorem (Section 4.4.1). (It should be noted that the three complex element formulation does not have the intuitively plausible basis that model resistances and capacities otherwise have. The values of resistances and capacities depend strongly on the assumed period of excitation.)

Déqué *et al.* (2000) describe a method (the 'grey box' model) to represent building envelopes using a limited number of parameters per thermal zone (10 or 50). The transient state is represented by a second-order model and increment to the main time constant (between 40 and 140 h) is taken to be proportional to wall surface area and inversely proportional to the total loss coefficient. The thermal model shown in Figure 1 of Mathews *et al.* (1997) has seven capacities, three of which represent an outer wall.

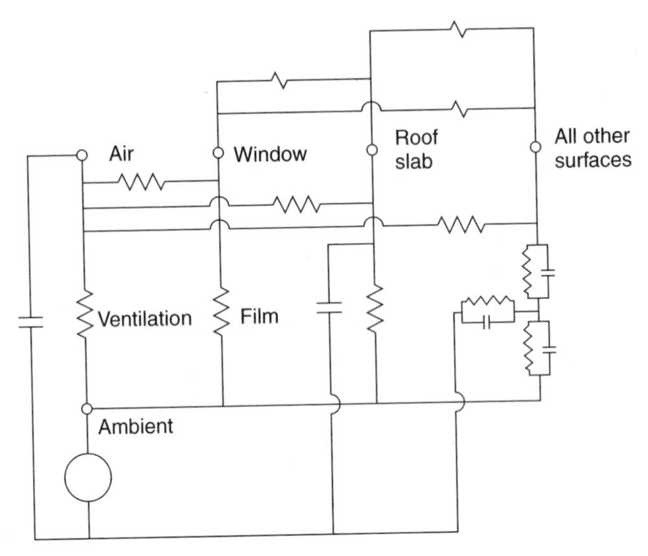

Figure 14.12 A room with two lumped capacities and one distributed capacity, modelled for sinusoidal variation of ambient temperature

14.4 DISCUSSION

The foregoing models have proved to be of value to their developers but have not gained wide acceptance as standard design methods. The information on models and algorithms to determine building response and heating and cooling loads is in fact very large. Spitler and Ferguson (1995) report that they uncovered a total of 1576 references – papers, reports, theses and books – in connection with preparation of an annotated guide to load calculation, of which 220 were selected for annotation and a further 176 were included as related references. Robinson (1996) states that more than 300 energy models had been reported, of varying complexity, function and ease of use. In his survey, 84% of the users of detailed models were engineers and 60% of the simplified models were architects. The historical development of methods to calculate building energy need in the US is discussed by Ayres and Stamper (1995) and Sowell and Hittle (1995).

The methods favoured by CIBSE and ASHRAE to model wall conduction are based on exact solutions to the Fourier continuity equation (implying, as we shall see, that a wall can be modelled by several discrete capacities).[4] Either the wall can be described by parameters which assume that the excitation to which the enclosure is subjected can be resolved into a steady state and a periodic component of period 24 h, and that it is in a steady-cyclic condition; the parameters are then the wall U value, admittance and cyclic transmittance. Alternatively, wall parameters in the form of sets of transfer coefficients describe the response of the wall due to excitation expressed at, say, hourly intervals.

[4]A number of available programs use the finite difference approach to model wall behaviour. The logic is broadly that leading to the 'hourly response' entry in Figure 14.13, but without the transfer coefficient or continuity equation entries since the finite difference method is based on the intuitively obvious principle that a wall can be represented approximately in discretised form.

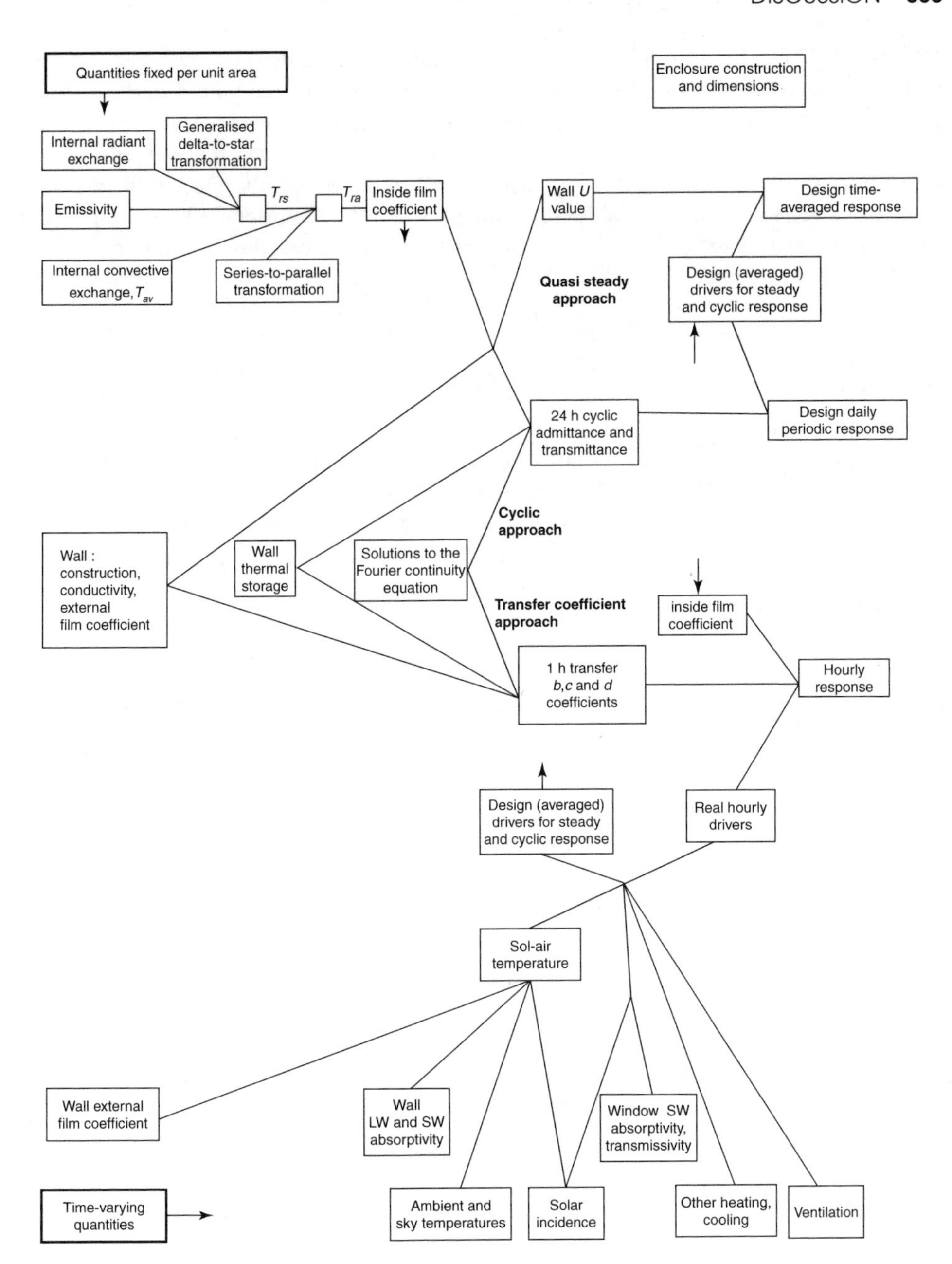

Figure 14.13 Room heat transfer parameters and the handling of excitation by cyclic and transfer coefficient methods. (Davies, 2001b, with permission from Elsevier)

The two approaches are indicated in Figure 14.13, a development of Figure 12.2. The leftmost column lists wall properties (layer conductivities and thicknesses, external film coefficient) related to unit area. It indicates how internal radiant and convective exchanges can be merged to arrive at an inside film coefficient. The wall U value derives from this information alone. For time-varying excitation, information on thermal capacity is required and is processed using one of the solutions to the Fourier continuity equation. If the periodic solution is used, we reach wall admittance and transmittance values for sinusoidal excitation of period 24 h (or other period). See Section 10.2 for the elementary one-capacity case and Chapter 15 for the general wall. If we adopt a Laplace transformation of the periodic solution, or suitable time-domain solutions as presented here, we arrive at the series of wall response factors and transfer coefficients, based on a sampling interval of 1 h (or other interval). They were introduced for a discretised wall in Chapter 11 and will be derived for the general wall in Chapter 17. These descriptors relate to unit area of the wall and in use are multiplied by the area concerned.

Time-varying drivers together with ventilation rate appear at the bottom of Figure 14.13 and those acting externally require processing according to whether they act through opaque or transparent wall elements. All excitation information has to be further processed for use with the various wall descriptors. Average values over a sufficiently long period are required for use with U values. We require some measure of the amplitude of sinusoidal variation of period 24 h about the daily mean over some representative period, perhaps several days, in connection with admittance and transmittance values (Chapter 16). For use with transfer coefficients, real (ongoing) values of the exciting variables are needed (Chapter 19).

15

Wall Parameters for Periodic Excitation

Chapter 1 discussed how the temperature in a room could be estimated when temperatures are unchanging, so that the thermal capacity of walls, external and internal, played no part. This technique, based upon the steady-state wall U value, served in earlier decades to estimate heat losses in winter and so to size heating systems. Traditional heavy brick buildings with small windows had sufficient thermal capacity to cope with relatively small solar gains, so large daily swings in temperature were not normally experienced. This situation changed in the middle of the twentieth century. Lightweight buildings provided less capacity and their larger glazed areas allowed large solar gains in sunny periods resulting in excessively hot areas within the building (and cold areas due to large losses in cold sunless periods).

This disturbing diurnal behaviour led to the development of a series of parameters to describe the thermal characteristics of a wall (taken here to include the floor and roof), similar to the U value, but including consideration of thermal capacity and time variation.[1] This chapter, based upon Davies (1994), considers parameters based on sinusoidal variation in time. In principle the approach is similar to finding a U value – the layer impedances have to be summed – but it is complicated by the need to describe travelling waves through the wall (involving trigonometric functions with magnitudes and phases), waves which are highly damped (introducing hyperbolic functions), and the fact that the thermal behaviour of the wall now has to be described by three complex parameters for each period P of excitation, in place of the single U value.

To gain a rough estimate of the peak temperature reached during some sunny spell, it may be sufficient to use the 24 h sinusoidal parameters alone. If these are known (and they are tabulated for a range of wall types), their use entails a simple hand calculation similar to the steady-state calculation. This enables the designer to run through a range of choices at the early design stage, regarding such matters as orientation, window size, weight of structure, effect of casual gains, etc., and he or she may thereby obtain a ready feel for the relation between cause and effect. The periodic method will also serve to find a closer estimate of daily variation of indoor temperature, but this will involve the designer finding

[1] The discussion in this book is confined to passive storage in walls and floors. Extensive use has been made, however, of the storage potential of floors, in particular, where ventilation air is passed through the structure so as to reduce the building's requirement for mechanical cooling. See for example Winwood *et al.* (1997).

Building Heat Transfer Morris G. Davies
© 2004 John Wiley & Sons, Ltd ISBN: 0-470-84731-X

the 12 h, 8 h, 6 h, etc., parameters for the walls (they are not tabulated), and preprocessing the meteorological data. The task is no longer a manual one.

15.1 THE FINITE-THICKNESS SLAB

The concept of the dynamic parameters **u** and **y** was introduced in Section 10.2.3 for a simplified wall whose thermal capacity was represented by a single lumped element. Walls however consist of layers in which the resistance and capacity are distributed. The basis for the periodic response of distributed systems was discussed in Section 12.2 and we go on here to find the **u** and **y** values, first for a homogeneous slab and later for a real wall.

Consider a slab of thickness X whose left surface ($x = 0$) experiences a sinusoidally varying temperature and flux of complex values $\mathbf{T_0}$ and $\mathbf{q_0}$, respectively, and values $\mathbf{T_1}$ and $\mathbf{q_1}$ at the right surface ($x = X$); see Section 15.8. They are related as

$$\mathbf{T_0} = \mathbf{T_1}\cosh(\tau + j\tau) \qquad + \mathbf{q_1}(\sinh(\tau + j\tau))/\mathbf{a}, \tag{15.1a}$$

$$\mathbf{q_0} = \mathbf{T_1}(\sinh(\tau + j\tau)) \times \mathbf{a} + \mathbf{q_1}\cosh(\tau + j\tau), \tag{15.1b}$$

where

$$\mathbf{a} = (2\pi\lambda\rho c_p/P)^{1/2}\exp(j\pi/4) \quad \text{and} \quad \tau = (\pi\rho c_p X^2/\lambda P)^{1/2}. \tag{15.2}$$

The magnitude of **a** is written as $a = (2\pi\lambda\rho c_p/P)^{1/2}$. Its phase is $\pi/4$ radians, 45°, 1/8 cycle or 3 h in a 24 h cycle. Values of the complex hyperbolic quantities are given in Table 15.1.

The single slab is not of interest in itself, but it illustrates certain properties needed for practical constructions. A cyclic *transmittance*[2] characteristic describes the heat flow at surface 1 due to excitation at surface 0 (or vice versa), and an *admittance*[3] or storage

Table 15.1 Values of the hyperbolic functions

	$\cosh(\tau + j\tau)$	$\sinh(\tau + j\tau)/\mathbf{a}$(m²K/W)	$\sinh(\tau + j\tau) \times \mathbf{a}$(W/m²K)
Cartesian form			
Real	$\cosh\tau\cos\tau$	$(\cosh\tau\sin\tau + \sinh\tau\cos\tau)/(a\sqrt{2})$	$(-\cosh\tau\sin\tau + \sinh\tau\cos\tau)a/\sqrt{2}$
Imaginary	$j\sinh\tau\sin\tau$	$j(\cosh\tau\sin\tau - \sinh\tau\cos\tau)/(a\sqrt{2})$	$j(\cosh\tau\sin\tau + \sinh\tau\cos\tau)a/\sqrt{2}$
Polar form			
Magnitude	$[(\cosh 2\tau + \cos 2\tau)/2]^{1/2}$	$[(\cosh 2\tau - \cos 2\tau)/2]^{1/2}/a$	$[(\cosh 2\tau - \cos 2\tau)/2]^{1/2}a$
Phase	$\arctan(\tan\tau\tanh\tau)$	$\arctan(\tan\tau/\tanh\tau) - \pi/4$	$\arctan(\tan\tau/\tanh\tau) + \pi/4$
Cartesian form when $\tau \to 0$			
Real	1.0	X/λ	0
Imaginary	j.0	j.0	$j2\pi\rho c_p X/P$
Polar form when τ is large			
Magnitude	$\frac{1}{2}\exp(\tau)$	$\frac{1}{2}\exp(\tau)/a$	$\frac{1}{2}\exp(\tau)a$
Phase	τ	$\tau - \pi/4$	$\tau + \pi/4$

[2]Also known as an alternating transmittance, or transfer admittance or cross-admittance.
[3]Also known as a self-admittance.

characteristic describes the response at surface 0 due to excitation at surface 0 itself. A similar quantity can be defined at surface 1, making three characteristics in all, but for a simple slab the two admittances are the same.

The cyclic transmittance $\mathbf{u}$ is

$$\mathbf{u} = \left(\frac{\mathbf{q_1}}{\mathbf{T_0}}\right)_{T_1=0} = \left(\frac{\mathbf{q_0}}{\mathbf{T_1}}\right)_{T_0=0} = \mathbf{a}\frac{1}{\sinh(\tau + \mathrm{j}\tau)}$$

$$= \underbrace{a\left(\frac{2}{\cosh(2\tau) - \cos(2\tau)}\right)^{1/2}}_{\text{magnitude}} \underbrace{\exp\left(\mathrm{j}\left(\frac{\pi}{4} - \arctan\left(\frac{\tan(\tau)}{\tanh(\tau)}\right)\right)\right)}_{\text{phase}}. \tag{15.3}$$

i ii iii iv

It has a magnitude and a phase and both of them have components deriving from $\mathbf{a}$ and from τ. Term i, a, is $(2\pi\lambda\rho c_p/P)^{1/2}$. It reflects the ability of the material both to conduct and to store heat. For unit variation in surface temperature, q_0 for example must on physical grounds decrease to zero as the imposed period P tends to infinity, and the expression confirms this. Term ii describes the effect that slab thickness has on the magnitude of $\mathbf{u}$. When τ is small $\mathbf{u}$ tends to X/λ. Its magnitude falls with increasing τ rather like a rectangular hyperbola, but it falls away more rapidly than this at high τ values; (τ is proportional to X and the steady-state transmittance λ/X is necessarily a hyperbola). Term iii, the phase shift $\pi/4$, is associated with $\mathbf{a}$ (3 h for 24 h periodicity) and the arctan term iv is the phase associated with thickness. The phase difference between $\mathbf{q_1}$ and $\mathbf{T_0}$ is nearly zero when τ is small, as must be the case. When τ is large, $\mathbf{q_1}$ lags behind $\mathbf{T_0}$, tending to a value $\tau - \pi/4$. The wave crest velocity is given as Λ/P, which is around 33 mm per hour for brick under diurnal excitation. The reciprocity relation may be noted:

$$(\mathbf{q_0}/\mathbf{T_1})_{T_0=0} = (\mathbf{q_1}/\mathbf{T_0})_{T_1=0}. \tag{15.4}$$

The ratio of $\mathbf{q_0}$ to $\mathbf{T_0}$, the admittance parameter, depends on whether surface 1 is taken to be adiabatic or isothermal. The adiabatic case is important in connection with room interior walls. Either side of a partition wall equally excited from both sides behaves as though the wall has a perfectly insulating surface at its midplane. The isothermal case leads to the discussion of an exterior wall.

Taking the adiabatic case first, the admittance

$$\mathbf{y_0} = \left(\frac{\mathbf{q_0}}{\mathbf{T_0}}\right)_{q_1=0} = \mathbf{a}\frac{\sinh(\tau + \mathrm{j}\tau)}{\cosh(\tau + \mathrm{j}\tau)}$$

$$= a\left(\frac{\cosh(2\tau) - \cos(2\tau)}{\cosh(2\tau) + \cos(2\tau)}\right)^{1/2} \exp\left(\mathrm{j}\left(\frac{\pi}{4} + \arctan\left(\frac{\sin(2\tau)}{\sinh(2\tau)}\right)\right)\right). \tag{15.5}$$

i ii iii iv

Both a and $\pi/4$ have the same interpretation as before. Term ii describes the effect of slab thickness on the magnitude of $\mathbf{y_0}$. It has values 0 and 1 when τ (or X) is zero and infinite respectively. It has a maximum value of 1.14 when $\tau = 1.18$ (values noted by Gruber and Toedtli in 1989). The arctan term is the phase shift due to τ and has values $\pi/4$ and 0 when τ is zero and infinite respectively. When the slab is adiabatic at its rear surface, the total phase shift varies between $\pi/2$ when $\tau = 0$ and $\pi/4$ when $\tau = \infty$, $\mathbf{q_0}$ always being ahead of $\mathbf{T_0}$ in phase. When τ is small (less than about 0.1), the slab can sustain only a small temperature difference between front and rear; thus it acts as an almost pure capacity $\rho c_p X$ and the phase difference of $\pi/2$ between $\mathbf{q_0}$ and $\mathbf{T_0}$ is similar to the phase difference between a sinusoidally varying voltage and current flowing into a condenser.

If the slab is isothermal at its rear surface, the admittance

$$\mathbf{y_0} = \left(\frac{\mathbf{q_0}}{\mathbf{T_0}}\right)_{T_1=0} = \mathbf{a}\frac{\cosh(\tau + \mathrm{j}\tau)}{\sinh(\tau + \mathrm{j}\tau)}$$

$$= \underset{\text{i}}{a} \underset{\text{ii}}{\left(\frac{\cosh(2\tau) + \cos(2\tau)}{\cosh(2\tau) - \cos(2\tau)}\right)^{1/2}} \exp\left(\mathrm{j}\left(\underset{\text{iii}}{\frac{\pi}{4}} - \underset{\text{iv}}{\arctan\left(\frac{\sin(2\tau)}{\sinh(2\tau)}\right)}\right)\right). \tag{15.6}$$

The effect of thickness is the complement of the adiabatic case. Term ii varies from infinity to 1 with a weak intervening minimum. The total phase difference varies from 0 to $\pi/4$, q_0 again being ahead of T_0 in phase. When τ is small, y_0 tends to X/λ with a phase difference of zero.

In a building, a wall of one or more layers exchanges heat by convection with the air at each exposed surface and also by radiation to the surroundings; the two processes can be combined to form a surface film transmittance h or resistance r. The external and internal spaces can be characterised by index temperatures, T_e and T_i, respectively, and the wall parameters **u** and **y** are best defined in terms of conditions imposed at T_e and T_i rather than at T_0 and T_1, (or T_n for a multilayer wall). An attempt to express **u** and **y** directly in terms of all primary layer quantities $(X, \lambda, \rho c_p)$ leads to bulky and uninformative algebraic forms and it is convenient to express the wall properties as a transmission matrix. Equation (15.1) can be written as

$$\begin{bmatrix} \mathbf{T_0} \\ \mathbf{q_0} \end{bmatrix} = \begin{bmatrix} \cosh(\tau + \mathrm{j}\tau) & (\sinh(\tau + \mathrm{j}\tau))/\mathbf{a} \\ (\sinh(\tau + \mathrm{j}\tau)) \times \mathbf{a} & \cosh(\tau + \mathrm{j}\tau) \end{bmatrix} \begin{bmatrix} \mathbf{T_1} \\ \mathbf{q_1} \end{bmatrix} = \begin{bmatrix} e_{11} & e_{12} \\ e_{21} & e_{22} \end{bmatrix} \begin{bmatrix} \mathbf{T_1} \\ \mathbf{q_1} \end{bmatrix}. \tag{15.7}$$

The matrix has the property that its real part is unity and its imaginary part is zero. For a homogeneous slab (and a symmetrical wall), $e_{22} = e_{11}$, but this is not true in general.

A matrix formulation for thermal purposes was first put forward by van Gorcum (1951), who expressed the elements as $e_{11} = e_{22} = \cos(\gamma X)$, $e_{12} = -\sin(\gamma X)/\lambda\gamma$, $e_{21} = \sin(\gamma X) \times \lambda\gamma$, where $\gamma^2 = -\mathrm{j}\omega\rho c_p/\lambda$ and $\omega = 2\pi/P$ and much subsequent work was later conducted in this notation. The γ, X, λ notation is exactly equivalent to the **a**, τ notation but is arguably not as suitable:

- It appears to suggest that the transmission matrix for a given period is a three-parameter operator rather than a two-parameters operator. The matrix is sometimes written as a *two-parameter* quantity retaining τ but expressing **a** as $(\tau + \mathrm{j}\tau)/r$, where r is the slab resistance X/λ.
- A term of form $\cos(u)$ suggests a quantity which oscillates with increasing u. Since γ is complex, $\sin(\gamma X)$ and $\cos(\gamma X)$ tend to infinity with increasing X, a property better expressed by a hyperbolic function.
- $(\tau + \mathrm{j}\tau)$ more clearly describes a complex quantity than does γX. Further, τ is readily interpretable.

- λ is already included in the definition of γ, so the single symbol **a** is more apt than $\lambda\gamma$.
- $j\omega$ notation is suitable in electrical and acoustic work where high frequencies are the rule. For building applications with a basic frequency of one cycle per day, it seems better to use the period P rather than angular frequency ω.

15.2 THE SLAB WITH FILMS

The heat transfer between the external index temperature $\mathbf{T}_e$ and the wall exterior surface $\mathbf{T}_0$ through the external film resistance r_e can be expressed as

$$\mathbf{T}_e = 1.\mathbf{T}_0 + r_e \cdot \mathbf{q}_0, \qquad (15.8)$$

$$\mathbf{q}_e = 0.\mathbf{T}_0 + 1 \cdot \mathbf{q}_0,$$

or

$$\begin{bmatrix} \mathbf{T}_e \\ \mathbf{q}_e \end{bmatrix} = \begin{bmatrix} 1 & r_e \\ 0 & 1 \end{bmatrix} \begin{bmatrix} \mathbf{T}_0 \\ \mathbf{q}_0 \end{bmatrix}. \qquad (15.9)$$

The inside film resistance r_i and any cavity resistance r_c can be formulated similarly. Conventional values are $r_e = 0.06$, $r_c = 0.18$ and $r_i = 0.12\,\text{m}^2\text{K/W}$. Since both temperature and heat flow are continuous at exposed or interfacial surfaces, the extreme values of T, q in a wall, taking account of the surface films and a homogeneous layer, can be written as

$$\begin{bmatrix} \mathbf{T}_e \\ \mathbf{q}_e \end{bmatrix} = \begin{bmatrix} 1 & r_e \\ 0 & 1 \end{bmatrix} \begin{bmatrix} e_{11} & e_{12} \\ e_{21} & e_{22} \end{bmatrix} \begin{bmatrix} 1 & r_i \\ 0 & 1 \end{bmatrix} \begin{bmatrix} \mathbf{T}_i \\ \mathbf{q}_i \end{bmatrix} = \begin{bmatrix} e'_{11} & e'_{12} \\ e'_{21} & e'_{22} \end{bmatrix} \begin{bmatrix} \mathbf{T}_i \\ \mathbf{q}_i \end{bmatrix}. \qquad (15.10)$$

A number of points may be noted:

- The sequence of matrices must be assembled in the order of the elements in the wall. The values of the elements e'_{jk} change if the positions of two of the matrices are reversed.
- In steady-state conditions, $e'_{11} = e'_{22} = 1$, $e'_{21} = 0$, and e'_{12} is formed as the sum of the thermal resistances such as r or X/λ. The conventional U value is $U = 1/e'_{12}$. In this case the order of the matrices does not matter.
- The real part of the determinant, $e'_{11}e'_{22} - e'_{21}e'_{12}$ is unity, the imaginary part is zero.
- The wall matrix consists of eight real numbers. Since there are two relations between them, it has only six degrees of freedom. These values summarise the behaviour of a wall of arbitrary (one-dimensional) structure.
- The behaviour of a wall driven at some period P has to be summarised by just six real or three complex parameters:

$$\text{outside admittance } (\mathbf{q}_e/\mathbf{T}_e)_{T_i=0} = e'_{22}/e'_{12}, \qquad (15.11a)$$

$$\text{inside admittance } (\mathbf{q}_i/\mathbf{T}_i)_{T_e=0} = e'_{11}/e'_{12'}. \qquad (15.11b)$$

Equation (15.10) gives this quantity a negative sign but that associates $\mathbf{T}_i$ with the phase of the *positively* directed value of $\mathbf{q}_i$, that is, from wall to room. Conventionally, however, $\mathbf{T}_i$ is linked to the phase of the flow from room to wall, so a positive sign is needed. Finally,

$$\text{transmittance } (\mathbf{q}_i/\mathbf{T}_e)_{T_i=0} = -(\mathbf{q}_e/\mathbf{T}_i)_{T_e=0} = 1/e'_{12}. \qquad (15.11c)$$

15.3 THERMAL PARAMETERS FOR AN EXTERNAL MULTILAYER WALL

As an illustration of these methods, we will find the parameters for an external lightweight wall, as described in Table 15.2, when excited sinusoidally with period 24 hours ($P =$ 86 400 s).

The steady transmittance or U value is $1/1.259 = 0.794\,\text{W/m}^2\text{K}$. The plaster has a small τ value (0.1264) and so behaves virtually as a lumped capacity acting at the centre of its lumped resistance. The concrete has a much larger τ value, so we expect considerable attenuation of a wave as it traverses the wall.

Computations will be reported to four places of decimals though it will be recognised that none of the input quantities is actually known accurately. It is convenient to define the quantities

$$A = \cosh(\tau)\cos(\tau), \qquad (15.12a) \qquad B = \sinh(\tau)\sin(\tau), \qquad (15.12b)$$

$$C = [\cosh(\tau)\sin(\tau) + \sinh(\tau)\cos(\tau)]/\sqrt{2}, \quad (15.12c) \qquad D = [\cosh(\tau)\sin(\tau) - \sinh(\tau)\cos(\tau)]/\sqrt{2}, \quad (15.12d)$$

so that $A^2 - B^2 + 2CD = 1$ and $2AB - C^2 + D^2 = 0$.

Then the transmission matrix of a single layer has the form

$$\begin{bmatrix} A + \mathrm{j}B & (C + \mathrm{j}D)/a \\ (-D + \mathrm{j}C)\cdot a & A + \mathrm{j}B \end{bmatrix}.$$

The plaster matrix is

$$\begin{bmatrix} 1.0000 + \mathrm{j} \times 0.0160 & 0.0260 + \mathrm{j} \times 0.00014 \\ & \text{m}^2\,\text{K/W} \\ -0.0065 + \mathrm{j} \times 1.2290 & 1.0000 + \mathrm{j} \times 0.0160 \\ \text{W/m}^2\,\text{K} & \end{bmatrix}.$$

Here the elements e_{111} and e_{221} are (nearly) unity. The element e_{121} (0.0260) is to this accuracy equal to the slab resistance ($0.026\,\text{m}^2\,\text{K/W}$) and the quantity $2\pi\rho c_p X/P$

Table 15.2 Details of wall construction

Element	X (m)	λ (W/m K)	ρ (kg/m^3)	c_p (J/kg K)	r_L (m^{2}K/W)	a_L (W/m^{2}K)	τ	A	B	C	D
Outer film	–	–	–	–	0.060	–	–				
Concrete	0.200	0.19	600	1000	1.053	2.8793	2.1431	−2.3406	3.5344	0.9588	4.1790
Plaster	0.013	0.50	1300	1000	0.026	6.8753	0.1264	1.000	0.0160	0.1788	0.0010
Inner film	–	–	–	–	0.120	–	–				
			Total resistance		1.259						

(1.2290W/m^2 K) is nearly equal to the element e_{212}. The other element in each pair is very small. All this is as expected for a thin slab.

The concrete matrix is

$$\begin{bmatrix} -2.3406 + \mathrm{j} \times 3.5344 & 0.3330 + \mathrm{j} \times 1.4514 \\ & \mathrm{m^2\,K/W} \\ -12.0326 + \mathrm{j} \times 2.7606 & -2.3406 + \mathrm{j} \times 3.5344 \\ \mathrm{W/m^2\,K} & \end{bmatrix}.$$

This slab is thermally thick and it is no longer possible to interpret the elements. The outside film matrix is simply

$$\begin{bmatrix} 1 & 0.060\ \mathrm{m^2\,K/W} \\ 0 & 1 \end{bmatrix}.$$

The product matrix, [outside film matrix] [concrete matrix] [plaster matrix], follows by routine multiplication[4]:

$$\begin{bmatrix} -5.1671 + \mathrm{j} \times 3.8767 & 0.0858 + \mathrm{j} \times 1.7622 \\ & \mathrm{m^2\,K/W} \\ -16.4045 + \mathrm{j} \times (-0.3314) & -2.7102 + \mathrm{j} \times 3.5669 \\ \mathrm{W/m^2\,K} & \end{bmatrix}.$$

Finally, this must be postmultiplied by the inside film matrix,

$$\begin{bmatrix} 1 & 0.120\ \mathrm{m^2\,K/W} \\ 0 & 1 \end{bmatrix},$$

to give

$$\begin{bmatrix} -5.1671 + \mathrm{j} \times 3.8767 & -0.5342 + \mathrm{j} \times 2.2274 \\ & \mathrm{m^2\,K/W} \\ -16.4045 + \mathrm{j} \times (-0.3314) & -4.6787 + \mathrm{j} \times 3.5272 \\ \mathrm{W/m^2\,K} & \end{bmatrix}.$$

This array has the structure

$$T_e = e'_{11} T_i + e'_{12} q_i, \tag{15.13a}$$

$$q_e = e'_{21} T_i + e'_{22} q_i, \tag{15.13b}$$

where of course

$$e'_{11} e'_{22} - e'_{12} e'_{21} = 1.$$

[4]This assumes that the plaster and concrete are everywhere in contact so there is no contact resistance and no step in temperature over the interface. The method of construction normally ensures this. Two preformed, rigid, plane slabs placed together only make contact over discrete areas. Transient flow in these circumstances is discussed by Barber (1989).

So the cyclic transmittance

$$\mathbf{u} = \left(\frac{\mathbf{q}_i}{\mathbf{T}_e}\right)_{T_i=0} = \frac{1}{e'_{12}} = \frac{1}{-0.5342 + \mathrm{j} \times 2.2274} = 0.437\ \mathrm{W/m^2K},\ 6.90\ \mathrm{h\ lag}.$$

This quantity is often expressed as a decrement factor, equal to u/U, so

$$\text{decrement factor} = 0.437/0.794,\ 6.90\,\mathrm{h} = 0.55,\ 6.9\,\mathrm{h}.$$

The inside admittance is

$$\mathbf{y_i} = \left(\frac{\mathbf{q}_i}{\mathbf{T}_i}\right)_{T_e=0} = \frac{-e'_{11}}{e'_{12}} = -\frac{-5.1671 + \mathrm{j} \times 3.8767}{-0.5342 + \mathrm{j} \times 2.2274} = 2.82\ \mathrm{W/m^2K},\ 2.64\ \mathrm{h\ lead}.$$

The outside admittance is

$$\mathbf{y_e} = \left(\frac{\mathbf{q}_e}{\mathbf{T}_e}\right)_{T_i=0} = \frac{e'_{22}}{e'_{12}} = \frac{-4.6787 + \mathrm{j} \times 3.5272}{-0.5342 + \mathrm{j} \times 2.2274} = 2.56\ \mathrm{W/m^2\,K},\ 2.63\ \mathrm{h\ lead}.$$

The outside admittance is not normally needed. Note that

$$\text{cyclic admittance} > U\,\text{value} > \text{cyclic transmittance}.$$

If a heat flow $\mathbf{q_s}$ acts at the exposed surface of the plaster (the node T_2 here), the heating effect it has within the enclosure is the same as the reduced flow $F\mathbf{q_s}$ acting at the room index temperature node T_i, where

$$F = \frac{\text{film transmittance}}{\text{film transmittance} + \text{surface admittance}}.$$

The surface admittance is found from the product matrix [outer film] [concrete] [plaster] given above:

$$\mathbf{y_{surface}} = \frac{-5.1671 + \mathrm{j} \times 3.8767}{0.0858 + \mathrm{j} \times 1.7622} = (2.0522 + \mathrm{j} \times 3.0321)\ \mathrm{W/m^2K},$$

so

$$F = \frac{1/0.120}{1/0.120 + (2.0522 + \mathrm{j} \times 3.0321)} = 0.770,\ 1.085\,\mathrm{h}.$$

The value of F is independent of all heat transfer paths between T_i and the exterior other than the one under discussion. This provides a useful simplification in manual calculations but the value at T_2 as estimated by taking $F\mathbf{q_s}$ to act at T_i is less than the correct value (found when $\mathbf{q_s}$ is properly taken to act at T_2) by an amount

$$\mathbf{q_s}/(\text{film transmittance} + \text{surface admittance}).$$

The values of $\mathbf{u}$ and of $\mathbf{y_i}$ depend on the period P chosen, 24 h here. Their values for the wall given in Table 15.2 are shown in Figure 15.1, in which the horizontal and vertical axes are the real and imaginary parts, respectively, the length of the vector is the magnitude, and the angle the vector makes with the horizontal is the phase angle. As the period increases, both $\mathbf{u}$ and $\mathbf{y_i}$ tend to the steady-state U value. As the period decreases,

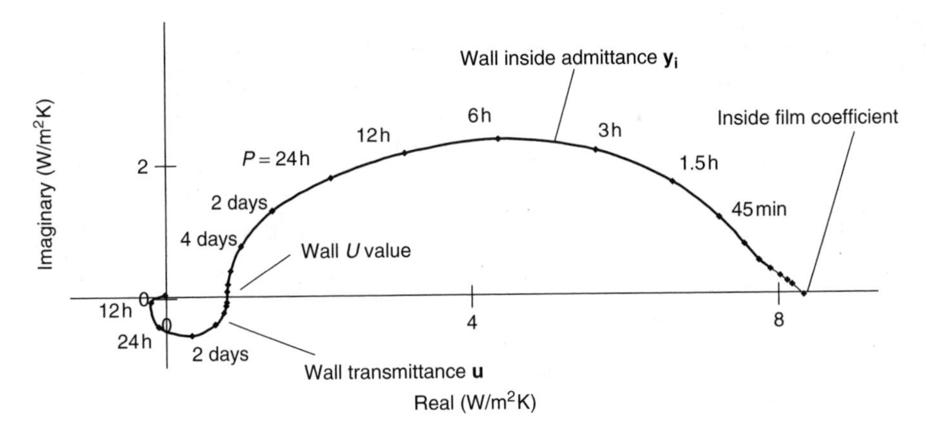

Figure 15.1 Variation of the transmittance and inside admittance of the wall of Table 15.2 with period P

$\mathbf{u}$ tends to zero and $\mathbf{y_i}$ tends to the inner film coefficient, $1/0.12 = 8.3\,\mathrm{W/m^2\,K}$ here. (As P decreases, the values of the elements in the layer and product matrices increase very rapidly and eventually exceed machine capability.) For a further example of $\mathbf{u}(P)$ and $\mathbf{y}(P)$ see Davies (1982b: Figure 3). Curves of this type can be expressed as the ratio of two complex polynomials (see Sanathanan and Koerner 1963). For the admittance,

$$Y(\mathrm{j}\omega) = \frac{A_0 + A_1(\mathrm{j}\omega) + A_2(\mathrm{j}\omega)^2 + A_3(\mathrm{j}\omega)^3 + \cdots}{1 + B_1(\mathrm{j}\omega) + B_2(\mathrm{j}\omega)^2 + B_3(\mathrm{j}\omega)^3 + \cdots}, \tag{15.14}$$

where $\omega = 2\pi/P$. A_0 is the wall U value.

Bode diagrams relate magnitude, phase and period, as does Figure 15.1. Bode diagrams given by Mao and Johannesson (1997) relate to an insulated roof treated as a one-dimensional construction and to the effect on its behaviour when considering the steel stiffening webs which form thermal bridges. The latter case, a two-dimensional situation, is investigated by dividing the construction into finite elements.

Table 15.3 gives a short selection of values taken from Book A of the 1986 *CIBSE Guide*. The resistances of the inside film, outside film and cavity where present, are 0.12, 0.06 and $0.18\,\mathrm{m^2K/W}$, respectively.

15.4 ADMITTANCE OF AN INTERNAL WALL

Although the steady-state heat losses from a room are normally associated with outside walls only, the periodic exchange of heat involves all surfaces. If we can assume that the daily pattern of temperature in an adjoining room is the same as in the room of interest, the heat flows into and out of the two surfaces of the partition wall are always the same. Thus although the wall midplane undergoes swings in temperature, no heat actually flows across it and it can be treated as an adiabatic surface. We are now concerned with only one parameter for the wall – its admittance.

This parameter could be found by determining first the transmission matrix for the wall semi-thickness; the surface admittance $\mathbf{y_s}$ is then e'_{21}/e'_{11}. To illustrate a rather different

Table 15.3 Wall parameters

Construction (outside to inside)	U value (W/m^{2}K)	Admittance		Decrement factor		Surface factor	
		Y (W/m^{2}K)	Lead (h)	f	Lag (h)	F	Lead (h)
Table A3.16							
3b 220 mm brickwork, 13 mm light plaster	1.9	3.6	1	0.46	7	0.61	1
7a 220 mm brickwork, 20 mm glass-fibre quilt, 10 mm plasterboard	1.0	1.4	1	0.34	7	0.84	0
10 200 mm heavyweight concrete block, 25 mm air gap, 10 mm plasterboard	1.8	2.5	1	0.35	7	0.64	0
15 200 mm lightweight concrete block, 25 mm air gap, 10 mm plasterboard	0.68	1.8	2	0.47	7	0.82	1
Table A3.17							
1a 105 mm brickwork, 25 mm air gap, 105 mm brickwork, 13 mm dense plaster	1.5	4.4	2	0.44	8	0.58	2
3a 105 mm brickwork, 50 mm urea-formaldehyde (UF) foam, 105 mm brickwork, 13 mm lightweight plaster	0.55	3.6	2	0.28	9	0.61	1
7 105 mm brickwork, 25 mm air gap, 100 mm medium concrete block, 13 mm lightweight plaster	1.3	3.4	1	0.40	8	0.64	1
9a 105 mm brickwork, 50 mm UF foam, 100 mm medium concrete block, 13 mm lightweight plaster	0.55	3.5	2	0.28	9	0.62	1

though equivalent approach, we will use equation (15.5) for the surface admittance for a slab which is adiabatic on its rear surface.

Take a lightweight concrete wall ($\lambda = 0.19$ W/m K, $\rho = 600$ kg/m^3 and $c_p = 1000$ J/kg K as before, so $a = 2.8793$ W/m^{2}K) and of full thickness 100 mm. Thus τ, based on the semi-thickness, is 0.5358. Using (15.5) it is found that $1/\mathbf{y_s} = 0.0875 - \mathrm{j} \times 0.4617$ m^2 K/W. Now the admittance $\mathbf{y_i}$ of the wall seen from the room index node is related to the surface admittance as

$$\frac{1}{\mathbf{y_i}} = \frac{1}{\mathbf{y_s}} + r_i, \tag{15.15}$$

as follows from continuity. With $r_i = 0.12$ m^2 K/W as before, $y_i = 1.975$ W/m^2 K, 4.39 h (or more realistically, 2 W/m^2 K, 4 h). A heavier and thicker wall will have a $\mathbf{y_i}$ value greater than this, but even though $\mathbf{y}_s$ itself is very large, the magnitude of $\mathbf{y}_i$ cannot exceed $1/r_i$ or 8.3 W/m^2 K in this case.

15.5 DISCUSSION

The steady-state U value depends on the thermal resistance of all elements in a wall, regardless of ordering. The cyclic transmittance $\mathbf{u}$ also depends on all elements, but because of the heat temporarily stored in the wall during the passage of a thermal wave, u is less than U. Small attenuation ($f = u/U$) is associated with a small time lag ϕ, and vice versa. The quoted values for (f, ϕ) in the 1986 *CIBSE Guide* for two contrasted walls, a 105 mm unplastered brick wall and for a 220 mm brick/cavity/220 mm brick wall are (0.88, 3 h) and (0.09, 15 h), respectively.

By contrast, the admittance, surface or index temperature-based, depends largely on the material nearest the surface and little on the construction remote from it. For masonry materials, the effective thickness for a periodicity of 24 h is around 0.1 m. Because of the temporary heat storage, the magnitude of $\mathbf{y}_i$ is always larger than U. The admittance of a construction having negligible thermal storage such as a window is the same as its U value. Magnitudes of $\mathbf{y}_i$ for walls vary comparatively little with construction; the smallest and largest values quoted in the *CIBSE Guide* are 2.1 and 4.3 W/m^2 K. Values less than 1 W/m^2 K are given for a flat roof with a fibreglass quilt beneath. Values of up to 6.5 W/m^2 K are suggested for internal partitions.

It was remarked earlier that a wall with $\tau = 1.18$ provided the largest magnitude of $\mathbf{y}_s(= 1.14a)$. Considerations of thermally 'optimally thick' wall have sometimes entered the design of passive solar buildings. It should be pointed out that when we take account of the film, the optimal magnitude of $\mathbf{y}_i$ is obtained with a smaller value of τ than 1.18. For example, if $a = 12$ W/m^2 K and $r_i = 0.12$ m^2 K/W, the optimal τ is about 0.8 (Davies 1982b: Table 7). The thermal penalty of making the wall thicker than this is hardly significant, but the thickness cannot be justified on thermal grounds.

15.6 AN EXACT CIRCUIT MODEL FOR A WALL

If a wall of any arbitrary one-dimensional construction is driven sinusoidally, it can be modelled exactly by six lumped elements. The representation differs in character from the intuitively plausible representation of a wall as a chain of RC elements (Chapter 10).

First consider the surface admittance of a semi-infinite slab. Its complex transmittance is

$$\frac{\mathbf{q}_0}{\mathbf{T}_0} = \mathbf{a} = a\exp(\mathrm{j}\pi/4) = \left(\frac{2\pi\lambda\rho c_p}{P}\right)^{1/2}\left(\frac{1}{\sqrt{2}} + \frac{\mathrm{j}}{\sqrt{2}}\right)$$

$$= \underbrace{\left(\frac{\pi\lambda\rho c_p}{P}\right)^{1/2}}_{\text{resistive}} + \mathrm{j}\times\underbrace{\left(\frac{\pi\lambda\rho c_p}{P}\right)^{1/2}}_{\text{reactive}}. \tag{15.16}$$

The complex transmittance of a semi-infinite resistive/capacitative medium can be expressed by two lumped elements: a purely resistive transmittance and a purely reactive transmittance. For lightweight concrete (Table 15.1), we have

$$\mathbf{q}_0/\mathbf{T}_0 = 2.0360 + \mathrm{j} \times 2.0360\ \mathrm{W/m^2\,K}$$

and it can be modelled as two transmittances in parallel. In resistance form,

$$\mathbf{T}_0/\mathbf{q}_0 = 0.2456 - \mathrm{j} \times 0.2456\ \mathrm{m^2\,K/W}$$

and it can be modelled as two elements in series. One is negative. These elements are purely equivalent constructs and have no physical reality. The real or resistive element includes a measure of the thermal capacity ρc_p of the slab and the imaginary or reactive element includes a measure of the thermal resistivity $1/\lambda$.

If the heat flow is to be expressed in terms of variation in T_i, the film resistance r_i (0.12 m^{2}K/W) must be included:

$$\mathbf{T}_i/\mathbf{q_0} = (0.2456 + 0.12) - \mathrm{j} \times 0.2456\ \mathrm{m^2\,K/W}.$$

The transmittance form is

$$\mathbf{q_0}/\mathbf{T}_i = 1.8848 + \mathrm{j} \times 1.2661\ \mathrm{W/m^2\,K}$$

Thus, a system comprising the effect of the semi-infinite slab together with convective/radiative exchange at its surface can be modelled by two lumped elements: two transmittance elements in parallel, or two resistive elements in series. There is little to choose between the two forms and the parallel form will be used here. The pair can be treated as a single complex quantity

For a finite-thickness wall including, if wanted, the convective/radiative exchange at its surfaces, we need a model with three complex elements. Consider the circuit of Figure 15.2 with three complex elements in T formation. This is described as a four-pole network, although two poles are identical. It can be shown that the input/output quantities are related as follows:

$$\begin{bmatrix} \mathbf{T}_e \\ \mathbf{q}_e \end{bmatrix} = \begin{bmatrix} 1 + \dfrac{\mathbf{c}}{\mathbf{a}} & \dfrac{1}{\mathbf{a}} + \dfrac{1}{\mathbf{b}} + \dfrac{\mathbf{c}}{\mathbf{ab}} \\ + \mathbf{c} & 1 + \dfrac{\mathbf{c}}{\mathbf{b}} \end{bmatrix} \begin{bmatrix} \mathbf{T}_i \\ \mathbf{q}_i \end{bmatrix}. \tag{15.17}$$

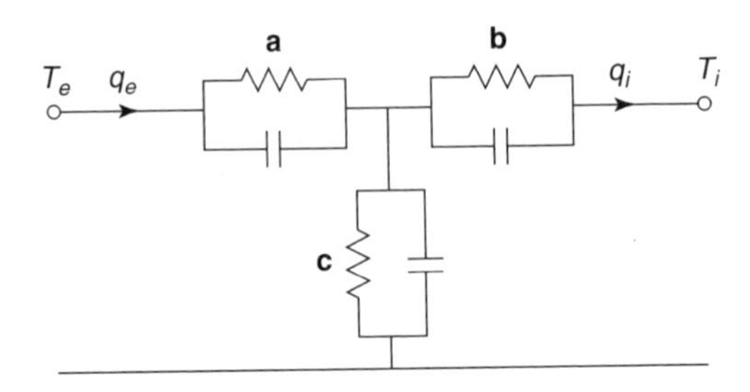

Figure 15.2 An assembly of three complex units to represent exactly the thermal behaviour of a wall when excited sinusoidally

The transmission matrix for a wall section of unit area is

$$\begin{bmatrix} \mathbf{T}_e \\ \mathbf{q}_e \end{bmatrix} = \begin{bmatrix} e'_{11} & e'_{12} \\ e'_{21} & e'_{22} \end{bmatrix} \begin{bmatrix} \mathbf{T}_i \\ \mathbf{q}_i \end{bmatrix}. \tag{15.18}$$

The wall matrix has six degrees of freedom; the T circuit is determined by the choice of six elements. Further, the wall matrix has a determinant of unity; the T circuit matrix also has this property. Thus the overall behaviour of the wall can be represented by a thermal circuit consisting of three complex elements in T formation. Values of **a, b** and **c** for a slab can be found from (15.7) (replacing **a** there by **a**$'$ to avoid ambiguous notation) and (15.17). It is easily checked that as P tends to infinity – we approach the steady state – **a** and **b** each tend to $2\lambda/X$, (together forming the steady-state transmittance λ/X) and **c** tends to $\mathrm{j}2\pi\rho c_p X/P$.

By identification the values for the concrete/plaster wall discussed earlier are

$$\mathbf{c} \equiv \quad e'_{21} \quad = -16.4045 + \mathrm{j} \times (-0.3314)\ \mathrm{W/m^2\,K}, \tag{15.19a}$$

$$\mathbf{a} \equiv \frac{e'_{21}}{e'_{11} - 1} = \quad 2.2531 + \mathrm{j} \times 2.2554\ \mathrm{W/m^2\,K}, \text{ coincidentally almost equal,} \tag{15.19b}$$

$$\mathbf{b} \equiv \frac{e'_{21}}{e'_{22} - 1} = \quad 1.8824 + \mathrm{j} \times 1.2370\ \mathrm{W/m^2\,K}. \tag{15.19c}$$

The cyclic transmittance and admittance can be expressed in terms of these circuit elements:

$$\mathbf{u} = \left(\frac{\mathbf{q}_i}{\mathbf{T}_e}\right)_{T_i=0} = \frac{\mathbf{ab}}{\mathbf{a}+\mathbf{b}+\mathbf{c}}, \tag{15.20a}$$

$$\mathbf{y}_\mathrm{i} = \left(\frac{\mathbf{q}_i}{\mathbf{T}_i}\right)_{T_e=0} = \frac{\mathbf{a}(\mathbf{b}+\mathbf{c})}{\mathbf{a}+\mathbf{b}+\mathbf{c}}. \tag{15.20b}$$

Although the equivalence of the six lumped elements in Figure 15.2 and the distributed resistance and capacity of the concrete/plaster wall is exact, the circuit does not provide a physical representation, for (i) the values of the 'resistive' elements depend on the wall capacity values as well as resistance values, and vice versa; all values depend on the period P (here 24 h) at which the wall response is required; (ii) two of its capacities are not linked to the reference temperature, as they must be in any physical analogue of thermal capacity.

An alternative representation is provided by a T-section with elements in series. Alternative representations are provided too by a Π-section with lumped elements either in series or in parallel. Section 16.1 illustrates the handling of a wall as three complex elements in conjunction with other room details.

15.7 OPTIMAL THREE-CAPACITY MODELLING OF A SLAB

Section 10.4 indicated how, on the basis of sinusoidal excitation, a slab could be modelled by a sequence of equal capacities located at the centre of equal resistances. It will be shown here that a rather better representation of slab behaviour is obtained if unequal resistances or capacities are adopted.

Consider a slab of resistance r and capacity c, which is to be modelled by three capacities. By symmetry, the outside capacities must be equal, so we choose the sequence as βc, $(1-2\beta)c$ and βc. The resistances are similarly αr, $(1-2\alpha)r/2$, $(1-2\alpha)r/2$ and αr. If the slab is divided into three equal T-sections, $\alpha = \frac{1}{6}$ and $\beta = \frac{1}{3}$. We have to write down the transmission matrix of the discretised slab and compare its elements with those of the slab itself.

The discretised product matrix is

$$\begin{bmatrix} 1 & \alpha r \\ 0 & 1 \end{bmatrix}\begin{bmatrix} 1 & 0 \\ \mathrm{j}\omega\beta c & 1 \end{bmatrix}\begin{bmatrix} 1 & \left(\frac{1}{2}-\alpha\right)r \\ 0 & 1 \end{bmatrix}\begin{bmatrix} 1 & 0 \\ \mathrm{j}\omega(1-2\beta)c & 1 \end{bmatrix}\begin{bmatrix} 1 & \left(\frac{1}{2}-\alpha\right)r \\ 0 & 1 \end{bmatrix}\begin{bmatrix} 1 & 0 \\ \mathrm{j}\omega\beta c & 1 \end{bmatrix}\begin{bmatrix} 1 & \alpha r \\ 0 & 1 \end{bmatrix}$$

$$= \begin{bmatrix} f_{111}+\mathrm{j}f_{112} & f_{121}+\mathrm{j}f_{122} \\ f_{211}+\mathrm{j}f_{212} & f_{221}+\mathrm{j}f_{222} \end{bmatrix}, \tag{15.21}$$

where $\omega = 2\pi/P$. We note that $r = X/\lambda$, $c = \rho c_p X$. Accordingly, the product matrix can be expressed in terms of $\tau = (\pi\rho c_p X^2/P\lambda)^{1/2}$ and $a = (2\pi\lambda\rho c_p/P)^{1/2}$:

$$\begin{aligned}
f_{111} = f_{221} &= 1 - 4\beta\left(\tfrac{1}{2}-\alpha\right)\left(\tfrac{1}{2}-\beta+\alpha\right)\tau^4, \\
f_{112} = f_{222} &= \tau^2 - 16\alpha\beta^2\left(\tfrac{1}{2}-\alpha\right)^2\left(\tfrac{1}{2}-\beta\right)\tau^6, \\
f_{121} &= \sqrt{2}\tau/a - 8\sqrt{2}\alpha\beta\left(\tfrac{1}{2}-\alpha\right)\left(\tfrac{1}{2}-\beta+\alpha\beta\right)\tau^5/a, \\
f_{122} &= 4\sqrt{2}\left[\alpha\beta(1-\alpha)+\tfrac{1}{4}\left(\tfrac{1}{2}-\beta\right)\right]\tau^3/a - 16\sqrt{2}\alpha^2\beta^2\left(\tfrac{1}{2}-\alpha\right)^2\left(\tfrac{1}{2}-\beta\right)\tau^7/a, \\
f_{211} &= 0 - 4\sqrt{2}\beta\left(\tfrac{1}{2}-\alpha\right)(1-\beta)\tau^3 a, \\
f_{212} &= \sqrt{2}\tau a - 8\sqrt{2}\beta^2\left(\tfrac{1}{2}-\alpha\right)^2\left(\tfrac{1}{2}-\beta\right)\tau^5 a.
\end{aligned} \tag{15.22}$$

Setting $a = 1$, the values for the slab (Table 15.1) are

$$\begin{aligned}
e_{111} = e_{221} &= \cosh\tau\cos\tau && \to 1, \\
e_{112} = e_{222} &= \sinh\tau\sin\tau && \to \tau^2, \\
e_{121} = e_{212} &= (\cosh\tau\sin\tau + \sinh\tau\cos\tau)/\sqrt{2} && \to \sqrt{2}\tau, \\
e_{122} = -e_{211} &= (\cosh\tau\sin\tau - \sinh\tau\cos\tau)/\sqrt{2} && \to (\sqrt{2}/3)\tau^3.
\end{aligned} \tag{15.23}$$

The coefficients tend to the values shown as τ tends to zero.

The overall difference between the e_{ijk} matrix representing the wall exactly and f_{ijk}, its discretised approximation, is the sum

$$\begin{aligned}
\Delta = {} & \frac{(f_{111}-e_{111})^2 + (f_{112}-e_{112})^2}{e_{111}^2 + e_{112}^2} + \frac{(f_{121}-e_{121})^2 + (f_{122}-e_{122})^2}{e_{121}^2 + e_{122}^2} \\
& + \frac{(f_{211}-e_{211})^2 + (f_{212}-e_{212})^2}{e_{211}^2 + e_{212}^2} + \frac{(f_{221}-e_{221})^2 + (f_{222}-e_{222})^2}{e_{221}^2 + e_{222}^2}.
\end{aligned} \tag{15.24}$$

(Quantities f_{11} and f_{22} are dimensionless, quantities f_{12} and f_{21} have dimensions of m^2 K/W and W/m^2 K, respectively. Δ is dimensionless and is independent of the characteristic admittance a. $(f_{111} - e_{111})^2 + (f_{112} - e_{112})^2$ is the square of the difference between the vectors $\mathbf{f}_{11}$ and $\mathbf{e}_{11}$). Δ provides a single overall measure of the difference in response of a parent wall and its discretised model. It would appear to be a more satisfactory means of comparing the two than attempting to compare magnitudes and phases of individual matrix elements – eight comparisons in all for a typical wall (Letherman 1977).

Δ is here largely proportional to τ^4. The useful quantities are the transmittance $\mathbf{u} = 1/(e_{121} + je_{122})$ and the admittance $\mathbf{y} = (e_{111} + je_{112})/(e_{121} + je_{122})$ or $\mathbf{y} = (e_{221} + je_{222})/(e_{121} + je_{122})$. Thus the fractional error in adopting values of $\mathbf{u}$ and $\mathbf{y}$ based on the f elements, (i.e. on the choice of α and β) is roughly proportional to $\sqrt{\Delta}/\tau^2$. Values for $\sqrt{\Delta}/\tau^2$ for even spacing ($\alpha = 1/6$ and $\beta = 1/3$) and for optimal spacing, found by variation of α and β, are listed in Table 15.4 together with the minimum value of Δ given by the values of α and β.

When τ tends to zero, only the $(f_{122} - e_{122})$ and the $(f_{211} - e_{211})$ terms in (15.24) are non-zero. This condition provides an explicit equation for Δ:

$$\Delta = \tau^4 \left\{ \left[4\left(\alpha\beta(1-\alpha) + \tfrac{1}{4}\left(\tfrac{1}{2} - \beta\right)\right) - \tfrac{1}{3} \right]^2 + \left[-4\beta\left(\tfrac{1}{2} - \alpha\right)(1-\beta) + \tfrac{1}{3} \right]^2 \right\}. \quad (15.25)$$

When $\partial\Delta/\partial\alpha = 0$ and $\partial\Delta/\partial\beta = 0$, $\alpha = 0.1366$ and $\beta = \frac{1}{3}$ exactly. This represents a significant change in α from its even-spacing value of 0.1667, whereas β has its even-spacing value. The minimised value of $\sqrt{\Delta}/\tau^2$, (0.0140), is substantially less than the even-spacing value (0.0414). As τ increases there are second-order changes in α and β. $\sqrt{\Delta}/\tau^2$ however increases somewhat. Thus $\sqrt{\Delta}$ and so the error in $\mathbf{u}$ or $\mathbf{y}$ increases rapidly with τ. Further, the usefulness of minimising Δ decreases. An increase in τ corresponds physically to an increase in slab thickness ($\tau \propto X$) or to an increase in the frequency of excitation, ($\tau \propto 1/\sqrt{P}$). The values are independent of the characteristic admittance a.

In the above treatment, the slab of resistance r and capacity c was assumed to have been first divided into three equal slices and each $\frac{1}{3}$ capacity placed *at the centre* of the corresponding resistance, forming three T-sections which, placed together, formed the seven-element chain $r/6, c/3, r/3, c/3, r/3, c/3, r/6$. Their values were then varied, keeping the total resistance and capacity constant. We could equally have supposed a capacity of $c/6$ to be placed *at the end* of each $r/3$ segment, forming three Π-sections, which when placed together form a seven-element chain – $c/6, r/3, c/3, r/3, c/3, r/3, c/6$. An analysis similar to the one above then indicates that, when optimised, the outer capacities decrease somewhat: when τ is small, the outer capacity is reduced from its even-spacing value ($0.1667c$) to $0.1366c$.

Table 15.4 Variation of $\sqrt{\Delta}/\tau^2$ with τ

τ	$\sqrt{\Delta}/\tau^2$ with $\alpha = \frac{1}{6}, \beta = \frac{1}{3}$	α	β	$\sqrt{\Delta}/\tau^2$ with listed α and β
0.01	0.0414	0.1366	0.3333	0.0140
0.5	0.0420	0.1361	0.3322	0.0159
1.0	0.0475	0.1350	0.3285	0.0288
2.0	0.0669	0.1351	0.3226	0.0590
3.0	0.0910	0.1426	0.3238	0.0886

This is not a useful analysis however as it stands. In practice, surface films at least have to be taken into account, and usually more than a single layer, in which case the a value and the τ value of each layer must be included. Davies (1983a) presents a method to optimally represent multilayer walls (with films) by two or three capacities for 24 h periodic excitation, but that is not fully satisfactory since the discretised wall does not have exactly the same resistance and capacity as the real wall. The analysis demonstrates, however, that provided there is an appropriate choice for the number of slices in the discretisation, improved accuracy in modelling can be achieved if finer division is made near exposed wall surfaces. It is intuitively clear that this must be so, since the admittance $\mathbf{y}$ of a wall depends much more on the wall construction near the relevant surface than on the construction remote from it. Section 17.8 outlines a procedure to discretise a wall in an optimum manner, based on transient wall response.

The full form of (15.24) may not be appropriate. Suppose that we have a wall of arbitrary construction and when investigating its dynamic behaviour, we are only interested in its admittance $\mathbf{y_I}$ as seen from inside (i.e. we are not interested in its external admittance). Now $\mathbf{y_i} = \mathbf{e}_{11}/\mathbf{e}_{12}$, so it would be appropriate here to form Δ from the 11 and 12 elements only of the transmission matrices of the real and discretised walls.

15.8 APPENDIX: COMPLEX QUANTITIES AND VECTOR REPRESENTATION

The quantity $\mathbf{T_0}$ is taken to denote $T_0 \exp(\mathrm{j}2\pi t/P)$. T_0 is the scalar amplitude of sinusoidal variation of the temperature T_0; $\exp(\mathrm{j}2\pi t/P)$ is a unit vector rotating anticlockwise with time t, period P. The observable value of T_0 at time t, the deviation from its central value, is the projection $T_0 \cos(2\pi t/P)$ on the horizontal. T_0 has its maximum value at $t = 0$.

In sinusoidal heat flow, the phases of temperatures and heat flows can be expressed in the dimensional term or the exponential term:

$$\begin{aligned}\mathbf{T}\exp(\mathrm{j}\omega t) &= (T_r + \mathrm{j}T_i)\exp(\mathrm{j}\omega t)\\ &= (T_r^2 + \mathrm{j}T_i^2)^{1/2}\exp(\mathrm{j}\arctan(T_i/T_r))\exp(\mathrm{j}\omega t)\\ &\equiv T\exp(\mathrm{j}(\omega t + \arctan(T_i/T_r))).\end{aligned} \tag{15.26}$$

If a wall is initially at zero and then experiences a heat flow into it, it initially registers little temperature. If the heat flow is sinusoidal, the maximum flow into the wall falls at a time before the maximum surface temperature it causes. The flow is written as $\mathbf{q_0}$, which denotes $q_0 \exp(\mathrm{j}(2\pi t/P + \phi))$, where q_0 is the scalar amplitude of variation and ϕ is its phase in relation to T_0. q_0 has its maximum value at the earlier time $t = -\phi P/2\pi$. In the expressions in this chapter, temperature and heat flow appear as vectors on both sides of the equations, and the time variation $\exp(\mathrm{j}2\pi t/P)$ can be dropped: only the phase difference between them is needed.

The product of two complex quantities in Cartesian form is

$$(a + \mathrm{j}b)(c + \mathrm{j}d) \equiv (ac - bd) + \mathrm{j}(ad + bc); \tag{15.27}$$

it is used in forming the product of wall layer matrices. The polar form is not needed.

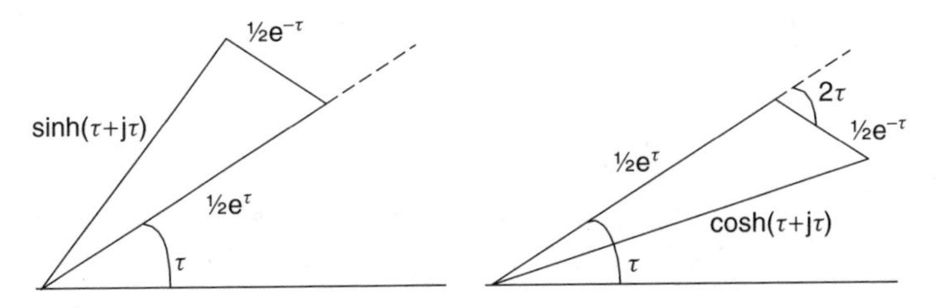

Figure 15.3 The vectors $\sinh(\tau+\mathrm{j}\tau)=\frac{1}{2}\mathrm{e}^{\tau}\mathrm{e}^{\mathrm{j}\tau}-\frac{1}{2}\mathrm{e}^{-\tau}\mathrm{e}^{-\mathrm{j}\tau}$ and $\cosh(\tau+\mathrm{j}\tau)=\frac{1}{2}\mathrm{e}^{\tau}\mathrm{e}^{\mathrm{j}\tau}+\frac{1}{2}\mathrm{e}^{-\tau}\mathrm{e}^{-\mathrm{j}\tau}$

The quotient is

$$\frac{a+\mathrm{j}b}{c+\mathrm{j}d}=\underbrace{\left(\frac{ac+bd}{c^2+d^2}\right)+\mathrm{j}\left(\frac{bc-ad}{c^2+d^2}\right)}_{\text{Cartesian form}}=\underbrace{\left(\frac{a^2+b^2}{c^2+d^2}\right)^{1/2}\exp\left(\mathrm{j}\arctan\left(\frac{bc-ad}{ac+bd}\right)\right)}_{\text{polar form}};\tag{15.28}$$

it is used in evaluating the admittance. Since the phase lies between 0 and $\pi/2$ radians (0 and 6 h when $P=24$ h), its evaluation is straightforward. The phase of the transmittance however may exceed $\pi/2$ radians and requires comment:

$$\mathbf{u}=\frac{\mathbf{q}}{\mathbf{T}}=\frac{1}{a+\mathrm{j}b}=\frac{1}{(a^2+b^2)^{1/2}}\exp(\mathrm{j}\arctan(-b/a)).\tag{15.29}$$

For a thin wall, a and b are positive numbers, so the phase difference $\phi=\arctan(-b/a)$ is negative, indicating that $\mathbf{q}$ lags behind $\mathbf{T}$ by less than $\pi/2$ radians or 6 h in a 24 h cycle. With increasing thickness, a becomes negative and the standard routine returns a positive value for ϕ. Thus if $a<0, \phi\to\phi-\pi$. Parameter b is initially positive but becomes negative. When a eventually becomes positive again, $\phi\to\phi-2\pi$. By this time the magnitude of $\mathbf{u}$, $1/(a^2+b^2)^{1/2}$, has become very small; see equation (10.59).

The Cartesian forms of $\cosh(\tau+\mathrm{j}\tau)$ and $\sinh(\tau+\mathrm{j}\tau)$ can be found using the usual expansions, noting that $\cosh(\mathrm{j}\tau)=\cos(\tau)$ and that $\sinh(\mathrm{j}\tau)=\mathrm{j}\sin(\tau)$. The polar forms can be demonstrated using a vector diagram. We note that

$$\cosh(\tau+\mathrm{j}\tau)=\tfrac{1}{2}\mathrm{e}^{\tau}\mathrm{e}^{\mathrm{j}\tau}+\tfrac{1}{2}\mathrm{e}^{-\tau}\mathrm{e}^{-\mathrm{j}\tau}\tag{15.30}$$

and can be represented as a vector of length $\frac{1}{2}\mathrm{e}^{\tau}$ making an angle of $+\tau$ with the horizontal, together with a vector $\frac{1}{2}\mathrm{e}^{-\tau}$ at an angle $-\tau$ with the horizontal (Figure 15.3). The magnitude (Table 15.1) follows from the cosine formula.

16

Frequency-Domain Models for Room Response

This chapter deals with room thermal models in which the thermal behaviour of the walls is handled using the period-based parameters of the last chapter. After a discussion of the basic principles underlying this approach, we discuss the case where just the 24 h period, together with the steady state are taken to describe the behaviour. This is followed by a more detailed analysis in which more harmonics are included.

16.1 BASIC PRINCIPLES

It is convenient to start with the thermal model shown in Figure 16.1, which is a development of the elementary enclosure of Figure 4.6. It corresponds to an enclosed space whose walls (and floor and ceiling) are of uniform massive construction so that its thermal behaviour can be represented by three complex elements **A, B** and **C**, as shown in Section 15.6. The resistance and capacity elements in this form of modelling are purely equivalent constructs and do not approximate to a physical model. The capacity elements are no longer necessarily connected to the reference temperature. The single surface temperature is T_3. The air temperature is T_a or T_2 and the ventilation conductance is V. There will be no radiant exchange since there is only one surface, but in order to generalise the model later, the rad-air model will be used. Thus there is a conductance H between T_3 and the rad-air temperature T_{ra} or T_1 representing a lumped convective/radiative exchange, and the conductance X, having no simple physical interpretation, between T_2 and T_1. In steady-state conditions the **A, B, C** configuration, together with H, reduces to the conductance L in Figure 4.6. Heat inputs are supposed to act at each node. In the present case, one source is redundant. If the enclosure were heated by a hot-body source, its radiant component would be represented as Q_3 acting at T_3 and its convective component Q_2 at T_2; Q_1 would be unnecessary. Alternatively, the radiation could be modelled as an augmented input Q_1 at T_1, in which case Q_2 represents the convective input as before, *less* the radiant excess; Q_3 is now superfluous but in a room with a window (not present in this simple model), it could represent absorbed solar radiation. The three separate inputs illustrate the structure of the solution.

The ventilation loss is driven by ambient temperature T_e. The heat flow inward through the shell is driven by ambient temperature together with solar gain absorbed at the wall

Building Heat Transfer Morris G. Davies
© 2004 John Wiley & Sons, Ltd ISBN: 0-470-84731-X

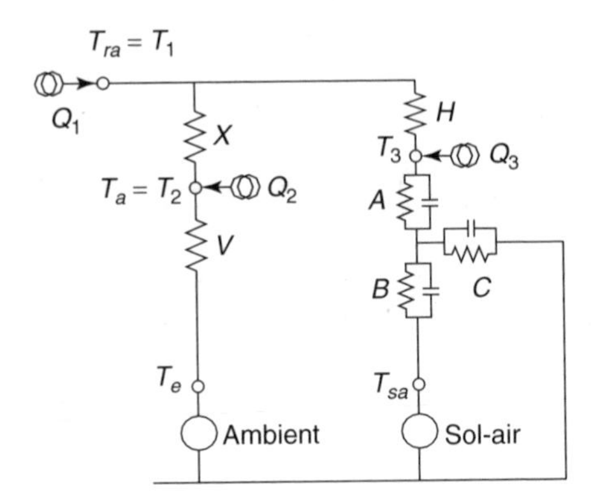

Figure 16.1 Thermal circuit to illustrate the basic principles of periodic response

exterior, and the two will be combined as sol-air temperature T_{sa}. All the drivers vary with time. Over a fixed period, such as a day or a week, the drivers at any instant t can be represented as a Fourier series, e.g., Q_1 can be represented as the real component of

$$Q_1(t) = Q_{10} + \mathbf{Q}_{11}\exp(\mathrm{j}\omega t) + \mathbf{Q}_{12}\exp(\mathrm{j}2\omega t) + \mathbf{Q}_{13}\exp(\mathrm{j}3\omega t) + \cdots, \tag{16.1}$$

where Q_{10} is the time-averaged value and $\mathbf{Q}_{11}$ is the complex amplitude of the first harmonic, etc. In the following analysis, the harmonic subscript will be dropped.

Heat continuity at nodes 1, 2, 3 and 4 (an equivalent node 'in' the wall) gives

$$(\mathbf{T}_1 - \mathbf{T}_2)X + (\mathbf{T}_1 - \mathbf{T}_3)H = \mathbf{Q}_1, \tag{16.2a}$$

$$(\mathbf{T}_2 - \mathbf{T}_1)X + (\mathbf{T}_2 - \mathbf{T}_e)V = \mathbf{Q}_2, \tag{16.2b}$$

$$(\mathbf{T}_3 - \mathbf{T}_1)H + (\mathbf{T}_3 - \mathbf{T}_4)\mathbf{A} = \mathbf{Q}_3, \tag{16.2c}$$

$$(\mathbf{T}_4 - \mathbf{T}_3)\mathbf{A} + (\mathbf{T}_4 - \mathbf{T}_{sa})\mathbf{B} + (\mathbf{T}_4 - 0)\mathbf{C} = 0. \tag{16.2d}$$

We wish to express the solution in terms of $\mathbf{L}_u$ and $\mathbf{L}_y$, where $\mathbf{L}_u = A\mathbf{u}$ and $\mathbf{L}_y = A\mathbf{y}$; A is the area of the massive walls, and $\mathbf{u}$ and $\mathbf{y}$ are the wall cyclic transmittance and admittance, respectively. For a period of 24 h, $\mathbf{u}$ and $\mathbf{y}$ are tabulated quantities; in the *CIBSE Guide* $\mathbf{u}$ is written as fU and $\mathbf{y}$ as Y.

$\mathbf{L}_u$ and $\mathbf{L}_y$ include the quantity $H(= Ah_i)$. The corresponding measures excluding H are $\mathbf{F}_u$ and $\mathbf{F}_y$, which are related to **A, B** and **C** as follows:

$$\mathbf{F}_u = \frac{\mathbf{AB}}{\mathbf{A} + \mathbf{B} + \mathbf{C}}, \tag{16.3a}$$

$$\mathbf{F}_y = \frac{\mathbf{A}(\mathbf{B} + \mathbf{C})}{\mathbf{A} + \mathbf{B} + \mathbf{C}}; \tag{16.3b}$$

see equation (15.20). They are related as

$$\mathbf{L}_u = \frac{H\mathbf{F}_u}{H + \mathbf{F}_y} \quad \mathbf{L}_y = \frac{H\mathbf{F}_y}{H + \mathbf{F}_y}. \tag{16.4}$$

The quantity C_v (*CIBSE Guide* notation) will be taken to combine X and V:

$$C_v = \frac{XV}{X+V}. \tag{16.5}$$

After some manipulation, the values for the temperatures resulting from all five independent drivers are given by the equation,

$$\begin{bmatrix} \mathbf{T}_1(\mathbf{L}_y + C_v) \\ \mathbf{T}_2(\mathbf{L}_y + C_v) \\ \mathbf{T}_3(\mathbf{L}_y + C_v) \end{bmatrix} = \begin{bmatrix} 1 & \frac{C_v}{V} & \frac{\mathbf{L}_y}{\mathbf{F}_y} & C_v & \mathbf{L}_u \\ \frac{C_v}{V} & \frac{\mathbf{L}_y + X}{X}\frac{C_v}{V} & \frac{\mathbf{L}_y}{\mathbf{F}_y}\frac{C_v}{V} & \frac{\mathbf{L}_y + X}{X}C_v & \frac{C_v}{V}\mathbf{L}_u \\ \frac{\mathbf{L}_y}{\mathbf{F}_y} & \frac{\mathbf{L}_y}{\mathbf{F}_y}\frac{C_v}{V} & \frac{C_v + H}{H}\frac{\mathbf{L}_y}{\mathbf{F}_y} & \frac{\mathbf{L}_y}{\mathbf{F}_y}C_v & \frac{C_v + H}{H}\mathbf{L}_u \end{bmatrix} \begin{bmatrix} \mathbf{Q}_1 \\ \mathbf{Q}_2 \\ \mathbf{Q}_3 \\ \mathbf{T}_e \\ \mathbf{T}_{sa} \end{bmatrix}. \tag{16.6}$$

We can make several points on its structure:

- It only differs from the corresponding steady-state equation in that a distinction must now be made regarding the wall's response to excitation from *inside* with $\mathbf{L}_y$ as factor, and excitation from *outside* by T_{sa} with $\mathbf{L}_u$ as factor. In the steady state, only the single scalar quantity $L(= AU)$ is needed.
- The simplest response due to heat input is that due to application of $\mathbf{Q}_1$ at node 1 when

$$\mathbf{T}_{11} = \frac{\mathbf{Q}_1}{\mathbf{L}_y + C_v}. \tag{16.7}$$

  Since $C_v/V < 1$ and $\mathbf{L}_y/\mathbf{F}_y < 1$, application of heat at node 2 or node 3 results in a smaller response at node 1, as is physically obvious.
- The reciprocity relationship

$$\mathbf{T}_{12}/\mathbf{Q}_2 = \mathbf{T}_{21}/\mathbf{Q}_1 \tag{16.8}$$

  is evident, and similarly for nodes 1 and 3 and nodes 2 and 3.
- The response at node 2 due to $\mathbf{T_e}$ is the same as that due to $\mathbf{Q}_2$ if $\mathbf{T_e}\mathrm{V} = \mathbf{Q}_2$ (as is evident from the continuity equation). There are five similar relations.
- The values of the temperatures in (16.6) are the components of the total temperatures at these nodes which vary sinusoidally with period P. These components can be represented as the projections on the horizontal of vectors rotating with period P. They, and their differences, can be presented in vector form as shown in Davies (1982a: Figure 11b). The heat flows through resistive elements are simply proportional to the differences, and the heat flow vector is parallel to the temperature-difference vector. There is a phase shift associated with heat flows into capacitative elements, so there is an angle between the flow vector and its associated temperature. These heat flows too can be shown in a vector diagram. If, for example, a node is the point of confluence of three flows, by continuity they map as a triangle.
- As formulated above, the equations estimate the response of a room to uncontrolled inputs. Now the net input $\mathbf{Q}_2$ to the air point, $\mathbf{T}_2$ or $\mathbf{T}_a$, is composed of the convective

fraction from internal heat sources, the withdrawal of excess radiation and the flow of heat or coolth from any air-conditioning unit. To find the heating/cooling load in a controlled environment, the steady-state component of air temperature T_2, say, is set to the required temperature and all T_2 harmonics are set to zero.

The scheme can be generalised. Figure 7.2 illustrates five distinct means of heat loss from a room: by ventilation from the air temperature (T_2); through the window (T_3), which is taken to have no thermal storage; through an outer wall (T_4); into a solid ground floor (T_5), where it may depend more on a constant deep ground temperature, possibly with yearly variation, than on daily variation in ambient, and through a partition wall (T_6) into an adjoining room. Figure 7.2(ii) illustrates modelling for periodic excitation (the element values depend on the period chosen). The outer wall may be expected to require use of the three complex elements shown, although a simple r-c-r element may be sufficient for a partition wall. If there are no significant periodic drives through the floor, its behaviour can be modelled by the single complex unit.

Figure 7.2(ii) for sinusoidally driven losses can be used in conjunction with any of the three models for internal exchange. If (c), the rad-air model, is selected, the model is similar to Figure 16.1 and a set of equations similar to (16.2) can be written down. One set is needed for each period taken. To find the response to some particular day's conditions, supposedly repeated indefinitely, we have to process the history of ambient temperature, solar incidence on each surface and internal heat production so we can find the amplitude and phase of each Fourier component, at $P = 24$, 24/2, 24/3, 24/4, etc. (P in hours). Between 5 and 10 harmonics may be needed. The author (Davies 1986 III: Figure 3) used this method to examine the response of a passive building to solar incidence. If the action of a sensor on an air-conditioning unit is to be examined, where a fast response is expected, correspondingly smaller values of P will be needed. The net response is the algebraic sum of all component responses.

The scheme can be further generalised to model say a week's meteorological data, in which case there may be minor contributions for a few components of type $P = 168$, 168/2, 168/3, etc., in addition to significant contributions for $P = 24$, 24/2, etc. (P in hours). In principle, several years' data could be examined in this way but unless a wall is very thick, its cyclic transmittance for a period of a year is near its steady-state value; if it is very thick, its transmittance is small and internal temperature will be largely determined by ventilation rate.

16.2 24 HOUR PERIODICITY: ADMITTANCE MODEL

In principle the above treatment applies to the steady state and any number of harmonics determined on the basis of a fixed period. The most natural choice of period is the daily period of 24 h and much useful analysis can be conducted when the steady state and first harmonic alone are considered. That is the topic of this section.

The bare essentials to find appropriate wall parameters, admittance $\mathbf{y}$ and cyclic transmittance $\mathbf{u}$, are described in Section 10.1. But as the analyses in Section 15.3 have shown, if a single layer of material with distributed capacity and resistance is substituted for a lumped capacity, the evaluation of wall parameters in pre-computer times, using mechanical calculating machines and tabulated hyperbolic functions, was a laborious task. It was correspondingly greater if two or more layers were considered. An interest developed in the 1930s. A large number of workers have been active in the area but their work is now of historical interest only; see the reviews of Stephenson (1962), Gupta (1970) and Gertis and Hauser (1975), in which there are many references to European work, and Davies (1983b). Work by the American authors Houghton *et al.* (1932) and Alford *et al.* (1939) probably led Mackey and Wright (1943, 1944, 1946b) to undertake their classical work, in effect, to find the transmittance **u** for use in cooling load calculations. They took fixed values for the wall inside and outside films, expressed a solid layer in terms of λ/X and $\lambda\rho c_p$ values and evaluated the magnitudes and phases of the quantity T_{si}/T_e, where T_{si} is the inside surface of the wall. In their graphs, the $m(T_{si}/T_e)$ axis is in logarithmic form, so that about half the area of the chart relates to values of 0.02 down to 0.001, which are rather small for use in cooling load calculations. There would have been merit in non-dimensionalising the units. To adapt these graphs to multilayer walls, they suggested using a composite $\lambda\rho c_p$ value. If r_i and c_i are the resistance and capacity respectively of layer i in a wall of n solid layers,

$$(\lambda\rho c_p)_{\text{composite}} = 1.1\left(\sum c_i\right) / \left(\sum r_i\right) - 0.1c_1/r_1,$$

summing over all layers and with layer 1 as the outermost layer. This expression assumes that r_1 is not small compared with the other resistances. With the pragmatically chosen values of 1.1 and 0.1, this is a convenient rule of thumb rather than a rational expression.

Significant advances were also made by Danter (1960), Pratt (1965a, 1965b) and Tavernier (1972). Danter started from Mackey and Wright's ideas but improved the presentation by use of the cyclic transmittance, non-dimensionalised as $\mathbf{f} = \mathbf{u}/U$; charts were given of the magnitude and time lag of **f** and are reproduced in Book A of the 1970 *IHVE Guide*, Book A p. 8–6.

The greater part of the work during this period was concerned with the *transmittance* property of a wall – the value of the heat flow q_i into the interior due to change in external temperature variation. There appears to have been much less effort devoted to the wall *admittance* property – variation in q_i when the internal temperature varies. Work by Shklover (1945), Muncey (1953), Barcs (1967) and Knabe (1971a, 1971b) may be noted. Values for $(\mathbf{q}_i/\mathbf{T}_i)$ when $T_e = 0$ were evaluated by Danter using an electrical analogue computer in the 1960s and are listed in Loudon (1970) and in Book A of the 1970 *IHVE Guide*.

The consequences of the continuity of temperature and heat flow at the interfaces of multilayer walls, if addressed directly by elementary algebra, lead to unwieldy expressions which are laborious to evaluate manually, although several authors tackled the problem; see Davies (1983b: 182). Evaluation is much better performed using matrix algebra.

The matrix method of handling transfer phenomena was discussed in detail by Pipes (1940) in a electrical context and for thermal purposes by van Gorcum (1951) and Vodicka (1956). Pipes (1957) independently presented an analysis of heat conduction problems in matrix terms but it was expressed in essentially electrical rather than thermal language and was not developed beyond analytical expressions in complex hyperbolic forms. There is no clear connection between these analyses and the cumbersome but practical working tools of Mackey and Wright, of direct help to the services engineer. The lists of transmittances and admittances listed in Book A of the 1986 *CIBSE Guide* (Tables A3.16 to A3.21) were taken from Milbank and Harrington-Lynn (1974), who used the matrix method.

The most advanced use of these sinusoidal wall parameters appears to have been in the 'admittance model', developed during the 1960s and 1970s at the UK Building Research Establishment (Loudon 1970; Danter 1960, 1974, 1983; Davies 1994.) The procedure was developed as a means of analysing the thermal behaviour of the lightweight, much-glazed buildings that had become popular after the Second World War and which, in the absence of air conditioning, had suffered severe overheating during prolonged sunny spells. In conjunction with the environmental temperature model, the admittance procedure forms the approach currently adopted by the UK Chartered Institute of Building Services Engineers (IHVE 1970; CIBSE 1986, 1999). Walsh and Delsante (1983) reviewed the frequency response approach at that time. In 1985 Oiry and Bardon reported a study on the thermal response of a simple building to periodic excitation, solar and ambient, conducted from first principles and without reference to the main body of relevant literature. In spite of the fact that computational tools are now sufficiently powerful and cheap to handle most problems in building response, Lebrun and Nusgens (2000) have argued that this simplified approach can help in detailed simulation.

The drivers were taken to be the daily mean air temperature and its amplitude of variation during the day, together with the daily mean solar radiation falling on some outer surface and its peak value. These were chosen to be almost extreme values. The procedure enabled the designer at the design stage to check the thermal consequences of the design factors over which they had control: window size, design and orientation, weight of structure and what effect the rate of air infiltration from outside might have on the peak temperature that might be reached during very warm periods. The calculation is conducted in two distinct stages: evaluation of the steady state reached in the building, and the swing of indoor temperature about its mean; the peak temperature reached is the sum of the two. The swing evaluation follows the same lines as the steady-state calculation but with some additional features. The procedure did not claim to be very accurate but it was quite quick to perform manually to examine the effect of some variation, perhaps the change from single glazing to double glazing. It provided excellent teaching material. There are three stages to setting up the model.

(a) Internal heat exchange

The separate processes of convective and radiative exchange within a room are lumped, so there is a conductance $A(\frac{6}{5}Eh_r + h_c) = A(\frac{6}{5} \times 0.9 \times 5.7{+}3.0) = 9.2\,A$ W/K between the room index temperature, environmental temperature T_{ei}, and the node representing the mean temperature of a surface of area A.[1] The mean air temperature is represented as a separate node T_a and there is the conductance of $h_a \sum A$ or X or $4.5 \sum A = (S + C)C/S$ between T_{ei} and T_a, where $S = \sum AEh_r$ and $C = \sum Ah_c$; (Section 7.3.1). Internal heat gains, variously due to the heating system, lighting, occupant and other casual gains have convective components totalling Q_c and radiant components Q_r. Q_c is taken to act at the air node T_a. Q_r could be taken as distributed over all the internal surfaces of the room, but it is more convenient to suppose instead that an augmented value $Q_e = (1 + C/S)Q_r$ is input at T_{ei} and the excess $(C/S)Q_r$ is withdrawn from T_a so that $Q_a = Q_c - (C/S)Q_r$. This follows from the series-parallel transform (Section 4.4.3). Solar gains are handled

[1]From CIBSE (1986: equation A5.100). A value of 8.3A W/K is suggested on page A3-8.

separately as noted below. As discussed in Section 7.3.3, the logic of setting up the model is flawed (Davies 1992a, 1996a), but it is satisfactory from a pragmatic viewpoint.

(b) External heat loss

The loss from the enclosure due to ventilation is $V = nsV_r$ where n is the number of air changes per second, s is the volumetric specific heat of air (1200 J/m^{3}K) and V_r is the room volume. The conduction conductance through a wall of area A is AU where $1/U = 1/(Eh_r + h_c) + \sum d/\lambda + 1/h_e$; h_e is the external film coefficient, dependent on wind speed but of order 18 W/m^{2}K. Their total is $\sum AU$. The model is shown in Figure 16.2a, after Figure A5.1 of the *CIBSE Guide*.

(c) Heat inputs and response

It is readily shown that the steady-state value of T_{ei} is

$$T_{ei} = T_e + \frac{\overline{Q}_e + \overline{Q}_{a.}C_v/V}{\sum AU + C_v} \approx \frac{\overline{Q}_e + \overline{Q}_a}{\sum AU + V}, \tag{16.9}$$

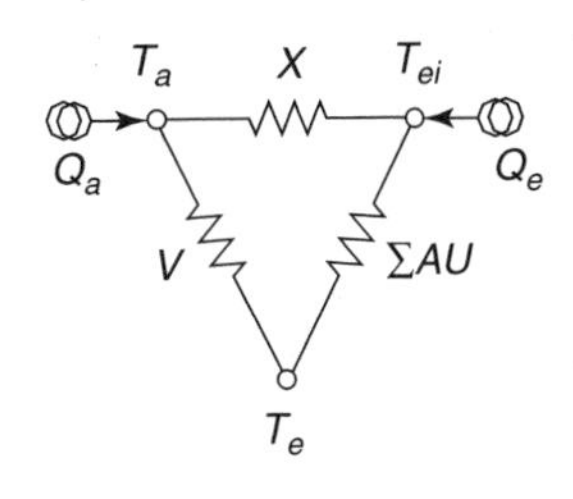

(a) The CIBSE steady-state environmental temperature model.

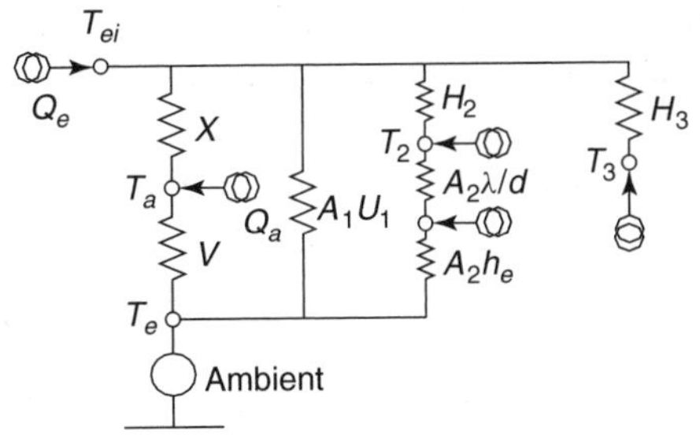

(b) The CIBSE model redrawn to show two representations of an outer wall and an inner wall.

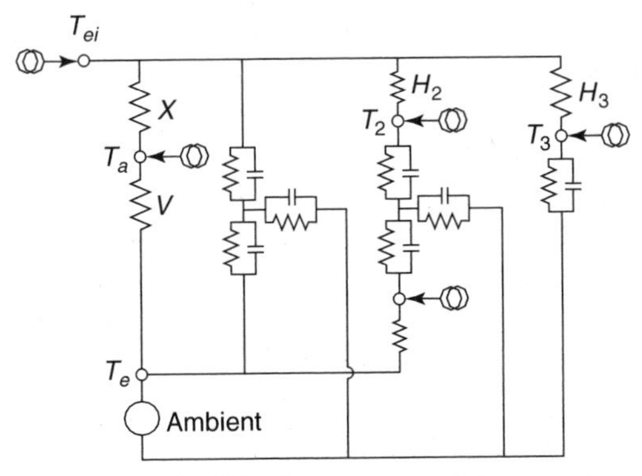

(c) Thermal circuit for the environmental temperature model for sinusoidal excitation, period 24 h.

Figure 16.2

where $1/C_v = 1/X + 1/V$ (equation 8.6 in the 1986 *CIBSE Guide*). (X tends to be numerically large compared with $\sum AU$ and V, and if it is neglected then T_{ei} and T_a coincide and we have the traditional simple building model.) It has been noted before that when evaluating T_{ei}, the effect of the full Q_a acting at T_a is the same as the reduced input $Q_a C_v / V$ acting at T_{ei}, as is evident generally in (16.5), although the value it leads to at T_a is lower than the correct value by an amount $Q_a/(V + X)$.

Figure 16.2b shows the model redrawn in the form adopted here for room thermal circuits, where the total conductive loss conductance $\sum AU$ is expanded to show three possible elements. The conductance for surface 1 is shown simply as a single element, $A_1 U_1$. Surface 2 is represented by its inside film $H_2 = A_2(Eh_r + h_c)$, the conductance $A_2 \lambda / d$ of a single layer, and the outer film $A_2 h_e$. Surface 3 is an internal adiabatic surface and does not contribute to $\sum AU$. Possible solar gains are indicated.

Solar gains due to direct and diffuse sunshine affect indoor temperatures through three mechanisms:

- Some of the radiation incident on external surfaces is absorbed there, much of it on dark surfaces and little on light surfaces. In the admittance procedure it can be handled by substituting the appropriate sol-air temperature for the surface concerned. The phase of the solar-generated component varies with the orientation of the surface.
- Some radiation is absorbed in the glass of a window. This is only a small fraction for clear glass but may be large for radiation-absorbing glasses. In the admittance procedure, the effect of window-absorbed gains is translated through use of solar gain factors into gains at the environmental and air temperature nodes.
- Radiation that is reflected outward from the window does not serve as a heat load. Transmitted radiation is partly absorbed, partly reflected at internal surfaces. Much of it can often be assumed to fall on the floor and furniture but since the pattern changes continuously through the day, any simple assumption regarding transmitted gains must be approximate. Thermal comfort is mainly determined by the total transmitted fraction; its distribution is of secondary importance. Some small fraction may be lost once more as short-wave radiation through the window. In the admittance procedure, transmitted solar radiation is handled using of the solar gain factors $\overline{S}_e$ and $\overline{S}_a$ defined below. A few typical values are listed in Table 16.2, evaluated for a south-west aspect at the latitude of London. They were found as described by Jones (1980) and are discussed in Chapter 9. Presumably they are for the summer period. In winter the sun is lower in the sky, leading to more normal incidence, less reflection and more transmission than in summer. Thus winter S values should be rather larger than summer values. Davies (1980: Tables 14 and 15) lists values for all months for a south-facing window. Not only do they take account of the hour-by-hour variation in beam radiation and the associated absorptivity and transmissivity of the relevant glazing, but they also include appropriate values of the inside and outside film coefficients. An example of their calculation is given in the 1999 *CIBSE Guide*.

$$\overline{S}_e = \frac{\text{mean solar gain, transmitted and absorbed, per square metre of glazing, acting at } T_{ei}}{\text{daily mean solar intensity on the facade}},$$

$$\overline{S}_a = \frac{\text{mean solar gain per square metre of glazing at } T_a}{\text{mean solar intensity on the facade}},$$

$$\tilde{S}_e = \frac{\text{total swing in gain per square metre acting at } T_e}{\text{swing in solar intensity on the facade}},$$

$$\tilde{S}_a = \frac{\text{total swing in gain per square metre acting at } T_a}{\text{swing in solar intensity on the facade}}. \qquad (16.10)$$

To represent the swing behaviour of an enclosure, the thermal capacity of the walls must be modelled as four-pole networks consisting of three parallel RC pairs (Figure 16.2c). The four-pole configuration can be designed to include the surface films, as for wall 1, or to model the solid construction only, as for wall 2, in which case the inside and outside films are made explicit. Locations for nodes are now available at which long-wave and solar radiation can be input. If an internal wall is in effect adiabatic at its rear surface (wall 3), a single RC pair is sufficient.

A circuit of this kind can be evaluated readily using standard circuit theory, inputting solar gains at the nodes where they act physically; a computer routine to solve simultaneous equations with complex coefficients is needed since the drivers have magnitude and phase, the wall elements are complex and the output temperatures (the swings in T_{ei} and T_a) have magnitude and phase.

This is too complicated for manual implementation. As a preliminary design tool, the wall is described instead by its complex **u** and **y** values for which the magnitude is of primary importance and the phase is of secondary importance. Sample values are given in Table 15.3. As noted earlier, **u** is written as fU, and the magnitude and time lag of f are listed. Also **y** is written as Y, and the magnitudes and time leads are listed.

Table 16.1 is a short list of solar gain factors taken from Tables A5.3 and A5.4 in Book A of the 1986 *CIBSE Guide*. The factors for use in the swing calculation depend on whether the building fabric can be described as 'heavy' or 'light', which is determined by the ratio

$$f_r = \frac{\sum AY + V}{\sum AU + V}. \qquad (16.11)$$

Table 16.1 Steady-state and alternating solar gain factors

	Steady state		Lightweight		Heavyweight	
	$\bar{S}_e$	$\bar{S}_a$	$\tilde{S}_e$	$\tilde{S}_a$	$\tilde{S}_e$	$\tilde{S}_a$
Clear 6 mm glazing	0.76	0	0.64	0	0.47	0
Clear 6 mm glazing with light horizontal slats	0.31	0.16	0.28	0.17	0.24	0.17
Body-tinted 6 mm glazing (BTG)	0.52	0	0.47	0	0.38	0
Body-tinted 6 mm glazing with light slatted blinds	0.19	0.20	0.18	0.22	0.17	0.22
Clear 6 mm + clear 6 mm	0.64	0	0.56	0	0.42	0
Clear 6 mm + clear 6 mm with light slatted blinds	0.26	0.19	0.25	0.21	0.21	0.21
BTG 10 mm + clear 6 mm	0.30	0	0.28	0	0.24	0
BTG 10 mm + clear 6 mm with light slatted blinds	0.10	0.13	0.10	0.14	0.09	0.14

It will be recalled that for a given wall, $Y > U$; in forming $\sum AY$, all room surfaces (six in a rectangular room) are included in the summation whereas $\sum AU$ involves an exterior wall only. Thus f_r is greater than 1. If $f_r > 6$ then the enclosure is described as heavy; if $f_r < 4$ then it is light. For an example of its use, see Simmonds (1991). (The factor f_r is called the room response factor, but it is unrelated to the series of coefficients, also called response factors, which describe the wall response at hourly intervals to a triangular temperature excitation.)

In implementing (16.9), $\overline{Q}_e$ is the average of the total heat input taken to act at T_{ei} summed over a 24 h period. If an allocation of the output from sources such as lighting, occupants and equipment can be made into convective and radiative components, the value taken to act at T_{ei} is the radiant fraction augmented by the factor $(1 + C/S)$. For the output from a heat appliance, Tables A9.1 to A9.7 of the 1986 *CIBSE Guide* suggest values for the radiant/convective fractions as listed in Table 16.2.

The solar gain calculation starts with the total radiation incident on some exterior surface area A_j during one day, W_j say, which may be of order 10×10^6 J/m^2. The daily mean value is $W_j/(24 \times 3600)$ in W/m^2. If it falls on a window of area A_j, the value to be input at T_{ei} is $\overline{S}_e(W_j/(24 \times 3600))A_j$. If it falls on an opaque outer surface j, ambient temperature T_e is replaced by the sol-air temperature $T_{sa,j} = T_e + \alpha_j[W_j/(24 \times 3600)]/h_{ej}$.

For a source whose radiant emissive fraction is C/S (around $\frac{2}{3}$), the whole input can be taken as input at T_{ei}. Equation (16.9) has to be rewritten as

$$C_v(\overline{T}_{ei} - \overline{T}_e) + \sum A_j U_j(\overline{T}_{ei} - \overline{T}_{sa,j}) = \sum Q_e + \sum Q_a C_v/V, \tag{16.12}$$

since surfaces of different orientations have different daily mean sol-air temperatures:

$$\overline{T}_{sa,j} = \overline{T}_e + \alpha_j[(H_j/24 \times 3600)]/h_{ej}, \tag{16.13}$$

where α_j is the short-wave absorptivity on the exterior of wall j and h_{ej} is the associated film coefficient. The value of $\overline{T}_{ei}$ follows.

The original task of the admittance procedure was to find the likely peak temperature $\hat{T}_{ei}$ in a room:

$$\hat{T}_{ei} = \overline{T}_{ei} + \tilde{T}_{ei}, \tag{16.14}$$

where $\tilde{T}_{ei}$ is the amplitude of the swing of T_{ei} about its mean value of $\overline{T}_{ei}$. The wall parameters **u** and **y** are exact for a sinusoidal driver of period 24 h, but not for other periods. Strictly, the response will only be correct if the drivers themselves are sinusoidal with period 24 h. In practice this cannot be and the response will be somewhat inaccurate.

Table 16.2 Long-wave radiant fraction from various heating appliances

Forced warm air heater	0%
Natural convectors and convector radiators	10%
Multi-column radiators	20%
Double- and treble-panel radiators and double-column radiators	30%
Single-column radiators, floor warming systems, block storage heaters	50%
Vertical and ceiling panel heaters	67%
High-temperature radiant systems	90%

The least problem is that of variation in ambient temperature which varies smoothly during periods of sustained stable conditions. The daily periodic variation in ambient temperature T_e can be resolved into a series of Fourier components, the first of which is by far the largest. Its amplitude can be estimated as

$$\tilde{T}_e = \tfrac{1}{2}(T_{e,\max} - T_{e,\min}). \tag{16.15}$$

Then the amplitude of energy flow due to the variation in ambient temperature through ventilation and by conduction through windows (area A_g, U value U_g) or other walls which have negligible thermal capacity is

$$\tilde{Q}_a = (C_v + A_g U_g)\tilde{T}_e. \tag{16.16}$$

Much of the daily variation in T_{ei} results from variation in the heat sources which are taken to act at T_{ei} and T_a. The variation $\tilde{T}_{ei}$ associated with a sinusoidally varying flow $\tilde{Q}_i$ into a room surface (of a room internal wall or a room external wall) of area A_w is

$$\tilde{T}_{ei} = \tilde{Q}_i/(A_w y). \tag{16.17}$$

No building heat sources vary sinusoidally with period 24 h. Accordingly, the fluctuating component of the internal gains, $\tilde{Q}_i$, is estimated as

$$\tilde{Q}_i \approx Q(\text{peak value during day}) - Q(\text{daily mean value}). \tag{16.18}$$

To see the consequences of this assumption, suppose that a source emitting w watts is switched on regularly for N hours a day. This square-topped input $Q(t)$ can be resolved into the Fourier series

$$\frac{Q(t)}{w} = \frac{N}{24} + \frac{2}{\pi}\sum \frac{(-1)^j}{j} \sin\frac{j\pi N}{24} \cos\frac{2j\pi t}{24}. \tag{16.19}$$

The first harmonic ($j = 1$) has an amplitude $w(2/\pi)\sin(\pi N/24)$. The table gives values of $\tilde{Q}$ given exactly and by the peak-to-mean estimate.

	Correct amplitude	Peak-to-mean approximation
$N = 12\,\text{h}$	$2\,w/\pi = 0.64\,w$	$0.5w$
N small	$(2N/24)w$	$(1 - N/24)\,w$

Thus the variation in a source which is on for a period of 12 h a day can be represented adequately by its peak-to-mean approximation (to the extent that $0.64 \approx 0.5$). The estimate becomes progressively worse as the source is on for a shorter period. The peak temperatures reached due to an input of 2 kW for one hour or to 1 kW for two hours will be nearly the same but according to the peak-to-mean expression, the swing due to the 2 kW input will approach double that of the 1 kW input.

The intensity of diffuse solar radiation rises gradually from almost zero at dawn to a maximum and back again. The incidence of direct radiation on a surface depends on the orientation of the surface and the time of year; see equation (9.7). In winter, at sunrise, sunshine falls directly on a south-facing wall of an open site and its intensity rises steeply; in summer the solar beam only falls on the wall some time after sunrise and its intensity increases slowly. In either case the incidence can be represented very approximately as a zeroth and first harmonic component with the daily average serving as the zeroth contribution and the peak-to-mean difference as the amplitude of 24 h sinusoidal variation. It leads, however, to negative values by night.

With this assumption, the fluctuating component of solar gain through a window acting at T_{ei} is

$$\tilde{Q}_s = \tilde{S}_e A_g (I_{\text{peak}} - I_{\text{daily mean}}), \tag{16.20}$$

where I is the intensity of radiation falling on the window and variously reflected, absorbed and transmitted there; A_g is the glazed area. The values of $\tilde{S}_e$ are less than $\overline{S}_e$ since they also take account of the fluctuating component of heat stored in the fabric. If the fabric had an infinite thermal capacity, it would absorb all the solar flux falling on it with no change in temperature, so the flux would have no effect at T_{ei} and $\tilde{S}_e$ would be zero. Note that heavyweight $\tilde{S}_e$ values are less than lightweight values.

The fluctuating component of sol-air temperature is

$$\tilde{T}_{sa} = \tilde{T}_e + \alpha (I_{\text{peak}} - I_{\text{daily mean}})/h_e, \tag{16.21}$$

where I is the intensity of radiation falling on the wall. The fluctuating component of the structural gain is then

$$\tilde{Q}_f = A_w \mathbf{u} \tilde{T}_{sa}, \tag{16.22}$$

where A_w is the area of the wall.

Summing the fluctuating gains, we have

$$\begin{pmatrix}\text{ventilation}\\ \text{gain}\end{pmatrix} + \begin{pmatrix}\text{casual}\\ \text{gain}\end{pmatrix} + \begin{pmatrix}\text{solar internally}\\ \text{absorbed gain}\end{pmatrix} + \begin{pmatrix}\text{solar externally}\\ \text{absorbed and}\\ \text{ambient gain}\end{pmatrix}$$
$$= \begin{pmatrix}\text{internal}\\ \text{temperature}\\ \text{variation}\end{pmatrix} \times \begin{pmatrix}\text{conductance}\\ \text{between}\\ T_{ei} \text{ and } T_e\end{pmatrix},$$

so the fluctuating component of the room index T_{ei} is

$$\tilde{T}_{ei} = \frac{\tilde{Q}_a + \tilde{Q}_i + \tilde{Q}_s + \tilde{Q}_f}{\sum A_j y_j + C_v}. \tag{16.23}$$

Worked examples illustrating the use of the admittance procedure are given in Section A5 of the 1999 *CIBSE Guide* and by Clarke (2001: 345); see also Campbell (1990). The approximate nature of much of the procedure will be apparent from this discussion. Basic theory requires the denominator of (16.23) to remain constant, that is, the ventilation rate should be constant throughout the 24 h period. Although equation (A5.85) of the 1986 *CIBSE Guide* gives an expression to handle varying rates (Harrington-Lynn 1974), it remains a limitation. The expression can be justified since in practice $\sum A_y$ for rooms is much larger that the ventilation term C_v and any variation in V is unimportant. It would be inappropriate for a tent, with negligible thermal storage and high and variable ventilation rates. Further, the theory assumes the building has been experiencing an exactly repeated pattern of excitation for several days and is in a steady-cyclic condition; the procedure cannot address transient change such as switching lights on and off. Its rather rough and ready nature however should not be allowed to obscure its strengths: it remains a valuable practical tool, easily implemented manually, to test out a choice of design parameters at an early stage in design.

The procedure was also used during estimation of air-conditioning load (Milbank and Harrington-Lynn, 1970). Since the aim of air conditioning is to maintain a largely unvarying indoor temperature, internal heat storage and its parameters are not as important here as when estimating mean and peak temperatures in an uncontrolled building. Other aspects of the procedure retain their significance. An extensive study has recently been completed (Spitler and Rees 1998; Rees *et al.* 1998, 2000a, 2000b). It compared air-conditioning load estimated by the admittance procedure and estimated by a more fundamental method based on hour-by-hour transfer coefficients (Chapter 19). The comparison involved over 7000 different combinations of zone type, internal loading and weather. The admittance procedure was implemented using the BRE-ADMIT program in Basic, which differed a little from the *CIBSE Guide* method. They concluded

> The BRE-ADMIT implementation of the admittance method tends to underpredict lighter weight zone cooling loads, and overpredict loads for heavyweight zones.
>
> The principal reason for the large range of errors for the admittance method appears to be due mainly to the poor treatment of solar gains from the reliance on solar gain factors in the BRE-ADMIT implementation of the method.
>
> A tendency to underpredict loads is introduced into the admittance method by the treatment of the radiant component of internal loads in that the interaction of this component of the load with the thermal storage of the fabric is only crudely represented. The effects of internal loads with a radiant component on peak cooling load are underestimated in both lightweight and heavyweight test cases.

16.3 SUBMULTIPLES OF 24 HOURS

When the thermal drivers are daily periodic but not sinusoidal, the accuracy of predicting the room response will be improved by using higher harmonics, $j = 2, 3, 4$, etc., or periods of 12, 8, 6, 4.8 hours, etc. Consider ambient temperature T_e. T_{ej} is the magnitude of the jth component of T_e and α_j is its phase relative to some arbitrary reference time. If noon is taken as reference and T_{e1} has its maximum value at 1400 hours, it has a phase angle α_1 of $2\pi(2/24)$. T_{ej} and α_j are found using standard Fourier analysis. Suppose that values of T_e are known at N regular instants during the period P. If T_e is known at half-hour intervals during a 24 h period, $P = 24$ h and $N = 48$. At time level i, $T_e(i)$ can be expressed as the Fourier series

$$T_e(i) = \sum_{j=0}^{N-1} T_{ej} \cos(2\pi ij/N - \alpha_j), \tag{16.24}$$

where

$$T_{ej} = (a_j{}^2 + b_j{}^2)^{1/2} \tag{16.25}$$

and

$$a_j = (1/N)\sum_{j=0}^{N-1} T_{ej}\cos(2\pi ij/N) \quad \text{and} \quad b_j = (1/N)\sum_{j=0}^{N-1} T_{ej}\sin(2\pi ij/N). \tag{16.26}$$

They have the property that $a_{N-j} = a_j$ and $b_{N-j} = -b_j$. Also,

- if $a_j \approx 0$ and $b_j > 0$ then $\alpha_j = \pi/2$;
- if $a_j \approx 0$ and $b_j < 0$ then $\alpha_j = -\pi/2$;
- if $a_j > 0$ and $b_j < 0$ then $\alpha_j = \arctan(b_j/a_j) - \pi$; (16.27)
- if $a_j > 0$ and $b_j > 0$ then $\alpha_j = \arctan(b_j/a_j)$;
- if $a_j < 0$ and $b_j > 0$ then $\alpha_j = \arctan(b_j/a_j) + \pi$.

The relation is exact provided that all N components are included in (16.24). But if the function is largely sinusoidal with some distortion, as is the case for ambient temperature during a steady-cyclic weather spell, a value of j up to 3 may be sufficient. $j = 0$ to 10 may serve for beam solar radiation; radiation should not become significantly negative by night.

The heat flow into a room due solely to variations in ambient temperature is

$$q_t = \sum m(\mathbf{u}_j)T_{ej}\cos(2\pi jt/24 - \alpha_j - p(\mathbf{u}_j)); \tag{16.28}$$

$m(\mathbf{u}_j)$ is the magnitude of the cyclic transmittance of period $24/j$ hours, and $p(\mathbf{u}_j)$ is its phase in radians.

The technique was used in pre-computer times. Mackey and Wright (1944) evaluated the magnitude and time lag of the quantity T_{si}/T_e for a homogeneous slab flanked by films, where T_{si} is the amplitude of sinusoidal variation of the inside surface and T_e is the amplitude of sinusoidal variation of period 24 h in ambient temperature. As noted earlier, $m(T_{si}/T_e)$ was plotted against λ/X with $\lambda\rho c_p$ as a parameter; they remarked that these values could be adapted for use with the jth harmonic by using the value $j\lambda\rho c_p$ in place of $\lambda\rho c_p$ and they supplied correspondingly high values. This was laborious to implement and they suggested a simplification for design purposes whereby the variation in T_{si} had the same shape as the variation in T_e but attenuated by the 24 h value of $m(T_{si}/T_e)$. This procedure somewhat overestimates the variation in T_{si}.

A number of studies using superposition of harmonic components were conducted around the 1960s; examples are Muncey (1953, 1963), Masuch (1966, 1969), Le Febve de Vivy (1966) and Knabe (1971a). Davies (1986 III: Figure 3) used the technique to estimate the response of a passively heated building to strong sunshine; the daily beam radiation was expressed as components of intensity I_0 to I_{10}. The study revealed the limitations of the harmonic method. It is comparatively easy to evaluate wall parameters for periodic excitation. But the weather data required to compute the thermal response are normally supplied as a time series, usually at hourly intervals. For use with the harmonic method, it has to be preprocessed into frequency-domain form and computations must conducted using complex quantities. This is not difficult but the computations are far removed from the simple manual calculations which made the harmonic approach initially so attractive. Although the method can be applied to passive buildings with constant ventilation rates, it is less valid for buildings with variable ventilation rates and cannot readily handle ongoing data.

16.4 FURTHER DEVELOPMENTS

Problems with the method were addressed by Athienitis and his colleagues, (Athienitis *et al.* 1985, 1986, 1987; Haghighat and Athienitis 1988; Athienitis 1993). It was assumed above that the pattern of excitation and use of a room over a period $P = 24\,\text{h}$ was repeated indefinitely, so the thermal behaviour could be expressed as the sum of components of period P, $P/2$, $P/3$, etc. However, a longer period may be chosen: a week, a month or a year, (only the day and the year are natural periods). A spectral analysis of ambient temperature and solar incidence indicates periods or frequencies of importance. Now the response of a room to unit excitation at such frequencies is readily found. Thus, by summing the products of driving forces and responses at certain discrete frequencies, we can obtain the total response. Once the driving data are available in frequency-domain form, the responses of series of rooms can be found more quickly in this way than by time-domain methods. This approach is well suited to analysing a series of rooms in a passive building. Iteration may sometimes be needed in time-domain methods, but it is not required in this approach, which also has a smaller storage requirement for weather data. It routinely handles the time delays in response due to the building fabric, of order hours, even a day. It can readily be extended to include the delays introduced by the HVAC system and the sensor control system, of order minutes, and transportation times in ducts and pipes, which may be less than 10 s. These in turn assist in stability studies and the response to variations in set point and load (Athientis 1993). For this purpose, using a least-squares fit, he represents the admittance of a wall in the form

$$1/y = (A_0 + A_1 s + A_2 s^2 + A_3 s^3)/(1 + B_1 s + B_2 s^2 + B_3 s^3 + B_4 s^4) \quad \text{where}$$
$$s = \mathrm{j}\omega = \mathrm{j}2\pi n/P \tag{16.29}$$

to 1% accuracy. $1/A_0$ is the steady-state U value.

The frequency-domain approach to finding room response assumes that the values of the room parameters do not vary in time (Section 10.6). The above authors, using thermal circuit notation, addressed the problem of variable ventilation rate or the addition of insulation to windows by night. The heat flow through a time-varying ventilation conductance, for example, may be replaced by an equivalent heat source composed of a series of sinusoidally varying components.

16.5 PERIODIC RESPONSE FOR A FLOOR SLAB

The steady-state response of a floor slab in contact with the ground was discussed in Chapter 3. A few remarks may be made about including heat storage of the soil. Soil thermal capacity is far too large to permit any significant response to daily fluctuations but there may be a significant relation between the slab and annual variation in soil temperature.[2] In some locations the soil temperature may be greater than the temperature of an air-conditioned building interior during summer.

The Fourier continuity equation in two dimensions is

$$\lambda\left(\frac{\partial^2 T(x,y,t)}{\partial x^2}+\frac{\partial^2 T(x,y,t)}{\partial y^2}\right)=\rho c_p\frac{\partial T(x,y,t)}{\partial t}. \tag{16.30}$$

Ambrose (1981) describes a two-dimensional finite difference analysis of the heat flow through a slab. A constant inside temperature of 17.5°C was assumed and monthly mean ambient temperatures between 3.1°C in January and 16.5°C in July were taken, together with measures of global radiation, long-wave radiant gains and evaporative loss. Soil properties were defined by the diffusivity $\kappa = \lambda/\rho c_p = 1.4/(1900 \times 1770) = 4.2 \times 10^{-7}\ \mathrm{m^2/s}$ (London clay). Average slab heat losses were estimated as follows.

	Winter loss	Summer loss (W/m^2)
Room 5.33 m × 5.33 m	11	2
Room 9.33 m × 9.33 m	7	2

These values were inferred by transforming a two-dimensional analysis to three dimensions. The model had been checked by comparing estimated monthly temperatures in soil with some measured values, showing generally satisfactory agreement. It is not clear how closely the London clay soil values corresponded to the values for the chalk soil of the site measurements. The author, however, calculated values of ground temperature using a value of $\lambda = 5$ W/m K, three times the 'correct' value. This led to substantially different estimates, too low in winter and too high in summer, as expected. Such estimates depend on the value of the diffusivity rather than its individual components, suggesting that the assumed and site κ values were in agreement.

Several authors have examined the problem of heat flow from floors. It will be recalled that, according to Macey's steady-state model, most of the loss of heat from an uninsulated slab laid on earth takes place near the walls. Spooner (1982) has shown that, as might be expected, the effect of a uniform layer of insulation is to smooth out this concentration (as well as reducing the loss). With 25 mm of insulation, the monthly heat losses per unit area from a peripheral area (0.6 m wide, 10 m^2 in area) and a central area (17 m^2) were broadly similar. However, the phase lags differed: the lowest air temperature in 1981 was around mid February; the highest heat loss from the peripheral area was mid March and from the central area roughly mid April. Kusada and Bean (1984) report earlier results of Lachenbruch and of Adamson for the heat loss from a large rectangular slab: in winter the heat flow paths are practically semicircular, centred on the edge of the slab but in summer they are directed downward; a steady-periodic condition beneath the slab

[2]The periodic penetration depth p_p in soil (Section 12.2) is less than 0.2 m for daily variation but around 3 m for annual variation.

is only established after more than three years. Kusada and Bean went on to calculate the positions of isotherms in the ground near a slab at monthly intervals throughout the year; the isotherms cluster strongly at the slab edges in winter (denoting large edge heat loss) and run more nearly horizontal beneath the slab in summer. Heat loss varied during the year but was around 5 W/m^2. Work by Mitalas (1983, 1987) and Perez Sanchez *et al.* (1988) may be noted. Delsante (1990) presents a comparison between his theory (Delsante *et al.* 1983) and measurements of floor losses made by Spooner (1982); it demonstrates satisfactory agreement.

Krarti and his colleagues (Krarti 1994a, 1994b, 1996; Krarti and Choi 1996; Krarti *et al.* 1988b, 1995) extended their steady-state analyses on heat losses from an insulated floor slab on soil to the response of the floor to yearly variation in ambient temperature. The solution to (16.30) for certain prescribed boundary conditions (constant floor temperature, constant water-table temperature and fluctuating soil surface temperature) remains laborious but is considerably simplified when the varying temperature, that of the soil surface, is taken to vary sinusoidally. In this case a solution for soil temperature can be found, made up of a steady-state component T_{ss} and the real part of a complex term $\Delta\mathbf{T}$:

$$T(x, y, t) = T_{ss}(x, y) + \mathrm{Re}[\Delta\mathbf{T}(x, y)\exp(\mathrm{j}2\pi t/P)]; \qquad (16.31)$$

P is 1 year. $\Delta\mathbf{T}$ depends on P; T_{ss} is independent of P. This approach is formally similar to the admittance procedure for daily excitation. They illustrate it for a slab of width 6 m at a constant temperature 18°C, a water table 5 m below it at 10°C and soil temperature varying sinusoidally between 1 and 15°C, that is, a mean value of 8°C and an amplitude of 7 K. Soil values of $\lambda = 1$ W/m K and $\kappa = 6.45 \times 10^{-7}$ m^2/s were assumed. Values of $T_{ss}(x, y)$ and of $\mathrm{Re}[\Delta\mathbf{T}(x, y)]$ are shown and the consequent distribution of summer and winter isotherms. For a poorly insulated floor, winter and summer losses of around 48 and 31 W per metre run were estimated. They were reduced when the slab was better insulated and increased when the depth of the water table was reduced to 1 m.

The 1994 pair of articles deal with vertically and horizontally placed insulation. Each is illustrated, however, with soil values of $\lambda = 1$ W/m K (as before) but a κ value now equal to 1.47×10^{-7} m^2/s. Given a value of $c_p = 840$ J/kg K, typical for inorganic porous materials generally, this implies a value of $\rho = 8100$ kg/m^3. This is much denser than any form of stone or soil material, so the temperature fluctuations may be underestimated. The results suggest that vertical insulation smooths the summer-winter variation in heat loss. Inner and outer horizontally placed insulation appears to be somewhat more effective than a uniform distribution. Since the effect of daily variation in temperature is so attenuated by the mass of the ground, Krarti and Choi (1996) have argued that the loss through the year from building to ground can be expressed as $Q_g(t) = Q_m + Q_s\cos(\omega t - \Phi)$. They illustrate the evaluation of these coefficients, taking into account soil thermal properties, foundation dimensions, placement of insulation, mean and swing of ground temperature and indoor air temperature, and the presence of a water table. In a later article (Chuangchid and Krarti 2001), ground isotherms are based on a value of $\rho c_p = 5.5 \times 10^6$ J/m^{3}K. This also appears to be too high; the highest value among common materials is for water, 4.2×10^6 J/m^{3}K.

A pair of articles (Claesson and Hagentoft 1991; Hagentoft and Claesson 1991) provide a comprehensive review of floor heat exchange and describe an approach which includes the steady-state, annual variation and the response to a step in temperature. Sobotka *et al.* (1995) report a study comparing the measured annual loss of heat from three basement rooms, two in Japan and one in Canada, with calculation methods due to Mitalas, European Standard, ASHRAE and a finite element method; the Mitalas method proved best in the three cases. The article by Davies *et al.* (1995) summarises a number of studies.

17

Wall Conduction Transfer Coefficients for a Layered System

There are three classes of parameter to describe the one-dimensional flow of heat through a wall: (i) the steady-state transmittance or U value, (ii) measures of the response to sinusoidal excitation, the **u** and **y** values for excitation of period 24 h and possibly other P values in the frequency-domain method, and (iii) measures of the response to time-wise excitation, response factors and the four vectors of transfer coefficients (**a, b, c, d**) in the time-domain method. The relative algebraic and computational effort associated with evaluating and implementing these measures is trivial, moderate and extensive, respectively.

Frequency-domain and time-domain methods have offered alternative means to determine the thermal behaviour of a building since the 1970s. Green and Ülge (1979) provided an early comparison of their respective merits. Broadly speaking, the **u, a** values are useful at the early design stage when typical site values for meteorological and usage measures are available or are assumed. More realistic values for room temperature variation and seasonal energy needs can be found from suitable (**a, b, c, d**) coefficients since they operate on hourly values of the driving functions and can take account of varying ventilation rates and extra insulation by night. Thus, like the finite difference method, they provide a means to examine the thermal behaviour of an enclosure in transient conditions. For a review, see Haghighat and Liang (1992).

U values were discussed in Chapter 1; **u, y** values were found in Chapter 10 for discretised walls and in Chapter 15 for a wall consisting of one or more layers where resistance and capacity is distributed. Chapter 11 described how the (**a, b, c, d**) parameters could be found for a discretised wall and the present chapter addresses the corresponding problem for real walls.

The concept of response factors goes back at least to Brisken and Reque (1956). They modelled a wall as a five-element discretised system as shown in Figure 10.4 and found the flux at hourly intervals due to a square-topped pulse of temperature. Ambient temperature was accordingly represented as a series of hourly square-topped pulses. Stephenson (1962) suggested the possibility of a triangular pulse in temperature and Stephenson and Mitalas (1967) and Mitalas and Stephenson (1967) found the response factors for a homogeneous slab, excited by a triangular pulse. They used the solution for the response of the slab to a ramp excitation given by Churchill (1958), who had deduced it using the Laplace transform; see equation (12.23). Mitalas and Arsenault (1967) developed a

Building Heat Transfer Morris G. Davies
© 2004 John Wiley & Sons, Ltd ISBN: 0-470-84731-X

computer program to evaluate such response factors for a multilayer wall using a matrix equation of Laplace transforms. Kusuda (1969) followed this method and presented series of response factors (written here as $\phi_{ee,j}, \phi_{ii,j}$ and $\phi_{ei,j} = \phi_{ie,j}$) for a sample two-layer wall with surface films.[1] A readable account of developments generally is given by Butler (1984). These and later authors used a time-consuming root-finding procedure, in effect, to find the decay times of the system, and in implementing it there was a danger of missing one when it lay close to neighbouring value, a possibility indicated in Section 14.4. To avoid this, Ouyang and Haghighat (1991) have proposed an alternative approach to finding response factors which avoids this problem.

In using conventional response factors or transfer coefficients to find the response of an enclosure, iteration of several days' data is needed before the effect of the assumed initial condition (e.g. that initially all room temperatures are zero) has faded and valid values are computed. To avoid this, Spitler *et al.* (1997) have developed another form of response factor for use in sizing air-conditioning systems. The building is assumed to be in a steady-cyclic (but non-sinusoidal) state and drive variables are assumed known at hourly intervals. The transmission behaviour is described by a set of 24 response factors. Their relation to transfer function methods is discussed in Spitler and Fisher (1999). This approach has the advantage that no iteration is needed. Such periodic response factors cannot be used to process seasonal behaviour.

Stephenson and Mitalas (1971) later developed the concept of the wall transfer coefficients, thereby summarising the information of the infinitely long series of ϕ values by the short series of **a, b, c** and **d** coefficients (as explained earlier) and provided a sample set of such coefficients. They used a Z transform approach, similar to the Laplace transform but applied to time variation sampled at regular intervals rather than treated as continuous. Later Peavy (1978) showed the connection between the d coefficients and (in effect) the wall decay times z and sampling interval δ (11.13). Mitalas (1978) presented a comparison between response factors and transfer coefficients (11.19). The series of b, d values for 42 roof and 41 wall constructions in Chapter 26 of the 1993 *ASHRAE Handbook of Fundamentals* were evaluated by Harris and McQuiston (1988).[2] As an example, their wall 7 consists of 100 mm of heavyweight concrete, 50 mm of insulation, surface finish and steel siding, and account is taken of surface films. The heat flow $q_{ie,i}$ at time level i into the room at zero due to hourly excitation by ambient temperature T_e is

$$\begin{aligned} q_{ie,i} = 0.00561T_{e,i} &+ 0.04748T_{e,i-1} + 0.02052T_{e,i-2} + 0.00039T_{e,i-3} \\ &+ 0.93970q_{ie,i-1} - 0.04664q_{ie,i-2} + 0.00000q_{ie,i-3}. \end{aligned} \quad (17.1)$$

[1] He recently gave an interesting account (Kusuda 2001) of his involvement in response factor and other building heat transfer studies, including the difficulties of electronic computing in its early days during the 1960s. He remarked that for him, thermal response factors – conduction transfer coefficients as they shortly became – provided the most exciting development in thermal physics, their mathematical background being precise and elegant. The method provided a pinnacle of computer applications for building thermal physics. Eigenfunction problems in general have held a certain fascination for many workers over two centuries. A notable result used here is the Sturm–Liouville theorem, dating from the 1830s.

[2] They only supplied the coefficients b and d, which provide means to estimate the heat flow at a wall interior due to temperature changes at the wall exterior, or vice versa. The c coefficients are needed to find the heat flow into the wall interior due to temperature changes at the wall interior (the a coefficients do the same but at the wall exterior, and they are not normally needed). Instead of the c coefficients, the 1993 *ASHRAE Handbook* used the further coefficients v and w as part of a calculation separate from the b, d calculation; in Chapter 19 the analyses of flows due to interior and exterior drives are conducted simultaneously. A textbook discussion of the ASHRAE method is given in Chapter 8 of McQuiston and Parker (1994).

Only the present value of temperature and the previous three values of temperature and flow are needed to find the present value of flow. Note that the coefficient of $q_{ie,i-2}$ is negative. Seem *et al.* (1989a) suggest this technique can be extended to flow in more than one dimension.

If the wall consists of n solid layers (or $n-1$ layers and a cavity) and is bounded between node 0 (exterior) and node n (interior), the equations have the form

$$q_{n0,i} = b_0 T_{0,i} + b_1 T_{0,i-1} + b_2 T_{0,i-2} + \cdots - d_1 q_{n0,i-1} - d_2 q_{n0,i-2} - \ldots \text{ with } T_n = 0, \tag{17.2a}$$

$$q_{nn,i} = c_0 T_{n,i} + c_1 T_{n,i-1} + c_2 T_{n,i-2} + \cdots - d_1 q_{nn,i-1} - d_2 q_{nn,i-2} - \ldots \text{ with } T_0 = 0, \tag{17.2b}$$

$$q_{00,i} = a_0 T_{0,i} + a_1 T_{0,i-1} + a_2 T_{0,i-2} + \cdots - d_1 q_{00,i-1} - d_2 q_{00,i-2} - \ldots \text{ with } T_n = 0, \tag{17.2c}$$

$$q_{0n,i} = b_0 T_{n,i} + b_1 T_{n,i-1} + b_2 T_{n,i-2} + \cdots - d_1 q_{0n,i-1} - d_2 q_{0n,i-2} - \ldots \text{ with } T_0 = 0. \tag{17.2d}$$

If both T_0 and T_n vary, the net flow $q_{n,i}$ out is

$$q_{n,i} = q_{nn,i} - q_{n0,i}. \tag{17.3}$$

If the wall is taken to include the outer film, ambient temperature T_e replaces T_0 above. In computing the values in (17.1), both outside and inside films were included so T_i, the room temperature, replaced T_n. It is more convenient, however, to include the inner film as part of the model for room internal exchange rather than as part of the wall model. As Seem *et al.* (1990) explain, equations of this form are often known as transfer functions. In control theory, a transfer function denotes the ratio of output/input at some frequency, such as the wall admittance and transmittance as defined in Chapter 15 for $P = 24$ h. In this case the quantities a, b, c and d will be called transfer coefficients but the form they take will not be called a transfer function.

We shall be concerned here with the evaluation of coefficients such as these, but using direct (time-domain) solutions of the Fourier continuity equation as in Chapter 11 rather than transform methods.[3] On the one hand, a slope solution is set up (the response to a steadily increasing temperature) which leads to a temperature profile through the wall; at $t = 0$, T_e and T_i are zero and the profile is wholly negative. On the other, a transient solution is developed which is equal and opposite to the slope solution at $t = 0$. The combined solutions provide the relations between the driving temperature T_e or T_i and the heat flows which they cause at these nodes. The general approach is shown in Figure 17.1. It can be illustrated without complication for a single layer.

[3]The coefficients can also be found through suitable processing of observational data on walls; see for example, Haghighat *et al.* (1988, 1991), Irving (1992), Dewson *et al.* (1993) and Hong *et al.* (1994). Burch *et al.* (1992a) have used finite difference methods to examine heat flows due to thermal bridging, from which they found sets of conduction transfer coefficients.

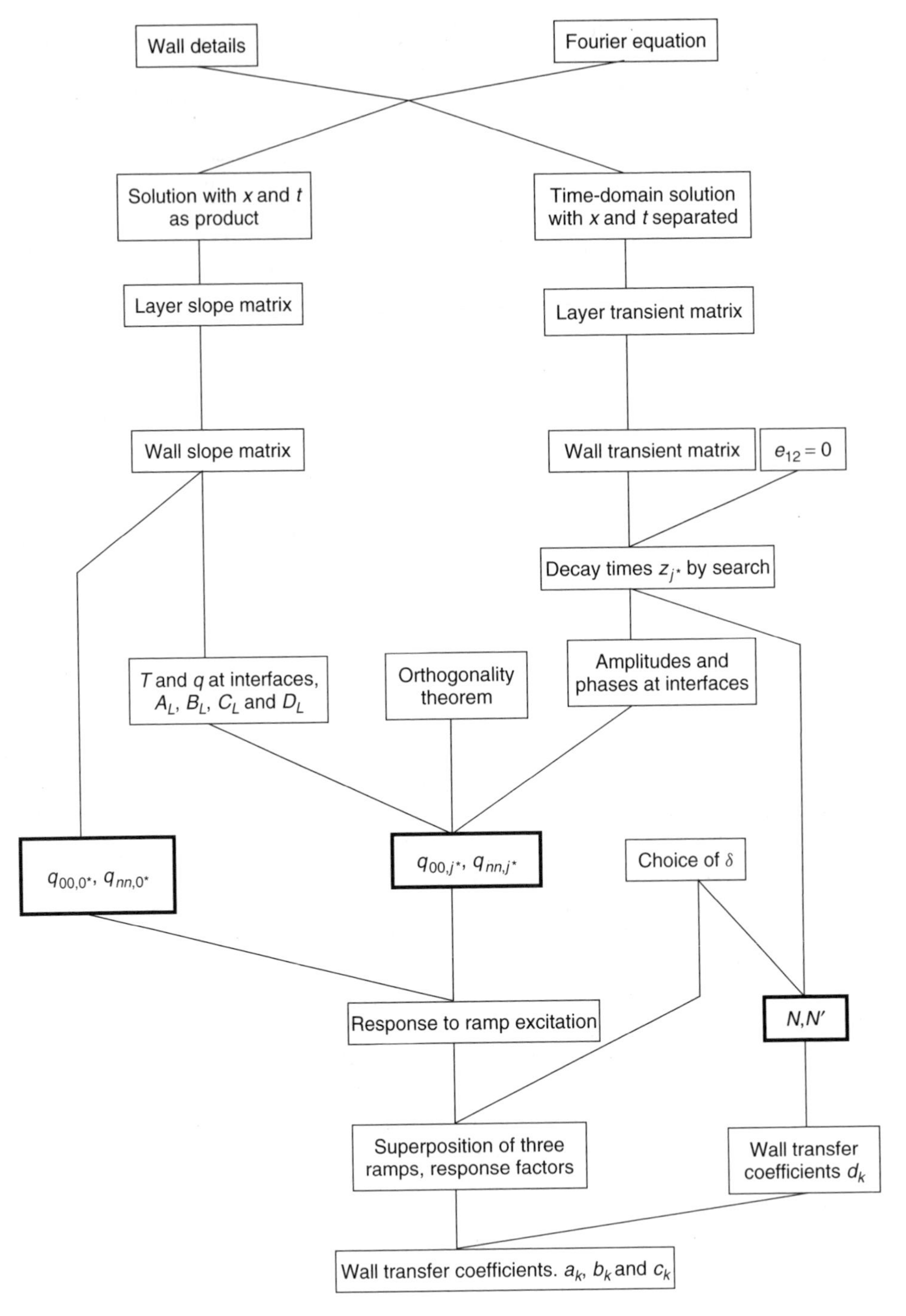

Figure 17.1 Scheme for calculation of wall transfer coefficients. The heavier boxes contain items that carry over into the discretisation procedure. (Davies, 2003, with permission from Elsevier Science)

17.1 THE SINGLE SLAB

Consider the slab $0 < x < X$, initially at zero, and where $T(X, t)$ is always zero but for $t > 0$, $T(0, t) = \theta t$. The subsequent heat flows $q(0, t)$ and $q(X, t)$ are given by (12.23). The response factors are the responses at $t = 0, \delta, 2\delta$, etc., to a triangular pulse in temperature applied at $x = 0$ or $x = X$, of base 2δ and height $\theta\delta = \Theta = 1$ K, centred at $t = 0$. It will be recalled that the pulse can be synthesised from three ramps, $+\theta$ acting from $t = -\delta$, -2θ acting from $t = 0$ and $+\theta$ acting from $t = \delta$.

As an example, we examine the response at $x = X$ due to excitation at $x = 0$. The zeroth response factor is due to the first ramp alone. From (12.23b),

$$\phi_{X0,0} = q(X, \delta)/\theta\delta = 1/r - \tfrac{1}{6}c/\delta - (2c/\delta)\sum[(-1)^j/(j^2\pi^2)]\exp(-\delta/z_j). \quad (17.4)$$

The factor for $t = \delta$ is composed of ramps 1 and 2:

$$\phi_{X0,\delta} = +\tfrac{1}{6}c/\delta - (2c/\delta)\sum[(-1)^j/(j^2\pi^2)][\exp(-2\delta/z_j) - 2\exp(-\delta/z_j)]. \quad (17.5)$$

All three ramps contribute to the factor at $t = 2\delta$:

$$\phi_{X0,2\delta} = -(2c/\delta)\sum[(-1)^j/(j^2\pi^2)][\exp(-3\delta/z_j) - 2\exp(-2\delta/z_j) + \exp(-\delta/z_j)]. \quad (17.6)$$

In general, at time level k

$$\phi_{X0,k\delta} = \phi_{X0,k} = -(2c/\delta)\sum[(-1)^j/(j^2\pi^2)][\exp(-\delta/z_j) - 2 + \exp(+\delta/z_j)]\exp(-k\delta/z_j). \quad (17.7)$$

When δ is small compared with z_1, the early terms in this series (small k) are near zero; see equation (12.24b). Later terms are positive, rise to a weak maximum and decrease to zero again. This expresses the physical observation that application of a temperature pulse at $x = 0$ has no immediate effect at $x = X$. After some delay, of order $F_0 = \lambda t/\rho c_p X^2 = t/rc \approx 0.2$ (Section 14.1), a perceptible response occurs there which rises and dies away again.

Values of $\phi_{00,k}$ can be written down similarly. $\phi_{00,0}$ is positive, $\phi_{00,1}$ is negative, and all subsequent values are negative, gradually tending again to zero. If $\delta << z_1$, $\phi_{00,0} \rightarrow (c/\pi r\delta)^{1/2}$; see equation (12.24a).

The values of the transfer coefficient d_j follow from the values of the decay times (12.14) and choice of δ. We recall that for a discretised wall, the heat flow at time level i can be expressed in terms of heat flows alone at previous levels, provided that the wall has not since been subjected to external excitation. Equation (11.10) generalised to a ladder of N capacities and $N + 1$ resistances becomes

$$q_{M0,i} = -d_1 q_{M0,i-1} - d_2 q_{M0,i-12} - d_3 q_{M0,i-3} - \cdots - d_N q_{M0,i-N}, \quad (17.8)$$

where $M = 2N + 1$ and denotes the right-hand node.

We assume that an equation of this form will serve too for the slab. Here we use (12.14) to obtain

$$z_j = cr/j^2\pi^2. \tag{17.9a}$$

As previously, β_j is defined as

$$\beta_j = \exp(-\delta/z_j) = \exp(-j^2\pi^2\delta/cr). \tag{17.9b}$$

Following the scheme in (11.13),

$$\begin{aligned}
d_1 &= -(\beta_1 + \beta_2 + \beta_3 + \beta_4 + \ldots),\\
d_2 &= +(\beta_1\beta_2 + \beta_1\beta_3 + \beta_1\beta_4 + \cdots + \beta_2\beta_3 + \beta_2\beta_4 + \cdots + \beta_3\beta_4 + \ldots),\\
d_3 &= -(\beta_1\beta_2\beta_3 + \beta_1\beta_2\beta_4 + \beta_1\beta_3\beta_4 + \beta_2\beta_3\beta_4 + \ldots),\\
d_4 &= +(\beta_1\beta_2\beta_3\beta_4 + \ldots),\\
&\text{etc.}
\end{aligned} \tag{17.10}$$

Since the slab has an infinite series of decay times, an infinite series of β values is in principle needed to evaluate d_1, etc. However, β_j values decrease more rapidly with j than do the z_j values. For a 200 mm brick wall with $\delta = 1$ h, or 3600 s, we have

$$\beta_1 = \exp(-\pi^2 \times 3600 \times 0.84/(800 \times 1700 \times 0.2^2)) = 0.577739.$$

Then

$$\beta_2 = 0.111\,410,\ \beta_3 = 0.007\,171,\ \beta_4 = 0.000\,154,\ \ \beta_5 = 0.000\,001,\ \text{ etc.}$$

Four-digit accuracy in d_1 is given here by taking the first four β values.

The argument must, however, be formalised. First we note from (11.36) that $\sum b_k/\sum d_k =$ the wall U value. Now it is important that the wall U value should be accurately represented so as to avoid the possibility of drift in seasonal response calculations. Thus $\sum d_k$ must be known accurately. Now, from (11.15),

$$\sum d_k = J = (1 - \beta_1)(1 - \beta_2)(1 - \beta_3)(1 - \beta_4)\ldots. \tag{17.11}$$

Since $1 - \beta_j$ tends to unity as j increases, the continued product J is easily computed. However, the d_k values alternate in sign and in order that their sum should be accurate, each individual d_k value must be accurate to some acceptable fraction ε of J (ε is of order 10^{-3} or 10^{-4}). Suppose that N of these d_k values are needed (so that d_{N+1} is negligible). N is determined when the magnitude of d_N, $|d_N| = \beta_1\beta_2\beta_3 \ldots \beta_N$, becomes less than εJ.

At this point, however, β_N itself is not less than εJ and since $d_1 = \sum \beta_j$, the summations indicated in (17.10) must be continued to $k = N'$, where $\beta_{N'} < \varepsilon J$.

As the slab becomes thicker, the values of N and N' increase and this has important implications. It is convenient to express the slab thickness in non-dimensional form:

$$V = (\rho c_p/4\pi\delta\lambda)^{1/2} X = (rc/4\pi\delta)^{1/2}, \tag{17.12}$$

Table 17.1 Parameters for a single-layer slab of thickness V. (Davies, 1997, with permission from Elsevier Science)

V	N	N'	$\sum d_k$	d_k(max)	NDP*
1	3	4	5.20×10^{-1}	$d_1 = -0.5$	3
2	5	7	7.63×10^{-2}	$d_1 = -1.5$	4
3	7	12	8.40×10^{-3}	$d_2 = +2.3$	5
4	10	17	8.21×10^{-4}	$d_2 = +5.0$	6
5	12	23	7.53×10^{-5}	$d_3 = -9.1$	7
6	14	30	6.63×10^{-6}	$d_3 = -18.4$	8
7	16	37	5.67×10^{-7}	$d_4 = +36.8$	9
8	18	45	4.76×10^{-8}	$d_4 = +72.1$	10
9	20	53	3.93×10^{-9}	$d_5 = -150$	11
10	22	61	3.20×10^{-10}	$d_6 = +301$	12
11	24	70	2.58×10^{-11}	$d_6 = +623$	13
12	26	79	2.07×10^{-12}	$d_7 = -1287$	14

*Number of places of decimals needed to express each d_j.

N the number of d_k values needed, N' the number of β_j values needed to find them together with $J = \sum d_k$ and the largest value of d_k in the series are listed in Table 17.1 as a function of V.

It will be seen that $\sum d_k$ is around 5×10^{-V} and d_k(max) is around 2.04^{V-2}. Theoretical expressions can be found for N and N':

- We note that for a slab,

$$\beta_j = \exp(-j^2\pi^2\lambda\delta/\rho c_p X^2) = \exp(-j^2\pi/4V^2), \tag{17.13}$$

so

$$\begin{aligned}\ln[\beta_1\beta_2\ldots\beta_N] &= \ln[\exp(-1^2\pi/4V^2)\exp(-2^2\pi/4V^2)\ldots\exp(-N^2\pi/4V^2)] \\ &= (-\pi/4V^2)(1^2+2^2+\cdots+N^2) \\ &= (-\pi/4V^2)N(N+1)(2N+1)/6. \end{aligned} \tag{17.14a}$$

This is to be approximately equal to

$$\ln\left(\varepsilon\sum d_k\right) = \ln(0.001 \times 5 \times 10^{-V}) = (-2.3 - V)/0.4343. \tag{17.14b}$$

So N is given approximately as

$$\frac{N(N+1)(2N+1)}{6} = \frac{4V^2(2.3+V)}{0.4343\pi}. \tag{17.15}$$

- N' is given when $\beta_{N'}$ is less than $\varepsilon\sum d_k$, leading to

$$N' \sim 1.7V\sqrt{2.3+V}. \tag{17.16}$$

- It can also be shown that if V is greater than about 1.5,

$$d_1 = V - \tfrac{1}{2}, \tag{17.17a}$$

$$d_2 = \tfrac{1}{2}V^2 - \tfrac{1}{2}(1 + 1/\sqrt{2})V + \tfrac{3}{8}, \tag{17.17b}$$

  d_3 is a polynomial in V^3, etc. The result comes about from an expression in elliptic integral theory and demonstrates the factor of 4π in the definition of V. Otherwise these relations do not appear to be of use.

Now $V \propto \sqrt{\text{resistance} \times \text{capacity}}$. This analysis demonstrates that as a wall resistance and capacity increase, $\sum d_k$ decreases as 10^{-V}, a very rapid rate. Since to preserve accuracy, each value of d_k must be known to an accuracy of $\varepsilon \sum d_k$ and so must be reported to a number of places of decimals (NDP) equal to $-\log_{10}(\varepsilon \sum d_k)$. NDP is listed in Table 17.1. When $V = 12$, d_k values must be specified to 14 places of decimals and since the largest of them is of order 10^3, each of the series of some $26 d_k$ values has to be reported to 18 significant places. This is going beyond the available machine precision.

To see the consequences of this, suppose that the response of a room with a concrete floor were wanted at 1 min intervals. The thickness of the floor for a value of $V = 12$ is $X = V\sqrt{4\pi\lambda\delta/\rho c_p} = 12\sqrt{4\pi \times 1.63 \times 60/2300 \times 1000} = 0.28$ m. On computational grounds, it would be inadvisable to attempt such modelling. On physical grounds too it would be inappropriate; it implies that we wish to determine the heat flow at 1 min intervals from one surface, notionally isothermal, due to changes at 1 min intervals at the opposite surface, a q_{01}-type flow. With so thick a slab, any minute-by-minute changes on the lower surface due to excitation on the upper surface would be far too small to be of consequence. According to the polynomial approximation for the penetration depth of a signal (12.73), a disturbance at the upper surface of the concrete block would only have penetrated to a distance $p = (1.63 \times 60/(2300 \times 1000 \times 0.0628))^{1/2}$ or 0.026 m, one-tenth the slab thickness, in 60 s. The flow of interest is the flow associated with temperature changes on the same side, a q_{00}-type flow. In the short term, this is independent of slab thickness and the heat flow can be found from the series of response factors, as shown.

The heat flow into the surface of a semi-infinite solid at a time t after imposition of a ramp excitation is

$$q(0, t) = \theta_0 t(\lambda\rho c_p/\pi t)^{1/2}. \tag{12.24a}$$

Suppose that $\theta_0 = 1$ K an hour or $1/\delta$ and set

$$a = (\lambda\rho c_p/\pi\delta)^{1/2}. \tag{17.18}$$

Then we follow the method described in (17.4) to (17.7). The $k = 0$ response factor results from a single ramp θ_0 imposed at $t = -\delta$:

$$\begin{aligned}\phi_{00,0} &= q(0, \delta)/\Theta = [(1/\delta)\delta](\lambda\rho c_p/\pi\delta)^{1/2}/\Theta \\ &= a[\sqrt{1}],\end{aligned} \tag{17.19a}$$

where $\Theta = 1$ K. The $k = 1$ factor results from the additional imposition of the ramp $-2\theta_0$ imposed at $t = 0$:

$$\phi_{00,1} = a[\sqrt{2} - 2\sqrt{1}]. \tag{17.19b}$$

The $k = 2$ factor results from the further imposition of the ramp $+\theta_0$ at $t = \delta$:

$$\phi_{00,2} = a[\sqrt{3} - 2\sqrt{2} + \sqrt{1}] \tag{17.19c}$$

so

$$\phi_{00,k} = a[\sqrt{k+1} - 2\sqrt{k} + \sqrt{k-1}] \rightarrow -\tfrac{1}{4}a/k^{3/2}. \tag{17.20}$$

The foregoing discussion has been concerned with finding response factors $\phi_{X0,k}$ and d_k values for a single slab. The transfer coefficients themselves follow using equations similar to (11.19). The argument must be extended to a multilayer wall.

17.2 SLOPE RESPONSE FOR A MULTILAYER WALL

Equation (12.6) gives the temperature profile through a single homogeneous slab with the boundary conditions $T(0, t) = \theta_0 t$ and $T(X, t) = \theta_1 t$ We have to determine the corresponding profile for a wall consisting of two or more layers in contact.

Equation (12.5) provides a steady-slope solution for one-dimensional heat flow in a homogeneous slab:

$$T(x, t) = A + B\frac{x}{X} + C\left(\frac{x^2}{X^2} + \frac{2t}{W}\right) + D\left(\frac{x^3}{X^3} + \frac{6xt}{XW}\right), \tag{17.21}$$

where $W = \rho c_p X^2/\lambda$ $(= cr)$ and has the units of time.

If two solid layers are placed in contact, there are two equations similar to (17.21) to describe the temperature profile through the pair corresponding to steady-slope excitation. The temperature and heat flow at the interface are continuous at all times. Suppose that temperatures are imposed at the two exposed surfaces so that $T_0 = \theta_0 t$ and $T_2 = \theta_2 t$. It is a straightforward but tedious task to express the layer constants, A_1 and A_2, etc., for slabs 1 and 2 in terms of r_1, c_1, r_2 and c_2, together with the values of θ_0 and θ_2.

For more than two layers the algebra becomes intractable. It is to be shown, however, that the transmission properties of a single layer can be expressed in matrix form and that the overall wall properties can then be found simply by multiplication of the layer matrices, so that we can find the steady-slope response of a wall consisting of any number of layers.

Suppose for convenience that the wall consists of three layers, 1, 2 and 3, bounded by surface 0 to the left and surface 3 to the right. In general, subscripts $L - 1$ and L denote the left and right surfaces respectively of layer L.

Suppose that layer 2 is initially at a uniform temperature throughout and then uniform rates of increase of temperature are imposed at its left and right surfaces $x = 0$ and $x = X_2$. When the transient response has died away,

$$T_2(x, t) = A_2 + B_2\frac{x}{X_2} + C_2\left(\frac{x^2}{X_2^2} + \frac{2t}{W_2}\right) + D_2\left(\frac{x^3}{X_2^3} + \frac{6xt}{X_2 W_2}\right). \tag{17.22}$$

The constants are determined from the conditions that

(i) at $x = 0$, $T_2(0, t) = T_1 + \theta_1 t$, and

(ii) at $x = X_2$, $T_2(X_2, t) = T_2 + \theta_2 t$,

where T_1 and T_2 are the $t = 0$ values, numerically negative, at the left and right surfaces respectively of layer 2. Then

$$T_2(x, t) = T_1 + \theta_1 t + \left((T_2 - T_1) - \left(\frac{\theta_1}{3} + \frac{\theta_2}{6} \right) W_2 - (\theta_1 - \theta_2) t \right) \frac{x}{X_2} + \left(\theta_1 \frac{W_2}{2} \right) \frac{x^2}{X_2^2} + \left((\theta_2 - \theta_1) \frac{W_2}{6} \right) \frac{x^3}{X_2^3}. \tag{17.23}$$

The heat flow $q_2(x, t) = -\lambda \partial T_2(x, t)/\partial x$, from which we have at surface 1 between layers 1 and 2,

$$q_1(X_1, t) = q_2(0, t) = \left(\frac{(T_1 - T_2) + (\theta_1 - \theta_2)t}{r_2} \right) + c_2 \left(\frac{\theta_1}{3} + \frac{\theta_2}{6} \right), \tag{17.24a}$$

and at surface 2 between layers 2 and 3,

$$q_2(X_2, t) = q_3(0, t) = \left(\frac{(T_1 - T_2) + (\theta_1 - \theta_2)t}{r_2} \right) + c_2 \left(\frac{\theta_1}{6} + \frac{\theta_2}{3} \right). \tag{17.24b}$$

The relations between q, T and θ at the two surfaces at the time $t = 0$ can be written in matrix form:

$$\begin{bmatrix} T_1 \\ q_1 \\ \theta_1 \\ \theta_1 - \theta_2 \end{bmatrix} = \begin{bmatrix} 1 & r_2 & c_2 r_2/2 & c_2 r_2/6 \\ 0 & 1 & c_2 & c_2/2 \\ 0 & 0 & 1 & 1 \\ 0 & 0 & 0 & 1 \end{bmatrix} \begin{bmatrix} T_2 \\ q_2 \\ \theta_2 \\ \theta_1 - \theta_2 \end{bmatrix}. \tag{17.25}$$

Layer 2 is flanked to the left by layer 1, whose left surface temperature at $t = 0$ is T_0, and to the right by layer 3, whose right surface is similarly at T_3. We suppose that these outside temperatures, together with the corresponding rates of rise, θ_0 and θ_3, are known; the intermediate temperatures and rise rates follow from them. Continuity at the interface between layers 1 and 2 requires that

$$\left(\frac{(T_0 - T_1) + (\theta_0 - \theta_1)t}{r_1} \right) + c_1 \left(\frac{\theta_0}{6} + \frac{\theta_1}{3} \right) = \left(\frac{(T_1 - T_2) + (\theta_1 - \theta_2)t}{r_2} \right) + c_2 \left(\frac{\theta_1}{3} + \frac{\theta_2}{6} \right). \tag{17.26}$$

This equation must hold for all values of time and there is a similar relation at the interface between layers 2 and 3. It follows that

$$\frac{(\theta_0 - \theta_1)}{r_1} = \frac{(\theta_1 - \theta_2)}{r_2} = \frac{(\theta_2 - \theta_3)}{r_3} \quad \text{which must equal} \quad \frac{(\theta_0 - \theta_3)}{r_1 + r_2 + r_3} \quad \text{or} \quad \frac{(\theta_0 - \theta_3)}{r}, \tag{17.27}$$

where $r = r_1 + r_2 + r_3$, the sum of the thermal resistances. Furthermore, θ_1 can be expressed incrementally as

$$\theta_1 = \theta_2 + \frac{r_2}{r}(\theta_0 - \theta_3). \tag{17.28}$$

A more useful form of the transmission matrix relating conditions at surfaces 1 and 2 of slab 2 then follows:

$$\begin{bmatrix} T_1 \\ q_1 \\ \theta_1 \\ \theta_0 - \theta_3 \end{bmatrix} = \begin{bmatrix} 1 & r_2 & c_2 r_2/2 & c_2 r_2^2/6r \\ 0 & 1 & c_2 & c_2 r_2/2r \\ 0 & 0 & 1 & r_2/r \\ 0 & 0 & 0 & 1 \end{bmatrix} \begin{bmatrix} T_2 \\ q_2 \\ \theta_2 \\ \theta_0 - \theta_3 \end{bmatrix}. \tag{17.29}$$

This matrix holds if the 'layer' is in fact a lumped resistance or a lumped capacity such as is discussed in Chapter 10. If the wall includes a ventilated cavity, the cavity can be represented by three matrices, where the non-diagonal elements of the middle one are zero except that $e_{21} = V$. In this case the chain of matrices must relate to the full wall area, not unit area, since V has units W/K; see equation (4.18).

r and $\theta_0 - \theta_3$ are globally known quantities. Thus by multiplication of the layer matrices in the order [layer 1] × [layer 2] × [layer 3], the extreme quantities can be related at time zero:

$$\begin{bmatrix} T_0 \\ q_0 \\ \theta_0 \\ \theta_0 - \theta_3 \end{bmatrix} = \begin{bmatrix} 1 & e_{12} & e_{13} & e_{14} \\ 0 & 1 & e_{23} & e_{24} \\ 0 & 0 & 1 & e_{34} \\ 0 & 0 & 0 & 1 \end{bmatrix} \begin{bmatrix} T_3 \\ q_3 \\ \theta_3 \\ \theta_0 - \theta_3 \end{bmatrix}. \tag{17.30}$$

Multiplication demonstrates the following properties of the elements:

$$e_{12} = r_1 + r_2 + r_3 = r, \text{ the overall thermal resistance of the construction,} \tag{17.31a}$$

$$e_{23} = c_1 + c_2 + c_3 = c, \text{ the overall thermal capacity of the construction,} \tag{17.31b}$$

$$e_{34} = 1, \tag{17.31c}$$

$$e_{12}e_{24} + e_{13} = rc. \tag{17.31d}$$

These properties are apparent in the elements for a single-slab matrix and they remain true for any number of layers. The slope response of a wall of arbitrary construction can be described by just four quantities: r, c, e_{13} and e_{14}.

In the current application of this theory, the initial extreme temperatures T_0 and T_3 are zero at $t = 0$; either the rise rate θ_0 or θ_3 too is to be zero. If θ_3 is zero, q_3 must be chosen as

$$q_3 = -e_{14}\theta_0/e_{12}. \tag{17.32}$$

Thus with T_3 and θ_3 equal to zero, the above value of q_3 and an imposed value of θ_0, values of T_2, q_2 and θ_2 at surface 2 can be found from the matrix of layer 3, and so on through the construction by successive multiplication; q_0 is the only quantity to be determined at surface 0. The values of T_1, q_1, T_2 and q_2 provide the values of A_2, B_2, C_2 and D_2 for layer 2:

$$A_2 = T_1, \tag{17.33a}$$

$$B_2 = -q_1 r_2, \tag{17.33b}$$

$$C_2 = -3T_1 + 3T_2 + (2q_1 + q_2)r_2, \tag{17.33c}$$

$$D_2 = 2T_1 - 2T_2 - (q_1 + q_2)r_2. \tag{17.33d}$$

These coefficients in turn determine the coefficients of the series of terms in the transient solution, as will be shown in Section 17.5.1.

For later use q_3 in (17.32) will be written as $q_{30,0^*}$. Subscript 1 indicates that the flow concerned is at node 3 and subscript 2 that it was due to excitation at node 0. The third subscript is introduced for consistency with transient values for q associated with eigenvalues $j = 1, 2, \ldots$; the slope solution is in a sense a $j = 0$ solution and is so indicated by the third subscript. The fourth subscript, *, denotes that the quantity concerned relates to a real wall. (It is replaced later by ^ in connection with an equivalent discretised wall.)

The procedure described above for three layers can be generalised to n layers in an obvious way. If the external and internal films are taken into account, the corresponding nodes are subscripted e and i, respectively. At the time when $T_e = T_i = 0$, the profile due to steady-slope excitation is everywhere negative, with a negative gradient at T_e and a positive gradient at T_i. It follows that

$$\frac{q_{ee}}{\theta_e} = e_{24} - \frac{e_{14}}{r}, \quad \frac{q_{ei}}{\theta_i} = \frac{e_{14}}{r}, \quad \frac{q_{ie}}{\theta_e} = -\frac{e_{14}}{r}, \quad \frac{q_{ii}}{\theta_i} = \frac{e_{14}}{r} - \frac{e_{13}}{r}; \tag{17.34}$$

q_{ie} and q_{ii} are numerically negative since they describe an *outward* flow.

The comparable quantities for steady-state excitation where T replaces θ_e or θ_i are

$$\frac{q_{ee}}{T_e} = \frac{1}{r}, \quad \frac{q_{ei}}{T_i} = -\frac{1}{r}, \quad \frac{q_{ie}}{T_e} = \frac{1}{r}, \quad \frac{q_{ii}}{T_i} = -\frac{1}{r}, \tag{17.35}$$

where $1/r$ is the wall U value.

Equations (17.34) provide four different measures for the slope response of the wall (which reduce to three since $q_{ei}/\theta_i = -q_{ie}/\theta_e$). They are fundamental measures for the wall and the simplest non-steady-state parameters possible. Use of (17.31d) shows that

$$(q_{ee} - q_{ii} + q_{ei} - q_{ie})/\theta = c. \tag{17.36}$$

Thus, taken as positive quantities, $\sum q/\theta$ is the wall total capacity. The slope parameters are not of use in themselves but together with a transient solution they lead to the response of a wall to ramp excitation.

Kossecka (1998: 71) provides explicit expressions for q_{ie}, etc., which are simpler to evaluate than by matrix multiplication. Thus in a wall consisting of n layers,

$$q_{ie}/\theta_e = (1/r^2)\sum_{L=1}^{n} c_L(-r_L^2/3 + r_L r/2 + r_{i-L} r_{L-e}), \tag{17.37}$$

where L denotes some layer, r_{i-L} is the total resistance between layer L and the interior, not including layer L, and r_{L-e} is similarly the resistance between r_L and the exterior. If the wall is taken to include outside and inside films, c_1 and c_n are zero. In the present application, however, once q_{ie}/θ_e is known, values of flux and temperature, needed at the interfaces, can be found and the matrix formulation provides a convenient general method to find them.

It can be shown that a multilayer wall, when undergoing steady-slope excitation, can be modelled exactly as a Π configuration of resistances αr, $(1-2\alpha)r$, αr and capacities βc and $(1-\beta)c$, where

$$\alpha - \alpha^2 = e_{14}/cr \tag{17.38a}$$

and

$$1 - \alpha - \beta + 2\alpha\beta = e_{13}/cr. \tag{17.38b}$$

It follows that for a single homogeneous slab where $e_{14} = cr/6, \alpha = 0.2113$ and not 0.25 as might have been expected. Stephenson and Starke (1959) showed that this value of α also serves approximately for the transmittance parameter of a homogeneous slab when excited sinusoidally. The present author showed, however, that if the admittance parameters too were to be taken into account, the optimum value for α was 0.193 (Davies 1983a: equation 34a).

17.3 TRANSIENT SOLUTION FOR A MULTILAYER WALL

The wall is supposed to consist of n layers. The temperature distribution for eigenvalue j in layer L at position x from the left surface of layer L and at time t can be expressed as

$$T_L(x, t, j) = q_{n0,j^*} E_{Lj} \sin(\phi_{Lj} + \alpha_{Lj}x).\exp(-t/z_j), \tag{17.39a}$$

where

$$\alpha_{Lj} = \sqrt{\rho_L c_{pL}/\lambda_L z_j}. \tag{17.39b}$$

The product $q_{n0,j^*} E_{Lj}$ is the amplitude of temperature in layer L at time zero. The factor q_{n0,j^*} denotes the heat flow out at T_n associated with eigenfunction j when the wall is subjected to a ramp excitation at node T_0 and is a descriptor of the wall as a whole; subscript 0 is not relevant until the slope and transient solutions are combined. The factor E_{Lj} relates to layer L alone. The constants are to be determined in the order z_j, E_{Lj} and ϕ_{Lj} together (in this section) and q_{n0,j^*} (Section 17.5.1).

The temperature and heat flow components for eigenfunction j and layer L at the two surfaces are related in (12.12) as

$$\begin{bmatrix} T_{L-1,j} \\ q_{L-1,j} \end{bmatrix} = \begin{bmatrix} \cos u_{Lj} & (\sin u_{Lj})/v_{Lj} \\ -(\sin u_{Lj}) \times v_{Lj} & \cos u_{Lj} \end{bmatrix} \begin{bmatrix} T_{L,j} \\ q_{L,j} \end{bmatrix}, \tag{17.40}$$

where

$$u_{Lj}^2 = \rho_L c_{pL} X_L^2/\lambda_L z_j = (c_L \times r_L)/z_j \tag{17.41a}$$

and

$$v_{Lj}^2 = \lambda_L \rho_L c_{pL}/z_j \quad = (c_L / r_L)/z_j. \tag{17.41b}$$

If the layer is purely resistive, the relation becomes

$$\begin{bmatrix} T_{L-1,j} \\ q_{L-1,j} \end{bmatrix} = \begin{bmatrix} 1 & r_L \\ 0 & 1 \end{bmatrix} \begin{bmatrix} T_{L,j} \\ q_{L,j} \end{bmatrix}. \tag{17.42}$$

This is the case for inside and outside film resistances and also a cavity resistance, if unventilated. If the layer is purely capacitative, we have

$$\begin{bmatrix} T_{L-1,j} \\ q_{L-1,j} \end{bmatrix} = \begin{bmatrix} 1 & 0 \\ -c_L/z_j & 1 \end{bmatrix} \begin{bmatrix} T_{L,j} \\ q_{L,j} \end{bmatrix}. \tag{17.43}$$

This form would be needed if a layer of some metal formed part of the wall, adding to its storage but negligibly to its resistance.

The values for the matrix elements can only be found for assumed z values. For an assumed z value, the wall product matrix

$$\begin{bmatrix} T_{0,j} \\ q_{0,j} \end{bmatrix} = \begin{bmatrix} e_{11j} & e_{12j} \\ e_{21j} & e_{22j} \end{bmatrix} \begin{bmatrix} T_{n,j} \\ q_{n,j} \end{bmatrix} \tag{17.44}$$

can readily be found as the product of n such layer matrices. Decay times are required for the condition that $T_{0,j}$ and $T_{n,j}$ should each be zero. Thus a search has to be made for values of z_j such that $e_{12j} = 0$ (Section 17.10). (Only when the boundary conditions have been fixed so that a series of values of z can be found to make one or other of the e_{ik} elements zero can we attach a subscript to z. Strictly speaking, the above equations should be drafted in terms of z, not z_j.)

The number of solutions needed for a multilayer wall follows as indicated for the single slab. The last decay time needed, the Nth value, is determined when $\beta_{N'} = \exp(-\delta/z_{N'})$ is less than $\varepsilon J = \varepsilon\Pi(1 - \beta_j)$.

The system of modes of decay in a multilayer wall is similar to the modes of vibration in a stretched string composed of several lengths having different masses per unit length; the tension is the same everywhere. A length of string having a negligible mass corresponds to a pure thermal resistance and a localised mass on the string corresponds to a pure capacity.

To find the amplitude $E_{L,j}$ for the jth eigenfunction temperature component in layer L and the corresponding phase $\phi_{L,j}$ in each layer, the value of $T_{n,j}$ is set equal to zero and $q_{n0,j}$ to 1 W/m^2. $T_{n-1,j}$ and $q_{n-1,j}$, etc., follow by back multiplication. (Thus $T_{L,j}$ here has units K/(W m^{-2}) and q has units (W m^{-2})/(W m^{-2}).) $E_{L,j}$ and $\phi_{L,j}$ are found from them as

$$E_{L,j} = (T^2_{L-1,j} + (q_{L-1,j}/\lambda_L\alpha_{L,j})^2)^{1/2}. \tag{17.45}$$

$$\text{If } q_{L-1,j} < 0,\ \phi_{L,j} = -\arctan\,(T_{L-1,j}/(q_{L-1,j}/\lambda_L\alpha_{L,j})), \tag{17.46a}$$

$$\text{If } q_{L-1,j} > 0,\ \phi_{L,j} = -\arctan\,(T_{L-1,j}/(q_{L-1,j}/\lambda_L\alpha_{L,j})) + \pi. \tag{17.46b}$$

The units of $E_{L,j}$ are m^{2}K/W.

The quantity q_{n,j^*} is now the only unknown. Its values are to be found from the n sets of coefficients A_L, B_L, C_L and D_L of the steady-slope solution (17.33) by using the orthogonality property of the eigenfunctions.

The wall transfer coefficients we wish to determine are based on the condition that the flanking temperatures T_e and T_n are prescribed, so we have to evaluate the elements e_{14} in the slope product matrix and e_{12} in the transient product matrix (to seek the values of z for which e_{12} is zero). Note that the structure of e_{12} is more complicated that the structure of e_{14}. For consider a seven-layer 'wall', where layers 1, 3, 5 and 7 are pure resistances, and layers 2, 4 and 6 are purely capacitative. It is easy to evaluate the elements and it is found that e_{14} consists simply of 10 terms of form $r_1c_2r_3$ whereas e_{12} consists of $(r_1 + r_3 + r_5 + r_7)$, the 10 terms of form $-r_1c_2r_3/z$, 6 terms of form $+r_1c_2r_3c_4r_5/z^2$ and a final term $-r_1c_2r_3c_4r_5c_6r_7/z^3$.

17.4 THE ORTHOGONALITY THEOREM

Equation (12.17) stated a well-known result in Fourier analysis and (13.12) gave similar results when the solid was in contact with a layer without capacity. Section 10.3.3 addressed the case of a succession of lumped capacities. The present discussion is the generalisation of these results. We have to demonstrate that the energy distribution throughout a wall associated with eigenfunction k is independent of eigenfunction j. This can be demonstrated analytically for a wall consisting of two layers, each of thickness X but with differing conductivities and volumetric specific heats, in contact at $x = 0$. $T_1(x, 0, j)$ is the temperature in slab 1 at position x and arbitrary time $t = 0$ for the jth eigenfunction temperature profile. $T_1(x, 0, j)(\rho_1 c_{p1})$ is the corresponding energy density. We have to show that

$$\int_{-X}^{0} T_1(x, 0, j)(\rho_1 c_{p1}).T_1(x, 0, k)\,\mathrm{d}x + \int_{0}^{X} T_2(x, 0, j)(\rho_2 c_{p2})T_2(x, 0, k)\,\mathrm{d}x = 0, \tag{17.47}$$

unless $k = j$, that is, the distributions $T_1(x, 0, j)$ and $T_1(x, 0, k)$ are orthogonal in slab 1 and similarly in slab 2.[4,5]

Now the standing wave general solution for eigenfunction j for a transient disturbance in the two layers can be written in the form

$$T_1(x, t, j) = q_j.E_{1j}\sin(\alpha_{1j}X + \alpha_{1j}x)\exp(-t/z_j), \qquad -X \le x \le 0, \tag{17.48a}$$

$$T_2(x, t, j) = q_j.E_{2j}\sin(\alpha_{2j}X - \alpha_{2j}x)\exp(-t/z_j), \qquad 0 \le x \le X, \tag{17.48b}$$

where E_{1j} and E_{2j} are defined below; q_j is the flow at the right surface. These expressions satisfy the Fourier continuity equation and ensure that the temperatures at the exposed surfaces ($x = -X$ and $x = +X$) are zero. Furthermore, temperature and heat flow are

[4]This is an example of the Sturm–Liouville boundary value result. (J.C.F. Sturm 1803–1855, F. Liouville 1809–1882). It is concerned with second-order differential equations applying over some interval with specified boundary conditions and has long been associated with the differential equations of Chebyshev, Bessel, Legendre, Hermite and Laguerre. It is usually drafted so as to make the ρc_p term, which here has a constant value within a slab, a continuous function of x. It is cited in this context by Özisik (1980: equation 14-3a), Mikhailov *et al.* (1983) and Mikhailov and Özisik (1984), where it is developed in great generality. A wide variety of thermal applications have been noted by Wirth and Rodin (1982), although they do not include studies of heat flow through building walls. One of the practical difficulties at that time was the difficulty of finding the possibly very long series of eigenvalues (in effect, the decay times) of the systems being analysed, something noted by Mikhailov and Vulchanov (1983). Many eigenvalues may be needed to model the early stages of some sudden change. In the early 1990s, the present author noted the orthogonality theorem independently and used it as an alternative to Laplace and Z transforms when evaluating transfer coefficients for the walls of buildings. In this case, with time steps of 1 h, only a few eigenvalues are needed and their evaluation presents no difficulty.

[5]The physical parameter in this equation associated with each layer is its thermal capacity ρc_p, but as equations 10.34a and b show, it could have equally well been drafted in terms of the layer resistivity $1/\lambda$. In a similar context, Salt (1983) equation 30 uses $1/\alpha$ or $\rho c_p/\lambda$.

continuous across the common surface $x = 0$. The relations provide values for the decay times z_j through solutions of the equation,

$$\frac{\sqrt{c_1/r_1}}{\tan\sqrt{c_1 r_1/z_j}} + \frac{\sqrt{c_2/r_2}}{\tan\sqrt{c_2 r_2/z_j}} = 0. \tag{17.49}$$

Also

$$E_{2j} = \left(\frac{r_2 z_j}{c_2}\right)^{1/2}, \; E_{1j} = E_{2j}\frac{\sin\sqrt{c_2 r_2/z_j}}{\sin\sqrt{c_1 r_1/z_j}}, \tag{17.50}$$

The first of the integrals can be written

$$I_1 = \frac{q_j q_k(\rho_1 c_{p1})\sqrt{z_j z_k}}{c_2/r_2}\frac{s_{2j}.s_{2k}}{s_{1j}.s_{1k}}\int_{-X}^{0}\sin(u_{1j}+\alpha_{1j}x)\sin(u_{1k}+\alpha_{1k}x)\,\mathrm{d}x, \tag{17.51}$$

where

$$u_{1j} = \alpha_{1j}X = \sqrt{c_1 r_1/z_j}, \; s_{1j} = \sin(u_{1j}). \tag{17.52}$$

Then

$$I_1 = \frac{q_j q_k}{c_2/r_2}\frac{z_j z_k}{z_k - z_j}s_{2j}s_{2k}\left(\frac{\sqrt{z_j}\sqrt{c_1/r_1}}{\tan u_{1k}} - \frac{\sqrt{z_k}\sqrt{c_1/r_1}}{\tan u_{1j}}\right). \tag{17.53a}$$

Similarly,

$$I_2 = \frac{q_j q_k}{c_2/r_2}\frac{z_j z_k}{z_k - z_j}s_{2j}s_{2k}\left(\frac{\sqrt{z_j}\sqrt{c_2/r_2}}{\tan u_{2k}} - \frac{\sqrt{z_k}\sqrt{c_2/r_2}}{\tan u_{2j}}\right). \tag{17.53b}$$

Adding these two terms and applying (17.49) to eigenfunctions j and k indicates that

$$I_1 + I_2 = 0,$$

demonstrating (17.47). Since none of the energy distribution of eigenfunction j 'leaks' into eigenfunction k as it decays, the distributions associated with j and k must be independent of each other and a result of this kind is to be expected. It is clear that the unweighted sum

$$\int T_1(x,t,j)T_1(x,t,k)\,\mathrm{d}x + \int T_2(x,t,j)T_2(x,t,k)\,\mathrm{d}x \tag{17.54}$$

itself cannot be zero, since if layer 1 were lightweight and layer 2 dense, T_1 values would be large and T_2 values small. It is not immediately obvious that the weighting factor should be the appropriate value of ρc_p rather than $(\rho c_p)^2$; the latter form would correlate the energy density of eigenfunction j with the energy density rather than just the temperature of eigenfunction k. However, the analysis in Section 10.3.3, where the wall consists of three resistances and two capacities, makes it clear that the quantity to be summed through the wall must be the product of the temperature and energy distributions; this is zero unless the two distributions relate to the same mode of decay.

It is easy to demonstrate the extension of the orthogonality theorem computationally to the case of three or more layers. Consider the integral for layer L in an assembly of n layers:

$$I_L = \int_0^X T_L(x,t,j)T_L(x,t,k)(\rho_L c_{pL})\,\mathrm{d}x \tag{17.55a}$$

$$= \int q_{n0,j^*}E_{Lj}\sin(\phi_{Lj}+\alpha_{Lj}x)q_{n0,k^*}E_{Lk}\sin(\phi_{Lk}+\alpha_{Lk}X)(\rho_L c_{pL})\,\mathrm{d}x. \tag{17.55b}$$

So

$$I_L = 0.5q_{n0,j^*}q_{n0,k^*}E_{Lj}E_{Lk}c_L\cdot\left(\frac{\sin(\phi_{Lj}-\phi_{Lk}+u_{Lj}-u_{Lk})-\sin(\phi_{Lj}-\phi_{Lk})}{u_{Lj}-u_{Lk}} - \frac{\sin(\phi_{Lj}+\phi_{Lk}+u_{Lj}+u_{Lk})-\sin(\phi_{Lj}+\phi_{Lk})}{u_{Lj}+u_{Lk}}\right). \tag{17.55c}$$

It turns out that I_L summed over the n layers is zero to the same order of accuracy as the terms e_{12j} and e_{12k} in the transient product matrix are zero; that is, it depends on the order of accuracy with which z_j and z_k have been determined.

17.5 HEAT FLOWS IN A MULTILAYER WALL

We proceed as previously in assuming that a suitable combination of transient solutions can be found such that, combined with the steady-slope solutions layer by layer, the net effect is to make the temperature zero everywhere in the wall at $t = 0$. If the temperature at the extreme left surface ($x = 0$ in layer 1) rises steadily at θ_0 (K/s) from $t = 0$ onward, the combined solution will provide the net temperatures within each layer of the wall and so the corresponding heat flows at the exposed wall surfaces. Thus we assume that

$$\sum T_L(x,0,j) = -T_{Lx}^S \text{ with } j = 1,2,\ldots,\infty, \tag{17.56a}$$

where the transient component

$$T_L(x,0,j) = q_{n0,j^*}E_{Lj}\sin(\phi_{Lj}+\alpha_{Lj}x) \tag{17.56b}$$

and T_{Lx}^S denotes the steady-slope solution at position x in layer L at $t = 0$:

$$T_{Lx}^S = A_L + B_L\frac{x}{X_L} + C_L\frac{x^2}{X_L^2} + D_L\frac{x^3}{X_L^3}. \tag{17.56c}$$

Each term for layer L on the left of (17.56a), together with its right-hand side, is to be multiplied by $T_L(x,0,k)\rho_L c_{pL}$ and integrated over x; $(0 < x < X_L)$. Then each side of

(17.56a) is summed over all n layers in the wall.[6]

$$\sum_{L=1}^{n} \int_{x=0}^{X_L} T_L(x,0,j)\rho_L c_{pL} T_L(x,0,k)\,\mathrm{d}x = \sum_{L=1}^{n} \int_{x=0}^{X_L} T_{Lx}^S \cdot \rho_L c_{pL} T_L(x,0,k)\,\mathrm{d}x. \quad (17.57)$$

According to the orthogonality theorem, all cross-product terms on the left are zero; only the term where $k = j$ is finite. On integration when $k = j$, we have for layer L that

$$F_{Lj} = \rho_L c_{pL} \int \sin^2(\phi_{Lj} + \alpha_{Lj} x)\,\mathrm{d}x = (c_L/2)(1 - (\mathrm{sn\ ks} - \mathrm{sf\ kf})/u), \quad (17.58a)$$

where

$$u = \alpha_{Lj} X_L = \sqrt{c_L r_L / z_j}, \quad (17.58b)$$

$$\mathrm{sn} = \sin(\phi_{Lj} + u), \quad (17.58c)$$

$$\mathrm{ks} = \cos(\phi_{Lj} + u), \quad (17.58d)$$

$$\mathrm{sf} = \sin\phi_{Lj} \quad (17.58e)$$

$$\mathrm{kf} = \cos\phi_{Lj}. \quad (17.58f)$$

Integration of the right-hand side gives

$$\begin{aligned} G_{Lj} &= \rho_L c_{pL} \int T_{Lx}^S \sin(\phi_{Lj} + \alpha_{Lj}.x)\,\mathrm{d}x \\ &= c_L[(A_L/u)(-\mathrm{ks} + \mathrm{kf}) \\ &\quad + (B_L/u^2)(\mathrm{sn} - u\ \mathrm{ks} - \mathrm{sf}) \\ &\quad + (C_L/u^3)(\mathrm{ks}(-u^2 + 2) + 2u\ \mathrm{sn} - 2\mathrm{kf}) \\ &\quad + (D_L/u^4)(\mathrm{ks}(-u^3 + 6u) + \mathrm{sn}(3u^2 - 6) + 6\mathrm{sf})]. \end{aligned} \quad (17.58g)$$

If r_L is zero, that is, if element L is a lumped capacity,

$$F_{Lj} = c_L \quad \text{and} \quad G_{Lj} = c_L A_L \sin\phi_{Lj}. \quad (17.58h)$$

[6]This procedure is similar to the classical representation of the function $f(x)$ in the homogeneous region $0 < x < X$ as, for example, $a_0/2 + \sum c_j \sin(\phi_j + 2j\pi x/X)$, in which the jth constant is found by integrating the product $f(x).\sin(\phi_j + 2j\pi x/X)$ between 0 and X. The integral indicates the extent of the correlation of the sine term with the function. This is valid since $\sin(\phi_j + 2j\pi x/X)$ and $\sin(\phi_k + 2k\pi x/X)$ are themselves uncorrelated over the region, i.e. they are orthogonal. Function $f(x)$ corresponds here to the steady-slope profile T_{Lx}^S at $t = 0$. Now in the sinusoidal term $T_L(x, 0, j)$, the argument $\alpha_{Lj} x = \sqrt{(c_L r_L / z_j)} x/X$ does not correspond closely to $2j\pi x/X$ since $z_1^{-1/2} : z_2^{-1/2} : z_3^{-1/2} : \ldots$ does not have the values 1:2:3: . . ., although they tend approximately to integral ratios as j increases. However, the jth and kth eigenfunctions are orthogonal, so the procedure is valid.

Then the jth component of the transient heat flux at the right wall surface at $t = 0$ due to excitation at the left surface of the wall is

$$q_{n0,j^*} = \frac{\sum E_{Lj} G_{Lj}}{\sum E_{Lj}^2 F_{Lj}} \tag{17.59a}$$

and the amplitudes of the left and right surface fluxes due to same-side excitation are

$$q_{00,j^*} = e_{22j} q_{n0,j^*}, \tag{17.59b}$$

$$q_{nn,j^*} = e_{11j} q_{n0,j^*}. \tag{17.59c}$$

Equation (17.59a) appears to be very similar to equation (26.2) of Vodicka (1955). However, it could not have been used numerically because of the difficulty in finding the eigenvalues of a wall of two or more layers.

The sums of q_{00,j^*} and q_{n0,j^*} (summing j from 1 to infinity) must equal the steady-slope fluxes $q_{00,0^*}$ and $q_{n0,0^*}$, respectively, at time $t = 0$. Convergence occurs faster at the right surface than at the left surface.

Thus the effect of imposing a ramp increase in temperature at the left layer of a wall, initially at zero temperature everywhere and whose right surface is subsequently maintained at zero, is to generate heat flows formed from the combined steady-slope and transient heat flows. At the left surface of layer 1, adjacent to the point of excitation, we have

$$q_{00,t} = \frac{\theta_0 t}{r} + q_{00,0^*} + \sum q_{00,j^*} \exp(-t/z_{j^*}). \tag{17.60a}$$

At the right surface of layer n, remote from the point of excitation,

$$q_{n0,t} = \frac{\theta_0 t}{r} + q_{n0,0^*} + \sum q_{n0,j^*} \exp(-t/z_{j^*}). \tag{17.60b}$$

The first two right-hand terms derive from the slope solution; the second from the wall slope matrix (17.30), together with (17.34).

At this point a choice has to be made of the sampling interval δ for which response factors and transfer coefficients have to be found. The heat flow into the room due to an externally imposed triangular pulse of height Θ[7] $= \theta\delta = 1$ K and base 2δ at time t is then

$$\phi_{n0}(t) = (q_{n0}(t+\delta) - 2q_{n0}(t) + q_{n0}(t-\delta))/\Theta \tag{17.61}$$

and the response factors $\phi_{n0}(0)$, $\phi_{n0}(1)$, $\phi_{n0}(2)$ are the values of $\phi_{n0,t}$ at times $t = 0\delta$, 1δ, 2δ. Conventionally, $\delta = 1$ h and $\theta_0 = 1$ K/hour or (1/3600) K/s. As noted earlier, the value of the first factor $\phi_{n0,0}$ derives from a single ramp and subsequent factors derive from the superposition of the three ramps so

$$\begin{aligned} \phi_{n0,k} &= \sum q_{n0,j^*}[\exp(+\delta/z_j) - 2 + \exp(-\delta/z_j)]\exp(-k\delta/z_j)/\Theta \\ &= \sum q_{n0,j^*}(1/\beta_j - 2 + \beta_j).(\beta_j)^k/\Theta. \end{aligned} \tag{17.62}$$

[7]Although Θ has the numerical value of unity, it is useful to include it for dimensional reasons; q has units W/m^2 and is proportional to θ with units K/s. Inclusion of Θ makes clear that ϕ has units W/m^{2}K.

The heat flow at time level i is

$$q_{n0,i} = \phi_{n0,0}T_{0,i} + \phi_{n0,1}T_{0,i-1} + \phi_{n0,2}T_{0,i-2} + \dots. \tag{17.63}$$

This is the flow due to cross-excitation. The corresponding values for same-side excitation are found by replacing subscripts $n0$ by 00 and nn.[8]

17.5.1 Same-Side and Cross Excitation

The steady-state transmittance of a wall is simply $1/r$ or U. It is convenient to comment here on the three forms of dynamic transmittance corresponding to cross-excitation that have been seen:

- For sinusoidal excitation,

$$(\mathbf{q}_i/\mathbf{T}_e)_{T_i=0} \quad = -(\mathbf{q}_e/\mathbf{T}_i)_{T_e=0} \quad = 1/e'_{12}. \tag{15.11c}$$

- For slope excitation,

$$q_{ie,0^*} = -q_{ei,0^*} = -\theta_i e_{14}/r. \tag{17.34}$$

- During transient decay, eigenvalue j (subscript e serving for 0 and subscript i for n)

$$q_{ie,j^*}/\theta_e = \pm q_{ei,j^*}/\theta_i. \tag{17.59a}$$

Thus in all cases there is a reciprocity relationship.

In the case of slope excitation, with $\theta_e = \theta_i = \theta$,

$$(q_{ee,0^*} - q_{ie,0^*}) - (q_{ii,0^*} - q_{ei,0^*}) = c\theta \tag{17.34}$$

so $q_{ie,0^*}$ or $q_{ei,0^*}$ can be found given $q_{ee,0^*}$ and $q_{ii,0^*}$.

Similarly, in the case of transient decay,

$$q_{ee,j^*} \cdot q_{ii,j^*} = \pm(q_{ie,j^*})^2 \tag{17.59}$$

since $e_{12,j} = 0$ and $e_{11j}e_{22j} = 1$. So again q_{ie,j^*} can be found given q_{ee,j^*} and q_{ii,j^*}. That is, in both cases, knowledge of the cross-excitation term does not add more information about the system.

In finding the sinusoidal parameters, however, the period P is an imposed, independent variable, $e_{12} \neq 0$ and the three sinusoidal parameters describing $(\mathbf{q}_e/\mathbf{T}_e)_{T_i=0}$, $(\mathbf{q}_i/\mathbf{T}_i)_{T_e=0}$ and $(\mathbf{q}_i/\mathbf{T}_e)_{T_i=0} = -(\mathbf{q}_e/\mathbf{T}_i)_{T_e=0}$ (equations (15.11)) are quasi-independent.

[8] Response factors are seen here simply as intermediate constructs, needed to find transfer coefficients. The choice of symbol ϕ to represent them is unimportant. However, they are sometimes seen as the end product of the argument and have been assigned the symbols X_m for $\phi_{nn,m}$, the internal response factors; Z_m for $\phi_{00,m}$, the external response factors; and Y_m for $\phi_{n0,m}$ or $\phi_{0n,m}$, the cross response factors. If the outside film resistance is included, subscript e replaces 0; if the inside film is to be included, subscript i replaces n.

17.5.2 Transfer Coefficients

The wall transfer d_k coefficients follow from the wall decay times and choice of δ (Section 11.2):

$$
\begin{aligned}
d_1 &= -(\beta_1 + \beta_2 + \beta_3 + \beta_4 + \ldots),\\
d_2 &= +(\beta_1\beta_2 + \beta_1\beta_3 + \beta_1\beta_4 + \cdots + \beta_2\beta_3 + \beta_2\beta_4 + \cdots + \beta_3\beta_4 + \ldots),\\
d_3 &= -(\beta_1\beta_2\beta_3 + \beta_1\beta_2\beta_4 + \beta_1\beta_3\beta_4 + \beta_2\beta_3\beta_4 + \ldots),\\
d_4 &= +(\beta_1\beta_2\beta_3\beta_4 + \ldots),\\
&\text{etc.,}
\end{aligned}
\tag{11.13}
$$

and the b_k coefficient follow as in (11.20):

$$
\begin{aligned}
b_0 &= d_0\phi_{n0,0},\\
b_1 &= d_0\phi_{n0,1} + d_1\phi_{n0,0},\\
b_2 &= d_0\phi_{n0,2} + d_1\phi_{n0,1} + d_2\phi_{n0,0},\\
b_3 &= d_0\phi_{n0,3} + d_1\phi_{n0,2} + d_2\phi_{n0,1} + d_3\phi_{n0,0},\\
b_4 &= d_0\phi_{n0,4} + d_1\phi_{n0,3} + d_2\phi_{n0,2} + d_3\phi_{n0,1} + d_4\phi_{n0,0}\\
&\text{etc. where } \beta_j = \exp(-\delta/z_j)
\end{aligned}
\tag{11.20}
$$

The a_k and c_k coefficients are found by replacing $\phi_{n0,k}$ by $\phi_{00,k}$ and $\phi_{nn,k}$ respectively.

There are various methods of finding the d_k values themselves.

(i) They can be evaluated directly using equations similar to (17.10). This is reliable since it involves only summation of positive quantities. For a thick and heavy wall, however, a long series of β_j values may be needed to achieve adequate accuracy for the early d_k values and the method is slow.

(ii) They are best found by successive evaluation:

- The zeroth iteration is

$$
d_0^{[0]} = 1. \tag{17.64}
$$

- The first iteration is

$$
\begin{aligned}
d_0^{[1]} &= 1,\\
d_1^{[1]} &= -\beta_1.
\end{aligned}
\tag{17.65}
$$

- The second iteration is

$$
\begin{aligned}
d_0^{[2]} &= 1,\\
d_1^{[2]} &= -(\beta_1 + \beta_2) = d_1^{[1]} - d_0^{[1]}\beta_2,\\
d_2^{[2]} &= +\beta_1\beta_2 = -d_1^{[1]}\beta_2.
\end{aligned}
\tag{17.66}
$$

- The third iteration is

$$
\begin{aligned}
d_0^{[3]} &= 1, \\
d_1^{[3]} &= -(\beta_1 + \beta_2 + \beta_3) = d_1^{[2]} - d_0^{[2]}\beta_3, \\
d_2^{[3]} &= +(\beta_1\beta_2 + \beta_1\beta_3 + \beta_2\beta_3) = d_2^{[2]} - d_1^{[2]}\beta_3, \\
d_3^{[3]} &= -\beta_1\beta_2\beta_3 = -d_2^{[2]}\beta_3,
\end{aligned}
\tag{17.67}
$$

and so on. In Pascal, where k denotes the iteration,

```
for k: = 0 to N' do d[k,0]:= 1;    d[1,1] = −β[1];
for k: = 2 to N' do
begin for j: = 1 to k-1 do d[k,j]:= d[k-1,j] - d[k-1, j-1]*β[k];
                           d[k,k]:=           - d[k-1, k-1]*β[k];
end;
```

(17.68)

Although this process must be continued N' times the values of d_k when $k > N$ are effectively zero. Since the process accumulates terms of the same sign, it is stable.

(iii) The d_j values can be found too from sums of powers of the β_k:

$$d_0 = 1, \tag{17.69a}$$

$$d_1 = -(d_0 s_1), \tag{17.69b}$$

$$2d_2 = -(d_0 s_2 + d_1 s_1), \tag{17.69c}$$

$$3d_3 = -(d_0 s_3 + d_1 s_2 + d_2 s_1), \tag{17.69d}$$

etc.

where

$$s_1 = \beta_1 + \beta_2 + \beta_3 + \ldots, \tag{17.70a}$$

$$s_2 = \beta_1^2 + \beta_2^2 + \beta_3^2 + \ldots, \tag{17.70b}$$

$$s_3 = \beta_1^3 + \beta_2^3 + \beta_3^3 + \ldots, \tag{17.70c}$$

etc.

Here values of d_0 and d_1 follow by definition. It follows from the identity

$$(a + b + c + \ldots)^2 = a^2 + b^2 + c^2 + 2(ab + ac + bc + \ldots) \tag{17.71a}$$

that

$$-d_1 s_1 = d_0 s_2 + 2d_2, \tag{17.71b}$$

which is (17.69c); d_k is the sum of alternately positive and negative terms, and machine precision places a limit on the resistance, capacity and time step parameters for the wall.

(iv) The N' values of b_j in (17.72) are roots of the equation

$$\beta_j^{N'} + d_1\beta_j^{N'-1} + d_2\beta_j^{N'-2} + \cdots + d_{N'} = 0. \quad (17.72)$$

Thus the d_k values can be found by writing this equation for $\beta_1, \beta_2, \ldots, \beta_{N'}$ and solving the set of simultaneous equations. In principle this is the same as the above methods and works satisfactorily for normal walls with a time interval of 1 h. However, the determinant of the matrix of β_k values has a factor $P = \beta_1\beta_2 \ldots \beta_{N'}$ and for walls of high resistance and capacity and for small values of the time step, P may be so small as to render the matrix effectively singular so that a solution becomes unattainable.

17.6 RESPONSE FACTORS AND TRANSFER COEFFICIENTS FOR AN EXAMPLE WALL

Although the wall transfer coefficients provide efficient means to estimate wall heat flows, they do not provide so interpretable an indication of the physical behaviour of the wall as do the first decay time and the three series of response factors. It is of interest therefore to look at these parameters before discussing the transfer coefficients themselves. We consider the wall described in Table 17.2, taken from Book A of the 1986 *CIBSE Guide*.

The total resistance is 0.546350 m^{2}K/W, so the U value is 1.830330. The total capacity is 372 800 J/m^{2}K.

17.6.1 Two-Layer Wall

It will be assumed first that the brickwork and concrete block are in contact and the three resistances will be ignored. Values of the decay times are given in Table 17.3.

If the two layers had identical λ and ρc_p. values, they would constitute a thermally homogeneous slab; if, further, the first decay time were 1.733 h, the decay times would decrease as

$$1/1^2 : 1/2^2 : 1/3^2 : 1/4^2 : 1/5^2 \quad \text{or} \quad 1.733 : 0.433 : 0.193 : 0.108 : 0.069.$$

Table 17.2 Details of a brick/cavity/block wall. (Davies, 1997, with permission from Elsevier Science)

	X (m)	λ (W/m K)	ρ (kg/m^3)	c_p (J/kg K)	c (J/m^{2}K)	r (m^{2}K/W)	$\lambda/\rho c_p$ (m^2/s)
Layer 1, outside film resistance					0	0.06	
Layer 2, brickwork	0.105	0.84	1700	800	142 800	0.125	6.2×10^{-7}
Layer 3, cavity resistance					0	0.18	
Layer 4, heavyweight concrete	0.100	1.63	2300	1000	230 000	0.06135	7.1×10^{-7}
Layer 5, inside film resistance					0	0.12	

Table 17.3 Decay times and left and right transient $t = 0$ heat flows for the brickwork/concrete block construction without film or cavity resistances. (Davies, 1997, with permission from Elsevier Science)

Eigen number	Decay time (h)	q_{00,j^*} (W/m^{2}K)	q_{20,j^*} (W/m^{2}K)	q_{22,j^*} (W/m^{2}K)
1	1.733	−10.69	19.26	−34.72
2	0.463	−4.91	−5.00	−5.10
3	0.193	−1.25	2.16	−3.74
4	0.116	−1.18	−1.27	−1.37
5	0.070	−0.49	0.79	−1.27

The two materials are not grossly different and their sequence of decay times follows this relation approximately.

The exposed brick and concrete surfaces are surfaces 0 and 2, respectively. Excitation consists in raising surface 0, for example, from zero to 1 K over an hour and returning to zero during a further hour while surface 2 is taken to be at zero throughout. By reversing the order of the layers, the effect of excitation at the right surface can be found. Table 17.3 lists the first five $t = 0$ heat flows. The flows decrease progressively with eigennumber and the tenth is around 1% of the first. The sum to infinity of such terms is exactly equal and opposite to the steady-slope flows at $t = 0$, although convergence of q_{00,j^*} with its uniform signs is slow; q_{20,j^*} with alternating signs converges faster.

The heat flows due to a triangular pulse are given in Table 17.4.

When a ramp temperature increase is imposed on the brick surface, we would expect the initial heat flow response there to be independent of the concrete and its boundary condition; this proves to be the case. Consider a semi-infinite slab, initially at zero temperature and excited from $t = 0$ onward so that its surface temperature $T(0, t) = \theta_0 t$. The temperature response is given in Carslaw and Jaeger (1959: 63, equation 4). It will be readily found that the heat flux at the surface is

$$q_{00,t} = 2\theta_0 t\sqrt{\lambda\rho c_p/\pi t}. \tag{17.73}$$

Thus 0.6 h after imposition of the first ramp excitation and with $\theta_0 = 1$ K/h, the heat flow should be

$$q = (2/3600) \times \sqrt{0.84 \times 1700 \times 800 \times 0.6 \times 3600/\pi} = 15.57\ \text{W/m}^2\,\text{K}.$$

This is the value noted in Table 17.4. At this time, $\phi_{20,t}$ is very small; thus the excitation imposed at the left surface is scarcely noticeable at the right, so the right boundary conditions have not yet affected the response at the left.

The quantity $\phi_{00,t}$ shows a rapid rise during the first hour, a rapid fall to negative values during the second hour and a slow return to zero. Since the first decay time is about 1.7 h, any disturbance will have largely died out after about four times this value, 7 h, and Table 17.4 shows this. At the other surface, $\phi_{20,t}$ initially remains at zero as noted, rises slowly to a weak maximum at 1.8 h and dies away again.

The sum to infinity of each of the three sets of response factors is equal to the wall U value. Once higher modes have died out, ϕ_{j+1}/ϕ_j has a constant value c less than unity.

Table 17.4 Heat flows (W/m^{2}K) into and out of the brick/concrete wall, following a triangular pulse excitation of height 1 K and base 2 h. (Davies, 1997, with permission from Elsevier Science)

		Brick/concrete alone			Wall with resistances		
Time (h)	Temp. (K)	ϕ_{00} (W/m^{2}K)	ϕ_{20} (W/m^{2}K)	ϕ_{22} (W/m^{2}K)	ϕ_{00} (W/m^{2}K)	ϕ_{20} (W/m^{2}K)	ϕ_{22} (W/m^{2}K)
−1.0	0.0	*	0.00	–	–	0.000	–
−0.8	0.2	8.99	0.00	16.28	2.516	0.000	1.532
−0.6	0.4	12.71	0.00	23.03	4.555	0.000	2.964
−0.4	0.6	15.57	0.00	28.20	6.362	0.000	4.337
−0.2	0.8	17.98	0.01	32.56	8.014	0.000	5.664
0.0	1.0	20.11	0.05	36.37	9.548	0.000	6.954
0.2	0.8	4.06	0.14	7.22	5.966	0.001	5.146
0.4	0.6	−1.60	0.29	−3.16	3.240	0.002	3.508
0.6	0.4	−5.64	0.51	−10.66	0.917	0.004	1.962
0.8	0.2	−8.87	0.78	−16.74	−1.157	0.008	0.479
1.0	0.0	−11.60	1.07	−21.91	−3.052	0.014	−0.952
1.2	0.0	−5.01	1.32	−10.18	−2.292	0.022	−0.810
1.4	0.0	−3.44	1.52	−7.49	−1.898	0.032	−0.731
1.6	0.0	−2.57	1.64	−5.99	−1.640	0.043	−0.677
1.8	0.0	−2.00	1.67	−4.97	−1.454	0.054	−0.638
2.0	0.0	−1.60	1.64	−4.23	−1.310	0.066	−0.607
3.0	0.0	−0.70	1.12	−2.16	−0.862	0.115	−0.500
4.0	0.0	−0.37	0.65	−1.19	−0.597	0.141	−0.423
6.0	0.0	−0.11	0.21	−0.37	−0.305	0.146	−0.308
10.0	0.0	−0.01	0.02	−0.04	−0.103	0.102	−0.170
15	0.0	−0.001	0.001	−0.002	−0.040	0.054	−0.083
20	0.0	−0.000	0.000	−0.000	−0.018	0.027	−0.041

*Asymptotically zero. 1000 terms summed to 0.017.

The sum beyond this point is proportional to $1 + c + c^2 + c^3 + \dots$ to infinity, which is equal to $1/(1 - c)$. Thus there is no convergence problem in demonstrating that $\sum \phi_{n0,j^*}$ for example is equal to U.

17.6.2 *Wall with Resistances*

A wall of brick and concrete in contact is not a practical example since excitation has to be taken as applied at ambient temperature T_e, rather than at some surface temperature as assumed above. Similarly, indoor temperature T_i – a mix of air and radiant temperatures – rather than indoor surface temperature is taken to be held constant. Furthermore, the solid elements are separated by a cavity in the wall we are considering. These extra resistances increase the decay times as Table 17.5 shows.

Consider the solution for the transient excitation only. By their construction, all temperature profiles have values of zero at the T_e and T_i nodes and consist of an integral number of half-waves. The profile through the wall associated with the first decay time z_1 is positive, say, everywhere. Although there may be some exchange of heat between the

Table 17.5 Decay times and left and right transient $t = 0$ heat flows for the brickwork/concrete block construction with the film and cavity resistances. (Davies, 1997, with permission from Elsevier Science)

Eigen number	Decay time (h)	q_{00,j^*} (W/m^{2}K)	q_{20,j^*} (W/m^{2}K)	q_{22,j^*} (W/m^{2}K)	β_j	$J = \Pi(1 - \beta_j)$
1	7.10	−15.06	22.81	−34.53	0.8686	0.131354
2	2.27	−12.60	−3.46	−0.95	0.6436	0.046817
3	0.36	−0.61	0.29	−0.14	0.0646	0.043793
4	0.32	−0.49	−0.22	−0.10	0.0439	0.041872
5	0.11	−0.15	0.01	−0.00	0.0001	0.041867
6	0.095	−0.00	−0.01	−0.02	0.0000	0.041866

brick and concrete, all the heat stored in the wall at $t = 0$ associated with the z_1 profile must eventually be lost slowly through the two flanking resistances to T_e and T_i. The value of z_1 is about 7 h. On the other hand, if the z_2 profile is positive in the brick, it is negative in the concrete. After $t = 0$ some heat in the brick is lost to T_e and some coolth in the concrete to T_i. There will now however be a substantial heat flow from brick to concrete across the cavity; on physical grounds, this must be faster-acting than the slump of the first eigenfunction, and z_2 is less than $\frac{1}{3}z_1$.

If the brick and concrete each had very high conductivities, the wall could be modelled as one composed of two lumped capacities (Figure 10.3) and would have only two decay times. Because of their moderate conductivities, they each form distributed systems, so there is an infinite series of decay times. It will be noted that the first two z values are large, corresponding roughly to a model having two lumped capacities; the further values are much smaller. The $j = 3$ and higher q_{00,j^*} values are much smaller than the first two.

The heat flows due to a triangular pulse are also shown in Table 17.4 for ease of comparison. As compared with the case of bare-wall excitation, the heat flows in the current case are smaller and of longer duration.

Equations (17.10) give the relation between d_k and the series of z_j together with choice of the interval δ, usually an hour. The values of $\beta_j = \exp(-\delta/z_j)$ and the continued product J (11.13) for the test wall are given in Table 17.5. It will be noted that, with a value of $\delta = 1$ hour, eigenvalues with decay times down to about 0.1 hour – about six in this case – are needed.

Although there is an indefinitely long series of decay times for this wall, only a limited number of values for z_j are useful in finding the d_k. No significant change in $J = \sum d_k$

Table 17.6 Transfer coefficients for the test wall detailed in Table 17.2. (Davies, 1997, with permission from Elsevier Science)

k	a_k	b_k	c_k	d_k
0	9.548397	0.000179	6.953625	1.0
1	−18.528113	0.013915	−12.223156	−1.620834
2	10.569717	0.043460	5.985915	0.726131
3	−1.575632	0.018036	−0.660046	−0.065025
4	0.062597	0.001034	0.020334	0.001594
5	−0.000339	0.000005	−0.000044	−0.000000
Sum	0.076628	0.076628	0.076628	0.041866

is seen beyond $N = 5$, so values up to d_5 are needed, although β_j values up to β_7 have to be included to compute them. Values of the transfer coefficients are given in Table 17.6

It will be seen that coefficients rise to some maximum magnitude and then decrease, effectively to zero. The $k = 0$ values of a_k, b_k and c_k are identical with the $t = 0$ response factors. The a_k, c_k and d_k values alternate in sign, whereas the b_k values are uniformly positive, as they must be on physical grounds. $\sum a_k$, $\sum b_k$ and $\sum c_k$ are all equal and $\sum b_k / \sum d_k = 1.830330\ \mathrm{W/m^2K}$, the wall U value. (Burch *et al.* (1992a: Table 3) present a series of values for a_k, b_k and c_k, called X, Y and Z in their notation. Of them, only a_2 and c_2 are negative. Values for d_k are not listed.)

Seem *et al.* (1990) describe a method of reducing the number of coefficients a_j etc. (such as the six in Table 17.6). They start with a wall where five b_j coefficients (a_j in their notation) are listed (so that the present response is found from the present value of temperature and $N = 4$ previous hourly values). They reduce this to a set of coefficients operating on $N = 2$ previous values, which preserves an accurate value for the wall transmittance; they demonstrate good agreement between the response to a step change in temperature found from the $N = 4$ set and the reduced $N = 2$ set. However, significant errors enter if the last two coefficients of the $N = 4$ set are simply dropped. The problem can be viewed the other way round. Table 6 of Davies (1996b) shows values of b_j and d_j for a sample wall where the number of d_j coefficients taken to represent the wall was increased from $N = 1$ (i.e. a value for d_1 alone since $d_0 = 1$ by definition) with corresponding values for b_0 and b_1, up to an $N = 5$ set of d and b values. In fact b_5 and d_5 were zero (to 6 places); an $N = 4$ set was sufficient and an $N = 3$ set would probably be adequate. For example, $b_2(N = 3)$ differed little from $b_2(N = 4)$. It was remarked earlier that b_0 derives from the ramp response alone and is the same whatever the choice of N.

17.6.3 Discussion

The main task ahead is to show how sets of such transfer coefficients can be used to estimate the response of an enclosure and this is presented in Chapter 19. The estimates cannot in general be exact and Chapter 18 compares estimates with values found from exact analytical solutions. Before doing so, it is convenient to examine what can be derived from transfer coefficients as they stand. In the following section, a check on internal consistency is illustrated, estimates of the 24 h frequency domain parameters found, together with values for the wall decay times and time lapse between excitation by an exterior temperature pulse and maximum flow at a wall interior. The general theory of transient flow can be used to find the optimal sizing of lumped resistive and capacitative elements for finite difference calculations (Section 17.8).

17.7 DERIVATIONS FROM TRANSFER COEFFICIENTS

Transfer coefficients form part of a set of interrelated quantities describing wall thermal behaviour, as discussed below.

17.7.1 Wall Thermal Capacity

A wall has two fundamental descriptors, its total resistance r and its total capacity c; r, whose reciprocal is U, the steady state transmittance, is of prime practical importance. The total capacity c is formally on the same footing as r but it is not of practical importance as such:

$$\sum d_k/\sum a_k = \sum d_k/\sum b_k = \sum d_k/\sum c_k = \sum r_j = 1/U, \qquad (17.74a)$$

$$-\sum^{N} k(a_k + c_k - 2b_k)\delta/\sum d_k = \sum c_j. \qquad (17.74b)$$

It may be recalled that the second expression is based on consideration of a wall initially at zero throughout (Section 11.6). At $t = 0$ the ambient temperature is raised to 1 K and remains there, while room temperature remains at zero. At first there is a large heat flow into the wall and a negligible flow out; the wall stores the difference. After a long time, the two flows become equal; as a function of progressive resistance, the wall has a uniform linear temperature gradient through it and it now stores no more heat. There are three independent computational methods to find the stored heat S'.

- As applied to the wall of Table 17.2,

$$S' = [(r_3 + r_5)c_2 + r_5 c_4]/(r_1 + r_3 + r_5), \qquad (17.75a)$$

  where r_1 is the sum of the outside film resistance and half that of the brickwork, r_3 that of the cavity, half brickwork and half concrete, and r_5 half concrete and inside film. With these values

$$S' = 174212.5\ \text{J/m}^2\ \text{K}.$$

- In Section 11.6 it was shown that when a value of a_k up to a_5 was significant,

$$S' = [5(a_0 - b_0) + 4(a_1 - b_1) + 3(a_2 - b_2) + 2(a_3 - b_3) + (a_4 - b_4)]\delta/\sum d. \qquad (17.75b)$$

  With the values in Table 17.6 and $\delta = 3600\,\text{s}$, we have

$$S' = 174210.9\ \text{J/m}^2\ \text{K}.$$

- At any time level i, the difference between the fluxes into and out of the wall, $q_{00,i}$ and $q_{n0,i}$ is given as

$$\Delta q(\text{external}, i) = \left[\sum a_k \times 1 - \sum d_k \times q_{00,i-k}\right] - \left[\sum b_k \times 1 - \sum d_k \times q_{n0,i-k}\right],$$

$$S' = \Delta q(\text{external})\delta. \qquad (17.75c)$$

  Thus the total heat stored up to time i due to external excitation is the sum of $\Delta q(\text{external}, i) \times 3600$ and, summing over a sufficiently long series of time levels i, tends to a steady value $\Delta q(\text{external}) \times 3600$. Computationally,

$$S' = 17\,4212.4\ \text{J/m}^2\ \text{K}.$$

These values are virtually the same.

If instead ambient temperature is held at zero and the room temperature is 1 K from $t = 0$ onward, the total heat stored in the steady state can be written as $\Delta q(\text{internal}) \times 3600$ (c_k replacing a_k) and this is similarly associated with a linear temperature gradient, computationally found as = 198 587.4 J/m^{2}K. If now the wall is subjected to these two excitations simultaneously, the total heat stored is $[\Delta q(\text{external}) + \Delta q(\text{internal})] \times 3600$, which is 372 799.8 J/m^{2}K. But the wall is now excited by steadily maintained temperatures of 1 K on each side, the temperature is everywhere 1 K and so the heat stored is by definition its total capacity c of 372 800 J/m^{2}K (Table 17.2).

It may be remarked that while $\sum^N k(a_k - b_k)$ and $\sum^N k(c_k - b_k)$ depend upon the film and layer resistances in the wall, their sum, $\sum^N k(a_k + c_k - 2b_k)$ does not. These tests provide a useful check on the full set of transfer coefficients.

17.7.2 Transfer Coefficients and Measures for Daily Sinusoidal Excitation

If a wall exterior is subjected to sinusoidal excitation of period 24 h, the heat flow into the room at zero per unit variation in ambient temperature is the cyclic transmittance **u**, a complex quantity with magnitude $m(u)$(W/m^{2}K) and a time lag $l(u)$ (h); see Section 15.2. The relation between **u** and the transfer coefficients is shown here.

Suppose that ambient temperature T_e varies sinusoidally by ±1 K and is given as

$$T_e(t) = 1 \times \exp(\mathrm{j}2\pi t/P) = 1 \times \exp(\mathrm{j}\omega t),$$

where t is in hours and P is the daily period of 24 h. The values of driving temperature and consequent heat flow into the room at present and previous times are as shown:

$$\begin{aligned} t &= 0, & T_{e,-0} &= \exp(\mathrm{j}\omega(-0)), & q_{ie,-0} &= u \times \exp(\mathrm{j}(\omega(-0) - \psi)), \\ t &= -1, & T_{e,-1} &= \exp(\mathrm{j}\omega(-1)), & q_{ie,-1} &= u \times \exp(\mathrm{j}(\omega(-1) - \psi)), \\ t &= -2, & T_{e,-2} &= \exp(\mathrm{j}\omega(-2)), & q_{ie,-2} &= u \times \exp(\mathrm{j}(\omega(-2) - \psi)), \\ &\text{etc.,} \end{aligned} \quad (17.76)$$

where ψ is the time lag of q_{ie} behind T_e, expressed in radians. Now the regression equation (11.1) can be written as

$$d_0 q_{ie,-0} + d_1 q_{ie,-1} + d_2 q_{ie,-2} + \ldots = b_0 T_{e,-0} + b_1 T_{e,-1} + b_2 T_{e,-2} + \ldots. \quad (17.77)$$

On substituting (17.76) into (17.77) and noting that $\exp(\mathrm{j}x) = \cos x + \mathrm{j} \sin x$, we have

$$u Dc \cos\psi - u Ds \sin\psi = Bc, \quad (17.78a)$$

$$u Dc \sin\psi + u Ds \cos\psi = Bs, \quad (17.78b)$$

where

$$\begin{aligned} Bc &= b_0 + b_1 \cos\omega + b_2 \cos 2\omega + \ldots, \\ Bs &= b_1 \sin\omega + b_2 \sin 2\omega + \ldots, \\ Dc &= d_0 + d_1 \cos\omega + d_2 \cos 2\omega + \ldots, \\ Ds &= d_1 \sin\omega + d_2 \sin 2\omega + \ldots. \end{aligned} \quad (17.79)$$

Thus the magnitude and time lag of the cyclic transmittance **u** are given as

$$m(u) = \left(\frac{Bc^2 + Bs^2}{Dc^2 + Ds^2}\right)^{1/2} \quad l(u) = \frac{24}{2\pi} \arctan\left(\frac{Bs \cdot Dc - Bc \cdot Ds}{Bc \cdot Dc + Bs \cdot Ds}\right). \tag{17.80}$$

The magnitude and time lead of the admittance **y** can similarly be found as

$$m(y) = \left(\frac{Cc^2 + Cs^2}{Dc^2 + Ds^2}\right)^{1/2} \quad l(y) = \frac{24}{2\pi} \arctan\left(\frac{Cs \cdot Dc - Cc \cdot Ds}{Cc \cdot Dc + Cs \cdot Ds}\right), \tag{17.81}$$

where Cc and Cs are defined similarly to Bc and Bs, with cs replacing bs. With the $N = 5$ values for b_k, c_k and d_k, the parameters found for the test wall are

$$m(u) = 0.735\ \text{W/m}^2\text{K}, l(u) = -4.600\ \text{h},$$
$$m(y) = 6.161\ \text{W/m}^2\text{K}, l(y) = \quad 1.440\ \text{h}.$$

They are close to the values found using the conventional frequency domain method:

$$m(u) = 0.739\ \text{W/m}^2\text{K}, l(u) = -4.600\ \text{h},$$
$$m(y) = 6.155\ \text{W/m}^2\text{K}, l(y) = \quad 1.450\ \text{h}.$$

The values from the transfer coefficients are based on a form of excitation in which temperature variation is made up of 24 linear segments per day, while the frequency domain method gives values corresponding to a strict sinusoidal variation in temperature; the two versions must therefore differ a little. Indeed, it is easy to show that in the case of a wall consisting simply of a lumped capacity c and outside film resistance r, so that $c_0 = 1/r + c/\delta, c_1 = -c/\delta, d_0 = 1$ and $d_1 = 0$ (11.22), we have

$$\frac{m^2(y) \text{ from transfer coefficients}}{m^2(y) \text{ exact}} = \frac{(1 + 4\pi^2 z^2/P^2 + 4\pi^2 z\delta/P^2)}{(1 + 4\pi^2 z^2/P^2)}, \tag{17.82}$$

so that $m(y)$ is slightly overestimated by the transfer coefficients. This proves computationally to be the case for all walls. The misestimation becomes worse for submultiples of $P = 24$ h.

Since knowledge of the exact values of the transfer coefficients b_k, d_k cannot lead to exact values for the cyclic transmittance $\mathbf{u_j}$ at period P_j, we would not expect that knowledge of the exact values of $\mathbf{u_j}$ at certain chosen values of P_j should lead to exact values of b_j, d_j. Davies (1997: 441) reports the unsatisfactory outcome of such an attempt where some values of b_j proved to be negative, which is physically unacceptable. Chen and Wang (2001) have recently attempted to find b_j and d_j values using a frequency-domain regression method. Their Table 2 shows the defects of the approach: for two of the walls illustrated, b_0 was negative and for no wall did the d_j values demonstrate the required alternating pattern of signs. Their further article (Wang and Chen 2003) demonstrates a_j, b_j, c_j values where $\sum a_j \neq \sum b_j \neq \sum c_j$, although they should be equal, and three different values for $\sum d_j$, none of which equals the correct $\sum d_j$ value for the wall studied. For hourly sampling of data, a wall has a unique set of d_j values and they are easy to find.

17.7.3 Transfer Coefficients, Decay Times and Time of Peak Flow

Two further descriptors of a wall can be found if its b_k, d_k values are supplied. According to (17.72) for a wall where $N = 3$, values of d_k are given by

$$\beta_j^3 + \beta_j^2 d_1 + \beta_j d_2 + d_3 = 0. \tag{17.83}$$

Solution of this equation gives three values β_j from which the first three decay times can be found as

$$z_j = -\delta / \ln \beta_j. \tag{17.84}$$

The d_k values in (17.83) must be such that there are, in this case, three positive fractional values of β_j. If $N = 2$, for example, this is so if $-2 < d_1 < 0$ and $d_1^2 - 4d_2 > 0$. The case of $d_1^2 - 4d_2 = 0$ corresponds to a symmetrical wall modelled as two lumped capacities with an infinite resistance between the capacities; the wall is now degenerate with two decoupled resistance/capacity components having the same decay times.

Davies (1996b: Table 8) describes the application of this procedure to the 41 wall constructions listed in the 1993 *ASHRAE Handbook*; it gives values for the decay times of between about 0.7 and 40 hours. For some walls solution of (17.83) gave complex values for the higher decay times. In two cases (wall groups 37 and 38, both with very low U values) even the first was complex. Thus the d_k values were defective; either the values were not quoted with sufficient precision or truncation of the series at $N = 6$ was inadequate. (It may be remarked though that the heat flow into a room through these walls due to ambient temperature fluctuations is so small that such defects are unlikely to be important.)

Secondly, using relations similar to (11.20), the response factors $\phi_{n0,k}$ can be found. The largest of them identifies approximately the time of largest heat flow after the apex of the triangular pulse of excitation and the time can be made a little more explicit by quadratic interpolation using the two adjacent factors. To judge from the values given for the ASHRAE walls, times of peak flow increase broadly with the time lag $l(u)$ for 24 h sinusoidal excitation; in most cases the peak flow time is a little less than $l(u)$ but for heavy walls, where $l(u)$ is around 13 h, it appears to exceed it a little.

17.7.4 Transfer Coefficients with and without Film Coefficients

The conventional wall U value, the cyclic parameters and the transfer coefficients normally include the values of the external and internal coefficients. In handling transmitted solar gains, however, it is conceptually simpler to suppose that they act at room surfaces themselves rather than as some scaled-down value at the room index node. The same applies in principle at wall exterior surfaces, but unless variation in the external convection coefficient is significant, the adoption of sol-air temperature as a driver is adequate.

Table 17.7 Parameters of a one-dimensional wall of total resistance r and total capacity c

Form of excitation	Number of thermal circuit elements needed	Parameters	Relationship
Steady state	1	U	$1/U = r$
Steady slope	5 (two equal)	$q_{ee}/\theta,\ q_{ii}/\theta,\ q_{ie}/\theta = -q_{ei}/\theta$	sum $\propto c$
Steady cyclic	6 (three pairs)	$\mathbf{u},\ \mathbf{y}_e,\ \mathbf{y}_i$	none
Ramp	$2N+1$,		
	N determined by	$a_0, a_1, \ldots, a_{N+1}$	$\sum d_k / \sum a_k = r$
	$\Pi_1^N \beta_j < \varepsilon \Pi_1{}^N (1-\beta_j)$,	$b_0, b_1, \ldots, b_{N+1}$	$\sum d_k / \sum b_k = r$
	$\varepsilon \sim 10^{-3}$	$c_0, c_1, \ldots, c_{N+1}$	$\sum d_k / \sum c_k = r$
		$d_0 = 1, d_1, \ldots, d_N$	$-\sum k(a_k + c_k - 2b_k)\delta / \sum d_k = c$

In all cases it is possible, given the value of the conventional parameter and the internal film coefficient h_i, to construct the corresponding parameter without the film coefficient. Thus if U' is the transmittance from the wall inside surface,

$$1/U' = 1/U - 1/h_i.$$

Values for the similar cyclic parameters $\mathbf{u}'$ and $\mathbf{y}'$ can be found along the lines of the numerical example in Section 15.3. Burch *et al.* (1992a) have demonstrated that without-film transfer coefficients too can be found from conventional values but it is computationally complicated; it might be simpler to evaluate the coefficients directly.

17.7.5 Summary of Modelling Parameters

Table 17.7 summarises the parameters relevant to various forms of wall excitation.

17.8 THE EQUIVALENT DISCRETISED WALL

To examine wall response using the finite difference method, each layer of the wall is supposed to be divided uniformly into a number of slices. Even spacing was discussed in Section 10.40 and non-uniform spacing in Section 15.7. Now using the Crank–Nicolson formulation, the time taken in solving the resulting set of N_{fd} equations, (N_{fd} is the total number of slices), is proportional to N_{fd}^3 and since in processing a year's weather data at hourly intervals the set must be solved 8760 times, it can result in long runtimes. Much effort has been spent on the question of nodal spacing so as to reconcile the conflicting criteria of accuracy and runtime. The problem has recently been reviewed and advanced by Tuomaala *et al.* (2000). It can be approached by making N_{fd} progressively larger until some measure, a heat flow at an isothermal boundary node, or temperature at an adiabatic boundary node when the wall is subjected to some excitation at the other bounding node, either yields negligible change, or better, sufficiently approximates to the value given by an analytical solution. This section (based on Davies (2003)) presents a different approach, making use of part of the theory of this chapter. It argues that there is a minimum number of nodes and this minimum depends on the series of decay times of the wall and the sampling interval of the driving data (the terms which lead to the series of d_k values). Once this number is known, a given wall construction is judiciously divided into the appropriate number of lumped resistances and capacities and its response to a ramp excitation is compared with the corresponding response of the parent wall, the difference being expressed non-dimensionally. By systematically changing the discretised resistances and capacities (but keeping the totals constant), the model response can be brought acceptably close to the response of the original wall.

First we note that, given the N values of d_k and the sampling interval δ for a real wall, approximate values for the first N values of z_j can be found. Now a ladder consisting of N capacities and $N+1$ resistances has just N decay times which result from the solution of $e_{12}=0$ in (17.44). It follows that a real wall described by N values of d_k should be representable fairly closely by a discretised wall of N capacities, as far as its response to excitation at time-discretised values of temperature is concerned. This value of N, determined from general time-domain considerations of wall heat flow, may be expected to be less than N_{fd}.

For a discretised wall of N capacities, we can express its node 0 response at time level k ($k=0$ denotes $t=0$) due to ramp excitation imposed at node T_0 as

$$q_{00,k^\wedge} = \theta k\delta/r + q_{00,0^\wedge} + \sum_{j=1}^{N} q_{00,j^\wedge} \exp(-k\delta/z_{j^\wedge}), \tag{17.85}$$

where the subscript $\wedge$ indicates that the quantities concerned are for the discretised wall. Summation is now performed only to $j=N$.

The difference between the exact flow $q_{00,1^*}$ at time level 1 and its estimate from the discretised wall $q_{00,1^\wedge}$ is found by subtraction of (17.85) and (17.86), which is (17.60a) with $t=k\delta$:

$$q_{00,k^*} = \theta k\delta/r + q_{00,0^*} + \sum_{j=1}^{N} q_{00,j^*} \exp(-k\delta/z_{j^*}). \tag{17.86}$$

The principle through which the elements of the model wall are sized is that the flows $q_{00,k}$ and $q_{nn,k}$ at the first time step ($k=1$) following imposition of a ramp input should be acceptably close. In order that spurious differences in sign should not cancel, we form the square of individual differences. Thus we form the sum of squares of differences:

$$\begin{aligned}\mathrm{SS} = {} & [(q_{00,0^\wedge} - q_{00,0^*})^2 + (q_{nn,0^\wedge} - q_{nn,0^*})^2] \\ & + \sum_{j=1}^{N-1} [\,(\underbrace{q_{00,j^\wedge}\exp(-\delta/z_{j^\wedge})}_{\text{model term}} - \underbrace{q_{00,j^*}\exp(-\delta/z_{j^*})}_{\text{parent term}})^2 + (\underbrace{q_{nn,j^\wedge}\exp(-\delta/z_{j^\wedge})}_{\text{model term}} - \underbrace{q_{nn,j^*}\exp(-\delta/z_{j^*})}_{\text{parent term}})^2\,].\end{aligned} \tag{17.87}$$

Consider the case $N=3$. There are $2N+1=7$ degrees of freedom. We have to find values of the four resistances r_1, r_3, r_5 and r_7 and the three capacities c_2, c_4 and c_6 such that their sums equal the respective resistances r and capacities c of the real wall. With this constraint, we vary them so as to minimise SS or reduce it to some acceptably low value. SS consists altogether of $(1+1)+2\times(3-1)=6$ squares so as to determine the remaining 5 degrees of freedom. Summation of j needs be conducted to $N - 1$ only since $\sum_{j=1}^{N} q_{00,j\wedge} = -q_{00,0\wedge}$ (similarly for q_{nn}), and inclusion of the term in $j = N$ provides no further information. $\sqrt{\text{SS}}$ has the units of flux (W/m^2) proportional to the assumed rate of temperature rise θ. It can be made independent of θ as the relative error

$$\Delta = \sqrt{\text{SS}/\text{SS}^*}, \tag{17.88}$$

where SS* is the sums of squares of all terms in (17.87) relating to the real wall only. SS* depends mainly on the values of $q_{00,0^*}$ and $q_{nn,0^*}$ since (i) each $q_{00j,*} < q_{00,0^*}$, (ii) each term is multiplied by a factor less than 1, and (iii) SS* consists of the squares of such terms. (We note that if $a + b = c$, $a^2 + b^2 < c^2$.)

The r_L, c_L values have to be systematically varied so as to reduce SS. The final value reached is a measure of how well the discretised wall matches the real wall. The scheme is shown in Figure 17.2. Note that in the usual method of designing a discretised wall by reducing the slice thickness, the capacity located at some node must be proportional to its adjacent resistances; the present approach is not restricted in this way.

Quantities shown in the heavier boxes of Figure 17.1 are carried over into the discretisation procedure. Calculations for the discretised wall follow a similar pattern and largely use the same code. The total number of elements is taken to be $2N + 1$. Values are assumed for $r_1, r_3, \ldots, r_{2N+1}$ totalling r, and $c_2, c_4, \ldots, c_{2N}$, totalling c. Since they are lumped elements, the steady-slope coefficients C_j and D_j are zero. Also, the decay times $z_{j\wedge}$ can be found by direct solution of a polynomial equation, a faster operation than the search that is needed for a wall of distributed layers. Using the values of $q_{00,0\wedge}$, etc., SS is found using (17.87).

Two minimisation procedures may be used to reduce SS. In procedure A, during the course of one iteration, each resistance r_j is associated with resistance r_k; $r_j \to r_j + r/\text{DV}$ and $r_k \to r_k - r/\text{DV}$, where the divisor DV is a large number, set initially to 1000 say. If this leads to a reduction in SS, these values are retained; if not, $r_j \to r_j - r/\text{DV}$ and $r_k \to r_k + r/\text{DV}$, and the procedure is repeated. If neither operation leads to a reduction in SS, DV is increased to DV$\cdot M$, where M is a multiplier (e.g. 1.02), for use in the next iteration. Similarly, with the capacities. This leads to a decrease in SS and thus convergence to an acceptable equivalent lumped wall. Procedure B is similar to this, but here one resistance r_j is increased to $r_j \pm r/\text{DV}$ and all the remaining resistances are changed so as to preserve a total of r. These procedures are crude and slow but the algebraic difficulties associated with finding first and second differential coefficients of SS preclude a more efficient procedure. Procedure A followed by procedure B 250 times (double iteration) led to a greater reduction of SS than 500 iterations of A alone or 500 iterations of B alone. Some results of their use follow.

17.8.1 Error and Wall Thickness

A homogeneous slab can be represented approximately by placing its total capacity c at a node in the centre of a resistance representing its total resistance r as shown in Figure 10.2b with $r_1 = r_3 = r/2$ and with $c_2 = c$. The model becomes progressively worse as the thickness X increases and the above ideas allow us to see the deterioration. It is

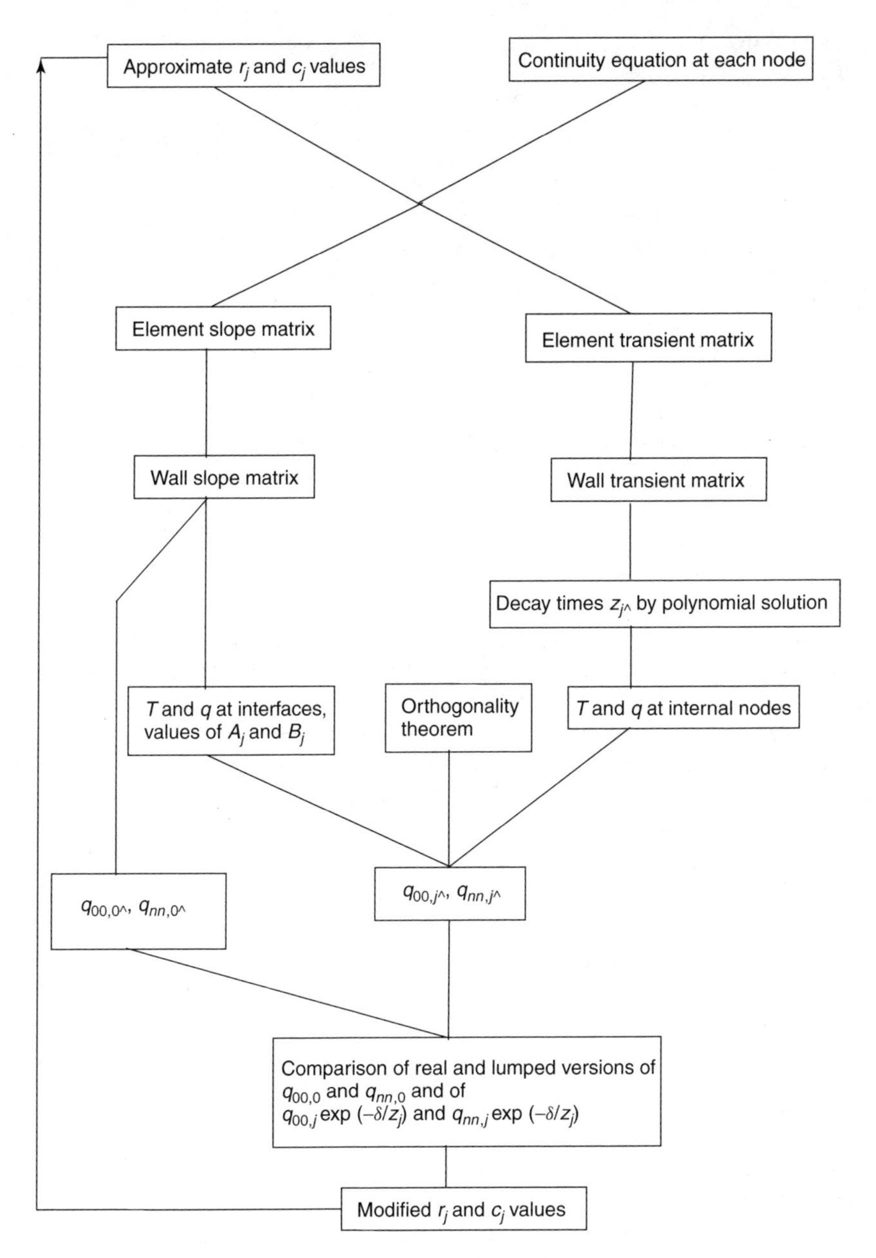

Figure 17.2 Scheme to evaluate the model wall elements. Note that depending on the context, subscript j denotes an eigennumber or a model element. (Davies, 2003, with permission from Elsevier Science)

convenient to summarise some of the equations. The general expression for heat flow into a wall due to same-side ramp excitation is

$$q_{00,t} = \frac{\theta_0 t}{r} + q_{00,0^*} + \sum q_{00,j^*} \exp(-t/z_{j^*}). \tag{17.60a}$$

For a single slab,

$$\frac{q(0,t)}{\theta_0} = \frac{\lambda t}{X} + \frac{\rho c_p X}{3} - 2\rho c_p X \sum \frac{1}{j^2\pi^2} \exp\left(-\frac{t}{z_j}\right), \tag{12.23a}$$

where

$$z_j = \rho c_p X^2/(j^2\pi^2\lambda) = cr/j^2\pi^2. \tag{12.14}$$

This can be written as

$$\frac{q_{00,t}}{\theta_0} = \frac{t}{r} + \frac{c}{3} - \frac{c}{\pi^2/2} \sum \exp\left(-\frac{j^2\pi^2 t}{cr}\right) /j^2. \tag{17.89}$$

The corresponding expression for a single capacity follows from (10.13), noting that $q_{00,t} = (\theta t - T_{2t})/\frac{1}{2}r$ and that $z = cr/4$:

$$\frac{q_{00,t}}{\theta_0} = \frac{t}{r} + \frac{c}{4} - \frac{c}{4} \exp\left(-\frac{4t}{cr}\right). \tag{17.90}$$

The similarity between the response of a slab and its T-section approximation will be noted. The slope term $c/3$ is replaced by $c/4$. The wall first decay time z_{1*} is cr/π^2 and the model wall decay time $z_\wedge$ is $cr/4$. Thus $z_\wedge$ is more than double z_{1*}. This implies that the model wall is much slower in its response than the real wall and this becomes important when the sampling interval δ is small compared to z_{1*}, that is, when the wall is thick. (Note that the $c/3$ term follows from the slope matrix in (17.25) together with q_{ee} (17.34). Further, $c/4$ follows from application of a matrix similar to that in (17.29) to the elements $r/2$, c and $r/2$. Also the transient terms in (17.89) sum to $-c/3$ at $t = 0$, as they must.) Thus the value of SS for the current problem is given as

$$\begin{aligned} SS/2\theta_0 = {} & [\quad c/4 \quad - \quad c/3 \quad]^2 \\ & + [(c/4)\exp(-4\delta/cr) - (c/(\pi^2/2))\exp(-\pi^2\delta/cr)]^2 \\ & + \sum_{j=2}^{N'} [\quad 0 \quad - (c/(j^2\alpha\pi^2/2))\exp(-j^2\pi^2\delta/cr)]^2. \end{aligned} \tag{17.91}$$

From symmetry, all q_{nn} and q_{00} terms are equal, so q_{nn} terms can be omitted, leading to SS/2 rather than SS.

To see the variation of SS with thickness X, consider a slab of material with the nominal values $\lambda = 1.0$ W/m K, $\rho = 1250$ kg/m^3 and $c_p = 800$ J/kg K (loosely representative of aerated cast concrete). A rate of temperature rise θ_0 of 1 K per hour or 1/3600 K/s will be used. Table 17.8 lists the values of N and N', $\sqrt{SS}$ and Δ for increasing values of thickness.

For a slab thickness less than about 50 mm the N, N' values (both 1) suggest that it can be adequately represented as an $r/2$, c, $r/2$ model. But if $X = 100$ mm then two capacities are needed, which accords with conventional wisdom. (It should be noted that the structure here is a simple slab. This implies that it may be excited by a heat source at one or other of its bounding surfaces at T_0 and T_1. However, if the slab forms part of a building wall,

Table 17.8 Modelling a slab by a single capacity. (Davies, 2003, with permission from Elsevier Science)

Thickness (m)	z_1 (h)	N	N'	$\sqrt{\text{SS}}$ (W/m^2)	$\Delta = \text{SS}/\sqrt{\text{SS}^*}$
0.05	0.07	1	1	1.16	0.25
0.10	0.28	2	2	2.75	0.30
0.2	1.13	4	4	6.86	0.36
0.3	2.53	4	6	9.47	0.32
0.5	7.0	6	10	14.8	0.28
1.0	28	12	23	30.2	0.28
2.0	112	22	58	61.6	0.28
5.0	704	52	215	155	0.28

it is necessarily flanked by inside and outside film resistances and index nodes T_e (excited by a temperature source) and T_i (excited by a heat source) replace the surface nodes. If no excitation is to be imposed at T_0 or T_1, representation of a thicker slab as a simple T-section may be satisfactory.)

If excitation is to act at T_0 or T_1, it is clear from the value of $\Delta(\geq 0.25)$ that the simple T-section can at best only provide a crude approximation to the transient response of a slab. This is intuitively obvious and the values of Δ are of little importance.

17.8.2 The Two-Capacity Homogeneous Wall

The simple one-capacity model discussed above cannot represent a real wall exactly, but a two-capacity $(r_1, c_2, r_3, c_4, r_5)$ can model it very closely if the wall is thermally 'thin'. To justify this, we note that the first decay time z_1 is the largest of the series and that when $t = 4.6z_1$, $\exp(-4.6z_1/z_1) = 1\%$, so transient effects have virtually died out after a time of around $4z_1$. If the sampling interval δ is larger than about $4z_1$, heat flows such as $q_{00,t}$ in (17.60a) are determined by the slope solution alone. Now in conditions of steady-slope excitation, the wall can be modelled exactly by a two-capacity model, with values determined by (17.38). Thus a wall whose first decay time is less than around $\delta/4$ can be represented almost exactly by a two-capacity model. The thinner the wall, the smaller its value of z_1 and if it is less than about $\delta/4$, the wall can be described as thermally thin.

It is instructive to see the contributions to SS of the slope and transient terms for a slab which is not thin – a homogeneous wall of 100 mm thickness with nominal values of λ, ρ and c_p given above. Here $z_1 = \rho c_p X^2/\pi^2\lambda = 1013$ s. If $\delta = 3600$ s then $N = 2$; that is, the wall should be modelled as a two-capacity wall. Now $r = 0.1$ m^{2}K/W and $c = 10^5$ J/m^{2}K so c_2 and c_4 should each be $c/2$ and $r_1 = r_5$. The total sum of squares of deviations (17.87) can be resolved into components based on the steady-slope terms $\sqrt{\text{SS}_\text{s}}$ and the components due to transient terms $\sqrt{\text{SS}_\text{tr}}$. They are shown in Table 17.9. $\Delta = \sqrt{\text{SS/SS}^*}$ where $\sqrt{\text{SS}^*} = 13.09$ W/m^2. $\sqrt{\text{SS}_\text{s}}$ has a low value when $r_1 = 0.021$ m^{2}K/W and is in fact zero when $r_1 = \alpha r$, where $\alpha = 0.2113$. $\sqrt{\text{SS}_\text{tr}}$ is zero when $r_1 = 0.191r$. However, SS_s varies more strongly with r_1 than does SS_tr and the overall minimum is near $r_1 = 0.021$.

Table 17.9 The homogeneous slab modelled by two capacities. (Davies, 2003, with permission from Elsevier Science)

r_1	Slope component, $\sqrt{SS_s}$	Transient component, $\sqrt{SS_{tr}}$	Total, $\sqrt{SS}$	$\Delta = \sqrt{SS}/\sqrt{SS^*}$
0.018	0.749	0.048	0.751	0.057
0.019	0.502	0.006	0.502	0.038
0.020	0.262	0.040	0.265	0.020
0.021	0.030	0.091	0.095	0.007
0.022	0.194	0.144	0.242	0.018
0.023	0.410	0.201	0.457	0.035
0.024	0.618	0.261	0.671	0.051
0.025	0.818	0.323	0.880	0.067

Conventional division would place c_2 and c_4 between two resistances of $r/4$ so that $r_1 = r_5 = r/4 = 0.025\,\mathrm{m^2K/W}$ and $r_3 = r/2 = 0.05\,\mathrm{m^2K/W}$. Both Stephenson and Starke (1959) and Davies (1983a) note that a better fit might result from a value of $r_1 = r_5 = 0.2113r$, based on a sinusoidal q_{n0}-like response (see p. 383). If both q_{n0}- and q_{00}-like responses were taken into account however, $r_1 = r_5$ should be about $0.193r$ (Davies 1983a: 26). Further it was shown above that the value $\alpha = 0.2113$, the solution of (17.38a) or $\alpha^2 - \alpha + \frac{1}{6} = 0$, is the condition that a two-capacity model should have exactly the same steady-slope response as a homogeneous wall. It is advantageous to put more resistance into the centre of the model than conventional modelling would suggest. Benard (1986) has addressed this problem.

17.8.3 The Real Wall Is Discretised

To test the minimisation procedure, it is useful to start with a 'real' wall which itself is made up of lumped elements. In this case the procedure should reproduce the original wall, at any rate approximately. Consider the three-capacity wall where $r_1 = 0.01$, $r_3 = 0.02$, $r_5 = 0.03$, $r_7 = 0.04$, totalling $0.10\,\mathrm{m^2K/W}$, and $c_2 = 20\,000\,\mathrm{J/m^2K}$, $c_4 = 40\,000$, $c_6 = 60\,000$, totalling $120\,000\,\mathrm{J/m^2/K}$. The parent values are shown in row 0 of Table 17.10.

Table 17.10 The real wall itself is lumped: approximations to its elements for different starting values. (Davies, (2003), with permission from Elsevier Science)

Row	r_1	c_2	r_3	c_4	r_5	c_6	r_7	Δ
0	0.01	20 000	0.02	40 000	0.03	60 000	0.04	–
1 Initial	0.025	40 000	0.025	40 000	0.025	40 000	0.025	0.2421
2 Final	0.012802	30 671	0.025613	37 374	0.022654	51 954	0.038912	0.0029
3 Initial	0.014	16 000	0.016	42 000	0.034	62 000	0.036	0.1146
4 Final	0.009799	19 467	0.019934	40 330	0.030233	60 203	0.040034	0.00004
5 After 50 000 iterations								
	0.010000	19 999	0.020000	40 000	0.030000	60 000	0.040000	–

The simplest starting values are the mean values $r_1 = r_3 = r_5 = r_7 = 0.10/4 = 0.025\,\text{m}^2\text{K/W}$ and $c_2 = c_4 = c_6 = 120\,000/3 = 40\,000\,\text{J/m}^2\text{K}$ (row 1). The approximations reached after 250 double iterations are shown in row 2. These approximations are poor. The c_L values are relatively bunched and the r_L values do not even have the correct rank order, since r_3 comes out greater than r_5.

The relative error Δ due to the initial choice is shown in the final column. After 250 iterations this was reduced to around 0.003 (and was still decreasing). This is a surprising result. It seems that although the model wall differs considerably from its parent, the two may be expected to have similar response characteristics since Δ expresses the total difference between the responses. The explanation is that, as indicated earlier, Δ depends mainly on the two steady-slope terms, $q_{00,0}$ and $q_{nn,0}$. Now the product steady-slope transmission matrix (17.30) has only four degrees of freedom and two quite dissimilar walls may have the same values for these elements; in the present case, the r and c values of parent and model wall are identical. The values of $q_{00,0}$ and $q_{nn,0}$ depend only on the values of the matrix elements, so two dissimilar walls could have identical $q_{00,0}$ and $q_{nn,0}$ values. Their series of $q_{00,j}$ and $q_{nn,j}$ values will of course differ. The minimisation procedure appears to have succeeded in largely matching model and parent walls as regards their steady-slope behaviour, but not their transient behaviour.

A better match is found when initial values are assumed which preserve the rank order of the real wall; see row 3. The r_L and c_L approximations reached after 250 double iterations are shown and they are now satisfactorily close to the parent values. Agreement improves with further iteration but by now it is very slow. These values depend somewhat on the initial choice and the chosen values of DV and M; here DV = 2000 and $M = 1.02$. The seven-dimensional topography of SS is complicated. For practical application, however, values for Δ of order 0.0001 are excellent; the accuracy with which basic information – the conductivity, density and specific heat of wall materials – differs fractionally by orders of magnitude from such Δ values. After sufficient iteration, the procedure leads back nearly exactly to the original wall. See row 5.

17.8.4 The Homogeneous Wall

Some insight into the effect of the procedure to reduce the deviation SS in (17.87) is provided by consideration of a homogeneous wall. Table 17.11 lists values for a wall of thickness $X = 0.3$ m with λ, ρ and c_p values as above.

Table 17.11 Flow amplitudes $q_{00,j}$, decay times z_j and deviations for the slab and its $N = 5$ model after 4000 iterations. (Davies, 2003, with permission from Elsevier Science)

j	q_{00,j^*} (W/m^{2}K)	$q_{00,j^\wedge}$ (W/m^{2}K)	z_{j^*} (h)	$z_{j^\wedge}$ (h)	$q_{00,j^*}\exp(-\delta/z_{j^*}) - q_{00,j^\wedge}\exp(-\delta/z_{j^\wedge})$ (W/m^{2}K)
0	27.7778	27.7836			−0.0058
1	−16.8869	−16.4835	2.533	2.702	0.0058
2	−4.2217	−3.2931	0.633	0.748	−0.0058
3	−1.8763	−0.7263	0.281	0.400	0.0060
4	−1.0554	−3.6571	0.158	0.158	0.0045
5	−0.6755	−3.6235	0.101	0.157	0.0063

The aim of the procedure is to reduce the deviations noted in (17.87). Values are listed in the final column; the procedure appears to reduce them fairly uniformly. We note that $\sum_{j=1}^{5} q_{00,j\wedge} = 27.7835\ \mathrm{W/m^2K}$ which equals $-q_{00,0\wedge}$, as it should. The model decay times are necessarily larger than the parent values. To justify this, we note that the z_j values are solutions to the relation $e_{12} = 0$ in (17.44) and might be written more explicitly as z_{12j} values. There are corresponding sets of z_{11j}, z_{21j} and z_{22j} values. Now it has been noted (10.28b) that for a real wall, we have

$$\sum_{j=1}^{\infty} z_{11j^*} + \sum_{j=1}^{\infty} z_{22j^*} = \text{wall resistance} \times \text{wall capacity}, \tag{17.92a}$$

although the sums of the series converge slowly, with a significant contribution for $j > 5$. For the model too, we have

$$\sum_{j=1}^{5} z_{11j\wedge} + \sum_{j=1}^{5} z_{22j\wedge} = \text{wall resistance} \times \text{wall capacity}, \tag{17.92b}$$

but with no contribution beyond $j = 5$. Thus the model z_{11j} values must be greater than the wall values, and this must be true for the z_{12j} values.

As wall thickness increases, the number N of capacities needed to model it increases. Values are shown in Table 17.12, rows 1 and 2, for a wall of thickness 0.4 m. Table 17.12 shows the error Δ for conventional modelling and also the values when the optimisation procedure (400 double iterations) is used. Since the wall is symmetrical, only half the element values, together with the central element, are given. The value of $\sqrt{SS^*}$ was $58.3\ \mathrm{W/m^2K}$. For this thickness $N = 6$.

The changes in the element values to optimise show that r_1(optimised) $< r_1$(raw) and correspondingly there is a greater concentration of resistance in the interior. Similarly c_2(optimised) $< c_2$(raw) with more capacity in the interior.

Conventional modelling gives a relative error of 2%, which is probably acceptable, but the optimisation procedure reduces it to 0.14%, so a smaller value of N might be chosen. By varying N, it was found that with $N = 4$ and with optimisation, $\Delta = 0.4\%$. A criterion that is sometimes used for sizing states that if $\lambda > 0.5\ \mathrm{W/mK}$, the maximum slice thickness should be 0.03 m. If this is adopted, for an $X = 0.4$ m slab, N should equal $0.4/0.03 = 13$ slices.

Table 17.12 Accuracy of modelling a homogeneous slab with uniform and optimal nodal placement. (Davies, (2003), with permission from Elsevier Science)

$X = 0.4$ m	r_1, r_{13}	c_2, c_{12}	r_3, r_{11}	c_4, c_{10}	r_5, r_9	c_6, c_8	r_7	Δ
1 Initial	0.033333	66 667	0.066667	66 667	0.066667	66 667	0.066667	0.019
2 Final	0.021944	54 591	0.070698	71 845	0.068735	73 564	0.077245	0.001

$X = 0.6$ m	r_1, r_{17}	c_2, c_{16}	r_3, r_{15}	c_4, c_{14}	r_5, r_{13}	c_6, c_{12}	r_7, r_{11}	c_8, c_{10}	r_9	Δ
3 Initial	0.037500	75 000	0.075000	75 000	0.075000	75 000	0.075000	75 000	0.075000	0.0132
4 Final	0.025514	61 604	0.076637	75 053	0.066287	68 028	0.079912	95 316	0.103302	0.0023

Rows 3 and 4 for $X = 0.6\,\text{m}$, $N = 8$, illustrate the same point. The value of Δ with even spacing, 1.3%, is nearly achieved with $N = 4$ optimised spacing, when $\Delta = 1.4\%$. But whereas reduced r_j and c_j values are found at the wall surfaces and large values in the centre, the progress from one to the other is not uniform.

17.8.5 Wall Modelling

The above discussion was concerned with a homogeneous wall without surface films. If these resistances are included, there is a decrease in the number of d_k values or the number of capacities needed to represent the wall. This will be clear qualitatively by considering the Groeber problem in Section 13.2. A slab of semi-thickness X and conductivity λ is adiabatic at one surface and in contact with an air film of transmittance h or resistance r_f and is initially at uniform temperature. At $t = 0$ the air temperature is reduced to zero and held there. The slab thickness is non-dimensionalised as the Biot number $B = hX/\lambda$ or $(X/\lambda)/r_f$ and time as $F_2 = h^2t/\lambda\rho c_p$ or $t/(r_f^2\lambda\rho c_p)$. Table 13.3 lists values of F_2 as a function of B and dimensionless surface temperature θ. For B values much less than 0.1, practically all the cooling regime is exponential cooling; that is, when the surface film resistance r_f is large compared with the slab resistance $X/\lambda, \theta \rightarrow \exp(-t/(r_f\rho c_p X))$ from (13.27). The wall behaves as a single lumped capacity $c = \rho c_p X$ with just a single decay time, so $N = 1$ independent of the slab resistance. For smaller values of film resistance, the slab thickness becomes significant.

Table 17.13 gives details of a real wall, including the outer film resistance. Finding the discretised equivalent is a two-stage process. First the details of the wall are input to determine the number of capacities N with which the wall should be modelled. Then by inspection of the resistances and capacities of each layer, initial values are selected for the $N + 1$ lumped resistances and the N capacities of the model. A large capacity with a small resistance (e.g. a heavy concrete slab) might be represented simply as a T-section. The initial values are given in Table 17.14. The value of $\sqrt{\text{SS}^*}$ was $79.76\,\text{W/m}^2$ and the values of $\sqrt{\text{SS}}$ resulting from the initial and optimised elements were 1.266 and 0.094, respectively, giving the values of $\Delta = \sqrt{\text{SS}/\text{SS}^*}$ noted.

Using the criterion that if $\lambda < 0.5\,\text{W/m K}$, $\Delta X_{\text{max}} = 0.02\,\text{m}$, if $\lambda > 0.5\,\text{W/m K}$, $\Delta X_{\text{max}} = 0.03\,\text{m}$, Table 17.13 indicates the number of slices into which each layer should be divided; the total number is 11. Each slice has a lumped resistance and a lumped capacity acting at a node representing the midpoint of the resistance. The total resistance

Table 17.13 Details of wall construction for wall 3.17 5a in the 1986 *CIBSE Handbook*. (Davies, (2003), with permission from Elsevier Science)

	X(m)	λ(W/m K)	ρ(kg/m^3)	c_p (J/kg K)	Resistance (m^{2}K/W)	Number of slices
Outer film	–				0.055	
Brickwork	0.105	0.840	1700	800		4
Air gap	–				0.180	
Phenolic foam	0.025	0.040	30	1400		2
Heavy concrete	0.100	1.630	2300	1000		4
Light plaster	0.013	0.160	600	1000		1

Table 17.14 Discretised elements for the sample wall in Table 17.13, $N = 5$. (Davies (2003) with permission from Elsevier Science.)

	r_1	c_2	r_3	c_4	r_5	c_6	r_7	c_8	r_9	c_{10}	r_{11}	Δ
Initial	0.085000	71 400	0.060000	71 400	0.810000	31 050	0.066300	200 000	0.066300	7800	0.040000	0.0159
Final	0.073013	52 847	0.068961	89 690	0.803451	25 871	0.074679	199 122	0.071093	14 121	0.036402	0.0012

between two nodes consists of two parts, equal when both lie in the same layer, unequal when they lie in different layers. The resistance between T_e and T_1 comprises the outer film resistance and the semi-slice resistance, and the resistance between two nodes separated by a cavity comprises the two semi-slice resistances and the cavity resistance. Thus division of the wall into N' slices implies $2N' + 1 = 23$ for this wall. Using the optimisation procedure, this can be reduced to 11, with an error of less than 1%. This and other examples are discussed in Davies (2003).

These values apply when the wall is randomly excited with data sampled at hourly intervals. For excitation at shorter intervals, more resistive/capacitative elements are needed.

17.9 TIME- AND FREQUENCY-DOMAIN METHODS COMPARED

Starting with wall details and the Fourier continuity equation, use of solutions in the time domain have led us to values for the wall transfer coefficients. The more usual way of performing this task is to use frequency-domain solutions (Section 12.5) and it is interesting to compare the two approaches.

The Laplace transformed transmission matrix for a single layer in a wall, resistance r, capacity c, can be written as

$$\begin{bmatrix} \cosh\sqrt{prc} & (\sinh\sqrt{prc})/[\sqrt{prc}/r] \\ (\sinh\sqrt{prc}) \times [\sqrt{prc}/r] & \cosh\sqrt{prc} \end{bmatrix} \quad \text{or} \quad \begin{bmatrix} A(p) & B(p) \\ C(p) & A(p) \end{bmatrix}. \tag{17.93}$$

This is identical with the form in (17.40) if $p = -1/z$. Thus the Laplace transformed heat flows at surfaces 0 and 1 can be expressed in terms of the Laplace transformed temperatures as

$$\begin{bmatrix} q_0(p) \\ q_1(p) \end{bmatrix} = \begin{bmatrix} A(p)/B(p) & -1/B(p) \\ 1/B(p) & -A(p)/B(p) \end{bmatrix} \begin{bmatrix} T_0(p) \\ T_1(p) \end{bmatrix}. \tag{17.94}$$

Suppose we wish to find the flux at surface 0 due to a ramp increase in temperature at surface 1. $T_0(p)$ is zero and $T_1(p)$ becomes $1/p^2$, the Laplace transform of a ramp (a rise in temperature of 1 K/s from $t = 0$). Thus the flux at surface 0 is the inverse of $q_0(p)$:

$$q_0(t) = L^{-1}[q_0(p)] = L^{-1}\left[\frac{1\sqrt{prc}}{p^2 r \sinh\sqrt{prc}}\right]. \tag{17.95}$$

According to complex variable theory, the inverse consists of the sums of residues at the poles of $q_0(p)e^{+pt}$; poles are defined as values of p where $q_0(p)$ is undefined, that is to say, infinite. Thus poles are located at $p = 0$ and $p = -j^2\pi^2/(rc)$.[9]

[9]Values of p prove to be negative, so to show this we write $p = -s$. Then $\sinh\sqrt{prc} = \sinh\sqrt{-src} = j\sin\sqrt{src}$, $(j = \sqrt{-1})$, which is zero when $\sqrt{src} = \sqrt{-prc} = j\pi$.

These residues were evaluated by Kusuda (1969: 250) and Stephenson and Mitalas (1971: 119) in the manner set out in Table 17.15. (They cite A/B, denoting $q_{nn,t}$, rather than the $1/B$ denoting $q_{0n,j}$ noted below.)

A further point of congruence follows from recognising that a series of values of driving temperatures (e.g. at hourly intervals) are in principle independent of each other. This leads to (11.19) in the time-domain method and to the equating of terms in z^{-n} in (12.63).

There is a formal similarity between the transmission matrix (17.93) and that for the heat loss from a fin (3.8). The diffusion of heat mainly between the above- and below-zero portions of some eigenfunction temperature distribution here corresponds to the external loss of heat from a fin; the term $p\rho c_p$ (p is the Laplace variable) corresponds to

Table 17.15 Comparison of time- and frequency domain methods

	Time-domain method	Frequency-domain method
Temperature boundary conditions imposed	Series of values of z_j found by search; interpreted as decay times	Series of poles p_j found by search; $p_j = -1/z_j$
	For a slab, $z_j = rc/j^2\pi^2$ (eq.12.2.8)	For a slab $p_j = -j^2\pi^2/\mathrm{rc}$
Temperature variation through the wall	Series of eigenfunction profiles: $j = 1$, profile everywhere positive $j = 2$, part positive, part negative $j = 3$, positive, negative, positive, etc.	Profile not made explicit
	Profiles decay as $\exp(-t/z_j)$	
Evaluation of terms in the equation for the response at node 0 due to ramp excitation at node n of unit slope, $\theta = 1$ K/s		
$q_{0n}(t) = t/r + q_{0n,0} + \sum q_{0n,j}\exp(-t/z_j)$:		
t/r	From elementary considerations; see Section 17.2	From Laplace transform theory, as $[1/B]_{p=0}$
$q_{0n,0}$	From slope matrix; e_{14} in equations (17.30) and (17.34)	From Laplace transform theory, as $[\mathrm{d}(1/B)/\mathrm{d}p]_{p=0}$
$q_{0n,j}$	From Fourier analysis using the slope and transient solutions; see (17.59b)	From Laplace transform theory, as $[1/(p^2\,\mathrm{d}B/\mathrm{d}p)]_{p=p(j)}$
		B here denotes the term in the matrix of a multilayer wall. $\mathrm{d}B/\mathrm{d}p$ is evaluated by the chain rule

the fin term hp/A (p here is the fin periphery). Equation (3.9) indicates that the lateral loss from a fin can be found through *differentiation* or through *integration*, and these alternative approaches are formally apparent from Table 17.15 in regard to the quantity $q_{0n,j}$: equation (17.59a) shows that $q_{0n,j}$ results from summation of layer integrals and the table shows that it can also be derived through differentiation of an element in the wall product transmission matrix.

The frequency domain formulation is elegant in that it expresses the elements of $q_0(t)$ explicitly in terms of the product matrix element B. The time-domain method, however, might be preferred for several reasons:

- It incorporates the requirement that $\sum q_{0n,j}$ should equal $-q_{0n,0}$, (since q_{0n} must be zero at $t = 0$).
- It is a simple extension of the well-known technique of Fourier analysis, in which some shape, such as a sawtooth, can be synthesised from a series of harmonics. The extension consists of including more than one layer, a situation addressed by the Sturm–Liouville theorem. The derivation remains physically interpretable at all stages. In particular, the integration is with respect to x, the distance through a layer, and its meaning is obvious. The frequency-domain method uses complex variable theory which rests on a lengthy exposition (Churchill *et al.* 1974) and is difficult to conceptualise. In particular, the differentiation is with respect to p (unlike the simple fin case, where differentiation is with respect to x): p_j (like its counterpart $1/z_j$) denotes a *rate of decay* of the transient components of temperature in each layer; at each value of $p = p_j$, $B(p_j)$ is zero but it is not obvious whether the value of $(\mathrm{d}B/\mathrm{d}p)$ at $p = p_j$ should have any physical significance. An expression to find the residue at a pole is given as the quantity b_1 on page 178 of Churchill *et al.* (1974). While the importance of complex variable theory in other fields is not in question, it is of little concern to workers in the field of building heat transfer.
- The time-domain approach draws attention to the status of the response of a wall to a steadily increasing temperature, a simpler response than is associated with periodic or transient excitation. As shown by (17.34) and (17.35), the slope response is little more complicated than the steady-state response. Further, the discretisation of a thin wall is largely determined through its slope characteristics; see equation (17.38) and Section 17.8.1.

17.10 APPENDIX: FINDING THE DECAY TIMES

A simple expression gives the values of the decay times for a single layer of material with specified boundary conditions. Even the addition of a film resistance as in the Groeber solution however makes it necessary to do a numerical search to find eigenvalues or decay times with adequate accuracy and this prevented the development of methods for examining multilayer behaviour before electronic computers became available. Indeed, in the 1960s, extraction of the decay times, or their equivalent, was a lengthy operation, but by the mid 1990s the time required had fallen to a few seconds, less than is needed to input the wall data.

A method frequently adopted is to take a sufficiently large value of z and evaluate the elements in (17.44), then successively reduce the value of z by some small fraction (e.g. to 0.99995 of its previous value) until the value of the element concerned (e_{12} here) changes sign. The first value of z to make $e_{12} = 0$ is z_1. The sign change in e_{12} locates z_1 approximately as z_1'. An improved value is given by the zero crossing of the tangent to $e_{12}(z_1')$ and further iteration. Alternatively, the two values of e_{12} before the sign change and the two values after the sign change can be used to fit a cubic equation to e_{12} over that range; its zero crossing gives z_1.[10]

Hittle and Bishop (1983) have drawn attention to circumstances where two consecutive roots are very close together and might remain undetected by an insufficiently small time step. They point out, however, that between two such close values, z_j and z_{j+1}, which result from $e_{12} = 0$, there must be a solution to $e_{11} = 0$. Thus z_j must be associated with a positive value of e_{11} and z_{j+1} with a negative value. If M denotes the number of sign changes in e_{11} since a value of z_j was detected, and M was found to be 2, then two z values must have been passed over, which should have been located either side of one of the intervening solutions to $e_{11} = 0$. Hittle and Bishop provide a proof of the result. An indication of the finding is given in the list of decay times listed in Section 10.3.1 for the three-capacity wall where $z_{11,1} > z_{12,1} > z_{11,2}$ and $z_{11,2} > z_{12,2} > z_{11,3}$. Since Hittle and Bishop's suggestion was put forward, computation has become very much faster and there little time penalty in taking small time steps; with a time interval of 1 h not many decay times are needed.

The construction below makes clear that solutions to $e_{11} = 0, e_{12} = 0$ follow consecutively as z decreases. Imagine a semi-infinite wave train representing temperature being compressed into the wall. Figure 17.3 illustrates this idea for the simplest case, the single

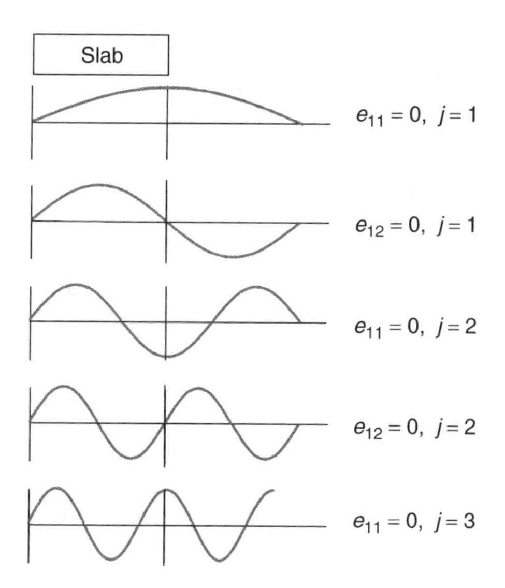

Figure 17.3 Sinusoidal temperature profiles in a slab, left-isothermal and alternatively right-adiabatic and right-isothermal

[10]The number of half-wavelengths of eigenfunction j in layer L is $\sqrt{c_L r_L / z_j}/\pi$. As j increases, z_j tends to approximately $\left(\sum_{L=1}^{n} \sqrt{c_L r_L}\right)^2 / (j^2 \pi^2)$. The expression neglects film resistances. It may serve as the starting point for a search to find the jth decay time. When two roots fall close together, the expression approximates one of the pair better than the other.

slab. In all cases the slab is left-isothermal. As the wave train compression progresses, the slab is successively right-adiabatic, corresponding to $e_{11} = 0$ and right-isothermal, corresponding to $e_{12} = 0$. There is no danger here that consecutive solutions to $e_{12} = 0$ should be close.

The danger does occur, however, when the wall consists of two concrete layers (say) separated by an insulating layer of zero capacity, as Hittle and Bishop note. Figure 17.4 shows two equal layers with distributed resistance and capacity separated by a purely resistive layer and indicates temperature profiles corresponding to the first four solutions to $e_{12} = 0$. In (i), where $j = 1$, the wall contains half a wavelength. In (ii), where $j = 2$, a half-wave has been pushed into the right layer. If the profile is plotted as a function of resistance, there is no change in gradient at the interfaces, so it may be seen that the $j = 1$ maximum in the left layer has moved a little from the interface itself to just inside to become the $j = 2$ maximum. When a further half-wave is pushed into the right layer so $j = 3$, the left layer maximum must move well to the left so that the minimum can be located at its interface, as shown in (iii). With a further half-wave ($j = 4$) the left layer maximum again moves only a little to the left, as shown in (iv).

This qualitative geometrical argument suggests that the solutions to $e_{12} = 0$ may fall into pairs: $j = 1$ and 2, $j = 3$ and 4, etc. Alternatively, it can be argued analytically.

The profiles are either symmetric or skew-symmetric, so we can consider the left half of the wall alone. The symmetrical solutions are independent of the resistive layer and

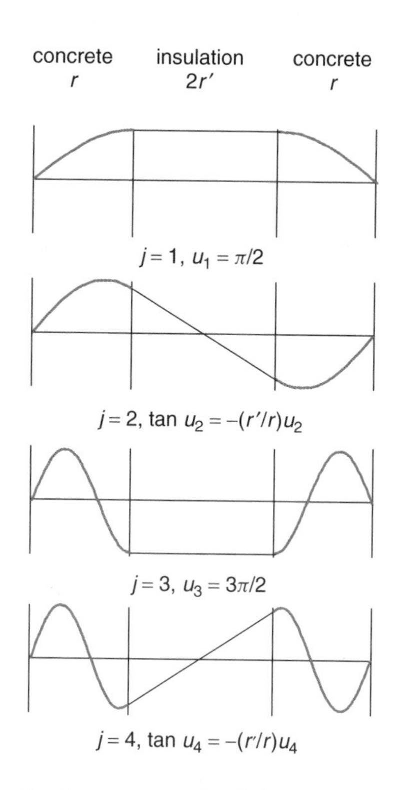

Figure 17.4 Temperature profiles in a concrete, insulation, concrete wall, resistances r, $2r'$, r, left- and right-isothermal

result from $e_{11} = 0$ solutions of the matrix

$$\begin{bmatrix} \cos u & (\sin u)/v \\ -(\sin u) \times v & \cos u \end{bmatrix}, \tag{17.96}$$

where $u^2 = cr/z$ and $v^2 = c/rz$. That is,

$$\cos u = 0 \text{ or } u_1 = \pi/2, u_3 = 3\pi/2, \text{ etc.}$$

The skew-symmetric solutions result from $e_{12} = 0$ solutions of the matrix product

$$\begin{bmatrix} \cos u & (\sin u)/v \\ -(\sin u) \times v & \cos u \end{bmatrix} \begin{bmatrix} 1 & r' \\ 0 & 1 \end{bmatrix}, \tag{17.97}$$

where r' is the *semi-resistance* of the middle layer. So

$$e_{12} = (\cos u)r' + (\sin u)/v = 0, \tag{17.98a}$$

or

$$\tan u = -r'v = -(r'/r)u, \tag{17.98b}$$

and solutions result from the intersection of $f_1 = \tan u$ and $f_2 = -(r'/r)u$. Figure 17.5 shows u_1, u_2, u_3 and u_4. It is clear that $u_2 - u_1 > u_4 - u_3$ and that as u increases, the difference between the pair becomes smaller, particularly when r'/r is large, that is, when the flanking layers are separated by a large resistance.

An alternative check that no eigenvalues have been missed is to trace each eigenfunction through the wall. The number of zero-crossings is $j - 1$. The exercise demonstrates that when j is large, most crossings are located in high-capacity low-resistance layers. Only one can fall in a cavity resistance.

Finally, this approach makes clear that for a succession of layers in series, all decay times must be separate, however close they may be. It confirms the argument in Section 13.5 for lumped resistances and capacities that even though consecutive roots may be close together, they cannot become exactly equal.

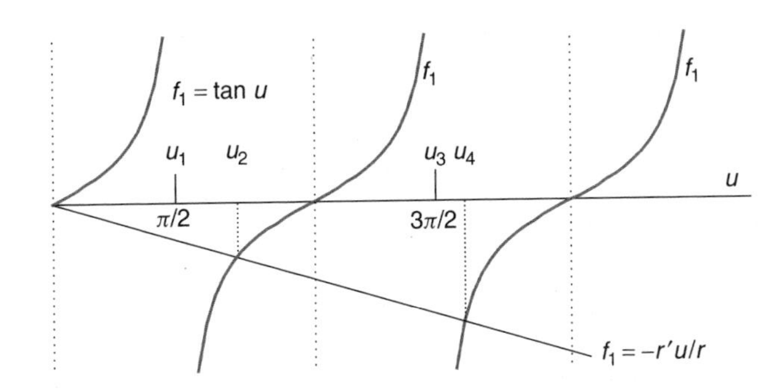

Figure 17.5 Construction to locate the skew-symmetric eigenfunction solutions for the concrete, insulation, concrete wall

17.11 APPENDIX: INCLUSION OF MOISTURE MOVEMENT

Some discussion of moisture movement in steady-state conditions was given in Chapter 8. If steady condensation (or evaporation) of rate Γ_x (kg/m^{3}s) occurs, it results in a latent heat input of $L\Gamma_x$ (W/m^2 section/m run) which is balanced by a change in the gradient $\lambda\, dT/dx$ (8.24). In non-steady conditions we must add a right-hand term of form $\rho c_p \partial T/\partial t$, which in principle takes account of the dry skeletal material, its liquid and its vapour content. The last contribution is negligible. For materials of low porosity the liquid-based term may be negligible and for materials such as fibreglass the skeletal material term may be negligible. The non-steady equation for continuity of mass flow must similarly include a right-hand term. The originators of the theory are Henry (1939), Krischer (1942), citing Henry but also his own work of 1938, and Luikov (1966) – in terms of physical variables (p. 247) and in terms of the mass transfer potential (p. 259); see Section 8.9. They are succinctly stated by Wyrwal (1988). The basic equations have been developed by many authors. The analysis of the equations in non-steady conditions is complicated and has not featured in routine design calculations. Developments in the 1990s, however, suggest that the situation may be changing and some brief account will be given.

In a lengthy article, Whitaker (1977) draws attention to a number of investigations from 1920 onward about the drying of various solid materials, attributing the drying mechanism either to diffusion based on the relation $\partial c/\partial t = D\partial^2 c/\partial x^2$, or in terms of surface tension or capillary action. (Henry does not cite this work.) An investigation in 1940 reported the observed moisture distribution at different times and at varying depth as a 2.5 cm layer of sand dried out. Theoretical distributions were calculated both on the basis of diffusion and capillary mechanisms; the capillary theory provided much the better fit. In the latter stages of drying, however, when the threads of moisture in the material had broken into globules, vapour diffusion provided the only transport mechanism. Whitaker states that Krischer was the first researcher to take account of these factors together with the transport of energy. (He does not cite Henry.) Whitaker goes on to present the formidable wealth of detail needed to model the drying process leading to 12 equations and 12 unknowns (Vafai and Whitaker 1986). Most workers since then have started with simpler formulations similar to the one sketched below. Building applications are of course as much concerned with the accumulation of moisture in walls as with its drying out.

Although the various formulations for the physics of the process are broadly the same, there is no agreement in the notation of the equations. We follow the discussion by Crank (1975: 356), itself based on Henry (1939). Here u denotes the moisture content as the ratio (kg of moisture absorbed at some cross section of wall material)/(kg of dry material); $\partial u/\partial t$ denotes the fractional rate of absorption or condensation; u (dimensionless) is related to the vapour concentration c (kg/m^3) and the temperature as

$$u = \text{constant} + \alpha_c c - \alpha_T T, \tag{17.99}$$

where the coefficients derive from experimental data such as in Figure 8.3 (the moisture isotherm function). The relation does not include any hysteresis in the absorption/desorption process and the coefficients can only be taken as constant over limited ranges. Moisture transport is assumed to take place only in the vapour phase, that is, condensate remains in situ. Molenda *et al.* (1993) report that it is necessary to take capillary hysteresis into account when the boundary conditions depend on time; it is not so important for steady boundary conditions.

Quantity v is the fraction of the total volume occupied by air and $1 - v$ is the fraction of the total volume occupied by solid material; the tortuosity τ (less than 1) is a measure of the resistance to diffusion of vapour within the porous material, reducing the in-air value of D_A to τD_A. Now the net amount of vapour diffusing into some element of the wall is balanced by the rates of increase of moisture in the air and moisture in solid material, so

$$\tau v D_A \frac{\partial^2 c}{\partial x^2} - (1 - v)\rho_s \frac{\partial u}{\partial t} = v \frac{\partial c}{\partial t}, \quad (17.100)$$

where ρ_s is the density of the material without pores and $\rho = (1 - v)\rho_s$ is the bulk density. Absorption or condensation results in a yield of latent heat L (J/kg); to this must be added the net conduction gain to balance the heat gain associated with the local rate of increase in temperature[11]:

$$\lambda \frac{\partial^2 T}{\partial x^2} + L\rho \frac{\partial u}{\partial t} = \rho c_p \frac{\partial T}{\partial t}. \quad (17.101)$$

The relations have been used in a variety of applications. Henry was concerned with the uptake of moisture by bales of cotton, Krischer with moisture in walls. Philip and de Vries (1957) and de Vries (1958, 1987) were concerned with soils but they did not include the effect of heat flow due to phase change, and Rossen and Hayakawa (1977) were concerned with foodstuffs. Fan *et al.* (2000) have recently applied them to a study of sorption and condensation in porous clothing. They note that wool can take up to 38% of its weight in moisture. Most applications of these equations to wall moisture transfer came some decades after their initial formulation.

One approach to the solution of these equations is based directly on Henry's own method. We can eliminate u by using (17.99) and the relations can be rewritten as

$$\frac{\tau v D_A}{v + \rho\alpha_c} \frac{\partial^2 c}{\partial x^2} + \frac{\rho\alpha_T}{v + \rho\alpha_c} \frac{\partial T}{\partial t} = \frac{\partial c}{\partial t}, \quad (17.102)$$

$$\frac{\lambda}{\rho(c_p + L\alpha_T)} \frac{\partial^2 T}{\partial x^2} + \frac{\rho L\alpha_c}{\rho(c_p + L\alpha_T)} \frac{\partial c}{\partial t} = \frac{\partial T}{\partial t}. \quad (17.103)$$

Equation (17.103) is similar to the Fourier continuity equation (12.1) except that a measure of the latent heat gain is added to the specific heat and, more importantly, it is coupled to the mass transfer equation through the heat of condensation or evaporation. The mass transfer equation can be similarly interpreted. For analytic purposes, the two equations can be redrafted using normalised coordinates into two mathematically independent relations:

$$\frac{\partial^2 (c + z_1 T)}{\partial x^2} = \mu_1 \frac{\partial (c + z_1 T)}{\partial t}, \quad (17.104a)$$

$$\frac{\partial^2 (c + z_2 T)}{\partial x^2} = \mu_2 \frac{\partial (c + z_2 T)}{\partial t}. \quad (17.104b)$$

[11] In his formulation of the pair of equations in terms of the mass transfer potential θ, Luikov (1966: equations 6.96a,b) uses $\partial^2\theta/\partial x^2$ where (17.101) has $\partial u/\partial t$.

The values of z and μ are bulky functions of the physical parameters in (17.102) and (17.103) but follow quite straightforwardly from the two solutions of a quadratic equation. Multiply (17.102) and (17.103) respectively by r and s, say, then add, giving a relation of form

$$\partial^2[Ac + BT]/\partial x^2 = \partial[Cc + DT]/\partial t; \tag{17.105}$$

set $A/C = B/D$ to achieve independence. This leads to a quadratic equation in $z = r/s$.

Thus the solutions to the Fourier equation, developed in Chapter 12 with T as variable and $\rho c_p/\lambda$ as parameter, can be used for the compound variable $c + zT$ with μ as parameter. Henry's interprets these equations as follows:

> Each diffusion 'wave' of vapour is accompanied by a temperature 'wave', proceeding at the same rate, whose magnitude is proportional to that of the vapour diffusion wave, the relation between the two depending only upon the properties of the materials. Similarly, the main temperature 'wave' is accompanied by a subsidiary vapour diffusion 'wave'. It will be seen that if one only of the external conditions alters – say the vapour concentration – there will nevertheless be the complete set of two vapour 'waves' and two temperature 'waves', though the latter may be small if the coupling is weak.

He cautions that the solutions are only strictly valid when the values of the parameters are constant; most are in fact variable. The dependence of conductivity on moisture content has already been noted. The results should be taken as applying quantitatively only to small changes in vapour concentration.

It was reported in Section 8.5 that Pedersen (1992) used Glaser's procedure (which omits the sorption properties of materials) to estimate condensation in a roof construction (felt, wood, expanded polystyrene, concrete) on the basis of monthly and of hour-by-hour data, leading to more reassuring conclusions with the more detailed data. The analysis was further improved by including the hygroscopicity of the wood and the concrete, leading to significant differences in the vapour pressure distribution of the concrete in January. He concludes that the combination of temperature and moisture content sets the wood fibreboard at risk for fungal growth in April and May.

Cunningham (1988) considers mass transfer in a timber-framed wall from the timber to the space between the timber (the y direction, if inside-outside is taken to be the x direction) and using a simple analytical model (without heat transfer) finds decay times of order tens of days. In a later paper (Cunningham 1990a) a finite difference scheme is described to evaluate the relations, with possible latent heat transfer. Each node has associated with it temperature, moisture concentration and vapour pressure; for both heat and mass transfer the space transfer between nodes consists of diffusion and convection components. Cunningham (1990b) presents values of the mean joist moisture content over a period of 40–100 days and a variety of imposed conditions (concrete roof, tiled roof, cold external conditions, warm external conditions). In each the water content was found from the analytical model of Cunningham (1988), the numerical model of Cunningham (1990a) and experimental values of Cunningham (1990b). The three methods mostly agreed well. In their study on walls, de Freitas *et al.* (1996) followed Philip and de Vries (1957) and omitted phase change, taking it to be negligible compared with transportation by diffusion.

Budaiwi *et al.* (1999) too have described a finite difference scheme to model transient behaviour in walls whose predictions are in satisfactory agreement with Swedish results and American results.

Kohonen (1984) had earlier developed a finite difference method to solve the equations and had compared its findings with an analytical solution based on the roots of a characteristic equation. The two approaches agreed satisfactorily; a small time step was needed to avoid oscillation.

Liu and Cheng (1991) adopt Luikov's formulation of (17.101) and apply it to the problem of the two-sided drying out of a sample of spruce, 24 mm thick, initially at 10°C, when placed in air at 110°C (a situation for which experimental data were available). The problem involved thermal and moisture boundary conditions as well as transfer in the solid. They used a further potential ϕ in which to express functions of T and θ, which led to a fourth-order differential equation in ϕ. This in turn led to eigenfunction solutions with a series of real eigenvalues (as seen in this chapter), but also a complex root, a possibility discussed by Lobo *et al.* (1987). Rossen and Hayakawa had noted the possibility of complex roots of the characteristic equation in 1977, though the physical significance of a complex root was not discussed. The initial conditions could not be satisfied using the series of real eigenvalues alone since the solution suggested that the moisture content should *increase* during the drying process. Inclusion of the complex root was essential to satisfy the initial conditions and to avoid this absurdity in the progress of drying. The process was largely complete after about 17 h. Pandey *et al.* (1999) have developed means of solving the Luikov equations by including many complex roots; their figures for the benchmark solutions show the need for including complex roots. Chang and Weng (2000) described further developments of this form of solution.[12]

Ilic and Turner (1989) describe a detailed numerical study of drying a 10 mm thick sample of wet brick initially at 47°C by air at 87°C; drying was complete after 140 min.

Burch and TenWolde (1993) have described a finite difference study of a lightweight wall used for mobile homes, hence without any load-bearing component. From inside to outside it consisted of 7.9 mm gypsum board (painted with latex), a vapour retarder, 89 mm of glass-fibre insulation, and a 10 mm waferboard siding (also painted with latex). Two nodes represented the gypsum board, 2 the vapour retarder, and 14 the siding; the insulation was treated as a non-storage storage layer. They used observed values for the moisture content (%) against relative humidity, and the permeance (kg/s m^2 Pa) against relative humidity for all these materials. Gypsum absorbs little moisture. Permeance values

[12]The physical meaning of a complex root is not made clear. Mention may be made however of a conduction study reported by Salt (1983) in which the imaginary roots enter quite straightforwardly. He considered a vertical stack of slabs (the x direction) having the same (horizontal) width (Y), infinitely long in the z direction but with differing thicknesses ($X_1, X_2, \ldots, X_{top}$) and different diffusivities ($\kappa_1, \kappa_2, \ldots$). The vertical sides of the stack and its lower surface were perfectly insulated and a convective loss took place from the top surface to ambient at zero after $t = 0$. The actual two-dimensional temperature distribution after $t = 0$ can be expressed in terms of eigenfunction components of temperature as discussed above for one dimension, components which decay in time in layer L (in the present notation) as $\exp[-(\lambda_L t/(\rho_L c_{pL} X_{top}^2))\tau_{L,j}^2]$, where $\tau_{L,j}^2 = (u_{vert,j,L}^2 + u_{horiz,j,L}^2)$ and the u values are the jth eigenvalues in the vertical and horizontal directions. Now, the $u_{horiz,j,L}$ values are simply proportional to $j\pi$ and are the same for all layers. It turns out, however, that some of the $u_{vert,j,L}$ values may prove to be imaginary. These are needed so that, following some change in ambient temperature, the future history of temperature in the stack can be found for an arbitrary initial condition. For a single layer and with no Y variation in temperature, we have the Groeber problem and the u_j values are given as $u_j \tan u_j = hX/\lambda$ (13.8).

can increase or decrease markedly with humidity: waferboard siding has its maximum value at 100% relative humidity, which is nearly 100 times its minimum value at 30%. They used the program to examine the behaviour of this construction, and some variations, to a cold climate and a hot, humid climate. In the coldest of the climates they considered (Madison, Wisconsin) and without the presence of the vapour barrier, the waferboard siding would become fully saturated during the winter months even though it would be too cold to lead to fungal growth. It dried out rapidly in the spring. The program was used to examine the hot, humid situation mentioned at the end of Section 8.5.

Liesen and Pederson (1999) base their analysis on the use of normalised coordinates. They point out that parameters for mass transfer – porosity, water vapour diffusivity and the coefficients representing moisture capacitance – are strong functions of temperature and moisture content itself. However, most buildings walls are in the pendular state, and the graphs they present show this to be valid over a large range of vapour density and temperature. They say that a transfer function of form $c + zT$ can be used with conduction transfer functions *within* a layer, but it cannot be used *between* layers; iteration is needed to conduct calculations for a multilayer wall. Their estimates of heat flows at the surface of a multilayer test wall agree well with finite difference solutions.

18
Accuracy of Temperature Estimates Using Transfer Coefficients

In the special circumstances that ambient temperature T_e varies linearly between its hourly values, and $T_n = 0$ at all times, the heat flows q_{ee} and q_{ne} found from the transfer coefficients a_k, b_k and d_k will be exact.[1] When the wall concerned forms part of a room, however, T_e can be taken as an independent variable, but T_n is neither isothermal nor adiabatic; its value depends on the thermal characteristics of the room and the way it is excited – it is a dependent variable. T_n can be calculated but its value will not be exact. In this chapter the alternative assumption will be made for T_n, namely that it is an adiabatic node, so that $q_{ne} = 0$ always. A linear excitation $\theta k \delta$ is to be imposed at T_e and the two consequent quantities q_{ee} and T_{ne} will be estimated using the wall transfer coefficients a_k, b_k, c_k and d_k, and their values will be compared with the corresponding values given by an exact solution. The errors so found provide an indication of the accuracy of practical calculations of room response. (If T_n represents the room index node, there is only a single T_n. If T_n represents a wall inner surface, there will be a T_n node for each wall and the notation would have to make this clear, but we confine ourselves to a single wall.) The examination is based on the study reported by Davies (2001a).

18.1 THE R-C MODEL

Figure 18.1a shows the elementary r-c model, a single lumped resistance and a single lumped capacity. In effect, the circuit is adiabatic at T_2. The model is too simple to be given a building interpretation and it will only be discussed analytically.

If $T_0 = \theta t$ from $t = 0$ onward, T_2 is given exactly by (10.9) as

$$T_2(\text{exact}) = \theta t - \theta z(1 - \exp(-t/z)), \tag{18.1a}$$

[1]It will be recalled that the transfer coefficients reproduce the wall resistance and thermal capacity to a high degree of accuracy; see equations (11.36) and (11.38)

Building Heat Transfer Morris G. Davies
© 2004 John Wiley & Sons, Ltd ISBN: 0-470-84731-X

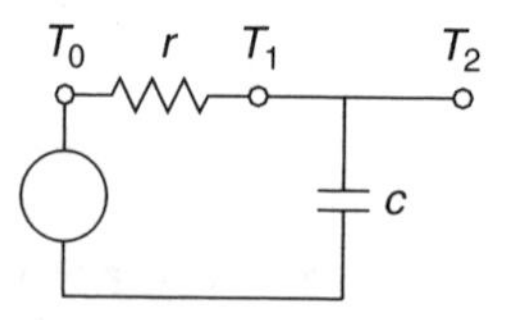

(a) The lumped resistance/capacity system.

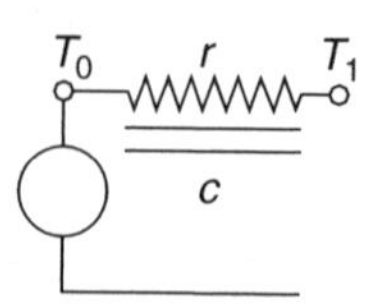

(b) The homogeneous slab, adiabatic at the right surface.

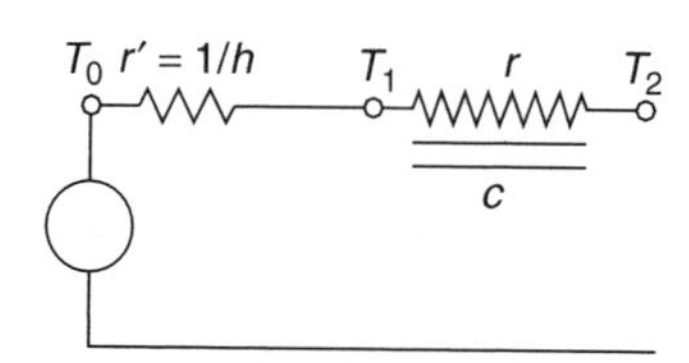

(c) The homogeneous slab with a film resistance.

Figure 18.1

where $z = rc$ and is the decay time, or time constant, of the circuit. The expression is the response of the system to ramp excitation in its simplest form: a linear term in t, a constant, and transient term in t. We are concerned with temperatures at discrete time steps of δ, so $t = k\delta$ and

$$T_{2,k}(\text{exact})/\theta z = k\delta/z - 1 + \exp(-k\delta/z). \tag{18.1b}$$

Now for this model (11.22) we have

$$b_0 = 1/r, b_1 = 0, c_0 = 1/r + c/\delta, c_1 = -c/\delta, d_0 = 1 \text{ (by definition) and } d_1 = 0. \tag{18.2}$$

So from equations similar to (11.2a) and (11.2b) we have

$$q_{20,k} = (1/r)T_{0,k} = (1/r)\theta k\delta, \tag{18.3a}$$

$$q_{22,k} = (1/r + c/\delta)T_{2,k} - (c/\delta)T_{2,k-1}. \tag{18.3b}$$

Since T_2 is taken to be adiabatic, these two flows are equal, so the transfer coefficient estimate of T_2 is

$$T_{2,k}(\text{TC})/\theta z = \frac{k\delta/z + (z/\delta)T_{2,k-1}/\theta z}{(1 + z/\delta)}. \tag{18.4}$$

Values for $z/\delta = 1$ are shown in Table 18.1.

Table 18.1 r-c circuit: Comparison of T_2 found from transfer coefficients and its exact value ($z/\delta = 1$)

Time level, k	1	2	3	4	5
$T_{2,k}(\text{TC})/\theta z$	0.500	1.250	2.125	3.062	4.031
$T_{2,k}(\text{exact})/\theta z$	0.368	1.135	2.050	3.018	4.007

Thus the estimates are larger than the exact values of temperature but they converge to the exact values. The same is true for other values of z/δ. It can be shown that for the first time step, the difference

$$T_{2,1}(\text{TC}) - T_{2,1}(\text{exact}) = \theta z[z/(\delta + z) - \exp(-\delta/z)] \tag{18.5}$$

is always positive. The TC estimates of temperature are slightly high.

18.2 THE SINGLE SLAB DRIVEN BY A RAMP

To extend the discussion, we consider a slab of uniformly distributed resistance and capacity (Figure 18.1b), thickness X, initially at zero temperature, adiabatic at $x = X$ and from $t = 0$ onward subjected to a rise of temperature $T_0 = \theta t$ at $x = 0$. An exact solution for $T(x, t)$ is given in Carslaw and Jaeger (1959: 104, equation 4). It can be shown from the solution that the heat flow into the slab at $x = 0$ is

$$q_{00}(\text{exact}) = \theta c \left\{1 - (8/\pi^2) \sum (2j - 1)^{-2} \exp(-(2j - 1)^2 \pi^2 t/4cr)\right\} \tag{18.6}$$

and that the temperature at $x = X$ is

$$T_{10}(\text{exact}) = \theta \left\{t - cr/2 + (16cr/\pi^3) \sum (-1)^{j-1} (2j - 1)^{-3} \exp(-(2j - 1)^2 \pi^2 t/4cr)\right\}, \tag{18.7}$$

where c is the slab thermal capacity $\rho c_p X$ and r is its resistance X/λ; j is to be summed from 1 to infinity. Equation (18.7) has the familiar structure: the first right-hand term is proportional to time, the second is the constant component of the slope response and the third is the sum of the transient components.

The heat flows at $x = X$ and at time level i as estimated by transfer coefficients are

$$q_{10,i} = b_0 T_{0,i} + b_1 T_{0,i-1} + b_2 T_{0,i-2} + \cdots - d_1 q_{10,i-1} - d_2 q_{10,i-2} - \cdots, \tag{18.8a}$$

$$q_{11,i} = c_0 T_{1,i} + c_1 T_{1,i-1} + c_2 T_{1,i-2} + \cdots - d_1 q_{11,i-1} - d_2 q_{11,i-2} - \cdots, \tag{18.8b}$$

and since the heat flows are equal and opposite, we have

$$T_{1,i} = \left\{\sum_{k=0}^{N} b_k T_{0,i-k} - \sum_{k=1}^{N} c_k T_{1,i-k}\right\} \Big/ c_0. \tag{18.9}$$

The heat flows components at $x = 0$ are

$$q_{00,i} = a_0 T_{0,i} + a_1 T_{0,i-1} + a_2 T_{0,i-2} + \cdots - d_1 q_{00,i-1} - d_2 q_{00,i-2} - \cdots, \quad (18.10a)$$

$$q_{01,i} = b_0 T_{1,i} + b_1 T_{1,i-1} + b_2 T_{1,i-2} + \cdots - d_1 q_{01,i-1} - d_2 q_{01,i-2} - \cdots, \quad (18.10b)$$

and the net flow at $x = 0$ and time level i is

$$q_{0,i} = q_{00,i} - q_{01,i}. \quad (18.11)$$

To illustrate these expressions, consider a concrete slab of thickness 0.17905 m, $\lambda = 1.63$ W/m K, $\rho = 2300$ kg/m^3 and $c_p = 1000$ J/kg K. The special value of thickness was chosen so that the slab should have a value of $V = (\rho c_p/4\pi\lambda\delta)^{1/2} X$ equal to 1.0 with a sampling time δ of 1 h. Except for a thin slab, $d_1 = \frac{1}{2} - V$ or -0.5 here. For a slab of this thickness, $N = 4$. It has decay times 1.27 h, 0.32 h, 0.14 h, ... Transfer coefficients are given in Table 18.2.

The slab is symmetrical, so $c_k = a_k$. Furthermore, $\sum a_k/\sum d_k = \sum b_k/\sum d_k = \sum c_k/\sum d_k = 4.73471/0.52011 = 9.10336 = 1.63/0.1790548 = \lambda/X$, the steady-state transmittance. The thermal capacity of the slab, as estimated by the coefficients, is also correct. It is supposed excited by a ramp increase of 1 K/h (so at $t = 12$ h the driving temperature is 12 K) and a comparison of heat flows and temperatures is given in Table 18.3

Table 18.2 Transfer coefficients for the slab

	a_k (W/m^{2}K)	b_k (W/m^{2}K)	c_k (W/m^{2}K)	d_k
0	36.41343	0.35841	36.41343	1.0
1	−39.52486	3.05280	−39.52486	−0.50001
2	8.03207	1.29267	8.03207	0.02013
3	−0.18604	0.03082	−0.18604	−0.00002
4	0.00011	0.00001	0.00011	0.00000

Table 18.3 Heat flows (W/m)^2 at $x = 0$ and temperatures (K) at $x = X$ in a slab ($V = 1$), adiabatic at $x = X$, due to increase of temperature at $x = 0$ of 1 K/h

Time	Heat flow at $x = 0$		Temperature at $x = X$	
	Exact	From TCs	Exact	From TCs
0	*	0.00	0.00	0.00
1	36.41	36.41	0.00	0.01
2	51.48	51.44	0.09	0.11
3	62.90	62.72	0.31	0.35
6	85.85	85.36	1.71	1.77
12	105.61	105.19	6.33	6.37
23	113.38	113.27	16.79	16.80
24	113.56	113.47	17.78	17.78

*Converges slowly to zero.

The transfer coefficient values are in excellent agreement with the exact values, although they marginally overestimate temperature. Some features may be noted:

- There is at first no perceptible change in temperature at $x = X$; this is as expected. Although no actual value can be attached to the speed with which a thermal signal passes through a material, a delay of an hour or two is to be expected for about 0.2 m of concrete.
- The slab has a first decay time of $z_1 = 1.27$ h (the first solution to $e_{12} = 0$, the slab being isothermal at $x = 0$ and $x = X$); this and the subsequent z_j values are used to find the transfer coefficients. In its response, however, being adiabatic at $x = X$, its first time constant is four times z_1, 5.1 h (the first solution to $e_{11} = 0$). Thus we should expect the transient excitation imposed at $t = 0$ to have largely died out at around four times this value, about 20 h. After 24 h the temperature at $x = X$ is increasing at virtually 1 K/h, the speed of excitation. The transient effect has by now almost died out and the temperature throughout the slab is increasing at nearly 1 K/h.
- The heat flow into the slab must initially be independent of its thickness and can then found by a simple exact solution. See Carslaw and Jaeger (1959: 63, equation 4), from which it follows that

$$q_{00,t} = 2\theta t \sqrt{\lambda \rho c_p / \pi t}, \tag{18.12}$$

giving a value of 36.41 W/m^2 after 1 h, the value noted in the table.

- When the transient excitation has died out, the constant heat flow is given as

$$q_{00,t} = \theta c = \theta \rho c_p X = (1/3600) \times (2300 \times 1000 \times 0.17905) = 114.39\ \text{W/m}^2 \tag{18.13}$$

and the flow is tending to this value.

- The error in estimation of surface temperature is virtually zero at first, rises to a maximum and then decreases again. The error varies with slab thickness and details are given in Table 18.4. The error is largest for a thin slab: for $V = 0.3$ it falls at $i = 1$ h and then decreases. This is consistent with the result noted above for the r-c system. The time of maximum error increases with slab thickness. The absolute error, however, decreases with thickness; as a fraction of the exact value, it decreases more.

Table 18.4 Variation of maximum error with slab thickness V

V	Time i (h)	$T_{1,i}$(exact) (K)	$T_{1,i}$(TC) − $T_{1,i}$(exact) (K)
0.3	1	0.50	0.0890
1.0	6	1.71	0.0538
2.0	22	5.68	0.0291
3.0	49	12.49	0.0198
4.0	83	19.94	0.0149

18.3 THE SINGLE SLAB DRIVEN BY A FLUX

The theory of response factors and transfer coefficients as used here presupposes that heat flows and temperatures result from a ramp increase in temperature. This is appropriate in finding the response of a building to varying ambient temperature and solar gains since they vary steadily in time and the response is based on a linear variation between given hourly values. However an internal source, in particular an electrical heater, when switched on, rapidly attains its steady-state value. It constitutes a step excitation and cannot be modelled exactly by the transfer coefficients. A 1 kW source switched on at say 1000 and off again at 2000 has to be handled as having values of 0, $\frac{1}{2}$ and 1 kW at 0900, 1000 and 1100 respectively, remaining at this value until 1900, then decreasing again to zero at 2100. It is implicit that the source increases linearly from 0900 to 1100 so that the total heat input is correct, but in handling it in this way, its effect is less abrupt. It is useful to test consequences of this procedure against an exact solution.

Carslaw and Jaeger (1959: 112, equation 3) provides an expression for the situation where a slab, initially at zero and insulated on its rear surface, is subjected to a constant flux F at its front surface from $t = 0$ onward. The temperature at the excited surface is

$$T_{00,t}(\text{exact}) = (F/c)\left\{t + rc/3 - (2rc/\pi^2)\sum[(1/j^2)\exp(-j^2\pi^2 t/rc)]\right\}, \tag{18.14}$$

and at the adiabatic surface it is

$$T_{10,t}(\text{exact}) = (F/c)\left\{t - rc/6 - (2rc/\pi^2)\sum[(-1)^j(1/j^2)\exp(-j^2\pi^2 t/rc)]\right\}, \tag{18.15}$$

summing j from 1 to ∞.

The transfer coefficient version of these values follows from (17.2). Noting that $q_{00,i} - q_{01,i} = F_i$, the imposed flux at time i, it is found that

$$a_0 T_{0,i} - b_0 T_{1,i} = \sum_{k=1}^{N}(-a_k T_{0,i-k} + b_k T_{1,i-k}) + \sum d_k F_{i-k}, \tag{18.16a}$$

$$-b_0 T_{0,i} + c_0 T_{1,i} = \sum_{k=1}^{N}(b_k T_{0,i-k} - c_k T_{1,i-k}), \tag{18.16b}$$

where F_i has the values 0, $\frac{1}{2}F$ and F for $i < 0$, $i = 0$ and $i > 0$, respectively. Solution of the equations gives the front and back temperatures (Table 18.5).

The TC estimate of the front temperature at $t = 0$ is in error, as it must be, and subsequent values are a little high. Rear surface temperatures are somewhat overestimated. In the steady-slope state, the hourly increase of either temperature is $F\Delta t/c = 100 \times 3600/(0.17905 \times 2300 \times 1000) = 0.87$ K, which is the difference between the 23 h and 24 h values. Furthermore, the steady-slope difference between front and back temperatures is $(F/c)(rc/3 - (-rc/6)) = Fr/2 = 100 \times 0.17905/(1.63 \times 2) = 5.49$ K. The 24 h values show this.

Table 18.5 Front and back temperatures (K) in a slab, adiabatic at $x = X$, driven by a heat flux of 100 W/m^2 at $x = 0$

Time (h)	Flux	temperature at $x = 0$		temperature at $x = X$	
		Exact	From TCs	Exact	From TCs
−1	0	–	0.00	–	0.00
0	50	*	1.37	0.00	0.01
1	100	3.50	3.55	0.03	0.16
2	100	4.95	4.97	0.38	0.57
3	100	6.07	6.12	1.00	1.19
6	100	8.89	8.98	3.43	3.57
23	100	23.77	23.88	18.27	18.39
24	100	24.64	24.76	19.15	19.26

*Converges slowly to zero.

18.4 THE SINGLE SLAB DRIVEN SINUSOIDALLY

If a wall is driven by a sinusoidal variation of ±1 K per 24 h, its response can be found from the three quasi-independent parameters,

$$\mathbf{y_0} = \mathbf{q_0}/\mathbf{T_0} \text{ when } T_1 = 0, \quad (18.17a)$$

$$\mathbf{y_1} = \mathbf{q_1}/\mathbf{T_1} \text{ when } T_0 = 0, \quad (18.17b)$$

$$\mathbf{u} = \mathbf{q_0}/\mathbf{T_1} \text{ when } T_0 = 0 \quad (18.17c)$$

$$= \mathbf{q_1}/\mathbf{T_0} \text{ when } T_1 = 0.$$

These are complex quantities with magnitudes W/m^2 K and the phase leads in $\mathbf{y_0}$ and $\mathbf{y_1}$ and the phase lag in **u** are conveniently expressed in hours. Parameters for the condition where instead a surface is maintained adiabatic can be similarly defined:

$$\mathbf{y_0}' = \mathbf{q_0}/\mathbf{T_0} \text{ when } q_1 = 0, \quad (18.18a)$$

$$\mathbf{r}' = \mathbf{T_1}/\mathbf{T_0} \text{ when } q_1 = 0. \quad (18.18b)$$

Values of the standard parameters were found directly from the transfer coefficients as in Section 17.7.2 and in their exact versions as in Section 15.3. For the slab in question,

$$\text{mag}(\mathbf{y_0})_{\text{TC}} = 14.820 \text{ W/m}^2\text{ K}, \qquad \text{lead}(\mathbf{y_0})_{\text{TC}} = 2.589 \text{ h},$$

$$\text{mag}(\mathbf{y_0})_{\text{exact}} = 14.515 \text{ W/m}^2\text{ K}, \qquad \text{lead}(\mathbf{y_0})_{\text{exact}} = 2.679 \text{ h},$$

$$\text{mag}(\mathbf{u})_{\text{TC}} = 8.549 \text{ W/m}^2\text{ K}, \qquad \text{lag}(\mathbf{u})_{\text{TC}} = 2.049 \text{ h},$$

$$\text{mag}(\mathbf{u})_{\text{exact}} = 8.597 \text{ W/m}^2\text{ K}, \qquad \text{lag}(\mathbf{u})_{\text{exact}} = 2.049 \text{ h}.$$

Since the slab is symmetrical, $\mathbf{y_1} = \mathbf{y_0}$. The exact values are found assuming strict sinusoidal variation in the driving temperature whereas the TC values are based on a 24-segment approximation to the sinusoid; strict agreement is not expected.

Transfer coefficients versions of $\mathbf{y_0}'$ and $\mathbf{r}'$ were evaluated from a time sequence using (18.8) and (18.10), convergence being comparatively slow, and the exact versions from Davies (1994). The values obtained are:

$$\begin{aligned}
\text{mag}(\mathbf{y_0}')_{\text{TC}} &= 18.83\ \text{W/m}^2\,\text{K}, & \text{lead}(\mathbf{y_0}')_{\text{TC}} &= 3.25\ \text{h},\\
\text{mag}(\mathbf{y_0}')_{\text{exact}} &= 18.78\ \text{W/m}^2\,\text{K}, & \text{lead}(\mathbf{y_0}')_{\text{exact}} &= 3.32\ \text{h},\\
\text{mag}(\mathbf{r}')_{\text{TC}} &= 0.577, & \text{lag}(\mathbf{r}')_{\text{TC}} &= 4.64\ \text{h},\\
\text{mag}(\mathbf{r}')_{\text{exact}} &= 0.592, & \text{lag}(\mathbf{r}')_{\text{exact}} &= 4.73\ \text{h}.
\end{aligned}$$

There is general agreement between the exact values of these adiabatic parameters and their estimates.

18.5 FILM AND SLAB DRIVEN BY A RAMP

As a final test, we consider the problem of a film and slab (Figure 18.1c). As before, the slab is adiabatic at its far surface but is wetted by an air film, convective coefficient $h = 8\ \text{W/m}^2\text{K}$ on the nearer surface; up to $t = 0$, the temperature $T(x, t)$ is zero but from $t = 0$ onward, the air temperature increases as θt The solution is given in Carslaw and Jaeger (1959: 127, equation 9):

$$\frac{T(x,t)}{\theta} = t + cr\left(\left(\frac{x^2}{2X^2} - \frac{1}{2} - \frac{1}{B}\right) + \sum_{j=1}^{\infty} \frac{2B}{u_j^2(B^2 + B + u_j^2)} \frac{\cos u_j x/X}{\cos u_j} \exp\left(-\frac{u_j^2 t}{cr}\right)\right). \tag{18.19}$$

Here B is the Biot number hX/λ and u_j is the jth eigenvalue found from solution of the equation $u \tan u = B$. In this case $B = 8 \times 0.17905/1.63 = 0.8788$, giving eigenvalues $u_1 = 0.8200$, $u_2 = 3.3949$, $u_3 = 6.4192$, etc., and decay times (h) of 18.7, 1.09, 0.31, etc.

Like the earlier exact solutions, this equation is structurally similar to (17.60) for flux. The first term is proportional to t, the second is the constant component of the slope term and the third is the sum of the transient components.

There are two approaches to this calculation using transfer coefficients. Either the slab can be described as above and the film considered as an addition (the 'separate' method), or the wall can be taken as the film and slab together (the 'combined' method). Each has its merits.

18.5.1 Film and Slab as Separate Entities

Certain heat inputs to an enclosure act directly on interior room surfaces. These include solar radiation transmitted as short wave through a window and the flux from a heated floor or cooled ceiling. Long-wave radiation from an internal heat source need not be handled in this way. If a flux acts at a surface node, the transfer coefficients for the wall must be based on the construction from the exterior up to that surface, but not including the inner film. The wall will respond to this flux and also to variation (in part concomitant) in the room temperature to which it is linked by its film. In this section, therefore, we examine the response of the above slab when driven by a ramp increase in temperature acting through a film, coefficient h. In Figure 18.1c T_0 must be interpreted as room temperature, T_1 as the wall interior surface node and T_2 as the wall exterior node. The previous equations apply with appropriate change of subscript:

$$q_{11,i} = a_0 T_{1,i} + a_1 T_{1,i-1} + a_2 T_{1,i-2} + \cdots - d_1 q_{11,i-1} - d_2 q_{11,i-2} - \cdots, \quad (18.20a)$$

$$q_{12,i} = b_0 T_{2,i} + b_1 T_{2,i-1} + b_2 T_{2,i-2} + \cdots - d_1 q_{12,i-1} - d_2 q_{12,i-2} - \cdots, \quad (18.20b)$$

$$q_{21,i} = b_0 T_{1,i} + b_1 T_{1,i-1} + b_2 T_{1,i-2} + \cdots - d_1 q_{21,i-1} - d_2 q_{21,i-2} - \cdots, \quad (18.20c)$$

$$q_{22,i} = c_0 T_{2,i} + c_1 T_{2,i-1} + c_2 T_{2,i-2} + \cdots - d_1 q_{22,i-1} - d_2 q_{22,i-2} - \cdots. \quad (18.20d)$$

As before, T_2 will be taken to be adiabatic so that q_{22} and q_{21} are equal and opposite at all times. The continuity equation at $x = 0$ is

$$(\theta i\delta - T_{1,i})h = q_{1,i} = q_{11,i} - q_{12,i}. \quad (18.21)$$

Thus we have a pair of simultaneous equations for T_1 and T_2:

$$(h + a_0)T_{1,i} - b_0 T_{2,i} = -\sum a_k T_{1,i-k} + \sum b_k T_{2,i-k} + \sum d_k (q_{11,i-k} - q_{12,i-k}) + \theta i\delta h, \quad (18.22a)$$

$$-b_0 T_{1,i} + c_0 T_{2,i} = +\sum b_k T_{1,i-k} - \sum c_k T_{2,i-k}; \quad (18.22b)$$

terms on the right are summed from $k = 1$ to N.

18.5.2 Film and Slab as a Combined Entity

Heat is driven into a wall exterior by ambient temperature T_e acting through an external film h and by solar gain I, of which $\alpha' I$ is absorbed at the exterior surface. These are independent but their action can be combined as the sol-air temperature $T_{sa} = T_e + \alpha' I/h$ whose value is supposed known. In this case the wall may be taken to consist of a film h (8 W/m^{2}K) and the slab as before. We now return to the former interpretation of Figure 18.1c: T_0 denotes sol-air temperature, T_1 is no longer of concern, and the transfer coefficients refer to nodes 0 and 2. Their values are given in Table 18.6.

Once again it will be found that $\sum a_k / \sum d_k = \sum b_k / \sum d_k = \sum c_k / \sum d_k = (1/h + X/\lambda)^{-1}$, the steady-state transmittance. The value of c_0 is the same as in Table 18.2, reflecting the fact that the heat flow at $x = X$ due to a ramp imposed at

Table 18.6 Exact calculation parameters and transfer coefficients for the film and slab

	a_k	b_k	c_k	d_k
0	6.72172	0.03416	36.41342	1.0
1	−6.80820	0.50768	−53.60080	−0.88595
2	1.07038	0.40255	19.46575	0.11461
3	−0.01327	0.02619	−1.31309	−0.00071
4	0.00002	0.00007	0.00537	0.00000

Table 18.7 A slab ($V = 1$) adiabatic at back surface, driven through a film at front surface

Time	Front heat flows (W/m^2)			Back surface temperatures (K)		
	Exact	TC, separate	TC, combined	Exact	TC, separate	TC, combined
0	0.00	0.00	0.00	0.00	0.00	0.00
1	6.72	6.56	6.72	0.00	0.00	0.00
6	32.35	32.16	32.31	0.45	0.48	0.47
12	54.88	54.63	54.77	2.32	2.37	2.35
24	83.08	82.81	82.93	9.15	9.21	9.19
48	105.73	105.58	105.64	29.01	29.03	29.02
167	114.38	114.38	114.38	146.42	146.42	146.42
168	114.38	114.38	114.38	147.42	147.42	147.42

$x = X$ for the first hour or so is independent of conditions at $x = 0$. Otherwise the values differ. In particular, the present value of a_0 is much less than the earlier value, since the heat flow at $x = 0$ due to a ramp of 1 K/h applied through a film has much less effect at $x = 0$ than when applied directly to an element of high conductivity and capacity.

Table 18.7 presents the values of q_0 and T_2 as found exactly, by treating the film separately, and in combination. It shows good agreement between the exact values of heat flows and temperature and their TC estimates by either method. The terminal heat flow of 114.38 W/m^2 is the same as when the slab surface itself is raised at 1 K/h (18.13) but it reaches this value much more slowly.

18.6 THE GENERAL WALL

These comparisons have been conducted for simple walls having well-known exact solutions for adiabatic surfaces. The theory of Chapter 17, however, can be readily adapted to find the exact solution for a wall of arbitrary (one-dimensional) construction. Suppose that the real wall consisted of three layers, e.g. an outer film S_1, concrete block S_2 and plaster S_3. Its set of transfer coefficients would be found in the normal way. To compute the exact response when the inside node T_3 is adiabatic, we set up a 'wall' of double construction $S_1, S_2, S_3, S_3, S_2, S_1 (n = 6)$ and suppose that both T_0 and T_6 have rise rates θ imposed upon them. Physically this amounts to symmetrical excitation of a symmetrical wall: the temperature gradient at midplane is zero, so T_3 is an adiabatic node. The relevant computer code provides sets of slope coefficients A_{L_1} etc., hence values for $q_{nO,j^{**}}$; the double asterisk denotes that they are for the double wall. When j is even, $q_{nO,j^{**}}$ is zero. The amplitudes of the transient components of response at midplane (T_3) are then $E_{3j}q_{nO,j^{**}}$ and the transient component of T_3 is found by summing terms given in (17.59a). The slope component of temperature is given straightforwardly by the slope matrix at its $n = 3$ stage of evaluation.

18.7 DISCUSSION

In all cases, transfer coefficient (TC) estimates of temperature and flux tend to their exact values with increasing time after the onset of transient change; this must be the case since the steady-slope components are common to both approaches and, although the transient components differ, both sets tend to zero with increasing time. Deviations will be seen in the earlier stages of change, when t is of order equal to the first decay time of the system. It has been pointed out that the 'system' here is the wall as it acts, one node being adiabatic, so its first decay time is larger than the first decay time of the wall used in finding the transfer coefficients.

The deviations, however, are small compared with the changes themselves and this must be seen in the context that we rarely know reliable values for the thermal constants of the wall layers. In conducting calculations on room response, the ingredient values must be taken as nominal and the estimates interpreted accordingly.[2] (This argument cannot be used however to justify approximating values for the a, b, c, d set. It will be recalled that

[2]In a study of the effect of rain on the heat gain through a building wall in tropical climates, Jayamaha *et al.* (1997) give the uncertainty in heat flux due to the estimated uncertainty in the parameters.

	Estimated uncertainty (%)	Change in heat flux (%)		Estimated uncertainty (%)	Change in heat flux (%)
Thermal conductivity (W/m K)	10	5.6	Moisture thermal diffusivity (kg/m s K)	15–20	2
Density (kg/m^3)	10	<1	Surface mass transfer coefficient (m/s)	15–20	2
Specific heat capacity (J/kgK)	20	2	Hydraulic conductivity (kg/Pa s m)	10	<1
Surface heat transfer coefficient (W/m^{2}K)	10	5	Vapour permeability (kg/Pa s m)	7.5	<1
Radiation intensity (W/m^2)	5	6	Retention curve	10	<1
Radiation absorptivity	5	6	Sorption isotherm	7.5	<1
Ambient temperature (°C)	5	4	Rainfall on wall	10	<1

These uncertainties are of course specific to their study. In particular, they remark that the variation in specific heat, 10–20%, is based on values for the dry and fully saturated state, presumably known reliably. The building analyst may not know reliably the value for the specific heat of some wall component but the table suggests that a 20% uncertainty in its value might lead to 2% uncertainty in an estimate of the heat flux. The authors' uncertainty for the heat flow predictions was computed 'based on the root-sum-square method using the individual uncertainties of the variables in the table and ... was estimated to be 12%. The uncertainty for the heat flow predictions made under dry conditions was found to be about 10%'.

the values of a, c, and d alternate in sign and may be large in magnitude. Calculations that involve summing terms which include them must be conducted to sufficient significant figures to avoid a build-up of errors.)

Finally, when T_e varies linearly between intervals of δ and $T_n = 0$, the flux q_{ne} is exact but when $q_n = 0$ T_{ne} has errors, as illustrated above. In practice however, when T_n is just one node among several which describe a room, it is neither isothermal nor adiabatic and we should expect errors to be somewhat less than these limiting values. Room internal radiant and convective exchanges involve no storage, so they impose no further computational errors. While we recognise that calculating room heat transfer is subject to a great many uncertainties, listed in the next chapter, the information can be handled reliably using transfer coefficients.

19

Room Thermal Response Using Transfer Coefficients

Although interest in the thermal performance of buildings emerged in the nineteenth century and dynamic behaviour in the 1930s, the current concern dates from the 1960s with the problems posed by overheating behind glazed facades, and more particularly from the energy crisis[1] and oil embargo of the 1970s. Only then did it become generally appreciated how much energy was consumed in the building sector. The rapid development of electronic computing by then had made advanced means of thermal analysis available, but it was very slow and cumbersome by current standards. While it is recognised that supplies of fossil fuels are limited, more concern had come by then to be expressed about the quality of the environment and the use of these fuels in particular.

Thus services engineers require to know the thermal response of buildings, either to estimate swings in temperature (most probably due to solar overheating) if it is uncontrolled, or to size the heating/cooling plant and make an estimate of energy requirements of complex multi-zone buildings over say a year when subjected to real weather and operating conditions. A model of their behaviour is 'one of the essential issues for the optimal operation, adaptive control, fault detection and diagnosis' (Chen and Athientis 2003).

A number of dynamic simulation programs have been developed to provide this information and Gough (1999) has provided a useful review of techniques that are available. The conductive contribution to the response can be estimated in a detailed way, either through the use of the finite difference technique (broadly speaking the current preferred European approach) or through the use of transfer coefficients (advanced by ASHRAE since 1977). In both cases it is normally assumed that information on air temperature and solar radiation is available at hourly intervals. Transfer coefficients provide the more efficient method of finding the response since the wall behaviour is summarised in a few, quickly processable parameters which relate to temperatures and heat flows at the wall surface only. Finite difference calculations involve estimation of temperatures throughout the wall, which are not of use, and may require solution of a set of simultaneous equations for each wall at each time step. This chapter demonstrates the use of transfer coefficients in finding the thermal response of a room.

[1] It is sometimes called the second energy crisis; the first crisis was shortly after the First World War.

Building Heat Transfer Morris G. Davies
© 2004 John Wiley & Sons, Ltd ISBN: 0-470-84731-X

19.1 SIMPLIFYING ASSUMPTIONS

Models commensurate with the ever-increasing speed and power of computation are constantly appearing and we shall discuss here only the basis of a model that might be used for design purposes. We make several assumptions about thermal models and these assumptions affect the accuracy of the estimates.

The model treats the enclosure globally, so that its internal temperature is described by two related global values. The model does not attempt to estimate the variation of temperature within the space. Thus the air temperature is supposed uniform throughout the room, or 'fully mixed'. There may, in fact, be several degrees difference between the air temperature at floor level and ceiling level in a kitchen, but any consideration of the difference is outside the scope of the model.

The models do not take account of the position of a radiant source within the enclosure, nor of the variation in local radiant temperature. Special consideration may be given to the radiant loss from the back of a radiator near a wall. The model is only concerned with the volume-averaged values of local air and radiant temperatures.

The space to be modelled is often taken to be empty, so the model ignores the effect that furnishings may have on the floor convective coefficient, radiant exchange near the floor and internal thermal capacity. The air capacity may also be ignored even though it is easy to include in the model.

There is no difficulty in modelling an hourly varying ventilation rate, but the actual air change rate (the number of volume air changes in an hour) may not be known reliably. Furthermore, any short-circuiting of the room due to a brisk flow between say a window and a door invalidates the assumption that the air is fully mixed and has a marked effect on heat loss. Everyone is familiar with the immediate effect that opening a window may have on room temperature.

It is assumed that the various surfaces of the envelope are isothermal. Logically, this is not possible, since the line of contact between two surfaces cannot be at two different temperatures; the steady change from one surface to the other is ignored. Again, while floor and ceiling temperatures may be different, we do not include the variation in temperature up the vertical walls. If there is a marked variation in floor temperature due to contact with an outer wall, it can be represented as two surfaces at different temperatures. No account is usually taken of the movement of beam radiation through a window falling on the floor and walls; its (hourly) average value is simply taken to act at the relevant temperature node. Similarly, the movement of shadows on external walls is usually neglected.

It is assumed that the thermophysical properties of the material composing the envelope are time-independent. The conductivity of masonry materials however varies with moisture content and so depends on relative humidity and wetting by rain. The cooling action due to the evaporation of rain is usually ignored.

The values assumed for λ, ρ and c_p normally have to be taken from tabulated data and these data may not correspond to values for the building they are intended to describe. Furthermore, sources may supply ambiguous information. The specific heat of low-density concrete, for example, is quoted in the 1993 *ASHRAE Handbook* (page 22-9) as 840 J/kg K, whereas lightweight concrete in the 1986 *CIBSE Guide* (Table A3.15) is given as 1000 J/kg K. ASHRAE's value for expanded polystyrene is 1210 J/kg K, whereas CIBSE's value for expanded polystyrene slab, phenolic foam, polyurethane board and

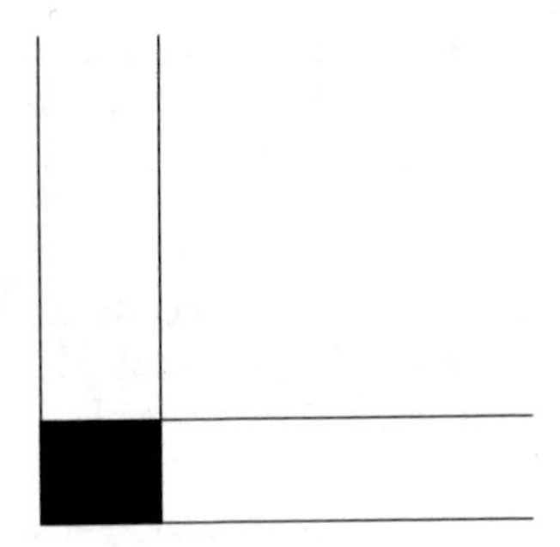

Figure 19.1 Junction of floor and wall

urea-formaldehyde foam is 1400 J/kg K. The *ASHRAE Handbook* suggests a range of values for the density of expanded polystyrene, 29–56 kg/m^3.

Heat conduction through the envelope is assumed to be one-dimensional. Thus the thermal conduction and storage of the black area in Figure 19.1 are ignored. Further, the performance of real walls is affected by thermal bridging and imperfections in construction which it may not be feasible to model.

Values must be assumed for the convective hear transfer coefficient at internal surfaces, e.g. $h_c = 3\,\text{W/m}^2\text{K}$, but they are not known with any certainty; details are given in Section 5.5. Provided the transfer is included in the room exchange model (i.e. not in the wall conduction model), there is no difficulty in making it dependent on temperature difference between air and surface and on the direction of heat flow at the floor and ceiling but it does not include variation from place to place on the surface, so it ignores low local h_c values in corners due to lower air speeds.

Values for h_c at external surfaces too must be assumed. They vary strongly with wind speed as seen in Section 5.7. For transfer at single glazing, this may have a marked influence on indoor temperatures but the uncertainty is less important as the wall resistance increases. The external convective resistance can be taken account of in finding the wall transfer coefficients, in which case the external driving temperature is sol-air temperature. If the convective resistance is not included in finding the transfer coefficients, the wall external node T_0 becomes explicit and must be included in the scheme of T values to be determined.

The long-wave radiative heat exchanges at the internal surfaces 1 and 2 depends upon the difference $T_1^4 - T_2^4$ and this can be linearised to $T_1 - T_2$ with reasonable accuracy when the difference is small. The exchange further depends on the surface emissivity ε of each surface. A value of 0.9 is often assumed but is rarely known reliably.

It is normally not possible to provide reliable information leading to values for the enclosure conductances. Indeed, the dimensions of the room, if rectangular, and the heat input, if electrical, are among the few ingredients whose values can be entered with confidence; most of the other drivers cannot be specified closely. Solar radiation often changes rapidly by large amounts, leading to large differences between hourly averages on sunny and cloudy days. Whatever the gain, it is often assumed to be distributed on the room surfaces in a manner independent of time; that is, the beam is not tracked. The designer may have little reliable information about casual heat gains due to industrial processes or about the metabolic output of an unknown group of occupants.

Against this, it must be recognised that occupants do not usually want an environment in which temperature does not vary with time or with location in the room. If they are provided with a broadly acceptable degree of thermal comfort, they can make minor personal adjustments through choice of clothing. It is neither possible nor desirable to attain high levels of accuracy in thermal modelling.

Sowell and Hittle (1995) have stated:

> Perhaps the most striking similarity behind nearly all building energy computer programs today is in the overall modeling strategy, commonly referred to the load, system, plant, and economics (LSPE) sequence. In this approach, which has withstood the test of time for nearly 30 years, heating and cooling loads are first calculated for all spaces, often for an entire year. Subsequently the secondary systems are simulated, thus calculating the required energy flows at the air handlers or other equipment supplied by the central plant. This is followed by simulation of the central plant, yielding source energy requirements. The final step calculates the cost of the source energy, sometimes introducing capital and other costs for a complete life-cycle economic analysis. In order to account for the effects of variable weather and occupancy conditions, each step is carried out hour by hour in the more sophisticated whole-building simulators.

The authors point out that this sequence is currently being replaced by an integrated approach. This chapter illustrates a realisation of the first of these stages, through use of transfer coefficients together with the model developed earlier to handle the internal exchange of heat by convection and radiation. The following sections from Davies (2001b) sketch a basic enclosure, examine more closely the response of a more detailed enclosure and indicate how it can be extended to include all surfaces.

19.2 A BASIC ENCLOSURE

We consider a basic enclosure consisting simply of a six-sided cubic shell (Figure 19.2). The walls are taken to be identical so that their internal temperatures are all the same and represented by one node, T_3. Similarly, one node represents all the external surfaces. In addition there are the internal air node, volume-averaged value T_{av} or T_2, and the ambient node T_e. The room is taken to be ventilated and driven by a hot-body source having a convective component which acts at T_2 and a radiative component which acts uniformly over all surfaces at the internal surface node T_3. Figure 19.2 also shows the possibility of a ducted air supply, heated or cooled; since we may expect this to be pressurised, the ventilation conductance V may be small or negligible. There is no net exchange of radiation between surfaces.

If one of the surfaces is singled out by having a different construction and surface temperature, leading to a radiative exchange, then we cannot use this model. Internal heat flow can in fact be modelled by variants of three methods indicated in Figure 7.2: (i) by independent convective and radiant transfer, with view factor and emissivity-based links for the radiant exchange, (ii) by independent convective and radiant exchange, assuming that the radiant flow takes place via a central (fictitious) radiant star node, and (iii) by a system where the convective and radiant exchanges are merged. The last possibility is usually adopted and the rad-air model, used here, provides a logical formulation.[2]

[2]The traditional UK approach distinguishes between the room air temperature and the room index temperature. The American approach does not, so the conductance X is infinite.

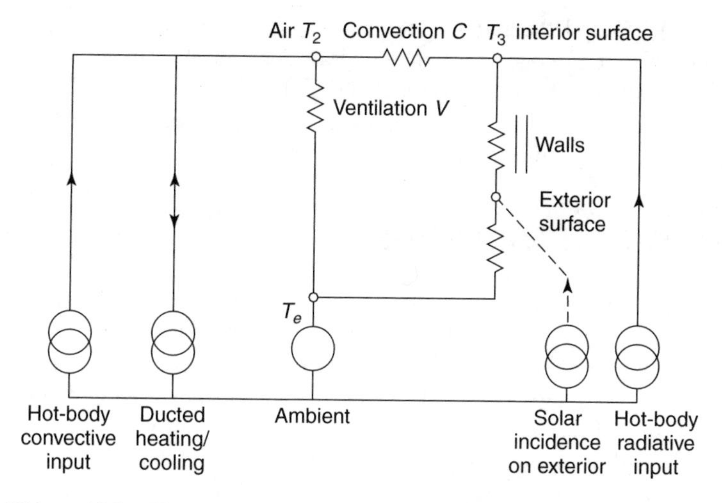

Figure 19.2 Thermal circuit of a basic enclosure with a uniform shell

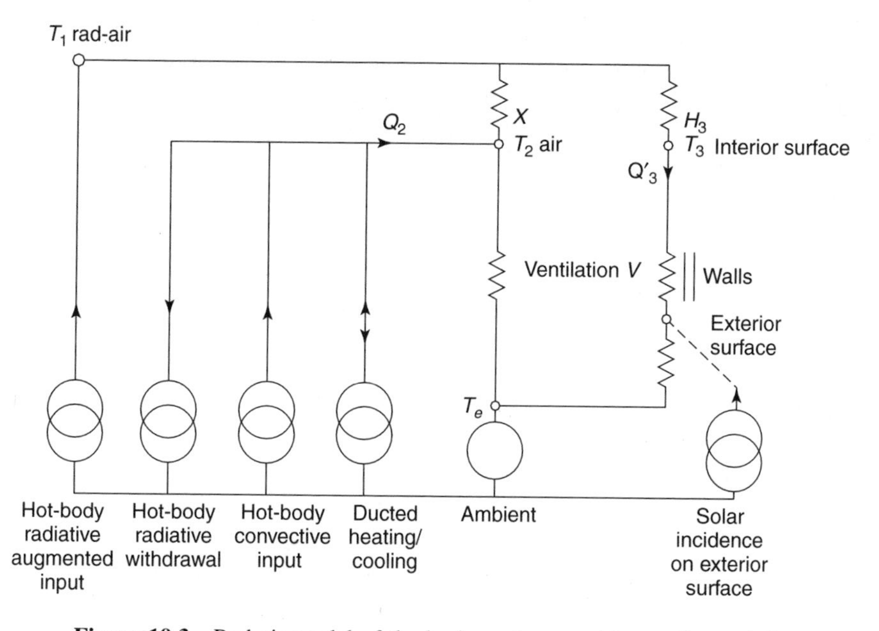

Figure 19.3 Rad-air model of the basic enclosure with a uniform shell

It is based on a radiant star network which optimally represents the exact radiant network (Section 6.5); the radiant and convective networks are then merged in a potentially exact manner (Section 7.1). The rad-air model is the appropriate model for room internal exchange when radiant and convective transfers are lumped.

The modification eventually needed to accommodate several different surfaces is shown in Figure 19.3. Most features preserve their identity, but the rad-air node, T_{ra} or T_1 appears, in effect, as a node placed on the convection conductance C (so that

$1/X + 1/H_3 = 1/C$) and the radiant component previously acting at T_3 is replaced by an augmented input at T_1 together with withdrawal of the excess from T_2, as explained earlier. The conductance H_3 is the sum of the convective conductances acting between T_{av} and T_3, together with the radiative conductances acting between the radiant star node and T_3. In this form, any number of surfaces can be modelled.

19.3 AN EXAMPLE ENCLOSURE

To illustrate the argument, a single room will be considered having internal floor dimensions of 4 m × 4 m and height $2\frac{1}{2}$ m (Figure 19.4). The north-facing wall consists of 10 m^2 of brick with thickness 220 mm, the south wall consists of 6 m^2 of similar brick together with a single-glazed window of area 4 m^2. The room will be taken as flanked above and below, to the east and the west by similar rooms with a floor/wall thickness of 110 mm of brick (so there are four surfaces with the common temperature T_4).

19.3.1 Internal Heat Transfer

The rad-air model is shown in Figure 19.5. T_1 denotes the rad-air temperature, T_{ra}, and T_2 denotes the volume-averaged air temperature, T_{av}. T_2 is linked to T_1 by the conductance X (which could be formally described as H_2) and to ambient by the ventilation conductance V, for which hourly values must be imposed. V = (room volume) × (number of air changes per hour/3600 s) × (1200 J/m^3K). Four separate surface temperatures, T_3 to T_6, require consideration.

We begin with the glass temperature T_3. Since the glass is thin and cannot sustain any significant temperature difference between its inside and outside surfaces, it is simply linked to the rad-air temperature by the conductance H_3 and to ambient by $F_3 = A_{window}/r_e$, where r_e is the outside film resistance (taken as 0.06 m^2 K/W here). If solar radiation of intensity I (W) per square metre of window falls on the glass, a fraction α of which is absorbed in the glass (Section 9.5), then a source $Q_3 = A_{window}\alpha I$ acts at T_3.

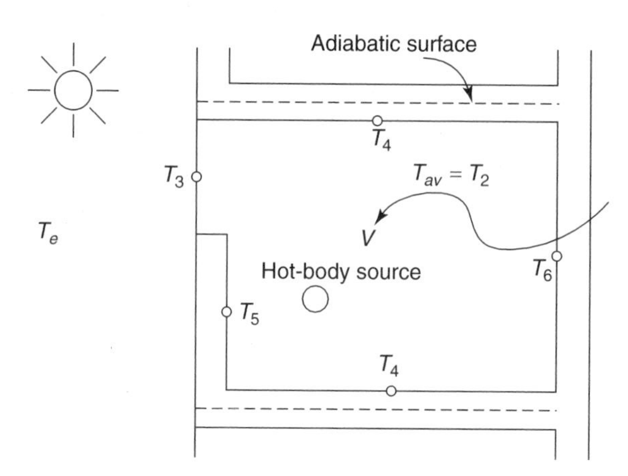

Figure 19.4 Elevation of the enclosure under investigation. (Davies, 2001b, with permission from Elsevier Science)

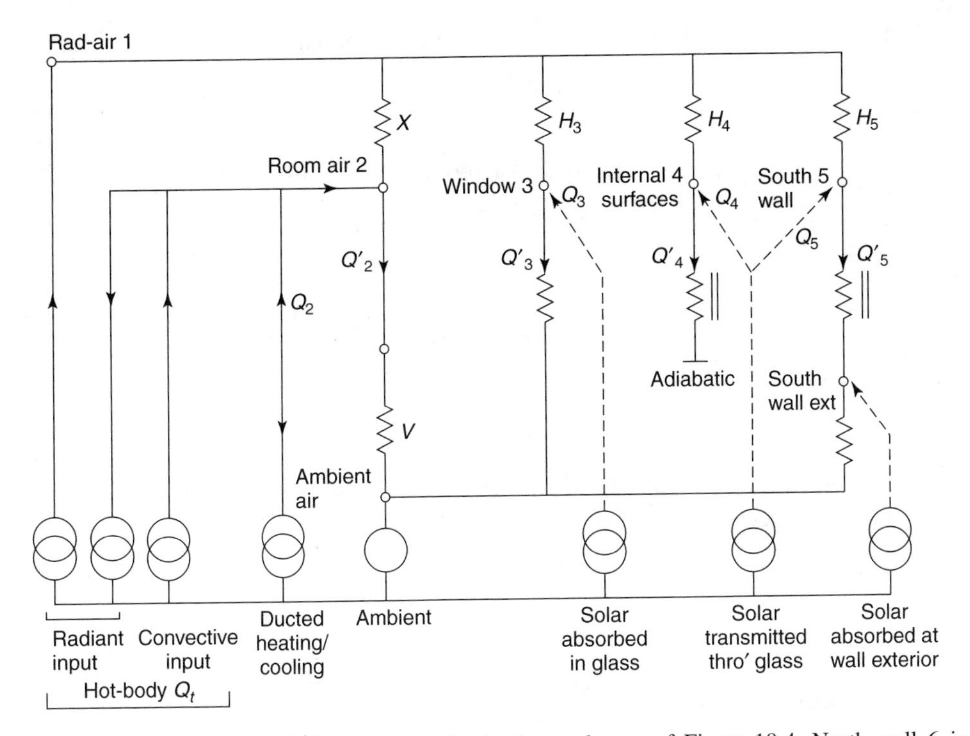

Figure 19.5 Thermal circuit for heat transfer in the enclosure of Figure 19.4. North wall 6 is similar to south wall 5. (Davies, 2001b, with permission from Elsevier Science)

To avoid some complication, the floor, ceiling and two flanking walls will be taken to have the common temperature T_4. (Figure 19.8 shows individual surfaces.) The transmitted fraction τ of the radiation on the window falls largely on the floor and constitutes a source $Q_4 = A_{window}\tau I$; much of it is likely to be absorbed there and the reflected fraction falling on and absorbed at the ceiling and flanking walls is included by the assumption of this common node. Some radiation will fall directly on the interior of the north wall T_6 and some of the radiation reflected at the floor will be absorbed on the interior surfaces of both north and south walls (T_6 and T_5). In supposing for simplicity that the transmitted short-wave flow is wholly input at T_4, these values will be in error to some extent but it will make little difference to the estimates of comfort temperatures. Q_4 is partly lost to T_{ra} and partly into the storage provided by the internal walls. Its handling is discussed later.

T_5 denotes the inside surface temperature of the south wall. The circuit detail for the north wall (surface temperature T_6) is similar and is not shown separately in Figure 19.5. Heat is driven from the exterior through each wall by the appropriate sol-air temperature.

The convective conductance C_w between surface w at T_w and the room air temperature T_2 is

$$C_w = A_w h_{cw}, \tag{19.1}$$

where A_w is the area of the surface and h_{cw} is its convective coefficient.

The radiative conductance S_w between T_w and the radiant star node T_{rs} is

$$S_w = A_w h_r E_w, \tag{19.2}$$

where h_r is the linearised radiative coefficient at 20°C, about 5.7 W/m²K, and

$$\frac{1}{E_w} = \frac{1-\varepsilon_w}{\varepsilon_w} + \beta_w, \tag{19.3}$$

from (6.51); ε_w is the surface emissivity of surface w. We also have equation (6.43)

$$\beta_w = 1 - b_w - 3.54\left(b_w^2 - \tfrac{1}{2}b_w\right) + 5.03\left(b_w^3 - \tfrac{1}{4}b_w\right) \tag{19.4a}$$

where

$$b_w = \frac{A_w}{\text{total enclosure area}} \quad \text{(Section 6.5.3)} \tag{19.4b}$$

The merged convective and radiative conductance, the inside film conductance, is

$$H_w = C_w + S_w. \tag{19.5}$$

Values are given in Table 19.1. Quantities are quoted to several digits to make clear the computational process. They are not known to this accuracy.

Thus

$$\alpha = C/S = 216/437.18 = 0.4941, \tag{19.6}$$

$$X = C(1+\alpha) = 216 \times (1 + 0.4941) = 322.72\,\text{W/K}. \tag{19.7}$$

19.3.2 Heat Flow through the Walls

The values of transfer coefficients b, c and d for the outer wall were found as explained in Chapter 17 and are listed in Table 19.2. They are based on a thickness $X = 220$ mm of brick, conductivity $\lambda = 0.84$ W/m K, density $\rho = 1700$ kg/m³ and specific heat $c_p = 800$ J/kg and a film resistance outside of $r_e = 0.06$ m²K/W. The inner film coefficient is not used in finding these values. They have of course the property that

$$\frac{\sum d_k}{\sum b_k} = \frac{\sum d_k}{\sum c_k} = \frac{X}{\lambda} + r_e. \tag{19.8}$$

Table 19.1 Room internal exchange parameters. (Davies, 2001b, with permission from Elsevier Science)

Node w	Surface	Dimensions (m)	Area A_w (m²)	Convective exchange		Radiative exchange				Surface films H_w (W/K)
				h_{cw} (W/m²K)	C_w (W/K)	ε_w	β_w	E_w	S_w (W/K)	
3	Window		4.0	3.0	12.0	0.9			23.06	35.06
5	South wall		6.0	3.0	18.0	0.9			34.60	52.60
6	North wall	4.0 × 2.5	10.0	3.0	30.0	0.9	0.8775	1.0115	57.66	87.66
4	Floor	4.0 × 4.0	16.0	3.0	48.0	0.9	0.7721	1.1323	103.27	
4	Ceiling	4.0 × 4.0	16.0	3.0	48.0	0.9	0.7721	1.1323	103.27	477.85
4	East wall	4.0 × 2.5	10.0	3.0	30.0	0.9	0.8775	1.0115	57.66	
4	West wall	4.0 × 2.5	10.0	3.0	30.0	0.9	0.8775	1.0115	57.66	
			$\sum A = 72.0$	$\sum C = 216.0$					$\sum S = 437.18$	$\sum H = 653.17$

Table 19.2 Wall transfer coefficients b_k, c_k (W/m^{2}K) and d_k (dimensionless). (Davies, 2001b, with permission from Elsevier Science)

	$k=0$	$k=1$	$k=2$	$k=3$	$k=4$	$k=5$
Outside walls						
b_k	0.003222	0.167770	0.335657	0.077515	0.001888	0.000003
c_k	20.100780	−32.660349	15.044552	−1.945390	0.046551	−0.000089
d_k	1.0	−1.039043	0.236217	−0.008542	0.000022	−0.000000
Inside partitions						
c_k	18.031627	−18.686151	0.654978	−0.000454		
d_k	1.0	−0.163762	0.000115	0.000000		

In finding the outside wall coefficients, either the node flanking the outer film (for the b_k) or the node at the inner surface (for the c_k) in the steady-slope solution was supposed to rise at a constant rate of 1 K/h or 1/3600 K/s; the other temperature was held at zero. This procedure has to be amended for the internal surfaces (floor, ceiling and flanking walls). The spaces surrounding the test enclosure will be considered to undergo the same thermal history as the enclosure itself. Accordingly, the steady-slope solution is based on a rate of rise of 1 K/h on *both* sides of the element (thickness 0.11 m, λ, ρ and c_p as above) and no film resistance. Thus the element is thermally symmetrical and is excited symmetrically. It behaves as though adiabatic at its midplane and the c_k, d_k values are defined accordingly. Values of c_k and d_k for the inner surfaces are listed in Table 19.2. They have the property that $\sum c_k = 0$.

The thermal capacity of the semi-thickness partition material is $(0.11/2)$ m $\times$ 1700 kg/m^3 $\times$ 800 J/kg K $=$ 74 800 J/m^{2}K. If it were treated as isothermal, (i.e. if it had an infinite conductivity), then $c_0 = -c_1 = 74\,800/3600 = 20.7$ W/m^{2}K, $c_2 = 0$, etc., and $d_1, d_2 = 0$, etc. The values in the table, which take account of conductivity, do not differ much from these values.

If we need N of the d_k values to describe a wall, it could be modelled by the finite difference method with N lumped capacities, optimally sized. In the model of Figure 19.4, no capacity is associated with nodes 1, 2 and 3, so $N = 0$ for each. For T_4 we have $N = 3$ and for T_5 and T_6 we have $N = 5$. The model might be seen therefore as a $3 + 5 + 5 = 13$ capacity model.

19.3.3 Thermal Response to Ambient Temperature and Heat Input

The enclosure can be excited by the three independently imposed variables, ambient air temperature, solar incidence and various internal heat inputs. Their effects will depend on the value of the ventilation rate which constitutes a further independent variable.

It will be supposed that meteorological data are available at hourly intervals. Values of air temperature T_e are known. The values for the example below were taken from Table A8.3 of the 1986 *CIBSE Guide* and are for May 22.

Values of solar irradiance I were taken from page A2-91 of the *CIBSE Guide* for a latitude of 55° and for May 22. Only values for direct incidence I on the south- and north-facing walls are included; diffuse radiation is omitted so as not to introduce too much detail.

For transmission through the north and south walls, ambient temperature was replaced by sol-air temperature as follows:

$$T_{saw,i} = T_{e,i} + \alpha'_w I_{w,i} r_e = T_{e,i} + 0.8 \times I_{w,i} \times 0.06, \tag{19.9}$$

where α' is the short-wave absorptivity of the outer wall surface (assumed to be 0.8). Angles of incidence of sun shining on the glass were calculated by standard means such as (9.3) and (9.6). The time-varying absorbed and transmitted fractions α and τ were then found. Values of α and τ were found for clear glass of thickness 4 mm, refractive index 1.526 and extinction coefficient 6.85 m^{-1}. The absorbed fraction in this case is very small but the transmitted fraction varies strongly with the angle of incidence.

If the room contains a hot-body source and/or casual sources due to occupants, lighting, etc., of total output Q_t with fractions of p and $1 - p$ in long-wave radiant and convective forms, respectively, the source is to be modelled as an input of $Q_t p(1 + \alpha)$ at T_1 and $-Q_t p\alpha + Q_t(1 - \alpha)$ at T_2 (Section 7.2). Convective heating or cooling Q_2 is to be applied at T_2.

19.3.4 The Continuity Equations

We have to equate the heat flow up to each of nodes 1 to 6 with the flow away. At time level i:

$$\begin{aligned}
&\text{at } T_1\text{: } (T_1 - T_2)X + (T_1 - T_3)H_3 + (T_1 - T_4)H_4 \\
&\qquad + (T_1 - T_5)H_5 + (T_1 - T_6)H_6 = Q_t p(1 + \alpha), \\
&\text{at } T_2\text{: } (T_2 - T_1)X + (T_2 - T_e)V = Q_2 - Q_t p\alpha + Q_t(1 - p), \\
&\text{at } T_3\text{: } (T_3 - T_1)H_3 + (T_3 - T_e)A_3/r_e = Q_3, \\
&\text{at } T_4\text{: } (T_4 - T_1)H_4 + Q'_4 = Q_4, \\
&\text{at } T_5\text{: } (T_5 - T_1)H_5 + Q'_5 = Q_5, \\
&\text{at } T_6\text{: } (T_6 - T_1)H_6 + Q'_6 = Q_6.
\end{aligned} \tag{19.10}$$

(There are no terms other than of form $T_j - T_1$ since the radiant exchange between surfaces has been transformed so that all of it passes via T_1.)

The equations can be resolved into a set of form

$$\mathbf{AT} = \mathbf{b}. \tag{19.11}$$

T is simply the vector of temperatures T_1 to T_6 at time level i, which result from solutions of the set. The vector **b** is made up of several components:

- It includes the quantities on the right-hand side of (19.10); Q_3 to Q_6 represent the solar radiation absorbed by and transmitted through the glass.
- T_e is known at time level i, so **b** includes the terms $V_i T_{e,i}$ and $(A_3/r_e)T_{e,i}$.

- The heat flows Q'_w into nodes on heat-storage elements have to be handled using the series of transfer coefficients. Denoting A_4c_{40} as C_{40}, for example, and including the time subscripts, the heat flow conducted into the floor, walls, ceiling node T_4 is

$$Q'_{4,i} = C_{40}T_{4,i} + \sum(C_{4k}T_{4,i-k} - d_{4k}Q'_{4,i-k}) \qquad (k = 1 \text{ to } N), \tag{19.12}$$

in which the value of the summed term is known at level i and is to be included in **b**. But $T_{4,i}$ is unknown, so its coefficient C_{40} is included in **A**.

- The outward flow at the interior surface of the south wall at T_5 is

$$\begin{aligned} Q'_{5,i} &= Q'_{5,out,i} - Q'_{5,in,i} \\ &= C_{50}T_{5,i} + \sum(C_{5k}T_{5,i-k} - d_{5k}Q'_{5,out,i-k}) \\ &\quad - B_{50}T_{sa,i} - \sum(B_{5k}T_{sa,i-k} - d_{5k}Q'_{5,in,i-k}) \qquad (k = 1 \text{ to } N). \end{aligned} \tag{19.13}$$

The summed terms and the term in T_{sa} (known at time level i) are included in **b** and C_{50} in **A**. Similarly, at the north wall.

Thus **AT** is

$$\begin{bmatrix} X+H_3+H_4+H_5+H_6 & -X & -H_3 & -H_4 & -H_5 & -H_6 \\ -X & X+V_i & & & & \\ -H_3 & & H_3+A_3/r_e & & & \\ -H_4 & & & H_4+C_{40} & & \\ -H_5 & & & & H_5+C_{50} & \\ -H_6 & & & & & H_6+C_{60} \end{bmatrix} \begin{bmatrix} T_{1,i} \\ T_{2,i} \\ T_{3,i} \\ T_{4,i} \\ T_{5,i} \\ T_{6,i} \end{bmatrix} \tag{19.14}$$

and the right-hand vector **b** is

$$\begin{bmatrix} Q_{t,i}p(1+\alpha) \\ Q_{2,i} - Q_{t,i}p\alpha + Q_{t,i}(1-p) + V_iT_{e,i} \\ Q_{3,i} \qquad + (A_3/r_e)T_{e,i} \\ Q_{4,i} - \sum_{k=1}^{N}(C_{4k}T_{4,i-k} - d_{4k}Q'_{4,out,i-k}) \\ Q_{5,i} - \sum_{k=1}^{N}(C_{5k}T_{5,i-k} - d_{5k}Q'_{5,out,i-k}) + \sum_{k=0}^{N} B_{5k}T_{sa5,i-k} - \sum_{k=1}^{N} d_{5k}Q'_{5,in,i-k} \\ Q_{6,i} - \sum_{k=1}^{N}(C_{6k}T_{6,i-k} - d_{6k}Q'_{6,out,i-k}) + \sum_{k=0}^{N} B_{6k}T_{sa6,i-k} - \sum_{k=1}^{N} d_{6k}Q'_{6,in,i-k} \end{bmatrix}, \tag{19.15}$$

where N is the number of values of d appropriate for the wall, 5 for the outer walls and 3 for the partitions.

The conductance matrix **A** has the usual pattern for a matrix of this kind in that the off-diagonal entry A_{wv} represents the negative of the conductance linking nodes w and v

but since all nodes are linked to T_1 only, X and H_j form the only off-diagonal elements. The diagonal element A_{ww} normally represents the sum of all the conductances linked to node w and this is true here for nodes 1, 2 and 3. Since nodes 4, 5 and 6 denote the surface temperatures of resistive/capacitative elements, we cannot assign a straightforward meaning to their 'conductances'. Physically speaking however, the quantity C_{w0} describes the heat flow into the wall after 1 h when its surface temperature is raised steadily at 1 K/h. (It will be recalled that the coefficients $C_{w1}, C_{w2}, \ldots$ depend further on the two additional ramp excitations, which taken together form a triangular pulse excitation of height 1 K and base 2 h.)

The matrix **A** has to be evaluated at each time step since some of its elements may vary in time. These include the ventilation rate V, subscripted as V_i. We might also include the value of the convective coefficient h_{cw} (and so H_w) at the floor and ceiling, which is higher for upward natural convection than for downward. Further, curtaining or other materials in place at night will reduce the value of H_{window}.

When more detail is specified than in the simple model of Figure 19.4, further off-diagonal terms may be included. Suppose that the floor, ceiling, east and west walls of this south-facing room were modelled separately and that T_7 and T_8 represent the temperatures of the east and west wall surfaces respectively. Since direct sunshine may fall first on the west wall, T_8 will tend to exceed T_7 early in the day, and vice versa. If the room is a module, flanked by rooms having the same thermal history, the temperature driving heat through the east wall to T_7 is the temperature of the west wall in the east adjacent room, which has the value T_8. Thus in the continuity equation for T_7, Q'_7 consists of a series of B_k terms with T_8 as driver, in addition to the series of C_k with T_7 as driver. The contribution at time i is $B_{80}T_{8,i}$, but since $T_{8,i}$ is unknown, its coefficient B_{80} has to be included as the off-diagonal term $A_{87} = -B_{80}$ in matrix **A**. Similarly, $A_{78} = -B_{70}$. Similar considerations apply to the heat flow through the floor and ceiling in a module room, with rooms above and below undergoing the same thermal history.

Finally, we note that the matrix **A** is symmetric. When the infiltration rate between rooms is included however, the quantity V enters unsymmetrically (Section 19.5).

Solution of the equations gives the values of T_1 to T_6 at time i. Of these, T_1 (the rad-air temperature T_{ra}) and T_2 (air temperature T_{av}) are required to find the radiant star temperature T_{rs}:

$$T_{rs} = T_{ra}(1 + \alpha) - T_{av}\alpha, \tag{19.16}$$

so the comfort temperature T_c is

$$T_c = 0.5T_{rs} + 0.5T_{av}, \tag{19.17a}$$

or perhaps

$$T_c = 0.4T_{rs} + 0.6T_{av} \tag{19.17b}$$

if there is perceptible air movement in the room for high ventilation rates. This estimate of T_c includes consideration of the long-wave radiation from a hot-body internal source which falls on furnishings, etc.

The current value of T_4, $T_{4,i}$, is needed so that $Q'_{4,i}$ can be found from (17.2b) for use at time level $i + 1$. $T_{5,i}$ and $T_{6,i}$ are needed to find the outward flows at the south and north walls.

19.3.5 Response of the Enclosure

The technique under discussion is intended for use with lists of hourly meteorological information. We will use the data for May 22. On starting a calculation, some time must elapse before valid values are returned, depending on the weight of the building.

Transient response

The simplest time-varying response that can be examined is the transition from one steady state to another. Suppose that the entire fabric is at 0°C up to a time of $t = -1$ h; at $t = -1$ h ambient temperature increases steadily to 10°C at $t = 0$ and then remains steady at this value. There is no other heat input. Figure 19.6 shows the variation of internal temperatures over the next few hours with an assumed ventilation rate V of zero. Since the window T_3 has no thermal storage and has a low-resistance link with ambient, it shows the most immediate rise in temperature and remains much the warmest element. Heat flows from window to T_1 and from T_1 on to T_4, T_5 and T_6. Since the south and north walls are of identical construction, T_6 is equal to T_5, and with $V = 0$ the air temperature $T_2 = T_1$. Heat flows into the internal storage (floor, flanking walls and ceiling), so T_4 remains less than T_1. Heat flows initially too into the external walls. Eventually the ramp change imposed at the outside wall surfaces makes itself apparent at the inside surfaces: after about 3 h, the interior temperatures of the two outside walls (T_5 and T_6) become greater than T_1, so the direction of heat flow changes. All temperatures slowly tend to ambient temperature, 10°C: T_5 changes from 9.99 to 10.00°C after 198 h.

If a finite ventilation rate is assumed, there is an additional means of warming the room and temperatures rise more quickly. Figure 19.7 shows the variation of comfort temperature for ventilation rates of zero and 5 air changes per hour. Ventilation leads to

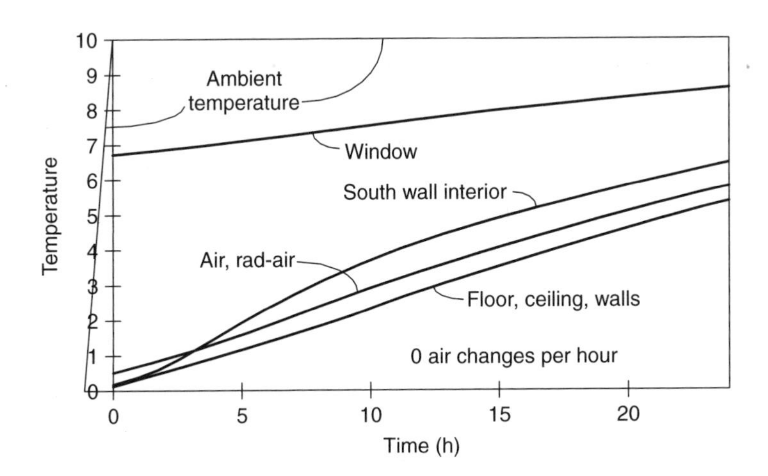

Figure 19.6 Response of the enclosure to a terminated ramp (0–10°C) in ambient temperature at zero ventilation rate (Reprinted from *Building and Environment*, vol. 36, M.G. Davies, Hourly estimation of temperature using wall transfer coefficients, 199–217, © 2001, with permission from Elsevier Science)

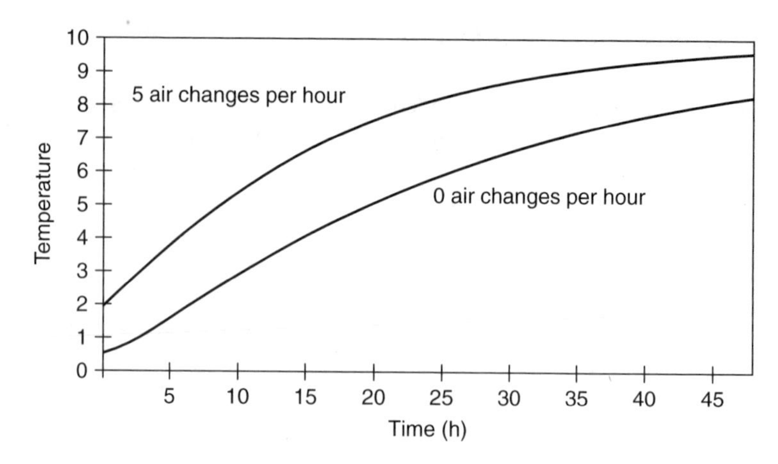

Figure 19.7 The effect of ventilation rate on the speed of response (comfort temperature) to the ramp (Reprinted from *Building and Environment*, vol. 36, M.G. Davies, Hourly estimation of temperature using wall transfer coefficients, 199–217, © 2001, with permission from Elsevier Science)

Table 19.3 Comfort temperature T_c due to a ramp input of 1000 W/h from $t = -1$ h for one hour, then a steady input of 1000 W. (Davies, 2001b, with permission from Elsevier Science)

Air changes	Time (h)																		
per hour	0	1	2	3	4	5	6	7	8	9	10	12	14	16	18	20	22	24	Final
0	2.7	3.3	3.8	4.4	4.8	5.3	5.8	6.2	6.6	7.0	7.4	8.1	8.8	9.4	9.9	10.5	11.0	11.4	19.2
5	2.1	2.5	2.9	3.2	3.5	3.8	4.1	4.3	4.5	4.7	4.9	5.3	5.6	5.9	6.1	6.4	6.6	6.7	7.9

higher internal temperatures and the direction of flow of heat into the walls now only reverses after 11 h. A time constant can be defined as the time taken for, say, comfort temperature to come within $1/e = 0.368$ of its terminal value, that is, $(1 - 1/e) \times 10 = 6.3$°C. The time constants for 0 and 5 air changes per hour are 28 and $13\frac{1}{2}$ h respectively.

The response to a heat input of 1000 W (300 W radiantly and 700 W convectively) is given in Table 19.3, which shows the values of comfort temperature T_c. With the higher ventilation rate, the final temperature reached is lower but convergence to this value is faster, as we saw for the temperature excitation.

Steady-cyclic response

To illustrate the response of the enclosure to a repeated pattern of daily variation, the program was run using the meteorological data noted for some days until $T_{1,i} - T_{1,i-24}$ was less than 0.01 K so that a steady-cyclic state had been established – about 6 days for zero ventilation and 4 days when 5 air changes per hour was assumed. Results in Table 19.4 indicate how peak temperatures may depend on variations in ambient temperature and solar gains associated with transparent and opaque parts of the fabric.

Table 19.4 Daily variation of comfort temperature due to ambient temperature and solar radiation, and the heating/cooling load to maintain comfort temperature at 20°C. (Davies, 2001b, with permission from Elsevier Science)

(1)	(2)	(3)	(4)	(5)	(6)	(7)	(8)	(9)	(10)	(11)	(12)	(13)	(14)	(15)
Hour	Ambient (°C)	I_S (W/m^2)	I_N (W/m^2)	0 ac/h, heat transfer via				5 ac/h, heat transfer via				Ventilation (all) by		Heating or cooling load (W)
				Air (°C)	Window (K)	Walls (K)	All (°C)	Air (°C)	Window (K)	Walls (K)	All (°C)	Night (°C)	Day (°C)	
0	11.0	0	0	14.6	6.1	2.0	22.7	14.3	2.8	1.0	18.1	21.4	21.8	124
1	10.0	0	0	14.5	5.9	1.9	22.3	14.0	2.6	1.0	17.6	21.1	21.4	193
2	9.0	0	0	14.3	5.6	1.9	21.9	13.6	2.4	0.9	17.0	18.9	21.0	260
3	9.0	0	0	14.2	5.4	1.8	21.5	13.4	2.3	0.9	16.5	18.4*	20.7	308
4	8.5	0	25	14.1	5.2	1.8	21.1	13.0	2.1	0.8	16.0	17.7*	20.3	361
5	9.0	0	135	13.9	5.0	1.7	20.7	12.9	2.0	0.8	15.7	17.3*	19.9	393
6	9.5	0	110	13.8	4.8	1.7	20.3	12.8	1.8	0.9	15.4	16.9*	19.6	420
7	10.5	0	0	13.7	4.7	1.7	20.0	12.9	1.7	0.7	15.3	16.6*	19.3	433
8	12.0	145	0	13.7	4.7	1.6	20.0	12.9	1.9	0.7	15.5	17.6*	19.3	361
9	13.5	285	0	13.7	5.1	1.6	20.4	13.1	2.2	0.7	16.0	18.1	19.7	172
10	15.0	395	0	13.7	5.6	1.6	21.1	13.4	2.8	0.7	16.9	18.9	20.5	−91
11	16.5	470	0	13.7	6.7	1.6	22.0	13.8	3.5	0.7	18.0	19.9	21.4	−377
12	18.0	495	0	13.8	7.5	1.6	23.0	14.2	4.3	0.8	19.2	21.0	22.4	−639
13	19.0	470	0	14.0	8.3	1.7	23.9	14.5	4.8	0.8	20.2	22.0	22.7*	−824
14	19.5	395	0	14.1	8.7	1.8	24.6	14.8	5.2	0.9	20.9	22.7	23.3*	−902
15	20.0	285	0	14.2	8.8	1.9	24.9	15.2	5.2	1.0	21.3	23.1	23.5*	−869
16	19.5	145	0	14.4	8.6	1.9	24.9	15.3	4.9	1.0	21.3	23.1	23.3*	−719
17	19.0	0	0	14.5	8.1	2.0	24.7	15.5	4.5	1.1	21.1	23.0	23.0*	−552
18	18.5	0	110	14.6	7.8	2.0	24.4	15.6	4.2	1.1	20.9	22.8	22.7*	−429
19	17.5	0	135	14.7	7.5	2.0	24.2	15.6	3.9	1.1	20.6	22.7	23.1	−316
20	16.0	0	25	14.7	7.2	2.1	24.0	15.4	3.7	1.1	20.1	22.5	22.9	−209
21	15.0	0	0	14.8	6.9	2.1	23.7	15.3	3.4	1.1	19.6	22.3	22.7	−120
22	13.5	0	0	14.7	6.6	2.0	23.4	15.0	3.2	1.1	19.2	22.0	22.4	−31
23	12.0	0	0	14.7	6.4	2.0	23.0	14.7	3.0	1.1	18.7	21.7	22.1	53
24	11.0	0	0	14.6	6.1	2.0	22.7	14.3	2.8	1.0	18.1	21.4	21.8	124

*Times of high ventilation.

Response to ambient temperature

Values for hourly ambient temperature are listed in column 2 of Table 19.4. Columns 5 and 9 list the variation in comfort temperature for ventilation rates of 0 and 5 air changes per hour (ac/h) assuming no solar or other gains. Mean indoor temperature must equal mean outdoor temperature. The variations are as follows:

- ambient: $20.0 - 8.5 = 11.5\,\mathrm{K}$,
- indoors with 0 ac/h: 1.1 K with time lags of 5 or 6 h,
- indoors with 5 ac/h: 2.6 K with time lags of 3 or 4 h.

Response to solar gains acting through the window

A large fraction of the solar radiation falling on the window is transmitted and falls on the floor, with some reflection to other surfaces; as explained, it is taken to act at T_4

A little radiation is absorbed in the glass and acts at T_3. Columns 6 and 10 show the increment that results. Peak values fall around 3 h after maximum insolation. Ventilation lowers temperature increments generally, but it has less effect on the variation: 4.2 K and 3.5 K for 0 and 5 ac/h, respectively.

Response to solar gains acting through the walls

Columns 7 and 11 show the increments to comfort temperature due to direct solar gain absorbed on the outside surfaces of the south and north walls. The increments are smaller than the other two components but are not negligible. The calculation indicates that the inside surface of the south wall (T_5) has its maximum value some 5 h after the time of maximum irradiation. The north wall temperature (T_6) varies very little but shows a weak maximum about 4 h after the morning burst of solar radiation falling on it. No evening maximum is apparent; presumably it is masked by the effect of the much larger gains through the opposite wall.

The computer estimates of net heat flows from and within the enclosure (T_1 to T_2, T_1 to T_3, etc.) also show some interesting detail. Most of the heat flow to the enclosure is through the south wall. The floor, etc., is a sink of heat from 1300 to 2200 and a source for the rest of the time. Heat is lost steadily through the window. With zero ventilation, heat is also lost through the north wall, except for a very small gain between 0800 and 1000. With 5 ac/h, heat is more readily lost, temperatures are generally lower, and heat is now lost through the north wall from 1400 to 2000.

Net response

The response to all three processes acting together is also given in Table 19.4, columns 8 and 12. It is simply the sum of the processes.

Variable ventilation

The above tests assumed fixed values of the ventilation rate but it is also possible to use hourly values. To demonstrate this, we examine the effect of imposing a period of 6 h of 5 ac/h, with 0 ac/h for the remainder of the period. These periods can be placed during the night so as to effect maximum cooling of the fabric, or during late afternoon to lower peak values. The results of the two strategies are shown in the final columns of the table. The asterisked values denote periods of high ventilation. All values in these columns necessarily fall between the 0 and 5 ac/h all-causes values. Night ventilation leads to marginally lower temperatures in the afternoon. With high ventilation during periods of occupation, comfort temperature should perhaps be more weighted on air temperature. It is straightforward to program this in. The effects of pre-cooling a building by night and at the weekend are described in a study by Ruud *et al.* (1990); their results showed an 18% reduction in cooling energy supplied during the daytime.

19.3.6 Heating or Cooling when Comfort Temperature Is Specified

We may regard the excitation due to ambient temperature and solar irradiation as imposed variables, and so far we have simply examined their consequence in the absence of any further heating or cooling. The procedure however allows some internal temperature or combination of temperatures to be fixed hourly, in which case the heating or cooling needed to achieve this value has to be determined (together with values of all the other temperatures). Suppose that comfort temperature T_c is fixed. Now

$$T_c = \tfrac{1}{2}T_{rs} + \tfrac{1}{2}T_{av}, \tag{19.18}$$

but

$$T_{ra} = (1/(1+\alpha))T_{rs} + (\alpha/(1+\alpha))T_{av}, \tag{19.19}$$

so

$$T_c = \tfrac{1}{2}(1+\alpha)T_{ra} + \tfrac{1}{2}(1-\alpha)T_{av} = \tfrac{1}{2}(1+\alpha)T_1 + \tfrac{1}{2}(1-\alpha)T_2. \tag{19.20}$$

It is convenient to implement this by giving the matrix **A** in (19.11) a seventh row $\left[\tfrac{1}{2}(1+\alpha)\ \tfrac{1}{2}(1-\alpha)\ 0\ 0\ 0\ 0\ 0\right]$ and a seventh column $[0\ \ -1\ 0\ 0\ 0\ 0\ 0]^{\mathrm{T}}$. The required value of T_c forms the final item in the right-hand column **b**. The heating or cooling load, input convectively at T_2, is given by the seventh item in the T vector (the units of course being W, not °C.)

Column 15 of the table shows the heating or cooling needed to hold comfort temperature at 20°C. Since this supply is likely to pressurise the system and prevent infiltration from ambient, V was taken to be zero. The maximum cooling load falls some 2 h after maximum solar gains.

It is a matter for further consideration whether the heating or cooling system should be operated so as to keep temperature constant. There is abundant anecdotal evidence to show that people welcome some variation of temperature in time (and location). Zmeureanu and Doramajian (1992) have demonstrated that by allowing the indoor environment to deviate from fixed comfort conditions, the energy consumption and costs in large office buildings can be reduced while maintaining generally acceptable conditions. In their study, by allowing temperature to increase after 1500 hours, there was an 11% saving of HVAC energy.

19.4 DEVELOPMENT OF THE MODEL

To show how the simple enclosure in Figure 19.4 with its combined surfaces can be extended to a more realistic enclosure, we will refer to the enclosure proposed by CEN (1995) as a standard against which to test computer programs for thermal response. The method is similar to the scheme used earlier by Walton (1981, 1983) – TARP (1988) and BLAST (1986) – and more recently by Pedersen *et al.* (1997). Seem *et al.* (1989b) used their star model combining internal radiant and convective exchange (Chapter 7) together with the transfer coefficients of each individual wall to derive a 'comprehensive room transfer function' for the room as a whole. It involved a procedure to reduce the number of transfer coefficients (i.e. to use less past information) than is formally necessary.

CEN (1995) gives details of the enclosure and how it is driven. Its floor area is 5.5 m × 3.6 m and its height is 2.8 m. Thermal details are given of all bounding surfaces and radiant and (temperature-dependent) convective exchange coefficients, together with hourly values of ambient temperature and direct and diffuse solar radiation, internal gains and ventilation rates. The long dimension is aligned E–W and there is a shaded window in the west elevation; shading details are supplied. Two sites are defined: site A at 40° and site B at 52°N. Appropriate hourly values of solar radiation and ambient air temperatures are supplied. In two sets of tests (A1,2 and B1,2) the enclosure is seen as a module, surrounded above and below, to the north and the south by enclosures undergoing an identical thermal history. Test 2 differs from test 1 only in that the ceiling provides more thermal storage. In a third set of tests (A3 and B3) the enclosure is seen as an attic room and so solar gains on the roof also have to be considered. For each of these conditions, three patterns of ventilation have to be taken: a constant low value, a constant high value, and one with high ventilation by night and low by day. Thus we have (2 latitudes) × (3 enclosure types) × (3 ventilation patterns) = 18 runs. For each run, the maximum, minimum and daily mean values of dry resultant temperature (or the air temperature) have to be found.

Figure 19.8 shows an appropriate thermal circuit for the attic version of the enclosure. Six individual surface nodes must be specified, together with those for a window surface, air and rad-air – a total of nine nodes. The circuit indicates the possibility of including the thermal capacity of the room air, significant in a large enclosure, and also the capacity of the window (less significant). CEN apparently left undefined the thermal conditions in the room below the attic room so the floor, node 6, is supposed here to be adiabatic at its lower surface. We have to assume that conditions in the adjoining attic rooms to the north and south undergo the same thermal behaviour as the test enclosure itself, so the enclosure north and south walls are handled as discussed earlier (introducing the off-diagonal terms A_{78} and A_{87}). Conditions in the space beyond the east wall too were undefined, so it was taken to be adiabatic at its midplane. Since the temperatures of the roof and the exterior surface of the west wall were not of interest, the drivers at these surfaces were the appropriate sol-air temperatures. CEN did not require the effect of any room furnishings to be included but were this the case, an additional element would be needed (element 10), lumped or distributed, adiabatic and linked to T_1 via H_{10}.

The test specification required that the convective coefficient at the floor and ceiling should be higher for upward flow than for downward flow. Thus the sign of $T_{air} - T_{surface}$ was tested every hour, leading to possible hourly variation in the H values. Furthermore, just as the ventilation rate can be supposed to vary, we can also take account of any insulation, curtaining or shutters applied to the window by night.

The numbers of d_k values associated with nodes 1 to 9 were respectively $N = 0$, 0 (or 1 with air capacity), 0 (or 1 with glass capacity), 6, 5, 8, 2, 2, 1, totalling 24, so that the model is equivalent to one consisting of 24 discrete capacities. The time taken was around 0.1 s per day, so that if the program were to process a year's similar data, the runtime (mid 1990s) would be some 30 s.

There are a large number of computational tools intended to provide design guidance on thermal and related matters. Balaras (1996) in an article focusing on the role of thermal mass as it affects cooling loads, reported that 128 programs had been identified that were able to calculate building cooling loads (54 calculated the variation in indoor temperature, 45 included mass and shading and 23 could simulate natural ventilation.) Of them he selected 17 simplified models and provided a useful table in which were succinctly

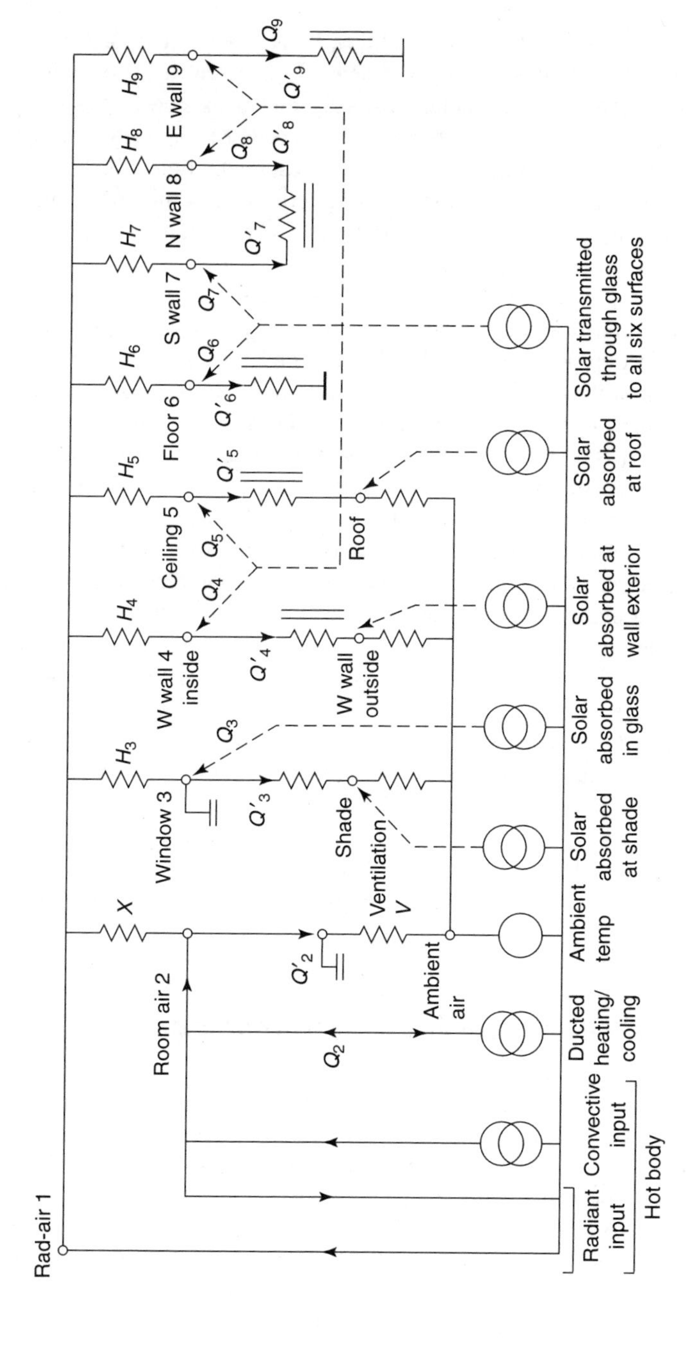

Figure 19.8 Thermal circuit for an attic room. (Davies, 2000b, with permission from Elsevier Science)

listed the extent of the input information to be supplied, the nature of the output, the parameter used to describe the thermal mass, any restrictions and whether software was available. Inputs included such matters as detailed information on the building geometry, construction and orientation, required by all programs. Other possible input data included ambient temperature (hourly, daily or monthly), solar radiation information (hourly or monthly), infiltration/ventilation inputs from sensible or latent heat loads and wind data (hourly or monthly). The outputs included hourly and peak cooling loads, hourly and peak air and surface temperatures. The parameters needed to describe the thermal mass took a variety of forms but included some of the measures that have been discussed here: elements in a thermal network, total thermal time constant, admittances and conduction transfer functions. A frequently noted 'restriction' was the omission of wind effects. A listing of this kind requires frequent updating.

How reliable are the predictions of a thermal model likely to be? The basic philosophy in the natural and applied sciences is that values of a quantity as estimated by some theory or procedure should accord with their observed values. There may be error and/or uncertainty in values as estimated and measured, and this poses severe problems for validation in building thermal response studies. Once a dynamic simulation program has been developed, it is a simple matter to run it to mimic some given history, but early experience showed that two programs ostensibly undertaking the same task (or indeed two workers using the same program) could, for many reasons, give substantially different results. Furthermore, observational or empirical data derive from a large number of independent variables (dimensions, materials, weather and solar incidence, patterns of use, heating/cooling plant and its control), and the many measurements needed may be costly and time-consuming, may be difficult to make or may demand considerable expertise. With uncertainty in estimated and observed versions of a variable, statistical means are needed to establish agreement. Irving (1988) describes validation studies in the 1980s. Methodologies have been discussed by Lomas and Eppel (1992) based on differential sensitivity analysis, Monte Carlo analysis and stochastic sensitivity analysis. A large and detailed validation study was reported by Lomas *et al.* (1997). Firm observational data were available for a series of enclosures. They were modelled by 25 independent simulation programs of European, American and Australian origin and the validation exercise succeeded in identifying those of them that performed well. Ahmad (1998) lists over 65 validation studies.

19.5 INFILTRATION BETWEEN ADJACENT ROOMS

Temperature-driven air movement between adjacent rooms was discussed in Section 5.6 but air may move between rooms due to cross-ventilation, the stack effect or by a fan; air movement transfers heat, moisture and possibly contaminants. Feustel and Kendon (1985) review infiltration models available at that time. A number of multi-zone models have been developed to estimate such interzonal flows, among them Etheridge (1988), Walton (1989), Axley and Grot (1989), Feustel and Sherman (1989), Riffat and Abdalla (1991), International Energy Agency (1992), Feustel (1992), El Diasty *et al.* (1993b), BREEZE 6.0 d (1993), Tuomaala (1993), Boyer *et al.* (1999), Feustel (1999), Clarke (2001), Li and Delsante (2001), Axley (2001), Plathner and Woloszyn (2002). Leakage measurement is described by Feustel (1990). Some aspects of these models are reviewed by Dascalaki and

Santamouris (1999). The heat transfer mechanism, however, differs in form from those due to conduction, convection and radiation. If two rooms share a common wall and room 1 is warmer than room 2, room 2 will gain heat by conduction. It will also gain heat by infiltration if there is air movement from 1 to 2. But if the air moves from 2 to 1, room 2 is uninfluenced by air temperature in room 1. Although the ventilation mechanism has so far been modelled by a conductance V on the same footing as the conduction/storage mechanism, this is clearly incorrect (Figure 4.2).

To analyse the situation in isolation, suppose that both rooms have massless adiabatic surfaces apart from their common partition. Thus no further surface nodes need be considered. Suppose too that room 1 is heated by a source Q whose radiant fraction p is equal to $S/(S+C)$, where S and C are respectively the total radiant and convective conductances in room 1. In this case, Q may be taken to act at the rad-air node T_{11} in room 1, so removing the complication of a withdrawal from the air node T_{12}. The thermal circuit is shown in Figure 19.9.[3] The partition has surface nodes T_{13} and T_{23}.

We suppose that air at ambient temperature T_e infiltrates into room 1 from ambient. If the thermal flow rate is V(W/K), the flow of internal energy into the room is VT_e and the outward flow is VT_{12}. Continuity at the six nodes requires that at time level i we have

$$\begin{aligned}
&(T_{11,i}-T_{12,i})X_1+(T_{11,i}-T_{13,i})H_{13}=Q_i,\\
&(T_{12,i}-T_{11,i})X_1 \qquad\qquad = V_iT_{e,i}-V_iT_{12,i},\\
&(T_{13,i}-T_{11,i})H_{13} \qquad\qquad = Q''_{1323,i}-Q'_{1313,i},\\
&(T_{23,i}-T_{21,i})H_{23} \qquad\qquad = Q''_{2313,i}-Q'_{2323,i},\\
&(T_{22,i}-T_{21,i})X_2 \qquad\qquad = V_iT_{12,i}-V_iT_{22,i},\\
&(T_{21,i}-T_{22,i})X_2+(T_{21,i}-T_{23,i})H_{23}=0,
\end{aligned} \tag{19.21}$$

where

$$\begin{aligned}
Q''_{1323,i} &= B_0T_{23,i}+\sum(B_kT_{23,i-k}-d_kQ''_{1323,i-k}) \qquad (k=1 \text{ to } N),\\
Q'_{1313,i} &= A_0T_{13,i}+\sum(A_kT_{13,i-k}-d_kQ'_{1313,i-k}),\\
Q''_{2313,i} &= B_0T_{13,i}+\sum(B_kT_{13,i-k}-d_kQ''_{2313,i-k}),\\
Q'_{2323,i} &= \underbrace{C_0T_{23,i}}_{\text{(i)}}+\underbrace{\sum(C_kT_{23,i-k}-d_kQ'_{2323,i-k})}_{\text{(ii)}}.
\end{aligned} \tag{19.22}$$

In forming the matrix equation $\mathbf{AT}=\mathbf{b}$, the values in (i) are current values and are not known, so their coefficients are included in the matrix $\mathbf{A}$. The values in (ii) have been

[3]The movement may be due to infiltration from outside, driven by wind forces or the stack effect (buoyancy), or by mechanical ventilation. The thermal flow rate V here is taken to be due to infiltration and has the same value at each stage of transfer in the circuit of Figure 19.9 but this will not normally be the case. An open doorway may allow a flow in both directions.

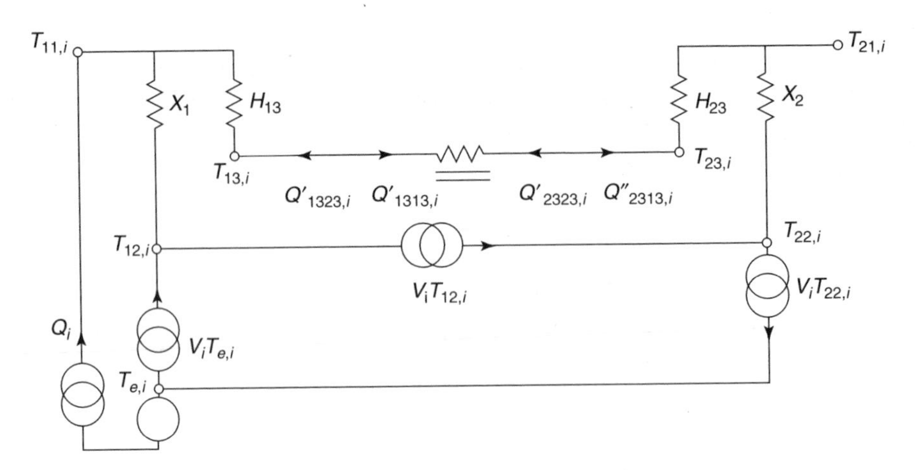

Figure 19.9 Thermal circuit for infiltration between rooms

evaluated at previous time steps and so form part of the vector **b**. Then

$$\begin{bmatrix} X_1+H_{13} & -X_1 & -H_{13} & & & \\ -X_1 & X_1+V_i & & & & \\ -H_{13} & & H_{13}+A_0 & -B_0 & & \\ & & -B_0 & H_{23}+C_0 & & -H_{23} \\ & -V_i & & & X_2+V_i & -X_2 \\ & & & -H_{23} & -X_2 & X_2+H_{23} \end{bmatrix} \begin{bmatrix} T_{11,i} \\ T_{12,i} \\ T_{13,i} \\ T_{23,i} \\ T_{22,i} \\ T_{21,i} \end{bmatrix}$$

$$= \begin{bmatrix} Q_i \\ V_i T_{e,i} \\ \sum(B_k T_{23,i-k} - d_k Q''_{1323,i-k}) - \sum(A_k T_{13,i-k} - d_k Q'_{1313,i-k}) \\ \sum(B_k T_{13,i-k} - d_k Q''_{2313,i-k}) - \sum(C_k T_{23,i-k} - d_k Q'_{2323,i-k}) \\ 0 \\ 0 \end{bmatrix}. \quad (19.23)$$

The matrix **A** is now no longer symmetrical since the element A_{52} is $-V_i$ while A_{25} is zero. Reversal of the direction of infiltration does not result in a change in the sign of V_i but by A_{25} becoming $-V_i$ and A_{52} zero. If infiltration takes place in both directions, for example through lower and upper apertures, both terms become finite.

19.6 DISCUSSION

This chapter has outlined a scheme using wall transfer coefficients to estimate room thermal response in a design context. It is a combination of two models, one to handle room internal radiant and convective exchange, and the other to compute wall conduction. They are independent of each other; their interface is the room internal surface.

It is clear that model estimates of temperature in a real building will correspond to their observed values only in so far as the values assumed for room parameters (convective coefficients, conductivity, etc.) and drives (ambient temperature, etc.) correspond to their actual values. It seems unlikely, however, that the computational process itself will add to the uncertainty. The internal checks in evaluating the transfer coefficients ensure they are free of error. As examined in Chapter 18, their use in evaluating an enclosure is not exact, but computational error is likely to be small.

Certain positive features of the scheme may be summarised:

- If temperature is allowed to float, comfort temperature can be determined. If comfort temperature is specified, the corresponding heating or cooling load is determined. The room could be allowed to float by night with temperature control by day.
- The procedure allows an hourly choice to be made of comfort temperature, ventilation rate, window insulation and convective coefficients (since convective flow at a horizontal surface depends on whether the air is warmer or cooler than the surface and they can be made temperature-dependent.)
- Transfer coefficients as defined here do not include an inner film since the film forms part of the rad-air model. As presented here, they include the effect of the outside film r_e so as to enable solar radiation absorbed on an outer surface to be handled as a simple sol-air temperature. They could however be defined so as to exclude this resistance, in which case hourly values of r_e too could be used. (Hourly values of r_e at a window can be specified without difficulty.)
- The scheme can be used for a suite of rooms provided that the directions of infiltration are specified.

The scheme has limitations.

- The rad-air model cannot take account of the *location* of internal heat sources. A radiator placed beneath a window will bring about a relatively warm floor and cool ceiling, but with all heat exchange taking place via a central node T_{ra}, this detail is lost. The full radiant exchange network would be needed. The scheme of the wall transfer coefficients remains unaltered, however.
- It assumes one-dimensional wall heat flow, so it ignores the extra conduction and storage in corners. It also assumes uniform mixing of room air.
- The procedure assumes that excitation variables change linearly between specified hourly values, so it cannot cope with the immediate effect of a step change in heat input. Without modification the technique is therefore unsuitable to estimate the short-term response of a room, perhaps to examine the action of a thermostat. It could be undertaken as discussed here but it would require b, c and d values based on a time interval shorter than 1 h and the series of these values would be longer.
- If wall properties change significantly, most probably due to variation of conductivity with water content, new coefficients will be needed.

A room thermal model could be improved by moving away from some of the assumptions listed earlier. Surfaces could be broken into smaller areas, furniture might be included, patches of direct sunlight might be tracked across furnishings and surfaces, account taken of two- and three-dimensional flow at edges and in corners, and use made of the Navier–Stokes equations for the room air. All this clearly increases the detail to be specified and runtimes in implementation by an order of magnitude while some of the other assumptions retain their force. As a design technique their use is exceptional.

19.7 CLOSURE

This book forms an attempt to bring together the elements of those factors determining building thermal response which relate to its design and materials of construction. As is usual with heat transfer texts, it does not touch on the associated technology – in this case the plant providing heating and cooling, together with the increasingly complex art of control. Furthermore, it provides an input at one end of the broad study of building simulation, discussed comprehensively by Clarke (1985, 2001). Spalding was urging in the mid 1980s that it was more cost-effective to promote the numerical approach to heat transfer through computer codes (which meet the designer's needs when needed) rather than in handbooks of ever-increasing size. The response of the fabric to weather excitation can be adequately examined by sampling at hourly intervals; at any rate, that is what has usually been assumed. Building simulation, however, requires inclusion of other factors: characteristics of the heating/cooling plant, the associated fluid flows and control element characteristics, which require shorter intervals. The possibilities of computer simulation have expanded to an extent that could hardly have been foreseen when digital methods overtook analogue computing in the early 1960s and there seems no slackening in the pace. Hong *et al.* (2000) review the wealth of facilities currently pursued in the field of building simulation. Murakami *et al.* (2001) describe computational fluid dynamic methods and details of a human model in connection with an air-conditioning system to assess thermal comfort. The possibilities of integrating computational fluid dynamics with building simulation have recently been examined by Zhai *et al.* (2002) and by Bartak *et al.* (2002) and Citherlet and Hand (2002) have extended the ESP-r simulation system, originally designed to model energy flow, to include considerations of lighting availability, thermal and visual aspects of user comfort, room acoustics and the environmental impacts related to construction over the lifespan of the building. I hope the topics discussed in this book may make some small contribution toward the cataloguing of this global endeavour.

Principal Notation

A list of many of the quantities needed in heat transfer, together with a recommended symbol and SI units, is given in the *American Society of Mechanical Engineers Journal of Heat Transfer* **122**, 416, May 2000. Most of the symbols below accord with this list. Symbols given in [ ] refer to standard usage elsewhere.

Physical constants

	Avogadro's number 6.0225×10^{26}	molecules/kmol
R	Universal gas constant 8.3144×10^{3}	J/kmol K
M	Molecular weight, usually of air (29) or water vapour (18)	kg/kmol
[R is often used to denote R/M]		
h	Planck's constant 6.62608×10^{-34}	J s
σ	Stefan–Boltzmann constant, 5.67051×10^{-8}	$W/(m^2 K^4)$

Heat flows

Q_t	Combined radiant and convective input from an internal source	W
Q_r	Radiant component from an internal source	W
Q_a	Convective component from an internal source	W
p	Q_r/Q_t	–
Q_2	Ducted heated or cooled air input acting at T_{av} or T_2	W
Q_w	Input at surface w due to solar and surface heating	W
Q'_w	Net conductive loss from T_w into the wall	W
q	Heat flow per unit area	W/m^2

Building Heat Transfer Morris G. Davies
© 2004 John Wiley & Sons, Ltd ISBN: 0-470-84731-X

Temperatures

[t is often used elsewhere, especially when its value is in relation to 0°C]

T_{ap}	Local or point air temperature	°C
T_a, T_{av}	Volume-averaged air temperature	°C
T_{rp}	Local or point observable radiant temperature	°C or K
T_{rv}	Volume-averaged observable radiant temperature	°C or K
T_{mp}	Local net observable temperature	°C
T_{rs}	Radiant star temperature, taken to be an estimate of T_{rv}	°C or K
T_{ra}	Rad-air temperature, composed of T_{av} and T_{rs}	
T_c	Dry resultant or volume-averaged comfort temperature, $\frac{1}{2}T_{av} + \frac{1}{2}T_{rv}$ or $\frac{1}{2}T_{av} + \frac{1}{2}T_{rs}$	°C
T_w	Area-averaged temperature of wall (or floor or ceiling) surface w	°C
T_{wb}	Black-body equivalent temperature of surface w	°C
T_e	Ambient (outdoor) temperature	°C
T_{saw}	Sol-air temperature for surface w	°C

Geometrical and physical parameters

t	Time	s

[τ is often used, especially when temperature is written as t]

δ	Time increment (usually 1 h)	s
δ	Convective film thickness, cavity width	m
P	Period of sinusoidal excitation, usually 24 h	s

Wall measures

A_w	Area of surface w	m^2
X_L	Thickness of layer L in a wall	m
n	Number of layers in the wall, including any cavity	–
λ_L	Thermal conductivity of layer L	W/m K

[k is often used for thermal conductivity]

r_L	Layer resistance X_L/λ_L	m^{2}K/W
r	Wall total resistance, $\sum X_L/\lambda_L$	m^{2}K/W
R	Wall resistance	K/W
C	Wall conductance, $1/R$	W/K
U_w	Steady-state transmittance of wall w (index-to-index temperature)	W/m^{2}K

[k is used for transmittance in European literature]

ρ_L	Density of layer L	kg/m^3
c_{pL}	Specific heat of layer L	J/kg K
c_L	Layer capacity $\rho_L c_{pL} X_L$	J/m^{2}K
c	Wall total capacity	J/m^{2}K
C	Wall total capacity	J/K

κ	Layer diffusivity $\lambda/\rho c_p$	m^2/s
[α is often used for diffusivity]		
$\mathbf{a}_L$	Characteristic admittance of layer L, $(2\pi\lambda_L\rho_L c_{pL}/P)^{1/2}$	W/m^2K and $\pi/4$ phase lead
τ	Cyclic thickness of layer L, $(\pi\rho_L C_{pL} X_L{}^2/\lambda_L P)^{1/2}$	–
V	Transient thermal thickness $(\rho_L C_{pL} X_L{}^2/4\pi\lambda_L\delta)^{1/2}$	–
$\mathbf{u}$	Dynamic or cyclic wall transmittance, usually for 24 h excitation	W/m^2K and phase lag
[fU is used in CIBSE literature]		
$\mathbf{y}$	Wall admittance, usually for 24 h excitation	W/m^2K and phase lead
[Y is used in CIBSE literature]		
$\phi_{n0,i}$	Response unit factor, heat flow at node n due to a triangular temperature pulse acting at node 0 at time level i	W/m^2K
N	Number of wall transfer coefficients needed, typically up to 6	–
a_k, c_k	Outside and inside admittance transfer coefficients, $0 \le k \le N$	W/m^2K
b_k	Transmittance transfer coefficient, $0 \le k \le N$	W/m^2K
d_k	Flux transfer coefficient, $1 \le k \le N$ $(d_0 = 1)$	–
J	$\sum d_k$	–

Room or rectangular space

L, W, H	Length, width, height	m
V_r	Room volume	m^3

(In a cavity when $W \ll H$, δ may be used in place of W)

Convective measures

μ	Viscosity	$[(kg\,m\,s^{-2}/m^2]/[(m/s)/m]$ or kg/m s
[η is often used to denote viscosity]		
ν	Kinematic viscosity μ/ρ	m^2/s
β	Coefficient of expansion	K^{-1}
λ	Thermal conductivity	W/m K
c_p	Specific heat at constant pressure	J/kg K
δ	Width of convective boundary layer	m
h_{wc}	Convective coefficient at inside enclosure surface w	W/m^2K
[α is often used for a film coefficient]		
r_{we}	Film resistance at the exterior surface of wall w	m^2K/W

Radiative measures

α_w	Absorptivity of surface w	–

h_r	Radiative heat transfer coefficient, $4\sigma T_{rs}{}^3$, about 5.7 indoors	W/m^{2}K
F_{jk}	View factor from surface j to surface k	–
G_{jk}	Geometrical conductance between surfaces j and k	m^2
b_w	A_w/total surface area of a rectangular room	–
β_w	$\dfrac{\text{Radiant conductance between } T_{wb} \text{ and an enveloping black-body surface}}{\text{Radiant conductance between } T_{wb} \text{ and the radiant star node } T_n}$ estimated as $1 - b_w - 3.54(b_w{}^2 - \frac{1}{2}b_w) + 5.03(b_w{}^3 - \frac{1}{4}b_w)$	
ε_w	Emissivity of surface w	–
E_w	$((1-\varepsilon_w)/\varepsilon_w + \beta_w)^{-1}$	–
α'_w	Short-wave absorptivity of surface w	–

Conductances

C_w	Convective conductance between T_w and T_{av}, equal to $A_w h_{cw}$	W/K
S_w	Radiant conductance between T_w and T_{rs}, equal to $A_w E_w h_r$	W/K
H_w	$C_w + S_w$, combined link between T_{ra} and T_w	W/K
X	Conductance between T_{ra} and T_{av}, equal to $(1+\alpha)C$	W/K
F_w	Fabric loss conductance between T_w and T_e	W/K
L_w	$1/(H_w{}^{-1} + F_w{}^{-1}) = A_w U_w$, conductance between T_{ra} and T_c	W/K
L	Total conduction loss conductance $\sum L_w = \sum A_w U_w$	W/K
$A_{w,k}$, $B_{w,k}$, $C_{w,k}$	Transfer coefficients for a wall of area A_w	W/K
C	$\sum C_w$	W/K
S	$\sum S_w$	W/K
α	C/S	–
V	Ventilation or infiltration conductance	W/K
j	$\sqrt{-1}$	

Subscripts

a	Air
c	Convection, comfort
e, o	Ambient
0	Outer surface of a wall
n	Inner surface of a wall, number of layers in the wall
i	Time level in hours, room interior
j	Eigenvalue number; element in a discretised wall
k	Transfer coefficient number, time lapse in units of δ
L	Number of layer in a wall of n layers
r	Radiation
p	Point value
v	Volume-averaged value
w	Number of wall, floor or ceiling surface

Dimensionless groups

The symbol L represents a measurable length which depends on the geometry of the physical situation such as gap width or layer thickness.
Non-dimensionalisation of time in conduction problems:

Fourier number	F_0	$\lambda t/(\rho c_p L^2)$, [usually written Fo]
	F_1	$ht/(\rho c_p L)$
	F_2	$h^2 t/(\lambda \rho c_p)$
Grashof number	Gr	$\rho^2 \beta g \Delta T L^3/\mu^2$ or $\rho^2 \beta g q L^4/\lambda \mu^2$ and is used for natural convection
Lewis number	Le	$\mathrm{Sc}/\mathrm{Pr} = \kappa/D$

(D is the diffusion coefficient in m^2/s)

Biot number	Bi	hL/λ, where λ is the conductivity of a solid layer
Nusselt number	Nu	hL/λ, where λ is the conductivity of a fluid.

For natural convection, h can also be non-dimensionalised as

$$h/[(\lambda \rho c_p)^{1/2} (\beta g \Delta T/L)^{1/4}]$$

Prandtl number	Pr	$\mu c_p/\lambda$
Rayleigh number	Ra	$\mathrm{Gr} \cdot \mathrm{Pr}$
Boussinesq number	Bo	$\mathrm{Gr} \cdot \mathrm{Pr}^2$
Richardson number	Ri	$\mathrm{Gr} \cdot \mathrm{Re}^{-2}$
Reynolds number	Re	$\rho u L/\mu$

where u is some characteristic velocity. Re is used for forced convection. There are several dimensionless groupings which include u.

References

Acharya, S. and Jetli, R., 1990, Heat transfer due to buoyancy in a partially divided square box, *International Journal of Heat and Mass Transfer* **33**(5), 931–942. 109

Adjali, M.H., Davies, M. and Littler, J., 1998, Earth-contact heat flows: review and application of design guidance predictions, *Building Services Engineering Research and Technology* **19**(3), 111–121. 49

Ahmad, Q.T., 1998, Validation of building thermal and energy models, *Building Services Engineering Research and Technology* **19**(2), 61–66. 454

Alford, J.S., Ryan, J.E. and Urban, F.O., 1939, Effect of heat storage and variation of outdoor temperature and solar intensity on heat transfer through walls, *ASHVE Trans.* **45**, 369–396. 357

Alifanov, O.M., 1994, *Inverse Heat Transfer Problems*, Springer, Berlin. 276, 330

Ambrose, C.W., 1981, Modelling losses from slab floors, *Building and Environment* **16**(4), 251–258. 368

Andersen, K.T., 2003, Theory of natural ventilation by thermal buoyancy in one zone with uniform temperature, *Building and Environment* **38**(11), 1281–1289. 109

Anderson, B.R., 1981, On the calculation of the U-value of walls containing slotted bricks or blocks, *Building and Environment* **16**(1), 41–50. 13

Anderson, B.R., 1991a, U-values of uninsulated ground floors: relationship with floor dimensions, *Building Services Engineering Research and Technology* **12**, 103–105. 44

Anderson, B.R., 1991b, Calculation of the steady state heat transfer through a slab-on-ground floor, *Building and Environment* **26**(4), 405–415. 44

Anderson, B.R., 1993, The effect of edge insulation on the steady-state heat loss through a slab-on-ground floor, *Building and Environment* **28**(3), 361–367. 50

Arens, E.A. and Baughman, A.V., 1996, Indoor humidity and human health, Part II: Buildings and their systems, *ASHRAE Trans.* **102**(1), 212–221. 168

Arfvidsson, J. and Cunningham, M.J., 2000, A transient technique for determining diffusion coefficients in hygroscopic materials, *Building and Environment* **35**(3), 239–249. 188

ASHRAE 1993, ASHRAE Handbook of Fundamentals, American Society of Heating, Refrigeration and Air-Conditioning Engineers, Atlanta, Georgia. many citations

Athienitis, A.K., 1993, A methodology for integrated building-HVAC system thermal analysis, *Building and Environment* **28**(4), 483–496. 323, 367

Athienitis, A.K., Chandrashekar, M. and Sullivan, H.F., 1985, Modelling and analysis of thermal networks through subnetworks for multizone passive solar buildings, *Applied Mathematical Modelling*, **9**, 109–116. 367

Athienitis, A.K., Sullivan, H.F. and Hollands, K.G.T., 1986, Analytical model, sensitivity analysis, and algorithm for temperature swings in direct gain room, *Solar Energy* **36**(4), 303–312. 367

Athienitis, A.K., Sullivan, H.F. and Hollands, K.G.T., 1987, Discrete Fourier series models for building auxiliary energy loads based on network formulation techniques, *Solar Energy* **39**(3), 203–210. 367

Awbi, A.W. and Hatton, H., 1999, Natural convection from heated room surfaces, *Energy and Buildings* **30**, 233–244. 103, 104

Building Heat Transfer Morris G. Davies

© 2004 John Wiley & Sons, Ltd

ISBN: 0-470-84731-X

Awbi, A.W. and Hatton, H., 2000, Mixed convection from heated room surfaces, *Energy and Buildings* **32**, 153–166. 104

Axley, J., 2001, Surface-drag flow relations for zonal modeling, *Building and Environment* **36**(7), 843–850. 454

Axley, J. and Grot, R., 1989, The coupled airflow and thermal analysis problem in building airflow system simulation, *ASHRAE Trans.* **95**, Part 2, 621–628. 454

Aydin, O., Ünal, A. and Ayhan, T., 1999, Natural convection in rectangular enclosures heated from one side and cooled from the ceiling, *International Journal of Heat and Mass Transfer* **24**(13), 2345–2355. 105

Ayres, J.M. and Stamper, E., 1995, Historical development of building energy calculations, *ASHRAE Trans.* **101**(1), 841–849. 332

Azizi, S., Moyne, C. and Degiovanni, A., 1988, Approche expérimentale et théorique de la conductivité thermique des milieux poreux humides: I. Expérimentation, II. Théorie, *International Journal of Heat and Mass Transfer* **31**(11), 2305–2317, 2319–2330. 31

Bachmann, H., 1938, *Tabeln über Abkühlungsvorgänge einfacher Körper*, Julius Springer, Berlin. 298

Von Baeyer, H.C., 1998, *Warmth Disperses and Time Passes*, Random House, New York. 33

Bahl, S. and Liburdy, J.A., 1991, Measurement of local convective heat transfer coefficients using three-dimensional interferometry, *International Journal of Heat and Mass Transfer* **34**(4/5), 949–960. 84

Bajorek, S.M. and Lloyd, J.R., 1982, Experimental investigation of natural convection in partitioned enclosures, *American Society of Mechanical Engineers: Journal of Heat Transfer* **104**, 527–532. 108

Baker, P.H., 2003, The thermal performance of a prototype dynamically insulated wall, *Building Services Engineering Research and Technology* **24**(1), 25–34. 39

Balaras, C.A., 1996, The role of thermal mass on the cooling load of buildings. An overview of computational methods, *Energy and Buildings* **24**(1), 1–10. 217, 452

Balcomb, J.D., 1984, Passive solar research and practice, *Energy and Buildings* **7**, 281–295. 216

Balcomb, S., 1984, Living in a passive solar home, *Energy and Buildings* **7**, 309–314. 216

Barber, J.R., 1989, An asymptotic solution for short-term transient heat conduction between two similar contacting bodies, *International Journal of Heat and Mass Transfer* **32**(5), 943–949. 341

Barcs, V., 1967, Wärmetechnische Bewertung von Räumen and Gebäuden, Heizung, Luftung, *Haustechnik* **18**, 415–419. 357

Barletta, A. and Zanchini, E., 1997, Hyperbolic heat conduction and local equilibrium: a second law analysis, *International Journal of Heat and Mass Transfer* **40**(5), 1007–1016. 25, 291

Bartak, M., Beausoleil-Morrison, I., Clarke, J.A., Denev, J., Drkal, F., Lain, M., Macdonald, I.A., Melikov, A., Popiolek, Z. and Stankov, P., 2002, Integrating CFD and building simulation, *Building and Environment* **37**(8/9), 865–871. 458

Batty, W.J., O'Callaghan, P.W. and Probert, S.D., 1981, Apparent thermal conductivity of glass-fibre insulant: effect of compression and moisture content, *Applied Energy* **9**, 55–76. 181

Baughman, A.V. and Arens, E.A., 1996, Indoor humidity and human health Part I: Literature review of health effects of humidity-influenced indoor pollutants, *ASHRAE Trans.* **102**(1), 193–211. 168

Bauman, F., Gadgil, A., Kammerud, R., Altmayer, E. and Nansteel, M., 1983, Convective heat transfer in buildings: recent research results, *ASHRAE Trans.* **89**(1A), 215–233. 102, 108

Beausoleil-Morrison, I., 2001, An algorithm for calculating convection coefficients for internal building surfaces for the case of mixed flow in rooms, *Energy and Buildings* **33**(4), 351–361. 105

Beausoleil-Morrison, I., 2002, The adaptive simulation of convective heat transfer at internal building surfaces, *Building and Environment* **35**(8–9), 791–806. 105

Beausoleil-Morrison, I. and Strachan, P., 1999, On the significance of modelling internal surface convection in dynamic whole-building simulation programs, *ASHRAE Trans.* **105**(2), 929–940. 104

Beck, J.V., 1970, Nonlinear estimation applied to the nonlinear inverse heat conduction problem, *International Journal of Heat and Mass Transfer* **13**, 703–716. 276

Beck, J.V., 1984, Green's function solutions for transient heat conduction problems, *International Journal of Heat and Mass Transfer* **27**(8), 1235–1244. 282

Beck, J.V. and Arnold, K.J., 1977, *Parameter Estimation in Engineering and Science*, John Wiley & Sons, Inc., New York. 276

Beck, J.V. and Litkouhi, B., 1988, Heat conduction numbering system for basic geometries, *International Journal of Heat and Mass Transfer* **31**(3), 505–515. 226

Beck, J.V., Blackwell, B. and St Clair, C.R., 1985, *Inverse Heat Conduction*, John Wiley & Sons, Inc., New York. 276

Beck, J.V., Blackwell, B. and Haji-sheikh, A., 1996, Comparison of some inverse heat conduction methods using experimental data, *International Journal of Heat and Mass Transfer* **39**(17), 3649–3657. 276

Beck, K., Al-Mukhtar, M., Rozenbaum, O. and Rautureau, M., 2003, Characterization, water transfer properties and deterioration in tuffeau: building material in the Loire valley, France, *Building and Environment* **38**(9/10), 1151–1162. 190

Becker, R., 1984, Condensation and mould growth in dwellings–parametric and field study, *Building and Environment* **19**(4), 243–250. 168

Bedford, T., 1964, *Basic Principles of Ventilation and Heating*, H.K. Lewis, *London*. 2

Bejan, A., 1980, A synthesis of analytical results for natural convection heat transfer across rectangular enclosures, *International Journal of Heat and Mass Transfer* **23**, 723–726. 94

Bejan, A., 1993, How to distribute a finite amount of insulation on a wall with nonuniform temperature, *International Journal of Heat and Mass Transfer* **36**(1), 49–56. 14

Bejan, A. and Lage, J.L., 1990, The Prandtl number effect on the transition in natural convection along a vertical surface, *American Society of Mechanical Engineers: Journal of Heat Transfer* **112**, 787–790. 91

Bellia, L. and Minichiello, F., 2003, A simple evaluator of building envelope moisture condensation according to a European Standard, *Building and Environment* **38**(3), 457–468. 177

Benard, C., 1986, Optimisation de la représentation réduite d'une paroi thermique, *International Journal of Heat and Mass Transfer* **29**(4), 529–538. 408

Bénard, C., Body, Y., Delisée, M., Depoid, C. and Gobin, D., 1990, Identification de l'erreur de mesure par thermocouple de la température d'une surface soumise a différentes conditions d'échanges, *International Journal of Heat and Mass Transfer* **33**(5), 785–796. 15

Bénard, H., 1901, Les tourbillons cellulaires dans une nappe liquide transportant de la chaleur par convection en régime permanent, *Annales de Chimie et de Physique*, **23**, 62–144. 95

Bernier, M.A. and Bourret, B., 1997, Effects of glass plate curvature on the *U*-factor of sealed insulated glazing units, *ASHRAE Trans.* **103**(1), 270–277. 100

Beuken, C.L., 1936, Wärmeverluste bei periodisch betriebenen elekrischen Öfen, Dissertation Bergakademie, Freiburg. 241

Bhattacharya, M.C., 1985, An explicit conditionally stable finite difference equation for heat conduction problems, *International Journal of Numerical Methods in Engineering*, **21**, 239–265. 238

Bhattacharya, M.C., 1993, Improved finite-difference solutions of the diffusion equation for heat conduction, *Communications in Numerical Methods in Engineering* **9**, 713–720. 238

Bhavnani, S.H. and Bergles, A.E., 1990, Effect of surface geometry and orientation on laminar natural convection heat transfer from a vertical flat place with transverse roughness elements, *International Journal of Heat and Mass Transfer* **33**(5), 965–981. 81

Billington, N.S., 1947, Solar heat gain through windows, *Journal of the Royal Institute of British Architects* **54**, 177–180. xviii

Billington, N.S., 1967, *Building Physics*, Pergamon, Oxford. 170

Billington, N.S., 1987, The evolution of environmental temperature, *Building and Environment* **22**(4), 241–249. 159

Blackwell, B.F., 1983, Some comments on Beck's solution of the inverse problem of heat conduction through the use of Duhamel's theorem, *International Journal of Heat and Mass Transfer* **26**(2), 302–305. 276

Bland, B.H., 1992, Conduction in dynamic thermal models: analytical tests for validation, *Building Services Engineering Research and Technology* **13**(4), 197–208. 294

BLAST (Building Loads and System Thermodynamics) Support Office, 1986, University of Illinois, *Urbana-Champaign*. 451

Bliss, R.W., 1961, Atmospheric radiation near the surface of the ground: a summary for engineers, *Solar Energy* **5**, 103–120. 58, 197

Boehrer, B., 1997, Convection in a long cavity with differentially heated end walls, *International Journal of Heat and Mass Transfer* **40**(17), 4105–4114. 101

Bohn, M.S. and Anderson, R., 1986, Temperature and heat flux distribution in a natural convection enclosure flow, *American Society of Mechanical Engineers: Journal of Heat Transfer* **108**, 471–475. 105

Bohn, M.S., Kirkpatrick, A.T. and Olson, D.A., 1984, Experimental study of three-dimensional natural convection high-Rayleigh number, *American Society of Mechanical Engineers: Journal of Heat Transfer* **106**, 339–345. 106

Boji'c, M.L. and Loveday, D.L., 1997, The influence on building thermal behavior of the insulation/masonry distribution in a three-layered construction, *Energy and Buildings* **26**, 153–157. 14

Boltzmann, L., 1884, Ableitung des Stefan'schen Gesetzes, betreffend die Abhängigkeit der Wärmestahlung von der Temperature aus der electromagnetischen Lichttheorie, *Wiedermanns Annalen* **22**, 291–294. 115

Bong, T.Y., Xue, H. and Liew, H.C., 1998, Predicting window condensation potential for a large viewing gallery, *Building and Environment* **33**(2/3), 143–150. 173

Bouzidi, M., 1991, Diffusion thermique non-stationnaire dans des milieux multicouches et problème aux valeurs propres, 1. Un paroi unique multicouches, *International Journal of Heat and Mass Transfer* **34**(4/5), 1259–1270. 280

Boyer, H., Lauret, A.P., Adelard, L. and Mara, T.A., 1999, Building ventilation: a pressure airflow model computer generation and elements of validation, *Energy and Buildings* **29**, 283–292. 107, 454

Brager, G.S. and de Dear, R.J., 1998, Thermal adaptation in the built environment: a literature survey, *Energy and Buildings* **27**(1), 83–96. 7

Braun, P.O., Goetzberger, A., Schmid, J. and Stahl, W., 1992, Transparent insulation of building facades – steps from research to commercial applications, *Solar Energy* **49**(5), 413–427. 16

BREEZE6.0d, 1993, User-Manual, Building Research Establishment (BRE). 454

Briggs, D.G. and Jones, D.N., 1985, Two-dimensional periodic natural convection in a rectangular enclosure of aspect ratio one, *American Society of Mechanical Engineers: Journal of Heat Transfer* **107**, 850–854. 101

Brinkworth, B.J., 1997, Collector sizing for solar heating of indoor swimming pools, *Building Services Engineering Research and Technology* **18**(4), 209–213. 39

Brisken, W.R. and Reque, S.G, 1956, Heat load calculations by thermal response, *ASHVE Trans.* **62**, 391–424. 371

Brown, W.G., 1962, Natural convection through rectangular openings in partitions: (2) horizontal partitions, *International Journal of Heat and Mass Transfer* **5**, 869–881. 109

Brown, W.G. and Solvason, K.R., 1962, Natural convection through rectangular openings in partitions: (1) vertical partitions, *International Journal of Heat and Mass Transfer* **5**, 859–868. 107

Brown, W.C. and Stephenson, D.G., 1993, A guarded hot box procedure for determining the dynamic response of full-scale wall specimens, part I, *ASHRAE Trans.* **99**(1), 632–642. 263

Bruckmayer, F., 1940, Die 'gleichspeichernde' Ziegeldicke, *Gesundheits-Ingenieur*, **63**, 61–65. 313, 314

Bruckmayer, F., 1951a, The cooling and warming of buildings, *Building Research Congress*, Division 3, Part II, pp. 66–74. 314

Bruckmayer, F., 1951b, Record of Discussion, *Building Research Congress*, p. 109. 314

Brush, S.G., 1976, *The Kind of Motion We Call Heat*, Book 1, North Holland, Amsterdam. 22

Buchberg, H., 1955, Electric analogue prediction of the thermal behaviour of an inhabitable enclosure, *ASHRAE Trans.* **61**, 339–386. 125

Buckingham, E., 1915, Model experiments and the forms of empirical equations, *Transactions of the American Society of Mechanical Engineers*, **37**, 263–296. 78

Budaiwi, I., El-Diasty, R. and Abdou, A., 1999, Modelling of moisture and thermal transient behaviour of multi-layer non-cavity walls, *Building and Environment* **34**(5), 537–551. 421

Budyko, M.I. 1958, *The Heat Balance of the Earth's Surface*. English translation by N. Stepanova, US Department of Commerce Weather Bureau, Washington DC. 206

Burberry, P.J., Letherman, K.M. and Valeri, D.A., 1979, Prediction techniques for seasonal energy consumption with special reference to control systems and installation performance, *Proceedings of the Second International CIB Symposium on Energy Consumption*, Copenhagen, May 1979. 242

Burch, D.M., 1993, An analysis of moisture accumulation in walls subject to hot and humid climates, *ASHRAE Trans.* **99**, Part 2, 1013–1022. 178

Burch, D.M. and TenWolde, A., 1993, A computer analysis of moisture accumulation in the walls of manufactured housing, *ASHRAE Trans.* **99**(2), 977–990. 178, 421

Burch, D.M., Seem, J.E., Walton, G.N. and Licitra, B.A., 1992a, Dynamic evaluation of thermal bridges in a typical office building, *ASHRAE Trans.* **98**, Part 1, 291–304. 373, 397, 402

Burch, D.M., Thomas, W.C. and Fanney, A.H., 1992b, Water vapor permeability measurements of common building materials, *ASHRAE Trans.* **98**(2), 486–494. 187

Burmeister, L.C., 1983, *Convective Heat Transfer*, John Wiley & Sons, Inc., New York. 86, 91, 95

Busch, J.F., 1992, A tale of two populations: thermal comfort in air-conditioned and naturally ventilated offices in Thailand, *Energy and Buildings* **18**(3–4), 235–249. 6

Butler, R., 1984, The computation of heat flows through multi-layer slabs, *Building and Environment* **19**(3), 197–206. 372

Cammerer, J.S., 1952, Die Berechnung der Wasserdampfdiffusion in den Wänden, *Gesundheits-Ingenieur*, **73**, 393. 170

Campbell, J., 1990, Calculation of heat requirements with intermittent heating, *ASHRAE Trans.* **96**(1), 120–123. 364

Candau, Y. and Piar, G., 1993, An application of spectral decomposition to model validation in the thermal analysis of buildings, *International Journal of Heat and Mass Transfer* **36**(3), 645–650. 330

Cannistraro, G., Franzitta, G., Giaconia, C. and Rizzo, G., 1992, Algorithms for the calculation of the view factors between human body and rectangular surfaces in parallelepiped environments, *Energy and Building* **19**(1), 51–60. 122

Carroll, J.A., 1980, An 'MRT method' of computing radiant energy exchange in rooms, *Proceedings of the Second Systems Simulation and Economic Analysis Conference*, San Diego, pp. 343–348. 132

Carroll, J.A., 1981, A comparison of radiant interchange algorithms, *Proceedings of the ASME Solar Energy Division Third Annual Conference on Systems Simulation, Economic Analysis/Solar Heating and Cooling Operational Results*, Reno, Nevada. 132, 133, 165

Carslaw, H.S., 1906, *Introduction to the Theory of Fourier's Series and Integrals*, Macmillan, London. 56

Carslaw, H.S. and Jaeger, J.C., 1959, *Conduction of Heat in Solids*, Clarendon Press, Oxford. xviii, 51, 265f, 287, 288, 299, 302, 305, 312, 317, 394

CEN, 1995, Thermal Performance of Buildings, Internal temperatures in summer of a room without mechanical cooling – General criteria and calculation procedures, ISO/DIS 13791:1995. 451, 452

CEN, 1998, Thermal Performance of Buildings, Revised CSTB proposal for an example of solution technique, CEN TC 89 WG6, Work item 39, November 1998. 328

Ceylan, T. and Myers, G.E., 1980, Long-term solutions to heat-conduction transients with time-dependent inputs, *American Society of Mechanical Engineers:, Journal of Heat Transfer* **102**(1), 115–120. (Ceylan's name is included in the list of technical papers but not on the article itself.) 226

Chadwick, M.L., Webb, B.W. and Heaton, H.S., 1991, Natural convection from two-dimensional discrete heat sources in a rectangular enclosure, *International Journal of Heat and Mass Transfer* **34**(7), 1679–1693. 106

Chang, W.-J. and Weng, C.-I. 2000, An analytical solution to coupled heat and moisture diffusion transfer in porous materials, *International Journal of Heat and Mass Transfer* **43**(19), 3621–3632. 421

Chapman, S. and Cowling, T.G., 1970, *The Mathematical Theory of Non-Uniform Gases*, 3rd edn, Cambridge University Press, London. 24, 27

Chen, F. and Wu, C.H., 1993, Unsteady convective flows in a vertical slot containing variable viscosity liquids, *International Journal of Heat and Mass Transfer* **36**(17), 4233–4266. 99

Chen, H.-T. and Lin, J.-Y., 1998, Simultaneous estimations of temperature-dependent thermal conductivity and heat capacity, *International Journal of Heat and Mass Transfer* **41**(14), 2237–2244. 276

Chen, P., and Pei, D.C.T., 1989, A mathematical model of drying processes, *International Journal of Heat and Mass Transfer* **32**(2), 297–310. 189

Chen, T.S. and Tzuoo, K.L., 1982, Vortex instability of free convection over horizontal and inclined surfaces, *American Society of Mechanical Engineers: Journal of Heat Transfer* **104**, 637–643. 83

Chen, T.S., Armaly, B.F. and Ramachandran, N., 1986, Correlations for laminar mixed convection flows on vertical, inclined, and horizontal flat plates, *American Society of Mechanical Engineers: Journal of Heat Transfer* **108**, 835–840. 113

Chen, T.Y. and Athienitis, A.K, 1993, Computer generation of semi-symbolic thermal network functions of buildings, *Building and Environment* **28**(3), 301–309. 331

Chen, T.Y and Athientis, A.K. 2003, Investigation of practical issues in building thermal parameter determination, *Building and Environment* **38**(8), 1027–1038. 435

Chen, Y. and Wang, S., 2001, Frequency-domain regression method for estimating CTF models of building multilayer constructions, *Applied Mathematical Modelling* **25**(7), 579–592. 400

Chi, D., Park, C.B., Kim, S.T. and Woo, S.B., 1991, The correction of water vapor pressure according to the International Temperature Scale of 1990, *ASHRAE Trans.* **97**(2), 293–297. 192

Choudhury, N.K.D. and Warsi, Z.U.A., 1964, Weighting function and transient thermal response of buildings. Part I: Homogeneous structure, *International Journal of Heat and Mass Transfer* **7**, 1309–1921. 307

Chuangchid, P. and Krarti, M., 2001, Foundation heat loss from heated concrete slab-on-grade floors, *Building and Environment* **36**(5), 637–655. 369

Churchill, R.V., 1958, *Operational Mathematics*, McGraw-Hill, New York. xix, 269, 371

Churchill, R.V., Brown, J.W. and Verhey, R.F., 1974, *Complex Variable and Applications*, McGraw-Hill, New York. 414

Churchill, S.W. and Chu, H.H.S., 1975, Correlating equations for laminar and turbulent free convection from a vertical plate, *International Journal of Heat and Mass Transfer* **18**, 1323–1329. 92

CIBSE, 1986, CIBSE Guide, Book A, Chartered Institution of Building Services Engineers, London. many citations

CIBSE, 1999, CIBSE Environmental Design Guide, Book A, Chartered Institution of Building Services Engineers, London. many citations

Ciofalo, M. and Karayiannis, T.G., 1991, Natural convection heat transfer in a partially – or completely – partitioned vertical rectangular enclosure, *International Journal of Heat and Mass Transfer* **34**(1), 167–179. 109

Citherlet, S. and Hand, J., 2002, Assessing energy, lighting, room acoustics, occupant comfort and environmental impact performance of buildings with a single simulation program, *Building and Environment* **37**(8/9), 845–856. 458

Claesson, J. and Hagentoft, C.-E. 1991, Heat loss to the ground from a building, I. General theory, *Building and Environment* **26**(2), 195–208. 40, 370

Clarke, J.A., 1985, *Energy Simulation in Building Design*, Adam Hilger, Bristol. xx, 148, 236, 458

Clarke, J.A., 2001, *Energy Simulation in Building Design*, 2nd edn, Butterworth-Heinemann, Oxford. xx, 19, 76, 104, 148, 167, 212, 364, 454, 458

Clarke, J.A., Johnstone, C.M., Kelly, N.J., McLean, R.C., Anderson, J.A., Rowan, N.J. and Smith, J.E., 1999, A technique for the prediction of the conditions leading to mould growth in buildings, *Building and Environment* **34**(4), 515–521. 167

Clausing, A.M. and Berton, J.J., 1989, An experimental investigation of natural convection from an isothermal horizontal plate, *American Society of Mechanical Engineers: Journal of Heat Transfer* **111**, 904–908. 85

Cleaveland, J.P. and Akridge, J.M., 1990, Slab-on-grade thermal loss in hot climates, *ASHRAE Trans.* **96**(1), 112–119. 50

Coelho, P.J., Goncalves, J.M., Carvalho, M.G. and Trivic, D.N., 1998, Modelling of radiative heat transfer in enclosures with obstacles, *International Journal of Heat and Mass Transfer* **41**(4/5), 745–756. 127

Cole, R.J., 1976, The longwave radiative environment around buildings, *Building and Environment* **11**, 3–13. 142

Cole, R.J., 1979, The longwave radiation incident upon inclined surfaces, *Solar Energy* **22**, 459–462. 59

Coley, D.A. and Penman, J.M., 1996, Simplified thermal response modelling in building energy management. Paper III: Demonstration of a working controller, *Building and Environment* **31**(2), 93–97. 330

Collins, K.J. and Hoinville, E., 1980, Temperature requirements in old age, *Building Services Engineering Research and Technology* **1**(4), 165–172. 5

Crabb, J.A., Murdoch, N. and Penman, J.M., 1987, A simplified thermal response model, *Building Services Engineering Research and Technology* **8**, 13–19. 330

Crank, J., 1975, *The Mathematics of Diffusion*, Clarendon Press, Oxford. 277, 418

Crank, J. and Nicolson, P., 1947, A practical method for numerical evaluation of solutions of partial differential equations of heat conduction type. *Proceedings of the Cambridge Philosophical Society*, **43**, 50–67. 238

Cunningham, M.J., 1983, A new analytical approach to the long time behaviour of moisture concentrations in building cavities: I. Non-condensing cavity, II Condensing cavity, *Building and Environment* **18**, 109–116 and 117–124. 178

Cunningham, M.J., 1984, Further analytical studies of building cavity moisture concentrations, *Building and Environment* **19**(1), 21–29. 178

Cunningham, M.J., 1988, The moisture performance of framed structures, a mathematical model, *Building and Environment* **23**(2), 123–135. 420

Cunningham, M.J., 1990a, Modelling of moisture transfer in structures: I. A description of a finite-difference nodal model, *Building and Environment* **25**(1), 55–61. 420

Cunningham, M.J., 1990b, Modelling of moisture transfer in structures: II. A comparison of a numerical model, an analytical model and some experimental results, *Building and Environment* **25**(2), 85–94. 420

Daian, J.-F. and Saliba, J., 1991, Determination d'un réseau aleatoire de pores pour modéliser la sorption et la migration d'humidité dans un mortier de ciment, *International Journal of Heat and Mass Transfer* **34**(8), 2081–2096. 171

Daniels, P.G. and Wang, P., 1994, Numerical study of thermal convection in tall laterally heated cavities, *International Journal of Heat and Mass Transfer* **37**(3), 375–386. 99

Danter, E. 1960, Periodic heat flow characteristics of simple walls and roofs, *Journal of the Institute of Heating and Ventilating Engineers* **28**, 136–146. 357, 358

Danter, E., 1974, Heat exchanges in a room and the definition of room temperature, *Building Services Engineering* **41**, 231–243. 158, 358

Danter, E., 1983, Room response according to the *CIBS Guide* procedures, *Building Services Engineering Research and Technology* **4**(2), 46–51. 358

Dascalaki, E. and Santamouris, M., 1999, Models for natural ventilation, *Building Services Engineering Research and Technology* **20**(2), B12–15. 454

Dascalaki, E., Santamouris, M., Balaras, C.A. and Asimakopoulos, D.N., 1994, Natural convection heat transfer coefficients from vertical and horizontal surfaces for building applications, *Energy and Buildings* **20**(3), 243–249. 81

Davenport, A.G., 1965, The relationship of wind structure to wind loading, *Proceedings of the National Physical Laboratory Symposium No. 16: Wind Effects on Buildings and Structures*, Vol. 1, pp. 54–115, HMSO, London. 113

Davies, A.D.M. and Davies, M.G., 1995, The adaptive model of thermal comfort: patterns of correlation, *Building Services Engineering Research and Technology* **16**(1), 51–53. 7

Davies, M., Tindale, A. and Littler, J. 1995, Importance of multi-dimensional conductive heat flows in and around buildings, *Building Services Engineering Research and Technology* **16**(2), 83–90. 370

Davies, M.G., 1978, Structure of the transient cooling of a slab, *Applied Energy* **4**, 87–126. 298

Davies, M.G., 1980, Useful solar gains through a south-facing window in the U.K. climate, *Building and Environment* **15**, 253–272. 216, 360

Davies, M.G., 1982a, Use of meteorological/constructional statistics to estimate building heat needs, *Building and Environment* **17**(4), 263–272. 216

Davies, M.G., 1982b, Transmission and storage characteristics of walls experiencing sinusoidal excitation, *Applied Energy* **12**, 269–316. 343, 345

Davies, M.G., 1983a, Optimum design of resistance and capacitance elements in modelling a sinusoidally excited building wall, *Building and Environment* **18**, 19–37. 316, 383, 408

Davies, M.G., 1983b, Transmission and storage characteristics of sinusoidally excited walls – a review, *Applied Energy* **15**, 167–231. 242, 357

Davies, M.G., 1983c, Optimal designs for star circuits for radiant exchange in a room, *Building and Environment* **18**(3), 135–150. 130, 152

Davies, M.G., 1984, The heat storage/loss ratio for a building and its response time, *Applied Energy* **18**, 179–238. 317, 319

Davies, M.G., 1985, Similarity between unsteady conduction and natural convection, *International Journal of Heat and Mass Transfer* **28**(12), 2385–2388. 82

Davies, M.G., 1986, 1987, The Passive Solar Heated School in Wallasey, seven articles in the *International Journal of Energy Research*: 216(I) Foreword and introduction, **10**, 101–120; (II) Background, preliminary analyses and patent specification, **10**, 121–136; (III) Model studies of the thermal response of a passive school building, **10**, 203–234; 302, 366(IV) An observational study of the thermal response of a passive school building, **10**, 305–332; 320(V) Energy requirements for a possible school building, **11**, 1–20; (VI) with Ann D. M. Davies, Thermal sensation and comfort in the classroom: a year-long study of the subjective and adaptive responses of children, **11**, 157–178; (VII) with Ann D. M. Davies, Window opening behaviour and the microclimate of a classroom, **11**, 315–326.

Davies, M.G., 1990, An idealised model for room radiant exchange, *Building and Environment* **25**, 375–378. 130

Davies, M.G., 1992a, Flaws in the environmental temperature model, *Building Services Engineering Research and Technology* **13**(4), 209–215. 161, 359

Davies, M.G., 1992b, The basis for a room global temperature, *Philosophical Transactions of the Royal Society of London Series A* **339**, 153–191. 126, 130 152

Davies, M.G., 1993, Heat loss from a solid ground floor, *Building and Environment* **28**(3), 347–359. 48, 56

Davies, M.G., 1994, The thermal response of an enclosure to periodic excitation – the CIBSE approach, *Building and Environment* **29**(2), 217–235. 335f

Davies, M.G., 1995, Solutions to Fourier's equation and unsteady heat flow through structures, *Building and Environment* **30**(3), 309–321. 263

Davies, M.G., 1996a, Comfort temperature: flawed status in the *CIBSE Guide*, *Building Services Engineering Research and Technology* **17**(3), 161–165. 161, 359

Davies, M.G., 1996b, A time-domain estimation of wall conduction transfer function coefficients, *ASHRAE Trans.* **102**, Part 1, 328–343. 397, 401

Davies, M.G., 1997, Wall transient heat flow using time-domain analysis, *Building and Environment* **32**, 427–446. 377, 393f

Davies, M.G., 1999, The time delay for a perceptible thermal disturbance in a slab, *American Society of Mechanical Engineers: Journal of Heat Transfer* **121**(4), 1072–1075. 287, 289, 296, 297

Davies, M.G., 2001a, Error analysis of wall heat flow using transfer coefficients, *Building and Environment* **36**(2), 189–198. 423

Davies, M.G., 2001b, Hourly estimation of temperature using wall transfer coefficients, *Building and Environment* **36**(2), 199–217. 333, 438f

Davies, M.G., 2003, A rationale for nodal placement for heat flow calculations in walls, *Building and Environment* **38**(2). 374, 403f

Davies, M.G., 2004, Wall thermal capacity and transfer coefficients, *Building and Environment* **39**(1), 109–112. 257

Davies, M.G. and Bhattacharya, M.C., 1984, Comments on Pratt's cooling solution, *International Journal of Heat and Mass Transfer* **27**, 1123. 305

Davies, M.G. and Message, P.J., 1992, The relation between the radiant star temperature in an enclosure and the mean observable temperature, *Building and Environment* **27**(1), 85–92. 137

Day, A.R. and Karayiannis, T.G., 1998, Degree-days: comparison of calculation methods, *Building Services Engineering Research and Technology* **19**(1), 7–13. 9

Day, A.R. and Karayiannis, T.G., 1999, Identification of the uncertainties in degree-day-based energy estimates, *Building Services Engineering Research and Technology* **20**(4), 165–172. 17

Dayan, A. and Gluekler, E.L., 1982, Heat and mass transfer within an intensely heated concrete slab, *International Journal of Heat and Mass Transfer* **25**(10), 1461–1467. 190, 193

de Dear, R.J., 1998, A global database of thermal comfort field experiments, *ASHRAE Trans.* **104**(1B) 1141–1152. 1

de Dear, R.J. and Brager, G.S., 1998, Developing an adaptive model of thermal comfort and preference, *ASHRAE Trans.* **104**(1A), 145–167. 6

de Freitas, V.P., Abrantes, V. and Crausse, P., 1996, Moisture migration in building walls – analysis of the interface phenomenon, *Building and Environment* **31**(2), 99–108. 420

de Graaf, J.G.A. and van der Held, E.F.M., 1953, The relation between the heat transfer and the convection phenomena in enclosed plane air layers, *Applied Science Research*, Martinus Nijhoff, The Hague, A3, pp. 393–409. 95, 96

Delaforce, S.R., Hitchin, E.R. and Watson, D.T.M., 1993, Convective heat transfer at internal surfaces, *Building and Environment* **28**(2), 211–220. 103

Delsante, A.E., 1988, Theoretical calculations of the steady-state heat losses through a slab-on-ground floor, *Building and Environment* **23**, 11–17. 44

Delsante, A.E., 1990, A comparison between measured and calculated heat losses thorough a slab-on-ground floor, *Building and Environment* **25**(1), 25–31. 369

Delsante, A.E., 1993, The effect of water table depth on steady-state heat transfer through a slab-on-ground floor, *Building and Environment* **28**(3), 369–372. 47

Delsante, A.E., Stokes, A.N. and Walsh, P.J., 1983, Application of Fourier transforms to periodic flow into the ground under a building, *International Journal of Heat and Mass Transfer*, **26**, 121–132. 44, 47, 48, 369

de Monte, F., 2000, Transient heat conduction in one-dimensional composite slab. A natural analytic approach, *International Journal of Heat and Mass Transfer* **43**(19), 3607–3619. 302

Déqué, F., Ollivier, F. and Poblador, A., 2000, Grey boxes used to represent buildings with a minimum number of geometric and thermal parameters, *Energy and Buildings* **31**(1), 29–35. 331

de Vahl Davis, G., 1968, Laminar natural convection in an enclosed rectangular cavity, *International Journal of Heat and Mass Transfer* **11**, 1675–1693. 101, 102

de Vahl Davis, G., 1983, Natural convection of air in a square cavity: a bench mark numerical solution, *International Journal for Numerical Methods in Fluids* **3**, 249–264. 102

de Vahl Davis, G. and Jones, I.P., 1983, Natural convection in a square cavity: a comparison exercise, *International Journal for Numerical Methods in Fluids* **3**, 227–248. 102

de Vries, D.A., 1958, Simultaneous transfer of heat and moisture transfer in porous media, *Transactions of the American Geophysical Union* **39**, 900–916. 419

de Vries, D.A., 1987, The theory of heat and moisture transfer in porous media revisited, *International Journal of Heat and Mass Transfer* **30**(7), 1343–1350. 170, 419

Dewson, T., Day, B. and Irving, A.D., 1993, Least squares parameter estimation of a reduced order thermal model of an experimental building, *Building and Environment* **28**(2), 127–137. 330, 373

Dines, W.H. and Dines, L.H.G., 1927, Monthly mean values of radiation from various parts of the sky at Benson, Oxfordshire, *Mem. Royal Meteorological Society*, **2**, 11. 142

Dinulescu, H.A. and Eckert, E.G.R., 1980, Analysis of the one-dimensional moisture migration caused by temperature gradients in a porous medium, *International Journal of Heat and Mass Transfer* **23**, 1069–1077. 189

Dixon, M. and Probert, S.D. 1975, Heat-transfer regimes in vertical, plane-walled, air-filled cavities, *International Journal of Heat and Mass Transfer* **18**, 709–710. 97

Duffie, J.A. and Beckman, W.A., 1980, *Solar Engineering of Thermal Processes*, John Wiley & Sons, Inc., New York. 58, 197, 199

Dufton, A.F., 1934, The warming of walls, *Journal of the Institution of Heating and Ventilating Engineers*, **2**, 416–417. 271

Dutt, G.S., 1979, Condensation in attics: are vapor barriers really the answer? *Energy and Buildings* **2**, 251–258. 173

Eames, I.W., Marr, N.J. and Sabir, H., 1997, The evaporation coefficient of water, *International Journal of Heat and Mass Transfer* **40**(12), 2963–2973. 176

Eckert, E.R.G., 1981, Pioneering contributions to our knowledge of convective heat transfer – one hundred years of heat transfer research, *American Society of Mechanical Engineers: Journal of Heat Transfer* **103**, 409–414. 78, 79

Eckert, E.R.G. and Drake, R.M., 1972, *Analysis of Heat and Mass Transfer*, McGraw-Hill, New York. 77, 95, 115, 184, 285, 290

Eckert, E.R.G. and Faghri, M., 1980, A general analysis of moisture migration caused by temperature differences in an unsaturated porous medium, *International Journal of Heat and Mass Transfer* **23**, 1613–1623. 31, 178, 188, 189, 195

Eckert, E.R.G. and Jackson, T.E., 1950, Analysis of turbulent free-convection boundary layer on flat plate, National Advisory Committee for Aeronautics, Technical Note 2207, also as Report 1015. 79, 91, 92, 94

Eckert, E.R.G. and Soehnghen, E., 1951, *Proceedings of the General Discussion on Heat Transfer*, Institution of Mechanical Engineers, London and American Society of Mechanical Engineers:, New York, pp. 321–323, 381, 387–388. 80

Ede, A.J., 1945, A new form of chart for determining temperatures in bodies of regular shape during heating or cooling, *Philosophical Magazine*, 7th series **36**, 845–851. 299

Ede, A.J., 1967a, Advances in Free Convection, in *Advances in Heat Transfer*, Vol. 4, pp. 1–64, Academic Press, New York, London. 75

Ede, A.J., 1967b, *An Introduction to Heat Transfer Principles and Calculations*, Pergamon, Oxford. 27, 91, 93, 94, 170, 174, 195

Elder, J.W., 1965, Laminar free convection in a vertical slot, *Journal of Fluid Mechanics* **23**, 77–98. 101

El Diasty, R., Fazio, P. and Budaiwi, I, 1993a, Dynamic modelling of moisture absorption and desorption in buildings, *Building and Environment* **28**(1), 21–32. 169

El Diasty, R., Fazio, P. and Budaiwi, I, 1993b, The dynamic modelling of air humidity: behaviour in a multi-zone space, *Building and Environment* **28**(1), 35–51. 454

ElSherbiny, S.M., Raithby, G.D. and Hollands, K.G.T., 1982a, Heat transfer by natural convection across vertical and inclined air layers, *American Society of Mechanical Engineers: Journal of Heat Transfer* **104**, 96–102. 99

ElSherbiny, S.M., Hollands, K.G.T. and Raithby, G.D., 1982b, Effect of thermal boundary conditions on natural convection in vertical and inclined air layers, *American Society of Mechanical Engineers: Journal of Heat Transfer* **104**, 515–520. 99

Erbs, D.G., Klein, S.A and Beckman, W.A., 1983, Estimation of degree-days and ambient temperature bin data from monthly average temperatures, *ASHRAE Journal*, June, 60–65. 9

Erhorn, H., 1990, Schimmelpilzanfälligkeit von Baumaterialien, Fraunhofer-Institut für Bauphysik, Mitteilung 196. 167

Erlandsson, M., Levin, P. and Myhre, L., 1997, Energy and environmental consequences of an additional wall insulation of a dwelling, *Building and Environment* **32**(2), 129–136. 15

Esser, W. and Krischer, O., 1930, *Die Berechnung der Anheizung und Auskühlung ebener und zylindrischer Wände*, Julius Springer, Berlin. 311

Etheridge, D.W., 1988, Modelling of air infiltration in single- and multi-cell buildings, *Energy and Buildings* **10**, 185–192. 454

Etheridge, D.W., 1998, Dynamic insulation and natural ventilation: feasibility study, *Building Services Engineering Research and Technology* **19**(4), 203–212. 39

Eto, J.H., 1988, On using degree-days to account for the effects of weather on annual energy use in office buildings, *Energy and Buildings* **12**, 113–127. 9

Fan, J., Luo, Z. and Li, Y., 2000, Heat and moisture transfer with sorption and condensation in porous clothing assemblies and numerical simulation, *International Journal of Heat and Mass Transfer* **43**, 2989–3000. 419

Fanger, P.O. 1970, *Thermal Comfort*, Danish Technical Press, Copenhagen. Reprinted 1972 by McGraw-Hill, New York. 2

Federspiel, C.C., 1998, Statistical analysis of unsolicited thermal sensation complaints in commercial buildings, *ASHRAE Trans.* **104**(1B), 912–923 7

Feustel, H., Zuercher, C., Diamond, R., Dickinson, B., Grimsrud, D. and Lipschutz, R., 1985, Temperature- and wind-induced air flow patterns in a staircase. Computer modelling and experimental verification, *Energy and Buildings* **8**, 105–122. 109

Feustel, H.E., 1990, Measurements of air permeability in multizone buildings, *Energy and Buildings* **14**(2), 103–116. 454

Feustel, H.E., 1992, Foreword to issue on COMIS (conjunction of multizone infiltration specialists model), *Energy and Buildings* **18**(2), 77. 454

Feustel, H.E., 1999, COMIS – an international multizone air-flow and contaminant transport model, *Energy and Buildings* **30**(1), 3–18. 454

Feustel, H.E. and Kendon, V.M., 1985, Infitration models for multicellular structures – a literature review, *Energy and Buildings* **8**, 123–126. 454

Feustel, H.E. and Sherman, M.H., 1989, A simplified model for predicting air flow in multizone structures, *Energy and Buildings* **13**(3), 217–230. 454

Fishenden, M. and Saunders, O.A., 1950, *An Introduction to Heat Transfer*, Clarendon Press, Oxford. 80, 299

Fisher, D.E. and Pedersen, C.O., 1997, Convective heat transfer in building energy and thermal load calculations, *ASHRAE Trans.* **103**(2), 137–148. 105

Fourier, J.B., 1822, *Théorie de la Chaleur*, Paris. Translated by A. Freeman as *The Analytical Theory of Heat*, Cambridge University Press, 1878. 263

Fusegi, T. and Farouk, B., 1990, A computational and experimental study of natural convection and surface/gas radiation in a square cavity, *American Society of Mechanical Engineers: Journal of Heat Transfer* **112**, 802–804. 102

Fusegi, T., Hyun, J.M., Kuwahara, K. and Farouk, B., 1991, A numerical study of three-dimensional natural convection in a differentially heated cubical enclosure, *International Journal of Heat and Mass Transfer* **34**(6), 1543–1557. 102

Galbraith, G.H. and McLean, R.C., 1990, Interstitial condensation and the vapour permeability of building materials, *Energy and Buildings* **14**(3), 193–196. 171, 194

Galbraith, G.H., McLean, R.C. and Tao, Z., 1993a, Vapour permeability: suitability and consistency of current test procedures, *Building Services Engineering Research and Technology* **14**(2), 67–70. 171

Galbraith, G.H., Tao, Z. and McLean, R.C., 1993b, Separation of moisture flow through porous building materials into vapour and liquid components, *Building Services Engineering Research and Technology* **14**(3), 107–113. 178

Galbraith, G.H., McLean, R.C. and Guo, J., 1998, Moisture permeability data: mathematical presentation, *Building Services Engineering Research and Technology* **19**(1), 31–36. 188

Ganzarolli, M.M. and Milanez, L.F., 1995, Natural convection in rectangular enclosures heated from below and symmetrically cooled from the sides, *International Journal of Heat and Mass Transfer* **38**(6), 1063–1073. 105

Gertis, K. and Hauser, G., 1975, Instationärer Wärmerschutz: (a) Der instationäre Wärmedurchgang durch Aussenbauteile – Grundlage und Vorschläge zur Normung. (b) Instationäre Berechnungsverfahren für den sommerlichen Wärmeschutz im Hochbau – eine zusammenfassende Darstellung auf Grund des vorliegenden Schrifttums. (c) Kenngrössen des instationären Wärmeschutzes von Aussenbauteilen – eine kritische Überprüfung der Kenngrössen – Eignung für die Neufassung von DIN 4108. *Berichte aus der Bauforschung*, no. 103, Verlag Wilhelm Ernst, Berlin. 357

Ghali, K., Jones, B. and Tracy, J., 1995, Modeling heat and mass transfer in fabrics, *International Journal of Heat and Mass Transfer* **38**(1), 13–21. 5

Gilly, B., Bontoux, P. and Roux, B., 1981, Influence des conditions thermiques de paroi sur la convection naturelle dans une cavité rectangulaire verticale, différentiellement chaufée, *International Journal of Heat and Mass Transfer* **24**(5), 829–841. 98, 101

Glaser, H., 1958a, Wärmeleitung und Feuchtigkeitsdurchgang durch Kühlraumisolierungen, *Kältetechnik* **10**, 86. 170

Glaser, H., 1958b, Temperature- und Dampfdruckverlauf in einer homogenen Wand bei Feuchtigkeitsausscheidung, *Kältetechnik* **10**, 174. 170

Glaser, H., 1958c, Vereinfachte Berechnung der Dampfdiffusion durch geschichtete Wände bei Ausscheidung von Wasser und Eis, *Kältetechnik* **10**, 358 and 386. 170

Glaser, H., 1959, Graphisches Verfahren zur Untersuchung von Diffusionsvorgängen, *Kältetechnik* **11**, 345. 170

van Gorcum, A.H., 1951, Theoretical considerations on the conduction of fluctuating heat flow, *Applied Science Research*, Martinus Nijhoff, The Hague, A2, pp. 272–280. 338, 357

Gorgolewski, M., 1996, Transparent insulating materials: steady-state model for assessing thermal performance, *Building Services Engineering Research and Technology* **17**(3), 141–146. 16

Goss, W.P. and Miller, R.G., 1989, Literature review of measurements and predictions of reflective building insulation performance, 1900–1989, *ASHRAE Trans.* **95**(2), 651–664. 11

Gouda, M.M., Danaher, S. and Underwood, C.P., 2002, Building thermal model reduction using nonlinear constrained optimisation, *Building and Environment* **37**(12), 1255–1265. 324

Gough, M., 1999, A review of new techniques in building energy and environmental modelling, U.K. Building Research Establishment, Contract BREA-42, 78pp. 435

Grant, C., Hunter, C.A., Flannigan, B. and Bravery, A.F. 1989, The moisture requirements of moulds isolated from domestic dwellings, *International Biodeterioration* **25**, 259–284. 167

Green, M.D. and Ülge, A., 1979, Frequency- and time-domain thermal response of dwellings, *Building and Environment* **14**, 107–118. 371

Greenberg, A., 1995, Development of HVAC&R for low- and high-rise buildings – one engineer's bird's eye view, *ASHRAE Trans.* **101**(1), 530–537. 1

Greenwood, J.A., 1991, Transient thermal contact resistance, *International Journal of Heat and Mass Transfer* **34**(9), 2287–2290. 275

Griffiths, I.D. and McIntyre, D.A., 1974, Sensitivity to temporal variation in thermal conditions, *Ergonomics* **17**(4), 499–507. 5

Groeber, H., 1925, Die Erwärmung und Abkühlung einfacher geometricher Körper, *Zeitschrift des Vereines Deutscher Ingenieure*, **69**, 705–711 xvi, 291.

Gruber, P. and Toedtli, J., 1989, On the optimal thermal storage capability of a homogeneous wall under sinusoidal excitations, *Energy and Buildings* **13**(3), 177–186 337.

Gryzagoridis, J., 1975, Combined free and forced convection from an isothermal vertical plate, *International Journal of Heat and Mass Transfer* **18**, 911–916. 92

Gupta, C.L., 1970, Heat transfer in buildings – a review, *Architectural Science Review*, 1–10. 357

Hagentoft, C.-E., 1988, Temperature under a house with variable insulation, *Building and Environment* **23**(3), 225–231. 50

Hagentoft, C.-E., 1996, Heat losses and temperature in the ground under a building with and without ground water flow, Building and Environment 31(1): I. Infinite ground water flow rate, 3–11, II. Finite ground water flow rate, 13–19. 50

Hagentoft, C.-E., 2002, Steady-state heat loss for an edge-insulated slab: part I, *Building and Environment* **37**(1), 19–25. 44, 51

Hagentoft, C.-E. and Claesson, J., 1991, Heat loss to the ground from a building: II. Slab on the ground, *Building and Environment* **26**(4), 395–403. 49, 50, 370

Haghighat, F. and Athienitis, A., 1988, Comparison between time domain and frequency domain computer program for building energy analysis, *Computer-Aided Design* **20**(9), 525–532. 367

Haghighat, F. and Liang, H., 1992, Determination of transient heat conduction through building envelopes - a review, *ASHRAE Trans.* **98** Pt1, 284–290. 371

Haghighat, F., Fazio, P. and Zmeureanu, R., 1988, A systematic approach for derivation of transfer function coefficients of buildings from experimental data, *Energy and Buildings* **12**, 101–111. 373

Haghighat, T., Jiang, Z. and Wang, J.C.Y., 1989, Natural convection and air flow pattern in a partitioned room with turbulent flow, *ASHRAE Trans.* **95**, Part 2, 600–610. 108

Haghighat, F., Sander, D.M. and Liang, H., 1991, An experimental procedure for deriving Z-transfer function coefficients of a building envelope, *ASHRAE Trans.* **97**, Part 2, 90–98. 373

Hagishima, A. and Tanimoto, J., 2003, Field measurements for estimating the convective heat transfer coefficient at building surfaces, *Building and Environment* **38**(7), 873–881. 112

Hall and colleagues., 1977–1995, Water movement in porous building materials, a series in *Building and Environment*: 168

Hall, C., 1977, I. Unsaturated flow theory and its applications, **12**, 117–125.

Gummerson, R.J., Hall, C. and Hoff, W.D., 1980, II. Hydraulic suction and sorptivity of brick and other masonry materials, **15**, 101–108.

Gummerson, R.J., Hall, C. and Hoff, W.D., 1981, III. A sorptivity test procedure for chemical injection damp proofing, **16**, 193–199.

Hall, C., 1981, IV. The initial surface absorption and the sorptivity, **16**, 201–207.

Hall, C., and Kalimeris, A.N., 1982, V. Absorption and shedding of rain by building surfaces, **17**, 257–262.

Hall, C., Hoff, W.D. and Nixon, M.R., 1984, VI. Evaporation and drying in brick and block materials, **19**(1), 13–20.

Hall, C. and Tse, K.-M., 1986, VII. The sorptivity of mortars, **21**(2), 113–118.

I'Anson, S.J. and Hoff, W.D., 1986, VIII. Effects of evaporative drying on height of capillary rise equilibrium in walls, **21**(3/4), 195–200.

Hall, C. and Yau, M.H.R., 1987, IX. The water absorption and sorptivity of concretes, **22**(1). 77–82.

Wilson, M.A., Hoff, W.D. and Hall, C., 1991, X. Absorption from a small cylindrical cavity, **26**(2), 143–152.

Wilson, M.A., Hoff, W.D. and Hall, C., 1994, XI. Capillary absorption from a hemispherical cavity, **29**(1), 99–104.

Wilson, M.A. and Hoff, W.D., 1994, XII, Absorption from a drilled hole with a hemispherical end, **29**(4), 537–544.

Wilson, M.A, Hoff, W.D. and Hall, C., 1995, XIII, Absorption into a two-layer composite, **30**(2), 209–219.

Wilson, M.A, Hoff, W.D. and Hall, C., 1995, XIV, Absorption into a two-layer composite ($S_A < S_B$), **30**(2), 221–227.

Hamady, F.J., Lloyd, J.R., Yang, H.Q. and Yang, K.T., 1989, Study of local natural convection heat transfer in an inclined enclosure, *International Journal of Heat and Mass Transfer* **32**(9), 1679–1708. 102

Hanby, V.I. and Dil, A.J., 1995, Stochastic modelling of building heating and cooling systems, *Building Services Engineering Research and Technology* **16**(4), 199–205. 330

Harmathy, T.Z., 1969, Simultaneous moisture and heat transfer in porous systems with particular reference to drying, *Industrial and Engineering Chemistry Fundamentals* **8**, 92–103. 184

Harrington-Lynn, J., 1974, The admittance procedure: variable ventilation, *Building Services Engineer* **42**, 199–200. 364

Harris, D.J., 1996, The moisture content of air in crawl spaces beneath suspended timber floors, *International Journal for Housing Science and its Applications* **20**(2), 109–120. 168

Harris, S.M. and McQuiston, F.C., 1988, A study to categorize walls and roofs on the basis of thermal response, *ASHRAE Trans.* **94**, Part 2, 688–715. 372

Hasan, A., 1999, Optimising insulation thickness for buildings using life cycle cost, *Applied Energy* **63**, 115–124. 14

Hastings, S.R., 1994, *Passive Solar Commercial and Institutional Buildings*, John Wiley & Sons, Ltd, Chichester. 212

Hatfield, D.W. and Edwards, D.K., 1981, Edge and aspect ratio effects on natural convection from the horizontal heated plate facing downwards, *International Journal of Heat and Mass Transfer* **24**(6), 1019–1024. 85

Hauf, W. and Grigull, U., 1970, Optical methods in heat transfer, *Advances in Heat Transfer*, Vol. 6, Academic Press, New York. 80

Häupl, P., Grunewald, J., Fechner, H. and Stopp, H., 1997, Coupled heat air and moisture transfer in building structures, *International Journal of Heat and Mass Transfer* **40**(7), 1633–1642. 190

Hawes, D.W., Feldman, D. and Banu, D., 1993, Latent heat storage in building materials, *Energy and Buildings* **20**(1), 77–86. 217

Heath, R., 1760, *Astronomia Accurata or the Royal Astronomer and Navigator*, London, Copy in the National Maritime Museum, London. 201

Heidt, F.D., Rabenstein, R. and Schepers, G., 1991, Performance and comparison of tracer gas methods for measuring airflows in two-zone buildings, *ASHRAE Trans.* **97**(2), 1078–1086. 169

Heisler, M.P., 1947, Temperature charts for induction and constant-temperature heating, *Transactions of the American Society of Mechanical Engineers* **69**, 227–236. 291, 299

Henkes, R.A.W.M. and Hoogendoorn, C.J., 1993, Scaling of the laminar natural-convection flow in a heated square cavity, *International Journal of Heat and Mass Transfer* **36**(11), 2913–2925. 101

Henry, P.S.H., 1939, Diffusion in absorbing media, *Proceedings of the Royal Society of London Series A*, **171**, 215–241. 180, 182, 418

Hens, H. and Fatin, A.M., 1995, Heat-air-moisture design of masonry cavity walls: theoretical and experimental results and practice, *ASHRAE Trans.* **101**(1), 607–626. 16, 66

Herring, H., 1999, Does energy efficiency save energy? The debate and its consequences, *Applied Energy* **63**, 209–226. xvi

Hess, C.F. and Henze, R.H., 1984, Experimental investigation of natural convection losses from open cavities, *American Society of Mechanical Engineers: Journal of Heat Transfer* **106**, 333–338. 110

Hetsroni, G., Yarin, L.P. and Kaftori, D., 1996, A mechanistic model for heat transfer from a wall to a fluid, *International Journal of Heat and Mass Transfer* **39**(7), 1475–1478. 94

Hill, P.G. and MacMillan, R.D.C., 1988, The properties of steam: current status, *American Society of Mechanical Engineers: Journal of Heat Transfer* **110**, 763–777. 193

Hitchin, E.R., 1981, Degree-days in Britain, *Building Services Engineering Research and Technology* **2**(2), 73–82. 9

Hitchin, E.R., 1983, Estimating monthly degree-days, *Building Services Engineering Research and Technology* **4**(4), 159–162. 9

Hitchin, E.R., 1990, Developments in degree-day methods of estimating energy use, *Building and Environment* **25**(1), 1–6. 9

Hittle, D.C. and Bishop, R., 1983, An improved root-finding procedure for use in calculating transient heat flow through multilayered slabs, *International Journal of Heat and Mass Transfer* **26**(11), 1685–1693. 415

Ho, C.J. and Chang, J.Y., 1994, A study of natural convection heat transfer in a vertical rectangular enclosure with two-dimensional discrete heating: effect of aspect ratio, *International Journal of Heat and Mass Transfer* **37**(6), 917–925. 98

Hoffman, M.E. and Feldman, M., 1981, Calculation of the thermal response of buildings by the total thermal time constant method, *Building and Environment* **16**, 71–85. 316

Hong, G., Irving, A.D., Dewson, T. and Day, B., 1994, Comparison of time series response factor estimators, *Energy and Buildings* **20**(3), 179–186. 373

Hong, T., Chou, S.K and Bong, T.Y., 2000, Building simulation: an overview of developments and information sources, *Building and Environment* **35**(4), 347–361. 458

Hosni, M.H, Sipes, J.M and Wallis, M.H., 1999a, Experimental results for diffusion and infiltration of moisture in concrete masonry walls exposed to hot and humid climates, *ASHRAE Trans.* **105**, Part 2, 191–203. 178

Hosni, M.H., Jones, B.W. and Xu, H., 1999b, Experimental results for heat gain and radiant/convective split from equipment in buildings, *ASHRAE Trans.* **105**(2), 525–539. 60

Hottel, H.C., 1976, A simple model for estimating the transmittance of direct solar radiation through clear atmospheres, *Solar Energy* **18**, 129–134. 204

Hottel, H.C. and Sarofim, A.F., 1967, *Radiative Transfer*, McGraw-Hill, New York. 120

Houghton, F.C., Blackshaw, J.L., Pugh, E.M. and McDermott, P., 1932, Heat transmission as influenced by heat capacity and solar radiation, *ASHVE Trans.* **38**, 231–284. 357

Howell, J., 2000, The millennium issue of JHT: Views by members of the ASME Heat Transfer Division on Heat Transfer in the new millennium, *American Society of Mechanical Engineers: Journal of Heat Transfer* **122**(1), 1–6. xx, 276

Howell, J.R., 1968, Application of Monte Carlo to heat transfer problems, *Advances in Heat Transfer*, Vol. 5, pp. 1–54, eds J.P. Hartnett and T. Irvine, Academic Press, San Diego. 122

Howell, J.R., 1998, The Monte Carlo method in radiative heat transfer, *American Society of Mechanical Engineers: Journal of Heat Transfer* **120**, 547–560. 122

Howieson, S.G., Lawson, A., McSharry, C., Morris, G., McKenzie, E. and Jackson, J., 2003, Domestic ventilation rates, indoor humidity and dust mite allergens: are our homes causing the asthma pandemic?, *Building Services Engineering Research and Technology* **24**(3), 137–147. 168

Hsieh, C.K. and Yang, S.L., 1984, A new theory on the critical thickness of insulation, *American Society of Mechanical Engineers: Journal of Heat Transfer* **106**, 648–652. 16

Huizenga, C., Hui, Z. and Arens, E., 2001, A model of human physiology and comfort for assessing complex thermal environments, *Building and Environment* **36**(6), 691–699. 6

Humphreys, M.A. and Nicol, J.F., 1998, Understanding the adaptive approach to thermal comfort, *ASHRAE Trans.* **104**, Part 1B, 991–1004. 6

Humphreys, M.A. and Nicol, J.F., 2000, Outdoor temperature and indoor thermal comfort: raising the precision of the relationship for the 1998 ASHRAE database of field studies, *ASHRAE Trans.* **106**, Part 2, 485–492. 7

IHVE, 1970, *Institution of Heating and Ventilating Engineers Guide Book A*, IHVE, London. (IHVE later became CIBSE). several citations

Ilic, M. and Turner, I.W., 1989, Convective drying of a consolidated slab of wet porous material, *International Journal of Heat and Mass Transfer* **32**(12), 2351–2362. 421

International Energy Agency, 1992, Annex XX. 454

International Energy Agency, 1997, *Solar Energy Houses: Strategies, Technologies, Examples*, edited by Hestnes, A.G., Hastings, R., and Saxhof, James and James, London. 212

Iqbal, M., 1983, *An Introduction to Solar Radiation*, Academic Press, Toronto. 197, 199

Irvine, T.F., 1963, Thermal radiation properties of solids, in *Modern Developments in Heat Transfer*, pp. 213–224, ed. W. Ibele, Academic Press, London. 118

Irving, A.D., 1988, Validation of dynamic thermal models, *Energy and Buildings* **10**, 213–220. 454

Irving, A.D., 1992, Dynamic response factor estimation: a point algebraic method, *ASHRAE Trans.* **98**(2), 79–85. 373

Ito, N., Kimura, K. and Oka, J., 1972, A field experiment study on the convective heat transfer coefficient on exterior surface of a building, *ASHRAE Trans.* **87**(1), 184–191. 111

Jakob, M., 1949, *Heat Transfer*, John Wiley & Sons, Inc., New York. 31

Jaeger, J.C., 1945, Conduction of heat in a slab in contact with well-stirred fluid, *Proceedings of the Cambridge Philosophical Society* **41**, 43–49. 302

Jannot, M. and Kunc, T., 1998, Onset of transition to turbulence in natural convection with gas along a vertical isotherm plane, *International Journal of Heat and Mass Transfer*, **41**(24), 4327–4340. 80

Jayamaha, S.E.G, Wijeysundera, N.E. and Chou, S.K., 1996, Measurement of the heat transfer coefficient for walls, *Building and Environment* **31**(5), 399–407. 110

Jayamaha, S.E.G, Wijeysundera, N.E. and Chou, S.K., 1997, Effect of rain on the heat gain through building walls in tropical climates, *Building and Environment* **32**(5), 465–477. 433

Jeans, J.H., 1954, *The Dynamical Theory of Gases*, Dover, New York. Reprint of the fourth edition from 1925. 24

Jeans, J.H., 1962, *An Introduction to the Kinetic Theory of Gases*, Cambridge University Press, London. Reprint of the first edition from 1940. 24

Jesperson, H.B., 1953, Thermal conductivity of moist materials and its measurement, *Journal of the Institute of Heating and Ventilating Engineers* **21**, 157–174. 31

Jones, R., 1993, Modelling water vapour conditions in buildings, *Building Services Engineering Research and Technology* **14**(3), 99–106. 169

Jones, R., 1995, Indoor humidity calculation procedures, *Building Services Engineering Research and Technology* **16**(3), 119–126. 169

Jones, R.H.L., 1980, Solar radiation through windows – theory and equations, *Building Services Engineering Research and Technology* **1**(2), 83–91. 208, 360

Jones, W.P., 1994, A review of CIBSE psychrometry, *Building Services Engineering Research and Technology* **15**(4), 189–198. 193

Jury, E.I., 1964, 1986, *Theory and Application of the Z-transform Method*, John Wiley & Sons, Inc., New York. 281

Kallel, F., Galanis, N., Perrin, B. and Javelas, R., 1993, Effects of moisture on temperature during drying of consolidated porous materials, *American Society of Mechanical Engineers: Journal of Heat Transfer* **115**, 724–733. 171, 189

Kamal, S. and Novak, P., 1991, Dynamic analysis of heat transfer in buildings with special emphasis on radiation, *Energy and Buildings* **17**(3), 231–241. 165

Kang, B.H. and Jaluria, Y., 1990, Natural convection heat transfer characteristics of a protruding thermal source located on horizontal and vertical surfaces, *International Journal of Heat and Mass Transfer* **33**(6), 1347–1357. 83, 84

Karlsson, K., Roos, A. and Karlsson, B., 2003, Building and climate influence on the balance temperature of buildings, *Building and Environment* **38**(1), 75–81. 9

Kaviany, M. and Mittal, M., 1987, Funicular state in drying of a porous slab, *International Journal of Heat and Mass Transfer* **30**(7), 1407–1418. 190

Khalifa, A.J.N and Marshall, R.H., 1990, Validation of heat transfer coefficients on interior building surfaces using a real-sized indoor test cell, *International Journal of Heat and Mass Transfer*, **33**(10), 2219–2236. 101

Khan, M.I., 2002, Factors affecting the thermal properties of concrete and applicability of its prediction models, *Building and Environment* **37**(6), 607–614. 180

Kimura, S. and Bejan, A., 1984, The boundary layer natural convection regime in a rectangular cavity with uniform heat flux from one side, *American Society of Mechanical Engineers: Journal of Heat Transfer* **106**, 98–103. 100

King, J.A. and Reible, D.D., 1991, Laminar natural convection heat transfer form inclined surfaces, *International Journal of Heat and Mass Transfer* **34**(7), 1901–1904. 84

Kirkpatrick, A.T. and Bohn, M., 1986, An experimental investigation of mixed cavity natural convection in the high Rayleigh number regime, *International Journal of Heat and Mass Transfer* **29**(1), 69–82. 106

Kitamura, K., Koike, M., Fukuoka, I. and Saito, T., 1985, Large eddy structure and heat transfer of turbulent natural convection along a vertical flat plate, *International Journal of Heat and Mass Transfer* **28**(4), 837–850. 81

Klobut, K. and Siren, K., 1994, Air flows measured in large openings in a horizontal partition, *Building and Environment* **29**(3), 325–335. 109

Knabe, G., 1971a, Frequenzverhalten ein- und mehrschichtiger Wände, *Luft- und Kältetechnik* **7**, 3–10. 357, 366

Knabe, G., 1971b, Optimales Speichervermögen von Wänden, *Luft- und Kältetechnik* **7**, 68–74. 357

Knudsen, M., 1950, *The Kinetic Theory of Gases*, Methuen, London. 176

Kobus, C.J. and Wedekind, G.L., 1996, Modeling the local and average heat transfer coefficient for an isothermal vertical plate with assisting and opposing combined forced and natural convection, *International Journal of Heat and Mass Transfer* **39**(13), 2723–2733. 92

Kohonen, R., 1984, Transient analysis of the thermal and moisture physical behaviour of building constructions, *Building and Environment* **19**(1), 1–11. 421

Kolesnikov, P.M., 1987, Generalized boundary conditions of the heat and mass transfer, *International Journal of Heat and Mass Transfer* **30**(1), 85–92. 225

Kollmar, A., 1950, *Die Strahlungsverhaltnisse im Beheizten Wohnraum*, Oldenbourg, Munich. 122

Kondratyev, K.Y., 1969, *Radiation in the Atmosphere*, Academic Press, New York. 199

Konev, S.V. and Mitrovic, J., 1986, An explanation for the augmentation of heat transfer during boiling in capillary structures, *International Journal of Heat and Mass Transfer* **29**(1), 91–94. 194

Korpela, S.A., Lee, Y. and Drummond, J.E., 1982, Heat transfer through a double pane window, *American Society of Mechanical Engineers: Journal of Heat Transfer* **104**, 539–544. 98

Kossecka, E., 1998, Relationships between structure factors, response factors, and Z-transform function coefficients for multilayer walls, *ASHRAE Trans.* **104**, Part 1A, 68–77. 382

Krarti, M., 1989, Steady-state heat transfer beneath partially insulated slab-on-grade floor, *International Journal of Heat and Mass Transfer* **32**(5), 961–969. 50

Krarti, M., 1993a, Steady-state heat transfer from horizontally insulated slabs, *International Journal of Heat and Mass Transfer* **32**(5), 2135–2145. 50

Krarti, M., 1993b, Steady-state heat transfer from slab-on-grade floors with vertical insulation, *International Journal of Heat and Mass Transfer* **32**(5), 2147–2155. 50

Krarti, M., 1994a, Time-varying heat transfer from slab-on-grade floors with vertical insulation, *Building and Environment* **29**(1), 55–61. 369

Krarti, M., 1994b, Time-varying heat transfer from horizontally insulated slab-on-grade floors, *Building and Environment* **29**(1), 63–71. 369

Krarti, M., 1996, Effect of spatial variation of soil thermal properties on slab-on-ground heat transfer, *Building and Environment* **31**(1), 51–57. 369

Krarti, M. and Choi, S., 1995, Optimum insulation for rectangular basements, *Energy and Buildings* **22**(2), 125–131. 50, 369

Krarti, M., Kreider, J.F. and Claridge, D.E., 1994, Schwarz–Christoffel transformation applied to steady-state ground-coupling problems, *Energy and Buildings* **20**, 193–203. 51, 54

Krarti, M. and Choi, S., 1996, Simplified method for foundation heat loss calculation, *ASHRAE Trans.* **102**(1), 140–152. 369

Krarti, M., Claridge, D.E. and Kreider, J.F., 1988a, The ITPE technique applied to steady-state ground-coupling problems, *International Journal of Heat and Mass Transfer* **31**(9), 1885–1898. 50

Krarti, M., Claridge, D.E. and Kreider, J.F., 1988b, ITPE technique applications to time-varying two-dimensional ground-coupling problems, *International Journal of Heat and Mass Transfer* **31**(9), 1899–1911. 369

Krarti, M., Claridge, D.E. and Kreider, J.F., 1995, Frequency response analysis of ground-coupled building envelope surfaces, *ASHRAE Trans.* **101**(1), 355–369. 369

Kreider, J.F. and Krieth, F., 1981, *Solar Energy Handbook*, McGraw-Hill, New York. 216

Krischer, O., 1942, Der Wärme- und Stoffaustausch im Trocknungsgut, *Verein deutscher Ingeneure*, **415**, 1–22. 312, 418

Kronberg, A.E., Benneker, A.H. and Westerterp, K.R., 1998, Notes on wave theory in heat conduction: a new boundary condition, *International Journal of Heat and Mass Transfer* **41**(1), 127–137. 291

Kumar, I.J., 1972, Recent mathematical methods in heat transfer, in *Advances in Heat Transfer*, Vol. 8, Academic Press, New York, pp. 1–91. 51

Künzel, H.M., 1998, The smart vapor retarder: an innovation inspired by computer simulations, *ASHRAE Trans.* **104**(2), 903–907. 177

Künzel, H.M. and Kiessl, K., 1996, Calculation of heat and moisture transfer in exposed building components, *International Journal of Heat and Mass Transfer* **40**(1), 159–167. 190

Kurnitski, J., 2000, Crawl space air change, heat and moisture behaviour, *Energy and Buildings* **32**(1), 19–39. 176

Kurnitski, J and Matilainen, M., 2000, Moisture conditions of outdoor air-ventilated crawl spaces in apartment buildings in a cold climate, *Energy and Buildings* **33**(1), 15–29. 176

Kusuda, T., 1969, Thermal response factors for multi-layer structures of various heat conduction systems, *ASHRAE Trans.*, **75**, Pt 1, 246–271. 372, 413

Kusuda, T., 2001, Building environment simulation before desk top computers in the USA through a personal memory, *Energy and Buildings* **33**(4), 291–302. 372

Kusuda, T. and Bean, J.W., 1984, Simplified methods for determining seasonal heat loss from uninsulated slab-on-grade floors, *ASHRAE Trans.* **90** Part 1B, 611–632. 368

Lakhal, E.K., Hasnaoui, M. and Vasseur, P., 1999, Etude numérique de la convection naturelle transitoire au sein d'une cavité chauffée périodiquement avec différents types d'excitations, *International Journal of Heat and Mass Transfer* **42**(21), 3927–3941. 106

Lam, S.W., Gani, R. and Symons, J.G., 1989, Experimental and numerical studies of natural convection in trapezoidal cavities, *American Society of Mechanical Engineers: Journal of Heat Transfer* **111**, 372–377. 100

Landman, K.A. and Delsante, A.E., 1986, Steady state heat losses from a building floor slab with vertical edge insulation I, *Building and Environment* **21**(3/4), 177–182. 49, 50

Landman, K.A. and Delsante, A.E., 1987a, Steady state heat losses from a building floor slab with vertical edge insulation II, *Building and Environment* **22**(1), 49–55. 49, 50

Landman, K.A. and Delsante, A.E., 1987b, Steady state heat losses from a building floor slab with horizontal edge insulation, *Building and Environment* **22**(1), 57–60. 49, 50

Laporthe, S., Virgone, J. and Castanet, S., 2001, A comparative study of two tracer gases: SF_6 and N_2O, *Building and Environment* **36**(3), 313–320. 169

Laret, L., 1980, Use of general models with a small number of parameters, Part 1: Theoretical analysis, *Proceedings of Conference Clima 2000*, Budapest, 263–276. 328, 330

Larsson, U and Moshfegh, B., 2002, Experimental investigation of downdraft from well-insulated windows, *Building and Environment* **37**(11), 1073–1082. 104

Lartigue, B., Lorente, S. and Bourret, B., 2000, Multicellular natural convection in a high aspect ratio cavity: experimental and numerical results, *International Journal of Heat and Mass Transfer* **43**(17), 3157–3170. 99

Launder, B.E. and Spalding, D.B., 1974, *Computer Methods in Applied Mechanics and Engineering*. National Academy Press, Washington DC. 75

Lebrun, J.J. and Nusgens, P.J., 2000, Simplified analysis of building thermal storage effects, *ASHRAE Trans.* **106**(1), 801–810. 358

Lee, S.C., 1989, Effect of fiber orientation on thermal radiation in fibrous media, *International Journal of Heat and Mass Transfer* **32**(2), 311–319. 181

Lee, Y. and Korpela, S.A., 1983, Multicellular convection in a vertical slot, *Journal of Fluid Mechanics* **126**, 91–121. 98

Le Febve de Vivy, D., 1966, Dynamique des températures d'une paroi. Fonctions de transfer de courbes de référence, *Rev. Gen. Thermique* **49**, 21–31. 366

Leong, W.H., Hollands, K.G.T. and Brunger, A.P., 1999, Experimental Nusselt numbers for a cubical-cavity benchmark problem in natural convection, *International Journal of Heat and Mass Transfer* **42**(11), 1979–1989. 102

Le Quéré, P., 1990, A note on multiple and unsteady solutions in two-dimensional convection in a tall cavity, *American Society of Mechanical Engineers: Journal of Heat Transfer* **112**, 965–974. 97

Letherman, K.M., 1977, A rational criterion for accuracy of modelling of periodic heat conduction in plane slabs, *Building and Environment* **12**, 127–130. 349

Letherman, K.M., 1988, Room air moisture content: dynamic effects of ventilation and vapour generation, *Building Services Engineering Research and Technology* **9**(2), 49–53. 168

Letherman, K.M., 1989, Condensation avoidance in layered structures: synthesis of design, *Building Services Engineering Research and Technology* **10**(1), 29–34. 177

Letherman, K.M. and Sarkis, B.L., 1984, The computation of two-dimensional steady-state temperature distributions using the Schwarz–Christoffel transformation, *Building and Environment* **19**(2), 101–109. 51

Levermore, G., 2002, Guest editorial [for a set of articles on climate change], *Building Services Engineering Research and Technology* **23**(4), 205–206. 10

Lewandowski, W.M., Radziemska, E., Buzuk, M. and Bieszk, H., 2000, Free convection heat transfer and fluid flow above horizontal rectangular plates, *Applied Energy* **66**(2), 177–197. 85

Lewis, R.H., 1995, Heating and air-conditioning systems – a historical overview and evolution, *ASHRAE Trans.* **101**(1), 525–527. 1

Lewis, W.K., 1922, The evaporation of a liquid into a gas, *Transactions of the American Society of Mechanical Engineers* **44**, 325–332. 195

Li, Y. and Delsante, A., 2001, Natural ventilation induced by combined wind and thermal forces, *Building and Environment* **36**, 59–71. 454

Liddament, M.W., 1998, Preface to special issue on optimum ventilation and air flow control in buildings, *Energy and Buildings* **27**(3), 221–222. 10

Liesen, R.J and Pedersen, C.O., 1997, An evaluation of inside surface heat balance models for cooling load calculations, *ASHRAE Trans.* **103**(2), 485–502. 125

Liesen, R.J and Pedersen, C.O., 1999, Modeling the energy effects of combined heat and mass transfer in building elements: Part 1, Theory. *ASHRAE Trans.* **105**, Part 2, 941–953. 170, 184, 422

Linthorst, S.J.M, Schinkel, W.M.M. and Hoogendoorn, C.J., 1981, Flow structure with natural convection in inclined air-filled enclosures, *American Society of Mechanical Engineers: Journal of Heat Transfer* **103**, 535–539. 98

Liu, J.Y. and Cheng, S., 1991, Solutions of Luikov equations of heat and mass transfer in capillary-porous bodies, *International Journal of Heat and Mass Transfer* **34**(7), 1747–1754. 421

Liu, M. and Claridge, D.E., 1995, Is the actual heat loss factor substantially smaller than you calculated? *ASHRAE Trans.* **101**(2), 3–13. 15

Lobo, P.D., Mikhailov, M.D. and Özisik, M.N., 1987, On the complex eigen-values of Luikov system of equations, *Drying Technology* **5**(2), 273–286. 421

Lock, G.S.H. and Zhao, L., 1992, Natural convection in honeycomb wall spaces, *International Journal of Heat and Mass Transfer* **35**(1), 155–164. 99

Lomas, K.J., 1996, The U.K. applicability study: an evaluation of thermal simulation programs for passive solar house design, *Building and Environment* **31**(3), 197–206. 104

Lomas, K.J. and Eppel, H., 1992, Sensitivity analysis techniques for building thermal simulation programs, *Energy and Buildings* **19**(1), 21–44. 454

Lomas, K.J., Eppel, H., Martin, C.J. and Bloomfield, D.P., 1997, Empirical validation of building energy simulation programs, *Energy and Buildings* **26**(3), 253–275. 454

Lombard, C., 1995, Heat-transfer for fins with two-port theory, *International Journal of Mechanical Engineering Education*, **24**, 61–72. 39

Lombard, C. and Mathews, E.H., 1992, Efficient, steady state solution of a time variable *RC* network, for building thermal analysis, *Building and Environment* **27**(3), 279–287. 324

Lombard, C. and Mathews, E.H., 1999, A two-port envelope model for building heat transfer, *Building and Environment* **34**(1), 19–30. 324

Lorenz, F. and Masey, G., 1982, Méthode d'évaluation de l'économie d'énergie apportée par l'intermittance de chauffage dans les bâtiments. Traitement par différences finies d'un modèle à deux constantes de temps. Report GM820130-01, Laboratoire de Physique du Bâtiment, Université de Liege. 324, 330

Loudon, A.G., 1967, The interpretation of solar radiation measurements for building problems, UK Building Research Station, Research Paper 73. Also in *Proceedings of CIE Conference*, Bouwcentrum, Rotterdam, pp. 111–118. 205

Loudon, A.G., 1970, Summertime temperatures in buildings without air conditioning, *Journal of the Institute of Heating and Ventilating Engineers* **37**, 280–292. 158, 357, 358

Loveday, D.L. and Craggs, C., 1993, Stochastic modelling of temperatures for a full-scale occupied building zone subject to natural random influences, *Applied Energy* **45**(4), 295–312. 322

Loveday, D.L. and Taki, A.H., 1996, Convective heat transfer coefficients at a plane surface on a full-scale building facade, *International Journal of Heat and Mass Transfer* **39**(8), 1729–1742. 112

Loveday, D.L. and Taki, A.H., 1998, Outside surface resistance: proposed new value for building design, *Building Services Research and Technology* **19**(1), 23–29. 113

Loveday, D.L, Taki, A.H. and Versteeg, H., 1994, Convection coefficients at disrupted building facades – laboratory and simulation studies, *International Journal of Ambient Energy* **15**(1), 17–26. 111, 112

Lu, W., Howarth, A.T. and Jeary, A.P., 1997, Prediction of airflow and temperature field in a room with convective heat source, *Building and Environment* **32**(6), 541–550. 76

Luikov, A.V., 1966, *Heat and Mass Transfer in Capillary-Porous Bodies*, Pergamon, Oxford. 27, 170, 179, 180, 195, 196, 418, 419

Luikov, A.V., 1975, Systems of differential equations of heat and mass transfer in capillary-porous bodies (review), *International Journal of Heat and Mass Transfer* **18**, 1–14. 180, 195

Macey, H.H., 1949, Heat loss through a solid floor, *Journal of the Institute of Fuel*, **2**, 369–371. 15, 40f, 55

Mackay, R.M. and Probert, S.D., 1996, Integrated policies for energy and the environment: options for the UK in a world context, *Applied Energy* **55**(3/4), 131–403. xv

Mackey, C.O. and Wright, L.T., 1943, Summer comfort factors as influenced by thermal properties of building materials, *ASHVE Trans.* **49**, 148. 64, 357

Mackey, C.O. and Wright, L.T., 1944, Periodic heat flow – homogeneous walls or roofs, *ASHVE Trans.* **50**, 293–312. 357, 366

Mackey, C.O. and Wright, L.T., 1946a, The sol-air thermometer – a new instrument, *ASHVE Trans.* **52**, 271–282. 64

Mackey, C.O. and Wright, L.T., 1946b, Periodic heat flow – composite walls or roofs, *ASHVE Trans.* **52**, 283. 357

Majumdar, A. and Tien, C.L., 1990, Effects of surface tension on film condensation in a porous medium, *American Society of Mechanical Engineers: Journal of Heat Transfer* **112**, 751–757. 194

Malalasekera, W.M.G and James, E.H., 1993, Thermal radiation in a room: numerical evaluation, *Building Services Engineering Research and Technology* **14**(4), 159–168. 127

Mao, G. and Johannesson, G., 1997, Dynamic calculation of thermal bridges, *Energy and Buildings* **26**(3), 233–240. 343

Markatos, N.C. and Pericleous, K.A., 1984, Laminar and turbulent natural convection in an enclosed cavity, *International Journal of Heat and Mass Transfer* **27**(5), 755–772. 101

Marquardt, W. and Auracher, H., 1990, An observer-based solution of inverse heat conduction problems, *International Journal of Heat and Mass Transfer* **33**(7), 1545–1562. 276

Martin, B.W., 1984, An appreciation of advances in natural convection along an isothermal vertical surface, *International Journal of Heat and Mass Transfer* **27**(9), 1583–1586. 86

Martin, T.J. and Dulikravich, G.S., 2000, Inverse determination of temperature-dependent thermal conductivity using steady surface data on arbitrary objects, *American Society of Mechanical Engineers: Journal of Heat Transfer* **122**(3), 450–459. 276

Masuch, J., 1966, Die Berechnung periodisch veränderlicher Wärmeströme durch zweischichtige Wände, *Gesundheits-Ingenieur* **87**, 315–325. 366

Masuch, J., 1969, Das thermische Verhalten mehrschaliger Bauteile unter einseitig periodischer Belastung, *Gesundheits-Ingenieur* **90**, 213–218. 366

Masmoudi, W. and Prat, M., 1991, Heat and mass transfer between a porous medium and a parallel external flow. Application to drying of capillary porous materials, *International Journal of Heat and Mass Transfer* **34**(8), 1975–1989. 188, 194

Mathews, E.H., 1986, Thermal analysis of naturally ventilated buildings, *Building and Environment* **21**(1), 35–39. 324

Mathews, E.H. and Richards, P.G., 1989, A tool for predicting hourly air temperatures and sensible energy loads in buildings at sketch design stage, *Energy and Buildings* **14**, 61–80. 324

Mathews, E.H. and Richards, P.G., 1993, An efficient tool for future building design, *Building and Environment* **28**(4), 409–417. 324

Mathews, E.H., Rousseau, P.G., Richards, P.G. and Lombard, C., 1991, A procedure to estimate the effective heat storage capability of a building, *Building and Environment* **26**(2), 179–188. 324

Mathews, E.H., Richards, P.G. and Lombard, C., 1994a, A first-order thermal model for building design, *Energy and Buildings* **21**, 133–145. 324

Mathews, E.H., Shuttleworth, A.G. and Rousseau, P.G., 1994b, Validation and further development of a novel thermal analysis method, *Building and Environment* **29**(2), 207–215. 324

Mathews, E.H., Etzion, Y., van Heerden, E., Weggelaar, S., Erell, E., Pearlmutter, D. and Meir, I.A., 1997, A novel thermal simulation model and its application on naturally ventilated desert buildings, *Building and Environment* **32**(5), 447–456. 152, 331

Mavroulakis, A. and Trombe, A., 1998, A new semianalytical algorithm for calculating diffuse plane view factors, *American Society of Mechanical Engineers: Journal of Heat Transfer* **120**, 279–282. 119

McAdams, W.H., 1954, *Heat Transmission*, McGraw-Hill, New York. 78, 80, 110, 116, 119, 121, 125

McClellan, T.M. and Pedersen, C.O., 1997, Investigation of outside heat balance models for use in a heat balance cooling load calculation procedure, *ASHRAE Trans.* **103**(2), 469–484. 59, 112, 145

McIntyre, D.A., 1980, *Indoor Climate*, Applied Science Publishers, London. 5

McLean, R.C. and Galbraith, G.H., 1988, Interstitial condensation: applicability of conventional vapour permeability values, *Building Services Engineering Research and Technology* **9**(1), 29–34. 171

McQuiston, F.C. and Parker, J.D., 1994, *Heating, Ventilating and Air Conditioning*, 4th edn, John Wiley & Sons, Inc., New York. 372

Meier, A., 1997, Editorial: special issue on urban heat islands and cool communities, *Energy and Buildings* **25**(2), 95–97. 9

Merrill, J.L. and TenWolde, A., 1989, Overview of moisture-related damage in one group of Wisconsin manufactured homes, *ASHRAE Trans.* **95**(1), 405–414. 177

Mikhailov, M.D. and Özisik, M.N. 1984, *Unified Analysis and Solutions of Heat and Mass Diffusion*, John Wiley & Sons, Inc., New York. 385

Mikhailov, M.D., Özisik, M.N. and Vulchanov, N.L., 1983, Diffusion in composite layers with automatic solution of the eigenvalue problem, *International Journal of Heat and Mass Transfer* **26**(8), 1131–1141. 385

Mikhailov, M.D. and Vulchanov, N.L., 1983, Computational procedure for Sturm–Liouville problems, *Journal of Computational Physics* **50**, 323–336. 385

Milbank, N.O. and Harrington-Lynn, J., 1970, Estimation of air-conditioning loads, Current Paper 13/70, (United Kingdom) Building Research Station. 365

Milbank, N.O. and Harrington-Lynn, J., 1974, Thermal response and the admittance procedure, *Building Services Engineering* **42**, 38–51. 357

Min, T.C., Schutrum, L.F., Parmlee, G.V and Vouris, J.D., 1956, Natural convection and radiation in a panel-heated room, *ASHRAE Trans.* **62**, 337–358. 103, 104

Mingfang, T. and Qigao, C., 1997, A simple expression for internal surface temperature distribution in corners of external walls, *Building and Environment* **32**(4), 313–316. 51

Mitalas, G.P., 1978, Comments on the Z-transfer function method for calculating heat transfer in buildings, *ASHRAE Trans.*, **84**, Part 1, 667–674. 372

Mitalas, G.P., 1983, Calculation of basement heat loss, *ASHRAE Trans.* **89**, Part 1, 420–437. 369

Mitalas, G.P., 1987, Calculation of below-grade residential heat loss: low-rise residential building, *ASHRAE Trans.* **93**(1), 743–783. 369

Mitalas, G.P. and Arseneault, J.G., 1971, Fortran IV program to calculate Z-transfer functions for the calculation of transient heat transfer through walls and roofs, NBS Building Science Series 39, October 1971. 371

Mitalas, G.P. and Stephenson, D.G., 1967, Room thermal response factors, *ASHRAE Trans.* **73**, Part 2, III. 2.1–10. xvii, xix, 308, 371

Mitchell, D.R., Tao, Y.-X. and Besant, R.W., 1995, Air filtration with moisture and frosting phase changes in fiberglass insulation: I. Experiment, II. Model validation, *International Journal of Heat and Mass Transfer* **38**(9), 1587–1596 and 1597–1604. 181, 185

Mitrovic, J. 1997, The Fick and Lagrange equations as a basis for the Maxwell–Stefan diffusion equations, *International Journal of Heat and Mass Transfer* **40**(10), 2373–2377. 26, 179

Molenda, C.H.A, Crausse, P. and Lemarchand, D., 1993, Heat and humidity transfer in non-saturated porous media: capillary hysteresis effects under cyclic thermal conditions, *International Journal of Heat and Mass Transfer* **36**(12), 3077–3088. 418

Moon, P., 1936, *Scientific Basis of Illuminating Engineering*, McGraw-Hill, New York. 123

Morgan, E.A., 1966, Improvements in solar heated buildings, United Kingdom Patent Specification 1 022 411, application date 6 April 1961, complete specification published 16 March 1966, The Patent Office, London. xviii, 216, 320

Motakef, S. and El-Masri, M.A., 1986, Simultaneous heat and mass transfer with phase change in a porous slab, *International Journal of Heat and Mass Transfer* **29**(10), 1503–1512. 182

Mull, W. and Reiher, H., 1930, Der Wärmeschütz von Luftschichten: seine experimentelle Bestimmung und graphische Berechnung, *Beiheft zum Gesundheits-Ingenieur*, Verlag Oldenbourg, Munich, 28, Reihe 1, pp. 1–26. 96

Muncey, R.W., 1953, The calculation of temperatures inside buildings having variable external conditions, *Australian Journal of Applied Science* **4**, 189–196. 357, 366

Muncey, R.W., 1963, The thermal response of a building to sudden changes of temperature or heat flow, *Australian Journal of Applied Science* **14**, 123–128. 366

Muneer, T. and Han, B., 1996, Multiple glazed windows: design charts, *Building Services Engineering Research and Technology* **17**(4), 223–229. 13

Murakami, S., Kato, S. and Zeng, J., 2000, Combined simulation of airflow, radiation and moisture transport for heat release from the human body, *Building and Environment* **35**(6), 489–500. 6

Murakami, S., Kato, S. and Kim, T., 2001, Indoor climate design based on CFD: coupled simulation of convection, radiation, and HVAC control for attaining a given PMV value, *Building and Environment* **36**(6), 701–709. 458

Murata, K., 1995, Heat and mass transfer with condensation in a fibrous insulation slab bounded on one side by a cold surface, *International Journal of Heat and Mass Transfer* **38**(17), 3253–3262. 182

Myers, G.E., 1971, *Analytical Methods in Conduction Heat Transfer*, McGraw-Hill, New York. 239

Neiswanger, L., Johnson, G.A. and Carey, V.P., 1987, An experimental study of high Rayleigh number mixed convection in a rectangular enclosure with restricted inlet and outlet openings, *American Society of Mechanical Engineers: Journal of Heat Transfer* **109**, 446–453. 105

Nelson, L.W., 1965, The analog computer as a product design tool, *ASHRAE Journal* **7**(11). 242

Neumann, G., 1990, Three-dimensional numerical simulation of buoyancy-driven convection in vertical cylinders heated from below, *Journal of Fluid Mechanics* **214**, 559–578. 95

Newsham, G.R. and Tiller, D.K., A field study of office thermal comfort using questionnaire software, *ASHRAE Trans.* **103**(2), 3–17. 2

Nicol, J.F. and Humphreys, M.A., 2002, Adaptive thermal comfort and sustainable thermal standards for buildings, *Energy and Buildings* **34**(6), 563–572. 6

Nielsen, P.V., 1998, The selection of turbulence models for prediction of room airflow, *ASHRAE Trans.* **104**(1B), 1119–1127. 76

Nottage, H.B. and Parmlee, G.V., 1954, 1955, Circuit analysis applied to load estimating, *ASHVE Trans.* **60**, 59–102 (part 1); **61**, 125–150 (part 2). 57, 242

Novak, M.H. and Nowak, E.S., 1993, Natural convection heat transfer in slender window cavities, *American Society of Mechanical Engineers: Journal of Heat Transfer* **115**, 476–479. 99

Ogniewicz, Y. and Tien, C.L., 1981, Analysis of condensation in porous insulation, *International Journal of Heat and Mass Transfer* **24**(3), 421–429. 182

Oiry, H. and Bardon, J.P., 1985, Comportement thermique de differents types d'habitation soumis a un ensoleillement et à une température extérieure périodiques, *International Journal of Heat and Mass Transfer* **28**(11), 1991–2004. 358

Onsager, L., 1931a,b, Reciprocal relations in irreversible processes, I and II, *Physical Review* **37**, 405–426 and **38**, 2265–2279. 30, 179

Oppenheim, A.K., 1956, Radiation analysis by the network method, *Transactions of the American Society of Mechanical Engineers* **78**, 725–735. 125

Oreszczyn, T., 1988, Cold bridging at corners: surface temperature and condensation risk, *Building Services Engineering Research and Technology* **9**(4), 167–175. 51

Oreszczyn, T. and Pretlove, S.E.C., 1999, Condensation Targeter II: modelling surface relative humidity to predict mould growth in dwellings, *Building Services Engineering Research and Technology* **20**(3), 143–153. 167

Ostrach, S., 1953, An analysis of laminar free-convection flow and heat transfer about a plate parallel to the direction of the generating body force, National Advisory Committee for Aeronautics (NACA), Report 1111. 86

Ostrach, S., 1972, Natural convection in enclosures, in *Advances in Heat Transfer*, Vol. 8, Academic Press, New York, pp. 161–227. 94

Ostrach, S., 1988, Natural convection in enclosures, *American Society of Mechanical Engineers: Journal of Heat Transfer* **110**(4B), 1175–1190. 94

Ouyang, K. and Haghighat, F., 1991, A procedure for calculating thermal response factors of multi-layer walls – state space method, *Building and Environment* **26**(2), 173–177. 372

Ozaki, A., Watanabe, T., Hayashi, T. and Ryu, Y., 2001, Systematic analysis on combined heat and water transfer through porous materials based on thermodynamic energy, *Energy and Buildings* **33**(4), 341–350. 31, 180, 196

Özisik, M.N., 1973, *Radiative Transfer*, John Wiley & Sons, Inc., New York. 123

Özisik, M.N., 1980, *Heat Conduction*, John Wiley & Sons, Inc., New York. 280, 385

Özisik, M.N., 1994, *Finite Difference Methods in Heat Transfer*, CRC Press. 239

Ozoe, H., Mouri, A., Hiramitsu, M., Churchill, S.W. and Lior, N., 1986, Numerical calculation of three-dimensional turbulent natural convection in a cubical enclosure using a two-equation model for turbulence, *American Society of Mechanical Engineers: Journal of Heat Transfer* **108**, 806–813. 103

Paláncz, B., 1987, Solution of the penetrating evaporation front model for finite porous medium using orthogonal collocation method, *International Journal of Heat and Mass Transfer* **30**(9), 1871–1878. 190

Pandey, R.N., Strivastava, S.K. and Mikhailov, M.D., 1999, Solutions of Luikov equations of heat and mass transfer in capillary porous bodies through matrix calculus: a new approach, *International Journal of Heat and Mass Transfer* **42**(14), 26449–2660. 421

Parmelee, G.V., 1945, The transmission of solar radiation through flat glass under summer conditions, *ASHVE Trans.* **51**, 317–350. 208

Paschkis, V., 1936, Elektrisches Modell zur Verfolgung von Wärmestrahlungsvorgängen, insbesondere in elektrischen Öfen, *Elektrotechnik und Maschinenbau* **52**, 617–621. 119, 125, 241

Paschkis, V. and Baker, H.D., 1942, A method of determining unsteady state heat transfer by means of an electrical analogy, *Transactions of the American Society of Mechanical Engineers* **64**, 105–112. 241

Pasqualetto, L., Zmeureanu, R. and Fazio, P., 1998, A case study of validation of an energy analysis program: MICRO-DOE2.1E, *Building and Environment* **33**(1), 21–41. 307

Paton, J., 1993, Dampness, mould growth and children's health, *Health and Hygiene* **14**, 141–4. 168

Pauker, M.T., Farley, B., Jeter, S.M. and Abdel-Khalik, S.I., 1995, An experimental investigation of water evaporation into low-velocity air currents, *ASHRAE Trans.* **101**(1), 90–96. 175

Peavy, B.A., 1978, A note on response factors and conduction transfer functions, *ASHRAE Trans.* **84**, Part 1, 688–690. 372

Pedersen, C.O., Fisher D.E. and Liesen, R.J., 1997, Development of a heat balance procedure for calculating cooling loads, *ASHRAE Trans.* **103**, Part 2, 459–468. 451

Pedersen, C.R., 1992, Prediction of moisture transfer in building constructions, *Building and Environment* **27**(3), 387–397. 177, 420

Pel, L., Broken, H. and Kopinga, K., 1996, Determination of moisture diffusivity in porous media using moisture concentration profiles, *International Journal of Heat and Mass Transfer* **39**(6), 1273–1280. 186, 187

Penman, J.M., 1990, Second-order system identification in the thermal response of a working school, *Building and Environment* **25**(2), 105–110. 330

Peppes, A.A., Santamouris, D.N. and Asimakopoulos, D.N., 2001, Buoyancy-driven flow through a stairwell, *Building and Environment* **36**(2), 167–180. 109

Pera, L. and Gebhart, B., 1973, On the stability of natural convection boundary layer flow over horizontal and slightly inclined surfaces, *International Journal of Heat and Mass Transfer* **16**, 1147–1163. 84

Perez Sanchez, M., Allard, F. and Achard, G., 1988, Thermal coupling between the ground and slab-on-grade buildings, *Building and Environment* **23**, 233–242. 369

Peuser, F.A., Remmers, K.-H. and Schnauss, M., 2002, *Solar Thermal Systems: Successful Planning and Construction*, James and James, London, and Solarpraxis, Berlin. 212

Pfrommer, P., Lomas, J.K. and Kupke, C., 1994, Influence of transmission models for special glazing on the predicted performance of commercial buildings, *Energy and Buildings* **21**(2), 101–110. 209

Philip, J.R. and De Vries, D.A., 1957, Moisture movement in porous materials under temperature gradients, *Transactions of the American Geophysical Union* **38**(2) 222–232. 419, 420

Phillips, J.R., 1996, Direct simulations of turbulent unstratified natural convection in a vertical slot for Pr = 0.71, *International Journal of Heat and Mass Transfer* **39**(12), 2485–2494. 99

Pierce, D.A. and Benner, S.M., 1986, Thermally induced hygroscopic mass transfer in a fibrous medium, *International Journal of Heat and Mass Transfer* **29**(11), 1683–1694. 180

Pipes, L.A., 1940, The matrix theory of four terminal networks, *Philosophical Magazine*, 7th, series **30**, 370–395. 357

Pipes, L.A., 1957, Matrix analysis of heat transfer problems, *Journal of the Franklin Institute*, **263**, 195–206. 357

Pitchumani, R. and Yao, S.C., 1991, Correlation of thermal conductivities of unidirectional fibrous composites using local fractal techniques, *American Society of Mechanical Engineers: Journal of Heat Transfer* **113**, 788–796. 29

Plathner, P and Woloszyn, M., 2002, Interzonal air and moisture transport in a test house: experiment and modelling, *Building and Environment* **37**(2), 189–199. 169, 454

Poulikakos, D. and Bejan, A., 1983, Natural convection experiments in a triangular enclosure, *American Society of Mechanical Engineers: Journal of Heat Transfer* **105**, 652–655. 100

Prasad, V. and Kulacki, F.A., 1984, Convective heat transfer in a rectangular porous cavity – effect of aspect ratio on flow structure and heat transfer, *American Society of Mechanical Engineers: Journal of Heat Transfer* **106**, 158–165. 100

Pratt, A.W., 1965a, Fundamentals of heat transmission through external walls of buildings, *Journal of Mechanical Engineering Science* **7**, 357–366. 357

Pratt, A.W., 1965b, Variable heat flow through walls of cavity construction, naturally exposed, *International Journal of Heat and Mass Transfer* **8**, 861–871. 357

Pratt, A.W. 1966, The thermal resistance of airspaces in building structures, *Journal of the Institution Heating and Ventilating Engineers* **34**, 133–145. 95

Pratt, A.W., 1981, *Heat Transmission in Buildings*, John Wiley & Sons, Ltd, Chichester. 304, 314

Pratt, A.W. and Ball, E.F, 1963, Transient cooling of a heated enclosure, *International Journal of Heat and Mass Transfer*, **6**, 703–718. 304, 305

Prins, G., 1992, On Condis and Coolth, *Energy and Buildings* **18**(3/4) 251–258. 6

Probert, D., 1997, Editorial, *Applied Energy* **58**(4), iii–v. xxi

Rao, V.R. and Sastri, V.M.K., 1996, Efficient evaluation of diffuse view factors for radiation, *International Journal of Heat and Mass Transfer* **39**(6), 1281–1286. 124

Raychaudhuri, B.C., 1965, Transient thermal response of enclosures: the integrated thermal time constant, *International Journal of Heat and Mass Transfer*, **8**, 1439–1449. 316

Raynaud, M., 1986, Some comments on the sensitivity to sensor location of inverse heat conduction problems using Beck's method, *International Journal of Heat and Mass Transfer* **29**(5), 815–817. 276

Rees, S.J., Spitler, J.D. and Haves, P., 1998, Quantitative comparison of North American and U.K. cooling load calculation procedures – results, *ASHRAE Trans.* **104**(2), 47–61. 365

Rees, S.J., Spitler, J.D., Davies, M.G. and Haves, P., 2000a, Qualitative comparison of North American and U.K. cooling load calculation procedures, *International Journal of Heating, Ventilating, Air-Conditioning and Refrigeration Research* **6**, 75–99. 365

Rees, S.J., Spitler, J.D., Holmes, M.J. and Haves, P., 2000b, Comparison of peak load predictions and treatment of solar gains in the admittance and heat balance load calculation procedures, *Building Services Engineering Research and Technology* **21**(2), 125–138. 365

Richards, R.F., Burch, D.M and Thomas, W.C., 1992, Water vapor sorption measurements of common building materials, *ASHRAE Trans.* **98**(2), 475–485. 187

Riffat, S.B., 1991, Algorithms for airflows through large internal and external openings, *Applied Energy* **40**(3), 171–188. 109

Riffat, S.B. and Abdalla, W.E., 1991, Airflow between two zones: comparison of experimental results with the MULTIC computer program, *Building Services Engineering Research and Technology* **12**(1), 79–82. 454

Riffat, S.B. and Kohal, J.S., 1994, Experimental study of interzonal natural convection through an aperture, *Applied Energy* **48**(4), 305–313. 109

Rios, J. de Mendoza, 1805, *A Complete Collection of Tables for Navigation and Nautical Astronomy*, printed by T. Bensley. 201

Roberts, B., 1997, *The Quest for Comfort*, Chartered Institution of Building Services Engineers, London. 1

Roberts, J.K., 1940, *Heat and Thermodynamics*, 3rd edn, Blackie, London. 176

Roberts, J.K. and Miller, A.R., 1960, *Heat and Thermodynamics*, Blackie, London. 24, 25

Robinson, D., 1996, Energy model usage in building energy design, *Building Services Engineering Research and Technology* **17**(2), 89–95. 332

Robinson, H.E., Powlitch, F.J. and Dill, R.J., 1954, The thermal insulating value of airspaces, US Housing and Home Finance Agency, Housing Research Paper 32, Washington. 97

Rossen, J.L and Hayakawa, K., 1977, Simultaneous heat and moisture transfer in dehydrated food: a review of theoretical models, *Symposium Series of the American Institute of Chemical Engineers* **163**, 71–81. 179, 419

Rowley, F.B., 1939, A theory covering the transfer of vapor through materials, *ASHVE Trans.* **45**, 545–560. 170

Ruud, M.D., Mitchel, J.W. and Klein, S.A., 1990, Use of building thermal mass to offset cooling loads, *ASHRAE Trans.* **96**(2), 820–835. 217

Salt, H., 1983, Transient conduction in a two-dimensional composite slab: I. Theoretical development of temperature modes, II. Physical interpretation of temperature modes, *International Journal of Heat and Mass Transfer* **26**(11), 1611–1616, 1617–1623. 385, 421

Sanathanan, C.K. and Koerner, J., 1963, Transfer function synthesis as a ratio of two complex polynomials, *IEEE Trans.* **AC-8**, 56–58. 343

Santamouris, M. and Wouters, P., 1994, Energy and indoor climate in Europe – past and present, *Proceedings of the European Conference on Energy Performance and Indoor Climate in Buildings*, Lyon. 6

Santamouris, M., Argiriou, A., Asimakopoulos, D., Klitsikas, N. and Dounis, A., 1995, Heat and mass transfer through large openings by natural convection, *Energy and Buildings* **23**(1), 1–8. 109

Sarkis, B.L. and Letherman, K.M., 1987, Heat flow rates and temperature distributions in corners of external walls of arbitrary shape, *Building and Environment* **22**, 251–258. 51

Sasaki, A., Aiba, S. and Fukuda, H., 1987, A study on the thermophysical properties of soil, *American Society of Mechanical Engineers: Journal of Heat Transfer* **109**, 232–237. 185

Saunders, O.A., 1936, The effect of pressure upon natural convection in air, *Proceedings of the Royal Society of London Series A* **157**, 278–291. 82, 83

Sayigh, A.A.M., 1979, *Solar Energy Applications in Buildings*, Academic Press, New York. 212

Schack, A., 1930, Zur Berechnung des Zeitlichen und örtlichen Temperaturverlaufs beim Glühvorgang, *Stahl und Eisen* **50**, 1290. 299

Schadler, N. and Kast, W., 1987, A complete model of the drying curve for porous bodies – experimental and theoretical studies, *International Journal of Heat and Mass Transfer* **30**(10), 2031–2044. 190

Scharmer, K. and Greif, J. (2000). *European Solar Radiation Atlas 2000*, 4th edn, Vol. 2, Presse de l'Ecole des Mines, Paris. 58, 59

Schinkel, W.M.M., Linthorst, S.J.M. and Hoogendoorn, C.J., 1983, The stratification in natural convection in vertical enclosures, *American Society of Mechanical Engineers: Journal of Heat Transfer* **105**, 267–272. 99

Schlichting, H., 1975, Tribute to Ludwig Prandtl, *International Journal of Heat and Mass Transfer*, **18**, 1333–1336. 77

Schmidt, E. and Beckmann, W., 1930, Das Temperature- und Geschwindigkeitsfeld vor einer Wārme abgebenden senkrechten Platte bei natürlicher Konvektion, *Forchungs-Ingenieur Wesen*, **1**, 391–406. 86, 87

Schoenau, G.J. and Kehrig, R.A., 1990, A method for calculating degree-days to any base temperature, *Energy and Buildings* **14**(4), 299–302. 9

Seem, J.E., Klein, S.A., Beckman, W.A. and Mitchell, J.W., 1989a, Transfer functions for efficient calculation of multidimensional transient heat transfer, *American Society of Mechanical Engineers: Journal of Heat Transfer* **111**, 5–12. 158, 373

Seem, J.E., Klein, S.A., Beckman, W.A. and Mitchell, J.W., 1989b, Comprehensive room transfer functions for efficient calculation of the transient heat transfer processes in buildings, *American Society of Mechanical Engineers: Journal of Heat Transfer* **111**, 264–273. 451

Seem, J.E., Klein, S.A., Beckman, W.A. and Mitchell, J.W., 1990, Model reduction of a transfer function using a dominant root method, *American Society of Mechanical Engineers: Journal of Heat Transfer* **112**, 547–554. 373, 397

Sefcik, D.M., Webb, B.W. and Heaton, H.S., 1991, Analysis of natural convection in vertically-vented enclosures, *International Journal of Heat and Mass Transfer* **34**(12), 3037–3046. 100

SEIS (Solar Energy Information Services), 1981, Volumes 1 and 2, US Department of Energy, PO Box 19475, Sacramento, California 95819. 216

Sezai, I. and Mohamad, A.A., 2000, Natural convection from a discrete heat source on the bottom of a horizontal enclosure, *International Journal of Heat and Mass Transfer* **43**(13), 2257–2266. 100

Shakun, W., 1992, The causes and control of mold and mildew in hot and humid climates, *ASHRAE Trans.* **98**(1), 1282–1291. 178

Shao, L. and Howarth, A.T., 1992, Air infiltration through background cracks due to temperature difference, *Building Services Engineering Research and Technology* **13**(1), 25–30. 109

Shapiro, A.B., 1985, Computer implementation, accuracy, and timing of radiation view factor analysis, *American Society of Mechanical Engineers: Journal of Heat Transfer* **107**, 730–732. 123

Shapiro, A.P. and Motakef, S., 1990, Unsteady heat and mass transfer with phase change in porous slabs: analytical solutions and experimental results, *International Journal of Heat and Mass Transfer* **33**(1), 163–173. 182

Sharples, S., 1984, Full scale measurement of convective energy losses from exterior building surfaces, *Building and Environment* **19**(1), 31–39. 111

Shavit, G., 1995, Short-time-step analysis and simulation of homes and buildings during the last 100 years, *ASHRAE Trans.* **101**(1), 856–867. 242

Shiralkar, G., Gadgil, A. and Tien, C.L., 1981, High Rayleigh number convection in shallow enclosures with different end temperatures, *International Journal of Heat and Mass Transfer* **24**(10), 1621–1629. 101

Shklover, A.M., 1945, Method of calculating heat transmission in buildings, Academy of Architecture, Moscow. Translated as Building Research Station Library Communication LC220, Garston, Herts, UK, 1947. 357

Siegel, R. and Howell, J.R., 1992, *Thermal Radiation Heat Transfer*, Hemisphere, Washington DC. 122

Simmonds, P., 1991, The utilisation and optimization of a building's thermal inertia in minimizing the overall energy use, *ASHRAE Trans.* **97**(2), 1031–1042. 362

Simmonds, P., 1992, A comparison of four European calculation methods, *Building Services Engineering Research and Technology* **13**(2), 85–94. 10

Simonson, C.J., Tao, Y.X. and Besant, R.W., 1993, Thermal hysteresis in fibrous insulation, *International Journal of Heat and Mass Transfer* **36**(18), 4433–4441. 180

Simonson, C.J., Tao, Y.X. and Besant, R.W., 1996, Simultaneous heat and moisture transfer in fiberglass insulation with transient boundary conditions, *ASHRAE Trans.* **102**(1), 315–327. 185

Simpson, A. and Stuckes, A.D., 1991, Thermal conductivity of porous materials: II. Theoretical treatment of radiative transfer, *Building Services Engineering Research and Technology* **11**(1), 13–19. 31

Simpson, A., O'Connor, D.E. and Stukes, A.D., 1991, Mineral fibre filled cavity wall: hygrothermal properties, *Building Services Engineering Research and Technology* **12**(4), 137–143. 177

Siviour, J.B., 1982, Thermal performance in practice of cavity walls insulated with urea-formaldehyde foam, *Building Services Engineering Research and Technology* **3**(2), 88–89. 15

Smith, C.C., Löf, G.O.G. and Jones, R.W., 1998, Rates of evaporation from swimming pools in active use, *ASHRAE Trans.* **104**(1A), 514–523. 175

Sobotka, P., Yoshino, H. and Matsumoto, S., 1995, The analysis of deep basement heat loss by measurements and calculations, *ASHRAE Trans.* **101**(2), 186–197. 370

Sonderegger, R.C., 1977, Harmonic analysis of building thermal response applied to the optimal location of insulation with the walls, *Energy and Buildings* **1**, 131–140. 14

Sowell, E.F. and Hittle, D.C., 1995, Evolution of building energy simulation methodology, *ASHRAE Trans.* **101**(1), 850–855. 332, 438

Söylemez, M.S., 1999, On the effective conductivity of building bricks, *Building and Environment* **34**(1), 1–5. 29

Sparrow, E.M., 1963, On the calculation of radiant interchange between surfaces, in *Modern Developments in Heat Transfer*, pp. 181–212, ed. W. Ibele, Academic Press, London. 125

Sparrow, E.M. and Bahrami, P.A., 1980, Experiments on natural convection from vertical parallel plates with either open or closed edges, *American Society of Mechanical Engineers: Journal of Heat Transfer* **102**, 221–227. 95

Sparrow, E.M. and Cess, R.D., 1978, *Radiation Heat Transfer*, Hemisphere, Washington DC. 122, 123

Sparrow, E.M., Kratz, G.K. and Schuerger, M.J., 1983, Evaporation of water from a horizontal surface by natural convection, *American Society of Mechanical Engineers: Journal of Heat Transfer* **105**, 469–475. 176

Spitler, J.D. and Ferguson, J.D, 1995, Overview of the ASHRAE annotated guide to load calculation models and algorithms, *ASHRAE Trans.* **101**(2), 260–264. 332

Spitler, J.D. and Fisher, D.E., 1999, On the relationship between the radiant time series and transfer function methods for design cooling load calculations, *Heating, Ventilation, Air-Conditioning and Refrigeration Research* **5**(2), 125–138. 372

Spitler, J.D. and Rees, S.J., 1998, Quantitative comparison of North American and U.K. cooling load calculation procedures – methodology, *ASHRAE Trans.* **104**(2), 36–46. 365

Spitler, J.D., Pederson, C.O. and Fisher, D.E., 1991, Interior convective heat transfer in buildings with large ventilative flow rates, *ASHRAE Trans.* **97**(1), 505–515. 105

Spitler, J.D., Fisher, D.E. and Pederson, C.O., 1997, The radiant time series cooling load calculation procedure, *ASHRAE Trans.* **103**, Part 2, 503–515. 372

Spooner, D.C., 1980, Results of a 'round robin' thermal conductivity test organized on behalf of the British Standards Institution, *Magazine of Concrete Research* **32**(111), 117–122. 30, 171

Spooner, D.C., 1982, Heat loss measurements through an insulated domestic ground floor, *Building Services Engineering Research and Technology* **3**, 147–151. 368, 369

Squire, H.B., 1953, Heat Transfer, in *Modern Developments in Fluid Dynamics*, Vol. II, pp. 801–810, eds L. Howarth, H.B. Squire and C.N.H. Lock, Clarendon Press, Oxford. 86

Stefan, J., 1879, Über die Beziehung zwischen der Wärmestrahlung und der Temperatur, Sitzung der akademischen Wissenschaft, Wien, *Mathematische und Naturwissenschaftklasse*, **79**, 391–428. 115

Stefanizzi, P., Wilson, A. and Pinney, A., 1990, Internal long-wave radiation in buildings: comparison of calculation methods: I. Review of algorithms, II. Testing of algorithms, *Building Services Engineering Research and Technology* **11**(3), 81–85 and 87–96. 115

Stephenson, D.G., 1962, Methods of determining non-steady-state heat flow through walls and roofs of buildings, *Journal of the Institution of Heating and Ventilating Engineers*, **30**, 64–73. 357, 371

Stephenson, D.G. and Mitalas, G.P., 1967, Cooling load calculations by thermal response factor method, *ASHRAE Trans.* **73**, Part 2, III.1.1–7. xvii, xix, 371

Stephenson, D.G. and Mitalas, G.P., 1971, Calculation of heat conduction transfer functions for multi-layer slabs, *ASHRAE Trans.* **77**, Part 2, 117–126. 281, 372, 413

Stephenson, D.G. and Starke, G.O., 1959, Design of a Π network for a heat flow analogue, *Journal of Applied Mechanics* **26**, 300–301. 383, 408

Swaid, H. and Hoffman, M.E., 1989, The prediction of impervious ground surface temperature by the surface thermal time constant (STTC) model, *Energy and Buildings* **13**(2), 149–157. 316

Szokolay, S.V., 1980, *World Solar Architecture*, Architectural Press, London. 216

Tahat, M.A., Babus'Haq, R.F. and O'Callaghan, P.W., 1993, Thermal energy storage, *Building Services Engineering Research and Technology* **14**(1), 1–11. 217

Taki, A.H. and Loveday, D.L., 1996, External convection coefficients for framed rectangular elements on building facades, *Energy and Buildings* **24**(2), 147–154. 111

Tang, D., 1997, Temperature distribution and heat transfer through a homogeneous corner, *Building and Environment* **32**(5), 457–463. 51

Tang, D.W. and Arak, N., 1996, Non-Fourier heat conduction in a finite medium under periodic surface thermal disturbance, *International Journal of Heat and Mass Transfer* **39**(8), 1585–1590. 291

Tang, R. and Etzion, Y., 2004, Comparative studies on the water evaporation rate from a wetted surface and that from a free water surface, Building and Environment **39**(1), 77–86. 175

Tao, W.Q. and Sparrow, E.M., 1985, Ambiguities related to the calculation of radiant heat exchange between a pair of surfaces, *International Journal of Heat and Mass Transfer* **28**(9), 1786–1787. 126

Tao, Y.-X., Besant, R.W. and Rezkallah, K.S., 1991, Unsteady heat and mass transfer with phase change in an insulation slab: frosting effects, *International Journal of Heat and Mass Transfer* **34**(7), 1593–1603. 185

Tao, Y.-X., Besant, R.W. and Rezkallah, K.S., 1992a, The transient thermal response of a glass-fiber insulation slab with hygroscopic effects, *International Journal of Heat and Mass Transfer* **35**(5), 1155–1167. 185

Tao, Y.-X., Besant, R.W. and Simonson, C.J., 1992b, Measurement of the heat of adsorption for a typical fibrous insulation, *ASHRAE Trans.* **98**(2), 495–501. 180

TARP, 1988, Loads Reference Manual, National Bureau of Standards. 451

Tavernier, E., 1972, Voorstel voor karakteristieken voor het thermisch gedrag in de zomer van bewoonde lokalen, *Warmte en Klimaat* **437**, 33–66. Available as translation 229, from the Heating and Ventilating Research Association. 357

Taylor, B.E., 1993, Interstitial condensation in building structures: revised model for identifying problem material combinations, *Building Services Engineering Research and Technology* **14**(1), 29–32. 178

Tewari, S.S. and Jaluria, Y. 1990, Mixed convection heat transfer from thermal sources mounted on horizontal and vertical surfaces, *American Society of Mechanical Engineers: Journal of Heat Transfer* **112**, 975–987. 83

Thekaekara, P.M., 1974, Data on incident solar energy, Supplement to the *Proceedings of the 20th Annual Meeting of the Institute for Environmental Science*, p. 21. 199

Thom, H.C.S., 1954, The rational relationship between heating degree-days and weather, *Monthly Weather Review*, **82**, 1–6. 9

Thomas, H.R. and Rees, S.W., 1999, The thermal performance of ground floor slabs – a full scale in-situ experiment, *Building and Environment* **34**(2), 139–164. 44

Thomas, W.C. and Burch, D.M., 1990, Experimental validation of a mathematical model for predicting water vapor sorption at interior building surfaces, *ASHRAE Trans.* **96**(1), 487–496. 187

Threlkeld, J.L., 1962, *Thermal Environmental Engineering*, Prentice Hall, Englewood Cliffs, New Jersey. 195

Tien, H.C. and Vafai, K., 1990, A synthesis of infiltration effects on an insulation matrix, *International Journal of Heat and Mass Transfer* **33**(6), 1263–1280. 185

Tindale, A., 1993, Third-order lumped-parameter simulation method, *Building Services Engineering Research and Technology* **14**(3), 87–97. 330

Tuomaala, P., 1993, New building air flow simulation model: theoretical basis, *Building Services Engineering Research and Technology* **14**(4), 151–157. 454

Tuomaala, P. and Piira, K., 2000, Thermal radiation in a room: an improved progressive refinement method, *Building Services Engineering Research and Technology* **21**(1), 9–17. 127

Tuomaala, P., Piira, K. and Vuolle, M., 2000, A rational method for the distribution of nodes in modelling of transient heat conduction in plane slabs, *Building and Environment* **35**(5), 397–406. 402

Tzou, D.Y., 1997, *Macro-to-Microscale Heat Transfer: The Lagging Behavior*, Taylor and Francis, Washington DC. 264

Udell, K.S., 1983, Heat transfer in porous media heated from above with evaporation, condensation and capillary effects, *American Society of Mechanical Engineers: Journal of Heat Transfer* **105**(3), 485–492. 194

Udell, K.S., 1985, Heat transfer in porous media considering phase change and capillarity – the heat pipe effect, *International Journal of Heat and Mass Transfer* **28**(2), 485–495. 191, 194

Uyttenbroeck, J., 1990, Building heat loss calculations: choice of internal temperature and of heat exchange coefficient h_i, *Building Services Engineering Research and Technology* **11**(2), 49–56. 164

Vafai, K. and Sarkar, S., 1986, Condensation effects in a fibrous insulation slab, *American Society of Mechanical Engineers: Journal of Heat Transfer* **108**, 667–675. 185

Vafai, K. and Whitaker, S., 1986, Simultaneous heat and mass transfer accompanied by phase change in porous insulation, *American Society of Mechanical Engineers: Journal of Heat Transfer* **108**, 132–140. 418

Van Straaten, J.F., 1967, *Thermal Performance of Buildings*, Elsevier, Amsterdam. 117

Vitharana, V.L. and Lykoudis, P.S., 1994, Criteria for predicting the transition to turbulence in natural convection along a vertical surface, *American Society of Mechanical Engineers: Journal of Heat Transfer* **116**, 633–638. 91

Vodicka, V., 1955, Eindimensionale Wärmeleitung in geschichteten Körpern, *Mathematische Nachrichten* **14**, 47–55. 389

Vodicka, V., 1956, Conduction of fluctuating heat flow in a wall consisting of many layers, *Applied Science Research*, Martinus Nijhoff, The Hague, A5, pp. 108–114. 357

Vozar, L. and Sramkova, T., 1997, Two data reduction methods for evaluation of thermal diffusivity from step-heating measurements, *International Journal of Heat and Mass Transfer* **40**(7), 1647–1655. 270

Waide, P.A. and Norton, B., 1995, Degree-hour steady-state temperature index, *Building Services Engineering Research and Technology* **16**(2), 107–113. 9

Wakitani, S., 1997, Development of multicellular solutions in natural convection in an air-filled vertical cavity, *American Society of Mechanical Engineers: Journal of Heat Transfer* **119**, 97–101. 99

Wallenten, P., 2001, Convective heat transfer coefficients in a full-scale room with and without furniture, *Building and Environment* **36**(6), 743–751. 104

Walraven, R., 1978, Calculating the position of the sun, *Solar Energy* **20**, 393–397. 203

Walsh, P.J. and Delsante, A.E., 1983, Calculation of the thermal behaviour of multi-zone buildings, *Energy and Buildings* **5**, 231–242. 358

Walton G.N., 1981, Passive solar extension of the building loads analysis and system thermodynamics (BLAST) program, Department of the Army Construction Engineering Research Laboratory, PO Box 4005, Champaign, Illinois 61820–1305. 165, 451

Walton G.N., 1983, Thermal Analysis Research Program (TARP) Refeence Manual, US Department of Commerce, National Bureau of Standards, Building Physics Division, Washington DC 20234. 451

Walton, G.N., 1989, Airflow network models for element-based building airflow modelling, *ASHRAE Trans.* **95** Part 2, 611–620. 454

Wang, S. and Chen, Y., 2003, Transient heat flow calculation for multilayer constructions using a frequency-domain regression method, *Building and Environment* **38**(1), 45–61. 400

Warsi, Z.U.A. and Choudhury, N.K.D., 1964, Weighting function and transient thermal response of buildings. Part II: Composite structure, *International Journal of Heat and Mass Transfer* **7**, 1323–1334. 307

Waters, J.R., 1981, An investigation of some errors due to the use of finite difference techniques for building heat transfer calculations, *Building Services Engineering Research and Technology* **2**(1), 51–59. 238

Waters, J.R. and Wright, A.J., 1985, Criteria for the distribution of nodes in multilayer walls in finite-difference thermal modelling, *Building and Environment*, **20**(3), 151–162. 236

Watmuff, J.H., Charters, W.W.F. and Proctor, D., 1977, Solar and wind induced external coefficients for solar collectors, Comples 2. 111

Wayner, P.C., 1982, Adsorption and capillary condensation at the contact line in change of phase heat transfer, *International Journal of Heat and Mass Transfer* **25**(5), 707–713. 187

Webb, R.L., 1991, Standard nomenclature for mass transfer processes, *ASHRAE Trans.* **97**, Part 2, 114–117. 194

Weber, C.F., 1981, Analysis and solution of the ill-posed inverse heat conduction problem, *International Journal of Heat and Mass Transfer* **24**(11), 1783–1792. 276

Weber, D. and Kearney, R.J., 1980, Natural convective heat transfer through an aperture in passive solar heated buildings, *Proceedings of the AS International Solar Energy Society Fifth Passive Solar Conference*, pp. 1037–1041. 108

Wenham, G.J., 1979, *The Book of Leviticus*, Eerdmans, Grand Rapids, Michigan. 196

Whitaker, S., 1977, Simultaneous heat, mass and momentum transfer in porous media: a theory of drying, in *Advances in Heat Transfer*, eds J.P. Hartnett and T.F. Irvine, Academic Press, New York, Vol. 13, pp. 119–203. 182, 418

Wickern, G., 1991, Mixed convection from an arbitrarily inclined semi-infinite plate, 1. The influence of the inclination angle, *International Journal of Heat and Mass Transfer* **34**(8), 1935–1945. 113

Wijeysundera, N.E. and Hawlader, M.N.E., 1992, Effects of condensation and liquid transport on the thermal performance of fibrous insulations, *International Journal of Heat and Mass Transfer* **35**(10), 2605–2616. 181

Wijeysundera, N.E. and Wilson, S.J., 1994, Transient heat transfer through an insulation slab with simultaneous moisture redistribution, *International Journal of Heat and Mass Transfer* **37**(16), 2391–2398. 180

Wijeysundera, N.E., Hawlader, M.N.E. and Tan, Y.T., 1989, Water vapour diffusion and condensation in fibrous insulations, *International Journal of Heat and Mass Transfer* **35**(10), 2605–2616. 181

Wijeysundera, N.E., Zheng, B.F., Iqbal, M. and Hauptmann, E.G., 1996, Numerical simulation of the transient moisture transfer through porous insulation, *International Journal of Heat and Mass Transfer* **39**(5), 995–1004. 181

Wilkes G.B. and Peterson, C.M.F., 1937, Radiation and convection across air spaces in frame construction, *Heating, Piping and Air Conditioning* **9**, 505–510. 95

Williams, L.O., 1994, Therapy for the Earth, *Applied Energy* **47**(2/3), 97–298. xv

Winters, K.H., 1987, Hopf bifurcation in the double-glazing problem with conducting boundaries, *American Society of Mechanical Engineers: Journal of Heat Transfer* **109**, 894–898. 101

Winwood, R., Benstead, R. and Edwards, R., 1997, Advanced fabric energy storage: I. Review, II. Computational fluid dynamics modelling, III. Theoretical analysis and whole-building simulation, IV. Experimental modelling, *Building Services Engineering Research and Technology* **18**(1), 1–30. 217, 335

Wirth, P.E. and Rodin, E.Y., 1982, A unified theory of linear diffusion in laminated media, *Advances in Heat Transfer*, pp. 283–330, Academic Press, New York. 385

Wirtz, R.A. and Stutzman, R.J., 1982, Experiments on free convection between vertical plates with symmetric heating, *American Society of Mechanical Engineers: Journal of Heat Transfer* **104**, 501–507. 100

Wong, S.P.W. and Wang, S.K., 1990, Fundamentals of simultaneous heat and moisture transfer between the building envelope and the conditioned space air, *ASHRAE Trans.* **96**(2), 73–83. 169, 179

Wood, A.S., 2001, A new look at the heat balance integral method, *Applied Mathematical Modelling* **25**(10), 815–824. 285

Wood, M. and Jesch, L.F., 1993, Effects of social costs on renewables: the case of transparent insulation, *International Journal of Ambient Energy* **14**(3), 123–146. 16

Woodbury, K.A., 1990, Effect of thermocouple sensor dynamics on surface heat flux predictions obtained via inverse heat transfer analysis, *International Journal of Heat and Mass Transfer* **33**(12), 2641–2649. 276

Woolf, H.M., 1968, On the computation of solar elevation angles and the determination of sunrise and sunset times, NASA-TM-X-1646, September 1968. 204

Wright, J.L. and Sullivan, H.F., 1994, A two-dimensional numerical model for natural convection in a vertical, rectangular window cavity, *ASHRAE Trans.* **100**(2), 1193–1206. 99

Wyrwal, J., 1988, Some theorems in Luikov's theory of heat and mass transfer in capillary-porous bodies, *International Journal of Heat and Mass Transfer* **31**(12), 2543–2546. 418

Wyrwal, J. and Marynowicz, A., 2002, Vapour condensation and moisture accumulation in porous building wall, *Building and Environment* **37**(3), 313–318. 184

Yang, C.-Y., 1998, A linear inverse model for the temperature-dependent thermal conductivity determination in one-dimensional problems, *Applied Mathematical Modelling* **22**(1/2), 1–9. 330

Yang, W.-M. and Leu, M.-C., 1993, Instability of radiation-induced flow in an inclined slot, *International Journal of Heat and Mass Transfer* **36**(12) 3089–3098. 100

Yewell, R., Poulikakos, D. and Bejan, A., 1982, Transient natural convection experiments in shallow enclosures, *American Society of Mechanical Engineers: Journal of Heat Transfer* **104**, 533–538. 106

Yousef, W.W., Tarasuk, J.D. and McKeen, W.J., 1982, Free convection heat transfer from upward-facing isothermal horizontal surfaces, *American Society of Mechanical Engineers: Journal of Heat Transfer* **104**, 493–500. 84

Yunnie, P.G., 1995, Early heating, ventilating, and air conditioning in the United Kingdom, *ASHRAE Trans.* **101**(1), 1129–1133. 1

Zaheer-uddin, M., 1990, Combined energy balance and recursive least squares method for the identification of system parameters, *ASHRAE Trans.* **96**(2), 239–244. 330

Zalewski, L., Lassue, S., Duthoit, B. and Butez, M., 2002, Study of solar walls – validating a simulation model, *Building and Environment* **37**(1), 109–121. 213

Zhai, Z., Chen, Q., Haves, P. and Klems, J.H., 2002, On approaches to couple energy simulation and computational fluid dynamics programs, *Building and Environment* **37**(8/9), 857–864. 458

Zhao, Y., Curcija, D. and Goss, W.P., 1997, Prediction of the multicellular flow regime of natural convection in fenestration glazing cavities, *ASHRAE Trans.* **103**, Part 1, 1009–1020. 97

Zhao, Y., Curcija, D. and Goss, W.P., 1999, Convective heat transfer correlations for fenestration glazing cavities: a review, *ASHRAE Trans.* **105**, Part 2, 900–908. 99

Zmeureanu, R. and Doramajian, A., 1992, Thermally acceptable temperature drifts can reduce the energy consumption for cooling in office buildings, *Building and Environment* **27**(4), 469–481. 451

Zmeureanu, R., Fazio, P. and Haghighat, F., 1987, Analytical and inter-program validation of a building thermal model, *Energy and Buildings* **10**, 121–133. 307

Zohrabian, A.S., Mokhtarzadeh-Dehghan, M.R., Reynolds, A.J. and Marriot, B.T.S., 1989, An experimental study of buoyancy-driven flow in a half-scale stairwell model, *Building and Environment* **24**(2), 141–148. 109

Zohrabian, A.S., Mokhtarzadeh-Dehghan, M.R. and Reynolds, A.J., 1990, Buoyancy-driven air flow in a stairwell model with through-flow, *Energy and Buildings* **14**(2), 133–142. 109

Bibliography

Billington, N.S., 1952, *Thermal Properties of Buildings*, Cleaver-Hume Press, London.
Billington, N.S. and Roberts, B.M., 1982, *Building Services Engineering: A Review of its Development*, Pergamon, Oxford.
Boyle, G., (editor), 1996, *Renewable Energy*, Oxford University Press, Oxford, with the Open University, Milton Keynes, U.K.
Burberry, P., 1983, *Practical Thermal Design in Buildings*, Batsford Academic and Educational Ltd., London.
Edwards, D.K., 1981, *Radiation Heat Transfer Notes*, Hemisphere, Washington DC.
Fisk, D.J., 1981, *Thermal Control of Buildings*, Applied Science Publishers, London.
Givoni, B., 1969, *Man, Climate and Architecture*, Elsevier, Amsterdam.
Hsu, S.T., 1963, *Engineering Heat Transfer*, Van Nostrand Company, Inc., Princeton, New Jersey.
Kimura, K.-I., 1979, *Scientific Basis of Air-Conditioning*, Applied Science Ltd., London.
Kreider, J.F. and Rabl, A., 1994, *Heating and Cooling of Buildings*, McGraw-Hill, Inc.
Markus, T.A. and Morris, E.N., 1980, *Buildings, Climate and Energy*, Pitman, London.
Muncey, R.W.R., 1979, *Heat Transfer Calculations for Buildings*, Applied Science Publishers, London.
Whitaker, S., 1977, *Fundamental Principles of Heat Transfer*, Pergamon Press Inc. New York, (reprinted 1983).

Building Heat Transfer Morris G. Davies
© 2004 John Wiley & Sons, Ltd ISBN: 0-470-84731-X

Index

Entries of form a are page numbers, a.b refer to equations, Fa.b to figures.

Building Heat Transfer Morris G. Davies

© 2004 John Wiley & Sons, Ltd

ISBN: 0-470-84731-X